Symbol	Term	Introduced in Chapter(s)
M^d	Money demand	5
M^s	Money supply	5
μ	Markup of prices over wages	15
N	Employment	15
N_n	Natural level of employment	15
P	GDP deflator/CPI/price level	2
$\$P_B$	Dollar price of a bond	5
π	Inflation	7
Π	Profit per unit of capital	8
Q	Imports	3
$\$Q$	Nominal stock price	9
R	Bank reserves	5
r	Real interest rate	7
S	Private saving	3
s	Private saving rate	22
T	Net taxes (taxes paid by consumers minus transfers)	3
θ	Reserve ratio of banks	5
U	Unemployment	2, 15
u	Unemployment rate	2, 15
$V(\$z)$	Present value of a sequence of nominal payments, z	7
$V(\$z^e)$	Present value of an expected sequence of nominal payments, z	7
u_n	Natural rate of unemployment	15
Y	Real GDP/output/supply of goods	2, 3
$\$Y$	Nominal GDP	2
Y_D	Disposable income	3
Y_L	Labour income	8
Y_N	Natural level of output	15
X	Exports	3
Z	Demand for goods	3
z	Sequence of payments	7
z	Factors that affect the wage, given unemployment	15

MACROECONOMICS

FIRST CANADIAN EDITION

Olivier Blanchard
Massachusetts Institute of Technology

Angelo Melino
University of Toronto

PRENTICE HALL CANADA INC.,
SCARBOROUGH, ONTARIO

Canadian Cataloguing in Publication Data

Blanchard, Olivier (Olivier J.)
  Macroeconomics

Includes index.
ISBN 0-13-095059-9

1. Macroeconomics. I. Melino, Angelo. II. Title.

HB172.5B556 1999 339 C98-932316-1

© 1999 Prentice-Hall Canada Inc., Scarborough, Ontario
A Division of Simon & Schuster/A Viacom Company

Original edition published by Prentice-Hall, Inc.
A Simon & Schuster Company
Upper Saddle River, New Jersey, USA
Copyright © 1998 by Prentice-Hall, Inc.

Prentice-Hall, Inc., Upper Saddle River, New Jersey
Prentice-Hall International (UK) Limited, London
Prentice-Hall of Australia, Pty. Limited, Sydney
Prentice-Hall Hispanoamericana, S.A., Mexico City
Prentice-Hall of India Private Limited, New Delhi
Prentice-Hall of Japan, Inc., Tokyo
Simon & Schuster Southeast Asia Private Limited, Singapore
Editora Prentice-Hall do Brasil, Ltda., Rio de Janeiro

ISBN 0-13-095059-9

Publisher: Patrick Ferrier
Acquisitions Editor: Sarah Kimball
Senior Marketing Manager: Ann Byford
Developmental Editor: Maurice Esses
Production Editor: Kelly Dickson
Copy Editor: Rodney Rawlings/Carol Fordyce
Production Coordinator: Jane Schell
Cover Design: Alex Li
Cover Image: Zigy Kaluzny/Tony Stone Images
Page Layout: Jack Steiner

1 2 3 4 5 03 02 01 00 99

Printed and bound in the USA

Visit the Prentice Hall Canada Web site! Send us your comments, browse our
catalogues, and more at **www.phcanada.com**. Or reach us through e-mail at
phcinfo_pubcanada@prenhall.com.

À Noelle
O.B.

To Effie
A.M.

ABOUT THE AUTHORS

OLIVIER BLANCHARD

Olivier Blanchard is the Class of 1941 Professor of Economics at MIT. He did his undergraduate work in France and received his Ph.D. in economics from MIT in 1977. He taught at Harvard from 1977 to 1982, and since 1983 has taught at MIT, where he has twice received the award for best teacher in the Department of Economics.

He has done research on many macroeconomic issues, including the effects of fiscal policy, the role of expectations, price rigidities, speculative bubbles, unemployment in Western Europe, and more recently transition in Eastern Europe. He has done work for many international organizations, including the World Bank, the IMF, the OECD, the Commission of the European Communities, and the European Bank for Reconstruction and Development. He has published over 150 articles and edited or written more than 10 books, including Lectures on Macroeconomics with Stanley Fischer.

He has been the editor of the Quarterly Journal of Economics since 1984. He is a research associate of the National Bureau of Economic Research, a fellow of the Econometric Society, and a vice-president of the American Economic Association.

Olivier Blanchard lives in Cambridge, Massachusetts, with his wife, Noelle. He has three daughters, Marie, Serena, and Giulia.

ANGELO MELINO

Angelo Melino was born and raised in Toronto. He received his B.A. from the University of Toronto in 1977 and his Ph.D. in economics from Harvard in 1983. Aside from one summer at the Bank of Canada and a few semesters in southern California, he has taught econometrics and macroeconomics at the University of Toronto since 1981.

He studied with Olivier Blanchard while at Harvard and the two have co-authored a paper on the behaviour of prices and quantities in the automobile industry. Angelo's research is best described as eclectic. He has written papers on issues in continuous-time econometrics, strikes, the term structure of interest rates, and pricing derivatives. In his spare time, he enjoys softball, squash, and ballroom dancing.

Angelo lives in Toronto with his wife, Effie. They have two daughters, Kate and Stacy.

BRIEF CONTENTS

*Optional chapter.

CONTENTS

*Indicates an optional chapter or section.

BOXES

PREFACE

This book has two main goals.

■ Our first goal is to help students see how macroeconomics can be used to try to understand some of the most important issues of the past thirty years, such as the transformation of Eastern Europe, high unemployment in Western Europe, rapid economic development in Asia, and widening income inequality in Canada and the United States.

To achieve this first goal, we set out to do two things.

First, we had to take on many issues not usually covered in intermediate macroeconomics textbooks. In particular, we give an increased emphasis to the role of expectations, to the openness of modern economies, and to the medium run (which typically falls in the void between business cycles and growth).

Second, we had to move back and forth between theory and facts. You will find many examples from around the world, both current and historical, in this textbook. These examples are treated in depth, not as footnotes, but as stories, interesting both in their own right and for what they tell us about macroeconomics in general. For example, German reunification is the topic of three detailed boxes. The first (in Chapter 6) looks at the short-run macroeconomic effects of reunification on Germany and provides a clear example of how monetary and fiscal policy must be used to help an economy adjust to such a large shock. The second (in Chapter 13) looks at the effects of reunification on Germany's partners in the European Union and shows the importance of interactions among today's

open economies. The third (in Chapter 26) focusses on the evolution of Eastern Germany since reunification, showing the nature of transition from central planning to a market economy and providing an example of the forces that determine the growth of output over time.

■ Our second goal is to give a sense of macroeconomics as an applied science, a science that deals with complex and often urgent issues, and a science that must maintain both internal consistency and external reference.

In the service of this second goal, the book is constructed around one underlying model that concentrates on the implications of equilibrium conditions in three sets of markets: goods, financial, and labour. Depending on the issue at hand, relevant parts of the model are developed in more detail and the others are simplified or left in the background. But the underlying model is always the same. We hope that, as a result, readers of this book will see macroeconomics as a coherent whole, not as a collection of models pulled from a hat.

At the same time, the book tries to convey the fact that macroeconomics is not an exact science but an applied one where ideas, theories, and models are constantly evaluated against the facts, and often modified or rejected. For example, our understanding of the relation between inflation and unemployment, and our assessment of the roles of fiscal and monetary policy, have changed dramatically over the past 30 years. They are likely to keep evolving. We try to give a sense of this process of construction and evolution, to show how macroeconomists build models, how they decide what aspects of reality to emphasize or leave aside, and how they estimate and test their models. Understanding this aspect of macroeconomics not only is essential to a full understanding of the field, but also makes macroeconomics less intimidating. This textbook does not present the "truth" but rather our best understanding of it at this time—an understanding based on observation and destruction, but also subject to failure and to change.

Organization

In building this book, we have made two main organizational choices:

■ The first has been to have more but shorter chapters than the typical macroeconomics text (30 chapters in all, at an average of about 20 pages each). Each chapter has a clear, single focus and can be covered in a lecture, or read and digested in a couple of hours. We have found that about 20 to 25 chapters can be covered in a semester course, between 15 and 20 in a quarter course.

■ The second has been to start with an analysis of the short run and then move to the medium run and long run. The reasons are twofold. It is our experience that it is easier for readers and for students to relate to short-run issues (such as the nature and the source of recessions) than to long-run issues (such as the role of capital accumulation in growth). And this organization allows a richer treatment of the medium and the long run, when they are covered later.

The book is divided into ten (short) parts.

■ Part One ("Introduction," Chapters 1 and 2) is an introduction to the basic facts and issues of macroeconomics. While Chapter 2 gives the basics of national income accounting, we have moved a detailed treatment of national income accounts to Appendix 2 at the end of the book. This both decreases the burden on the reader new to economic theory and allows for a more thorough treatment in the appendix.

The next three parts focus on the short run.

■ Part Two ("The Basics," Chapters 3 to 6) focusses on equilibrium in goods and financial markets and introduces the basic short-run model used in macroeconomics, the *IS-LM* model.

■ Part Three ("Expectations," Chapters 7 to 10) introduces and develops one of the major themes of the book, the role of expectations in determining decisions and macroeconomic outcomes.

■ Part Four ("Openness," Chapters 11 to 14) introduces and develops another major theme of the book, the importance of the increasing openness of modern economies.

The next three parts move from the short run to the medium and long run.

■ Part Five ("The Supply Side," Chapters 15 to 19) introduces equilibrium in the labour market and develops a model of aggregate supply and aggregate demand that describes output and price movements from the short run to the long run.

■ Part Six ("Pathologies," Chapters 20 and 21) looks at times when something in the economy goes very wrong. Chapter 20 looks at periods of high unemployment. Chapter 21 looks at episodes of hyperinflation.

■ Part Seven ("The Long Run," Chapters 22 to 24) looks at growth (the steady increase in output over time) and focusses on the roles of capital accumulation and technological progress.

The last three parts focus on change and transition, policy, and macroeconomics past and present.

■ Part Eight ("Change and Transition," Chapters 25 and 26) shows how we can use what we have learned in earlier chapters to examine two of the most important economic issues today. Chapter 25 studies the relations among technological progress, the level of unemployment, and the distribution of wages. Chapter 26 looks at transition from central planning to market economies in Eastern Europe.

■ Most of the first 26 chapters discuss macroeconomic policy in one form or another. Part Nine ("Policy," Chapters 27 to 29) ties all of the threads together. Chapter 27 looks at the role and the limits of macroeconomic policy in general. Chapters 28 and 29 review monetary and fiscal policy respectively.

■ Part Ten ("Epilogue," Chapter 30) concludes by putting macroeconomics in historical perspective, showing the evolution of macroeconomics in the past 50 years and discussing current directions of research.

Alternative Course Outlines

Within the book's broad organization, there is plenty of room for alternative course organizations. Here are six alternative course outlines.

A Very Short Core Course

1. The core of the book is composed of 10 chapters: Chapters 1 and 2, the description of the *IS-LM* model in Chapters 3 to 6, the description of the model of aggregate supply and aggregate demand in Chapters 15 to 17, and Chapter 27 on policy.

   A short course can be based on these chapters, using some of the boxes in other chapters as additional examples.

Three Courses with an Emphasis on the Short Run and the Medium Run

2. A short course can be organized around the core, plus Chapters 7 and 10 on expectations, and Chapters 11 to 13 on the open economy (15 chapters in all).

3. A longer course can be organized around the core plus Chapters 7 to 10 on expectations, Chapters 11 to 14 on the open economy, and a combination of:

   ■ Chapters 18 and 19 on inflation, output, and exchange rates

   ■ Chapters 20 and 21 on pathologies

   ■ Chapters 28 and 29 on fiscal and monetary policy

4. Some instructors may want to go directly from the *IS-LM* model to the model of aggregate supply and aggregate demand (that is, before looking at the role of expectations and the implications of openness). We have written Chapters 7 to 17 in such a way as to make this possible. An alternative outline for a longer course would begin with Chapters 1 to 6, directly followed by Chapters 15 to 17, then followed by Chapters 7 to 10 on expectations and Chapters 11 to 14 on the open economy, plus a combination of:

   ■ Chapters 18 and 19 on inflation, output, and exchange rates

   ■ Chapters 20 and 21 on pathologies

   ■ Chapters 28 and 29 on fiscal and monetary policy

Two Courses with an Emphasis on the Medium Run and the Long Run

5. A short course can be organized around the core, plus Chapters 22 to 24 on growth, and Chapters 25 and 26 (15 chapters in all).

6. A longer course can be organized around the core plus Chapters 22 to 25 on growth, and a combination of:

   ■ Chapters 20 and 21 on pathologies

   ■ Chapters 25 and 26 on current issues

   ■ Chapters 28 and 29 on fiscal and monetary policy

In addition, we have flagged nine chapters as optional. Optional chapters fall into two categories. Two of them, Chapter 14 (on the behaviour of exchange rates) and Chapter 19 (on the interactions among inflation, interest rates, and exchange rates), are conceptually somewhat more difficult than other chapters. The other seven (Chapters 20 and 21, 25 and 26, 28 and 29, and 30) deal with specific issues and can be skipped without loss of continuity.

Features

In presenting each argument in this book, we have made sure to tell it in three different ways:

- Using algebra. Algebra forces us to be logical. (The book uses only basic algebra. A refresher on the required mathematics is given in Appendix 3 at the end of the book.)

- Using graphs. Graphs are less precise than algebra, but for many of us they are more intuitive.

- Using only words and stating the intuition behind the results.

We have also made sure never to present a theoretical result without relating it to the real world. In addition to discussions of facts in the text itself, we have introduced three types of boxes: **Focus** boxes, which expand on a point made in the text; **In Depth** boxes, which look at a particular macroeconomic episode in detail; and **Global Macro** boxes, which look at macroeconomic events from around the world.

Each chapter ends with three ways of making sure that the material in the chapter has been thoroughly understood: a **summary** of the chapter's main points; a list of **key terms** (with a glossary at the end of the book); and a series of **end-of-chapter exercises**.

Finally, for students who want to explore macroeconomics further, we have introduced the following three feature: **Digging Deeper** footnotes, which expand on an argument in the text, often by indicating how an implicit assumption in the text can be relaxed and what the implications would be; **short appendices** to some chapters, which show how a proposition in the text can be derived more rigorously or expanded; and a **further reading** section at the end of each chapter, indicating where to find more information.

Furthermore, Appendix 1 at the end of the book indicates where to find data on nearly any macroeconomic variable, for nearly any country in the world. When available, Internet addresses are given, indicated with this symbol:

The Teaching and Learning Package

The book comes with a number of supplements.

- **Instructor's Manual.** The instructor's manual discusses pedagogical choices, alternative ways of presenting the material, and ways of reinforcing students' understanding. For each chapter of the book, the manual has seven sections: objectives, how it fits in, logical structure (summary), key tools, alternative perspectives, extensions, and firming up and testing concepts (conceptual observations, empirical observations, additional exercises, and answers). The instructor's manual also includes the answers to all end-of-chapter questions and exercises.

- **Study Guide.** Each chapter of the student-friendly guide begins with a presentation of objectives and review, organized in the form of a tutorial and covering the important points of the chapter, with learning suggestions along the way. The tutorial is followed by quick self-test questions, review problems, and multiple-choice questions. Solutions are provided for all the study guide problems.

- **Test Item File and Test Manager (for Windows).** The printed test bank includes at least 25 multiple-choice questions per chapter. The electronic version of the test item file, known as Test Manager, is a computerized package that allows users to custom-design, save, and generate classroom tests. Custom Test allows professors to edit and add or delete questions from the test item file, edit existing graphics, and create new graphics. It also has a gradebook feature and allows online testing.

- **Transparency Masters.** A complete set of black and white transparency masters for all figures in the text is available to adopters upon request.

Acknowledgements and Thanks

This book owes much to many.

We have benefited from the comments and encouragement of many colleagues and friends, in both the United States and Canada. The list is long, but ot hose thanked in the U.S. edition we would like to add a few more: the wise voices that regularly contribute to the departments's lunch table at the University of Toronto for their sage advice; various memabers of the Economics Department and the Faculty of Management at the University of Toronto who found the first U.S. edition of sufficient interest that we were encouraged to adapt the book for a Canadian audience; and six anonymous reviewers whose comments helped shape and focus the Canadian edition.

We have many people to thank at Prentice Hall Canada. Sarah Kimball convinced Angelo to undertake this Canadian edition; for this she has earned the thanks of at least one of us. Maurice Esses provided initial editorial guidance and was a reliable and much-valued sounding board. We owe a special debt to Carol Fordyce; her sharp eye alerted us to some embarrassing errors and her good taste spared the reader from our unintended attacks on the English language. Kelly Dickson and her staff pulled together the final product.

We have benefited from the comments and suggestions of many reviewers, including Alastair Robertson, Wilfrid Laurier University, Apostolos Serletis, The University of Calgary, Gary Tompkins, University of Regina, Christian Zimmerman, University of Quebec at Montreal.

Mark Flanagan collected data, checked calculations, and spent many hours with Angelo talking about the book and suggesting changes.

Our daughters did not ignore our pleas for assistance.

Our wives kept us sane.

Olivier Blanchard   Angelo Melino
Cambridge, Massachusetts   Toronto, Ontario

1

A TOUR OF THE WORLD

1-1 WHAT IS MACROECONOMICS?

How does macroeconomics differ from microeconomics? The answer is given by the Greek roots: *macro* means large, and *micro* means small. **Macroeconomics** studies aggregate economic variables, such as production for the economy as a whole (aggregate output) or the average price of all goods (the aggregate price level). In contrast, **microeconomics** studies production and prices in specific markets.

This answer raises another question. Isn't macroeconomics just a branch of microeconomics? Isn't an economy just a collection of specific markets, so that once you understand each of them, you understand the entire economy? The answer to this question is: yes in principle, no in practice.

In principle, one can think of writing the equilibrium condition that demand equals supply for each of the millions of markets that compose a modern economy, listing all the variables that affect demand and supply in each market, and using a powerful computer to solve for the conditions under which demand equals supply simultaneously in all markets.

In practice, macroeconomists' knowledge of the economy, and of all the interactions in it, is just not good enough for this strategy to be feasible. And even if they succeeded in building such a gigantic model and solving it with a computer, the model would be just as complicated as the economy, and nearly as hard to understand.

THE SIMPLIFICATIONS OF MACROECONOMICS

Given the field's complexity, the task of macroeconomics is to find ways of simplifying in order to explain the behaviour of aggregate variables. Macroeconomists do this by pretending that the world is much simpler than it really is. For example, instead of keeping track of the many goods and markets in which goods are sold, they usually rely on the fiction that there is only one good, with only one demand curve and one supply curve, trading in one market. Then,

I

instead of keeping track of the millions of demand and supply curves that exist in reality, they need to think only of the demand curve and the supply curve for "the" good.

Having made such simplifications, they can construct simple structures to think about and interpret the economy. These structures are called **models.** Macroeconomic models are logical, internally consistent ways of describing the workings of an economy. Sometimes they are just descriptions in words; more often they rely on mathematics. One of the advantages of using mathematics is that it forces one to make sure that a model does not suffer from flaws in logic.

How do macroeconomists come to choose particular simplifications? First, by thinking hard about their implications and the dangers of making them. Some simplifications are benign for some purposes, dangerous for others. To take an example from outside economics, booksellers may safely ignore the fact that their customers differ in height, but it would be very unwise for suitmakers to make the same assumption.

Second, by continually comparing the simple models they construct to the reality they observe. Today's world, with close to 190 countries, provides them with many experiences to explain.[1] History provides them with many more. Thus macroeconomists are continually challenged to ask: Can we explain these experiences? If we cannot, how should we change our thinking? Do we have to change the structure of our models radically, or simply change some of our simplifications?

Once in a while, an event indeed causes major rethinking. This was the case during the Great Depression, the period of high unemployment that lasted for more than a decade before World War II. The existing body of macroeconomics in the 1930s proved unable to explain the depth and the length of the Depression. When, in 1936, the English economist John Maynard Keynes offered a new way of thinking about output determination and a way of explaining the Depression, his approach quickly swept the field. Within a decade, Keynes's ideas had completely transformed macroeconomics.

Such defining moments aside, many events lead macroeconomists to realize they have left a crucial element out of their models, or misunderstood another. This was the case in the 1970s, when most countries experienced nearly a decade of *stagflation* (a clumsy term coined by economists to denote a period of simultaneous *stagnation* and *inflation*). The occurrence of high unemployment and high inflation did not fit well the then-prevailing notion that countries could have either high unemployment or high inflation, but not both at the same time. After some study, it became clear that the shocks hitting the economy at the time—namely, the very large increases in the price of oil imposed by OPEC, the cartel of oil producers— were different from those that had happened in earlier decades. Because they hadn't seen such shocks before, macroeconomists had not included them in their models. Within a few years, however, the effects of shocks such as oil price increases were much better understood, and cost shocks came to be included in models. Stagflation was no longer a puzzle.

---

[1]*It is not easy to say how many countries there are in the world. The United Nations has 185 members, but some countries, such as Taiwan, are not members. The telephone system uses 190 different country codes. The International Organization for Standardization, which assigns codes to country names, has 239 different codes, but the list includes some, such as Hong Kong (the former British colony, now part of China) and the French Islands of St. Pierre and Miquelon, that don't fit the usual notion of a country.*

WHY MACROECONOMISTS SOMETIMES DISAGREE

Macroeconomics is thus the result of a sustained process of construction, of an interaction between ideas and events. What macroeconomists believe today is the result of an evolutionary process in which they have eliminated those ideas that failed and kept those that appear to explain reality well.

This does not mean that macroeconomics today is "right." Surely, new events will lead macroeconomists to question some of their thinking; some may even lead to radical rethinking. Nor does it imply that the lessons of history and the interactive process between ideas and events are so strong that all macroeconomists agree on everything. They disagree on many issues, although often less so than is commonly perceived. When they disagree, they do so for two very different reasons.

First, even when they share the same view of the way the economy works, they often disagree on the weight they assign to different objectives. Some economists are willing to reduce income inequality even if some of the means needed to achieve this goal, such as high tax rates, have adverse effects on aggregate activity. Others believe that income inequality should be accepted and that high aggregate activity is more important. Some economists put more weight on fighting high unemployment than on fighting inflation because they see unemployment as a major social evil. Others put more weight on fighting inflation, which they see as more dangerous to society. Often, lines of disagreement run along political lines. Economists with left-wing leanings usually care more about income inequality and unemployment; economists with right-wing leanings usually care more about growth and fighting inflation. The measures that both groups recommend differ accordingly. As long as people (and thus economists) have different values, these disagreements will remain.

Second, reality often does not speak strongly enough to make all economists agree. In contrast to researchers in most other applied sciences, economists cannot do **controlled experiments.** When an engineer wants to find out how the temperature affects a material's conductivity, she builds an experiment in which she changes the temperature, making sure that everything else remains the same, and looks at the change in conductivity. But macroeconomists who want to find out, for example, how changes in the money supply affect aggregate activity cannot perform such experiments; they cannot make the world stop while they ask the central bank to change the money supply. Typically, changes in the money supply coincide with a myriad of other events, ranging from changes in tax legislation, to strikes, to unusual weather, and so on. Thus, to isolate the effects of the change in the money supply on output, economists must, when they look at data, control for the other variables that moved at the same time. This is difficult enough, and involves enough judgements about what to control for and how to do so, that different economists looking at the same episode can reach different conclusions. Looking at the same episode, one economist can see a strong effect of money on activity, while another sees a weaker effect.[2]

---

[2]*Even if they were feasible, it is not clear that society would be willing to put up with controlled experiments.* **Randomized experiments**, *where the path of the money supply is chosen on the basis of a purely random device, such as a roll of dice, are feasible and could be used to uncover causal relationships. Most economists, however, are reluctant to submit their neighbours to a purely random monetary policy.*

The availability and the study of more and more episodes, and the use of better and better techniques to examine the data, narrow such differences of opinion over time. For example, there is large agreement about the effects of money on economic activity, if not about the specific channels through which these effects take place. But disagreements do and will remain. This book focusses on the common core of macroeconomics, on what the majority of macroeconomists agree on and how they use it to study the world. But along the way, it will also point out where and why they disagree. ⁻

1-2 LOOKING AT THE WORLD

A quick look at the performance of and problems facing economies around the world provides a nice introduction to the sort of questions that concern macroeconomists. Keeping an eye on developments elsewhere is also of particular interest to Canadians.

Canada is a very open economy. Exports and imports of goods and services both amount to about 40% of our national output. The bulk of our foreign trade (about four-fifths) is with the United States, but trade flows with Europe and the Pacific are also large. Economic developments around the rest of the world, particularly those affecting our neighbours to the south, are an important source of shocks to the Canadian economy. Our trading partners also provide a useful benchmark to gauge our own economic performance. How do we measure up?

In Canada, at the time of this writing, not all is well, but there are some reasons for cautious optimism. After a very deep recession in the early 1990s, economic growth was weak and uneven, but the prospects for the last few years of the decade look good. Inflation is low and shows no signs of heating up. Governments at all levels have made deep cuts but finally appear to have their budgets on a sustainable path. In the short term, job creation remains the biggest headache. Employment growth is improving, but the unemployment rate remains stubbornly high, particularly for young people.

In contrast, Americans are enjoying an enviable economic boom. Inflation in the United States is running a bit higher than in Canada, but it is low by postwar standards. Growth has been robust and the unemployment rate has dropped to levels not seen in 30 years. The long expansion has allowed attention to shift to long-run problems. In particular, the underlying growth rate of the U.S. economy appears to have declined since the 1970s. Low growth, combined with widening wage inequality, is leading to declining income for those at the bottom of the skill distribution.

The economic outlook in the world's other rich countries is slightly bleaker than in Canada. In Europe, the good news is that most countries are experiencing both output growth and low inflation. But many European countries have unemployment rates that are even higher than those in Canada and show even fewer signs of returning to low levels any time soon. The news from Japan is worse. After sustaining high growth for most of the postwar period, Japanese growth fell to very low levels in the 1990s. Growth in 1996 was surprisingly strong but it is unlikely to be sustained. A financial crisis in Asia is making policy makers all over the world nervous, but the risks are especially high for Japan. Getting the Japanese economy growing again is the major worry of Japanese macroeconomists and policy makers at this point.

In the rest of this chapter, we will look more closely at what is happening in these diverse parts of the world, placing current events in the larger context of these countries' economic performance since 1960. Read the rest of this chapter as you would read an article in a newspaper. Do not worry about the exact meaning of the words or about understanding all the arguments in detail. The purpose is to give you some background and to introduce you to the issues of macroeconomics; the words will be defined and the arguments articulated later in the book. Indeed, once you have read the book, you might come back to this chapter, see where you stand on the various issues, and evaluate how much progress you have made in your study of macroeconomics.

CANADA

When macroeconomists look at an economy, they focus first on three measures. The first is *aggregate output* and its rate of growth. The second is the *unemployment rate*, the proportion of workers in the economy who are not employed and are looking for a job. The third is the *inflation rate*, the rate at which the average price of goods in the economy is increasing over time.

By at least one of these measures, Canada (Figure 1-1) is doing well. Table 1-1 gives the numbers for the three measures. The first column gives the average since 1960; the next four columns give the numbers for 1994 to 1997. The numbers for 1997 are preliminary, as of December 1997.

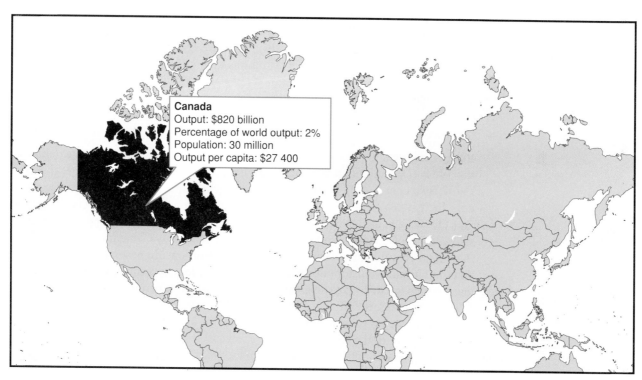

Canada
Output: $820 billion
Percentage of world output: 2%
Population: 30 million
Output per capita: $27 400

FIGURE 1-1
Canada, 1996

Source: International Monetary Fund, *International Financial Statistics*.

TABLE 1-1

GROWTH, UNEMPLOYMENT, AND INFLATION
IN CANADA, 1960–1997

	1960–1996	1994	1995	1996	1997 (PRELIMINARY)
Output growth rate	3.8	4.1	2.3	1.5	3.6
Unemployment rate	7.6	10.4	9.5	9.7	9.4
Inflation rate	4.7	0.7	1.5	1.3	1.1

Growth rate: Annual rate of GDP growth
Unemployment rate: Average over the year
Inflation rate: Annual rate of change of the GDP deflator

Source: OECD Economic Outlook, December 1997.

Canada started the 1990s with a recession—that is, a decline in aggregate output. Output growth turned positive in 1992 and has remained positive since, but the expansion has been weak. Except for 1994, output growth has been well below the postwar average. High interest rates in late 1994 and early 1995 were followed by very low growth in 1996, but growth appears to have jumped sharply in 1997 and the prospects for the next few years look bright.

Inflation is low and is expected to stay low. The average rate of increase of prices was only 1.3% in 1996, well below the average since 1960.

What keeps policy makers and macroeconomists awake at night is the unemployment rate (Figure 1-2). Canada's unemployment rate has been trending upwards since the 1970s, but even after taking this into account the performance of the last decade has been disappointing. In 1992, the unemployment rate for all workers in Canada peaked at 11.3%; although it has fallen, the decline has been excruciatingly slow and in January 1998 it stands just below 9.0%. As usual, the incidence of unemployment is unevenly spread across the economy. Job prospects are very good in Alberta, but Newfoundland has an unemployment rate more than twice the national average. Young workers aged 20–24 face an unemployment rate of over 14%.

Budget deficits. Getting deficits under control has dominated economic debate in the 1990s. The federal government had been running large deficits since the mid-1970s. By contrast, until 1990, provincial governments were able to more or less balance their books. But both federal and provincial governments (especially Ontario and Quebec) began the decade facing large and rapidly increasing deficits. Spiralling costs of financing this mountain of debt left many Canadians worried about the legacy that they were leaving to future generations. In response to these fears, and to a shift in public opinion towards a taste for less government, politicians at all levels began to cut both expenditures and transfers. The results have been dramatic. In February 1998, the Minister of Finance, Paul Martin, announced that the federal government would achieve a balanced budget for its 1997 fiscal year and he forecast a zero deficit until the end of the decade.

Government deficits in Canada are measured on both a public accounts basis (prepared by individual governments) and a national accounts basis (prepared by Statistics Canada). There are a number of differences, the most important being that loans from certain trust funds controlled by the federal government are not included in the national accounts estimates of the deficit. Basically, Statistics Canada

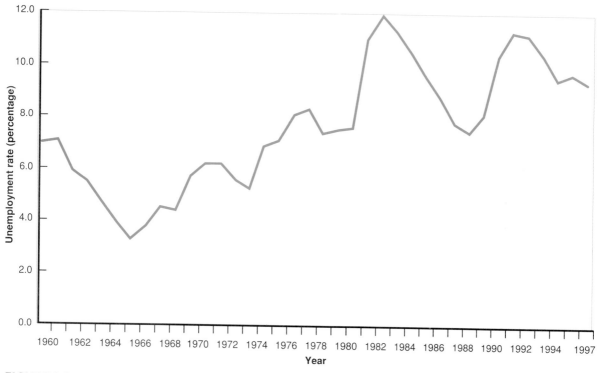

FIGURE 1-2

The Unemployment Rate in Canada, 1960–1997

The unemployment rate in Canada has been trending upwards over time.

Sources: Statistics Canada, CANSIM Series D44950 and *Historical Statistics of Canada.*

measures government expenditures when they are made, whereas the public accounts attempts to measure expenditure obligations (such as pensions for civil servants) when they are incurred. In recent years, the two measures of the deficit have diverged considerably. Statistics Canada is currently reviewing the way it constructs the national accounts estimate of the deficit with an eye to reducing the discrepancy. By the time you read this, it may be safe to ignore the difference, but at the present time the deficits measured on a national accounts basis are running several billions of dollars below their public accounts counterpart.[3]

In 1996, the federal government had a deficit of about $16 billion, measured on a national accounts basis. This amounted to about 2% of GDP. Figure 1-3 shows the evolution of the federal deficit since 1960. Although small at first, deficits grew sharply in the early 1980s. In 1985, the federal deficit ballooned to over 6.5% of GDP. Economic growth reduced the deficit, but it was still very high when the next recession hit a few years later, and it jumped to almost 5% of GDP in 1993. The Liberal government elected in 1993 set out deficit targets and pledged to reduce the public accounts deficit to 3% of GDP by 1996. Finance Minister Paul Martin comfortably hit his target and it appears that the federal government achieved a small surplus in 1997.

---

[3]*From the public accounts, the federal deficit is sometimes reported on a financial requirements basis—literally a measure of the amount of borrowing in the market that is done to manage the government's cash flow. This measure is used less often in Canada but it comes closest to the way that the U.S. federal deficit is reported.*

FIGURE 1-3

Canadian Government Deficits as a Proportion of Output since 1960

Large federal government deficits date back to the mid-1970s.

Source: Statistics Canada, CANSIM Series D10000, D10171, and D10193.

The sharp improvement in the finances of the federal government has opened up a new debate. Some argue that we should cut government spending more deeply. They point out that a string of large deficits has left Canadians with a large debt that makes us vulnerable to an increase in world interest rates and puts us in a poor position to handle a downturn in the economy. They also argue that individual taxpayers are better at spending their own money than government bureaucrats. They propose using the savings from a smaller government to pay down the national debt and to cut taxes. Others argue that fiscal restraint has been excessive. They believe that the cutbacks have hurt the economic recovery and have had detrimental long-term consequences. They also believe that restraint has gutted many important government programs and that we should seize the opportunity to restore funding in important areas like health care and education. The debate is heated, but everyone agrees that it is nice to finally have a choice to make.

THE UNITED STATES

With strong growth, unemployment well below average, and low inflation (Table 1-2), it would appear that U.S. macroeconomic policy makers have little to worry about. This is not quite the case. While business cycle issues still dominate Canadian policy discussion, the excellent current performance of the U.S. economy (Figure 1-4) has shifted the attention down south to long-run issues. Although they were

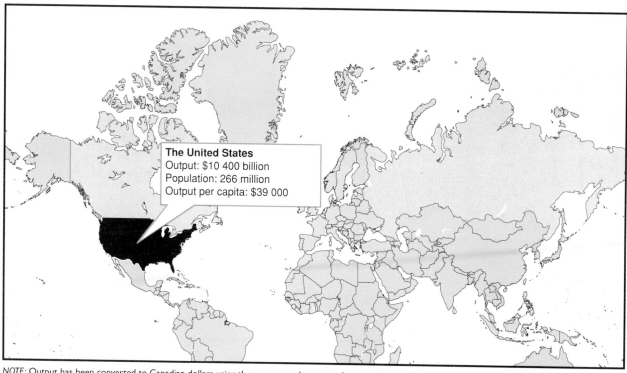

NOTE: Output has been converted to Canadian dollars using the average exchange rate between the U.S. dollar and the Canadian dollar in 1996.

FIGURE 1-4

The United States, 1996

Source: International Monetary Fund, *International Financial Statistics*.

....

TABLE 1-2

GROWTH, UNEMPLOYMENT, AND INFLATION IN THE UNITED STATES, 1960–1997

	1960–1996	1994	1995	1996	1997 (PRELIMINARY)
Output growth rate	3.0	3.5	2.0	2.8	3.8
Unemployment rate	6.1	6.1	5.6	5.4	5.0
Inflation rate	4.4	2.4	2.5	2.3	2.0

Growth rate: Annual rate of GDP growth
Unemployment rate: Average over the year
Inflation rate: Annual rate of change of the GDP deflator

Source: OECD Economic Outlook, December 1997.

never as bad as in Canada, budget deficits in the U.S. are being brought under control. A smaller debt load has given policy makers in the U.S. more room to manoeuver. The budget agreement struck between President Clinton and the Congress in 1997 aims at a slower elimination of the deficit but more tax reductions than Canadians can expect. With the deficit under control, the main issues of concern are a decrease in the country's underlying growth rate and increasing wage inequality.

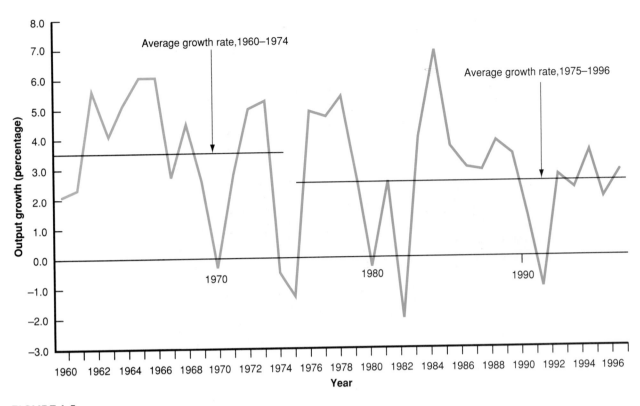

FIGURE 1-5

U.S. Output Growth since 1960

The underlying U.S. growth rate appears to have decreased since the mid-1970s.

Source: U.S. Department of Commerce, Bureau of Economic Analysis, National Income and Product Accounts.

The slowdown in growth. While growth rates vary from year to year, the evidence suggests that the average growth rate in the United States has decreased in the last 20 years. You can see this in Figure 1-5 which plots the annual rate of output growth since 1960 together with the average growth rate from 1960 to 1975 and from 1975 to 1996. The first horizontal line shows a growth rate of 3.5% for the period 1960–1974; the second horizontal line shows the lower rate of 2.6% for the period 1975–1996. While the range of annual growth rates is large—note the negative growth rates, or recessions, in 1974-1975, 1979, 1981, and 1991—the figure shows that the movements in growth rates have taken place around a lower average since the mid-1970s.

A decrease in the average growth rate of 0.9% per year—the difference between the average rate for 1960–1974 and the average rate for 1975-1996—may not strike you as very consequential. But it is. One way of thinking about it is this: If the average growth rate after 1975 had remained equal to the average from 1960 to 1974, U.S. output would be 22% higher today; income per capita (measured in Canadian dollars) would be $47 500 instead of its actual value of $39 000.

Why has growth slowed? The growth rates of most rich countries, including Canada, appear to be affected, so one should not look for explanations specific to the United States. Some economists argue that rich countries have lost their edge, that the research process is less productive than it used to be. Others point to low investment and low capital accumulation. Still others claim that the slowdown is largely a figment of data construction, that official measures of aggregate output underestimate the increase in the sophistication of new products and thus underestimate the growth rate of output. At this point all explanations are tentative, but the question clearly is one of the most important in macroeconomics.

Increasing wage inequality. Since the early 1980s, wage inequality has increased in the United States. Workers with fewer skills and less education have seen their wages decline relative to the average wage. Combined with low growth, this increase in wage inequality has led to an absolute wage decline for many types of workers. Since 1980, the average wage of workers who have not finished high school has *decreased* by roughly 1% per year. Although the Canadian experience has been different in many respects, we too have seen a sharp increase in wage inequality in recent decades, particularly among male workers.

Where does this increase in wage inequality come from? Most economists see two main causes. The first is international trade. Unskilled workers are increasingly competing with workers from countries with very low wages, and this competition is pulling down their wages. The second is the nature of technological progress. The argument here is that new technology requires increasingly skilled workers to operate it. Thus, the relative demand for skilled workers is steadily increasing, the relative demand for unskilled workers is steadily decreasing, and these two evolutions are reflected in the relative wages of skilled and unskilled workers. There is disagreement about the relative importance of these two causes, however. Some economists believe that international trade is the main culprit; others disagree. Identifying the causes of increasing wage inequality is one of the most active research topics in economics today.

THE EUROPEAN UNION

In 1957, six nations—Belgium, France, Germany, Italy, Luxembourg, and the Netherlands—decided to form a common European market. Since then, nine more —Austria, Denmark, Finland, Greece, Ireland, Portugal, Spain, Sweden, and the United Kingdom—have joined. This union is now known as the **European Union**, or **EU**. (Until a few years ago, the official name was the *European Community*, or *EC*. You are likely to encounter both names in your readings.) Not only has the number of members increased, but the ties among them have tightened. Together, they form a formidable power. Their combined output slightly exceeds that of the United States and accounts for about one-quarter of the world's output. As Figure 1-6 shows, many of them have a standard of living (measured by their output per capita) higher than that of Canada.[4]

Although the economic performance of the EU countries has been disappointing in the 1990s, the situation is slowly improving. Most countries are now growing again: EU growth was 1.6% in 1996 and is estimated to be 2.6% in 1997, still below the average of 3.2% since 1960.

While growth is positive, unemployment is still very high. In 1996, the average unemployment rate was 11.3%, and it appears to have decreased only slightly in 1997. The numbers for some countries are astounding: In Spain, for example, the unemployment rate stood at 23% in 1996.

---

[4]*In Chapter 22, we will discuss better measures for comparing standards of living across countries.*

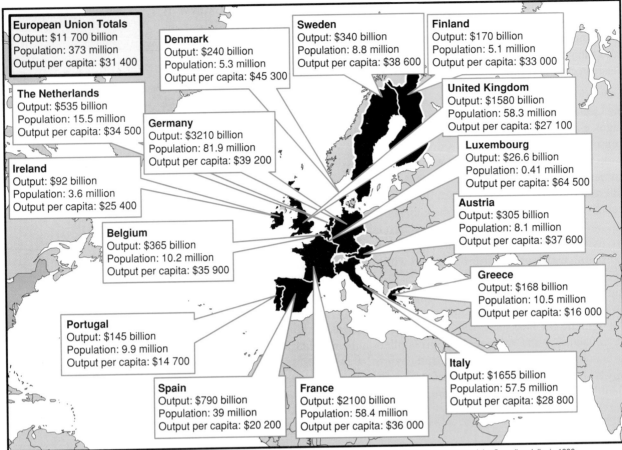

European Union Totals
Output: $11 700 billion
Population: 373 million
Output per capita: $31 400

Denmark
Output: $240 billion
Population: 5.3 million
Output per capita: $45 300

Sweden
Output: $340 billion
Population: 8.8 million
Output per capita: $38 600

Finland
Output: $170 billion
Population: 5.1 million
Output per capita: $33 000

The Netherlands
Output: $535 billion
Population: 15.5 million
Output per capita: $34 500

Germany
Output: $3210 billion
Population: 81.9 million
Output per capita: $39 200

United Kingdom
Output: $1580 billion
Population: 58.3 million
Output per capita: $27 100

Luxembourg
Output: $26.6 billion
Population: 0.41 million
Output per capita: $64 500

Ireland
Output: $92 billion
Population: 3.6 million
Output per capita: $25 400

Austria
Output: $305 billion
Population: 8.1 million
Output per capita: $37 600

Belgium
Output: $365 billion
Population: 10.2 million
Output per capita: $35 900

Greece
Output: $168 billion
Population: 10.5 million
Output per capita: $16 000

Portugal
Output: $145 billion
Population: 9.9 million
Output per capita: $14 700

Italy
Output: $1655 billion
Population: 57.5 million
Output per capita: $28 800

Spain
Output: $790 billion
Population: 39 million
Output per capita: $20 200

France
Output: $2100 billion
Population: 58.4 million
Output per capita: $36 000

NOTE: Output for each country is quoted in 1996 Canadian dollars, using the average exchange rate between that country's currency and the Canadian dollar in 1996.

FIGURE 1-6
The European Union, 1996

Source: International Monetary Fund, *International Financial Statistics.*

················
TABLE 1-3
GROWTH, UNEMPLOYMENT, AND INFLATION IN THE EUROPEAN
UNION, 1960–1997

	1960–1996	1994	1995	1996	1997 (PRELIMINARY)
Output growth rate	3.2	2.9	2.5	1.7	2.6
Unemployment rate	6.0	11.5	11.2	11.4	11.3
Inflation rate	6.2	2.6	2.9	2.4	1.8

Growth rate: Annual rate of GDP growth
Unemployment rate: Average over the year
Inflation rate: Annual rate of change of the GDP deflator

Source: OECD Economic Outlook, December 1997.

The good news concerns inflation. Just as in Canada, inflation is low in the European Union, running at an annual rate of 2.4% in 1996, much lower than the average of 6.2% since 1960.

The European Union is confronting two major macroeconomic issues. The first is, rather obviously, high unemployment. The second is the speed at which to achieve economic and monetary integration.

High unemployment. High unemployment is not a European tradition. Figure 1-7, which plots the evolution of unemployment rates in the European Union and the United States, shows how low the European unemployment rate was in the 1960s. Indeed, at the time, the talk in the United States was about the European "unemployment miracle," and U.S. macroeconomists went to Europe in the hope of discovering the secrets of that miracle. But by the late 1970s the miracle had vanished. And since the early 1980s, the unemployment rate in Europe has been higher than that in Canada and much higher than that in the United States.

There are many views as to why European unemployment is high and what should be done to decrease it. At one end of the spectrum are those who point to high levels of worker protection as the main cause. In many European countries it is difficult and costly for firms to lay off workers. Although such protection may

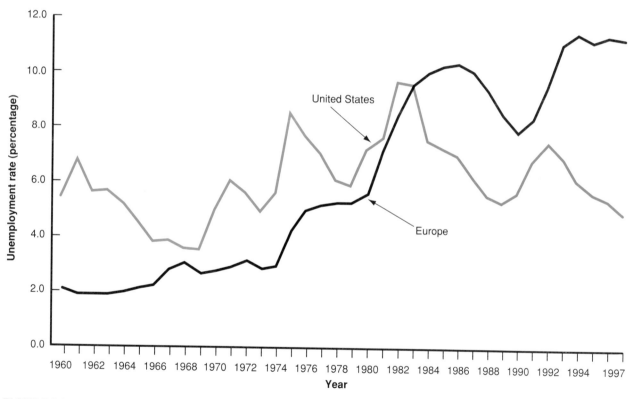

FIGURE 1-7

Unemployment Rates: Europe versus the United States, since 1960

The European unemployment rate has gone from being much lower to being much higher than that of the United States.

Sources: U.S. Department of Labor, Bureau of Labor Statistics; *OECD Economic Outlook,* various issues.

have been warranted when economies were growing fast and international competition was much weaker, they argue, in today's world the EU countries cannot afford to maintain the level of worker protection they used to offer. They point to the Canadian experience, where labour market institutions and the unemployment rate fall between those of the United States and the EU, as supporting their view. At the other end of the spectrum are those who argue that labour costs and worker protection in the European Union are for the most part reasonable, and that high unemployment comes instead from primarily misguided macroeconomic policies. They argue that an expansion of demand—through lower interest rates, for example—could decrease unemployment rates to levels similar to those of the United States.

Most macroeconomists find themselves somewhere between these two extreme views, and believe that both changes in the labour market and some demand expansion are needed. The debate is far from settled, but its outcome is crucial to Europe's future.

European integration. The European Union has eliminated barriers to trade in goods among its member countries (a feat that has so far eluded Canada's provinces). Barriers to the movement of people and capital are also being steadily reduced. One of the next steps is the creation of a common currency to replace national currencies. The current plan is to have a common currency in place by the end of the decade, but the debate about the pros and cons of such a move is intense.

Those in favour argue that it is of enormous symbolic importance for Europe to have a common money and that different currencies are a major hindrance to trade.[5] Those against the common currency point out that a common currency implies a common monetary policy and that individual member countries will have to give up their national monetary policies. Giving up such an important policy instrument may have large macroeconomic costs, larger than any of the benefits that a common currency will bring. At this point, the most likely outcome appears to be that eleven of the fifteen EU countries—including Germany, France, and Italy—will adopt a common currency, and a few others — such as the United Kingdom and Denmark — will wait for a while.

JAPAN

In 1960, Japan (Figure 1-8) would not have been included in our tour of rich countries. Its output per capita was much lower than that of Canada. But things are very different today. As Table 1-4 shows, Japan's average annual growth rate since 1960 has been 6% compared to 3.8% in Canada. Japan's GDP is almost eight times that of Canada's, and its output per capita is almost twice Canadian output per capita.

Although it is the richest, Japan is not the only East Asian country that has been growing fast. Following close behind are the so-called *four tigers:* Singapore, Hong Kong, South Korea, and Taiwan. Behind them, poorer but also growing fast, are three economies of Southeast Asia: Malaysia, Thailand, and Indonesia. China, which is still very poor, is now growing at close to 10% a year. However, a financial crisis in Asia that began in Thailand in June 1997 threatens a sharp reduction in growth for the entire region that may spread to the rest of the world.

Japan's economic performance since 1960 is extremely impressive, but its economic performance in the 1990s has been dismal. Output growth has been far below

---

[5]*Quebec separatists appear to share this view: They propose that a sovereign Quebec will continue to use the Canadian dollar as its national currency.*

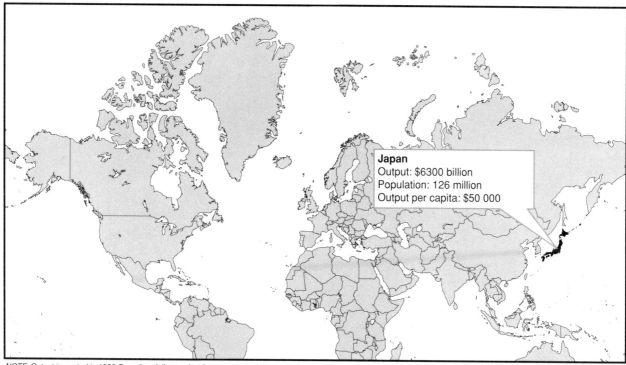

NOTE: Output is quoted in 1996 Canadian dollars, using the average exchange rate between the yen and the Canadian dollar during 1996.

FIGURE 1-8

Japan, 1996

Source: International Monetary Fund, *International Financial Statistics*.

· · · · · · · · · · · · · · ·

TABLE 1-4

GROWTH, UNEMPLOYMENT, AND INFLATION IN JAPAN, 1960–1997

	1960–1996	1994	1995	1996	1997 (PRELIMINARY)
Output growth rate	5.9	0.6	1.4	3.5	0.5
Unemployment rate	2.0	2.9	3.1	3.4	3.4
Inflation rate	4.4	0.2	−0.6	0.0	1.1

Growth rate: Annual rate of GDP growth
Unemployment rate: Average over the year
Inflation rate: Annual rate of change of the GDP deflator

Source: *OECD Economic Outlook*, December 1997.

the 6.0% average growth rate since 1960. The unemployment rate—which traditionally has been very low in Japan— has increased to record highs. The unemployment rate for 1996 was 3.4%, a number that Canadians can only dream of, but a postwar record high for Japan. The good news, if any, is about inflation. Inflation in Japan today is very close to zero.

Japan's current major macroeconomic issue is clearly whether and when it will return to the high growth rates of the past. Another issue, which has led over the years to tensions with its trading partners, is the size of its trade surplus.

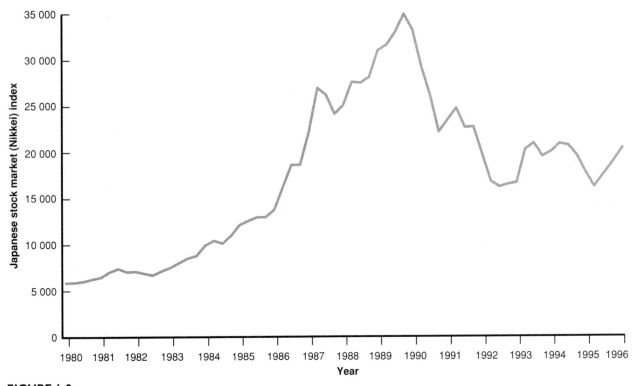

FIGURE 1-9

The Japanese Stock Market (Nikkei) Index since 1980

The large increase in the index in the 1980s was followed by a very sharp decrease in the early 1990s.

Source: U.S. Department of Commerce, Bureau of Economic Analysis.

The recent slump. What caused the slump in Japanese activity in the first half of the 1990s? Many economists attribute it in large part to the movement in asset prices since the mid-1980s.

Figure 1-9 shows the behaviour of the Nikkei index, an index of stock prices in the Japanese stock market, since 1980. From 1985 to 1989, the Nikkei increased from about 13 000 to 33 000; in other words, the average price of a share in the Japanese stock market nearly tripled in price in less than five years. This sharp increase was followed in the early 1990s by an equally sharp decrease: In less than three years, from 1990 to 1992, the Nikkei went back down from 33 000 to 17 000. In late 1997, the index stood at 17 500, about 45% below its 1989 value.

Many economists interpret the Nikkei's rise and fall as a speculative bubble, an excessive increase in stock prices, followed by a crash and a return to reality. They point to similar movements in other asset prices, such as land prices. They argue that the bubble's result was a boom in demand and output in the late 1980s, followed by a sharp decrease in demand in the early 1990s.[6] The question now is: How can Japan return to higher growth rates? In an effort to increase demand, the Japanese Central Bank has decreased interest rates to very low levels. The interest rate in Japan has

---

[6]*Crashes in asset values are not uncommon. In the summer of 1997, a foreign exchange and stock market crisis swept across Southeast Asia and engulfed Thailand, Indonesia, the Phillipines, Singapore, and Malaysia. Stock market values there fell by as much as 40% in a few months.*

been hovering around 2%. However, a series of events in 1997 have made a quick return to high growth unlikely. An increase in Japan's value-added tax, similar to Canada's GST, in April 1997 from 2% to 5% appears to have put the brakes on economic growth. Japanese citizens scurried to purchase big-ticket items before the tax came into place, and consumer spending plummeted after April. The financial crisis in Asia led the government to change course and stimulate the economy once again through large deficits. But Japan is not expected to grow much in 1998. Japanese financial institutions, already vulnerable because of the bad loans used to finance the real estate speculation of the 1980s, have been further weakened by the Asian crisis and several have already declared themselves insolvent. Those remaining are reluctant to make loans until their balance sheets improve.

The trade surplus. Japan exports far more than it imports. The trade surplus for 1996 was equal to 3% of output. Japan's trading partners have argued that the trade surplus reflects restrictions on imports of foreign goods in Japan. Such trade barriers, they argue, prevent foreign firms from entering Japanese markets, and should be dismantled. But the Japanese government argues that the trade surplus has little to do with trade barriers and that Japan should indeed have a trade surplus. One way of thinking about the Japanese trade surplus is that Japan, as a country, is saving. Just as someone who spends less than he earns is saving, a country that buys less from abroad is also saving. There may be good reasons for Japan to want to save. Japan will soon have one of the largest proportions of retirees to workers in the world; it may indeed be wise to save in anticipation of such times.

1-3 LOOKING AHEAD

This concludes our tour. There are many other countries and regions worth looking at, from the nations of Eastern Europe (which are shifting from central planning to a market system), to Latin America (which alternates between high and low inflation), to Africa (which seems unable to grow). But there is a limit to how much should be presented in this first chapter. Think of the issues to which you have been exposed already:

■ Why do growth rates differ so much from country to country, and over long periods of time? Why has Japan grown so much faster than North America and Europe over the last 40 years? Why has growth slowed down since the mid-1970s in most rich countries?

■ Why was Japan able to maintain such low unemployment for so many years? Why has Europe gone from low to high unemployment over the last two decades?

■ Are budget deficits too costly? If they are, how fast should they be reduced? Can a trade surplus be equally undesirable? Should Japan aim at reducing its trade surplus?

■ Why is inflation so much lower in the 1990s than in earlier decades? What is so bad about inflation? Should countries aim for zero inflation?

The purpose of this book is to help you think about these questions and their answers.

Throughout this chapter we have examined data from around the world. But where do these data come from? For example, where does one find the number for inflation in Germany over the last two decades? Forty years ago, the answer would have been to learn German, find a library with German publications, find the page where inflation numbers were given, write them down, and plot them on a nice clean sheet of paper. Today, improvements in the collection of data and the development of computers and electronic databases make the task much easier.

International organizations now collect data from many countries. For the richest countries, the most useful source is the **Organization for Economic Cooperation and Development (OECD)**, which is based in Paris. Most of the world's rich countries belong to the OECD. The complete list of 29 countries now consists of Austria, Australia, Belgium, Canada, the Czech Republic, Denmark, Finland, France, Germany, Greece, Hungary, Iceland, Ireland, Italy, Japan, Korea, Luxembourg, Mexico, New Zealand, the Netherlands, Norway, Poland, Portugal, Spain, Sweden, Switzerland, Turkey, the United Kingdom, and the United States. Together, these countries account for about 70% of world output. *OECD Economic Outlook,* which is published twice yearly, gives basic data on output growth, inflation, unemployment, and other major aggregate variables for member countries, as well as an assessment of their recent macroeconomic performance. The data, going back to 1960, are available on diskette; they are on many macroeconomists' hard disks.

For those countries that are not members of the OECD, information is available from other international organizations. The main world economic organization, a sort of world economic club, is **the International Monetary Fund (IMF)**. It now has 182 members. The IMF publishes monthly *International Financial Statistics (IFS)*, which contains basic macroeconomic data for all IMF members. Twice a year, it also publishes *World Economic Outlook*, an assessment of macroeconomic developments in the various parts of the world. Although their language is sometimes stilted, both *World Economic Outlook* and *OECD Economic Outlook* are precious sources of information.

Because these publications sometimes do not contain sufficient detail, one may need to turn to country publications. Major countries now have remarkably clear statistical publications, often with an English translation available. Statistics Canada has a user-friendly web site at <www.statcan.ca>. It is evolving, but it currently gives access to recent values for a number of major macroeconomic time series. Historical data can be obtained from the CANSIM database or from publications such as *Canadian Economic Observer*.

A longer list of data sources, both for Canada and for the rest of the world, as well as how to access some of them through the Internet, is given in Appendix 1 at the end of the book.

SUMMARY

■ Macroeconomics studies aggregate economic variables, such as production for the economy as a whole (aggregate output) or the average price of all goods (the aggregate price level). In contrast, microeconomics studies production and prices in specific markets.

■ One of the reasons that macroeconomists disagree is because they assign different weights to different objectives. For example, some care more about unemployment, others about inflation; their recommendations will differ accordingly.

■ Macroeconomists also sometimes disagree about the interpretation of facts. One reason is that they cannot do controlled experiments. If they want to find out, for example, how changes in the money supply affect aggregate activity, they cannot make the world stop while they ask the central bank to change the money supply. Looking at the same episode, one economist can see a strong effect of money on activity, while another sees a weaker effect.

■ The availability and the study of more and more episodes narrow differences of opinion over time. For example, there is large agreement about the effects of money on economic activity. This book focusses on the common core of macroeconomics, on what the majority of macroeconomists agree on and how they use it to study the world.

It is harder to summarize our economic tour of the world. You may want to remember the following main points:

■ In the early 1990s, most rich countries (among them Canada, the United States, the members of the European Union, and Japan) experienced a recession—that is, a decline in output. Most are now experiencing an expansion, with output growth rates at or above their historical average. This growth has been accompanied by low inflation, lower than the average since 1960.

■ All is not perfect, however. Unemployment in Canada and Europe is still very high. And in most rich countries, average output growth appears to be lower than it was before the mid-1970s.

KEY TERMS

....

■ macroeconomics, 1
■ microeconomics, 1
■ models, 2
■ controlled experiments, 3

■ European Union (EU), 11
■ Organization for Economic Cooperation and Development (OECD), 18
■ International Monetary Fund (IMF), 18

QUESTIONS AND PROBLEMS

....

1. Using the tables and graphs in this chapter, label each of the following statements "true" or "false." Explain briefly.
   a. The Canadian federal government has run a deficit every year since 1960.
   b. The Japanese trade surplus has grown larger every year since 1980.
   c. Since 1960, the Canadian unemployment rate has consistently exceeded the unemployment rate of the European Union.
   d. From 1994 to 1996, U.S. output growth consistently exceeded output growth in Europe.
   e. Recently, inflation in both Canada and the United States has been below its historical average (1960–1996), but inflation in Europe and Japan has been higher than its historical average.
   f. Europe has made significant progress bringing down unemployment over the last three decades.

*2. If Canadian output had grown from 1960 to 1996 at the same rate as Japan's output grew, how much greater would Canadian output have been in 1996 compared to its actual level in 1996? (Express your answer as a proportion of the actual level of Canadian output in 1996.)

3. In 1994, output in the European Union increased, and yet the unemployment rate rose. Since more output ordinarily requires more workers, how can this be?

*4. In 1996, the output of the People's Republic of China was about $1.1 trillion, and was growing at an annual rate of about 10%. If China's economy continues to grow at 10% per year, while U.S. output grows at 2% per year, when will China have a GDP larger than that of the United States?

5. "The budget deficit will shrink as a proportion of output only if government spending decreases or tax revenues increase." Is this statement correct? Explain.

6. Update Table 1-3 in the text, using the most recent issue of *OECD Economic Outlook*. Using your update:
   a. Were the 1997 preliminary estimates for output, unemployment, and inflation reasonably accurate?
   b. Has output growth increased or decreased since 1996? Do all of the EU countries have an output growth rate that is close to the average, or is there wide variation?

7. Discuss how new technology has increased the relative demand for skilled workers over the last 15 years in each of the following industries:
   a. Retail food
   b. Hollywood films
   c. The post office and express mail services
   d. Banking
   e. Temporary secretarial assistance

──────────

Problems marked with an asterisk are more challenging.

FURTHER READING

To learn about current economic events and issues, you can do no better than read *The Economist,* a weekly magazine published in England. The trademark of articles in *The Economist* is that they are well-informed, well-written, witty, and opinionated.

A Tour of the Book

Terms such as *output, unemployment, inflation, budget deficits,* and *trade deficits* have become part of the modern vocabulary. They appear daily in newspapers and on the evening news, and when we used them in Chapter 1 you knew roughly what we were talking about. We now need to define them more precisely. This is what we do in the first part of this chapter. Having done so, we can then give you a tour of the book, with a description of the major sights.

2-1 Aggregate output

Economists in the nineteenth century and during the Great Depression had no aggregate measure of activity on which to rely. They had to put together bits and pieces of information, such as the production of pig iron or sales at department stores, to infer what was happening to the economy as a whole.

It was not until the end of World War II that **national income and expenditure accounts** were put together in major countries.[1] These measures of aggregate output have been published on a regular basis in Canada since 1947. (You will find measures of aggregate output for earlier times, but these have been constructed retrospectively.)

Like any accounting system, the national income accounts define the concepts they use, indicate how to construct corresponding measures, and show how these measures relate to one another. One needs only to look at statistics from countries

---

[1] *Putting them together was a gigantic intellectual achievement. Two economists have been given the Nobel Prize for their contributions to the development of the national income accounts: Simon Kuznets, from Harvard University, in 1971, and Richard Stone, from Cambridge University, in 1984.*

that have not yet developed such systems to realize how crucial such precision and consistency are. Without them, numbers that should add up do not add up; one does not quite know what variables the published numbers correspond to (an exercise not unlike trying to balance someone else's chequebook). We shall not burden you with the details of national income accounting here. But because you will occasionally need to know the definition of a variable and how variables relate to each other, Appendix 2 at the end of the book gives you the basic accounting framework used in Canada (and, with minor variations, in most other countries). You will find it useful whenever you want to look at economic data on your own.

GDP, VALUE ADDED, AND INCOME

The measure of aggregate output in the national income accounts is **gross domestic product,** or **GDP** for short.[2] There are three ways of thinking about an economy's GDP. Let's examine each in turn.

(1) GDP is the value of the final goods and services produced in the economy during a given period. The important word is *final.* An example will help here. Suppose that the economy is composed of just two firms. Firm 1 produces steel, employing workers and using machines. It then sells the steel for $100 to firm 2, which produces cars. Firm 1 pays its workers $75 and keeps what remains, $25, as profit. Firm 2 buys the steel and uses that steel, together with workers and machines, to produce cars. Revenues from car sales are $210. Of the $210, $100 goes to pay for steel and $65 goes to workers in the firm, leaving $45 in profit. We can summarize all this information in a table:

Steel Company		
Revenues from sales		$100
Expenses (wages)		75
Profit		25
Car Company		
Revenues from sales		$210
Expenses:		170
Wages	$ 65	
Steel purchases	100	
Profit		45

What is GDP in this economy? Is it the sum of the values of all production in the economy—the sum of $100 from the production of steel and $210 from the production of cars, thus $310? Or is it the value of the production of final goods, here cars, thus $210?

A moment of thought suggests that the right answer is $210. Why? Because steel is an **intermediate good** that is used in the production of the final good, cars, and thus should not be counted in GDP—which is the value of *final* output. We can look at this example in another way. Suppose the two firms merge, so that the sale of steel takes place inside the new firm and is no longer recorded. All one would see is one firm sell-

---

[2] *You will also encounter the term* **gross national product,** *or* **GNP.** *There is a subtle difference between "domestic" and "national," and thus between GDP and GNP, which we examine in Chapter 11 (see also Appendix 2). For the moment, you can ignore the difference between the two.*

ing cars for $210, paying workers $75 + $65 = $140, and making $25 + $45 = $70 in profits. The $210 measure would remain unchanged, as it indeed should.

This example suggests constructing GDP by recording and adding up the production of final goods, and this is indeed roughly the way that actual GDP numbers are put together. But the example also suggests an alternative way of thinking about and constructing GDP:

(2) GDP is the sum of value added in the economy during a given period.   The term **value added** means exactly what it suggests. The value a firm adds in the production process is equal to the value of its production minus the value of the intermediate goods it uses in production. In our two-firm example, value added by the steel company is $100 because the firm does not use intermediate goods. Value added by the car company is equal to revenues minus the value of intermediate goods, $210 − $100 = $110. Total value added in the economy, or GDP, is thus equal to $100 + $110 = $210. Note that aggregate value added would remain the same if the steel and car firms merged and became one firm.

This definition gives us a second way of thinking about GDP. Put together, the two definitions imply that the value of final goods and services—the first definition of GDP—can always be thought of as the sum of the value added by all firms along the chain of production of those final goods—the second definition of GDP.

In reality, GDP is made up of much more than just steel and cars. What does the composition of Canadian GDP actually look like? The answer from the national income accounts for 1961 and 1996 is summarized in Table 2-1. The striking fact in the table is the large and increasing share of services. Today, more than half of GDP—66% to be precise—is accounted for by services, from health care to phone services, up from 59% in 1960. The production of goods accounts for only 34 percent.[3]

(3) GDP is the sum of incomes in the economy during a given period. A third way of looking at GDP is from the income side. The difference between the value of a firm's production and the value of intermediate goods must go to one of three places: to workers as labour income, to firms as profit, or to the government in the form of indirect taxes, such as sales taxes collected from the proceeds of final sales.

In our example, we ignore depreciation and indirect taxes. Of the $100 of value added by the steel manufacturer, $75 goes to labour income and the remaining $25 to

....
TABLE 2-1
THE COMPOSITION OF
CANADIAN GDP, 1961 AND 1996

	1961	1996
Goods	41%	34%
Services	59%	66%

Source: Statistics Canada, CANSIM Matrix 4670.

----

[3]*The steadily increasing proportion of services has prompted Alan Greenspan, the Chairman of the Federal Reserve System (the U.S. central bank) to remark that the physical weight of GDP is steadily declining, and to muse about the implications of that fact for international trade and macroeconomic policy. As you go through the book, you may find it fun to wonder about the same issues.*

THE COMPOSITION OF GDP BY TYPE
OF INCOME, 1961 AND 1996

	1961	1996
Labour income	51%	52%
Capital income	25%	22%
Depreciation	12%	13%
Indirect taxes	12%	13%

Source: Statistics Canada, CANSIM Matrix 6547.

profit. Of the $110 of value added by the car manufacturer, $65 goes to labour income and $45 to profit. Thus, for the economy as a whole, value added is $210, of which $140 goes to labour income and $70 goes to profit.

In our example, labour income accounts for 67% of GDP, profit for 33%, and depreciation and indirect taxes for 0%. Table 2-2 shows the breakdown of value added among the different types of income in Canada in 1961 and 1996. The table shows that except for the presence of depreciation and indirect taxes (which our example does not have), the proportions we have been using in our example are roughly those of the Canadian economy. Labour income accounts for 52% of GDP. Capital income (which includes not only profit but also rental income and interest payments paid by firms), accounts for 22%. Depreciation (also called **capital cost allowance**) is the value of the capital stock that is "used up" in the production process. It amounts to 13% of GDP. Indirect taxes account for the remaining 13%.

Nominal and Real GDP

If you look at Statistics Canada data in CANSIM,[4] you will find that Canadian GDP was equal to $820 billion in 1996, compared with $41 billion in 1961. Is it really the case that Canadian output in 1996 was 20 times as high as in 1961? The answer is no, and this leads us to the distinction between nominal GDP and real GDP.

Nominal GDP is simply the sum of the quantities of final goods produced times their current price. A warning is in order here. People often use the word *nominal* to denote small amounts. Economists use *nominal* for variables expressed in dollars (or in units of the currency of the relevant country). And economists surely do not refer to small amounts. The numbers you will see in this book are typically expressed in billions of dollars.

Nominal GDP increases over time for two reasons. The first is that the production of most goods increases over time. The second is that the dollar price of most goods also increases over time. We produce more and more cars each year, and their dollar price increases each year as well. If our intention is to measure production and its change over time, we need to eliminate the effect of increasing prices. For this purpose, economists focus on *real* rather than *nominal* GDP.

To construct **real GDP,** they first choose a base year. They then construct real GDP in any year as the sum of quantities produced times their price in the base year.

---

[4]*For a description of CANSIM, see Appendix 1.*

An example will help here. Suppose that an economy produces two goods, potatoes and cars. In year 0—which we shall take as the base year—it produces 100 000 pounds of potatoes and sells them at $1 a pound, and 10 cars that sell for $10 000 a car. One year later, in year 1, it produces and sells 100 000 pounds of potatoes at a price of $1.20 a pound, and 11 cars at $10 000 a car. Nominal GDP in year 0 is thus equal to $200 000, and nominal GDP in year 1 equal to $230 000. This information is summarized in Table 2-3.

The increase in nominal GDP from year 0 to year 1 is equal to $30 000/$200 000 = 15%. But what is the increase in *real* GDP? Let's take year 0 as the base year—that is, let's add quantities in both years 0 and 1 using year 0 prices for potatoes and cars. Because we take year 0 as the base year, real GDP is equal to nominal GDP in year 0; real and nominal GDP are always equal in the base year. In year 1, real GDP is constructed by using year 1 quantities and year 0 prices, so that it is equal to $(100\ 000 \times \$1) + (11 \times \$10\ 000) = \$210\ 000$. The increase in real GDP is thus equal to $10 000/$200 000, or 5%.

Instead of using year 0 as the base year, we could have used year 1, or indeed any other year. The choice of the base year will typically affect the measure of real GDP growth. For example, if we had used year 1 as the base year, real GDP in year 0 would be equal to $[(100\ 000 \times \$1.20) + (10 \times \$10\ 000)] = \$220\ 000$. By construction, real and nominal GDP would be the same in year 1, both equal to $230 000. The increase in real GDP would be equal to $10 000/$220 000, thus 4.5%. It would thus be smaller than the increase in real GDP we obtained using year 0 as the base year.

In most countries the practice has been to use a base year and change it infrequently, say every five years or so. For example, in Canada the base year used from 1992 to 1997 was 1986. That is, measures of real GDP published in 1997 both for 1997 and for earlier years were all constructed using 1986 prices. In 1998, national income accounts will shift to 1992 as a base year; measures of real GDP for all past years will be recalculated using 1992 prices.[5]

GDP numbers change over time for a variety of reasons. For example, they are revised regularly as more and better data become available. In addition, accounting

....

TABLE 2-3

NOMINAL GDP IN YEAR 0 AND IN YEAR 1

	Year 0					
	Quantity	×	$ Price	=	$ Value	
Potatoes	100 000		1		100 000	
Cars	10		10 000		100 000	
	Nominal GDP				**$200 000**	
	Year 1					
	Quantity	×	$ Price	=	$ Value	
Potatoes	100 000		1.20		120 000	
Cars	11		10 000		110 000	
Nominal GDP					**$230 000**	

[5] *One way around this problem is to use what's known as a* **chained index**. *The United States moved to computing real GDP in this way in December 1995. Although Statistics Canada also computes a chained index, the method is not as prominent in Canada.*

conventions change. In December 1997, Statistics Canada plans to move from the 1968 SNA (System of National Accounts) recommended by the United Nations to the 1993 SNA.

At the time this book is being written, the 1992 base year numbers are unavailable. Thus, the numbers for real GDP and its components presented in this book are those computed using 1986 as the base year.

Figure 2-1 plots the evolution of both nominal and real GDP in Canada since 1960. Note that the two are the same in 1986, the base year. Real GDP in 1996 stood at 3.8 times its level of 1960—a considerable increase, but clearly much less than the twentyfold increase in nominal GDP. The difference between the two is due to price increases over the period.

The terms *nominal GDP* and *real GDP* each have many synonyms, and you are likely to encounter them in your readings:

■ Nominal GDP is also called **dollar GDP** or **GDP in current dollars.**

■ Real GDP is also called **GDP in terms of goods, GDP in constant dollars, GDP adjusted for inflation,** or **GDP in 1986 dollars**—if the base year is 1986.

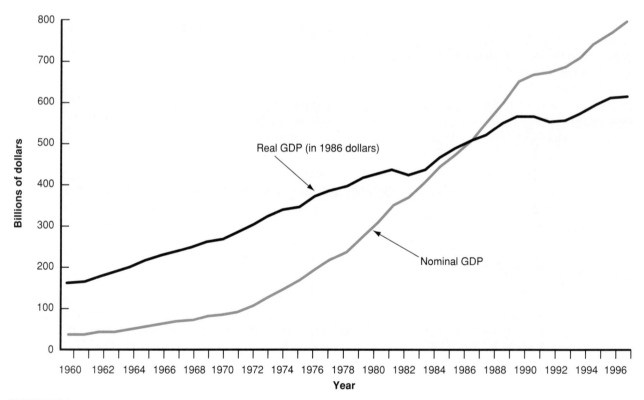

FIGURE 2-1
Nominal and Real Canadian GDP, 1960–1996
From 1960 to 1996, Canadian nominal GDP increased by a factor of 20. Real GDP increased by a factor of 3.7.

Source: Statistics Canada, CANSIM Series D10000 and D10373.

This concludes our discussion of the main macroeconomic variable, GDP. In the chapters that follow, unless we indicate otherwise, GDP will refer to real GDP and Y_t will denote real GDP in year t. Nominal GDP, and variables measured in current dollars in general, will be denoted by a dollar sign in front of the term—for example, $\$Y_t$ for nominal GDP.

Similarly, **GDP growth** in year t will refer to the growth rate of real GDP in year t; GDP growth equals $(Y_t - Y_{t-1})/Y_{t-1}$. Periods of positive GDP growth are called **expansions.** Periods of negative GDP growth are called **recessions.** To avoid calling a single quarter of negative growth a recession, macroeconomists usually use the word only if the economy experiences at least two consecutive quarters of negative growth. The Canadian recession of 1990–1991, for example, was characterized by four consecutive quarters of negative growth, the last three quarters of 1990 and the first quarter of 1991.

REAL GDP, TECHNOLOGICAL PROGRESS, AND THE PRICE OF COMPUTERS

Computing real GDP from nominal GDP may seem relatively easy, as indeed it would be if all goods were like potatoes. All we would need to do to get real GDP in a given year would be to multiply the quantity of potatoes sold in that year by the price of potatoes in the base year.

Things become more complicated when we deal with new goods—goods that are produced today but were not produced in the base year—or when we deal with goods that change from year to year, in reality the large majority of goods.

One of the most difficult cases is that of computers. In 1987, the average price of a personal computer was $3500. In 1997, the average price was roughly the same. Should we assume that a computer in 1997 is the same beast as a computer produced ten years earlier? This would clearly be absurd: For roughly the same price, one is clearly buying much more computing in 1997 than one was in 1987. The difficult question, however, is: How much more? Does a 1997 computer provide twice, 10 times, or 100 times the computing services of a 1987 computer? How should we take account of improvements in internal speed, in the size of the RAM or of the hard disk, the fact that computers can be used as fax machines, that they weigh less, and so on?

Economists adjust for these improvements by looking at how the market values computers with different characteristics in a given year. An example will help. Suppose that the evidence shows that the market is willing to pay 10% more for a computer with a speed of 200 rather than 166 megahertz. Suppose that all new computers this year have a speed of 200 megahertz compared with a speed of 166 megahertz for all new computers last year. Suppose also that the price of new computers is the same as last year. Then, economists in charge of computing the adjusted price of computers will assume that new computers are in fact 10% cheaper than last year.

This approach, which treats goods as providing a collection of characteristics—here speed, memory, and so on—each with an implicit price, is called **hedonic pricing** (*hedone* means pleasure in Greek). It is used to estimate changes in the price of complex and fast-changing goods, such as automobiles and computers. Using this approach, the U.S. Department of Commerce estimates that computers have indeed provided about 10% more services each year since 1987. Equivalently, it estimates that while the dollar price of computers has remained roughly constant, the adjusted price of computers has fallen at an average rate of 10% per year since 1987.

2-2 THE OTHER MAJOR MACROECONOMIC VARIABLES

GDP is the main macroeconomic variable, but it is not the only one. Others, from unemployment to inflation to trade or budget deficits, tell us about important aspects of the economy. In this section we define these variables and briefly discuss why we care about them.

THE UNEMPLOYMENT RATE

The **labour force** is defined as the sum of those employed and those unemployed:

$$L = N + U$$
$$\text{labour force} = \text{employed} + \text{unemployed}$$

The **unemployment rate** in turn is defined as the ratio of the number of unemployed to the labour force:

$$u = \frac{U}{L}$$

$$\text{unemployment rate} = \frac{\text{unemployed}}{\text{labour force}}$$

What determines whether a worker is defined as unemployed or not? Until 1945 in Canada, and until more recently in most other countries, the number of people registered at unemployment offices was the only available source of data on unemployment, and only those workers who were registered in unemployment offices were counted as unemployed. This system led to a poor measure of unemployment. How many of the truly unemployed actually registered varied both across countries and across time. Those who had no incentive to register—for example, those who had exhausted their unemployment benefits—were unlikely to take the time to come to the unemployment office and thus were not counted. Countries with less generous benefit systems were likely to have fewer unemployed registering, and thus smaller measured unemployment rates.

Today, most countries rely on large surveys of households to compute the unemployment rate. In Canada, this survey is called the **Labour Force Survey (LFS),** and it relies on interviews of about 50 000 households every month. The survey classifies somebody as employed if he or she has a job at the time of the interview; it classifies somebody as unemployed if he or she does not have a job and has been looking for work in the last four weeks. Most other countries use a similar concept of unemployment, although the definition of what "looking for work" means exactly varies across countries. In Canada in 1996, estimates based on the LFS survey showed that on average over the year, 13.68 million people were employed and 1.47 million people were unemployed. The unemployment rate was thus 1.47/(13.68 + 1.47) or 9.7%.

Note an important characteristic of the definition of the unemployment rate. Only those looking for work are counted as unemployed; those who are not looking are counted as **not in the labour force.** But when unemployment is high, many of those without jobs simply give up looking for work and thus are no longer counted as unemployed. These people are known as **discouraged workers.** To take an extreme case, if all workers without a job gave up looking, the unemployment rate would equal zero, and the unemployment rate would be a very poor indicator of what is happening in the labour market. This extreme case does not hold, but a milder version is present: Typically, high unemployment is associated with many workers dropping out of the labour force. Equivalently, a high unemployment rate is typically associated with a low **participation rate,** defined as the ratio of the labour force to the total population of working age. In Canada, we see dramatic differences in the unemployment and participation rates across regions. In 1996, Alberta's robust economy enjoyed an unemployment rate of 7% and a participation rate of 72%. Newfoundland suffered an unemployment rate of 19.4% and a participation rate of 52%.

Why do macroeconomists care about unemployment? There are two main reasons. First, the unemployment rate tells us something about whether an economy is operating above or below its normal level. Second, unemployment has important social consequences. Let's look at both reasons in turn.

Unemployment and activity. In most countries there is a reliable relation between GDP growth and the change in the unemployment rate. This relation is known as **Okun's law,** after the economist Arthur Okun, who first identified and interpreted it in the 1960s. The relation between these two variables in Canada since 1960 is plotted in Figure 2-2, which shows the change in the unemployment rate on the vertical axis and the rate of GDP growth on the horizontal axis. Each point in the figure shows the growth rate and the change in the unemployment rate for a given year. Figures such as Figure 2-2, which plot one variable against another over time, are called **scatter diagrams.**

The figure shows that high output growth is typically associated with a decrease in the unemployment rate, and conversely that low growth is associated with an increase in the unemployment rate. This makes good sense: High output growth leads to high employment growth, as firms have to hire more workers to produce more. High employment growth in turn leads to a decrease in unemployment.

The relation has a simple implication. If the current unemployment rate is too high—what constitutes *too high,* or *too low,* or *about right* will be the topic of many chapters later; for the moment we shall remain purposely vague—it will take a period of faster growth to reduce it. If, instead, the unemployment rate is about right, then output should grow at the rate that is consistent with an unchanged unemployment rate. Thus, the unemployment rate provides macroeconomists with a signal of where the economy stands and what growth rate is desirable. If unemployment is too high, higher output growth is desirable; if too low, slower growth is needed.

Social implications of unemployment. Macroeconomists also care about unemployment because of its direct effects on the welfare of the unemployed. Although unemployment benefits are much more generous today than they were during the Great Depression, unemployment is still often associated with substantial financial

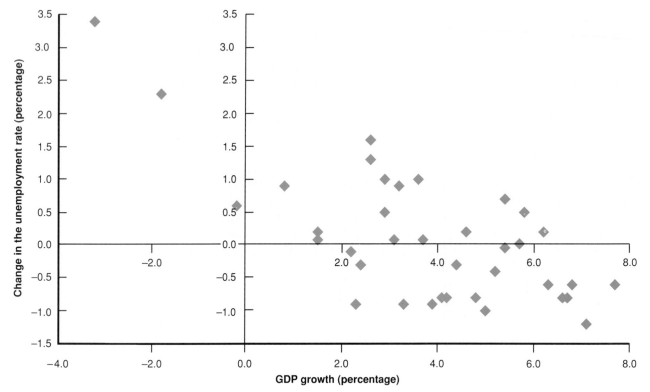

FIGURE 2-2

Change in the Canadian Unemployment Rate versus GDP Growth, 1960–1996

High output growth is typically associated with a decrease in the unemployment rate.

Sources: Statistics Canada, CANSIM Series D20463 and D44950 and *Historical Statistics of Canada.*

DOES SPAIN REALLY HAVE A 23% UNEMPLOYMENT RATE?

In 1995, the official unemployment rate in Spain was an astonishing 23%. Yet there were no riots and Spain looked nothing like Canada or the United States during the Great Depression; there were few homeless, and most cities looked prosperous. Can one really believe that nearly one-fourth of the Spanish labour force was looking for work?

To answer this question, we must examine how the Spanish unemployment number is put together. Much as in Canada, it comes from a large survey of 60 000 households. People are classified as unemployed if they indicate that they are not working, but that they are seeking work.

However, can one be sure that people are telling the truth? No. Although there is no obvious incentive to lie—answers to the survey are confidential and are not used to determine whether people are eligible for unemployment benefits—those who are working in the underground economy may prefer to play it safe and report that they are unemployed instead.

The size of the **underground economy**—that part of economic activity that is not measured in official statistics, either because the activity is illegal or (more important here) because firms and workers would rather not report it and thus not pay taxes—is indeed an old issue in Spain. Because of that, we know more about it than in many other countries. In 1985 the Spanish government tried to find out more and organized a detailed survey of 60 000 individuals. To try to elicit the truth from those interviewed, the questionnaire asked interviewees for an extremely precise account of the use of their time, making it difficult to lie undetected. The answers were indeed interesting. The underground economy in Spain—defined as the number of people working without declaring it to the social security administration—accounted for 10% to 15% of employment. But it was composed mostly of people who already had a job and were taking a second or a third job. The best estimate from the survey was that only about 15% of the unemployed were in fact working. This implied that the unemployment rate, which at the time was officially equal to 21%, was in fact closer to 18%, still a very high number. In short, the Spanish underground economy is significant, but it does not explain the high unemployment rate.

Do the unemployed survive because unemployment benefits are unusually generous in Spain? The answer is no. Except for unusually generous unemployment systems in two regions, Andalusia and Extremadura—which have even higher unemployment than the rest of the country—unemployment benefits are very much in line with those in other OECD countries. Benefits are typically equal to 70% of the wage for the first six months, 60% after. They are paid for a period of 4 to 24 months, depending on how long people have worked before. The 30% of unemployed who have been unemployed for more than two years do not, at this point, receive unemployment benefits.

So how do the unemployed survive? The answer lies in the Spanish family structure. The unemployment rate is highest among the young: It exceeds 40% for those between 16 and 19, and is around 40% for those between 20 and 24. The young typically stay at home until their late 20s, and have indeed increasingly done so as unemployment has increased. Looking at households rather than at individuals, the proportion of households where nobody was employed in 1994 was less than 10%; the proportion of households that received neither wage income nor unemployment benefits was around 3%. In short, it is the family structure that has allowed many of the unemployed to survive.

and psychological suffering. How much suffering depends on the nature of unemployment. (See the Global Macro box entitled "Does Spain Really Have a 23% Unemployment Rate?") One image of the unemployed is that of a stagnant pool of people remaining unemployed for long periods of time. As we shall see later in the book, this image is wrong most of the time. Typically, the flows in and out of unemployment are large. Each month, many people become unemployed, and many of the unemployed find jobs. But some groups (often the young, some ethnic minorities, and the unskilled) suffer disproportionately, remaining chronically unemployed and being most vulnerable to unemployment when the unemployment rate increases.

THE INFLATION RATE

Inflation is a sustained rise in the general level of prices, in the **price level.** The **inflation rate** is the rate at which the price level increases.

The practical issue is how to define this price level. Macroeconomists typically look at two measures of the price level, at two *price indexes:* the GDP deflator and the consumer price index.

The GDP deflator.   Suppose that nominal GDP, $\$Y_t$, increases, but that real GDP, Y_t, remains unchanged. Then, clearly the increase in nominal GDP must be due to the increase in prices. This motivates the definition of the **GDP deflator,** which gives the average price of the final goods produced in the economy. The GDP deflator in year t, P_t, is defined as the ratio of nominal GDP to real GDP in year t:

$$P_t = \frac{\$Y_t}{Y_t}$$

To see what this definition implies, let's return to our earlier car–potato economy example. In year 0, the base year, nominal GDP and real GDP are equal by construction. Thus, the GDP deflator in year 0 is equal to 1 by construction. This is a point worth emphasizing: The GDP deflator is what is called an **index number.** Its level is chosen arbitrarily—here it is equal to 1 in the base year—and thus has no economic interpretation.[6] But its rate of change is well defined and has an economic interpretation.

In year 1, the GDP deflator is equal to the ratio of nominal to real GDP in year 1, \$230 000/\$210 000, or about 1.10. If we define the inflation rate as the rate of change of the GDP deflator, $(P_t - P_{t-1})/P_{t-1}$, the inflation rate is equal to $(1.10 - 1.00)/1.00 = 10$ percent.

Note that the inflation rate is a weighted average of the rate of price increase for each of the two goods. Potato inflation—the rate of change of the price of potatoes—is equal to 20%; car inflation is equal to 0%. The average of the two is 10%. This result is more general than our example: The inflation rate obtained by using the GDP deflator can be thought of as a weighted average of the rate of change of individual prices.

One of the main advantages of the GDP deflator is that it is closely related to our measure of real GDP. Indeed, reorganizing the previous equation gives

$$\$Y_t = P_t Y_t$$

Nominal GDP is equal to real GDP times the GDP deflator.

The consumer price index. The GDP deflator gives the average price of the goods included in GDP, thus of the final goods *produced* in the economy. However, consumers care about the average price of the goods they *consume.* The two prices need not be the same, because the set of goods produced in the economy is not the same as the set of goods bought by consumers. This is true for two reasons. Some of the goods in GDP are sold not to consumers but to firms (machine tools, for example), to the government, or to foreigners. And some of the goods bought by consumers are not produced at home, but rather imported from abroad.

Thus, to measure the average price of consumption, or equivalently, the **cost of living,** macroeconomists look at an alternative index, the **consumer price index (CPI).**[7] The CPI has been in existence since the early 1900s, and comes out monthly. In contrast, GDP numbers and the GDP deflator come out only quarterly.[8]

The CPI gives the cost in dollars of a given list of goods and services over time. The list, which is based on a detailed study of consumer spending, originally attempted to replicate the consumption basket of a typical urban consumer, but, since 1995, it reflects the spending patterns of all households in Canada. It is revised ap-

---

[6]*Published indexes are often set equal to 100 rather than 1 in the base year. The number 100 is short for 100%, which in decimal terms is equal to 1.*

[7]*The CPI should not be confused with the **industrial product price index (IPPI),** which is an index of prices of domestically produced goods in manufacturing.*

[8]*Statistics Canada also publishes an estimate of GDP at factor cost each month, but it does not calculate a monthly estimate of the deflator. See Appendix 2 for the details.*

proximately every four or five years. Each month, Statistics Canada employees visit stores to find out what has happened to the prices of the goods on the list. Over 60 000 price quotations are collected from supermarkets, department stores, garages, and so on from all across Canada. These prices are then used to construct the consumer price index.

Like the GDP deflator, the CPI is an index. It is set equal to 100 (see footnote 5) for a given year called the *time base*. In Canada, from 1990 to 1998, the time base was 1986, so the average value of the CPI in 1986 was set to 100.[9] The CPI is what economists call a *Laspeyres price index;* it is a weighted average of the ratio of current to base year prices. Statistics Canada updates the weights every four or five years; as of late 1997, they are based on 1992 expenditure patterns. In 1996, the CPI stood at 136; this means the cost of purchasing a typical basket of goods and services was 36% higher in 1996 than it was in 1986.

You may wonder how the rate of inflation differs depending on whether one uses the GDP deflator or the CPI. The answer is given in Figure 2-3, which plots both inflation rates since 1960 for Canada. The figure yields two conclusions:

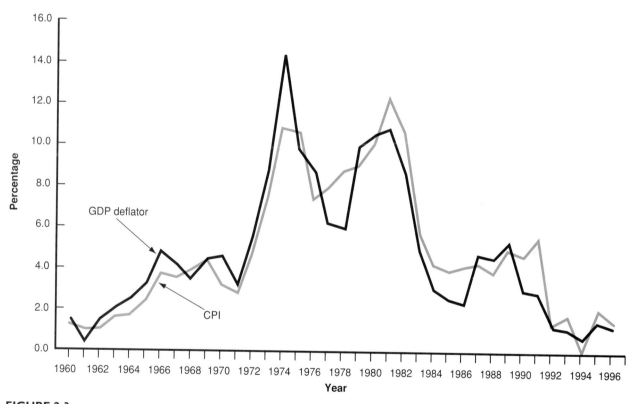

FIGURE 2-3

Canadian Inflation Rate, Using the CPI and the GDP Deflator, 1960–1996

The inflation rates computed using either the CPI or the GDP deflator are largely similar.

Source: Statistics Canada, CANSIM Series D20556 and P700000.

---

[9]*Beginning with the January 1998 CPI, Statistics Canada changed the time base to 1992 and shifted to 1996 expenditure weights.*

■ The CPI and the GDP deflator move together most of the time. In most years, the two inflation rates differ by less than 1%.

■ But there are clear exceptions. The reason is not hard to find. Recall that the GDP deflator is the price of goods *produced* in Canada, and the CPI is the price of goods *consumed* in Canada. Thus, when the price of imported goods increases relative to the price of goods produced in Canada, the CPI increases faster than the GDP deflator.

In what follows, we do not distinguish between the two indexes unless we want to focus precisely on their difference. Thus, we simply talk about *the price level,* and denote it by P_t, often without indicating whether we have the CPI or the GDP deflator in mind.

Inflation and unemployment.   Is there a relation between inflation and either output or unemployment, or does inflation have a life of its own? There is indeed a relation, but it is far from mechanical; it varies across time and country.

The relation between unemployment and inflation in Canada is shown in Figure 2-4. The *change* in the inflation rate (using the CPI)—that is, the inflation rate

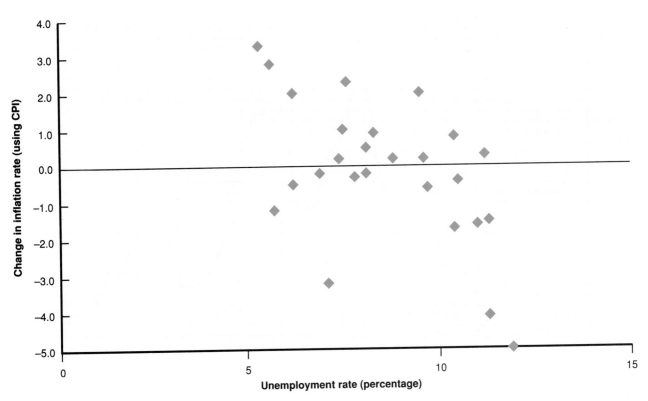

FIGURE 2-4

Change in the Canadian Inflation Rate versus the Canadian Unemployment Rate, 1970–1996

When the unemployment rate is low, inflation tends to increase. When the unemployment rate is high, inflation tends to decrease.

Sources: Statistics Canada, CANSIM Series D20463, D44950, D20556, and P700000 and *Historical Statistics of Canada.*

this year minus the inflation rate last year—is plotted on the vertical axis. The unemployment rate is plotted on the horizontal axis. The figure gives the combinations of unemployment rates and changes in inflation rates for each year since 1970.

The figure shows a clear negative relation between the unemployment rate and the change in inflation. When the unemployment rate is low, inflation tends to increase. When the unemployment rate is high, inflation tends to decrease. This negative relation is called the *Phillips relation*, and the curve that fits the set of points best is called the **Phillips curve,** named for the economist who first documented the relation between unemployment and inflation.[10] Where this relation comes from, why it changes through time and place, and what it implies will be the focus of many later chapters.

Why do economists care about inflation?   If higher inflation meant just a faster proportional increase in all prices and wages—a phenomenon known as *pure inflation*—inflation would be only a minor inconvenience. Relative prices would not be affected by inflation. Take, for example, the *real wage*—the wage measured in terms of goods rather than in dollars—received by workers. In an economy with 10% inflation, prices would increase by 10% a year but so would wages. Real wages would thus remain the same.[11] Inflation would not be entirely irrelevant; people would have to keep track of the increase in prices and wages in making decisions, but this would be a small burden, hardly justifying making control of the inflation rate one of the major goals of macroeconomic policy.

So why do economists care about inflation? Precisely because there is no such thing as pure inflation. During periods of inflation, not all prices and wages go up proportionately. Thus inflation affects income distribution. For example, retirees in many countries receive payments that do not keep up with the price level, and thus lose relative to other groups when inflation is high. Canada Pension Plan benefits automatically go up with the CPI, protecting retirees from inflation. But during the very high inflation that took place in Russia in the early 1990s, retirement pensions did not keep up with inflation, and many retirees are literally starving.

Inflation also leads to distortions. Some prices, which are fixed by law or by regulation, lag behind the others. Inflation thus leads to changes in relative prices. Taxation interacts with inflation to create more distortions. If tax brackets are not adjusted for inflation, for example, people move into higher and higher tax brackets as their nominal income increases, even if their real income remains the same. Variations in relative prices lead also to more uncertainty, making it harder for firms to make decisions about the future, such as investment decisions.

In short, economists see high inflation as affecting income distribution, and creating both distortions and uncertainty. How important these problems are, and whether they justify trying to achieve and maintain zero inflation, are much-debated questions, and we take them up later in the book.

---

[10] *As we shall see in Chapter 17, the nature of the Phillips curve has changed since Phillips first documented it in 1958, but the name is still used.*

[11] *This abstracts from changes in real wages that would occur even if there were no inflation. A more accurate statement is that, under pure inflation, the change in real wages would be independent of the inflation rate.*

BUDGET AND TRADE DEFICITS

Finally, let's look at **budget deficits** (the excess of government expenditures over government revenues) and **trade deficits** (the excess of imports from the rest of the world over exports to the rest of the world).

Why do economists care about budget and trade deficits? The answer is: for the same reason that you care about your own deficit—although you do not usually call it that—namely, the excess of your spending over your income. When your spending exceeds your income and you are accumulating debt, you know that sooner or later you will have to repay what you've borrowed, and thus decrease your spending then. It may still make sense for you to borrow (for example, to finance your college education) if you know that your income will be higher in the future. But whether borrowing is a good idea must be assessed case by case. Very much the same reasoning applies to governments and to countries.

■ A government that runs a deficit accumulates debt over time. Higher debt means higher interest payments on the debt. To finance the higher interest payments, the government must either increase taxes or lower spending elsewhere. It may still make sense for a government to borrow if expenditures are unusually high, such as during a war or after an earthquake. Otherwise, deficits may be unwise.

■ A country that runs a trade deficit is buying more from abroad than it is selling abroad, and is thus accumulating debt vis-à-vis the rest of the world. Again, this may make sense: Borrowing to finance investment, which in turn leads to higher output later, can easily be justified. But running a trade deficit to finance a consumption binge may be as unwise for a country as it is for an individual.

Thus, the reason economists look at and worry about budget deficits and trade deficits is that they may signal a need for painful adjustments in the future.

2-3 A ROAD MAP

Having defined the main concepts, let's now turn to the central question of macro-economics. What determines the level of aggregate output?

A reading of daily newspapers suggests one answer: Movements in output come from movements in the demand for goods. You have all read news stories that begin: "Production and sales of automobiles were higher last month, apparently because of a surge in consumer confidence, which drove consumers to showrooms in record numbers." Such explanations point out the role of demand in determining aggregate output, as well as factors ranging from consumer confidence to taxes and interest rates.

But a longer view, based on the steady increase in real GDP since 1960 shown in Figure 2-1 or on the large differences in output per capita across countries, suggests a different answer. In that view it is the supply side that determines output: How much an economy produces depends on the size of its labour force, how many machines the country has, and how sophisticated its technology is.

Both views are correct, but each applies over a different time horizon. Movements in output from one year to the next, or over a few years at most—what macroeconomists call the **short run**—depend primarily on movements in demand. Decreases in demand, which can arise from changes in consumer confidence or any other source, can lead to a decrease in output (a recession). But, over long periods of time, say decades or more—what macroeconomists call the **long run**—output is determined by supply factors, from the labour force to the capital stock to the state of technology. For periods of time in between—which, not surprisingly, macroeconomists call the **medium run**—both demand and supply factors play a role.

This way of thinking about movements in output underlies this book's organization. We focus first on short-run fluctuations, and thus on the role of demand. We then move on to the medium run and the long run. The book is organized as follows (the organization is also summarized visually in Figure 2-5):

■ Chapters 3 and 4 look at the *goods markets*[12] in a closed economy and develop a preliminary model of economic activity. The model is simple, but it is useful. It provides a user-friendly way of introducing many of the basic tools of macroeconomics. However, the model is appropriate only for the short run: It assumes that firms are willing to supply any quantity at a given price. The focus is thus on the demand for goods as the determinant of output.

■ Chapters 5 and 6 introduce *financial markets,* and then look at equilibrium in both goods and financial markets. The resulting framework is known as the basic *IS-LM* model. Developed in the late 1930s, the basic *IS-LM* model provides a simple way of thinking about the joint determination of output and interest rates in the short run, and it remains a basic building block of macroeconomics. It also allows for a first pass at the role of *fiscal policy* and *monetary policy* in affecting output and interest rates.

■ The basic *IS-LM* model ignores *expectations*. But expectations play an essential role in macroeconomics. Nearly all the economic decisions people and firms make—whether to buy bonds or stocks, whether or not to buy a machine—depend on their expectations of future profits, of future interest rates, and so on. Fiscal and monetary policy affect activity not only through their direct effects, but also through their effect on expectations. Chapters 7 to 10 focus on the role of expectations, and their implications for fiscal and monetary policy.

■ The basic *IS-LM* model takes the economy as *closed,* ignoring its interactions with the other economies of the world. Canadians have never been able to ignore the rest of the world. But the increase in trade, in both goods and financial assets, has made it clear that all countries are more and more interdependent. The nature of this interdependence and the implications for fiscal and monetary policy are the topics of Chapters 11 to 14.

■ The basic *IS-LM* model assumes that firms supply any quantity at a given price. But in fact changes in output, which lead also to changes in unemployment, lead to changes in wages and prices over time. The determination of wages in the *labour markets* and their relation to unemployment are the topics

---

[12]*From now on,* goods *will stand for both goods and services.*

FIGURE 2-5
The Organization of the Book

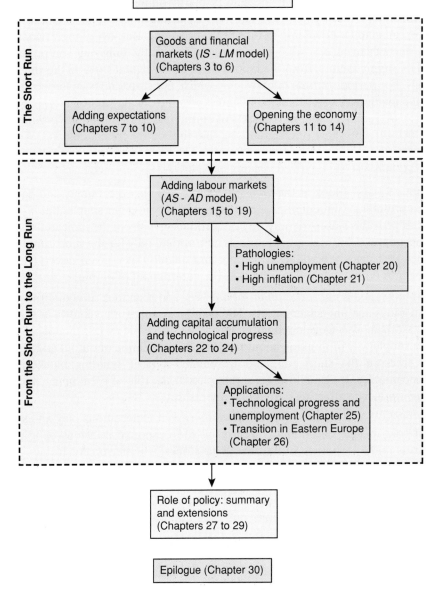

of Chapter 15. Chapters 16 to 19 then look at movements in output, unemployment, and prices, taking into account equilibrium in goods, financial, and labour markets. The model we develop in these chapters is known as the *aggregate supply–aggregate demand* (*AS-AD*) model of output. Chapters 16 to 19 show how it can be used to think about the short, medium, and long runs and apply it to a wide range of examples drawn from what has happened in the world over the last two decades.

■ Sometimes the word *fluctuations* does not accurately capture what is happening in the economy. Sometimes something goes very wrong. Inflation reaches extremely high rates. Or unemployment remains very high for very long, as during the Great Depression. (As we saw in Chapter 1, high and persistent unemployment is also the case in Canada and many European countries today.) These *pathologies* are the topics of Chapters 20 and 21.

■ The next five chapters focus on the medium run and the long run, on what makes economies rich or poor, grow fast or stagnate. Two factors become essential here: capital accumulation and technological progress. These are the focus of Chapters 22 to 24. Chapters 25 and 26 then apply what we have learned to two current issues. The first is the relation between technological progress and unemployment. The second is the transition in Eastern Europe from central planning to a market system.

■ With the framework fully in place, the next three chapters return to the role of macroeconomic policy. Chapter 27 focusses on general issues of policy, such as whether macroeconomists know enough to use policy at all, and whether policy makers can be trusted to do what is right. Building on previous chapters, Chapters 28 and 29 assess the role of monetary and fiscal policy, respectively.

■ From the outside, macroeconomics sometimes looks like a field divided between schools—Keynesians, monetarists, new classicals, supply siders, and so on—hurling arguments at each other. The actual process of research is more orderly and more productive than this image suggests. In the final chapter, Chapter 30, we look at the recent history of macroeconomics and how we have come to believe what we believe today. We identify the main differences between macroeconomists, and state the set of propositions that defines the core of macroeconomics today.

SUMMARY

....

■ One can think of GDP, the measure of a country's aggregate activity, in three equivalent ways: (1) GDP is the value of the final goods and services produced in the economy during a given period; (2) GDP is the sum of value added in the economy during a given period; and (3) GDP is the sum of incomes in the economy during a given period.

■ Nominal GDP is equal to the sum of the quantities of final goods produced times their current prices. Real GDP is the sum of quantities produced times their prices in the base year.

■ The labour force is defined as the sum of those employed and those unemployed. The unemployment rate is defined as the ratio of the number of unemployed to the labour force. Somebody is classified as unemployed if he or she does not have a job and has been looking for work in the last four weeks.

■ Okun's law is an empirical relation between GDP growth and the change in the unemployment rate. It shows that high output growth is associated with a decrease in the unemployment rate and, conversely, that low growth is associated with an increase in the unemployment rate.

■ Inflation is a rise in the general level of prices, a rise in the price level. The inflation rate is the rate at which the price level increases. Macroeconomists look at two measures of the price level. The first is the GDP deflator, which gives the average price of goods produced in the economy. The second is the consumer price index (CPI), which gives the average price of goods consumed in the economy.

■ The Phillips curve is the empirical relation between the unemployment rate and the change in the inflation rate. This relation has changed over time and also varies across countries. In Canada today, it takes the following form: When the unemployment rate is low, inflation tends to increase; when the unemployment rate is high, inflation tends to decrease.

■ Economists see high inflation as creating changes in income distribution, distortions, and uncertainty.

■ Governments that spend more than their revenues run budget deficits. Countries that import more than they export run trade deficits. Macroeconomists look at and worry about budget deficits and trade deficits because they may signal the need for painful adjustments later.

It would not make sense to summarize the road map here. But you should remember the basic principle behind the book's organization:

■ Movements in output over short periods of time—the short run—depend primarily on movements in demand. Over long periods of time—the long run—movements in output depend instead on supply factors, from the labour force, to the capital stock, to the state of technology. For periods of time in between—the medium run—both demand and supply factors play a role.

KEY TERMS

- national income and expenditure accounts, 21
- gross domestic product (GDP), 22
- gross national product (GNP), 22
- intermediate good, 22
- value added, 23
- nominal GDP, dollar GDP, or GDP in current dollars, 24, 26
- real GDP, GDP in terms of goods, GDP in constant dollars, GDP adjusted for inflation, or GDP in 1986 dollars, 24, 26
- GDP growth, 27
- expansions, 27
- recessions, 27
- hedonic pricing, 27
- labour force, 28
- unemployment rate, 28
- Labour Force Survey (LFS), 28
- not in the labour force, 29

- discouraged workers, 29
- participation rate, 29
- Okun's law, 29
- scatter diagram, 29
- inflation, 31
- price level, 31
- inflation rate, 31
- GDP deflator, 31
- index number, 32
- cost of living, 32
- consumer price index (CPI), 32
- industrial product price index (IPPI), 32
- Phillips curve, 35
- budget deficit, 36
- trade deficit, 36
- short, long, and medium run, 37

QUESTIONS AND PROBLEMS

1. During a given year, the following activities occur:
   (i) A silver mining company pays its workers $75 000 to mine 25 kilograms of silver, which it sells to a jewellery manufacturer for $100 000.
   (ii) The jewellery manufacturer pays its workers $50 000 to make silver necklaces, which it sells directly to households for $400 000.
   a. Using the "production of final goods" approach, what is GDP?
   b. What is the value added at each stage of production? Using the value-added approach, what is GDP?
   c. What are the total wages and profit earned from this activity? Using the income approach, what is GDP?

2. An economy produces three goods: books, bread, and beans. Production and prices in 1998 and 1999 are as follows:

	1998		1999	
	Quantity	Price	Quantity	Price
Books	100	$10.00	110	$10.00
Bread (loaves)	200	1.00	200	1.50
Beans (kilograms)	500	0.50	450	1.00

a. What is nominal GDP in 1998?
b. What is nominal GDP in 1999?
c. Using 1998 as the base year, what is real GDP in 1998 and 1999? By what percentage has real GDP changed between 1998 and 1999?
d. Using 1999 as the base year, what is real GDP in 1998 and 1999? By what percentage has real GDP changed between 1998 and 1999?

e. "The growth rate we obtain for real GDP depends on which base year's prices we use to measure real GDP." True or false?

3. Using the data in problem 2, and 1998 as the base year, calculate:
   a. The GDP deflator in 1998 and 1999.
   b. The rate of inflation over this period.

4. Suppose that, in a given month in Canada, there are 16 million working-age people. Of these, only 14 million have jobs. Of the remainder, 2 million are looking for work, 1.5 million have given up looking for work, and 0.5 million do not want to work.
   a. What is the labour force?
   b. What is the labour-force participation rate?
   c. What is the official unemployment rate?
   d. If all discouraged workers were counted as unemployed, what would be the unemployment rate?

5. Suppose that you are asked to compute a hedonic price index for each of the following goods:

   (i) automobiles
   (ii) video cameras
   (iii) computer printers
   (iv) cable television
   For each good:
   a. Which characteristics would you separate out for implicit pricing?
   b. How would hedonic pricing affect our estimate of the good's change in price over the past decade?

6. Look up the value of "real gross domestic product" on the Statistics Canada website.
   a. Do the changes over the most recent four quarters suggest that the economy was in a recession? An expansion? Neither? Explain briefly.
   b. For each of the most recent two years, compute the percentage of total GDP consisting of consumption, investment, exports and imports.

FURTHER READING

If you want to know more about the many economic indicators that are regularly reported on the news—from the help-wanted index to the retail sales index—a good reference is:

> John Grant, *A Handbook of Economic Indicators,* (Toronto: University of Toronto Press, 1992).

For more about the LFS or the CPI, see:

> Statistics Canada, *Guide to the Labour Force Survey*, January 1997.
> Statistics Canada, *Your Guide to the Consumer Price Index*, No. 62-557-XPB, December 1996.

Both publications can be downloaded from the Statistics Canada web site.

3

THE GOODS MARKET

Central to economists' thinking about year-to-year movements in economic activity is the interaction among *aggregate production, income,* and *demand.* Changes in the demand for goods lead to changes in production. Changes in production lead to changes in income, and, in turn, to changes in the demand for goods. The cartoon on the next page makes the point. A more formal way of capturing this interaction is with a simple diagram, Figure 3-1.

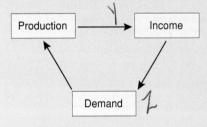

FIGURE 3-1
Production, Income, and the Demand for Goods

The purpose of this chapter and the next is to look at this interaction and its implications.

3-1 THE COMPOSITION OF GDP

Firms' purchases of machines are likely to depend on quite different factors than people's purchases of food, or the federal government's purchases of new airplanes. Thus, a good starting point is to look at aggregate production (GDP) from the point of view of the different goods being produced and the different types of buyers for these goods. The decomposition of GDP typically used by macroeconomists is given in Table 3-1. (A more detailed version, with more formal definitions, is given in Appendix 2.)

Cartoon by Tom Toles, *The Buffalo News,* October 1991; by permission.

■ The first component of GDP is **consumption,** which we shall denote by **C.** These are the goods and services purchased by consumers, ranging from food to airline tickets, to vacations, to new cars, and so on. Consumption is by far the largest component of GDP, accounting for 58% of GDP in 1996.

■ The second component is **investment (I)**, sometimes called **fixed investment** to distinguish it from inventory investment, described on page 45. Investment is the sum of two components. The first, **nonresidential investment,** is the

....

TABLE 3-1

THE COMPOSITION OF CANADIAN GDP, 1996

		BILLIONS OF DOLLARS*		PERCENTAGE OF GDP	
	GDP (Y)	820		100	
1	Consumption (C)	478		58	
2	Investment (I)	122		15	
	Nonresidential		82		10
	Residential		40		5
3	Government spending (G)	188		23	
4	Net exports	32		4	
	Exports (X)		321		39
	Imports (Q)		−289		−35
5	Inventory investment (I_S)	1		0	

*Detail does not add to total GDP because of rounding.

Source: Statistics Canada, *CANSIM Matrix 6548.*

purchase of new plants or new machines—from turbines to computers—by firms. The second, **residential investment,** is the purchase of new houses or apartments by people. The two types of investment and the decisions behind them have more in common than might first appear. Firms buy machines or plants to get more output in the future. People buy houses or apartments to get *housing services* in the future. This is the justification for lumping both under the same heading, "investment." Together, the two components of investment accounted for 15% of GDP in 1996.

Note that economists use the word *investment* in a narrower way than does the person on the street or the financial press. In common parlance, "investment" refers to the purchase of any asset, such as gold or shares of General Motors. Economists reserve the term to refer to the purchase of *new capital goods,* such as machines, buildings, or houses. When referring to the purchase of financial assets, economists use the term *financial investment*.

■ The third component is **government spending** on goods and services (G). These are the goods and services purchased by the federal, provincial, and local governments. The goods range from airplanes to office equipment. The services include spending on health and education, and the services provided by government employees. In effect, the national income accounts treat the government as buying the services provided by government employees, and then providing these services to the public.

Note that G does not include **government transfers,** such as Employment Insurance (E.I.) payments or the Canada Pension Plan (C.P.P.) benefits, nor interest payments on the government debt. Although these are clearly government expenditures, they are not purchases of goods and services. Thus, the number for government spending on goods and services in Table 3-1, about 23% of GDP, is smaller than the number for total government spending including transfers and interest payments, which was equal to roughly 45% of GDP in 1996.

■ The sum of lines 1 to 3 in the table gives us the purchases of goods and services by Canadian consumers, Canadian firms, and the Canadian governments. To get to the total purchases of Canadian goods and services, we must take two more steps.

First, we must exclude **imports** (Q), the purchases of foreign goods and services by Canadian consumers, Canadian firms, and the Canadian governments. Second, we must add **exports** (X), the purchases of Canadian goods and services by foreigners. The difference between exports and imports is called **net exports,** or the **trade balance.** If exports exceed imports, a country is said to run a **trade surplus.** A trade surplus means that the trade balance is positive, a **trade deficit** that the trade balance is negative. In 1996, exports accounted for 39% of GDP. Imports were equal to 35% of GDP, so Canada was running a trade surplus of 4% of GDP.

■ The sum of lines 1 to 4 in the table gives the purchases of Canadian goods and services in 1996. To get to Canadian production in 1996, we need to take one further step. Some of the goods produced in a given year may not be sold in that year, but sold in later years. And some of the goods sold in a given year may have been produced in an earlier year. The difference between production and sales is called **inventory investment** and is denoted I_S (the subscript S stands for **stocks,** another term for inventories). If production exceeds sales, inventories of goods increase: Inventory investment is positive. If production is less than sales, inventories decrease: Inventory investment is negative. Inventory investment is typically small, positive in some years, negative in others. In 1996, inventory investment was little more than 0.1% of GDP.

With the help of this decomposition of GDP, we can now turn to our first model of output determination.

3-2 THE DETERMINATION OF DEMAND

Let's denote the demand for goods by Z. From above, we can write Z as

$$Z \equiv C + I + G + X - Q$$

Demand is the sum of consumption, plus investment, plus government spending, plus exports minus imports. Note that this equation holds by definition: It defines the demand for goods, Z. Such an equation, called an **identity,** is written using the symbol "$\equiv$" rather than an equal sign.

Assume now that there are only three sources of demand: consumption, investment, and government spending. (A model nearly always starts with the word "assume"; this is an indication that reality is about to be simplified in order to focus on the issue at hand.) In other words, ignore exports and imports by pretending that the economy is *closed,* that it does not trade with the rest of the world. (We know that this is asking you to make a very big assumption for the Canadian economy, but we ask you to be patient. For pedagogical reasons, we think it helps to follow the road map that we sketched in Chapter 2. We will first introduce the basic model with as few actors and markets as possible. After establishing some important ideas, we will augment the model to deal with some of its more glaring deficiencies. The closed economy assumption is less heroic when applied to the U.S. economy, so we will rely rather heavily on U.S. historical episodes in the next few chapters.) In Chapter 11 we shall "open" the economy, and reintroduce exports and imports.

Assume next that all firms produce the same good, which can be used by consumers for consumption, by firms for investment, or by the government.[1] With this assumption, we need to look at only one market, the market for "the" good (thus the title of this chapter, "The Goods Market," rather than "The Goods Markets"). All that we need to do now is think about what determines supply and demand in that market.

Assume further that firms are willing to supply any amount of the good at a given price—call it P. In other words, ignore the fact that as firms supply more their costs may go up, forcing them to raise prices. This assumption will allow us to focus on the role of demand in the determination of output. But as we shall see later in the book, it is an assumption that is approximately correct only in the short run.

---

[1] *Such a good, which can either be eaten (like a consumption good) or produce other goods (like a machine), is often called a* shmoo, *for the Li'l Abner cartoon character. Other goods that would qualify are cows and rabbits.*

Thus, the model we are about to develop will be most relevant for movements in output in the short run. To think about what determines output over, say, a decade, we shall need to relax this assumption; we do so starting in Chapter 15.

We can now concentrate on what determines the demand for goods. Under our assumptions, it is the sum of consumption, investment, and government spending:

$$Z \equiv C + I + G$$

Let's discuss each component in turn.

CONSUMPTION (C)

The main determinant of consumption is surely income, or more precisely **disposable income,** the income that remains once consumers have received transfers from the government and paid their taxes. When their disposable income goes up, people buy more goods; when it goes down, they buy fewer goods. Other variables affect consumption, but for the moment we shall ignore them.

Let C denote consumption and Y_D denote disposable income. We can write

$$C = C(Y_D)$$
$$(+)$$

This is just a formal way of stating that consumption is a function of disposable income. The function $C(Y_D)$ is called the **consumption function.** The positive sign below Y_D means a positive relation between disposable income and consumption: It captures the fact that, when disposable income increases, so does consumption. Economists call such an equation a **behavioural equation,** to indicate that the equation captures some aspect of behaviour—in this case, the behaviour of consumers.

We shall use functions in this book as a simple but formal way of capturing relations between variables. What you need to know about functions—which is very little—is described in Appendix 3 at the end of the book. The appendix develops the mathematics you need to go through this book. Do not worry: We shall always describe a function in words when we introduce it for the first time.

It is often useful to be more specific about the form of the function—for example, to assume that the function is linear. Here is such a case. It is reasonable to assume that the relation between consumption and disposable income is given by

$$C = c_0 + c_1 Y_D \qquad (3.1)$$

We are assuming now that the function is a **linear relation;** it is characterized by two **parameters,** c_0 and c_1. Let's look at each in turn.

The parameter c_1 is called the **marginal propensity to consume.** It gives the effect on consumption of an additional dollar of disposable income. If c_1 is equal to 0.7, then an additional dollar of disposable income increases consumption by $1 \times 0.7 = 70$ cents. A natural restriction on c_1 is that it must be positive: An increase in disposable income is likely to lead to an increase in consumption. Another natural restriction is that c_1 must be less than 1: People are likely to consume only part of any increase in income, and to save the rest.

The parameter c_0 has a simple interpretation. It is what people would consume if their disposable income in the current year were equal to zero: If Y_D equals

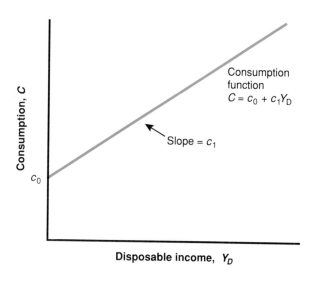

Consumption
function
$C = c_0 + c_1 Y_D$

Slope $= c_1$

c_0

Disposable income, Y_D

Consumption, C

FIGURE 3-2
Consumption and
Disposable Income
Consumption increases
with disposable income, but
less than one for one.

zero in equation (3.1), then $C = c_0$. A natural restriction is that if current income is equal to zero, consumption is still positive: People must eat! This implies that c_0 is positive. How can people have positive consumption if their income is equal to zero? The answer is: by dissaving—by drawing down their assets, or by borrowing.

The relation between consumption and disposable income implied by equation (3.1) is drawn in Figure 3-2. Because it is a linear relation, it is represented by a straight line. Its intercept with the vertical axis is equal to c_0; its slope is equal to c_1. Because c_1 is less than 1, the slope of the line is less than 1: The line is flatter than a 45-degree line. (A refresher on graphs, slopes, and intercepts is given in Appendix 3.)

Next, we need to define disposable income. The exact relation of disposable income to aggregate income is given in Appendix 2. Let's assume here that disposable income is given simply by

$$Y_D \equiv Y - T$$

where Y is aggregate income and T is taxes paid minus transfers received by consumers. Note that this equation is again an identity, thus the use of the symbol "$\equiv$." For short, we shall refer to T simply as taxes, but remember that it is equal to taxes minus transfers.

Replacing Y_D in equation (3.1) for consumption gives

$$C = c_0 + c_1(Y - T) \tag{3.2}$$

Consumption is a function of income and taxes. Higher income increases consumption, although less than one for one. Higher taxes decrease consumption, also less than one for one.

INVESTMENT (I)

Models have two types of variables. Some variables depend on other variables in the model and are therefore explained within the model. Such variables are called **endogenous.** This is the case for consumption here. Other variables are not explained within the model but are instead taken as given. Such variables are called

exogenous. This is how we shall treat investment here. We shall take investment as given and write

$$I = \bar{I} \qquad (3.3)$$

Putting a bar on investment will remind us that we take investment as given.

The reason for taking investment as given is to keep our model simple. But the assumption is not innocuous. It implies that, when we look at the effects of changes in production below, we shall do so under the assumption that investment does not respond to such changes in production. It is not hard to see that this implication may be quite wrong as a description of reality: Firms that experience an increase in production may well decide that they need more machines, and thus increase their investment. We leave this mechanism out of the model for the moment, but we shall introduce a more realistic treatment of investment in Chapter 5.

GOVERNMENT SPENDING (G)

The third component of demand in our model is government spending, G. Together with taxes (T), G describes the government's **fiscal policy**—that is, the government's choice of taxes and spending. Just as we just did for investment, we shall take G and T as exogenous. The rationale for doing so is a bit different than for investment. It is based on two considerations.

First, governments do not behave with the same regularity as consumers or firms, so there is no reliable rule that we could write for G or T corresponding to the rule we wrote for consumption. This first reason is not fully convincing, however. Even if governments do not follow simple behavioural equations as consumers do, a good part of their behaviour is predictable. We look at this issue later, in Chapters 27 to 29, but we leave it aside until then.

The second consideration is the more important one. Part of the task of macro-economists is to advise governments on spending and tax decisions. Thus, we do not want to look at a model in which we have already assumed something about their behaviour. We want to be able to say: "If you were to choose these values for G and T, this is what would happen." The approach in this book will thus typically be to treat G and T as variables chosen by the government, and not try to explain them.[2]

3-3 THE DETERMINATION OF EQUILIBRIUM OUTPUT

Let's collect the pieces we have introduced so far. Under our assumption that exports and imports are both equal to zero, the demand for goods is the sum of consumption, investment, and government spending:

$$Z \equiv C + I + G$$

If we replace C and I by their expressions from equations (3.2) and (3.3), we get

$$Z = c_0 + c_1(Y - T) + \bar{I} + G \qquad (3.4)$$

---

[2]*Because we shall nearly always take G and T as exogenous, we do not use a bar to denote their value. This keeps the notation lighter.*

The demand for goods (Z) depends on income (Y) and taxes (T), which both affect consumption spending, as well as on investment ($\bar{I}$) and government spending (G), which we take as exogenous.

Let's now turn to **equilibrium** in the goods market. Assume that firms do not hold inventories, so that inventory investment is equal to zero. (Think, for example, of the economy producing only services. Because of the very nature of services, firms cannot hold inventories of services.) Then, *equilibrium in the goods market is simply the condition that the supply of goods (Y) be equal to the demand for goods (Z):*

$$Y = Z \tag{3.5}$$

This equation is called an **equilibrium equation.** Models are composed of three types of equations: behavioural equations, identities, and equilibrium conditions. We now have seen examples of each.

Replacing demand (Z) by its expression from equation (3.4) gives

$$Y = c_0 + c_1(Y - T) + \bar{I} + G \tag{3.6}$$

Equation (3.6) captures the mechanism that we described informally at the beginning of the chapter. Production, Y, must be equal to demand, Z. Demand in turn depends on income, Y. Note that we are using the same symbol, Y, for production and income. This is no accident. As we saw in Chapter 2, income and production are identically equal: They are the two ways of looking at GDP—one from the production side, the other from the income side.

Having constructed a model, we can now *solve it* to look at what determines the level of output and how output changes in response to, say, a change in government spending. Solving a model means not only solving it algebraically but also understanding why the results are what they are. Thus, in this book, solving a model will also mean characterizing the results using graphs—sometimes skipping the algebra altogether—and then describing the results and the mechanisms in words. These three steps are also those that macroeconomists use in their research: algebra to make sure that the logic is right, graphs to build the intuition, and words to explain the results.

THE ALGEBRA

Rewrite the equilibrium equation (3.6) as

$$Y = c_0 + c_1 Y - c_1 T + \bar{I} + G$$

Subtract $c_1 Y$ from both sides and reorganize the right side to get

$$(1 - c_1)Y = c_0 + \bar{I} + G - c_1 T$$

Finally, divide both sides by $(1 - c_1)$:

$$Y = \frac{1}{1 - c_1}(c_0 + \bar{I} + G - c_1 T) \tag{3.7}$$

This equation characterizes equilibrium output. Let's look at the first and second terms on the right-hand side of this expression, taking them in reverse order.

The second term, $(c_0 + \bar{I} + G - c_1 T)$, has a simple interpretation. It is what the demand for goods would be if output were equal to zero. If output were equal to zero, we know from equation (3.2) that consumption would be equal to $c_0 - c_1 T$. Investment and government spending, which by assumption do not depend on the level of output, would still be equal to $\bar{I}$ and G, respectively. Putting things together and reorganizing, demand would be equal to $(c_0 + \bar{I} + G - c_1 T)$. This term is called **autonomous spending,** to capture the idea that it is the component of the demand for goods that does not depend on the level of output.

Can we be sure that autonomous spending is positive? We cannot, but it is very likely to be. Suppose, for example, that the government is running a **balanced budget,** that taxes are equal to government spending. If $T = G$ and the marginal propensity to consume (c_1) is less than 1 (as we have assumed), then $(G - c_1 T)$ is positive and so is autonomous spending. Only if the government ran a large budget surplus—if taxes were much larger than government spending—could autonomous spending be negative. We shall ignore this unlikely case here.

Now consider the first term, $1/(1 - c_1)$. Because the marginal propensity to consume (c_1) is between zero and one, $1/(1 - c_1)$ is a number greater than 1. This number, which multiplies the effect of autonomous spending, is called the **multiplier.** The closer c_1 is to 1, the larger the multiplier.

What does the multiplier imply? Suppose that, at their initial level of income, consumers suddenly decide to consume more. More specifically, assume that c_0 in equation (3.2) increases by \$1 billion. Equation (3.7) tells us that output increases by more than \$1 billion. For example, if c_1 is equal to 0.6, the multiplier is equal to $1/(1 - 0.6) = 2.5$, so that output increases by $2.5 \times \$1$ billion = \$2.5 billion. We have looked here at an increase in consumption, but clearly any increase in autonomous spending, from increases in investment to increases in government spending or reductions in taxes, will have the same qualitative effect: It will increase output by more than its direct effect on autonomous spending.

Where does the multiplier effect come from? Looking at equation (3.6) gives the beginning of a clue. An increase in c_0 increases demand. The increase in demand then leads to an increase in production and income. But the increase in income increases consumption further, which increases demand further, and so on. The best way to strengthen and refine this intuition is to use a graphical approach.

A GRAPH

Equilibrium requires that the production of goods (Y) be equal to the demand for goods (Z). Figure 3-3 plots both production and demand as functions of income; the equilibrium is the point at which production and demand are equal.

The first step is to plot production as a function of income. Production is measured on the vertical axis and income on the horizontal axis. Plotting production as a function of income is straightforward, as production and income are always equal. Thus, the relation between the two is simply the 45-degree line, the line with a slope equal to 1, in Figure 3-3.

The second step is to plot demand, Z, as a function of income. The relation between demand and income is given by equation (3.4). Let's rewrite it here for convenience, regrouping the terms for autonomous spending together in the term in parentheses on the right:

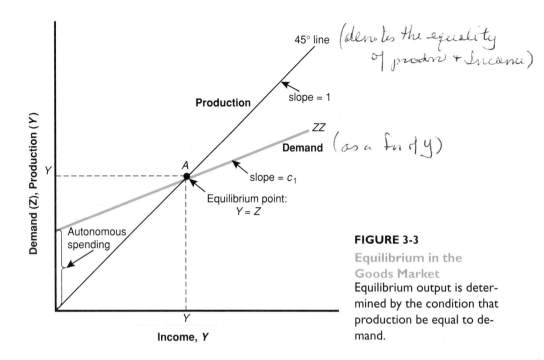

Handwritten annotations on figure:
45° line (denotes the equality of prodrn + Income)
Demand (as a fn of y)

FIGURE 3-3

Equilibrium in the Goods Market
Equilibrium output is determined by the condition that production be equal to demand.

$$Z = c_1 Y + (c_0 + \bar{I} + G - c_1 T)$$

Demand depends on income, through its effect on consumption, and on autonomous spending. The relation between demand and income is drawn as the line denoted ZZ in the figure. The intercept—the value of demand when income is equal to zero—is equal to autonomous spending. The slope of the line is equal to the marginal propensity to consume, c_1. Under the restriction that the slope term c_1 is positive but less than 1, the line is upward-sloping but with slope less than 1.

Equilibrium holds when production is equal to demand. Thus, equilibrium output, Y, is given by the intersection of the 45-degree line and the demand relation, ZZ, at point A. To the left of A, demand exceeds production; to the right, production exceeds demand. Only at A are the two equal.

Now return to the example we looked at earlier. Suppose that at a given level of income, consumers increase their consumption: c_0 increases by \$1 billion. Figure 3-4, which builds on Figure 3-3, shows what happens as a result. For any value of income, demand is higher by \$1 billion. Thus, if the relation between demand and income was given earlier by the line ZZ, the new relation is given by the line ZZ', which is parallel to ZZ but higher by \$1 billion. In other words, the demand relation shifts up by \$1 billion. The new equilibrium is at the intersection of the 45-degree line and the new demand relation, thus at point A'. Equilibrium output increases from Y to Y'. It is clear that the increase in output, $Y' - Y$, which we can measure on either the horizontal or the vertical axis, is larger than the initial increase in consumption of \$1 billion. This is the multiplier effect.

With the help of the graph, it becomes easier to tell how and why the economy moves from A to A'. The initial increase in consumption leads to an increase in demand. At the initial level of income, Y, the level of demand is now given by point B:

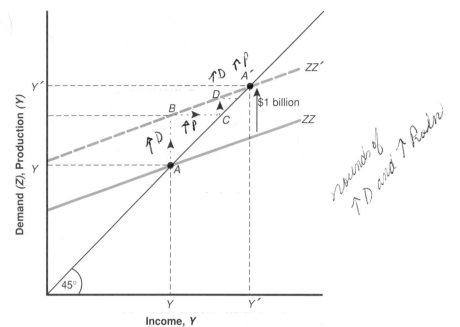

FIGURE 3-4

The Effects of an Increase in Autonomous Spending on Output
An increase in autonomous spending has more than a one-for-one effect on equilibrium output.

Demand is $1 billion higher. To satisfy this higher level of demand, firms increase production by $1 billion. The economy moves to point *C,* with both demand and production higher by $1 billion. But this is not the end of the story. The higher level of production leads to a further increase in demand, so that demand is now given by point *D. D* in turn leads to a higher level of production, and so on, until the economy is at *A′,* where production and demand are again equal, and which is therefore the new equilibrium.[3]

IN WORDS

How can we summarize our findings in words?

Production depends on demand, which in turn depends on income, which is itself equal to production. An increase in demand, such as an increase in consumer spending, leads to an increase in production and an increase in income. This increase in income leads to a further increase in demand, which leads to a further increase in production, and so on. The end result is an increase in output that is larger than the initial shift in demand, by a factor equal to the multiplier.

We have focussed on *increases* in demand. But the mechanism is symmetric: Decreases in demand lead to decreases in output. The 1990–1991 recession in the United States was indeed due largely to a sudden drop in consumer confidence, leading to a sharp decrease in consumption demand and, in turn, to a sharp decline in output. The origins of the 1990–1991 recession are examined in detail in the In Depth box entitled "Consumer Confidence and the 1990–1991 Recession in the United States."

---

[3]*This explanation suggests an adjustment of output that takes place over a period of time, rather than instantaneously as we are assuming here. We explore this time dimension explicitly in Chapter 4.*

In the third quarter of 1990, after the invasion of Kuwait by Iraq but before the beginning of the Persian Gulf War, U.S. GDP growth turned negative and remained negative for the next two quarters. As we saw in Chapter 2, economists use the word *recession* to denote at least two consecutive quarterly decreases in GDP, so this episode is known as the 1990–1991 recession.

The second column of the accompanying table gives real GDP—in billions of 1987 U.S. dollars—from the second quarter of 1990 to the second quarter of 1991. The third column gives the change in GDP from one quarter to the next. Note that GDP is measured at "an annual rate"; thus, the numbers are equal to four times their true value for the quarter. Reporting monthly or quarterly variables at an annual rate may at first appear confusing. But reporting all variables, whether daily, monthly, or quarterly, at an annual rate—and thus a common rate—makes comparison among them easier. In 1990:3, 1990:4, and 1991:1, the change in GDP is negative. This is the 1990–1991 recession.

Macroeconomists did not predict the 1990–1991 recession. To a large extent, it was due to the effects of shifts in spending that hadn't been or couldn't be anticipated. This is clear from columns (4) and (5) of the table. Column (4) gives forecasts made for each quarter during the preceding quarter. For example, the forecast of GDP made in the second quarter of 1990 for the third quarter was $4931 billion. Column (5) gives the **forecast error,** the difference between the actual value of GDP and the forecast. A positive forecast error indicates that actual GDP turned out to be higher than expected; a negative forecast error indicates that actual GDP turned out lower than expected. To pursue our example: Because the actual value of GDP in the third quarter of 1990 was $4898 billion, the forecast error for that quarter was 4898 − 4931, or −$33 billion.

As you can see, the forecast errors are negative during all three quarters of the recession. They are actually larger than the actual declines in GDP in each of the first two quarters of the recession. Put another way, at the beginning of these two quarters, the forecasts were of positive growth, although growth actually turned out to be negative. For example, the forecast made in the second quarter for the third quarter was for an increase in GDP of 4931 − 4915 = $16 billion. The actual movement in GDP was a decrease of $17 billion.

Where did these forecast errors come from? In terms of equation (3.7), which of the determinants of

··············
TABLE 1

U.S. GDP, CONSUMPTION, AND FORECAST ERRORS, 1990–1991

(1) QUARTER	(2) REAL GDP	(3) CHANGE IN GDP	(4) FORECAST OF GDP	(5) FORECAST ERROR FOR GDP	(6) FORECAST ERROR FOR c_0	(7) CONSUMER CONFIDENCE
1990:2	4915	8	4913	2	1	105
1990:3	4898	−17	4931	−33	1	90
1990:4	4861	−37	4906	−45	−24*	61
1991:1	4822	−39	4841	−19	−18*	65
1991:2	4831	10	4798	33	6	77

Notes for headings:
Column (2)—Real GDP: Billions of 1987 U.S. dollars, at an annual rate.
Column (3)—Change in GDP: change from last quarter, at annual rate, in billions of U.S. dollars.
Column (4)—Forecast of GDP for the current quarter, as of the preceding quarter, in billions of U.S. dollars.
Column (5)—Forecast error for GDP: GDP minus forecast of GDP, in billions of U.S. dollars.
Column (6)—Forecast error for c_0 from article cited at the end of the box, in billions of U.S. dollars.
Column (7)—Consumer confidence index
Note: Asterisks in column (6) denote an unusually large error; errors of this magnitude happen only 1% of the time.

spending was the main culprit? Was it c_0, or $\bar{I}$, or G, or T? Research looking at the evolution of each of these components suggests that the main cause, for the last two quarters of the recession, was an adverse shift in consumption, an unexpected decrease in c_0. Forecast errors of c_0 are given in column (6) of the table. There are two large negative errors for the last two quarters of the recession.

Why did consumption decrease so much, given income, in late 1990 and early 1991? The direct cause is shown in the last column of the table, which gives the value of the **consumer confidence index.** This index is computed from a monthly survey of about 5000 households; the survey asks consumers how confident they are about both current and future economic conditions, from job opportunities to their expected family income six months ahead. As you can see, there was a dramatic decrease—and one very unusual in size—in the index in the fourth quarter of 1990. Consumers lost confidence, leading them to cut

on consumption given income, and triggering the recession.

This brings us to the last question: Why did consumers lose confidence in late 1990? Why did they become more pessimistic about the future? The truth is that, even today, economists are not sure. It is more than likely that this mood change was related to the increasing probability of a war in the Middle East—a war that indeed started in early 1991, after the beginning of the recession. People worried that the United States might get involved in a prolonged and costly war. They also worried that a war in the Middle East could lead to a large increase in oil prices and to a recession: The two previous large increases in oil prices in the 1970s had both been associated with recessions. Whatever the reason, the decrease in consumer confidence was a major factor behind the 1990–1991 recession.

Source: Olivier Blanchard, "Consumption and the Recession of 1990–1991," *American Economic Review*, May 1993.

3-4 INVESTMENT EQUALS SAVING: AN ALTERNATIVE WAY OF THINKING ABOUT GOODS-MARKET EQUILIBRIUM

Thus far we have thought about equilibrium in terms of equality between the production of and the demand for goods. An alternative—but equivalent—way of thinking about equilibrium focusses on investment and saving instead. This is how John Maynard Keynes first articulated this model in 1936, in *The General Theory of Employment, Interest and Money*.

By definition, **saving (S)** by consumers is equal to their disposable income minus their consumption:

$$S \equiv Y_D - C$$

Using the definition of disposable income, we can rewrite saving as income minus taxes minus consumption:

$$S \equiv Y - T - C$$

Now return to the equation for equilibrium in the goods market. Production must be equal to demand, which in turn is the sum of consumption, investment, and government spending:

$$Y = C + I + G$$

Take $Y = C + I + G - T$

Subtract taxes (T) from both sides and move consumption to the left-hand side of the equation to get

$$Y - T - C = I + G - T$$

Note that the left-hand side of this equation is simply equal to saving (S), so that we can rewrite

$$S = I + G - T \quad (public)$$

or, equivalently,

$$I = S + (T - G) \tag{3.8}$$

Equation (3.8) gives us another way of looking at equilibrium in the goods market. On the left-hand side is investment. The first term on the right-hand side is saving by consumers, which we can call **private saving.** The second term is equal to taxes minus government spending. If it is positive, the government is running a budget surplus, or, equivalently, **public saving** is positive. If it is negative—which, as we saw in Chapter 2, has been the case in Canada for most of the period since 1960—the government is running a budget deficit.

Thus, equation (3.8) states, equilibrium in the goods market requires that investment equal the sum of private and public saving. This way of looking at equilibrium is why the equilibrium condition for the goods market is called the **IS relation,** for "**I**nvestment equals **S**aving."

To strengthen your intuition for equation (3.8), think of an economy where there is only one person, who has to decide how much to consume, invest, and save—a "Robinson Crusoe" economy. For Robinson Crusoe, the saving decision and the investment decision are one and the same: What he invests (say, by keeping animals for reproduction rather than eating them), he automatically saves. In a modern economy, however, investment decisions are made by firms, while saving decisions are made by consumers and the government. In equilibrium, equation (3.8) tells us, all those decisions have to be consistent: Investment must be equal to saving.

We can study the characteristics of the equilibrium using equation (3.8) and the behavioural equations for saving and investment. Note first that *consumption and saving decisions are one and the same:* Once consumers have chosen consumption, their saving is determined, and vice versa. The way we specified consumption behaviour implies that private saving is given by

$$S \equiv Y - T - C$$
$$= Y - T - c_0 - c_1(Y - T),$$

$1Y - 1T - C, Y + C_1 T$ *is the equivalent of:*

Rearranging, we get

$$S = -c_0 + (1 - c_1)(Y - T) \leftarrow$$

mps

In the same way that we called c_1 the marginal propensity to consume, we can call $(1 - c_1)$ the **marginal propensity to save.** The marginal propensity to save tells us how much people save out of an additional unit of income. The assumption we made earlier that the marginal propensity to consume (c_1) is between zero and one implies that the marginal propensity to save ($1 - c_1$) is also between zero and one. Private saving increases with disposable income, but less than one for one.

In equilibrium, investment must be equal to saving, the sum of private and public saving. Replacing private saving by its expression from above in equation (3.8) gives

$$I = -c_0 + (1 - c_1)(Y - T) + (T - G)$$

Solving for output, as we did earlier, gives

$$Y = \frac{1}{1 - c_1}(c_0 + \bar{I} + G - c_1 T)$$

(3.7)

This is exactly the same expression as equation (3.7). This should come as no surprise. We are looking at the same model, just looking at it in a different way. This way will prove useful at various points later in the book.

3-5 THE PARADOX OF SAVING

As we grow up, we are told of the virtues of thrift. Those who spend all their income are condemned to end up poor, or so the story goes. Those who save are promised a happy life. Similarly, governments often tell us that an economy that saves is an economy that will grow strong and prosper. Equation (3.7), however, tells us a different and quite surprising story.

Suppose that, at a given level of income, consumers decide to save more. In terms of equation (3.2), the equation describing consumption, they decrease c_0, thus decreasing consumption and increasing saving at a given level of income. What happens to output and to saving?

Equation (3.7) makes it clear that equilibrium output decreases when c_0 decreases. As people save more at their initial level of income, they decrease their consumption. But this decreased consumption in turn decreases demand, which decreases production.

Can we tell what happens to saving? On the one hand, consumers are saving more at any level of income; this tends to increase saving. But, on the other hand, income is now lower, which tends to decrease saving. The net effect would seem to be ambiguous. In fact, we can tell which way it goes. Remember that we can think of the equilibrium condition as the condition that saving equals investment. By assumption, investment does not change. So, the equilibrium condition tells us, saving does not change. While people want to save more at a given level of income, income decreases by an amount such that saving is unchanged. This means that attempts by people to save more lead both to a decline in output and to unchanged saving. This surprising pair of results is known as the **paradox of saving.**[4]

So should you forget the old wisdom? Should the government tell people to be less thrifty? No. The results of this simple model are of much relevance in the *short run*. It is indeed the case that the consumers' desire to save more led to the 1990–1991 recession in the United States (see the In Depth box earlier in this chapter). But as we

---

[4]**Digging deeper.** *You may think that the result that saving is unchanged is really the result of one of the less appealing assumptions we have made so far, namely that investment is fixed. In fact, if we were to allow investment to move with output, the result would be even more dramatic. Attempts by people to save more would still lead to a decrease in output. And the decrease in output would lead to a decrease in investment, and thus to a decrease in saving.*

(handwritten margin notes:)

$(3.2)\ C = C_0 + C_1(Y-T)$

Suppose $C_0 \downarrow$ $\$1000$

$Y = \dfrac{C_0 + \bar{I} + G - C_1 T}{1 - C_1}$ $\downarrow -1000 \Rightarrow \uparrow S \,\$1000$ $.2\,mps$

$Y \downarrow -5000 \times .2\,mps$

$= -1000$

$= 0\ net\ Svg.$

$\uparrow Y \downarrow$ $\uparrow S \downarrow S$

Paradox of Svg. $\therefore$ One $t\,I$

$\Rightarrow$ recession 1990-91

Short Run concept.

In Long Run, $\uparrow S, \uparrow I,$ $\uparrow$ Growth.

Assumes $S \ne I$, but $I \ne f(y)$

shall see later in this book when we make the model more realistic, other mechanisms come into play and an increase in the saving rate is likely to lead to *higher saving and higher income in the long run.* An important warning remains, however: Policies that encourage saving may be good in the longer run, but may nevertheless lead to a recession in the short run.

3-6 IS THE GOVERNMENT REALLY OMNIPOTENT? A WARNING

Equation (3.7) implies that, by choosing the level of spending (G) or the level of taxes (T), the government can affect output as it wants. If it wants output to increase by, say, $1 billion, all it needs to do is increase G by $(1 - c_1)$ billion dollars; this increase in government spending, in theory, will lead to an output increase of $1 billion. But can governments really choose the level of output they wish? The existence of recessions makes it clear that the answer is no.

The answer is no because there are many aspects of reality that we have not yet incorporated in our model. We shall do so in due time. But it is worth briefly listing some of them here:

■ Changing government spending or taxes may be far from easy. Tax changes have to wait for the annual Budget announcement. Although federal and provincial governments can change spending fairly quickly, it can be difficult to get the various levels of government to cooperate. Fiscal choices by one government can be offset by the actions of another (see Chapters 27 and 29).

■ The effects of spending and taxes on demand are much less mechanical than equation (3.7) makes them appear. They may happen slowly, consumers and firms may be scared of the budget deficit and change their behaviour, and so on (see Chapter 10).

■ Maintaining a desired level of output may come with unpleasant side-effects. For example, trying to achieve too high a level of output may lead to accelerating inflation (see Chapter 17).

■ Cutting taxes or increasing government spending may lead to large budget deficits and an accumulation of public debt. Such debt may have adverse implications in the longer run (see Chapters 23 and 29). Also, in an open economy, the short-run impact can be greatly reduced (see Chapter 12).

As we refine our analysis, the role of government will become increasingly difficult. Governments will never again have it so good as in this chapter.

SUMMARY

What you should remember about the components of GDP:

■ GDP is the sum of consumption, plus investment, plus government spending, plus exports, minus imports, plus inventory investment.

■ Consumption (C) is the purchase of goods and services by consumers. Consumption is the largest component of demand.

■ Investment (I) is the sum of nonresidential investment (the purchase of new plants and new machines by firms) and of residential investment (the purchase of new houses or apartments by people).

- Government spending (G) is the purchase of goods and services by federal, provincial, and local governments.
- Exports (X) are purchases of Canadian goods by foreigners. Imports (Q) are purchases of foreign goods by Canadian consumers, Canadian firms, and the Canadian government.
- Inventory investment (I_S) is the difference between production and sales. It can be positive or negative.

What you should remember about our first model of output determination:

- In the short run, production is determined by demand, which in turn depends on production and income.
- The consumption function shows how consumption depends on disposable income. The marginal propensity to consume describes how consumption increases for a given increase in disposable income.
- Equilibrium output is the point at which production is equal to demand. In equilibrium, output is equal to autonomous spending times the multiplier. Autonomous spending is that part of demand that does not depend on income. The multiplier is equal to $1/(1 - c_1)$, where c_1 is the marginal propensity to consume.
- Increases in consumer confidence, investment demand, or government spending, or decreases in taxes, all increase equilibrium output in the short run.
- An alternative way of stating the goods-market equilibrium condition is that investment must be equal to saving, the sum of private and public saving. For this reason, the equilibrium condition is called the IS relation.
- In the short run, an attempt by people to save more, or by the government to reduce its deficit, has no effect on saving but leads to a decrease in output. This is called the paradox of saving. In the longer run, however, other factors come into play, and both saving and output are likely to increase instead.

KEY TERMS

- consumption (C), 43
- investment (I), or fixed investment, 43
- residential and nonresidential investment, 43–44
- government spending (G), 44
- government transfers, 44
- imports (Q), 44
- exports (X), 44
- net exports ($X - Q$), or trade balance, 44
- trade surplus, 44
- trade deficit, 44
- inventory investment (I_S), 45
- stocks, 45
- identity, 45
- disposable income, 46
- consumption function, 46
- behavioural equation, 46
- linear relation, 46
- parameters, 46

- marginal propensity to consume, 46
- endogenous variables, 47
- exogenous variables, 48
- fiscal policy, 48
- equilibrium, 49
- equilibrium equation, 49
- autonomous spending, 50
- balanced budget, 50
- multiplier, 50
- forecast error, 53
- consumer confidence index, 54
- saving (S), 54
- private saving, 55
- public saving (S), 55
- *IS* relation, 55
- marginal propensity to save, 55
- paradox of saving, 56

QUESTIONS AND PROBLEMS

1. Suppose that an economy is characterized by the following behavioural equations:

$$C = 100 + 0.6Y_D$$
$$I = 50$$
$$G = 250$$
$$T = 100$$

Solve for:
a. Equilibrium GDP (Y)
b. Disposable income (Y_D)
c. Consumption spending (C)
d. Private saving
e. Public saving
f. The multiplier

2. For the economy in question 1, verify that, in equilibrium:
  a. Production equals demand.
  b. Total saving equals investment.

3. Suppose that the government wishes to increase equilibrium GDP by 100.
  a. What change in government spending is required? [*Hint:* What is the value of the multiplier?]
  b. If government spending cannot change, what change in taxes is required? [*Hint:* The answer is different from the answer to part a.]

4. To keep our model simple, we have assumed that taxes are exogenous. In reality, we know that taxes tend to rise and fall with income. Suppose that taxes depend linearly on income, according to the equation

$$T = T_0 + t_1 Y$$

where t_1 is the tax rate and is between 0 and 1. All other behavioural equations are as specified in the chapter.
  a. Find the equation for equilibrium GDP. [*Hint:* It will be similar, but not identical to, equation (3.4).]
  b. Find the expression for the multiplier.
  c. Is the multiplier when taxes are endogenous larger than, smaller than, or the same as the multiplier when taxes are exogenous?

5. A student is overheard to say, "I don't get macroeconomics. Sometimes, changes in income seem to cause changes in consumption. Other times, changes in consumption seem to cause changes in income. Which is it?" Help this student resolve this confusion. [*Hint:* Distinguish between changes in c_0 and changes in $c_1 Y_D$.]

6. Although the marginal propensity to consume states the relation between *aggregate* income and *aggregate* consumption, it can also apply to the relation between income and consumption for an individual.
  a. Determine your own marginal propensity to consume out of current income. [*Hint:* For raw data, pose some hypothetical questions. For example, if you are already working, how much would your spending increase if you were given a 20% raise? If you are not working, how much would your spending increase if you began a job at a salary of, say, $30 000 per year?]
  b. Is your marginal propensity to consume consistent with the two "natural restrictions" stated in the text? If not, explain.

FURTHER READING

For a lively—and somewhat controversial— assessment of the cause of Canada's recession in the early 1990s, see Pierre Fortin's presidential address to the Canadian Economics Association, "The Great Canadian Slump," *Canadian Journal of Economics,* November 1996. Fortin places a lot of the responsibility for the 1990–1991 recession and the subsequent slow recovery in Canada on the Bank of Canada. You may want to look at his paper again after you have had a chance to study monetary policy in an open economy (Chapter 13).

THE GOODS MARKET: DYNAMICS

A consumer who gets a raise at work may not adjust her consumption right away. A firm that faces an increase in demand may decide to wait before adjusting its production, relying meanwhile on its inventories to satisfy demand. A government that wants to increase spending to fight a recession needs to draft a plan and get it through Parliament before it can start spending the money.

This time dimension was absent from the model we built in Chapter 3. But it is very important in practice. Thus the first section of this chapter builds a model in which firms do not adjust their production instantaneously to changes in output, and shows the model's implications for the **dynamics** (the term economists use for "movements over time") of output in response to changes in spending. The second section shows how one can use **econometrics**—statistical methods applied to economics—to get estimates of the model's important parameters. The third section then simulates the estimated model and characterizes the dynamic effects of changes in autonomous spending on output.

4-1 PRODUCTION, SALES, AND INVENTORY INVESTMENT

We assumed in Chapter 3 that inventory investment was always equal to zero. Put another way, we assumed that firms reacted to a change in demand with an instantaneous and equal change in production. We now want to relax this assumption and capture the notion that, faced with a change in demand, firms may not adjust production right away. Production schedules are typically set in advance and are costly to change, and increasing the number of shifts and getting additional workers on short notice may be costly and difficult.[1]

DIVIDING TIME

Let us first divide time into discrete periods. Depending on the question at hand, the right period to use may be a month, a quarter, a year, or longer. Here, let's think of a period as roughly the time it takes for firms to adjust production. A quarter is a reasonable time period in this case.

As we shall refer to variables in different quarters, we must identify them by the quarter to which they correspond. For example, Y_{t+1} refers to production during quarter $t + 1$, Y_t refers to production during quarter t, Y_{t-1} to production in quarter $t - 1$, and so on.[2] It is often convenient to refer to the value of the variable in the preceding period as the **lagged value** of that variable; we shall, for example, refer to Y_{t-1} as the lagged value of Y_t, or to Y_t as the lagged value of Y_{t+1}. To go light on the use of time indexes in the text, we shall refer to quarter t as the *current* quarter, to $t + 1$ as *next* quarter, and to $t - 1$ as *last* quarter.

We shall use time indexes throughout this book except when they can be omitted without risk of confusion. In Chapter 3, for example, we could have used a time index for all variables. But, because we were looking at just one period in time, they would all have been indexed by the same index t. Not much would have been gained except a heavier notation. So we just ignored time indexes. Given our focus on movements in the variables here, we cannot do the same in this chapter.

PRODUCTION AND SALES

With the technicalities of time notation out of the way, let's now concentrate on the issue at hand, the movement in output over time when firms do not adjust production right away in response to changes in demand.

Assume that firms set their level of production at the beginning of each quarter, before they know the level of demand for that quarter. Once set, production cannot be adjusted during the quarter. If demand exceeds production, firms satisfy demand by drawing down their inventories. (The assumption that firms always satisfy demand—rather than telling their customers that they have run out of goods—implies that sales are equal to demand. We shall thus use *demand* and *sales* interchangeably in what follows.) Inventories decline or, equivalently, inventory investment is negative. If, instead, sales are less than production, firms accumulate inventories: Inventory investment is positive.

This first assumption raises the issue of the level at which firms set production at the beginning of each quarter. A reasonable assumption is that they set production at the level of sales they expect for the quarter. How do they form those expectations? Let's assume that they expect this quarter's sales to be equal to last quarter's sales, and

---

[1] **Digging deeper.** *Another argument is that firms may not want to do major adjustments in production if they expect the increase in demand to be temporary. Following this argument would force us to think about the role of expectations in decisions, a central theme of this book, but one that we defer until Chapter 7.*

[2] *If we had divided time in years rather than quarters, then Y_t would refer to production in* year t. *This is what we did in Chapter 2, for example, where we used Y_t to denote real GDP in* year t.

thus set production equal to last quarter's sales. Expectations will be the focus of Chapters 7 to 10; we defer further discussion of how expectations are formed until then.

Denote sales—equivalently, demand—by Z, and production by Y. Together, our assumptions imply that firms set next quarter's production at a level equal to this quarter's sales:[3]

$$Y_{t+1} = Z_t \tag{4.1}$$

This quarter's sales, in turn, are equal to consumption plus investment plus government spending.[4] At time t, Z_t is thus given by

$$Z_t \equiv C_t + I_t + G_t \tag{4.2}$$

Let's make the same assumptions about consumption, investment, and government spending as in Chapter 3. Consumption depends on current disposable income (income minus taxes):

$$C_t = c_0 + c_1(Y_t - T_t)$$

The only change from Chapter 3 is the explicit use of time indexes. Investment, government spending, and taxes are exogenous. Let's assume that they are constant through time, so that there is no need to include a time index:

$$I_t = \bar{I}$$
$$G_t = G$$
$$T_t = T$$

Replacing these equations in equation (4.2) gives sales in quarter t:

$$Z_t = \underbrace{C_t}_{} + \underbrace{I_t + G_t}_{}$$
$$= c_0 + c_1(Y_t - T) + \bar{I} + G$$

Putting together all the terms that do not depend on income in a term in parentheses gives

$$Z_t = c_1 Y_t + (c_0 + \bar{I} + G - c_1 T) \tag{4.3}$$

Sales this quarter depend on income this quarter—which affects sales through consumption—and on autonomous spending. There is nothing new here: Both the effect of income on demand (given by the marginal propensity to consume, c_1) and autonomous spending (the term in parentheses) are the same as in the model of Chapter 3.

EQUILIBRIUM IN THE GOODS MARKET REVISITED

Equation (4.1) tells us that firms set next period's production, Y_{t+1}, equal to current sales, Z_t. Equation (4.3) tells us that current sales, Z_t, depend in turn on current production, Y_t, and on autonomous spending. Putting the two equations together gives

$$Y_{t+1} = c_1 Y_t + (c_0 + \bar{I} + G - c_1 T) \tag{4.4}$$

---

[3]*Equations indexed by time, such as (4.1), appear often in macroeconomics. You should think of (4.1) as a collection of equations, one for each value of the time index* t.

[4]*We are still ignoring imports and exports.*

The central feature of equation (4.4) is that production next quarter, Y_{t+1}, depends on current production, Y_t. Where does this dependence come from? Production next quarter depends on sales this quarter. Sales this quarter depend in turn on income and thus production this quarter.

Suppose that government spending, taxes, and investment all remain constant, and that we follow the economy through time. It is plausible that output will eventually settle to a constant level as well. So let us start by asking: To what equilibrium level will output settle?

When the economy settles to a constant level of output—call it Y without a time index, to indicate that it is indeed constant—Y_{t+1} is then equal to Y_t and both are equal to their common value, Y. Replacing Y_{t+1} and Y_t by Y in equation (4.4), we get

$$Y = c_1 Y + (c_0 + \bar{I} + G - c_1 T)$$

Solving for Y yields

$$\underbrace{Y}_{\text{equilibrium output}} = \underbrace{\frac{1}{1 - c_1}}_{\text{multiplier}} \underbrace{(c_0 + \bar{I} + G - c_1 T)}_{\substack{\text{autonomous} \\ \text{spending}}}$$

This is exactly the same expression as equation (3.7). Equilibrium output is equal to the multiplier times autonomous spending. So, although we have introduced a time dimension in our model, output still eventually settles to the same level as before. Shifts in consumption, investment, or fiscal policy still affect output through a multiplier effect.

No subscript t in equil.

The difference is that output does not adjust to this new level right away. We must now examine how this adjustment takes place.

THE DYNAMIC EFFECTS OF AN INCREASE IN GOVERNMENT SPENDING

In Chapter 3 we looked at the effects of a shift in consumption, a shift in c_0. For a change, let's look instead at the effects of an increase in G.

First we have to define carefully the experiment we want to look at. The use of the word **experiment** is not accidental. In the same way that a scientist conducts an experiment by changing, say, the temperature of a material and looking at the effect on the material's resistance to electricity, we want to change government spending in our model and trace the effects of this change on production over time. Let's define the experiment here as a *permanent* increase in government spending. Suppose that the values of the variables in this economy have not changed for some time. Suppose also that, in the first quarter of the year 2000, the level of government spending is increased by $1 billion and remains at this new higher level forever after. We want to trace the effects of this change on production in the first, second, and third quarters of 2000, and so on. What we are doing here—using the model to look at the effects of a change in an exogenous variable on the variables in the model—is called a **simulation** of the model.

Before tracing those effects let's make one more assumption, to simplify notation and avoid carrying the parameter c_1 around. Let's assume that c_1 is equal to 0.5:

An increase in current disposable income of $1 billion leads to an increase in current consumption of $0.5 billion.

Table 4-1 shows what happens over time. The top half of the table gives the quarter-to-quarter changes in production, government spending, consumption, and sales. The bottom half gives the cumulative changes in the same variables, cumulated from 2000:1, the quarter during which government spending is increased. The best way to understand the table is to take it one line at a time, starting with the top half.

In the first quarter of 2000, production is set before the change in government spending and thus does not change. Government spending increases by $1 billion. Because production is unchanged, income remains unchanged and so does consumption. Thus, sales increase by $1 billion, the increase in government spending.

In the second quarter of 2000, firms, having observed an increase in sales in the first quarter, increase production by $1 billion. Government spending remains at its new level and therefore no longer changes. As production and thus income have increased by $1 billion, consumption increases by $0.5 billion. The increase in sales over the preceding quarter is thus equal to the increase in consumption spending, $0.5 billion.

In the third quarter, firms, having observed an increase in sales in the second quarter of $0.5 billion, further increase production by $0.5 billion. Government spending remains at its new higher level, so the change in government spending is equal to zero. As production and income have further increased by $0.5 billion, consumption increases by 0.5 times the change in income, thus $0.5 \times \$0.5$ billion $= \$0.5^2$ billion. The increase in sales over the previous quarter is thus equal to the further increase in consumption spending, $\$0.5^2$ billion.

What happens in the following quarters? Each increase in production leads to an increase in income, thus to a further increase in consumption, thus to a further increase in sales, thus to an increase in production in the following quarter.

The top half of the table records the quarter-to-quarter increases in production and its components. If we want to know the total increase in production that results from the increase in government spending, we must add—equivalently,

TABLE 4-1

THE EFFECTS OF A CHANGE IN GOVERNMENT SPENDING

QUARTER	(1) PRODUCTION	(2) GOVERNMENT SPENDING	(3) CONSUMPTION	(4) SALES
		Quarter-to-quarter increases		
2000:1	0.0	1.0	0.0	1.0
2000:2	1.0	0.0	0.5	0.5
2000:3	0.5	0.0	0.5^2	0.5^2
2000:4	0.5^2	...	...	...
		Cumulative increases		
2000:1	0.0	1.0	0.0	1.0
2000:2	1.0	1.0	0.5	$1.0 + 0.5$
2000:3	$1.0 + 0.5$	1.0	$0.5 + 0.5^2$	$1.0 + 0.5 + 0.5^2$
2000:4	$1.0 + 0.5 + 0.5^2$	...	...	...

All numbers are in billions of dollars.

cumulate—these quarter-to-quarter increases. The bottom half of the table gives the cumulative increases in production and its components. Look at column (1). In 2000:4, for example, the cumulative increase in production is equal to 1.0 (the increase in 2000:2) plus 0.5 (the increase in 2000:3) plus 0.5^2 (the increase in 2000:4).

Note that as time passes production keeps increasing. This does not mean that it increases without bounds, however. The cumulative increase in output $n + 1$ quarters after the increase in government spending is given by

$$1 + (0.5) + (0.5)^2 + \cdots + (0.5)^n$$

Or, more generally, if c_1 is the marginal propensity to consume, the cumulative increase in output is given by a **geometric series,** a sum of the form

$$1 + c_1 + c_1^2 + \cdots + c_1^n$$

Geometric series will come up often in this book. A refresher on their properties is given in Appendix 3 at the end of the book. The main property of such series is that, when c_1 is less than 1 (which is the case here) and as n gets larger and larger, the sum approaches a limit that is given by $1/(1 - c_1)$. Thus, the eventual increase in output is given by $1/(1 - c_1)$. This expression should by now be quite familiar: It is simply our old multiplier, derived another way. Taking $c_1 = 0.5$, the eventual increase in output is $1 billion times $1/(1 - 0.5) = 2$ billion, twice the increase in government spending.

USING A GRAPH

A graph can again strengthen the intuition. In Figure 4-1, let production (equivalently, income) in quarter t be measured on the horizontal axis, and demand (equivalently, sales) in quarter t be measured on the vertical axis. The relation of demand

See Cameron Math notes

IS curve =

45° diagram shows the law of motion

$$\frac{dy}{dt} = \lambda (E - Y) \qquad \lambda > 0$$

& IS is where

$$\frac{dy}{dt} = 0$$

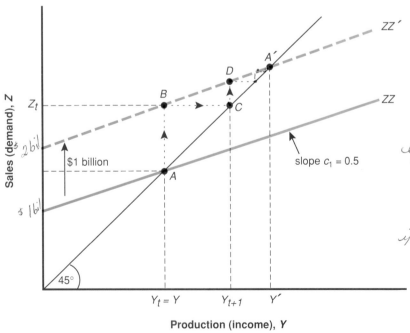

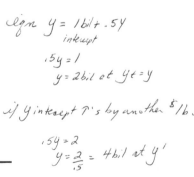

eqn $y = 1 bil + .5Y$
        intercept

$.5y = 1$
$y = 2 bil$ at $Yt = Y$

if Y intercept ↑'s by another $1 b.

$.5y = 2$
$y = \frac{2}{.5} = 4 bil$ at y'

FIGURE 4-1
The Dynamic Effects of an Increase in Government Spending
An increase in government spending increases output over time.

to income is given by equation (4.3). Under our assumptions about the marginal propensity to consume, demand is represented by an upward-sloping line, with slope of 0.5.

The demand relation that holds before the change in government spending is drawn as line ZZ, and the initial equilibrium is thus given by point A, with production equal to sales and output equal to Y. After the increase in government spending by $1 billion in quarter t, demand is higher, at any level of production, by $1 billion. The new demand relation is given by line ZZ', and the point at which production is eventually again equal to sales is A'.

How does the economy go from A to A'? In quarter t, production, Y_t, is unchanged and thus still given by Y. As government spending increases by $1 billion, demand increases to point B, which is $1 billion higher than it was at point A. The economy is thus at point B, with higher sales but the same level of production. In quarter t + 1, firms set production equal to demand in quarter t, thus at $Y_{t+1} = Z_t$. With production equal to Y_{t+1}, demand is given by Z_{t+1}, so that the economy is now at point D, and so on. Output eventually increases from Y to Y', corresponding to point A'.

Figure 4-2 presents the same results in another way. It plots the evolution of production and sales, measured on the vertical axis, against time, measured on the horizontal axis. Sales increase by $1 billion in the first quarter, but production does not increase until the second quarter. From then on, sales and production keep in-

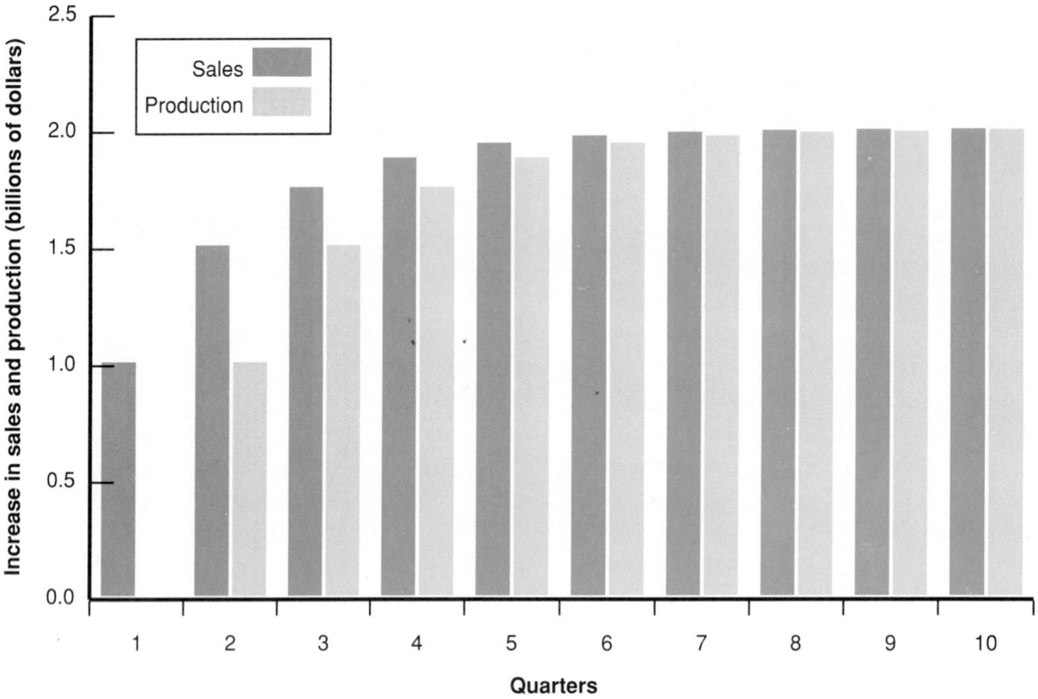

FIGURE 4-2

The Effects of an Increase in Government Spending on Sales and Production

Both sales and production increase over time after a one-time increase in government spending, until they eventually reach the same level.

creasing, with production lagging behind. The two eventually become equal to each other at their new higher equilibrium value. Note how similar this story is to the informal story we told in Chapter 3; what we did there was in effect describe the adjustment process informally.[5]

Telling *how* the economy adjusts over time not only adds to our model's realism, but also strengthens the intuition as to *why* the economy goes from one equilibrium to another. For that very reason, we shall often tell stories of how the economy adjusts over time, although we shall usually not write down the dynamic equations as precisely as we have in this section.

*4-2 GOING EMPIRICAL

What is the actual value of the marginal propensity to consume? Is it the case that consumers fully respond to changes in disposable income within the quarter—as we have assumed so far—or do they actually respond with a lag? To answer these questions, we must look at the data and use econometrics, the set of statistical techniques designed for use on the problems confronting economists. Econometrics can get fairly mathematical, but the basic principles behind econometric techniques are simple. Our purpose in this section is to show you these basic principles.

ESTIMATING THE MARGINAL PROPENSITY TO CONSUME

Remember that the marginal propensity to consume tells us by how much consumption changes when disposable income changes. Thus, to get at the marginal propensity to consume, let's first construct changes in consumption and changes in disposable income, say from quarter to quarter, since 1960. Because changes in variables will be used often in this book, it is worth introducing a symbol to denote them. For a given variable, x, we shall denote $x_t - x_{t-1}$ (that is, the change in the variable x from period $t - 1$ to period t) by Δx_t. Thus, here, we shall denote changes in consumption and changes in disposable income by ΔC_t and ΔY_{Dt} respectively. Both consumption and disposable income are measured in 1986 dollars, at annual rates.

Figure 4-3 plots changes in consumption against changes in disposable income. The vertical axis measures the change in consumption in each quarter since 1960, as a deviation from the mean (average) change. More formally, the variable measured on the vertical axis is constructed as $\Delta C_t - \overline{\Delta C}$, where $\overline{\Delta C}$ is the average quarterly increase in consumption since 1960. Similarly, the horizontal axis measures the change in disposable income minus its mean value for the period 1960–1996, $\Delta Y_{Dt} - \overline{\Delta Y_D}$. Thus, a particular point in the figure gives the deviations of the change in consumption and disposable income from their respective means for a particular quarter since 1960.

---

[5]**Digging deeper.** *You might notice, however, something slightly unpleasant about the adjustment process that we have just described. During every quarter along the adjustment process, production is smaller than sales, so that inventory investment is negative. Put another way, inventories decrease steadily and are lower in the new equilibrium than in the original one. A better specification would allow firms to increase production so as to return to their initial, and presumably preferred, level of inventories. This is very much worth doing but would lead to more complicated dynamics of output. Thus we do not consider it here.*

*An asterisk preceding a chapter or section number indicates that the chapter or section is optional.

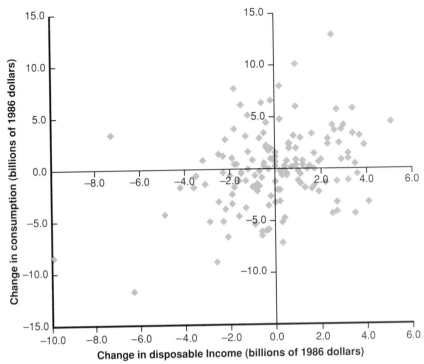

FIGURE 4-3

Changes in Consumption versus Changes in Disposable Income, 1960–1996

There is a clear positive relationship between changes in consumption and changes in disposable income. But the relation is far from tight.

Source: Statistics Canada, CANSIM Series D20111, D20464, and D20557.

Two main conclusions can be drawn from Figure 4-3. First, there is a clear positive relation between changes in consumption and changes in disposable income. The majority of points lie in the northeast and southwest quadrants of the figure: When disposable income increases by more than average, consumption also typically increases by more than average; when disposable income increases by less than average, so typically does consumption. Second, the relation between the two variables is far from tight. In particular, some points lie in the northwest and southeast quadrants: These correspond to quarters when larger-than-average changes in disposable income were associated with smaller-than-average changes in consumption, or the reverse.

The use of econometrics allows us to state these two conclusions more precisely and to get an estimate of the marginal propensity to consume. Using an econometrics software package, we can find the line that fits the cloud of points best. This line-fitting process is called **ordinary least squares (OLS)**.[6] The estimated equation corresponding to the line is called a **regression,** and the line itself is called the **regression line.**

---

[6]*The term* least squares *comes from the fact that the line has the property that it minimizes the sum of the squared distances of the points to the line—thus gives the "least" "squares." The word* ordinary *comes from the fact that distances are measured in the "ordinary" way.*

In our case, the estimated equation is given by

$$\Delta C_t - \overline{\Delta C} = 0.20(\Delta Y_{Dt} - \overline{\Delta Y_D}) + \text{residual} \quad R^2 = 0.11 \qquad (4.5)$$

The regression line corresponding to this estimated equation is drawn in Figure 4-4.

Equation (4.5) reports two important numbers. (Econometrics packages actually give more information than reported above. A typical printout, together with further explanation, is given in the Focus box entitled "A Guide to Understanding Econometric Results" on page 71.)

■ The first is the estimated marginal propensity to consume. The estimated equation tells us that an increase in disposable income of $1 billion above normal is typically associated with an increase in consumption of $0.20 billion above normal. In other words, the estimated marginal propensity to consume is 0.20. It is positive but much smaller than 1.

■ The second important number is R^2, which is a measure of how well the regression line fits.

Having estimated the effect of disposable income on consumption, we can decompose the change in consumption for each quarter into that part that is due to the change in disposable income—the first term on the right in equation (4.5)—and the rest, which is called the **residual.**

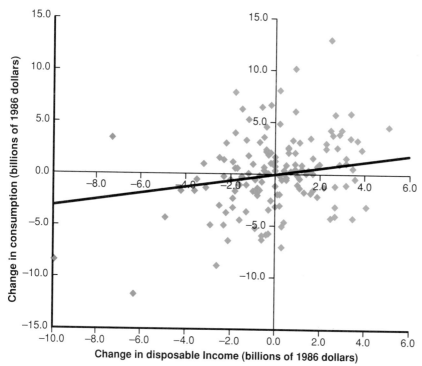

FIGURE 4-4

Changes in Consumption and Changes in Disposable Income:
The Regression Line

The regression line is the line that fits the scatter of points best.

Source: Statistics Canada, CANSIM Series D20111, D20464, and D20557.

If all the points in Figure 4-4 were exactly on the estimated line, all residuals would be equal to zero; all changes in consumption would be explained by changes in disposable income. As you can see, however, this is far from the case. R^2 is a statistic that tells us how well the line fits. R^2 is always between zero and one. A value of 1 would imply that the relation between the two variables is perfect, that all points are exactly on the regression line. A value of zero would imply that the computer could see no relation between the two variables. The R^2 value of 0.11 in equation (4.5) is positive but quite low. It tells us what we already knew from looking at Figure 4-3. While movements in disposable income clearly affect consumption, there is also a lot of movement in consumption that cannot be explained by current movements in disposable income.

ALLOWING FOR LAGS

We have assumed so far that the response of consumption to a change in disposable income happened within the same quarter. This was more for convenience than for realism. It is more than likely that, faced with, say, an unusual increase in disposable income, consumers take a few quarters to adjust their consumption.

In practice, how long does it take consumers to adjust? Put another way, how much does consumption depend on current versus lagged values of disposable income? This is a question that can be answered only by looking at the data. The visual approach we used in Figure 4-3 to look at the relation between current consumption and current disposable income no longer works, however: We would now have to look in three dimensions or more to find the best fit among consumption and current and lagged values of disposable income. But the increase in the number of variables is no problem for econometrics. Using an econometrics package, we can easily find the relation among changes in consumption, current changes in disposable income, and lagged changes in disposable income that fits the data best. It is given by

$$\Delta C_t - \overline{\Delta C} = 0.21(\Delta Y_{Dt} - \overline{\Delta Y_D}) + 0.08(\Delta Y_{Dt-1} - \overline{\Delta Y_D}) \qquad (4.6)$$
$$+ \text{residual} \qquad R^2 = 0.13$$

The computer tells us that, indeed, the response of consumption to changes in disposable income takes some time. The estimated parameters imply that an increase in disposable income this quarter of $1 billion leads to an increase in consumption of $0.21 billion this quarter, and a further increase of $0.08 billion next quarter. Thus, the total increase in consumption due to an increase in disposable income of $1 billion is (0.21 + 0.08) = $0.29 billion. In other words, the marginal propensity to consume is equal to 0.29—a bit higher than what we estimated earlier when we ignored lagged disposable income in equation (4.5); but most of the change in consumption takes place within the same quarter as the increase in disposable income.

How much better do we explain movements in consumption when we allow for a role of both current and lagged disposable income? The answer is given by the value of R^2. R^2 goes up from 0.11 in equation (4.5) to 0.13 here. Thus, allowing for lags of disposable income leads to a small but significantly better fit.

In your readings, you may run across results of estimation using econometrics. Here is a guide, which uses the slightly simplified but otherwise untouched computer output for equation (4.5).

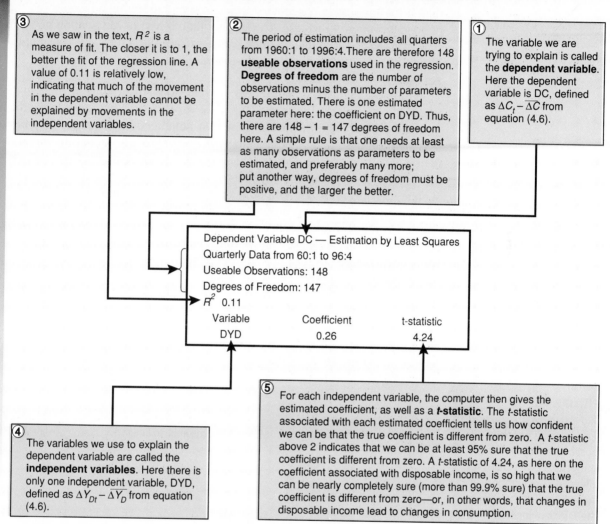

③ As we saw in the text, R^2 is a measure of fit. The closer it is to 1, the better the fit of the regression line. A value of 0.11 is relatively low, indicating that much of the movement in the dependent variable cannot be explained by movements in the independent variables.

② The period of estimation includes all quarters from 1960:1 to 1996:4. There are therefore 148 **useable observations** used in the regression. **Degrees of freedom** are the number of observations minus the number of parameters to be estimated. There is one estimated parameter here: the coefficient on DYD. Thus, there are 148 − 1 = 147 degrees of freedom here. A simple rule is that one needs at least as many observations as parameters to be estimated, and preferably many more; put another way, degrees of freedom must be positive, and the larger the better.

① The variable we are trying to explain is called the **dependent variable**. Here the dependent variable is DC, defined as $\Delta C_t - \overline{\Delta C}$ from equation (4.6).

Dependent Variable DC — Estimation by Least Squares
Quarterly Data from 60:1 to 96:4
Useable Observations: 148
Degrees of Freedom: 147
R^2 0.11

Variable	Coefficient	t-statistic
DYD	0.26	4.24

④ The variables we use to explain the dependent variable are called the **independent variables**. Here there is only one independent variable, DYD, defined as $\Delta Y_{Dt} - \overline{\Delta Y_D}$ from equation (4.6).

⑤ For each independent variable, the computer then gives the estimated coefficient, as well as a **t-statistic**. The t-statistic associated with each estimated coefficient tells us how confident we can be that the true coefficient is different from zero. A t-statistic above 2 indicates that we can be at least 95% sure that the true coefficient is different from zero. A t-statistic of 4.24, as here on the coefficient associated with disposable income, is so high that we can be nearly completely sure (more than 99.9% sure) that the true coefficient is different from zero—or, in other words, that changes in disposable income lead to changes in consumption.

CORRELATION VERSUS CAUSALITY

What we have established so far is that consumption and disposable income typically move together. More formally, we have seen that there is a positive **correlation**—the technical term for "co-relation"—between quarterly movements in consumption and quarterly movements in disposable income. We have interpreted this relation as showing **causality**—that an increase in disposable income *causes* an increase in consumption.

We need to think again about this interpretation. A positive relation between consumption and disposable income may indeed reflect the effect of disposable income on consumption. But it may also reflect the effect of consumption on disposable income. Indeed, the model we have developed in this and Chapter 3 tells us that, if for any reason consumers decide to spend more, then income and thus disposable income will increase. This implies that, if part of the relation between consumption and disposable income comes from the effect of consumption on disposable income, then any conclusion based on equation (4.6) will not be right. An example will help here.

Suppose that consumption does not depend on disposable income, so that the true value of c_1 is equal to zero. (This is not very realistic, but it will make the point most clearly.) Thus, draw the consumption function as a horizontal line in Figure 4-5. Next, suppose that disposable income is equal to Y_D, so that the initial combination of consumption and disposable income is given by point A.

Now suppose that for any reason, say improved consumer confidence, consumers increase their consumption, so that the consumption line shifts up. If demand affects output, then income and in turn disposable income increase, so that the new combination of consumption and disposable income will be given by, say, point B. If, instead, consumers become more pessimistic, the consumption line shifts down, and so in turn does output, leading to a combination of consumption and disposable income given by, say, point D.

If we look at this economy, we observe points A, B, and D. If, as we did above, we then draw the best-fitting line through these points, we estimate an upward-sloping line, such as CC', and thus estimate a positive value for the marginal propensity to consume, c_1. Remember, however, that the true value of c_1 is zero. Why do we get the wrong answer? Because we interpret the positive relation between dis-

FIGURE 4-5

A Misleading Regression
The relation between disposable income and consumption comes from the effect of consumption on income rather than from the effect of income on consumption.

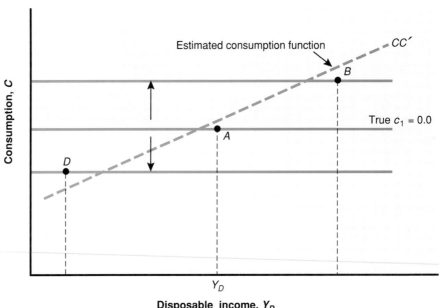

posable income and consumption as showing the effect of disposable income on consumption, where in fact the causality goes the other way: The relation reflects the effect of consumption on disposable income.

There is an important lesson here, *the difference between correlation and causality.* The fact that two variables move together does not imply that the first variable causes the second to move. Perhaps the causality runs the other way. Or perhaps, as is likely here, the causality runs *both* ways: Disposable income affects consumption *and* consumption affects disposable income.

Is there a way out of the correlation-versus-causality trap? If we are interested—as we indeed are—in the effect of disposable income on consumption, can we still learn that from the data? The answer is yes, but only by using more information. Suppose that we *knew* that some changes in disposable income were not triggered by changes in consumption in the first place. Then, by looking at the reaction of consumption to *those* changes in disposable income, we could learn how consumption responds to disposable income; we could estimate the marginal propensity to consume.

This answer would simply seem to assume away the problem: How can we *know* that a change in disposable income is not due to a change in consumption? In fact, we sometimes can. Suppose, for example, that the government embarks on a major increase in defence spending, leading to an increase in demand and, in turn, an increase in output. In that case, if we see both disposable income and consumption increase, we can safely assume that the movement in consumption reflects the effect of disposable income on consumption, and thus estimate the marginal propensity to consume.

This example suggests a general strategy. First find exogenous variables—that is, variables that affect output but are not in turn affected by it. Then look at the movement in consumption in response not to all movements in disposable income—as we did in our earlier regressions—but to those movements in disposable income that can be explained by movements in these exogenous variables. By doing so, we can be confident that what we are estimating is indeed the effect of disposable income on consumption, and not the other way around.

The problem of finding such exogenous variables is known as the **identification problem** in econometrics. These exogenous variables, when they can be found, are called **instruments.** Methods of estimation that rely on the use of such instruments are called **instrumental variable methods.**[7]

When equation (4.6) is estimated using an instrumental variable method—using changes in government spending as the instrument—rather than ordinary least squares as we did earlier, the estimated equation becomes

$$\Delta C_t - \overline{\Delta C} = 0.16(\Delta Y_{Dt} - \overline{\Delta Y_D}) + 0.07(\Delta Y_{Dt-1} - \overline{\Delta Y_D}) \qquad (4.7)$$
$$+ \text{residual} \quad R^2 = 0.12$$

Note that the sum of coefficients on current and lagged disposable income is smaller than it was in equation (4.6). The response of consumption to an increase in disposable income of $1 billion is now only $(0.16 + 0.07) = \$0.23$ billion. In other words, the marginal propensity to consume is 0.23, compared with 0.29 in equation (4.6). This

---

[7]*There is no free lunch. With purely observational data, such as we have in economics, statistical methods alone can never distinguish causation from correlation. To do that, we must exploit economic theory.*

decrease in the estimated marginal propensity to consume is exactly what one would expect. Our earlier estimate in equation (4.6) reflected not only the effect of disposable income on consumption, but also the effect of consumption back on disposable income. The use of instruments eliminates this second effect and thus leads to a smaller estimated effect of disposable income.

*4-3 SIMULATING THE ESTIMATED MODEL

Now that we have estimated a consumption function in the form of equation (4.7), let's return to the dynamic effects of a change in government spending. If we maintain the assumption that current production is equal to last quarter's sales, our model is now given by

$$Y_{t+1} = Z_t$$
$$Z_t \equiv C_t + I_t + G_t$$
$$C_t = c_0 + 0.16Y_{Dt} + 0.07Y_{Dt-1}$$
$$Y_{Dt} \equiv Y_t - T_t$$
$$I_t = \bar{I}$$
$$G_t = G$$

Note that this model contains three types of equations. The first equation is the equilibrium condition in the goods market. The second and fourth equations are identities, which define demand and disposable income respectively. The others are behavioural equations. The only change from the model of Section 4-1 is the specification of the consumption function, where we now allow for lags of disposable income and use the coefficients estimated in equation (4.7) and assume the residual to be equal to zero.[8]

We can simulate this estimated model to look at the effects of, say, a permanent increase in government spending in the first quarter of 2001 on output and sales over time. Solving the model by hand is unwieldy but is no problem for the computer. Figure 4-6 plots the response of sales and production over time. If you compare the response of production in this case to the response of production in Figure 4-2, you will notice two differences.

The first is that the ultimate effect on output is smaller. In Figure 4-2 the increase in government spending of $1 billion eventually led to an increase in output of $2 billion. Here, the eventual effect is only of $1.3 billion. The reason is that the estimated marginal propensity to consume, 0.23, is smaller than the value we had assumed in Figure 4-2, namely 0.5. Thus, the multiplier is smaller.

The second is that the adjustment is slightly slower. This is no surprise: The slower adjustment of consumption to disposable income implies a slower effect of income on demand, and thus back on output. But it gives a first hint about the difficulty of using fiscal policy, say to take the economy out of a recession. Even if the govern-

---

[8]***Digging deeper.*** *Note that, whereas we estimated equation (4.7) by looking at the relation between changes in consumption and changes in disposable income, we state the relation here in terms of levels. Also, because we estimated the relation in terms of changes, we did not get an estimate for c_0, but c_0 plays no role in the effect of government spending on output.*

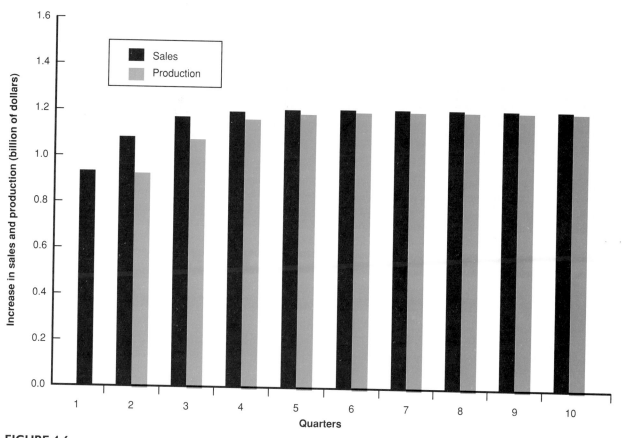

FIGURE 4-6

Estimated Effects of an Increase in Government Spending on Sales and Production

When we allow for lags in the response of consumption to disposable income, the effect of government spending on output is slower. Compare this graph with Figure 4-2.

ment can increase spending right away, the effect of an increase in spending takes a while to build up. By the time it actually has a large effect on output, the recession may already be over. The issue of lagged effects of policy is one to which we shall often return.

4-4 CONCLUSIONS

We have covered a lot of ground in this chapter, especially in the last part of it. We wanted, however, to give you a feeling for how economists build models. The model we have developed is simple, indeed too simple: It obviously leaves out important aspects of reality that we shall introduce in future chapters. But it already captures a central macroeconomic mechanism, the interaction among production, income, and demand. It is actually quite similar to the first macroeconometric model of the U.S. economy ever built, a model built by Lawrence Klein, from the University of Pennsylvania, in the 1950s—a time when computers were in their infancy and even simple regressions required considerable time and effort.[9] Macroeconometric models today are much larger and much more complex. But the principles underlying their construction are the same as those that we have seen in this chapter: specification of the relations in the model, estimation using econometrics, and simulation of the model to look at the effects of shocks or changes in policy.

SUMMARY

■ Most decisions in economics have a time dimension. A consumer who gets a raise at work may not adjust her consumption right away. A firm that faces an increase in demand may decide to wait before adjusting its production, relying meanwhile on its inventories to satisfy demand.

■ When firms adjust production to demand with a lag, changes in autonomous spending have dynamic effects on output. In response to an increase in autonomous spending, output increases over time to a new, higher level. This new level is given by the multiplier times the change in autonomous spending.

■ Macroeconomists use econometrics to estimate relations among variables. For example, econometrics can be used to estimate the marginal propensity to consume, as

well as the dynamic response of consumption to changes in disposable income.

■ One of the difficulties in interpreting data is the need to distinguish between correlation and causality. The fact that two variables move together does not imply that a change in the first causes the second to move. To find causal relations, such as, for example, the effect of changes in disposable income on consumption, macroeconomists use instrumental variable techniques.

■ Macroeconomic models are collections of equations. Some equations are identities. Some are equilibrium conditions. Some are behavioural relations estimated using econometrics. These models are then simulated to study the effects of shocks or policies on the economy.

KEY TERMS

■ dynamics, 60
■ econometrics, 60
■ lagged value, 61
■ experiment, 63
■ simulation, 63
■ geometric series, 65
■ ordinary least squares (OLS), 68
■ regression, regression line, 68
■ residual, 69
■ R^2, 71

■ dependent variable, 71
■ independent variable, 71
■ useable observations, 71
■ degrees of freedom, 71
■ t-statistic, 71
■ correlation, 71
■ causality, 72
■ identification problem, 73
■ instruments, 73
■ instrumental variable methods, 73

QUESTIONS AND PROBLEMS

1. Consider the following dynamic model of the goods market:

$$C_t = 0.50 + 0.75(Y_t - T)$$
$$I = 25$$
$$G = 150$$
$$T = 100$$
$$Z_t = C_t + I + G$$
$$Y_{t+1} = Z_t$$

a. Solve for equilibrium GDP, under the assumption that GDP is constant.

b. Suppose that the economy is in equilibrium, and then government spending decreases by 100 in period t. Construct a table showing what happens to consumption (C), demand (Z), and output (Y) in periods t, $t + 1$, and $t + 2$.

c. Plot income and demand over time, from t to $t + 2$.

---

[9]The main difference between Klein's model and the one presented here is that Klein's included an equation relating investment to sales, whereas we have not tried to explain investment here. Klein was given the Nobel Prize in 1980 for his contributions to macroeconometric modelling.

d. When output eventually returns to a new constant value, what are the values of:
   (i) Income?
   (ii) Demand?
   (iii) Consumption?
e. How many periods must pass before 75% of the ultimate decrease in output has occurred?

2. In the early 1980s, several studies showed that children with poor eyesight tend to have higher IQs, and that the worse the eyesight, the greater the IQ. One study concluded that IQ and poor eyesight are inherited through the same gene.
   a. Give at least two alternative explanations for the data.
   b. How does this question relate to the material in this chapter? Explain briefly.

3. For each of the following statements, answer "true" or "false" and explain briefly.
   a. In a dynamic model of the goods market, there is no multiplier.
   b. In a dynamic model of the goods market, firms maintain constant levels of inventories from period to period.
   c. In a dynamic model of the goods market, the only way to calculate equilibrium GDP is to run a simulation and add up the changes to GDP in each period.

*4. Suppose that we have the following dynamic model of the goods market:

$$C_t = 100 + 0.5(Y_{t-1} - T)$$
$$I_t = 200 + 0.25Y_{t-1}$$
$$G = 100$$

$$T = 100$$
$$Z_t = C_t + I_t + G$$
$$Y_{t+1} = Z_t$$

Note that in this model investment spending is endogenous, and both investment spending and consumption spending depend on *lagged* output.
   a. Solve for equilibrium output, under the assumption that output is constant.
   b. Suppose that the economy is in equilibrium in period 1 and then government spending rises from 100 to 200 in period 2. What will happen to C_t, I_t, Z_t, and Y_t in periods 2, 3, 4, and 5? Summarize your results in a table.
   c. Determine the ultimate impact of the rise in G on equilibrium GDP. [*Hint:* What geometric series do you see in your table?]
   d. Based on your answer in part c, what is the value of the multiplier in this model? Does the presence of endogenous investment make the multiplier larger or smaller?

5. The consumption equations in two alternative dynamic models are as follows:

   Model A: $C_t = c_0 + 0.25Y_{Dt} + 0.15Y_{Dt-1}$
   Model B: $C_t = c_0 + 0.20Y_{Dt} + 0.15Y_{Dt-1} + 0.05Y_{Dt-2}$

   a. What is the marginal propensity to consume in model A? In model B?
   b. What is the multiplier in model A? In model B?
   c. Would a policy maker need to know which of these two models describes consumption spending in the economy? Why or why not?

FURTHER READING

A good introduction to understanding econometrics is given in Peter Kennedy, *A Guide to Econometrics, 4th ed.* (Cambridge: MIT Press, 1998).

FINANCIAL MARKETS

Barely a day goes by without newspaper discussions of whether Canada's central bank, the **Bank of Canada,** or its U.S. counterpart, the **Federal Reserve Bank** (the **Fed**), is going to increase/decrease interest rates and what this is likely to do to the economy. The model of economic activity we developed in Chapters 3 and 4 did not include interest rates. This was a strong simplification, and it is time to relax it. As we shall see, interest rates play an important role in the determination of the demand for goods, and thus the determination of output. And interest rates are determined in part by our central bank, implying an important role for monetary policy in the determination of output.[1]

Interest rates, their determination, and their effects on economic activity are the focus of this and the next chapter. This chapter focusses on the determination of interest rates in **financial markets**—the markets where financial assets are bought and sold—and the role of monetary policy in their determination. The next chapter looks at financial and goods markets simultaneously, focussing on the interaction between interest rates and output, and on the roles of fiscal and monetary policy.

---

[1] *The role played by the Fed (and other central banks around the world) in determining Canadian interest rates is extremely important but will have to wait until Chapter 13.*

5-1 MONEY VERSUS BONDS

In countries with modern financial markets, such as Canada, people have the choice among literally thousands of different financial assets, from money, to bonds, to stocks, to mutual funds.

To focus on the basic mechanism by which interest rates are determined, we shall begin by ignoring most of that complexity. We shall think of an economy in which there are only two financial assets:

- **Money,** which can be used for transactions and pays zero interest
- **Bonds,** which cannot be used for transactions but pay a positive interest rate, which we denote by i

Later in this chapter, we shall distinguish between different forms of money: **currency** (coins and bills issued by the central bank) and **chequable deposits** (deposits at banks and other financial institutions on which one can write cheques). And in later chapters (starting in Chapter 7), we shall distinguish among short-term bonds, long-term bonds, and stocks. For simplicity's sake, the precise definitions of these different assets are best left until later.

In our simplified economy, people thus face one financial choice: how much of their wealth to hold in money versus how much to hold in bonds. In this section we

SEMANTIC TRAPS: MONEY, INCOME, AND WEALTH

In everyday life we use the word "money" to denote many things. We use it as a synonym for income: "making money." We use it as a synonym for wealth: "she has a lot of money." In economics, you must be more careful. Here is a basic guide to some terms and their precise meanings.

Income is what you earn—from working, or from rental income, or from interest and dividends. It is a **flow**—that is, it is expressed per unit of time: weekly income, monthly income, or yearly income. J. Paul Getty was once asked what his income was. To the surprise of the person who had asked the question, Getty answered: "$1000." He meant but did not say: per minute.

Saving is that part of after-tax income that is not consumed. It is also a flow. For example, if you save 10% of your income and your income is $3000 per month, you save $300 per month. **Savings** (the plural form) is sometimes used as a synonym for wealth, the value of what you have accumulated over time. To avoid potential confusion, we shall not use it in this book.

Your **financial wealth,** or simply **wealth** for short, is the value of all your financial assets minus all your financial liabilities. In contrast with income or saving, which are flow variables, financial wealth is a **stock** variable. It gives the value of wealth at a given moment of time. At a given moment of time, you cannot change the total amount of your financial wealth. You can do this only over time, as you save or dissave, or/and as the values of your assets change. But you can change the composition of your wealth; you can, for example, change the proportion of stocks versus bonds in your portfolio.

Those financial assets that can be used directly to buy goods are called *money*. Money includes currency and chequable deposits, deposits against which one can write cheques. Money is also a stock. Somebody can have a large wealth but small money holdings—for example, $1 000 000 worth of stocks but only $500 in his chequing account. Or somebody can have a large income but small money holdings—for example, be paid $10 000 a month but have a very small positive balance in her chequing account.

Investment is a term that economists reserve for the purchase of new capital goods, from machines to plants to office buildings. When you refer to the purchase of shares of stock or other financial assets, you should use the expression **financial investment** instead.

The bottom line: Learn how to be economically correct. Do not say "Mary is making a lot of money"; say "Mary receives a high income." Do not say "Joe has a lot of money"; say "Joe is very wealthy."

focus on this choice. In the next two sections we look at both the demand for and the supply of money, the determination of the interest rate, and the role of the central bank in that determination. Before we do so, however, it is useful to clarify some of the terms we shall be using as we go along. This is done in the Focus box entitled "Semantic Traps: Money, Income, and Wealth."

THE BASIC CHOICE

Suppose that as a result of having steadily saved part of your income in the past, your financial wealth today is $50 000. You may intend to keep saving in the future and increase your wealth further, but its value today is given. The choice you have to make today is how to allocate this $50 000 between money and bonds.

Think of buying or selling bonds as implying some cost: for example, a phone call to a broker and the payment of a transaction fee. What should your portfolio choice be? More precisely, how much of your $50 000 should you hold in money, and how much should you hold in bonds?

Holding all your wealth in money completely avoids the need to call your broker or pay transaction fees. But it also means receiving no interest income on your wealth. In contrast, holding all your wealth in bonds implies receiving interest on all your wealth, but having to call your broker whenever you need to take the subway or pay for a cup of coffee—a rather inconvenient and unpleasant way of going through life.

Clearly, you should hold both money and bonds. But in what proportions? Your decision will depend on two main variables:

■ The first is your *level of transactions.* You want to have enough money on average so as to avoid going through bond sales too often. Say that you typically spend $5000 a month. You may want to have on average, say, two months' worth of spending on hand, or $10 000 in money, and thus $50 000 − $10 000 = $40 000 in bonds. If, instead, you typically spend $6000 a month, you may want to have $12 000 in money and thus only $38 000 in bonds.

■ The second is the *interest rate on bonds.* The only reason to hold any of your wealth in bonds is that they pay interest. If bonds paid no interest, you would hold all of your wealth in money, simply because money can be used for transactions and is therefore more convenient. The higher the interest rate, the more you will be willing to incur the hassle and the costs associated with buying and selling bonds. If the interest rate is very high, you may decide to squeeze your money holdings to an average of only two weeks' worth of spending, or $2500 (assuming that your monthly spending is $5000). This means that you will be able to keep on average $47 500 in bonds, getting more interest as a result.

Let's make this last point more concrete. Most of you do not hold bonds directly; few of you, we suspect, have a broker. But many of you may hold bonds indirectly, through a money market account. **Money market funds** receive funds from people and firms and use these funds to buy bonds, typically government bonds. Money market funds pay an interest rate close to the interest rate on the bonds that they hold, the difference coming from the administrative costs of running these funds and from their profit margin. In the early 1980s, with the interest rate on money market funds reaching 14% per year, people who had previously kept all their financial wealth in their chequing accounts (which paid no interest) realized how much interest they could earn by holding part of those funds in a money market account instead. Chequing account balances dwindled. Since then, however, the interest rate has decreased. In 1997, the interest rate paid by money market funds was down to around 2.5 percent. This is better than zero—the rate paid on many chequing accounts—but clearly much less attractive than the rate in 1981. As a result, most people are less careful about putting as much as they can in their money market fund. Put another way, for a given level of transactions, they tend to keep more on average in their chequing account than they did in 1981.

THE DEMAND FOR MONEY

Let's formalize our discussion of the demand for money as follows.

Let "$Wealth" represent the financial wealth of households in dollars—thus, the dollar sign. At a point in time, financial wealth is given. Households must choose how much to hold in money and how much to hold in bonds. Let their demand for money be denoted by M^d and their demand for bonds by B^d. (The superscript d stands for *demand*.) Whatever they choose, their decisions must be such that money and bond holdings add up to their wealth:

$$M^d + B^d = \$Wealth \tag{5.1}$$

We just saw that an individual's money demand depends primarily on two variables, his level of transactions and the interest rate. This suggests that money demand for the economy as a whole depends on the overall level of transactions in the economy and on the interest rate. The overall level of transactions is hard to measure. But it is reasonable to assume that it is roughly proportional to nominal income—equivalently nominal output—which we denoted by $Y in Chapter 2. If nominal income increases by, say, 10%, it is reasonable to think that the amount of transactions in the economy will also increase by roughly 10%. Thus, we write

$$M^d = \$YL(i) \tag{5.2}$$
$$(-)$$

Price level
real GDP
financial Innovations

Read this equation as "Money demand is equal to nominal income times a function of the interest rate, denoted $L(i)$." This equation summarizes what we have said so far:

■ First, the demand for money increases in proportion to nominal income. If

FIGURE 5-1

The Demand for Money

For a given level of nominal income, the demand for money is a decreasing function of the interest rate.

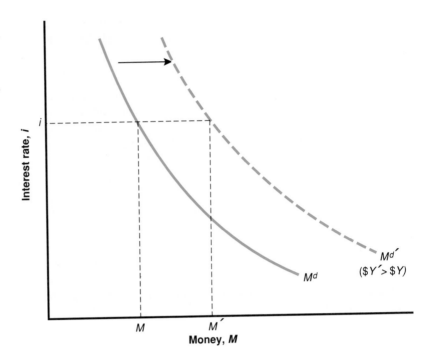

income doubles, increasing from $\$Y$ to $\$2Y$, then the demand for money increases from $\$YL(i)$ to $\$2YL(i)$; thus it also doubles.

■ Second, the demand for money depends negatively on the interest rate. This relation is captured by the function $L(i)$ and the negative sign under the interest rate, which indicates a negative relationship: The demand for money decreases when the interest rate increases.

The relation between the demand for money and the interest rate implied by equation (5.2) is represented in Figure 5-1. The interest rate, i, is measured on the vertical axis. Money, M, is measured on the horizontal axis. The M^d curve represents the demand for money for a given level of nominal income, $\$Y$. It is downward-sloping, because a lower interest rate leads to a higher demand for money. The $M^{d'}$ curve gives the demand for money for a higher level of nominal income, $\$Y'$. At a given interest rate, say i, an increase in nominal income from $\$Y$ to $\$Y'$ increases the demand for money from, say, M to M'. Put another way, the demand for money shifts to the right, from M^d to $M^{d'}$. At any interest rate, the demand for money is larger than before the increase in nominal income.

THE DEMAND FOR BONDS

Remember that the demand for money and the demand for bonds are not independent decisions but that the two have to add up to financial wealth. We can thus look at the demand for bonds implied by the demand for money. From equations (5.1) and (5.2), the demand for bonds is given by

$$B^d = \$\text{Wealth} - M^d \qquad (5.3)$$
$$= \$\text{Wealth} - \$YL(i)$$

An increase in wealth leads to a one-for-one increase in the demand for bonds. This conclusion comes from our assumption that the demand for money depends on income and the interest rate, not on wealth. Thus, an increase in wealth goes into higher bond holdings rather than into higher money holdings. An increase in income leads to an increase in the demand for money, and thus to a decrease in the demand for bonds.[2] And an increase in the interest rate, which makes bonds more attractive, leads to an increase in the demand for bonds.

Throughout this chapter we shall work with the demand for money and think of the equilibrium condition in financial markets as the condition that the demand for money equals the supply of money. But we could work instead with the demand for bonds, and look at the equilibrium condition as the condition that the demand for bonds is equal to the supply of bonds. These are two ways of looking at the same condition.

MONEY DEMAND AND THE INTEREST RATE: THE EVIDENCE

Before moving on, we must ask the same question we asked of our consumption function in Chapter 4. How well does our equation for the demand for money, equation (5.2), fit the facts? In particular, how much does the demand for money actually respond to movements in the interest rate?

To get at the answer, note first that dividing both sides of equation (5.2) by $\$Y$, we can write

$$\frac{M^d}{\$Y} = L(i)$$

This equation says that the ratio of money demand to nominal income is a decreasing function of the interest rate. This relation provides the motivation for Figure 5-2, which plots the ratio of money to nominal income and the interest rate against time for the period 1960 to 1996.

The ratio of money to nominal income is constructed as follows. Money, M, is constructed as the sum of currency (Canadian coins and bills held outside banks), and chequable deposits (deposits against which cheques can be written) less the value of cheques in transit.[3] This measure of money, which is constructed by the Bank of Canada, is called **M1.** The Bank also constructs other "monetary aggregates," called $M1B$, $M2$, $M3$, and so on. It does so because, in the real world, the boundary between money and other assets is fuzzier than in our simple economy with money and bonds alone. For example, money market funds (which are not included in $M1$) usually allow for cheques to be written against them, but the cheques have to exceed a certain amount, often $500. Savings deposits are not included in

---

[2]***Digging deeper.*** *This result may seem counterintuitive. Your intuition may have been that higher income leads to higher saving, and thus to an increase in bond holdings over time. If so, your intuition is right: As they save more and their wealth increases over time, people are indeed likely to hold more bonds. But we are looking at what happens at the moment when income increases and wealth has not yet changed. At that moment,* given wealth, *the only way to increase money holdings is to decrease bond holdings.*

[3]***Digging deeper.*** *While we have assumed that money does not pay interest, some chequable deposits actually pay a positive interest rate. But the interest rate is substantially lower than the interest rate on bonds, so that our assumption that money pays no interest is not misleading. You may want to think about how equation (5.2) would change if we allowed money to pay a positive, but low, interest rate.*

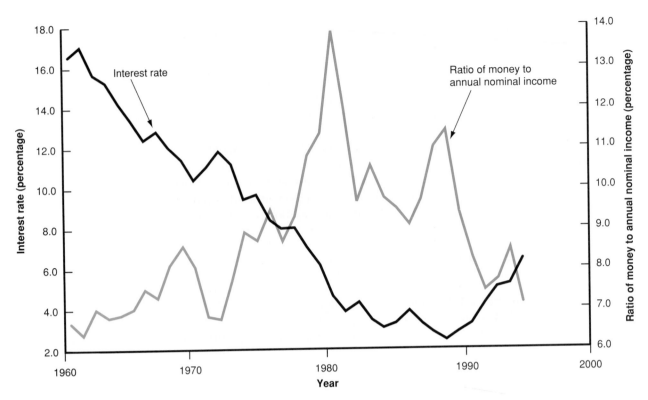

FIGURE 5-2

The Ratio of Money to Nominal Income and the Interest Rate, 1960–1996

The ratio of money to nominal income has decreased over time. Leaving aside this trend, the interest rate and the ratio of money to nominal income move roughly in opposite directions.

Source: Statistics Canada, CANSIM Series B14001, B1627, and D10000.

$M1$, but funds can be shifted from savings accounts to chequing accounts at nearly no cost and in nearly no time. We shall return to the construction of monetary aggregates in Chapter 28. For the time being, we shall use $M1$ as our basic measure of money.

Nominal income is measured by nominal GDP (at an annual rate), $\$Y$. The interest rate, i, is measured by the average interest rate paid by government bonds during each year. (More specifically, it is the average over the year of the three-month Treasury-bill rate. We shall defer a precise definition of Treasury bills until Section 5-2).

Figure 5-2 has two striking characteristics:

(1) The first is the large decline in the ratio of money to nominal income since 1960. The interest rate was roughly the same in 1996 as it was in the early 1960s. Yet the ratio of money to nominal income declined from nearly 13% in 1960 to 8% in 1996.

Economists sometimes refer to the inverse of this ratio—that is, to the *ratio of nominal income to money*—as **velocity.** The word comes from the intuitive idea that when the ratio of nominal income to money is higher the number of transactions for a given quantity of money is higher, and money is thus changing hands faster; in other words, the *velocity* of money is higher. Therefore, another—equivalent—way of stating the first characteristic of Figure 5-2 is that velocity increased from about 7.7 (1/.13) in 1960 to about 12.5 (1/.08) in 1996.

Why has velocity increased so much over the last 40 years? The reason is not hard to guess. Many innovations in financial markets have made it possible to hold lower money balances for a given amount of transactions. Perhaps the most important development is the increased use of credit cards. At first glance, credit cards would appear to be money: When we go to a store, aren't we asked whether we want to pay with cash, cheque, or credit card? But, despite appearances, credit cards are not money. You actually do not pay when you use your credit card at the store; you pay when you receive your bill and send your monthly payment. What credit cards allow you to do is concentrate many of your payments in one day, and thus decrease the average amount of money you need to have during the rest of the month.[4] One would therefore expect the introduction of credit cards to reduce money demand steadily in relation to nominal income over time. Figure 5-2 shows that this indeed has been the case.

(2) The second characteristic of Figure 5-2 is the negative relation between the ratio of money to nominal income and the interest rate. For example, note how the ratio has increased as the interest rate has decreased since 1981. Note also how the increase in the interest rate in the late 1980s is reflected in a decrease in the ratio, the decrease in the interest rate in the early 1990s is reflected in an increase in the ratio, and so on.

The trend evolution of the ratio of money to nominal income in Figure 5-2 makes it difficult to see the relation between this ratio and the interest rate. To show the relation more clearly, the scatter diagram in Figure 5-3 plots *changes* in the ratio versus *changes* in the interest rate from year to year. Changes in the interest rate are measured on the vertical axis. Changes in the ratio of money to nominal income are measured on the horizontal axis. Each point in the figure corresponds to a given year. The figure gives a clearer picture than Figure 5-2. It shows a clear negative relation between year-to-year changes in the interest rate and changes in the ratio. If we were to fit a regression line, it would not fit perfectly; this is again a reminder that behavioural equations such as equation (5.2) never fit exactly. But the line would clearly be downward-sloping, as predicted by our money demand equation.

5-2 THE DETERMINATION OF THE INTEREST RATE: I

Let's turn to the determination of the interest rate. (There are many interest rates in the real world. But for the moment we are assuming that there is only one type of bond, and thus only one interest rate. We shall relax this assumption in Chapter 7.) We have looked at the demand for money—and the implied demand for bonds. We now need to introduce the supply side. We shall assume in this section that money is supplied only by the central bank, that all money is currency.[5] In Section 5-3 we introduce banks and look at the mechanics of money supply in an economy with both currency and chequable deposits.

---

[4]Some, but not all, credit cards also allow you to defer payments, and thus to borrow, often at a high interest rate. This is a separate function and not the one that is relevant here.

[5]Strictly speaking, the Bank of Canada supplies bank notes but not coins; we shall ignore this distinction.

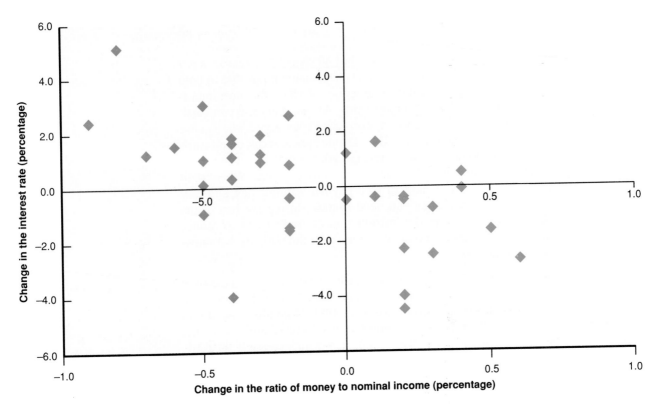

FIGURE 5-3

Changes in the Interest Rate versus Changes in the Ratio of Money to
Nominal Income, 1960–1996

Increases in the interest rate are typically associated with a decrease in the ratio of
money to nominal income, decreases in the interest rate with an increase in that ratio.

Source: Statistics Canada, CANSIM Series B14001, B1627, and D10000.

MONEY DEMAND, MONEY SUPPLY, AND THE EQUILIBRIUM INTEREST RATE

Let's start with the simplest possible assumptions about supply: Assume that the stocks of money and bonds are both given. Let's denote them by M and B respectively. Thus, we can write financial wealth in the economy as

$$\text{\$Wealth} = M + B \tag{5.4}$$

Financial wealth is equal to the sum of the stock of money (that is, the money supply) and the stock of bonds.

Financial markets are in equilibrium if the supply of money is equal to the demand for money, or—equivalently—if the supply of bonds is equal to the demand for bonds. Looking at the supply and demand for money, and using equation (5.2) for money demand, the equilibrium condition is thus given by

$$\text{money supply} = \text{money demand}$$
$$M = \$YL(i) \tag{5.5}$$

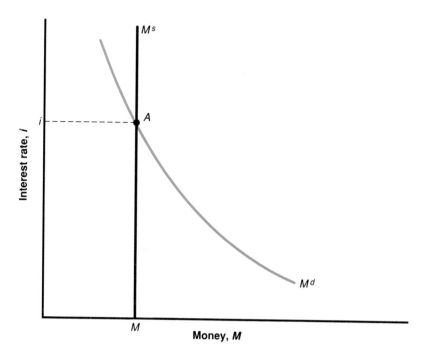

FIGURE 5-4

The Determination of the Interest Rate

The equilibrium interest rate is such that the supply of money is equal to the demand for money.

This equation tells us that the interest rate must be such that people are willing to hold an amount of money equal to the existing money supply. This equilibrium relation is called the **LM relation.**[6]

Before we explore what equation (5.5) implies for the reaction of the interest rate to changes in nominal income and to changes in the money supply, let's check that the condition that the supply of bonds be equal to the demand for bonds indeed gives the same relation as equation (5.5). From equation (5.3), the demand for bonds is given by

$$B^d = \$Wealth - M^d$$

From equation (5.4), the supply of bonds and the supply of money are related by

$$supply \ of \ bonds = demand \ for \ bonds$$
$$B = \$Wealth - M$$

If $M = M^d$, it follows that $B = B^d$. If the supply of and the demand for money are equal, so are the supply of and demand for bonds. Thus, we can think of equation (5.5) as either the condition that the supply of and demand for money be equal, or that the supply of and demand for bonds be equal.

The equilibrium condition in equation (5.5) is characterized in Figure 5-4. Just as in Figure 5-1, money is measured on the horizontal axis, and the interest rate is measured on the vertical axis. The demand for money, M^d, is drawn for a given level

---

[6]**Digging deeper.** *The letters* L *and* M *in* LM *stand for "liquidity" and "money." Economists use liquidity as a measure of how easily and how cheaply an asset can be exchanged for money. Thus, money is fully liquid, other assets less so, and one can think of the demand for money as a demand for liquidity. This is the origin of the letter "L" in* LM. *The letter "M" stands for money: The demand for liquidity must be equal to the supply of money. (The relation between liquidity and money is explored at more length in Chapter 28.)*

of nominal income. It is downward-sloping: A higher interest rate implies a lower demand for money. The supply of money is fixed and thus does not depend on the interest rate. It is a vertical line at M; the line is denoted by M^s in the figure. Equilibrium is thus at point A, with interest rate i.

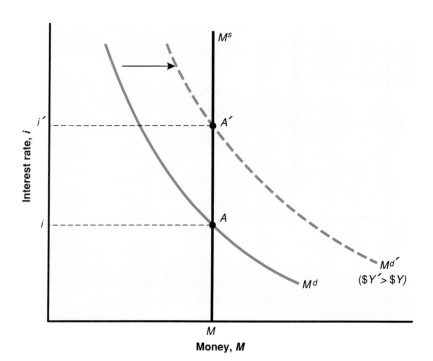

FIGURE 5-5

The Effects of an Increase in Nominal Income on the Interest Rate

An increase in nominal income leads to an increase in the interest rate.

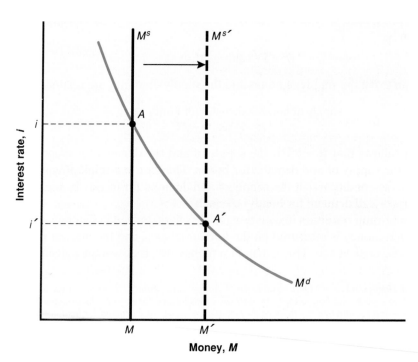

FIGURE 5-6

The Effects of an Increase in the Money Supply on the Interest Rate

An increase in the money supply leads to a decrease in the interest rate.

Figure 5-5 shows the effects of an increase in nominal income on the interest rate. The figure replicates Figure 5-4, and the initial equilibrium is at point A. An increase in nominal income increases the level of transactions and thus increases the demand for money at any interest rate. The demand curve shifts to the right, from M^d to $M^{d'}$. The equilibrium moves from A to A', and the equilibrium interest rate increases from i to i'. Thus, *an increase in nominal income leads to an increase in the interest rate.* The reason is clear: At the initial interest rate, the demand for money exceeds the (unchanged) supply. An increase in the interest rate is needed to decrease the amount of money people want to hold and to re-establish equilibrium.

Figure 5-6 shows the effects of an increase in the money supply on the interest rate. The initial equilibrium is at point A, with interest rate i. An increase in the money supply, from M to M', leads to a shift of the supply curve to the right, from M^s to $M^{s'}$. The equilibrium moves from A to A', and the interest rate decreases from i to i'. Thus, *an increase in the supply of money leads to a decrease in the interest rate.* The decrease in the interest rate is needed to increase the demand for money so that it is equal to the larger money supply.

MONETARY POLICY AND OPEN-MARKET OPERATIONS

We can get a better intuition for the results we saw in Figures 5-5 and 5-6 by looking more closely at how the central bank actually changes the money supply and what happens in financial markets when it does so.

Think of our economy as having a market where people buy and sell bonds in exchange for money. Those who want to increase the proportion of bonds in their portfolio buy bonds. Those who want to decrease it sell bonds. In equilibrium, the interest rate is such that the demand for bonds is equal to the supply of bonds, or, equivalently, the supply of money is equal to the demand for money, and the equilibrium condition (5.5) holds.

Now, think of the central bank as changing the stock of money in the economy by buying and selling bonds in the bond market. If it wants to increase the stock of money, the central bank buys bonds and pays for them by creating money. If it wants to decrease the stock of money, the central bank sells bonds and removes from circulation the money it receives in exchange for the bonds. Such operations are called **open market operations,** so called because they take place in the "open market" for bonds.[7]

The balance sheet of the central bank in this economy is given in Figure 5-7. The central bank's assets are the bonds that it holds in its portfolio. Its liabilities are the stock of money in the economy. Open market operations lead to equal changes in assets and liabilities. If the central bank buys, say, $1 million worth of bonds, the amount of bonds it holds is higher by $1 million and so is the amount of money in circulation. If it sells $1 million worth of bonds, both the amount of bonds held by the central bank and the amount of money in the economy are lower by $1 million.

---

[7]*Open-market operations also include the use of purchase and resale agreements. These and other mechanisms used to "fine-tune" the money supply are discussed in Chapter 28.*

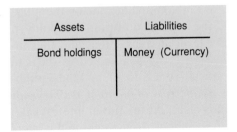

Assets	Liabilities
Bond holdings	Money (Currency)

FIGURE 5-7

The Balance Sheet of the Central Bank

Open market operations change both the assets and the liabilities of the central bank.

Let us now be more specific about the way the bond market works in this economy. We have assumed that the bond market determines the interest rate on the bonds. In fact, bond markets typically determine not the interest rate but rather the price of bonds. The interest rate can then be inferred from the price. Let us look at the relation between interest rate and price more closely.

Let the bonds be one-year bonds that promise payment of $100 a year hence. In Canada, such bonds, when issued by the government and promising payment in a year or less, are called **Treasury bills,** or simply **T-bills.** Thus, you can think of the bonds in our economy as one-year T-bills. Let their price today be $\$P_B$, where B stands for "bond." If you buy the bond today and hold it for a year, the rate of return on holding the bond for a year is equal to $(\$100 - \$P_B)/\$P_B$ (what you get for the bond at the end of the year minus what you pay for the bond today, divided by the price of the bond today). Thus, the interest rate on the bond is defined by

$$i = \frac{\$100 - \$P_B}{\$P_B}$$

For example, if $\$P_B$ is equal to $95, the interest rate is equal to $5/$95, or 5.3%. If $\$P_B$ is $90, the interest rate is 11.1%. *The higher the price of the bond, the lower the interest rate.*

Equivalently, if we are given the interest rate, we can infer the price of the bond. Reorganizing the formula above, the price of a one-year bond is given by

$$\$P_B = \frac{\$100}{1 + i}$$

The price of the bond is equal to the final payment divided by 1 plus the interest rate. Thus, if the interest rate is positive, the price of the bond is less than the final payment. And the higher the interest rate, the lower the price today. When newspapers write that "bond markets went up today," they mean that the prices of bonds went up and therefore that interest rates went down.

Let's now look at the effects of an open-market operation in which the central bank increases the supply of money—an **expansionary open-market operation.** In such a transaction, the central bank buys bonds in the bond market and pays for them by creating money. As it buys bonds, the demand for bonds goes up and thus the price of bonds goes up. Equivalently, the interest rate on bonds goes down. When, instead, the central bank wants to decrease the supply of money—and thus does a **contractionary open-market operation**—it sells bonds. This leads to a decrease in their price, and thus to an increase in the interest rate.

↑ ms ⟹ Central bank must buy up bonds; BD↑ which ↑P Bonds ↓ ↓ in i

↓ ms ⟹ C.B. must sell bonds ↑ Bs ↓P Bonds ↑r

MONETARY POLICY AND THE INTEREST RATE

Our economy with its two assets, money and bonds, is a much-simplified version of actual economies and their many financial assets and many financial markets. But the basic lesson we learn here is general: The interest rate is determined by the equality of the supply of money and the demand for money. The central bank can affect the interest rate. It does so through open-market operations in bond markets. Open-market operations in which the central bank increases the money supply by buying bonds lead to an increase in the price of bonds, and thus to a decrease in the interest rate. Open-market operations in which the central bank decreases the money supply by selling bonds lead to a decrease in the price of bonds, and an increase in the interest rate. Open-market operations are indeed one of the basic tools used by most modern central banks to affect interest rates.

We have made one simplification in this section, however, that we must now relax. We have assumed that all money is currency, supplied by the central bank. Most money is actually supplied by private banks. In Canada, only 20% of $M1$ is currency; the rest, 80%, is chequable deposits. Thus, we must have a second look at the supply of money, taking into account the money supplied by banks. This is what we do in the next section. That section is optional. For those of you who decide to skip it, let us state its basic conclusion. In an economy in which money is in part currency and in part chequable deposits, the central bank no longer controls the total stock of money directly. However, it controls it indirectly. It can still use open market operations—purchases and sales of bonds—to increase or decrease the total stock of money and thus affect the interest rate. In Chapter 28, we will look in more detail at how the Bank currently influences the money supply.

*5-3 THE DETERMINATION OF THE INTEREST RATE: II

We start this section by looking at what banks do. We then return to the determination of the money supply and the effects of open-market operations.

THE ROLE OF BANKS

Banks are **financial intermediaries:** They receive funds from people and firms and use these funds to make loans and to buy bonds. What makes them different from other financial intermediaries is that they receive those funds by offering deposits—chequable deposits—that allow depositors to write cheques or get cash on demand, up to an amount equal to the account balance. (For this reason, these accounts are also called **demand deposits.**) Because cheques can be used to pay for transactions, these deposits are money. The balance sheet of banks is given in Figure 5-8(b). On the liability side are chequable deposits.[8] Chequable deposits are liabilities of the banks. On the asset side are reserves, loans, and bonds.

---

[8]***Digging deeper.*** *In fact, banks also offer other types of deposits, such as savings and time deposits. These cannot be used directly in transactions and thus are not money. We shall ignore this part of banks' activity, which is not central for our purposes. Also, chartered banks are the main but not the only financial institutions to offer chequable deposits; trust companies and credit unions also offer such deposits. We also ignore this complication here and use "banks" to denote all suppliers of chequable deposits.*

(a) Central Bank				(b) Banks		
Assets	Liabilities			Assets	Liabilities	
Bonds	Central bank money = Reserves + Currency	(1) (2) (3)		Reserves Loans Bonds	Chequable deposits	(4)

FIGURE 5-8
The Balance Sheet of the Central Bank Revisited,
and the Balance Sheet of the Banks

Why do banks hold reserves? On any given day, some depositors withdraw cash from their chequing accounts, while others deposit cash into their accounts. There is no reason for the inflows and outflows of cash to be equal, so a bank must keep some cash on hand. In the same way, on any given day, what the bank owes to other banks (as a result of cheques written by people with accounts at the bank) may be greater or smaller than what other banks owe to the bank (as a result of cheques received and deposited by people with accounts at the bank). Thus, for both reasons, banks want to keep some reserves even if they are not required to do so.

In many countries, banks are required to hold reserves in some proportion to their deposits—that is, a certain **reserve ratio**. Canada removed all such legal requirements in June 1994 and in practice reserve requirements had disappeared a few years earlier. Chartered banks and a few other financial institutions continued to maintain deposits at the Bank of Canada in order to facilitate cheque-clearing. We will continue to refer to deposits held by the banks at the Bank of Canada as *reserves*, although a more accurate description would be *clearing* or *settlement balances*.

Less than 0.5% of the banks' assets are held as reserves. Banks use their funds primarily to make loans to firms and consumers. Loans and foreign assets represent roughly 90% of banks' nonreserve assets. Government bonds account for the rest. In what follows, we shall assume for simplicity that banks do not make loans and thus hold only reserves and bonds on the asset side. The distinction between bonds and loans is unimportant for our purpose, which is to understand the determination of the money supply. It is, however, important for other purposes, from the likelihood of bank runs to the role of federal deposit insurance. These topics are explored in the Focus box entitled "Bank Runs."

THE DETERMINATION OF THE MONEY SUPPLY

Let's now turn to Figure 5-8(a), the balance sheet of the central bank. It is very similar to the balance sheet we saw earlier in Figure 5-7. The liabilities of the central bank are the money it has issued, **central bank money.** Part of central bank money is now held as reserves by banks, and the rest is held as currency by the public. The assets of the central bank are the bonds it holds in its portfolio.

Making a loan to a firm and buying a government bond are more similar than they may seem at first. In both cases, what is involved is the payment of an initial amount of money (the loan, or the payment for the bond) in exchange for a promise of payment of principal plus interest later. This is why, for simplicity, we have assumed in this text that banks held only bonds.

But in one respect, making a loan is very different from buying a bond. Bonds, especially government bonds, are very liquid; in case of need, they can be sold quickly and cheaply in the bond market. Loans, in contrast, are often not liquid at all. Calling them back may be impossible: The firm, which has used the loan to buy inventories or a new machine, no longer has the cash. Selling the loan itself to a third party may be very difficult, because potential buyers know little about how reliable the borrower is.

This fact has one important implication. Take a healthy bank, a bank with a good portfolio of loans. Now suppose that rumours start that the bank is not doing well and that some loans will not be paid back. On the belief that the bank may fail, people with deposits at the bank will want to close their accounts and get the cash. If enough people do so, the bank will run out of reserves. Given that the loans cannot be called back, the bank will not be able to satisfy the demand for cash, and it will indeed have to close.

Thus, the belief that a bank may close may lead it to close, even if all its loans are good. The financial history of the United States up to the 1930s is full of such **bank runs.** The typical story is one in which one bank fails for the right reasons (that is, bad loans),

leading depositors at other banks to get scared and run on their own banks, thus forcing them to close, whether or not their loans are good. You may remember *It's a Wonderful Life,* an old movie with James Stewart that runs on TV every year around Christmas. Because of the failure of another bank in town, depositors at the savings and loan of which James Stewart is the manager get scared and come to get their money back. It takes all of James Stewart's persuasion to avoid closure. The movie has a happy ending. In contrast, most bank runs didn't.

What can be done to avoid bank runs? The United States has dealt with this problem since 1934 with federal deposit insurance. The **Canada Deposit Insurance Corporation** has done the same since 1968 for Canadians.

Many observers have pointed out, however, that federal deposit insurance leads to problems of its own. Depositors who do not have to worry about their deposits no longer look at the activities of the banks in which they have their deposits, and banks may misbehave. Banks may make dangerous loans that they would not have made in the absence of the insurance. Both countries saw first-hand the consequences in the 1980s. In the United States, deposit insurance led to a crisis in the savings and loan industry that cost the Treasury over $50 billion. Canada had not had a bank failure since 1924. But the availability of deposit insurance led to the birth of many new trust companies. Several ran into financial difficulty and had to be absorbed by the larger banks, and two failed. The CDIC and Canadian taxpayers are still paying the bill.

In Section 5-2, where we ignored the presence of banks, the money supply was simply equal to central bank money. This is no longer the case. Central bank money is equal to reserves plus currency, lines (2) and (3) in Figure 5-8. But the money supply—that is, the amount of money held by the public for transaction purposes—is equal to currency plus chequable deposits, lines (3) and (4) in Figure 5-8.

The question we now take up is that of the relation between central bank money and the money supply. We shall see that the money supply is given by

$$\text{money supply} = \text{money multiplier} \times \text{central bank money}$$

You can think of what we do now as deriving the value of the **money multiplier.** To do so, let's first introduce some definitions and notation.

The central bank controls central bank money. Central bank money is held either as currency by the public, or as reserves by the banks:

$$H = CU + R \tag{5.6}$$

Central bank money is denoted by H. The more usual name for central bank money is the **monetary base,** and this is the name we shall generally use from now on. (The letter H stands for **high-powered money,** yet another name for central bank money. The origin of the expression is that an increase in central bank money leads, because of the money multiplier, to a more than one-for-one increase in the money supply, and thus is "high-powered.") CU stands for currency held by the public. R stands for reserves held by the banks.

The money supply (M) is the sum of currency (CU) and chequable deposits (D):

$$M = CU + D \tag{5.7}$$

What is the relation between the monetary base (H), which is controlled by the central bank, and M, the money supply? Let's answer this question in two steps.

The multiplier in an economy with no currency.   Suppose first that people hold all their money in the form of chequable deposits (people do not hold currency and pay for all transactions by cheque). This is an unrealistic assumption, but it will help us understand the mechanics at work, and we shall relax it soon. In this case, $CU = 0$ by assumption. Thus, from equations (5.6) and (5.7), it follows that

$$H = R$$
$$M = D$$

All central bank money is held as reserves by banks, and the money supply is equal to chequable deposits.

Assume now that the reserve ratio, the ratio of reserves to chequable deposits, is equal to θ (the Greek lowercase letter theta), so that

$$R = \theta D$$

As we saw earlier, θ is a decision variable of banks. We shall not try to explain how banks chose θ; we shall simply take θ as a parameter here. Dividing both sides of the previous equation by θ, we can rewrite the relation between reserves and chequable deposits as

$$D = \frac{1}{\theta}R \tag{5.8}$$

If the reserve ratio is equal to θ, then the supply of chequable deposits by banks is equal to $(1/\theta)$ times the amount of reserves held by banks. If $\theta = 5\%$, for example, then the supply of chequable deposits is equal to 20 times reserves.

Now use the facts that $H = R$ (that all central bank money is held as reserves by banks) and that $M = D$ (that the money supply is composed only of chequable deposits), and substitute in equation (5.8) to get

Central Bank

Assets	Liabilities
Bonds	(H) Central Bank $(1)

= Reserves (2)
+ Currency (3)

Banks

Assets	Liabilities
Reserves	(D) Chequable deposits (4)
Loans	
Bonds	

$$M = \frac{1}{\theta}H \qquad (5.9)$$

The money supply is equal to the *money multiplier* (the first term on the right-hand side of the equation) times the *monetary base, H.* The money multiplier is equal to $1/\theta$, the inverse of the reserve ratio. If, for example, $\theta = 5\%$, the money supply is equal to 20 times the monetary base. The reason is simple: The entire monetary base is held as reserves by banks. For every dollar of reserves, the banking system can create 20 dollars of deposits.

We have, however, derived this result under the assumption that people do not hold currency. This assumption is clearly counterfactual, and we must now relax it.

The multiplier in an economy with currency and chequable deposits. Suppose that people hold both currency and chequable deposits. What is the relation between central bank money and the money supply?

Let's assume that the demand for currency is a proportion, *c*, of the demand for chequable deposits:

$$CU = cD \qquad (5.10)$$

If people want to hold only chequable deposits, the parameter *c* is equal to zero. If people want to hold mostly currency, then *c* is very large. The value of *c*, which we shall take as a parameter, depends mainly on two sets of factors. First, it depends on the type of transactions that people engage in. Currency is more convenient for small transactions (also for illegal transactions, but we shall ignore these here); cheques are more convenient for large transactions. Second, it depends on the cost of getting cash from your chequing account. The development of automated banking machines (ABMs) has made it easier to take cash from your account and thus reduce the average amount of cash you need to keep.

Let's now derive the money multiplier. From equation (5.6) and our assumptions that the ratio of currency to deposits is equal to *c* and the reserve ratio of banks is equal to θ, we get

$$\begin{aligned} H &= CU + R \\ &= cD + \theta D \\ &= (c + \theta)D \end{aligned}$$

or, equivalently,

$$D = \frac{1}{(c + \theta)}H$$

Chequable deposits are equal to a multiple, $1/(c + \theta)$, of the monetary base. The multiple is equal to 1 over the sum of the ratio of currency to deposits and the reserve ratio.

From equation (5.7) and our assumption that the ratio of currency to chequable deposits is equal to *c*, we get

$$\begin{aligned} M &= CU + D \\ &= cD + D \\ &= (1 + c)D \end{aligned}$$

Eliminating D between the preceding two equations gives

$$M = \frac{1+c}{c+\theta} H \tag{5.11}$$

As in equation (5.9), the money supply is equal to *the money multiplier* times *the monetary base*. But the money multiplier in equation (5.11) is a more complicated expression than before. Let's look at it more closely.

Consider first the case where people hold no currency, only chequable deposits. In that case, $c = 0$ and the money multiplier is $1/\theta$; this is the case we examined above and that led to equation (5.9). As people hold more in the form of currency, as c increases, the multiplier decreases. Consider the extreme case where people hold mostly currency, so that the ratio of currency to chequable deposits, c, is very large. If c is very large compared to 1 and θ, then $(1 + c)/(c + \theta)$ is very close to c/c and thus the multiplier is very close to 1. This is a result that should be familiar from Figure 5-8 and Section 5-2: If people hold only currency (chequable deposits are equal to zero), then all money is central bank money; that is, the money multiplier must be 1.

What is the size of the money multiplier in Canada? We saw earlier that people hold 20% of money in the form of currency, 80% in the form of chequable deposits. Thus c, the ratio of currency to deposits, is equal to $0.2/0.8 = 0.25$. The ratio of reserves to chequable deposits is equal to 5%. Thus, the money multiplier is equal to $1.25/0.3 = 4.17$. An increase in the monetary base of $1 increases the money supply by $4.17.

We have derived the relation of the money supply to the monetary base in an economy in which there are both currency and chequable deposits. But the reasoning may still feel a bit abstract. A good way to strengthen the intuition for how an increase in the monetary base leads to a larger increase in the money supply is to follow through the effects of an actual open-market operation. Let's do that now.

THE EFFECTS OF AN OPEN-MARKET OPERATION

For computational simplicity, let's assume that people hold only chequable deposits (so that $c = 0$) and that the reserve ratio is $\theta = 0.05$. Let's trace the effects of an open-market operation by the Bank of Canada:

- Suppose that the Bank of Canada buys $100 worth of bonds in an open market operation. It pays the seller—call her seller 1—$100 and thus creates $100 in central bank money. Thus, the increase in the monetary base is $100.

  When we looked earlier at the effects of an open-market operation in an economy in which there were no banks, this was the end of the story. Here it is just the beginning.

- Seller 1 (who, we have assumed, does not want to hold any currency) deposits the $100 in a chequing account at her bank—call it bank A. This leads to an increase in chequable deposits of $100.

- Bank A keeps $100 times $0.05 = \$5$ in reserves and buys bonds with the rest, $100 times $0.95 = \$95$. It thus pays $95 to the seller of those bonds—call him seller 2.

- Seller 2 deposits $95 in a chequing account in his bank—call it bank B. This leads to an increase in chequable deposits of $95.

- Bank B keeps $95 times 0.05 = $4.75 in reserves and buys bonds with the rest, $95 times 0.95 = $90.25. It pays $90.25 to the seller of those bonds—call him seller 3.

- Seller 3 deposits $90.25 in a chequing account in his bank—call it bank C. And so on.

By now, the chain of events should be clear. What is the eventual increase in the money supply? The increase in chequable deposits is $100 when seller 1 deposits the proceeds on her bond sale in bank A, plus $95 when seller 2 deposits the proceeds of his bond sale in bank B, plus $90.25 when seller 3 does the same, and so on. Let's write the sum as

$$\$100(1 + 0.95 + 0.95^2 + \cdots)$$

The series in parentheses is a geometric series, so its sum is equal to $1/(1 - 0.95) = 20$. Thus, the money supply increases by $2000—20 times the initial increase in the monetary base.[9]

The fact that the money multiplier is equal to 20 is no surprise. We derived earlier in this section the result that, when people do not hold currency, the money multiplier is equal to the inverse of the reserve ratio, which is equal to 0.05 here. But the derivation gives us another and useful way of thinking about the money multiplier. We can think of the ultimate increase in the money supply as the result of *successive rounds of open-market operations*—of purchases of bonds—the first by the Bank of Canada, the following ones by banks. Each successive round leads to an increase in the money supply; eventually, the increase in the money supply is equal to 20 times the initial increase in the monetary base.[10]

Our specific derivation of the chain of open-market operations was based on the assumption that people did not hold currency. But this was just for mathematical simplicity. Make sure that you can derive the chain, and the resulting value of the money multiplier, when people hold both currency and demand deposits, when c is different from zero. In that case, in each round, sellers of bonds keep some of the proceeds in currency and some in their chequing accounts. Thus, in each round, chequable deposits increase by less than in our example, banks buy fewer bonds, and the multiplier is accordingly smaller.

CONCLUSIONS: MONEY DEMAND AND MONEY SUPPLY

Let's put together what we have learned in this section.

We had seen in the first section that the demand for money is a function of the level of transactions (which we proxy by nominal income) and of the interest rate:

---

[9] *If needed, see Appendix 3 for a refresher on geometric series.*

[10] **Digging deeper.** *There is a clear parallel between our interpretation of the money multiplier as the result of successive purchases of bonds and the interpretation of the goods-market multiplier in Chapter 4 as the result of successive rounds of spending. There is indeed a general principle here: Multipliers can often be derived as the sum of a geometric series, and be interpreted as the result of successive rounds of decisions. This interpretation often gives a better intuition for the process at work.*

$$LM \text{ relation:} \qquad M^d = \$YL(i)$$

We have learned in this section that the supply of money, M, is equal to the money multiplier times the monetary base:

$$M = \frac{1+c}{c+\theta}H$$

where c is the ratio of currency to chequable deposits and θ is the ratio of reserves to deposits.

Equilibrium in financial markets requires that the supply of money and the demand for money be equal:

$$\frac{1+c}{c+\theta}H = \$YL(i)$$

An increase in the monetary base, H, leads to a larger increase in the money supply, in a proportion given by the money multiplier. At a given level of nominal income, this increase in the money supply leads to a decrease in the interest rate.

SUMMARY

■ The demand for money depends positively on the level of transactions in the economy, and negatively on the interest rate.

■ Given the supply of money, an increase in income leads to an increase in the demand for money and an increase in the interest rate. An increase in the money supply leads to a decrease in the interest rate.

■ Central banks affect the interest rate primarily through open market operations. A purchase of bonds (equivalently an increase in the money supply) leads to an increase in the price of bonds, equivalently a decrease in the interest rate. A sale of bonds (equivalently a decrease in the money supply) leads to a decrease in the price of bonds, equivalently an increase in the interest rate.

■ The money supply is equal to the money multiplier times central bank money (the monetary base). The money multiplier depends on both the ratio of currency to chequable deposits and on the ratio of reserves to chequable deposits.

KEY TERMS

■ Bank of Canada, 78
■ Federal Reserve Bank (Fed), 78
■ financial markets, 78
■ money, 79
■ bonds, 79
■ currency, 79
■ chequable deposits, 79
■ income, 79
■ flow, 79
■ saving, 79
■ savings, 79
■ financial wealth, or wealth, 79
■ stock, 79
■ investment, 80
■ financial investment, 80
■ money market funds, 81

■ M1, 83
■ velocity, 84
■ LM relation, 87
■ open-market operations, 89
■ Treasury bill, or T-bill, 90
■ expansionary and contractionary open-market operations, 90
■ financial intermediary, 91
■ demand deposits, 91
■ reserve ratio, 92
■ central bank money, or monetary base, or high-powered money, 92
■ bank run, 93
■ Canada Deposit Insurance Corporation (CDIC), 93
■ money multiplier, 93

QUESTIONS AND PROBLEMS

1. Automated banking machines have no doubt shifted the public's preferences away from cash and toward deposits. What do you think they have done to the demand for money as a whole?

2. Suppose that a person with wealth of $25 000 and a yearly income of $50 000 has the following demand-for-money function:

$$M^d = \$Y(.5 - i)$$

   a. What is the person's demand for money when the interest rate is 5%? 10%?
   b. What is the person's demand for bonds when the interest rate is 5%? 10%?
   c. Summarize your results by stating the impact of a rise in the interest rate on the demand for money and the demand for bonds.

3. Using the information in problem 2 and assuming that the demand for money is equal to the supply of money:
   a. Find an expression for velocity at any given interest rate.
   b. Use the expression to determine the impact on velocity of a rise in the interest rate.

4. A bond will pay $1000 in one year.
   a. What is the interest rate on the bond if the price today is
      (i) $700?
      (ii) $800?
      (iii) $900?
   b. Do your answers suggest a positive or negative relationship between the price of a bond and the interest rate on that bond?
   c. What would the price have to be for the bond to pay an interest rate of 10%?

5. Suppose the following:
   (1) The public holds no currency.
   (2) The ratio of reserves to deposits is 0.2.
   (3) The demand for money is given by the following equation:

$$M^d = \$Y(0.2 - 0.8i)$$

   Initially, the monetary base is $100 billion and nominal income is $5000 billion.
   a. Determine the value of the money supply.
   b. Determine the equilibrium interest rate. [*Hint:* The money market must be in equilibrium, so equate money demand and money supply.]
   c. Determine the impact on the interest rate if the central bank increases the stock of high-powered money to $150 billion.
   d. With the original money supply, determine the impact on the interest rate if nominal income increases from $5000 billion to $6250 billion.

*6. Look in the most recent *Bank of Canada Review* or go to their Web site, get the *Weekly Financial Statistics,* and find the table entitled, "Selected Monetary Aggregates and Their Components."
   a. Determine the value of the *M*1 measure of the money supply in the most recent month and one year earlier.
   b. By what percentage has the money supply increased or decreased over the period?
   c. Now, use the consumer price index to compute the *real* money supply in the most recent month and 12 months earlier.
   d. By what percentage has the real money supply increased or decreased over the period?
   e. Which of the two measures—real money supply or nominal money supply—is the better one to use when gauging the effects of monetary policy?

FURTHER READING

For a more detailed description of financial markets and institutions, you should read a textbook on money and banking. A good one is *Money, Banking and Financial Institutions: Canada and the Global Environment,* 2nd ed., by Pierre Siklos (Toronto: McGraw-Hill Ryerson, 1997).

GOODS AND FINANCIAL MARKETS: THE *IS-LM*

We spent Chapters 3 and 4 looking at the goods market, then Chapter 5 looking at financial markets. We now look at goods and financial markets together. By the end of this chapter you will have a conceptual framework to think about how output and the interest rate are determined in a closed economy, and about the effects of monetary and fiscal policies on the economy.

In developing this framework we shall follow a path first traced by two economists, John Hicks and Alvin Hansen, in the late 1930s and the early 1940s. When Keynes's *General Theory* was published in 1936, there was much agreement that the book was both fundamental and nearly impenetrable. (You may want to have a look at it to convince yourself of that.) There were many debates about what Keynes "really meant." In 1937, John Hicks summarized what he saw as one of Keynes's main contributions: the joint description of goods and financial markets. His analysis was later extended by Alvin Hansen. Hicks and Hansen called their formalization the *IS-LM* model. For obvious reasons, it is also called the Hicks-Hansen model.

Macroeconomics has made substantial progress since the early 1940s. This is why the *IS-LM* model is treated in Chapter 6 rather than in Chapter 30 of this book. (Think of it: If you had taken this course 40 years ago, you would be nearly done.) But to most economists, the *IS-LM* model still represents an essential building block—one that, despite its simplicity, captures much of what happens in the economy in the short and medium run. This is why the *IS-LM* model is still taught and used today.

6-1 THE GOODS MARKET AND THE *IS* RELATION

Let's first take stock of what we learned in Chapter 3. We characterized equilibrium in the goods market as the condition that production, *Y*, is equal to demand, *Z*. We called this condition the *IS* relation, because it can be reinterpreted as the condition that investment is equal to saving.

We defined demand as the sum of consumption, investment, and government spending. We assumed that consumption was a function of disposable income (income minus taxes), and took investment spending, government spending, and taxes as given. The equilibrium condition was thus given by

$$Y = C(Y - T) + \bar{I} + G$$

Using this equilibrium condition, we then looked at the factors that changed equilibrium output. We looked in particular at the effects of changes in government spending and of shifts in consumption demand.

Perhaps the main simplification of this first model was that the interest rate did not affect the demand for goods. Our first task in this chapter is thus to introduce the interest rate in our model of goods-market equilibrium. For the time being, we shall focus only on its effect on investment and leave until later its effect on the other components of demand.

INVESTMENT, SALES, AND THE INTEREST RATE

In our first model of output determination, investment was left unexplained and thus assumed constant in the face of movements in output. Let's now relax this assumption. Investment—spending on new machines and plants by firms—depends primarily on two factors:[1]

---

[1] *We focus here on nonresidential fixed investment by firms and ignore* residential investment, *the purchase of new housing by people.*

■ The first is the level of sales. Firms faced with high sales and the need to increase production will typically want to buy additional machines, to build additional plants. Firms that face low sales will feel no such need and will spend little if anything on investment.

■ The second is the interest rate. Consider a firm that is deciding whether or not to buy a new machine. Suppose that to buy the new machine the firm must borrow, either by taking a loan from a bank or by issuing bonds. The higher the interest rate, the less likely the firm is to borrow and buy the machine. At a high enough interest rate, the additional profits from the new machine will just not cover interest payments and the new machine will not be worth buying.

To capture these two effects, we write the investment relation as follows:

$$I = I(Y, i) \qquad (6.1)$$
$$(+, -)$$

Equation (6.1) states that investment depends on two variables: production, Y, and the interest rate, i. While our discussion suggests that sales may be a more appropriate variable, we shall assume that sales and production are equal—in other words, we shall assume that inventory investment is always equal to zero—and use production instead. The positive sign under Y indicates a positive relation: An increase in production leads to an increase in investment. The negative sign under the interest rate i indicates a negative relation: An increase in the interest rate leads to a decrease in investment.

THE *IS* CURVE

Taking into account the investment relation (6.1), the equilibrium condition in the goods market becomes

$$\underbrace{Y}_{\text{production}} = \underbrace{C(Y - T) + I(Y, i) + G}_{\text{demand}} \qquad (6.2)$$

This is our expanded *IS relation*. We can now look at what happens to output when the interest rate changes.

The first step is taken in Figure 6-1. Demand—the right-hand side of equation (6.2)—is measured on the vertical axis. Output (equivalently production, or income; remember that production is a synonym for output, and that production and income are always equal) is measured on the horizontal axis.

The curve ZZ plots demand as a function of output for a given value of the interest rate, i. As output, and thus income, increases, so does consumption; we studied this relation at length in Chapter 3. And as output increases investment increases; this is the relation between investment and production that we have introduced in this chapter. Thus, through its effects on both consumption and investment, an increase in output leads to an increase in demand: ZZ is upward-sloping.[2]

---

[2]*Since we have not assumed that the consumption and investment relations in equation (6.2) are linear, ZZ is in general a curve rather than a line. Thus, we draw it as a curve in Figure 6-1. But all the arguments that follow would apply if we assumed instead that the consumption and investment relations were linear and that ZZ was a line instead.*

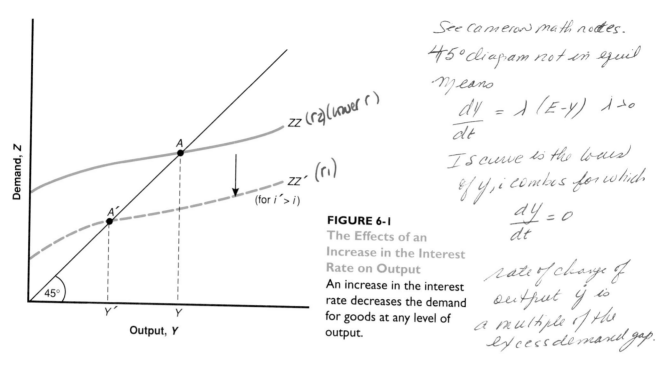

Handwritten notes (right margin):

See Cameron math notes.

45° diagram not in equil means

$$\frac{dY}{dt} = \lambda(E - Y) \quad \lambda > 0$$

IS curve is the locus of Y, i combos for which

$$\frac{dY}{dt} = 0$$

rate of change of output Y is a multiple of the excess demand gap.

FIGURE 6-1

The Effects of an Increase in the Interest Rate on Output

An increase in the interest rate decreases the demand for goods at any level of output.

Note that we have drawn ZZ so that it is flatter than the 45-degree line. Put another way, we have assumed that an increase in output leads to a less than one-for-one increase in demand. In Chapter 3, where investment was constant, this restriction naturally followed from the restriction that consumers spend only part of their additional income on consumption. But now that we allow investment to respond to production, this restriction may no longer hold. When output increases, the sum of the increase in consumption and investment could exceed the initial increase in output. Although this is a theoretical possibility, the empirical evidence suggests that it is not the case in practice. Thus, we shall assume that the response of demand to output is indeed less than one for one and draw ZZ flatter than the 45-degree line.

Equilibrium is reached at the point where demand is equal to production—thus at point A, the intersection of ZZ and the 45-degree line. The equilibrium level of output is given by Y.

We have drawn the demand relation, ZZ, for a given value of the interest rate. Suppose now that the interest rate increases from its initial value i to a new higher value i'. At any level of output, investment decreases. Thus, the demand curve ZZ shifts down to ZZ'. The new equilibrium is at the intersection of the new demand curve ZZ' and the 45-degree line, thus at point A'. The equilibrium level of output is now given by Y'.

Let's explain what happens in words. An increase in the interest rate decreases investment. The decrease in investment leads to a decrease in output, which in turn further decreases consumption and investment. In other words, the initial decrease in investment leads to a larger decrease in output through the multiplier effect.

Using Figure 6-1, we can find the equilibrium value of output associated with any value of the interest rate. The relation between equilibrium output and the interest rate is derived in Figure 6-2. Figure 6-2(a) reproduces Figure 6-1. The interest

FIGURE 6-2

The Derivation of the *IS* Curve

Equilibrium in the goods market implies that output is a decreasing function of the interest rate. The *IS* curve is downward-sloping.

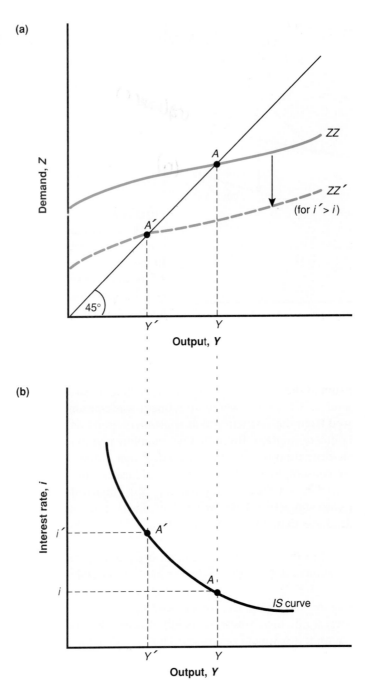

rate *i* implies a level of equilibrium output equal to *Y*. The higher interest rate *i'* implies a lower equilibrium level of output *Y'*. Figure 6-2(b) plots equilibrium output *Y* on the horizontal axis against the interest rate on the vertical axis. Point *A* in Figure 6-2(a) corresponds to point *A* in Figure 6-2(b), and point *A'* in Figure 6-2(a) corresponds to *A'* in Figure 6-2(b). More generally, equilibrium in the goods market implies that the higher the interest rate, the lower the equilibrium level of output.

This relation between the interest rate and output is represented by the downward-sloping curve in Figure 6-2(b). This curve is called the **IS curve**.[3]

SHIFTS IN THE *IS* CURVE

Note that we have derived the *IS* curve in Figure 6-2 for given values of taxes, *T*, and government spending, *G*. Changes in either *T* or *G* will shift the *IS* curve.

To see how, consider Figure 6-3. The *IS* curve gives the equilibrium level of output as a function of the interest rate. It is drawn for given values of taxes and government spending. Now consider an increase in taxes, from *T* to *T'*. At a given interest rate, say *i*, consumption decreases, leading to a decrease in the demand for goods and, through the multiplier, to a decrease in equilibrium output. The equilibrium level of output decreases, from, say, *Y* to *Y'*. Put another way, the *IS* curve shifts to the left: At any interest rate, the equilibrium level of output is now lower than before the increase in taxes.

More generally: Any factor that for a given interest rate decreases the equilibrium level of output leads the *IS* curve to shift to the left. We have looked at an increase in taxes. But the same would hold for a decrease in government spending, or a decrease in consumer confidence (which decreases consumption given disposable income). In contrast, any factor that for a given interest rate increases the equilibrium level of output—a decrease in taxes, an increase in government spending, an increase in consumer confidence—leads the *IS* curve to shift to the right.

Let's summarize the main results of this first section. Equilibrium in the goods market implies that output is a decreasing function of the interest rate. This relation

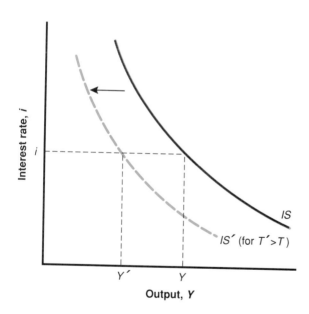

Output, *Y*

FIGURE 6-3

Shifts in the *IS* Curve

An increase in taxes shifts the *IS* curve to the left.

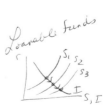

---

[3]**Digging deeper.** *You may want to think about what happens to investment and saving as we move down the* IS *curve. As we move down, the interest rate decreases and production increases; both factors increase investment. Also, as we move down, income increases, and thus saving increases. Thus, as we move down the* IS *curve, both investment and saving increase; indeed, by the construction of the* IS *curve, they increase by the same amount, so that investment remains equal to saving.*

is represented by the *IS* curve. Changes in factors that decrease or increase the demand for goods given the interest rate shift the *IS* curve to the left or to the right.

6-2 FINANCIAL MARKETS AND THE *LM* RELATION

Let's now turn to financial markets. We saw in Chapter 5 that the interest rate is determined by the equality of the supply of and the demand for money:

$$M = \$YL(i)$$

The variable *M* on the left-hand side is the nominal money stock. We shall ignore here the details of the money-supply process and simply think of the central bank as controlling *M* directly. The right-hand side gives the demand for money, which is a function of nominal income, $\$Y$, and of the nominal interest rate, *i*. An increase in nominal income increases the demand for money. An increase in the interest rate decreases the demand for money. Equilibrium requires that money supply (the left-hand side of the equation) be equal to money demand (the right-hand side of the equation).

REAL MONEY, REAL INCOME, AND THE INTEREST RATE

The foregoing equation gives a relation between money, nominal income, and the interest rate. It will prove more convenient here to rewrite it as a relation between real money (that is, money in terms of goods), real income, and the interest rate.

Recall from Chapter 2 that nominal income divided by the price level is equal to real income, *Y*. Thus, dividing both sides of the equation by the price level *P* (which we take as given here) gives

$$\frac{M}{P} = YL(i) \tag{6.3}$$

Thus, we can restate our equilibrium condition as the condition that the *real money supply*—that is, the money stock in terms of goods, not dollars—be equal to the *real money demand,* which in turn depends on real income *Y* and the interest rate *i*. The notion of a "real" demand for money may feel a bit abstract, so an example may help here. Think not of your demand for money in general but just of your demand for coins. Suppose that you like to have coins in your pocket to buy four cups of coffee during the day. If a cup costs 90 cents, you will want to keep about $3.60 in coins; this is your nominal demand for coins. Equivalently, you want to keep enough in your pocket to buy four cups of coffee. This is your demand for coins in terms of goods, here in terms of cups of coffee.

From now on, we shall refer to equation (6.3) as the *LM relation.* The advantage of writing things this way is that *real income, Y,* now appears on the right-hand side of the equation in place of *nominal income $Y.* And real income is the variable we focus on when looking at equilibrium in the goods market. To make the reading lighter, we shall refer to the right- and left-hand sides of equation (6.3) simply as "money supply" and "money demand" rather than the more accurate but heavier "real money supply" and "real money demand." Similarly, we shall refer to income rather than "real income."

THE *LM* CURVE

To see the relation between output and the interest rate implied by equation (6.3), let's start with Figure 6-4. Let the interest rate be measured on the vertical axis, and (real) money be measured on the horizontal axis. Money supply is given by the vertical line at M/P, and is denoted M^s. For a given level of income, Y, money demand is a decreasing function of the interest rate. It is drawn as the downward-sloping curve denoted M^d. Except for the fact that we measure real money rather than nominal money on the horizontal axis, the figure is the same as Figure 5-4. The equilibrium is at point A, where money supply is equal to money demand and the interest rate is equal to i.

Now consider an increase in income from Y to Y', which leads people to increase their demand for money at any given interest rate. Money demand shifts to the right, to $M^{d'}$. The new equilibrium is at A', with a higher interest rate, i'. Thus, an increase in income leads to an increase in the interest rate. Why is this so? When income increases, money demand increases. But the money supply is given. Thus, the interest rate must go up until the two opposite effects on the demand for money—the increase in income that leads people to want to hold more money, and the increase in the interest rate that leads people to want to hold less money—cancel each other. At that point, the demand for money is equal to the unchanged money supply, and financial markets are again in equilibrium.

Using Figure 6-4, we can find out what value of the interest rate is associated with *any* value of income for a given money supply. The relation is derived in Figure 6-5. Figure 6-5(a) reproduces Figure 6-4. When income is equal to Y, money demand is given by M^d and the equilibrium interest rate is equal to i. When income is equal to the higher value Y', money demand is given by $M^{d'}$ and the equilibrium interest rate

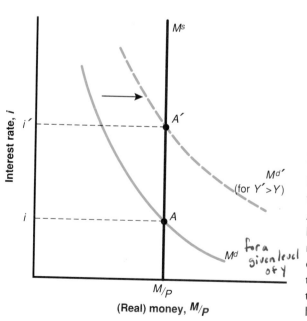

See Cameron math notes

Law of motion re LM curve

$$\frac{di}{dt} = \mu \, (L - m) \ + $$

LM curve is locus
of steady state
i, y combos for
which di/dt = 0

FIGURE 6-4

The Effects of an Increase in Income on the Interest Rate

An increase in income leads, at a given interest rate, to an increase in the demand for money. Given the money supply, this leads to an increase in the equilibrium interest rate.

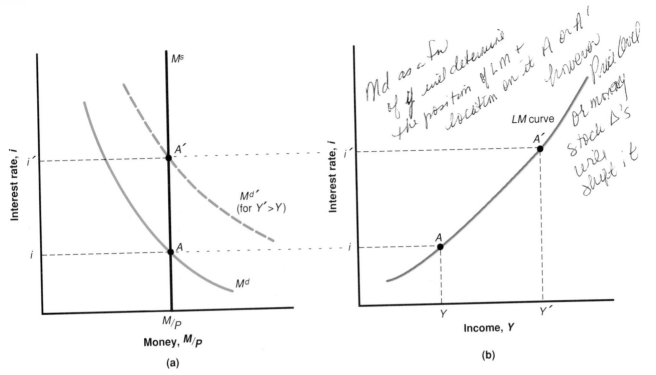

Handwritten notes in figure: "Md as a fn of Y will determine the position of LM + location on it A on A¹ however Price level of money stock Δ's uries shift it"

FIGURE 6-5

The Derivation of the *LM* Curve

Equilibrium in financial markets implies that the interest rate is an increasing function of the level of income. The *LM* curve is upward-sloping.

is equal to i'. Figure 6-5(b) plots the equilibrium interest rate i on the vertical axis against income on the horizontal axis. Point A in Figure 6-5(b) corresponds to point A in Figure 6-5(a), and point A' in Figure 6-5(b) corresponds to point A' in Figure 6-5(a). More generally, equilibrium in financial markets implies that the higher the level of output, the higher the demand for money, and thus the higher the equilibrium interest rate. This relation between output and the interest rate is represented by the upward-sloping curve in Figure 6-5(b). This curve is called the ***LM* curve.**

SHIFTS IN THE *LM* CURVE

We have derived the *LM* curve in Figure 6-5 taking both the nominal money stock, M, and the price level, P—and thus the real money stock, M/P—as given. Changes in M/P, whether they come from changes in the nominal money stock, M, or from changes in the price level, P, will shift the *LM* curve.

To see how, consider Figure 6-6. The *LM* curve gives the interest rate as a function of the level of income. It is drawn for a given value of M/P. Now consider an increase in the nominal money supply, from M to M', so that, at an unchanged price level, the real money supply increases from M/P to M'/P. At a given level of income, Y, this increase in money leads to a decrease in the equilibrium interest rate from i to i'. Put another way, the *LM* curve shifts down; at any level of income, an increase in money leads to a decrease in the equilibrium interest rate. By the same reasoning, at any level of income, a decrease in money leads to an increase in the interest rate. Thus, a decrease in money leads the *LM* curve to shift up.

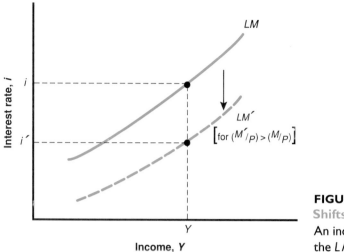

FIGURE 6-6
Shifts in the *LM* Curve
An increase in money leads
the *LM* curve to shift down.

Let's summarize what we have learned in this section. Equilibrium in financial markets implies that the interest rate is an increasing function of the level of income. This relation is represented by the *LM* curve. Increases in money shift the *LM* curve down; decreases in money shift the *LM* curve up.

6-3 THE *IS-LM* MODEL: EXERCISES

We can now put the *IS* and *LM* relations together. At any point in time, the supply of and the demand for goods must be equal. And the same must hold for the supply of and the demand for money. Thus, both the *IS* and *LM* relations must hold:

$$IS \text{ relation:} \qquad Y = C(Y - T) + I(Y, i) + G$$

$$LM \text{ relation:} \qquad \frac{M}{P} = YL(i)$$

Figure 6-7 on the next page plots both the *IS* curve and the *LM* curve on one graph. Output—equivalently production or income—is measured on the horizontal axis. The interest rate is measured on the vertical axis.

Any point on the downward-sloping *IS* curve corresponds to equilibrium in the goods market. Any point on the upward-sloping *LM* curve corresponds to equilibrium in financial markets. Only at point *A* are both equilibrium conditions satisfied. Thus point *A*, with associated levels of output *Y* and interest rate *i*, is the overall equilibrium, the point at which there is equilibrium in both the goods market and the financial markets.

The *IS* and *LM* relations that underlie Figure 6-7 contain a lot of information about consumption, investment, money demand, and equilibrium conditions. But you may well be asking at this point: So what if the equilibrium is at point *A?* How does this fact translate into anything directly useful about the world? Don't despair: Figure 6-7 does in fact hold the answer to many central questions in macroeconomics. Used properly, it allows us to study what happens to output and the interest rate when the central bank decides to increase the money supply, or when the government decides to increase

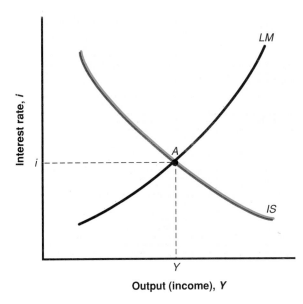

FIGURE 6-7
The *IS-LM* Model
Equilibrium in the goods market implies that output is a decreasing function of the interest rate. Equilibrium in financial markets implies that the interest rate is an increasing function of output. Only at point *A* are both goods and financial markets in equilibrium.

taxes, or when consumers become more pessimistic about the future, and so on. We will now look at what the *IS-LM* model can do.

FISCAL POLICY, ACTIVITY, AND THE INTEREST RATE

Suppose that the government concludes that the budget deficit is too large. A decision is made to increase taxes while keeping government spending unchanged. Such a policy, aimed at reducing the budget deficit, is often called a **fiscal contraction.** (In contrast, an *increase* in the deficit, due either to an increase in spending or to a decrease in taxes, is called a **fiscal expansion.**) What are the effects of such a fiscal contraction on output, on its components, and on the interest rate?

In answering this or any question about the effects of changes in policy, always go through the following three steps:

- *Step 1:* Ask how this change affects goods and financial markets equilibrium relations, how it shifts the *IS* and/or the *LM* curve.
- *Step 2:* Characterize the effects of these shifts on the equilibrium.
- *Step 3:* Describe the effects in words.

With time and experience, you will often be able to go directly to step 3; by then you will be ready to give an instant commentary on the economic events of the day. But until you get to that level of expertise, go step by step.

Going through step 1, the first question to tackle is how the increase in taxes affects equilibrium in the goods market—that is, how it affects the *IS* curve. Let's draw, in Figure 6-8(a), the *IS* curve corresponding to equilibrium in the goods market before the increase in taxes. Take an arbitrary point, *B*, on this *IS* curve. By construction of the *IS* curve, output Y_B and the corresponding interest rate i_B are such that the supply of goods is equal to the demand for goods.

Now, at the interest rate i_B, ask what happens to output if taxes increase from *T* to *T'*. We saw the answer in Section 6-1. Because people have less disposable in-

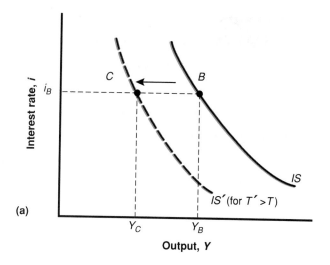

(a)

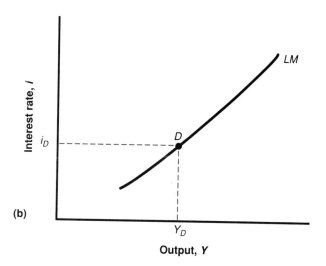

(b)

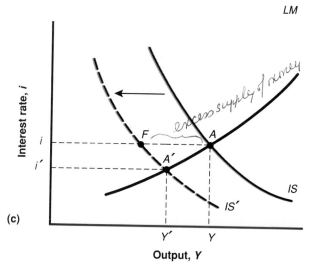

(c)

FIGURE 6-8

The Effects of an Increase in Taxes

An increase in taxes shifts the *IS* curve to the left, and leads to a decrease in the equilibrium level of output and the equilibrium interest rate.

come, the increase in taxes decreases consumption, and through the multiplier decreases output. Thus, at interest rate i_B, output decreases from Y_B to Y_C. More generally, at *any* interest rate, higher taxes lead to lower output: The *IS* curve shifts to the left from *IS* to *IS'*.

Next, let's see if anything happens to the *LM* curve. Figure 6-8(b) draws the *LM* curve corresponding to financial-markets equilibrium before the increase in taxes. Take an arbitrary point, *D*, on this *LM* curve. By construction of the *LM* curve, the interest rate i_D and income Y_D are such that the supply of money is equal to the demand for money.

What happens to the *LM* curve when taxes are increased? The answer is: nothing. At the given level of income Y_D, the interest rate at which the supply of money is equal to the demand for money is the same as before, namely i_D. In other words, because taxes do not appear in the *LM* relation, they do not affect the equilibrium condition. They do not affect the *LM* curve.

Note the general principle here: *A curve shifts in response to a change in an exogenous variable only if this variable appears directly in the equation represented by that curve.* Taxes enter equation (6.2), so the *IS* curve shifts. But taxes do not enter equation (6.3), so the *LM* curve does not shift.

Now let's consider the second step, the determination of the equilibrium. Let the initial equilibrium in Figure 6-8(c) be at point *A*, at the intersection between the initial *IS* curve and the *LM* curve. After the increase in taxes, the *IS* curve shifts to the left, and the new equilibrium is at the intersection of the new *IS* curve and the unchanged *LM* curve, at point *A'*. Output decreases from *Y* to *Y'*. The interest rate decreases from *i* to *i'*. Thus, as the *IS* curve *shifts,* the economy *moves along* the *LM* curve, from *A* to *A'*. The reason these words are italicized is that it is always important to distinguish *shifts in* curves (here the *IS* curve) and *movements along* a curve (here the *LM* curve). Most mistakes and confusion come from not distinguishing between the two.

The third and final step is to tell the story in words. The story goes like this: The increase in taxes leads to lower disposable income, which causes people to consume less. The result through the multiplier effect is a decrease in output and income. The decrease in income reduces the demand for money, leading to a decrease in the interest rate. The decline in the interest rate mitigates but does not completely offset the effect of higher taxes on the demand for goods. [If the interest rate did not decline, the economy would go from point *A* to point *F* in Figure 6-8(c). But as a result of the decline in the interest rate—which stimulates investment—the decline in activity is only to point *A'*.]

What happens to the components of demand? By assumption, government spending remains unchanged: We have assumed that the fiscal consolidation takes place through an increase in taxes. Consumption surely goes down, both because taxes go up and because income goes down. But what happens to investment? On the one hand, lower output means lower sales and lower investment. On the other, a lower interest rate leads to higher investment. Without knowing more about the exact form of the investment relation, equation (6.1), we cannot tell which effect dominates. If investment depends only on the interest rate, then investment surely increases; if investment depends only on sales, then investment surely decreases. In general, we cannot tell. Contrary to what is often stated by politicians, a reduction in the budget deficit does not necessarily lead to an increase in investment. (The

DEFICIT REDUCTION: GOOD OR BAD FOR INVESTMENT?

You may have heard the argument before: "Private saving goes toward either financing the budget deficit or financing investment. Thus, it does not take a genius to conclude that reducing the budget deficit leaves more saving available for investment, and thus increases investment."

This argument sounds simple and convincing. How do we reconcile it with what we just saw in this text, that deficit reduction may decrease rather than increase investment?

Remember from Chapter 3 that we can also think of goods-market equilibrium as the condition that

$$\underset{\text{investment}}{I} = \underset{\text{private saving}}{\left(\dfrac{Y-T-C}{S}\right)} - \underset{\text{budget deficit}}{(G-T)}$$

Investment must be equal to total saving—that is, private saving minus the budget deficit (which is just dis-

saving by the government). And it is indeed true that, *given private saving*, if the government reduces its deficit—either by increasing taxes or reducing government spending—investment must go up.

The crucial part of this statement, however, is "given private saving." And a fiscal contraction affects private saving as well: A fiscal contraction leads to lower output, lower income; since consumption goes down by less than the decrease in income, private saving also goes down. And it may go down by more than the reduction in the budget deficit, leading to a decrease rather than an increase in investment. To sum up, a fiscal contraction may indeed decrease investment. Or, looking at the reverse case, a fiscal expansion—that is, a decrease in taxes or an increase in spending—may actually increase investment.

Focus box entitled "Deficit Reduction: Good or Bad for Investment?" discusses this result at more length.) We shall return to the relation between fiscal policy and investment many times in this book, and we shall qualify this first answer in many ways. But the result that *deficit reduction may decrease investment,* at least in the short run, will remain.

MONETARY POLICY, ACTIVITY, AND THE INTEREST RATE

An increase in the money supply is called a **monetary expansion.** A decrease in the money supply is called a **monetary contraction** or **monetary tightening.**

Let's take the case of a monetary expansion here. Suppose that the central bank increases nominal money, M, through an open-market operation. Given our assumption that the price level is fixed, this increase in nominal money leads to a one-for-one increase (in percentage terms) in real money, M/P. Let us denote the initial real money supply by M/P and the new higher one by M'/P, and trace the effects of the money supply increase on output and the interest rate.

The first step is again to see whether and how the IS and the LM curves shift. Let's look at the IS curve first. The money supply does not affect *directly* either the supply of or the demand for goods. In other words, M does not appear in the IS relation. Thus, a change in M does not shift the IS curve.

Money enters the LM relation, however, so that the LM curve shifts when the money supply changes. As we saw in Section 6-2, an increase in money shifts the LM down: At a given level of income, an increase in money leads to a decrease in the interest rate.

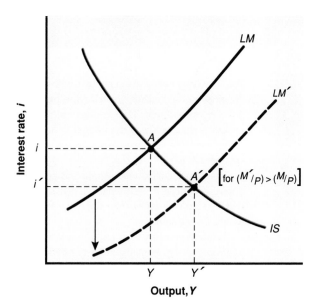

FIGURE 6-9

The Effects of a
Monetary Expansion

A monetary expansion
leads to higher output and a
lower interest rate.

Putting things together, a monetary expansion shifts the *LM* curve down and does not affect the *IS* curve. Thus, in Figure 6-9, the economy moves along the *IS* curve, and the equilibrium changes from point *A* to point *A'*. Output increases from *Y* to *Y'*, and the interest rate decreases from *i* to *i'*. In other words, the increase in money leads to a lower interest rate. The lower interest rate leads to an increase in investment and, through the multiplier, to an increase in demand and output.

In the case of a monetary expansion, as opposed to the case of fiscal contraction we looked at earlier, we can tell exactly what happens to the various components of demand. With higher income and unchanged taxes, consumption goes up. With higher sales and a lower interest rate, investment also unambiguously goes up. A monetary expansion is thus more investment-friendly than a fiscal expansion.

USING A POLICY MIX

We have looked so far at fiscal and monetary policy in isolation. Our purpose was to show how each one worked. In practice, the two are often used together. The combination of monetary and fiscal policies is known as the **monetary-fiscal policy mix,** or simply as the **policy mix.**

Sometimes monetary and fiscal policies are used for a common goal. For example, from 1993 to 1997 federal and provincial governments reduced their deficits while the Bank of Canada chose to partially counteract the adverse effects of a fiscal contraction on economic activity with a more expansionary monetary policy. In terms of the *IS-LM* diagram in Figure 6-10, the decrease in government spending and increase in taxes (leading to a leftward shift of the *IS* curve from *IS* to *IS'*) was partially offset by a shift in the *LM* curve down (from *LM* to *LMε*). The Bank's response was to lead the economy from a point A to a point such as Aε rather to a point such as B.

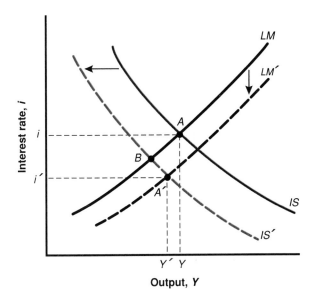

FIGURE 6-10
Fiscal Contraction and Monetary Expansion
Combining a monetary expansion with a fiscal contraction allows for a decrease in the interest rate with little or no decrease in output.

Sometimes the fiscal-monetary mix emerges from tensions or even disagreements between the government (which is in charge of fiscal policy) and the central bank (which is in charge of monetary policy). A typical scenario is one in which the central bank, disagreeing with what it considers a dangerous fiscal expansion, embarks on a course of monetary contraction to offset some of the fiscal expansion's effects on activity. A recent example of such a tug-of-war is that of Germany after unification in the early 1990s. It is described in the In Depth box entitled "German Unification and the German Fiscal-Monetary Policy Mix."

You should remember the method we have developed in this section to look at changes in activity and the interest rate. We have used it to look at the effects of changes in policy; Table 6-1 summarizes what we have learned. But you can use it to look at other changes as well. For example, you may want to trace the effects of a decrease in consumer confidence through its effect on consumption demand.

TABLE 6-1
THE EFFECTS OF FISCAL AND MONETARY POLICY

	SHIFT IN *IS*	SHIFT IN *LM*	MOVEMENT IN OUTPUT	MOVEMENT IN INTEREST RATE
Increase in taxes	left	none	down	down
Decrease in taxes	right	none	up	up
Increase in government spending	right	none	up	up
Decrease in government spending	left	none	down	down
Increase in money	none	down	up	down
Decrease in money	none	up	down	up

In 1990 East Germany and West Germany became one country again. Whereas the two parts had been at a roughly comparable level of economic development before World War 11, this was no longer the case by 1990. West Germany was far richer and far more productive than East Germany. The economic consequences of unification were many; the focus here will be just on the implications of unification for fiscal and monetary policy. (A full discussion of the transition from central planning to a market economy in East Germany, and Eastern Europe in general, will have to wait until Chapter 26, when we will have developed the required tools.)

Upon unification, it became clear that most firms in the Eastern Lander (as the former German Democratic Republic is now known) were just not competitive. Many simply had to be closed in part or in total, and the others needed new and more modern equipment. It soon became obvious that transition would require large increases in government spending on new infrastructure, on cleaning up environmental damage, on unemployment benefits to workers losing their jobs, and on transfers to firms to keep them operating until they turned around.

Faced with this large increase in transfers and spending, the German government decided to rely partly on increased taxes and partly on a larger deficit. Table 1 gives basic numbers on some of the major macroeconomic variables from 1988 to 1991 (for West Germany only).

The numbers show that, even before unification, Germany was experiencing a strong expansion. GDP growth in 1988 and 1989 was close to 4%. Investment was booming. And, because tax revenues depend on economic activity, the strong growth in GDP was the source of high government revenues in 1989, leading to a fiscal surplus of 0.2% of GDP in 1989.

The effects of unification were to increase demand further. In 1990, the rate of investment growth was even higher than in 1989. Because of the increase in spending and transfers due to unification, West Germany's fiscal position went from a budget surplus in 1989 to a budget deficit of 1.8% of GDP in 1990. In terms of the *IS-LM* model, 1990 was thus characterized by a sharp increase in government spending, a large shift of the *IS* to the right, from *IS* to *IS'* in Figure 1.

The Berlin Wall, which had separated the two halves of Berlin since 1961, was dismantled in November 1989.

Seeing these developments, the German central bank (Bundesbank) worried that growth was too strong, that the economy was operating at too a high a level of activity, and that the result would be inflation (a mechanism we explore later in the book). The central bank concluded that growth should be slowed. Thus, even though the interest rate had already increased from 4.3% in 1988 to 7.1% in 1989, the Bundesbank decided on a policy of tight money; it let the interest rate go even higher, to 9.2% in 1991. In terms of the *IS-LM* in Figure 1, the central bank decided to shift the *LM* up, in order to slow down economic activity.

Thus, one of the effects of German unification was fiscal expansion combined with tight monetary policy. The result was fast growth (from the fiscal expansion) and high interest rates (from tight money). These high interest rates had important implications not only for Germany, but for all of Europe. Indeed, they have been accused of being one of the main causes of the recession in Europe in the early 1990s. We discuss this argument in Chapter 13.

....

TABLE I

SELECTED MACRO VARIABLES FOR WEST GERMANY, 1988–1991

	1988	1989	1990	1991
GDP growth (%)	3.7	3.8	4.5	3.1
Investment growth (%)	5.9	8.5	10.5	6.7
Budget surplus (% of GDP)				
(minus sign: deficit)	−2.1	0.2	−1.8	−2.9
Interest rate (short term)	4.3	7.1	8.5	9.2

"Investment" refers to nonresidential investment.

Source: OECD Economic Outlook, June 1992.

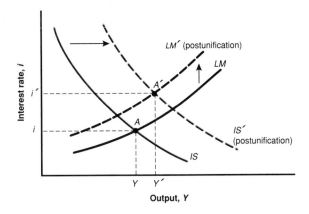

FIGURE I
The Fiscal-Monetary Policy Mix
in Postunification Germany

6-4 ADDING DYNAMICS

In Chapter 4 we added dynamics to our description of the goods market and were able to describe the adjustment of output both more realistically and more intuitively. We do the same here, indeed building on what we learned in Chapter 4.

Let's return first to the *IS* curve and examine the effects of a tax increase. As we have seen, a tax increase shifts the *IS* curve to the left. In Figure 6-11(a), the *IS* curve shifts from *IS* to *IS'*. At a given interest rate, say i_A, the equilibrium level of output decreases from Y_A to Y_B.

Will output really decline instantaneously from Y_A to Y_B? No. We saw some reasons in Chapter 4: It takes a while for production to respond to the decrease in sales and for consumers to respond to the decrease in income. We can now add to that list the fact that it is also likely to take a while for firms to revise their investment plans in light of a decrease in sales. For all these reasons, the decline in output will happen only over time. In terms of Figure 6-11(a), output will decrease slowly from Y_A to Y_B.

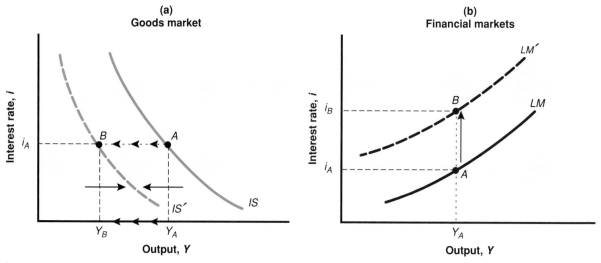

(a)
Goods market

(b)
Financial markets

FIGURE 6-11

Introducing Dynamics in the *IS-LM* Model

When output is above the level implied by the *IS* relation, it adjusts slowly to that level. In contrast, interest rates adjust quickly, so that the *LM* relation is always satisfied.

More generally, it is reasonable to assume that when output is to the right of the equilibrium curve—which, in our case, after the tax increase, is *IS′*—output decreases slowly; when it is to the left of the equilibrium curve, output increases slowly. This basic conclusion is represented by the two large arrows on each side of *IS′* in Figure 6-11(a).

Now let's look at the *LM* curve and the effects of a monetary contraction. As we have seen, a monetary contraction shifts the *LM* curve up. In Figure 6-11(b) the *LM* curve shifts from *LM* to *LM′*: At a given level of income, say Y_A, the interest rate increases from i_A to i_B. How long does it take for the interest rate to adjust? The answer is: practically no time. Interest rates adjust very quickly to changes in supply and demand in most financial markets. The market for government bonds (in which the Bank of Canada buys and sells) is one of the most efficient markets in the world and clears within seconds of changes in demand or supply. Thus, the right assumption is that the decrease in the money supply causes the interest rate to increase instantaneously from i_A to i_B in Figure 6-11(b). For the rest of the book, we shall assume that the adjustment of the interest rate to any change in the demand for or the supply of money is so fast that *the economy is always on the LM curve.*

Equipped with these dynamics, let's reexamine the effects of a monetary contraction on activity and the interest rate. The adjustment is shown in Figure 6-12. Before the decrease in the money supply, the economy is at point *A,* with output *Y* and interest rate *i.* When the central bank decreases the money supply, the *LM* curve shifts from *LM* to *LM′.* The economy jumps to point *A″*: Output does not change right away, and the interest rate must do all the adjustment, increasing from *i* to *i″.* Over time, the higher interest rate leads to lower investment, a lower demand for goods, and lower output, so that output slowly decreases from its initial level. The economy moves along *LM′,* and eventually reaches point *A′.* At *A′,* the interest rate

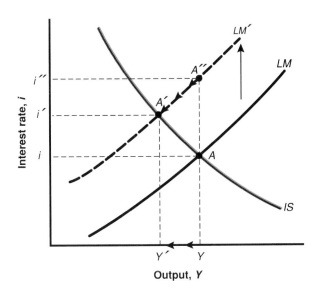

FIGURE 6-12

The Dynamic Effects of a Monetary Contraction

A monetary contraction leads to an increase in the interest rate. The higher interest rate leads, over time, to a decline in output.

is equal to i' and output is equal to Y'. Note that the eventual increase in the interest rate is smaller than the initial increase. This is because, as output contracts, so does the demand for money, which puts some pressure on the interest rate to decrease.

It is easy to describe the adjustment in words. The monetary contraction leads initially to a sharp increase in the interest rate. Over time, this increase leads to a decrease in output. This time dimension is important, and there is a general lesson about policy to be drawn from it. Monetary policy can affect the interest rate quickly but cannot affect output right away. Thus, the central bank must be careful not to be fighting the last battle. For example, there is no point in fighting a recession through a lower interest rate if the recession is already over when the lower interest rate starts affecting economic activity.

We shall let you look at the dynamic effects of a change in fiscal policy on your own. And, from now on, we shall often rely on these dynamics to tell more realistic stories of how changes in policy or behaviour affect economic activity.

6-5 DOES THE *IS-LM* MODEL ACTUALLY CAPTURE WHAT HAPPENS IN THE ECONOMY?

The *IS-LM* model gives us a way of thinking about the determination of output and the interest rate. But it is a model based on many assumptions and many simplifications. How do we know that we have made the right simplifications? How much should we believe the answers given by the *IS-LM* model?

These are the questions facing any theory, whether in macroeconomics or anywhere else. To pass muster, a theory must pass two tests.

■ First, the assumptions and the simplifications must be reasonable. But what "reasonable" means is not entirely clear. Surely the assumptions we have made, such as that there is only one type of good in the economy, that there is no foreign sector, and so forth, are factually wrong, but we may still call the model a reasonable simplification of reality if taking into account that these extensions led to a more complicated model, but to roughly the same results about aggregate activity, the interest rate, and so on.

■ Second, the major implications of the theory must be roughly consistent with what we actually see in the world. This is easier to check. Using econometrics, we can trace the effects of changes in monetary policy and fiscal policy and see how closely the effects correspond to the predictions of the *IS-LM* model. And it turns out that for an essentially closed economy, like the United States, the *IS-LM* model does quite well indeed.

Figure 6-13 makes this point nicely. It shows the results of a recent econometric study of the effects of changes in monetary policy on activity, using data from the United States from 1960 to 1990.[4] The authors of the study focus on the effects of movements in the **federal funds rate,** a short-term interest rate that is most directly affected by changes in U.S. monetary policy (you can think of it as the American counterpart to the Bank of Canada rate). They then trace the typical effects of such a change on activity. These effects are shown in the figure.

Figure 6-13(a) shows the effects of an increase in the federal funds rate of 1% on output over time. The percentage change in output is plotted on the vertical axis; time, measured in quarters, on the horizontal axis. The figure plots three lines. The best estimate of the effect of the change in the interest rate on output is given by the blue line. But, as we saw in Chapter 4, when using econometrics there is no such thing as learning the exact value of a coefficient or the exact effect of one variable on another. Rather, econometrics provides a best estimate—here, the blue line—and a measure of confidence we should have in the estimate. The true value of the effect lies within the two dashed lines with 60% probability. For this reason, the space between the two lines is called a **confidence band.**

Focussing on the best estimate—the blue line—we see that, following an increase in the federal funds rate of 1%, output slowly declines over time. The largest decrease, −0.7%, is achieved after eight quarters. This slow decline is very much consistent with the result predicted by our dynamic version of the *IS-LM* model in Figure 6-12.

The other parts of Figure 6-13 give more detail. Figure 6-13(b) shows how the increase in the federal funds rate leads to lower employment: As firms cut production, they also cut employment. As with output, the decline in employment is slow and steady, reaching −0.5% after eight quarters. The decline in employment is reflected in an increase in unemployment, shown in Figure 6-13(c).

Figures 6-13(d) and (e) show that the lagged response of production to sales—which we looked at in Chapter 4—is indeed present in the data. Figure 6-13(d) shows that sales decrease more quickly than production, and Figure 6-13(e) shows the implication for the behaviour of inventories. As sales decline faster than production, inventories initially increase. After three quarters, however, firms cut production by more than sales, and inventories start decreasing. Over time, inventories not only go back to normal but actually keep declining. The fact that eventually production declines more than sales—so that inventories decline—was not present

---

[4]*Lawrence Christiano, Martin Eichenbaum, and Charles Evans, "The Effects of Monetary Policy Shocks: Evidence from the Flow of Funds,"* WP 94-2, *Federal Reserve Bank of Chicago, 1994.*

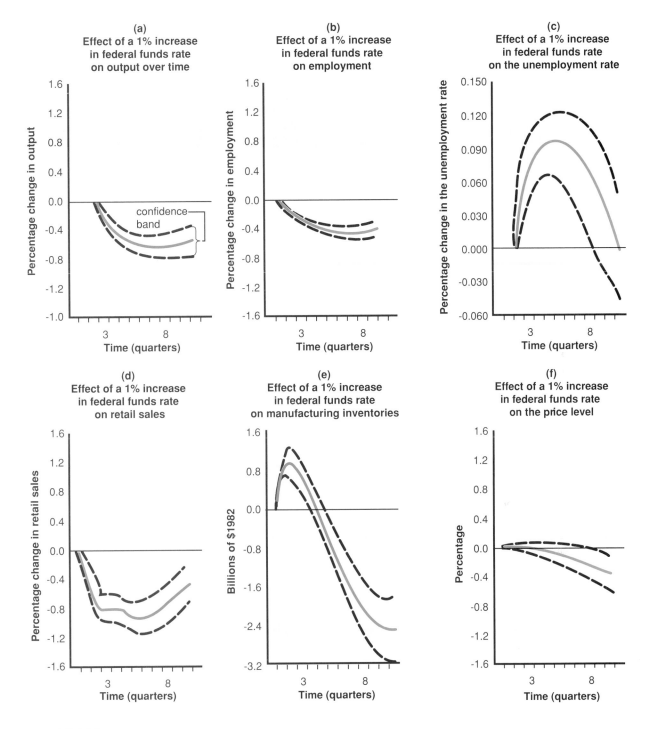

(a)
Effect of a 1% increase in federal funds rate on output over time

(b)
Effect of a 1% increase in federal funds rate on employment

(c)
Effect of a 1% increase in federal funds rate on the unemployment rate

(d)
Effect of a 1% increase in federal funds rate on retail sales

(e)
Effect of a 1% increase in federal funds rate on manufacturing inventories

(f)
Effect of a 1% increase in federal funds rate on the price level

FIGURE 6-13

The Empirical Effects of an Increase in the Federal Funds Rate on U.S. Variables

An increase in the federal funds rate leads, over time, to a decrease in output and to an increase in unemployment, but has little effect on the price level.

Source: Lawrence Christiano, Martin Eichenbaum, and Charles Evans, "The Effects of Monetary Policy Shocks: Evidence from the Flow of Funds," *WP 94-2*, Federal Reserve Bank of Chicago, 1994.

in the model of Chapter 4, and suggests the need for a richer model than the one we developed in that chapter.

Finally, Figure 6-13(f) looks at the behaviour of the price level. Remember that one of the assumptions of the *IS-LM* model is that the price level is given and thus does not change in response to changes in demand. Figure 6-13(f) shows that the assumption is not a bad approximation of reality in the short run. The price level is nearly unchanged for the first five quarters or so. Only after the first five quarters does the price level appear to decline. This gives a strong hint as to why the *IS-LM* model becomes less reliable as we look at the medium run. One can no longer assume that the price level is given, and movements in the price level become important.

Figure 6-13 is comforting. It shows that the implications of the *IS-LM* model are consistent with what we observe in the world's largest economy. Even though the events of the past 30 years have convinced macroeconomists that several important features are missing, the *IS-LM* model looks like a solid basis on which to build. In the next four chapters, we build by looking at the role of expectations in determining activity.

SUMMARY

■ The *IS-LM* model characterizes the implications of simultaneous equilibrium in the goods and in the financial markets.

■ The *IS* relation and the *IS* curve show the combinations of the interest rate and the level of output that are consistent with equilibrium in the goods market. An increase in the interest rate leads to a decline in output.

■ The *LM* relation and the *LM* curve show the combinations of the interest rate and the level of output consistent with equilibrium in financial markets. Given the real money supply, an increase in output leads to an increase in the interest rate.

■ A fiscal expansion shifts the *IS* curve to the right, leading to an increase in output and an increase in the interest rate. A monetary expansion shifts the *LM* curve down, leading to an increase in output and a decrease in the interest rate.

■ The combination of monetary and fiscal policies is known as the monetary-fiscal policy mix, or simply as the policy mix. Sometimes monetary and fiscal policies are used for a common goal. But the monetary-fiscal mix occasionally emerges from tensions or even disagreements between the government (which is in charge of fiscal policy) and the central bank (which is in charge of monetary policy).

■ The *IS-LM* model appears to describe well the behaviour of the U.S. economy in the short run. In particular, the effects of monetary policy appear to be similar to those implied by the *IS-LM* model once dynamics are introduced in the model. An increase in the interest rate due to a monetary contraction leads to a steady decrease in output, with the maximum effect taking place after about eight quarters.

KEY TERMS

■ *IS* curve, 105
■ *LM* curve, 108
■ fiscal contraction, or fiscal consolidation, 110
■ fiscal expansion, 110
■ monetary expansion, 113

■ monetary contraction, or monetary tightening, 113
■ monetary-fiscal policy mix, or policy mix, 114
■ federal funds rate, 120
■ confidence band, 120

QUESTIONS AND PROBLEMS

1. One reason why higher interest rates discourage investment is that many firms must borrow funds to purchase plant and equipment. But what about investment projects financed from a firm's own *retained earnings*—income from profits that is kept within the firm? Since no borrowing occurs, will higher interest rates discourage investment in this case? Why or why not?

2. As discussed in the text, a decrease in the budget deficit may cause investment spending to increase or decrease. If you were asked by the prime minister to determine which way investment would change after a deficit reduction, what specific information would you need to give him an answer?

3. Consider the following numerical version of the *IS-LM* model:

$$C = 400 + 0.5Y_D$$
$$I = 700 - 4000i + 0.1Y$$
$$G = 200$$
$$T = 200$$

Real money demand: $(M/P)^d = 0.5Y - 7,500i$
Real money supply: $(M/P)^s = 500$

(Note that in this problem money demand is assumed to be linear to make the mathematics easier.)
Try to keep decimal fractions (like 0.0667) in the form of simple fractions (2/30) until calculating your final values for Y, i, and so forth.

a. Find the equation for the *IS* curve. [*Hint:* Goods-market equilibrium. You want an equation with Y on the left-hand side, all else on the right.]

b. Find the equation for the *LM* curve. [*Hint:* It will be convenient for later use to write the equation with i on the left-hand side, all else on the right.]

c. Solve for the equilibrium real output (Y). [*Hint:* Substitute the expression for the interest rate (given in the *LM* equation) into the *IS* equation, and solve for Y.]

d. Solve for the equilibrium interest rate (i). [*Hint:* Substitute the value you obtained for Y above into either the *LM* equation or the *IS* equation, and solve for i. You can substitute into both equations to check your work.]

e. Solve for the equilibrium values of consumption spending and investment spending, and verify the value you obtained for Y by adding up C, I, and G.

f. Now suppose that government spending increases by 500 to 700. Solve again for Y, i, C, and I, and once again verify that $Y = C + I + G$ in equilibrium.

g. Summarize the effects of the expansionary fiscal policy in part **f** by stating what has happened to Y, i, C, and I.

h. Set all variables back to their initial values. Now, suppose that the money supply increases by 500. Solve again for Y, i, C, and I. Once again, verify that $Y = C + I + G$ in equilibrium.

i. Summarize the effects of the expansionary monetary policy in part **h** by stating what has happened to Y, i, C, and I.

4. Suppose that policy makers want to decrease the deficit while guaranteeing that there will be no decrease in either output or investment spending. Is there any monetary-fiscal policy mix that can achieve this goal?

5. Using the dynamic assumptions of the text, what sequence of events can we expect after an expansionary fiscal policy?

EXPECTATIONS: THE BASIC TOOLS

The consumer who decides whether or not to buy a new car must ask himself: Can I safely take out a new car loan? How much of a wage raise can I expect over the next few years? How safe is my job? The manager who observes an increase in current sales must ask herself: Is this a temporary boom that I should meet with the existing production capacity? Or does this upswing reflect a permanent increase in sales, in which case I should probably order new machines? How much additional profit can I expect to make if I buy a new machine? A pension fund manager who observes a boom in the stock market must ask himself: Are stock prices going to increase further, or is the boom likely to fizzle? Does this increase in prices reflect expectations of higher profits by firms in the future? Do I share those expectations? Should I reallocate some of my funds and put them into the stock market?

These examples make it clear that many economic decisions depend not only on what is happening today but also on expectations of the future. Indeed, one can push the argument further. Some decisions should depend very little on what is happening today. For example, why should an increase in sales today, if it is not accompanied by expectations of higher sales in the future, lead a firm to alter its investment plans? The new machines may not be in operation before sales have gone back to normal. By then, they might well sit idle, gathering dust until they are discarded.

Until now, we have not paid much attention to the role of expectations. When looking at the goods market, we assumed that consumption depended on current income and that investment depended on current sales. When looking at financial markets, we lumped assets together and called them "bonds"; we then focussed on the choice between bonds and money, and ignored the choice between bonds and stocks, between short-term bonds and long-term bonds, and so on. We introduced these simplifications to build intuition. But it is now time to consider the role of expectations. Indeed, a recurring theme of this book will be that expectations matter very much in modern economies. Consumers, firms, and participants in financial markets spend much of their time peering into the future. Pessimism about the future, justified or not, can trigger a decline in spending and a recession; optimism, justified or not, can trigger an expansion. Policies work not only because of their direct effects, but also because of and sometimes mainly through their effects on expectations. Prime ministers, premiers, and finance ministers go on TV to explain their economic programs, hoping to shape perceptions of what the programs will do and to increase the chance that the programs will work.

Introducing expectations complicates our task. Current developments depend on expectations of the future, and in turn expectations of the future depend on current developments. To keep things manageable, we must develop a few key concepts and tools. This is our goal in this chapter, which introduces the concepts of *real* versus *nominal interest rates* and of *expected present discounted value*. The returns from this investment in tools come in the following chapters. Chapter 8 looks at the role of expectations in consumption and investment decisions. Chapter 9 looks at the role of expectations in financial markets. Chapter 10 extends our *IS-LM* model to allow for the role of expectations and takes another look at the role and the limits of policy.[1]

7-1 NOMINAL VERSUS REAL INTEREST RATES

In 1980, the *one-year T-bill rate*—the interest rate on one-year government bonds—stood at 12.2 percent. In late 1997, the one-year T-bill rate stood at 4.2 percent. Although most of us cannot borrow at the same interest rate as the government, the interest rates we face as consumers have also gone down substantially since 1980. Borrowing was clearly much cheaper in 1997 than it was in 1980.

---

[1]You can go directly from Chapter 7 to Chapter 10. What you need to know from Chapters 8 and 9 is summarized at the beginning of Chapter 10.

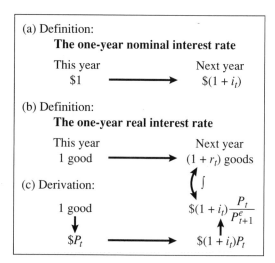

FIGURE 7-1

Nominal and Real
Interest Rates

(a) Definition:

The one-year nominal interest rate

This year   Next year

$1   ⟶ $(1 + i_t)$

(b) Definition:

The one-year real interest rate

This year   Next year

1 good   ⟶ $(1 + r_t)$ goods

(c) Derivation:

1 good $\$(1 + i_t)\dfrac{P_t}{P^e_{t+1}}$

$\$P_t$ ⟶ $\$(1 + i_t)P_t$

Or was it? In 1980, inflation was running above 12%. In 1997, inflation was running at about 2%. This information is very relevant. The interest rate tells us how many dollars we shall have to pay in the future in exchange for having one more dollar today. But we do not consume dollars; we consume goods. Thus what we really want to know when we borrow is how many goods we shall have to give up in the future in exchange for getting one more good today. Conversely, when we lend, we want to know how many goods—not how many dollars—we shall get in the future if we give up one good today. In the presence of inflation, the distinction is important. What is the point of receiving high interest payments in the future if inflation between now and then is so high that we shall be able to buy few goods with the proceeds?

Let us introduce two definitions at this point. Let us refer to interest rates in terms of dollars (or, more generally, in units of the national currency) as **nominal interest rates.** The interest rates printed in the financial pages of newspapers are nominal interest rates. For example, when we say that the one-year T-bill rate is 4.2%, we mean that for every dollar the government borrows by issuing one-year T-bills it promises to pay 1.042 dollars a year from now. More generally, if the nominal interest rate for year t is i_t, borrowing one dollar this year requires you to pay $1 + i_t$ dollars next year. This relation is represented in Figure 7-1(a): One dollar this year corresponds to $1 + i_t$ dollars next year.

Let us refer to interest rates expressed *in terms of goods* as **real interest rates.** Thus, if we denote the real interest rate for year t by r_t, then, by definition, borrowing the equivalent of one good this year requires you to pay the equivalent of $1 + r_t$ goods next year. This relation is represented in Figure 7-1(b): One good this year corresponds to $1 + r_t$ goods next year.

The practical problem we face at this point is that the real interest rate is not published in newspapers. The next question is therefore how we compute it.

COMPUTING THE REAL INTEREST RATE

Suppose that there is only one good in the economy, say bread (we shall relax this assumption later). If you borrow enough to eat one more kilogram of bread this year, how much will you have to repay, in terms of kilograms of bread, next year?

Figure 7-1(c) helps us derive the answer.

■ Suppose that the price of a kilogram of bread this year is P_t dollars. Thus, to eat one more kilogram of bread, you must borrow P_t dollars. This step is captured by the arrow pointing down in Figure 7-1(c).

■ Let i_t be the one-year nominal interest rate, the interest rate in terms of dollars. Thus, if you borrow P_t dollars, you will have to repay $(1 + i_t)P_t$ dollars next year. This step is captured by the horizontal arrow from left to right at the bottom of Figure 7-1(c).

■ What you care about, however, is not dollars, but kilograms of bread. Thus, the last step involves converting dollars to kilograms of bread next year. Let P_{t+1}^e be the price of bread you expect to prevail next year. The superscript "e" indicates that this is an expectation; you do not know yet what the price of bread will be next year. How much you expect to repay next year, in terms of kilograms of bread, is thus equal to $(1 + i_t)P_t/P_{t+1}^e$. This last step is captured by the arrow pointing up in Figure 7-1(c).

Putting together parts (b) and (c) of Figure 7-1, it follows that 1 plus the one-year real interest rate r_t is defined by

$$1 + r_t \equiv (1 + i_t)\frac{P_t}{P_{t+1}^e} \tag{7.1}$$

This expression is somewhat intimidating. Two simple manipulations make it look simpler.

Denote expected inflation by π_t^e. Given that we are assuming there is only one good—bread—the expected rate of inflation is defined as the expected change in the dollar price of bread between this year and next year, divided by the dollar price of bread this year:

$$\pi_t^e \equiv \frac{P_{t+1}^e - P_t}{P_t} \tag{7.2}$$

Given equation (7.2), note that we can rewrite P_t/P_{t+1}^e in equation (7.1) as $1/(1+\pi_t^e)$ —add 1 to both sides in equation (7.2) and take the inverse on both sides— so that we can rewrite equation (7.1) as

$$1 + r_t = \frac{1 + i_t}{1 + \pi_t^e} \tag{7.3}$$

One plus the real interest rate is equal to the ratio of 1 plus the nominal interest rate to 1 plus the expected rate of inflation.

Equation (7.3) gives us the exact definition of the real interest rate. However, when the nominal rate and expected inflation are not too large—say, less than 20% per year—a close approximation to this equation is given by the simpler relation[2]

$$r_t \approx i_t - \pi_t^e \tag{7.4}$$

----

[2]*This approximation is derived in proposition 6, Appendix 3. To see how close the approximation is, suppose that the nominal interest rate is 10% and expected inflation is 5%. Using the exact formula (7.3) gives $r_t = 4.8\%$. The approximation given by equation (7.4) is 5%, which is indeed close. The approximation becomes worse when nominal interest rates and expected inflation are very high. When nominal interest rates and expected inflation are equal to 100% and 80% respectively, for example, the exact formula gives a real interest rate of 11% while the approximation yields 20%.*

Equation (7.4) is a simple expression, and one you should remember. It says that *the real interest rate is (approximately) equal to the nominal interest rate minus expected inflation.* It has a number of important implications.

Only when expected inflation is equal to zero are the nominal interest rate and real interest rate equal. Because expected inflation is typically positive, the real interest rate is typically lower than the nominal interest rate. The higher the expected rate of inflation, the lower the real interest rate.

The case where expected inflation happens to be equal to the nominal rate is worth looking at more closely. Suppose, for example, that the nominal interest rate and expected inflation are both equal to 10%. Look at things first from the borrower's point of view. While it is true that for every dollar you borrow, you will have to repay 1.10 dollars next year, dollars will be worth 10% less in terms of goods next year. Thus, if you borrow the equivalent of one good, you will have to repay the equivalent of one good next year; the real cost of borrowing—the real interest rate—is equal to zero. Now look at things from the lender's point of view. True, every dollar you lend will yield 1.10 dollars next year, and this looks attractive; but dollars next year will be worth 10% less in terms of goods. If you lend the equivalent of one good, you will get the equivalent of one good next year. Despite a 10% nominal interest rate, the real interest rate is equal to zero.

We have assumed so far that there is only one good—bread. But what we have done easily generalizes. All we need to do is substitute the *price level*—the average price of goods—for the price of bread. If we use the consumer price index (the CPI) to measure the price level, the real interest rate tells us how much consumption we must give up next year to consume more today.

NOMINAL AND REAL INTEREST RATES IN CANADA SINCE 1980

Let us return to the question that started this section. We can now rephrase it more generally as follows: What has happened to the real interest rate in Canada since 1980?

A partial answer is given in Figure 7-2, which plots both nominal and **ex post real interest rates** since 1980. The ex post real interest rate is the difference between the nominal interest rate and the rate of inflation that actually occurred. For each year, the nominal interest rate is the one-year T-bill rate at the beginning of the year. To construct the (ex ante) real interest rate, we need a measure for expected inflation—more precisely, for the rate of inflation expected as of the beginning of each year. Such data are hard to find for Canada, which is why we've settled for the ex post real interest rate.

Figure 7-2 shows the importance of adjusting for inflation. While the nominal interest rate was much lower in 1996 than it was in 1980, the ex post real interest rate was actually *higher* in 1996 than it was in 1980, 2.6% in 1996 versus 0.1% in 1980. This comes from the fact that inflation has steadily declined since the early 1980s. Most economists expect inflation in 1998 to be around 2%, so the (ex ante) real interest rate in late 1997 is just over 2%. Looking at nominal rates only can clearly send the wrong signal about the true cost of borrowing. This theme is ex-

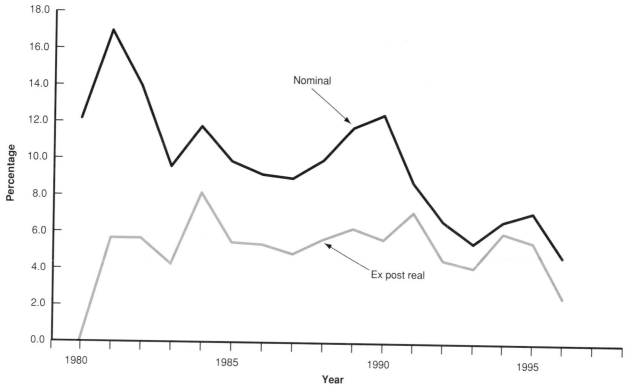

FIGURE 7-2

Nominal and Ex Post Real One-Year T-Bill Rates in the Canada, 1980–1996

While the nominal interest rate was much lower in 1996 than in 1980, the ex post real interest rate was actually higher in 1996.

Source: Statistics Canada, CANSIM Series B14062 and P700000.

plored further in the Global Macro box entitled "Nominal and Real Interest Rates around the World," which looks at differences in nominal and real interest rates across countries rather than at variations over time as we have just done.

NOMINAL AND REAL INTEREST RATES AROUND THE WORLD

Every month or so, the magazine *The Economist* gives both actual values and forecasts of the main macro-economic variables for the major OECD countries. These forecasts are usually averages of commercial forecasts published in each country by professional forecasters. Using the numbers from the November 11, 1995, issue of *The Economist,* Table 1 gives the nominal short-term interest rates, together with the forecasts of inflation for 1996 as of November 1995, for a

number of countries. It then computes the real short-term interest rate for each country as the difference between the nominal rate and the forecast of inflation.

In November 1995, nominal interest rates varied from 10.9% in Italy to 0.5% in Japan, a 10.4% difference. But real interest rates varied less, from 5.8% in Italy to 0.3% in Japan, a 5.5% difference. This is because countries with high nominal interest rates were also typically countries with higher inflation rates. (As we shall see in Chapter 19, this is a general proposition. Countries with higher inflation tend to have higher nominal interest rates, so that differences in real interest rates are typically smaller than differences in nominal interest rates.)

In most OECD countries today inflation and expected inflation are low, so that nominal interest rates are not very different from real interest rates. The picture can be sharply different when inflation is high. Take, for example, Brazil in 1993 (inflation has decreased since then). In September 1993, the Brazilian monthly nominal interest rate was 36.9 percent. [Note that this was the monthly rate; the annual rate was equal to $(1 + .369)^{12} = 4335\%$! When interest rates get so high, it becomes more convenient to compute and report them monthly.] But the picture was very different for the real interest rate. Brazil was suffering from very high inflation: The Brazilian CPI, normalized to be equal to 100 in 1985, was equal to 4 539 000 in September 1993, up from 3 382 000 in August 1993. Thus the monthly rate of inflation in September was equal to $(4\,539\,000 - 3\,382\,000)/(3\,382\,000) = 34.2\%$. We do not know what Brazilian markets expected inflation to be in the future as of September 1993. If we assume that they expected future inflation to be roughly the same as the rate of inflation in September, the real interest rate was 36.9% − 34.2%, thus only 2.7%. (Note that "only" should probably be in quotes. A monthly real interest rate of 2.7% corresponds to a yearly real interest rate of 37.7%, still a very high real interest rate.)

TABLE I

NOMINAL AND REAL INTEREST RATES FOR SELECTED COUNTRIES, NOVEMBER 1995

	NOMINAL RATE (%)	EXPECTED INFLATION (%)	REAL RATE (%)
Belgium	4.1	2.2	1.9
Germany	4.0	2.2	1.8
Italy	10.9	5.1	5.8
Japan	0.5	0.2	0.3
Spain	9.6	4.5	5.1

Source: The Economist, November 11, 1995, 106.

7-2 EXPECTED PRESENT DISCOUNTED VALUES

Let's now turn to the second important concept we introduce in this chapter, the concept of expected present discounted value.

To see why the concept is helpful, let's return to the example of the manager who considers whether or not to buy a new machine. On the one hand, buying and installing the machine involves a cost today. On the other hand, the machine will allow for higher production, higher sales, and thus higher profits in the future. The question facing the manager is thus whether the value of these expected profits is higher than the cost of buying and installing the machine. This is where the concept of expected present discounted value comes in handy: The **expected present discounted value** (or **present discounted value** or **present value**) of a sequence of payments is the value today of this expected sequence of payments. Thus, once the manager has computed the expected present discounted value of the sequence of profits, her problem becomes simple. If this value exceeds the initial cost, she should go ahead and buy the machine. If it does not, she should not.

As in the case of the real interest rate in Section 7-1, the practical problem is that expected present discounted values are not directly observable. They must be constructed from information about the sequence of expected payments and interest rates. Let's first look at the mechanics of construction.

COMPUTING EXPECTED PRESENT DISCOUNTED VALUES

If the one-year nominal interest rate is i_t, lending one dollar today yields $1 + i_t$ dollars next year. Equivalently, borrowing one dollar today implies paying back $1 + i_t$ dollars next year. In that sense, one dollar today is worth $1 + i_t$ dollars next year. This relation is captured in Figure 7-3(a).

Let's turn the argument around and ask: How many dollars is one dollar *next year* worth today? The answer, shown in Figure 7-3(b), is clearly $1/(1 + i_t)$ dollars today. Think of it this way: If you lend $1/(1 + i_t)$ dollars today, you will receive $1/(1 + i_t)$ times $(1 + i_t) = 1$ dollar next year. Equivalently, if you borrow $1/(1 + i_t)$ dollars today, you will have to repay exactly one dollar next year.

Thus, one dollar next year is worth $1/(1 + i_t)$ dollars today. More formally, we say that $1/(1 + i_t)$ is the *present discounted value* of one dollar next year. The term "present" comes from the fact that we are looking at the value in terms of dollars *today*. The term "discounted" comes from the fact that the value next year is discounted, with $1/(1 + i_t)$ being the **discount factor** (i_t, the one-year nominal interest rate, is sometimes called the **discount rate**). Note that because the nominal interest rate is always positive, the discount factor is always less than 1. Having a dollar next year is worth less than having a dollar today. The higher the nominal interest rate, the lower the value today of a dollar next year. For example, if $i = 5\%$, the value today of a dollar next year is $1/1.05 \approx 95$ cents; if $i = 10\%$, the value today of a dollar next year is $1/1.10 \approx 91$ cents.

Let's now apply the same logic to the value today of a dollar two years from now. Let's for the moment ignore uncertainty and assume that current and future one-year nominal interest rates are known with certainty. Let i_t be the nominal interest rate this year, and let i_{t+1} be the one-year nominal interest rate next year.

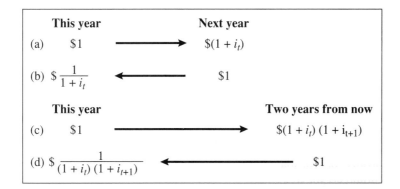

FIGURE 7-3
Computing Present Discounted Values

If you lend one dollar for two years, you will get $(1 + i_t)(1 + i_{t+1})$ dollars two years from now. Put another way, one dollar today is worth $(1 + i_t)(1 + i_{t+1})$ dollars two years from now. This relation is represented in Figure 7-3(c).

What is one dollar two years from now worth today? By the same logic as before, the answer is $1/(1 + i_t)(1 + i_{t+1})$ dollars. If you lend $1/(1 + i_t)(1 + i_{t+1})$ dollars today, you will get exactly one dollar in two years. More formally, the *present discounted value of a dollar two years from now* is equal to $1/(1 + i_t)(1 + i_{t+1})$ dollars. This relation is shown in Figure 7-3(d). If, for example, the one-year nominal interest rate is constant and equal to 5%, so that $i_t = i_{t+1} = 5\%$, then the present value of a dollar in two years is equal to $1/(1.05)^2$ or about 91 cents today.

A general formula.   Having gone through these steps, it is now easy to derive the present discounted value for the general case.

Consider a sequence of payments in dollars, now and in the future. Assume for the moment that future payments are known with certainty. Denote the current payment by $\$z_t$, the payment next year by $\$z_{t+1}$, the payment two years from now by $\$z_{t+2}$, and so on.

The present discounted value of this sequence of payments—the value in today's dollars of the sequence of payments, which we will call $\$V_t$—is then given by

$$\$V_t = \$z_t + \frac{1}{1 + i_t}\ \$z_{t+1} + \frac{1}{(1 + i_t)(1 + i_{t+1})}\ \$z_{t+2} + \cdots$$

Each payment in the future is multiplied by its proper discount factor. The more distant the payment, the smaller the discount factor, and thus the smaller the value of the payment today. In other words, future payments are discounted more heavily, so their present value is lower.

We have assumed so far that both future payments and future interest rates are known with certainty. But actual decisions are typically made under uncertainty and thus have to be based on expectations rather than actual values. In our example, the manager cannot be sure how much profit the new machine will actually bring; nor can she be sure of what interest rates will be. The best she can do is get the best forecasts she can, and then compute the *expected present discounted value* of profits, based on these forecasts.

How does one compute an expected present discounted value? Exactly in the same way that we just did, but replacing *known* future payments and *known* interest rates by *expected* future payments and *expected* interest rates. Formally, denote expected payments next year by $\$z_{t+1}^e$, expected payments two years from now by $\$z_{t+2}^e$, and so on. Similarly, denote the expected one-year nominal interest rate next year by i_{t+1}^e, and so on. (The one-year nominal interest rate this year, i_t, is known as of today and thus does not need a superscript *e*.) The expected present discounted value of this expected sequence of payments is then given by

$$\$V_t = \$z_t + \frac{1}{1 + i_t}\ \$z_{t+1}^e + \frac{1}{\phantom{(1 + i_t)(1 + i_{t+1}^e)}}\ \$z_{t+2}^e + \cdots \tag{7.5}$$

"Expected present discounted value" is a heavy expression to carry; we shall often use, for short, just *present value*. Also, it will be convenient to have a shorthand way of writing expressions like equation (7.5). To denote the present value of an expected sequence for $z, we shall write $V(\$z_t)$, or just $V(\$z)$.

APPLICATIONS

A simpler notation does not remove the fact that equation (7.5) is not simple. On the one hand, the present value depends positively on expectations of future payments. An increase in any future z^e leads to an increase in the present value. On the other hand, the present value depends negatively on interest rates, current and expected. An increase in i or in future i^e leads to a decrease in the present value. Intuition for these effects, and what they imply, is best built by going through a number of examples.

Constant interest rates. To focus on the effects of the sequence of payments on the present value, assume that interest rates are expected to be constant over time, so that $i_t = i^e_{t+1} = \ldots$, and denote their common value by i. The present value formula—equation (7.5)—becomes

$$\$V_t = \$z_t + \frac{1}{1+i}\,\$z^e_{t+1} + \frac{1}{(1+i)^2}\,\$z^e_{t+2} + \cdots \tag{7.6}$$

In this case, the present value is a *weighted sum* of current and expected future payments: The weights decline *geometrically* through time. That is, the weight on a payment today is 1, the weight on the payment n years from now is $[1/(1+i)^n]$. With a positive interest rate, the weights get closer and closer to zero as we look further and further into the future. For example, with an interest rate equal to 10%, the weight on a payment in 10 years is equal to $1/(1 + 0.10)^{10} = 0.386$, so a payment of \$1000 in 10 years is worth \$386 today; the weight on a payment in 30 years is $1/(1 + 0.10)^{30} = 0.057$, so a payment of \$1000 in 30 years is worth only \$57 today!

Constant interest rates and payments. In some cases, the sequences of payments for which we want to compute the present value are particularly simple. For example, a typical fixed-rate five-year car loan requires constant dollar payments over five years. Thus, consider a sequence of equal payments—call them $z without a time index—over n years, including the current year. In this case, the present value formula in equation (7.6) simplifies further to

$$\$V_t = \$z\left[1 + \frac{1}{1+i} + \cdots + \frac{1}{(1+i)^{n-1}}\right]$$

Because the terms in the expression in brackets represent a geometric series, we can compute the sum of the series, and get[3]

$$\$V_t = \$z\,\frac{1 - [1/(1+i)^n]}{1 - 1/i}$$

Suppose that you have just won a million dollars in a lottery and have been presented with a six-foot \$1 000 000 cheque on TV. When the TV presentation is

---

[3]*By now, you have seen enough geometric series that they should hold no secrets. But if they still do, see Appendix 3.*

over, you are told that, to protect you from your worst spending instincts as well as from your now many friends, you will receive the million dollars in equal yearly installments of $50 000 over the next 20 years. What is the present value of your prize? Taking an interest rate of 6%, the foregoing equation gives $V =$ $50 000(0.688/.0566) = about $608 000. This is not bad, but winning the prize did not make you a millionaire.

Constant interest rates and payments, going on forever.   Let's go one step further and assume that payments are not only constant but go on forever. Real-world examples are harder to come by for this case, but one comes from nineteenth-century England, when the government issued *consols,* bonds paying a fixed yearly amount forever.[4] Let z be the constant payment. Assume that payments start next year rather than right away as in the preceding example (this makes for slightly simpler algebra). From equation (7.6), we have

$$\$V_t = \frac{1}{1+i} \$z + \frac{1}{(1+i)^2} \$z + \cdots$$

$$= \frac{1}{1+i} \left[1 + \frac{1}{(1+i)} + \cdots\right] \$z$$

where the second line follows by factoring out $1/(1 + i)$. The reason for doing so should be clear from looking at the term in brackets: It is an infinite geometric sum, so we can use the property of geometric sums to rewrite the present value as

$$\$V_t = \left(\frac{1}{1+i}\right) \frac{1}{1 - [1/(1+i)]} \$z$$

Or, simplifying,

$$\$V_t = \frac{\$z}{i}$$

The present value of a constant sequence of payments of $z is thus simply equal to the ratio of $z to the interest rate i. If, for example, the interest rate is expected to be equal to 5% forever, the present value of a consol that promises $10 per year forever is equal to $10/.05 = $200. If the interest rate increases and is now expected to be equal to 10% forever, the present value of the consol decreases to $10/0.10 = $100.

Zero interest rates.   Because of discounting, computing present discounted values typically requires the use of a calculator. There is, however, a special case worth keeping in mind where computations simplify. This is the case where the interest rate is equal to zero. Because the interest rate is usually positive, this is only an approximation, but it is a very useful one. The reason is obvious from equation (7.6). In that case, $1/(1 + i)$ is just equal to 1, and so is $1/(1 + i)^n$ for any power n. For that reason, the present discounted value of a sequence of expected payments at a zero interest rate is then just the *sum* of those expected payments.

---

[4]*Many of the consols were bought back by the British government at the end of the nineteenth century and early in the twentieth century. But some are still around.*

Nominal versus Real Interest Rates, and Present Values

We have so far computed the present value of a sequence of dollar payments by using interest rates in terms of dollars—nominal interest rates. Specifically, we have written the following relation, equation (7.5):

$$\$V_t = \$z_t + \frac{1}{1+i_t}\,\$z^e_{t+1} + \frac{1}{(1+i_t)(1+i^e_{t+1})}\,\$z^e_{t+2} + \cdots$$

where i_t, i^e_{t+1}, ... is the sequence of current and expected future nominal interest rates, and $\$z_t$, $\$z^e_{t+1}$, $\$z^e_{t+2}$, ... is the sequence of current and expected future dollar payments.

Suppose that we want to compute instead the present value of a sequence of *real* payments—that is, payments in terms of goods rather than in terms of dollars. Following the same logic as before, what we need to do is use the proper interest rates for this case, namely, interest rates in terms of goods—*real interest rates.* Specifically, we can write the present value of a sequence of real payments as

$$V_t = z_t + \frac{1}{1+r_t}\,z^e_{t+1} + \frac{1}{(1+r_t)(1+r^e_{t+1})}\,z^e_{t+2} + \cdots \qquad (7.7)$$

where r_t, r^e_{t+1}, ... is the sequence of current and expected future real interest rates, z_t, z^e_{t+1}, z^e_{t+2}, ... is the sequence of current and expected future real payments, and $V_t \equiv \$V_t/P_t$ is the real present value of future payments.

These two ways of writing the present value are equivalent.[5] That is, we can compute the present value as (1) the present value of the sequence of payments expressed in dollars, discounted using nominal interest rates, or (2) the present value of payments expressed in real terms, discounted using real interest rates.

So why present the two formulas? Because which one is more helpful depends on the context. Bonds typically are claims to a sequence of nominal payments over a period of years. For example, a 10-year bond may promise $50 a year for 10 years, plus a final payment of $1000 in the last year. Thus, when we look at the pricing of bonds in Chapter 9, we shall rely on equation (7.5) rather than on equation (7.7).

But sometimes we have more precise expectations of future real values than expected dollar values. You may, for example, have little idea of what your dollar income will be in 20 years; this value depends very much on what happens to inflation between now and then. But you may be fairly confident that your nominal income will increase at least as much as inflation—equivalently, that your real income will not decrease. In this case, using equation (7.5), which requires you to form expectations of future dollar income, may be difficult; using equation (7.7), which requires you to form expectations of future real income, will be easier. For that reason, when we discuss consumption and investment decisions in Chapter 8, we shall rely on equation (7.7) rather than on equation (7.5).

---

[5]*The proof is given in the appendix to this chapter. You may want to go through it just to test your understanding of the concepts introduced in this chapter: real versus nominal interest rates, and present values.*

NOMINAL AND REAL INTEREST RATES, AND THE *IS-LM* MODEL

We shall spend the next three chapters using the tools we have just developed to explore the role of expectations in determining activity. In this section we take a first step, by introducing the distinction between real and nominal interest rates into the basic *IS-LM* model and exploring some of the implications.

In the *IS-LM* model developed in Chapter 6, we saw that the interest rate entered in two places: It affected investment in the *IS* relation, and it affected the choice between money and bonds in the *LM* relation. Which interest rate were we talking about in each case?

Take the *IS* relation first. Our discussion earlier in this chapter should make it clear that, in deciding how much (if any) investment to undertake, firms care about the *real interest rate:* They want to know how much they will have to repay, not in terms of dollars but in terms of goods. So what belongs in the *IS* relation is the real interest rate. Let r denote the real interest rate (we shall drop time subscripts in this section). The *IS* relation must therefore be rewritten as

$$Y = C(Y - T) + I(Y, r) + G$$

Investment spending, and thus the demand for goods, depends on the real interest rate.

Now turn to the *LM* relation. In deriving the *LM* relation, we argued that the demand for money depends on the interest rate. Were we referring to the nominal interest rate or the real interest rate?

The answer is the *nominal interest rate.* Remember why the interest rate affects the demand for money. When thinking about whether to hold money or bonds, people take into account the opportunity cost of holding money rather than bonds. Money pays a zero nominal interest rate. Bonds pay a nominal interest rate of i. Thus, the opportunity cost of holding money is equal to the difference between the two interest rates, $i - 0 = i$, which is just the nominal interest rate.[6]

Therefore, the *LM* relation is still given by

$$\frac{M}{P} = YL(i)$$

Collecting the two equations and the relation between the real and the nominal interest rates, the *IS-LM* model is now given by

---

[6]*That, given the nominal interest rate, expected inflation does not affect portfolio choice and thus does not enter the* LM *relation may still feel odd. To clarify, let's look at it in another way, in terms of the real rates of return on money and on bonds. Because money pays a zero nominal interest rate—a zero nominal rate of return—the real rate of return on money is equal to $0 - \pi^e = -\pi^e$. Thus, higher expected inflation implies a more negative real rate of return on money. For example, an expected rate of inflation of 10% implies losing 10% of the real value of your money holdings over a year. Now think about the alternative, holding bonds. For a given nominal interest rate, i, the real interest rate on bonds, $r = i - \pi^e$, also goes down when expected inflation goes up. Thus, real rates of return on both money and bonds go down. But the difference between the two remains equal to the interest rate [the real rate of return on bonds, $i - \pi^e$, minus the real rate of return on money, $-\pi^e$, is equal to $i - \pi^e - (-\pi^e) = i$], and is not affected by expected inflation.*

IS:
$$Y = C(Y - T) + I(Y, r) + G$$

LM:
$$\frac{M}{P} = YL(i)$$

Real interest rate:
$$r \approx i - \pi^e$$

The *IS-LM* model is now composed of three equations and determines three variables: output *Y*, the nominal interest rate *i*, and the real interest rate *r*. We can easily reduce it, however, to a model with two variables, output and the real interest rate. Note that from the third equation, we can write the nominal rate as the sum of the real interest rate and expected inflation: $i = r + \pi^e$. Replacing the nominal interest rate *i* by $r + \pi^e$ in the *IS-LM* relation gives

IS:
$$Y = C(Y - T) + I(Y, r) + G$$

LM:
$$\frac{M}{P} = YL(r + \pi^e)$$

Investment depends on the real interest rate. The demand for money depends on the nominal interest rate, which is equal to the sum of the real interest rate and expected inflation.

Let's draw this modified *IS-LM* model in Figure 7-4. Let's measure the real interest rate on the vertical axis and output on the horizontal axis. To begin with, assume that expected inflation is equal to zero, so that the real and the nominal interest rates are the same. In this case the *IS-LM* model is the same as before. An increase in the interest rate (real or nominal, as the two are the same if expected inflation is equal to zero) decreases investment and equilibrium output in the goods market: The *IS* curve is downward-sloping. An increase in income increases the demand for money and leads to an increase in the interest rate consistent with equilibrium in financial markets: The *LM* curve is upward-sloping.

Consider now an increase in expected inflation from zero to some positive number, perhaps as the result of a public perception that inflation has been positive in the recent past. What happens to the nominal interest rate, to the real interest rate, and to output? To answer these questions, we need to see whether and how each of the two curves in Figure 7-4 shifts.

The *IS* curve does not shift, because expected inflation (π^e) does not enter into the *IS* relation. At a given real interest rate, investment remains the same, and so does output.

The *LM* curve shifts down by an amount equal to the increase in expected inflation. To see why, note that, at a given level of income and for a given money supply, equilibrium in financial markets determines the nominal interest rate. The real interest rate, which we are measuring on the vertical axis, is equal to the nominal interest rate minus expected inflation. Thus, it decreases one for one with an increase in expected inflation. A numerical example may help here. Suppose that, at a given level of income, the nominal interest rate consistent with equilibrium in financial markets is 8%, and that expected inflation increases from 0% to 5%. The real interest rate will decrease from 8% to 3%: The *LM* curve will shift down, so that at a given level of income the real interest rate corresponding to financial market equilibrium is now lower by 5%. The new *LM* curve is denoted by *LM'*.

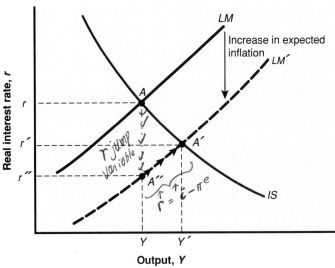

FIGURE 7-4

The Effects of an Increase in Expected Inflation on Output and the Real Interest Rate

An increase in expected inflation leads to a decrease in the real interest rate and an increase in output.

Handwritten notes (left margin):

if $i = r$, then ↑r ↓Investment and ↓Y (movement along IS)

- if there is inflation: no Δ en IS curve
- Lm: just a movement up & down it . with nominal i ✓
- but if r is on vertical axis, then Lm shifts to right; r decreases one for one with π^e ↑
$$r = i - \pi^e$$
↓

Note: over time movement from A″ to A′ ✓
$$\hat{r} = \hat{i} - \pi^e$$
i ↑ due to ↑y causing ↑ md i [graph: md_e(y'), md_i(y), md/ms]

Main text:

Equilibrium therefore moves from point A to point A′, with higher output, Y′, and a lower real interest rate, r′. The nominal interest rate goes up, however. To see why, note that the decrease in the real interest rate from r to r′ must be smaller than the vertical shift in the LM curve—which is itself equal to the increase in expected inflation. If the real interest rate decreases by less than the increase in expected inflation, then the nominal rate, which is the sum of the real interest rate and expected inflation, must go up.

Why does higher expected inflation lead to an increase in output? The best way to tell the story is to introduce dynamics. Assume, as we did in earlier chapters, that financial markets adjust instantaneously but that it takes time for a change in the real interest rate to affect output. Now suppose that expectations of inflation increase. Given output, the demand for money does not change. With an unchanged demand for money and an unchanged supply of money, the nominal interest rate does not change initially and the real interest rate decreases by the increase in expected inflation. The economy thus goes from A to A″. The real interest rate decreases from r to r″, by an amount equal to the increase in the expected inflation.

This is just the beginning of the story, however. The lower real interest rate leads over time to an increase in investment, to an increase in the demand for goods, and to an increase in output. As output increases, the economy moves along LM′ from A″ to A′. As output increases, so does the demand for money, and this in turn leads to an increase in the nominal interest rate. In the end, higher expected inflation leads to higher output, a lower real interest rate, and a higher nominal interest rate.

This exercise is useful as it forces you to think hard about the distinction between nominal and real interest rates. However, you should not conclude that because higher expected inflation leads to more output in Figure 7-4, inflation is good for output. What we have described is only one of many interactions between output and inflation; the complete picture will have to wait until later.

Summary

- The nominal interest rate tells us how many dollars one has to repay in the future in exchange for one dollar today.
- The real interest rate tells us how many goods one has to repay in the future in exchange for one good today.
- The real interest rate is approximately equal to the nominal interest rate minus expected inflation.
- The expected present discounted value of a sequence of payments is the value today of this expected sequence of payments. It depends positively on current and future expected payments. It depends negatively on current and future expected interest rates.
- In discounting a sequence of current and expected fu-

ture nominal payments, one should use current and expected future nominal interest rates. In discounting a sequence of current and expected future real payments, one should use current and expected future real interest rates.
- Investment decisions depend on the real interest rate. The choice between money and bonds depends instead on the nominal interest rate. Thus, the real interest rate enters the *IS* relation, while the nominal interest rate enters the *LM* relation.
- In the *IS-LM* model, an increase in expected inflation leads to an increase in output, an increase in the nominal interest rate, and a decrease in the real interest rate.

Key Terms

- nominal interest rate, 126
- real interest rate, 126
- ex post real interest rate, 128
- expected present discounted value, or present discounted value, or present value, 130

- discount factor, 131
- discount rate, 131

Questions and Problems

1. Prove the assertion in the text that $P_t/P^e_{t+1} = 1/(1 + \pi^e_t)$.
2. For each of the following, calculate:
   (i) The *exact* real interest rate [using equation (7.3)]
   (ii) The *approximate* real interest rate [using equation (7.4)]
   a. $i_t = 6\%$; $\pi^e_t = 1\%$
   b. $i_t = 10\%$; $\pi^e_t = 5\%$
   c. $i_t = 50\%$; $\pi^e_t = 45\%$
3. Can the real interest rate ever be negative? Under what circumstances? Explain in words what this implies about borrowing and lending.
4. Suppose that the *monthly* inflation rate in a country is expected to remain constant at 30%.
   a. What is the rate of inflation per *year?*
   b. If the yearly nominal interest rate is 2340%, calculate the yearly real interest rate using:
      (i) The exact formula
      (ii) The approximation formula
5. Suppose that you win the lottery and will receive $100 000 now, another $100 000 one year from now, and a final $100 000 two years from now. Calculate the present discounted value of these payments when the interest rate is expected to remain constant at

   a. 0%
   b. 5%
   c. 10%
6. For which of the following problems would you want to use *real* payments and *real* interest rates, and for which would you want to use *nominal* payments and *nominal* interest rates to calculate the expected present discounted value? In each case, why?
   a. Estimating the present discounted value of your lifetime income
   b. Trying to decide whether to continue renting your apartment or to buy a house
7. Using the *IS-LM* model, determine the impact on *C*, *I*, *Y*, *i* (nominal interest rate), and *r* (real interest rate) for each of the following:
   a. A decrease in expected inflation
   b. A decrease in expected inflation combined with a contractionary monetary policy
8. Find the most recent issue of *The Economist* magazine, and look at the tables in the back ("Economic Indicators" and "Financial Indicators"). Use the three-month money market interest rate (at an annual rate) as the nominal interest rate, and the three-month change in consumer prices (at an annual rate) as the expected rate of inflation.

a. What country has the highest nominal interest rate, and what country has the lowest?
b. What country has the highest real interest rate, and what country has the lowest?

9. Suppose you are about to enter into a business venture from which you expect to earn $10 000 in the current year, and the same amount next year and the year after. The one-year nominal interest rate is 5%, and you expect it to remain unchanged in the future.
   a. What is the expected present value of your future earnings?
   b. Calculate the new expected present value of future earnings under each of the following changes:

(i) A transitory (current year only) increase in earnings from $10 000 to $20 000
(ii) An increase in earnings from $10 000 to $20 000, expected to be permanent (all three periods)
(iii) A transitory (current year only) increase in the nominal interest rate from 5% to 10%
(iv) An increase in the current nominal interest rate from 5% to 10%, expected to be permanent

APPENDIX

DERIVING THE PRESENT DISCOUNTED VALUE USING REAL OR NOMINAL INTEREST RATES

The purpose of this appendix is to show that the two ways of expressing a present discounted value, equations (7.5) and (7.7) in the text, are equivalent.

Let's first rewrite the two equations for convenience.

Equation (7.5) gives the present value as the sum of current and future expected nominal payments, discounted using current and future expected nominal interest rates:

$$\$V_t = \$z_t + \frac{1}{1+i_t} \$z^e_{t+1} + \frac{1}{(1+i_t)(1+i^e_{t+1})} \$z^e_{t+2} + \cdots \quad (7.5)$$

Equation (7.7) gives the present value as the sum of current and future expected real payments, discounted using current and future expected real interest rates:

$$V_t = z_t + \frac{1}{1+r_t} z^e_{t+1} + \frac{1}{(1+r_t)(1+r^e_{t+1})} z^e_{t+2} + \cdots \quad (7.7)$$

Divide both sides of equation (7.5) by the current price level, P_t. The left-hand side becomes $\$V_t/P_t = V_t$, the real present discounted value, the same as the left-hand side of equation (7.7).

Now take each term on the right-hand side of equation (7.5) in turn.

The first term becomes $\$z_t/P_t = z_t$, the current payment in real terms. This term is the same as the first term on the right-hand side of equation (7.7).

The second term is given by $[1/(1+i_t)](\$z^e_{t+1}/P_t)$. Multiplying top and bottom by P^e_{t+1}, the price level expected for next year, gives

$$\frac{1}{1+i_t} \frac{P^e_{t+1}}{P_t} \frac{\$z^e_{t+1}}{P^e_{t+1}}$$

The third fraction is the expected real payment at time $t+1$. Consider the second fraction. Note that (P^e_{t+1}/P_t) can be rewritten as $1 + [(P^e_{t+1} - P_t)/P_t]$—thus, using the definition of expected inflation, as $(1 + \pi^e_t)$. This gives

$$\frac{1+\pi^e_t}{1+i_t} z^e_{t+1}$$

Finally, using the definition of the real interest rate in equation (7.3) earlier gives

$$\frac{1}{1+r_t} z^e_{t+1}$$

This is the same as the second term on the right-hand side of equation (7.7).

The same method applies to the other terms; make sure that you can derive the next one. It follows that equations (7.5) and (7.7) are equivalent ways of stating and deriving the expected present discounted value of a sequence of payments.

EXPECTATIONS, CONSUMPTION, AND INVESTMENT

We now return to two of the questions that motivated the hard work of the previous chapter. What determines consumption and investment decisions, and how do expectations affect these decisions? This chapter starts with the study of consumption, then turns to investment. This characterization of consumption and investment decisions will provide one of the building blocks for our expanded model of output determination in Chapter 10.

8-1 CONSUMPTION

How do people decide how much to consume and how much to save? In Chapter 3 we simply assumed that consumption and saving depended on current income. By now, you do not need to be convinced that they depend on much more, particularly on expectations of the future. We now explore how those expectations affect the consumption decision.

The theory of consumption on which this section is based was developed independently in the 1950s by Milton Friedman of the University of Chicago, who called it the **permanent income theory of consumption,** and by Franco Modigliani of MIT, who called it the **life cycle theory of consumption.**[1] Both Friedman and Modigliani

[1] *Milton Friedman received the Nobel Prize in 1976; Franco Modigliani received it in 1985.*

Panel data sets are data sets that give the value of one or more variables for many individuals or many firms over time. In Canada, panel data sets are only a few years old. The Survey of Labour and Income Dynamics (SLID) starts in 1993. Because of delays, the second wave of data (that is, for 1994) wasn't released until late 1997, so Canadian researchers have not yet had much opportunity to exploit the panel structure.

U.S. researchers are more blessed. The Panel Study of Income Dynamics, or PSID, was started in 1968, with approximately 4800 families. Interviews of these families have been conducted every year since, and are continuing. The survey has grown as new individuals have joined the original families, either by marriage or by birth. Every year, interviewers ask people in the survey questions about their income, wage rate, number of hours worked, health, and food consumption.*

By giving more than 25 years of information about individuals and about extended families, the survey has allowed economists to ask and answer questions for which there was previously only anecdotal evidence.

Among the many questions to which the PSID has been applied in the recent past are:

■ How much does consumption respond to transitory movements in income: for example, the loss of income from a period of unemployment?

■ How much risk-sharing is there within families? For example, when a family member becomes sick or unemployed, how much help does he or she get from other family members?

■ How much do people care about staying geographically close to their families? When somebody becomes unemployed, for example, how does the probability that he will migrate to another city depend on how many family members live in the city in which he currently lives?

The focus on food consumption comes from the fact that one of the survey's initial aims was to understand better the living conditions of poor families. The survey would be much more useful if it asked about all of consumption rather than food consumption. Unfortunately, it does not.

chose these labels carefully. By choosing "permanent income," Friedman emphasized that consumers look beyond current income. By choosing "life cycle," Modigliani emphasized that consumers' natural planning horizon is their entire lifetime.

The behaviour of aggregate consumption has remained a hot area of research ever since, for two reasons. The first is simply the sheer size of consumption in GDP, and thus the importance of understanding movements in consumption: Remember from Chapter 3 that consumption spending accounts for 61% of total spending in Canada. The second is the increasing availability of large surveys of individual consumers, such as the PSID described in the Focus box entitled "Up Close and Personal: Learning from Panel Data Sets." These surveys, which were not available when Friedman and Modigliani developed the theory of consumption, have allowed economists to steadily improve their understanding of how consumers actually behave.

THE VERY FORESIGHTED CONSUMER

Let's start our discussion with an assumption that will surely—and rightly—strike you as too extreme, but that will serve as a convenient benchmark. We'll call it the theory of the *very foresighted consumer*. How would a very foresighted consumer decide how much to consume? He would proceed in two steps.

First, he would estimate his total wealth by adding up the value of the stocks and bonds he owns, the value of his chequing and savings accounts, the value of the house he owns minus the mortgage still due, and so on. This would give him a notion of his financial and housing wealth.[2] But this is only part of his wealth. Indeed, for many consumers, the major part of wealth is neither financial nor housing wealth, but rather the present value of the after-tax labour income they expect to receive throughout their working lives. Thus, the foresighted consumer would sit down, estimate what his after-tax labour income is likely to be, and compute its present value (which he would have learned to compute from Chapter 7). Economists call the labour-income component of wealth **human wealth,** and call financial and housing wealth **nonhuman wealth.** Adding his human and nonhuman wealth, he would end with an estimate of his **total wealth.**

[handwritten margin note: lifetime wages = h.w. assets: house etc nonh.w.]

He would then decide how much of this total wealth to consume this year. A reasonable assumption is that he would decide to consume a proportion of total wealth such as to maintain roughly the same level of consumption each year throughout his life. If that level of consumption was higher than his current income, he would borrow the difference. If it was lower than his current income, he would save the difference.

Let's write this relation more formally. Let Y_{Lt} denote labour income in year t. Let T_t denote taxes (net of transfers). Using the notation we introduced in the previous chapter, let $V(Y_{Lt}^e - T_t^e)$ denote the expected present value of after-tax labour income. What we have described is a consumption decision of the form

[handwritten margin note: PV^e of net (after tax) wages.]

$$C_t = C(\text{total wealth}_t)$$
$$(\quad + \quad)$$

(8.1)

where

$$(\text{total wealth})_t = (\text{nonhuman wealth})_t + (\text{human wealth})_t$$
$$= (\text{financial + housing wealth})_t + V(Y_{Lt}^e - T_t^e)$$

[handwritten margin note: expect Y - expected taxes]

Consumption is an increasing function of total wealth; this relation is indicated by the positive sign under "total wealth" in the consumption equation. Total wealth in turn is the sum of financial wealth, housing wealth, and the present value of expected after-tax labour income.

We have just sketched a simple consumption rule: The very foresighted consumer computes his total wealth and then consumes some fraction of it each period. This description clearly contains some truth. We surely do think about our wealth and about our future expected labour income in deciding how much to consume today. But one cannot help but think that it assumes too much computation and foresight on the part of the typical consumer.

To get a better sense of what the description implies and what is wrong with it, let's apply this decision process to the problem facing a typical university student.[3]

---

[2]*"Housing wealth" is a bit of a misnomer, as we have in mind not only housing but also the other goods that he may own, from cars to paintings and so on.*

[3]*One of the nice aspects of consumption theory is that each of us is a consumer. Thus, each of us can use introspection as a way of checking on the plausibility of a particular theory. However, introspection is not without pitfalls: Economists are often warned that they do not think like other people.*

AN EXAMPLE

Let's assume that you are 21 years old, with three more years of university before you take your first job. Based on what we know today, your starting salary should be around $40 000 (in 2001 dollars) and will increase by an average of 3% a year in real terms until your retirement at age 60. About 33% of your income will go to taxes. Some of you may be in debt today, having borrowed to go to university; some of you may own a car and a few other worldly possessions. For the sake of simplicity, we shall assume that your debt and your possessions roughly offset each other, so that your nonhuman wealth is equal to zero.

Your *total wealth* is thus equal to your *human wealth,* the present value of your expected after-tax labour income. Building on what we saw in Chapter 7, let's compute this present value as the value of real expected after-tax labour income, discounted using real interest rates [equation (7.7)]. To make the computation simpler, let's assume that the real interest rate is equal to zero. This implies that your wealth is equal to the sum of real expected after-tax labour income over the 36 years of your working life (you will start earning income at age 25, and work until age 60), or

$$V(Y_L^e - T^e) = 0.67[1 + (1.03) + (1.03)^2 + \cdots + (1.03)^{35}](40\ 000)$$

The first term (0.67) comes from the fact that one dollar in income leaves you only 67 cents after tax. The second term $[1 + (1.03) + (1.03)^2 + \cdots + (1.03)^{35}]$ reflects the fact that you expect your real income to increase at 3% a year until retirement. The third term (40 000) is the initial level of labour income, in 2001 dollars. Using the properties of geometric series to solve for the sum in brackets gives

$$V(Y_L^e - T^e) = 0.67(63.3)(40\ 000) = 1\ 696\ 440$$

Thus, your total wealth today, the expected value of your lifetime after-tax labour income, is around $1.7 million.

How much should you consume? You can expect to live about 20 years after retirement, so that your expected remaining life today is 59 years. You may want to consume roughly the same amount every year. The constant level of consumption that you can afford to maintain is thus equal to your total wealth divided by your expected remaining life, or $1 696 440/59 = $28 753 a year. Given that your income until you get your first job is equal to zero, this implies borrowing $28 753 a year for the next three years, and starting to save when you get your first job.

TOWARD A MORE REALISTIC DESCRIPTION

Your first reaction to this computation may be that this is a rather stark and sinister way of summarizing your life prospects. Your second reaction may be that, although you agree with most of the ingredients that went into the computation, you surely do not intend to borrow $86 259 over the next three years. It is worth thinking about why. There are likely to be four reasons:

1. You may not want to plan for constant consumption over your lifetime and may be quite happy with deferring higher consumption until later. Student life usually does not leave much time for expensive activities. You may want to

defer memberships in golf clubs and trips to the Galápagos to later in life. You also have to think about the additional expenses that will come with having children, sending them to nursery school, summer camp, college, and so on.

2. You may find that the amount of computation and foresight involved in the computation we just went through far exceeds the amount you use or need right now. You may never have thought until now about exactly how much income you are going to make, and for how many years. You may feel that most consumption decisions are made in a simpler, less forward-looking fashion.

3. The computation of total wealth is based on forecasts of what can reasonably be expected to happen. But things can turn out better or worse. What happens if you are unlucky and you become unemployed or sick? How will you pay back what you borrowed? You may well want to be prudent, make sure that you can adequately survive even the worst outcomes, and thus borrow much less than $86 259.

4. Even if you decided to borrow $86 259, you are likely to find the bank from which you try to borrow that amount to be rather unreceptive. Why? The bank may worry that you are taking on a commitment you will not be able to afford if times turn bad, and that you may not be able or willing to repay the loan. Thus, the bank is unlikely to lend you the full amount.

These reasons are good ones. They imply that if our purpose is to characterize consumers' actual behaviour, we must modify the description we gave earlier. The last three reasons in particular suggest a consumption function in which consumption depends not only on total wealth but also on current income.

Take the second reason. You may, because it is a simple rule, decide to let your consumption follow your income and not think about what your wealth might be. In that case consumption will depend on current income, not on your wealth. Now take the third reason. It implies that a safe rule may be to consume no more than your income. This way, you do not run the risk of accumulating debt that you could not repay if times were to turn bad. Or take the fourth reason. It implies that you may have little choice anyway. Even if you wanted to consume more than your income, you may be unable to do so, since no bank will give you a loan.[4]

If we want to allow for a direct effect of current income on consumption, what measure of current income should we use? A convenient variable is after-tax labour income, which we introduced earlier in defining human wealth. This leads to a modified consumption function of the form

$$C_t = C(\text{total wealth}_t, Y_{Lt} - T_t) \qquad (8.2)$$
$$(\quad + \quad , \quad + \quad)$$

Current after-tax income.

where again

$$(\text{total wealth})_t = (\text{nonhuman wealth})_t + (\text{human wealth})_t$$
$$= (\text{financial + housing wealth})_t + V(Y^e_{Lt} - T^e_t)$$

---

[4] *Market imperfections, such as the inability to borrow against future labour earnings, are often at least partially offset by other institutions (the family or government). But unless the offset is complete, our point still stands.*

The question of how much consumption depends on current income versus expected future income is not easy to answer. This is because, most of the time, expectations of future income move very much with current income. If we get promoted and receive a raise, not only does our current income go up, but so typically does the income we can expect to receive in future years. Whether or not we are very foresighted, our consumption will typically move closely with our current income.

What can economists do to disentangle the effects of current versus future income? They must look for times and events where current income and expected future income move in different ways, and then look at what happens to consumption. Such events are called **natural experiments.** The word "experiment" comes from the fact that, like laboratory experiments, these events allow us to test a theory or to get a better estimate of an important parameter. The word "natural" comes from the fact that, unlike researchers in the physical sciences, economists typically cannot run experiments themselves. They must rely on experiments given by nature—or, as we shall see in our second example below, created by policy makers.

Here are two examples from recent research on consumption.

(1) RETIREMENT

Retirement implies a clear, predictable change in labour income: Labour income drops from some positive number to zero. By looking at how people save for retirement, we can in principle find out whether, when, and by how much people take into account the predictable decline in their future labour income.

A recent study by Steven Venti and David Wise, based on a U.S. panel data set called the "Survey of Income and Program Participation," sheds some light on retirement behaviour. Table 1, taken from their study, shows the mean level and the composition of wealth for people between 65 and 69 years old in 1991.

A mean wealth of $313 807 is a substantial number (for reference, U.S. per capita personal disposable income in 1991 was $16 205). The picture suggested by

....
TABLE 1

MEAN WEALTH OF PEOPLE, AGE 65–69, IN 1991 (IN CURRENT U.S. DOLLARS)

Social Security pension	$ 99 682
Employer-provided pension	62 305
Personal retirement assets	10 992
Other financial assets	42 018
Home equity	64 955
Other equity	33 855
Total	$313 807

Source: Venti and Wise, Table A1.

the table is thus one of forward-looking individuals making careful saving decisions and thus retiring with enough wealth to enjoy a comfortable retirement.

However, a closer look at the table, and at differences between individuals, suggests two caveats.

■ Note that the largest component of wealth is the present value of Social Security benefits, an amount over which workers have no control. Indeed, one of the main motivations behind the introduction of the Social Security program in the United States was to make sure that people contributed to their retirement, whether or not they would have done so on their own. Note that the third-largest component is an employer-provided pension, again a component over which workers have limited control. The only components that clearly reflect an individual saving decision (personal retirement assets + other financial assets) account only for $53 010, about 17% of total wealth. Thus, one can also read the evidence as suggesting that people save enough for retirement only because they are forced to do so, through Social Security and other contributions.

■ The numbers reported in the table are averages and hide substantial differences across individuals. The same study shows that most people retire with little more than their Social Security pensions. More generally, studies of retirement saving give the following picture: Most people appear to give little thought to retirement saving until some time during their 40s. At that point, many start saving for retirement. But many also save little and rely exclusively on their government-provided pensions when they retire.

(2) ANNOUNCED TAX CUTS

In 1981 the Reagan administration designed a fiscal package with phased-in tax cuts over 1981–1983. Income tax rates were to be reduced in three steps: 5% in 1981, 10% in 1982, and 8% in 1983, implying a cumulative reduction of 23%, a very large amount indeed. Congress passed the package in July 1981 and it became law in August 1981.

This period of U.S. history provides us with a natural experiment. The experiment is a change in future expected after-tax labour income coming from an anticipated decrease in taxes. The question we want to answer is a simple one: Did consumers react in 1981 to the expected decrease in taxes in 1982 and 1983, and, if so, by how much?

This is exactly the question James Poterba, from MIT, asked in a recent article. Using econometrics, Poterba looked for evidence of an unusual increase in consumption, given disposable income, in the summer of 1981 (the time when Congress passed the package). He found no evidence of such an increase.

Is this conclusive evidence that consumers do not take account of future expected income in their consumption decision? Not necessarily. One can think of a number of alternative interpretations of the facts. People may have believed that Congress would change its mind, leading them to take a wait-and-see attitude and wait for the actual decreases in taxes to adjust their consumption. Or maybe people do not take into account expected changes in taxes, but do take into account other expected changes in their income (say, an expected promotion or the coming of retirement). These arguments cannot be dismissed. But what can safely be said is that the evidence from that particular natural experiment does not provide evidence for a strong effect of expected future tax changes on consumption.

References

On retirement: Steven Venti and David Wise, "The Wealth of Cohorts: Retirement and Saving and the Changing Assets of Older Americans," mimeo, Kennedy School, Harvard University, October 1993.

On the Reagan tax cuts: James Poterba, "Are Consumers Forward Looking? Evidence from Fiscal Experiments," *American Economic Review,* May 1988, 413–418.

Or in words: *Consumption is an increasing function of total wealth and of current after-tax labour income. Total wealth is the sum of financial wealth, housing wealth, and the present value of expected after-tax labour income.*

The main issue then becomes how much consumption actually depends on total wealth (and thus on expectations of future income), and how much it depends on current income. Some consumers, especially those who have temporarily low income and poor access to credit, are likely to consume just their current income, regardless of what they expect will happen to them in the future. A worker who becomes unemployed may have a hard time borrowing to maintain her level of consumption, even if she is fairly confident that she will soon find another job. Consumers who are richer and have easier access to credit are likely to give more weight to the expected future and to try to maintain roughly constant consumption through time.

The relative importance of wealth and income can be settled only by looking at the empirical evidence. This is not so easy to do, and the In Depth box entitled "How Much Do Expectations Matter? Looking for Natural Experiments" explains why. But even if some of the details still need to be filled in, the basic evidence is clear and unsurprising: Both total wealth and current income affect consumption.

PUTTING THINGS TOGETHER: CURRENT INCOME, EXPECTATIONS, AND CONSUMPTION

Let's go back to what motivates this chapter, the importance of expectations in the determination of spending. For this purpose, the description of consumption behaviour we have just gone through has two main implications.

First, *consumption is likely to respond less than one for one to movements in current income.* In thinking about how much they should consume, consumers look at more than current income. If they conclude that a decrease in income is permanent, consumers may decrease consumption one for one with the decrease in income. But if they conclude that the decrease in current income is transitory, they will adjust their consumption by less.

Thus, in a recession, consumption adjusts less than one for one to decreases in income. This is because consumers know that recessions typically do not last for more than a few quarters. The converse is true in expansions. Faced with an unusually rapid increase in income, consumers are unlikely to increase consumption by as much as income. They are likely to assume that the boom is transitory and that things will soon return to normal.

Second, *consumption may move even if current income does not change.* In late 1996 and early 1997, consumption climbed although disposable income was virtually stagnant. In part this was due to the sharp increases in the stock market. Capital gains are not included in measures of current income, but they clearly affect the level of nonhuman wealth. But another factor that helped boost spending was the sharp increase in consumer optimism. In early 1997, people became more optimistic about the future, and their own future in particular. According to surveys, many felt that it was a good time to buy big-ticket items like cars and houses, even though current incomes were not changing much. Some of this upbeat mood reflected real improvements in the economy, but many of these improvements were widely anticipated: It is hard to see what could have changed so much in a few months. To this day economists do not understand the wide swings in optimism and pessimism that often play a large role in the business cycle.

8-2 INVESTMENT

Let's now turn to investment. The natural starting point here is the assumption that a firm's objective is to make profit. Thus, when deciding whether to buy a new machine, the firm must think about how much profit the machine will generate over its productive life. If the value today of the sequence of expected profits is larger than the cost of buying and installing the machine, the firm will move ahead and invest. Otherwise, it will not. This, in a nutshell, is how economists think about firms' investment decisions.

INVESTMENT AND EXPECTATIONS OF PROFIT

Let's look at the steps a firm must take to determine whether to buy a new machine. (While we refer to a machine for concreteness, the same reasoning applies to the other components of investment—for example, the building of a new factory, the renovation of an office complex, and so on.)

(1) *The firm must estimate how long the machine will last.* Most machines are like cars. They can last nearly forever; but as time passes they become less and less reliable, more and more expensive to maintain. A simple way of capturing this depreciation is to assume that a machine loses its usefulness at rate δ (the Greek lowercase letter delta) per year. That is, a machine that is new this year is worth only $(1 - \delta)$ machines next year, $(1 - \delta)^2$ machines in two years, and so on. The parameter δ, called the **depreciation rate,** measures how much usefulness the machine loses from one year to the next.[5] What are reasonable values for δ? This is a question that the statisticians in charge of computing how the capital stock changes over time have had to answer. Based on their studies of depreciation of specific machines and buildings, they use numbers between 4% and 15% for machines, and between 2% and 4% for buildings and factories.

(2) *The firm must compute the present discounted value of profits.* As a way of capturing the fact that it takes some time to put machines in place (and even more time to build a factory or an office building), assume that a machine bought in year t becomes operational—and starts depreciating—only one year later, in year $t + 1$.

Denote profit per machine in real terms by Π. (This is an uppercase pi, as opposed to the lowercase pi, which we use to denote inflation.) If the firm purchases a machine in year t, the machine generates its first expected profit in year $t + 1$; denote this expected profit by Π_{t+1}^e. The present value, in year t, of this expected profit in year $t + 1$, is given by

$$\frac{1}{1 + r_t} \Pi_{t+1}^e$$

The construction of this term is represented by the arrow pointing left in the first line of Figure 8-1. Note that because we are measuring profit in real terms, we are using real interest rates to discount future profits. This is one of the lessons we learned in Chapter 7.

Denote expected profit per machine in year $t + 2$ by Π_{t+2}^e. Because of depreciation, only $(1 - \delta)$ of the machine bought in year t is left in year $t + 2$, so that the expected profit from the machine is equal to $(1 - \delta)\Pi_{t+2}^e$. The present value of this expected profit as of year t is equal to

$$\frac{1}{(1 + r_t)(1 + r_{t+1}^e)} (1 - \delta)\Pi_{t+2}^e$$

This computation is represented by the arrow pointing left in the second line of Figure 8-1.

The same reasoning applies to expected profit in following years. Putting the pieces together gives us *the present value of expected profits* from buying the machine in year t, call it $V(\Pi_t^e)$:

$$V(\Pi_t^e) = \frac{1}{1 + r_t} \Pi_{t+1}^e + \frac{1}{(1 + r_t)(1 + r_{t+1}^e)} (1 - \delta)\Pi_{t+2}^e + \cdots \tag{8.3}$$

---

[5] *If we think of a large number of machines rather than one machine, we can interpret δ differently. We can think of δ as the proportion of machines that die every year. Thus, if the firm starts the year with K working machines and does not buy new ones, it has only K(1 −) machines left one year later, and so on.*

FIGURE 8-1

Computing the Present
Value of Expected
Profits

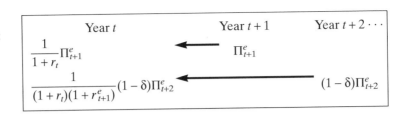

The expected present value is equal to the discounted value of expected profit next year, plus the discounted value of expected profit two years from now (taking into account the depreciation of the machine), and so on.

(3) *The firm must then decide whether to buy the machine.* This decision clearly depends on the relation between the present value of expected profits and the price of a machine. To simplify notation, let's assume that the real price of a machine—that is, the machine's price in terms of the basket of goods produced in the economy—is equal to 1. What the firm must then do is compare the present value of profits with 1.

If the present value is less than 1, the firm should not buy the machine. If it did so, it would be paying more for the machine than it expects to get back in profits later. If the present value exceeds 1, then the firm has an incentive to buy the new machine.[6]

We have so far looked at the decision of an individual firm. Let's now jump from our one-firm, one-machine example to investment in the economy as a whole. Let I_t denote aggregate investment. Denote profit per machine, or more generally, profit per unit of capital—where capital includes not only machines, but also factories, office buildings, and so on—for the economy as a whole by Π_t. Denote the expected present value of profit per unit of capital by $V(\Pi_t^e)$, defined as in equation (8.3). Our discussion suggests an investment function of the form

$$I_t = I(V(\Pi_t^e)) \qquad (8.4)$$
$$(\ +\ )$$

Investment depends positively on the expected present value of future profits (per unit of capital). The higher current or expected profits, the higher the level of investment.

A CONVENIENT SPECIAL CASE

There is a special case where the relation we have just derived between investment, profit, and (real) interest rates becomes very simple. Suppose that firms expect both future profits (per unit of capital) and future interest rates to remain at the same level as today, so that $\Pi_{t+1}^e = \Pi_{t+2}^e = \cdots = \Pi_t$, and $r_{t+1}^e = r_{t+2}^e = \cdots = r_t$. Under these assumptions, equation (8.3) becomes (the derivation is given in the appendix to this chapter):

---

[6]*This way of relating investment to the present value of expected profits was first presented by James Tobin of Yale University. He denoted the ratio of the present value of profits to the price of a machine by Q. This is why the theory of investment presented here is known as the "Q theory" (not quite as good a name as those chosen by Friedman and Modigliani for their theories of consumption). Tobin received the Nobel Prize in 1981, for this and many other contributions.*

$$V(\Pi_t^e) = \frac{\Pi_t}{r_t + \delta} \quad\text{(8.5)}$$

Profit (handwritten, pointing to numerator)

→ these are the rental cost of K (handwritten, pointing to denominator) = 'r' (handwritten)

The present value of expected profits is simply equal to the ratio of profit to the sum of the real interest rate and the depreciation rate.

Replacing (8.5) in equation (8.4), investment is given in turn by

$$I_t = I\left(\frac{\Pi_t}{r_t + \delta}\right) \quad\text{(8.6)}$$

Let's look more closely at the expression in equation (8.5). The denominator—the sum of the real interest rate and the depreciation rate—is called the **user cost or rental cost of capital.** To see why, suppose that instead of buying the machine, the firm rented it by the year from a rental agency.[7] How much would the rental agency charge? Even if the machine did not depreciate, the agency would have to charge an interest charge equal to r_t times the price of the machine (which we have assumed to be 1 in real terms): The agency has to get at least as much from renting the machine as it would from holding bonds. In addition, the rental agency would have to charge for depreciation: δ times the price of the machine (which, again, is assumed to be 1). Thus, the rental cost would be equal to $r_t + \delta$.[8] Even though firms typically do not rent their machines, $r_t + \delta$ still captures the implicit cost—sometimes called the *shadow cost*—to the firm of using the machine for one year.

The investment function given by equation (8.6) thus has a simple interpretation: *Investment depends on the ratio of profit to the user cost.* The higher the profit compared to the user cost, the higher the level of investment. The higher the user cost, the lower the level of investment.

(handwritten marginal note: key)

This relation between profit, the real interest rate, and investment relies on a strong assumption: that the future is expected to be the same as the present. It is nevertheless a useful one to remember, and one that macroeconomists keep handy in their toolbox.

CURRENT VERSUS EXPECTED PROFIT

The theory we have developed so far implies that investment should be forward-looking and depend primarily on *expected future profits.* One striking empirical fact about investment, however, is how strongly it moves with movements in *current profit.*

This relation is shown in Figure 8-2, which plots yearly changes in investment and proft since 1960 for the Canadian economy. Investment is measured as fixed nonresidential investment in 1986 dollars. Profit is constructed as the sum of after-tax profits plus interest payments and then converted to 1986 dollars using the implicit price deflator for fixed nonresidential investment.[9]

The positive relation between changes in investment and changes in current profit is clear to the eye in Figure 8-2. Is this relation inconsistent with the theory we have just developed, which holds that investment should be related to the present value of expected future profits rather than to current profit? It need not be. If firms

(handwritten marginal note: current vs expected profit)

---

[7]*Such arrangements actually exist. Many firms lease the cars they need from car leasing companies.*

[8]**Digging deeper.** *If we had allowed the price of a unit of capital in terms of goods to be, say, P_{Kt} rather than 1, the user cost would be given by $P_{Kt}(r_t + \delta)$ instead.*

[9]*Definitions of these terms are given in Appendix 2.*

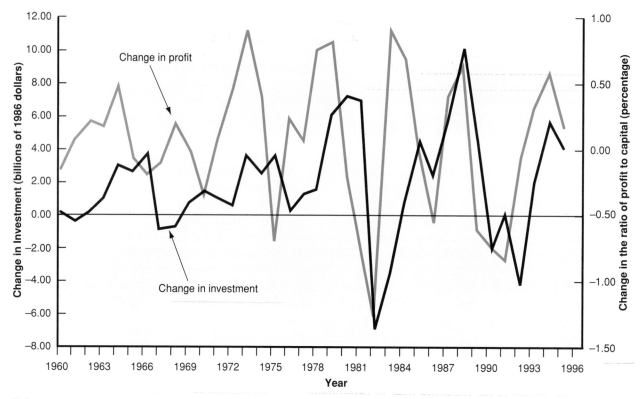

FIGURE 8-2

Changes in Investment and Changes in Profit in Canada, 1960–1996

Investment and profit move very much together.

Source: Statistics Canada, CANSIM Series D11000, D11004, D12101, D14458, D14459, D14486, D14488, D14489, and D884008.

expect future profits to move very much like current profit, then the present value of profits will move very much like current profit, and so will investment.

Economists who have looked at the question more closely have concluded, however, that the effect of current profit on investment is stronger than the theory we have developed would predict. How they gathered some of the evidence is described in the Focus box entitled "Profitability versus Cash Flow." On the one hand, some firms with highly profitable investment projects but low current profits appear to be investing too little. On the other hand, some firms that have high current profit appear sometimes to invest in projects of doubtful profitability. In short, current profit appears to affect investment, even after controlling for the expected present value of profits.

Why does current profit play a role in the investment decision? Our earlier discussion of why consumption may depend directly on current income is relevant here. Many of the reasons we used to explain the behaviour of consumers also apply to firms:

■ First, if current profit is low, a firm that wants to buy new machines can only get the funds it needs by borrowing. It may be reluctant to do so; while expected profits may look good, things may turn bad, leaving the firm unable to repay the debt. But if current profit is high, the firm may be able to finance its investment just by retaining some of its earnings and without having to borrow. Thus, higher current profit may lead the firm to invest more.

152   EXPECTATIONS

■ Second, even if the firm wants to invest, it may find it difficult to borrow. Potential lenders may not be convinced that the project is as good as the firm says, and they may be worried that the firm will be unable to repay. If the firm has large current profits, it does not have to borrow and thus does not need to convince potential lenders. It can proceed and invest as it pleases, and thus it is more likely to do so.

In summary, to fit the investment behaviour we observe, the investment equation is better written as

$$I_t = I(V(\Pi_t^e), \Pi_t)$$
$$(\ +\ ,\ +)$$

(8.7)

Key relationships

Investment is a function of the expected present value of profits as well as the current level of profit.

PROFITABILITY VERSUS CASH FLOW

How much does investment depend on the expected present value of profits, and how much does it depend on current profit? Economists often refer to the question as that of the respective roles of **profitability** (the expected present discounted value of profits) versus **cash flow** (current profit, the net flow of cash the firm is receiving) in investment decisions.

The problem in answering this question is very similar to the problem of identifying the relative importance of current and expected future income on consumption, a problem we discussed in the preceding box. Most of the time, cash flow and profitability are likely to move together. Firms that do well typically have both large cash flows and good future prospects. Firms that have losses often have poor future prospects as well.

As in the case of consumption, the best way to isolate the effects is to identify times or events when cash flow and profitability move in different directions, and then look at what happens to investment. This is the approach taken in a recent paper by Owen Lamont, an economist at the University of Chicago. An example will help you understand Lamont's strategy.

Think of two firms. The first one, A, is involved only in steel production. The second one, B, is composed of two parts. The first is steel production, and the other is petroleum exploration.

Suppose now that there is a sharp drop in the price of oil, leading to losses in oil exploration. This shock decreases firm B's cash flow. Indeed, if the losses in oil exploration are large enough to offset the profits from steel production, firm B may show an overall loss.

The question we can now ask is: As a result of the decrease in the price of oil, will firm B invest less in its steel operation than firm A does? If only *profitability* in steel production matters, there is no reason for firm B to invest less in its steel operation than firm A. But if current *cash flow* also matters, the fact that firm B has a lower cash flow may prevent it from investing as much as firm A in its steel operation. Thus, looking at investment in the steel operations of the two firms can tell us how much investment depends on cash flow versus profitability.

This is the empirical strategy that Lamont follows. He focusses on what happened in the mid-1980s, when the price of oil in the United States dropped by 50%, leading to large losses in oil-related activities. He then looks at whether firms that had substantial oil activities cut investment in their non-oil activities relatively more than other firms in the same non-oil activities. He concludes that they did. He finds that for every $1 decrease in cash flow due to the decrease in the price of oil, investment spending in non-oil activities was reduced by 10 to 20 cents. Current cash flow indeed matters.

Huntley Schaller, from Carleton University, uses another strategy to estimate the impact of cash flows on Canadian investment. He compares firms that belong to a tightly-knit web of corporate directorships, a sort of Canadian analogue to the Japanese keiretsu, with those that are largely independent and on their own. He estimates that one dollar of after-tax cash flow can have very different impacts on the investment decisions of the two kinds of firm. For firms that are members of an industrial group the increase may be as little as 5 cents, while for firms that are largely on their own the impact may be as large as 60 cents.

References

Owen Lamont, "Financial Constraints and Investment: Evidence from Internal Capital Markets," mimeo, MIT, 1994.

Huntley Schaller, "Asymmetric Information, Liquidity Constraints, and Canadian Investment," *Canadian Journal of Economics*, August 1993.

A general review of studies along these lines is given by R. Glenn Hubbard in "Capital-Market Imperfections and Investment," *Journal of Economic Literature*, 1995.

PROFIT AND SALES

We have argued that investment depends on both current and expected profit. We need to take one last step, and ask: What in turn determines profit? The answer is, primarily two factors: (1) the level of sales and (2) the existing capital stock. If current sales are low or if the capital stock is already high, profit per unit of capital is likely to be low.

Let's write this more formally, ignoring the distinction between sales and output, and let Y_t denote output. Let K_t denote the capital stock at time t. Our discussion suggests the following relation:

$$\Pi_t = \Pi\left(\frac{Y_t}{K_t}\right) \qquad (8.8)$$
$$(\ +\ )$$

Profit per unit of capital is an increasing function of the ratio of sales to the capital stock. Given the capital stock, the higher the sales, the higher the profit. Given sales, the higher the capital stock, the lower the profit.

How does this relation hold in practice? Figure 8-3 plots yearly changes in profit per unit of capital and changes in the ratio of output to capital, for Canada, since 1960. Profit per unit of capital is again defined as the sum of after-tax profits plus interest payments divided by the capital stock. The ratio of output to capital is constructed as the ratio of GDP to the aggregate capital stock, both measured in current dollars.

The figure shows indeed a tight relation between changes in profit and changes in the ratio of output to capital. Given that most of the year-to-year changes in the ratio of output to capital come from movements in output (capital, which is a stock, moves slowly over time), we can state the relation as follows: Profit decreases in recessions, and increases in expansions.

Why is this relation between output and profit relevant here? Because it implies a link between *current and expected output* on the one hand, and *investment* on the other. For example, the anticipation of a long, sustained economic expansion will lead firms to anticipate sustained profits, now and for some time in the future. These expectations in turn will lead to higher investment. The effect of current and expected output on investment, together with the effect of investment back on demand and output, will play a crucial role when we return to the determination of output in Chapter 10.

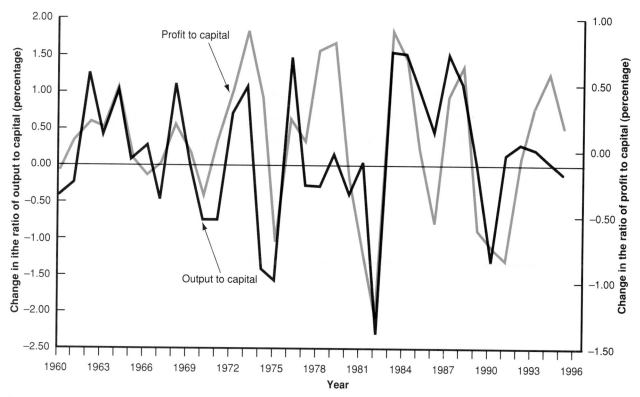

FIGURE 8-3

Changes in the Ratio of Output to Capital and Changes in Profit, 1960–1996
Profit and output move largely together.

Source: Statistics Canada, CANSIM Series D11000, D11004, D12101, D14458, D14459, D14486, D14488, D14489, and D884008.

8-3 THE VOLATILITY OF CONSUMPTION AND INVESTMENT

You will surely have noticed the many similarities between our treatment of consumption and of investment in Sections 8-1 and 8-2. Whether consumers perceive current movements in income to be transitory or permanent affects their consumption decisions. In the same way, whether firms perceive current movements in sales to be transitory or permanent affects their investment decisions. The less they expect a current increase in sales to last, the less they revise their assessment of the present value of profits, and thus the less likely they are to buy new machines or build new factories or new offices. This is why, for example, the boom in sales that happens every year between late November and Christmas (in Canada, retail sales are typically 20% higher in December than in other months[10]) does not lead to a boom in investment every year in December. Firms fully understand that this boom is transitory.

---

[10]*In France and Italy, sales are 60% higher in December. These numbers and other facts about such seasonal cycles come from J. Joseph Beaulieu and Jeffrey Miron, "A Cross Country Comparison of Seasonal Cycles and Business Cycles,"* Economic Journal, *July 1992, 772–778.*

But there are also important differences between consumption and investment decisions that you must keep in mind. The main implication of these differences is that investment is more volatile than consumption. To understand why, consider the responses of consumption and investment to permanent changes in income and sales.

The theory of consumption we developed implies that, when faced with an increase in income they perceive as permanent, consumers respond with *at most* an equal increase in consumption. The permanent nature of the increase in income implies that they can afford to increase consumption now and in the future by the same amount as the increase in income. Increasing consumption more than one for one would require cuts in consumption later, and there is no reason for consumers to want to plan consumption this way.

Now consider the behaviour of firms faced with an increase in sales they believe to be permanent. The present value of expected profits increases, leading to an increase in investment. In contrast to consumption, there is no implication that the increase in investment should be no greater than the increase in sales. Indeed, once a firm has decided that an increase in sales justifies the purchase of a new machine or the building of a new factory, it may want to proceed quickly, leading to a large but short-lived increase in investment spending. This increase may exceed the increase in sales.

More concretely, take a firm that has a ratio of capital to its annual sales of, say, 3. An increase in sales of $10 million this year, if expected to be permanent, requires the firm to spend $30 million on additional capital if it wants to maintain the same ratio of capital to output. If the firm buys the additional capital right away, the increase in investment spending this year will be equal to *3 times* the increase in sales. Once the capital stock has adjusted, the firm will return to its normal pattern of investment. This example is extreme, because firms are unlikely to adjust their capital stock right away. But even if they do adjust their capital stock more slowly, say over a few years, the increase in investment may still exceed the increase in sales for a while.

We can tell the same story in terms of equation (8.8). As we make no distinction here between output and sales, the initial increase in sales leads to an equal increase in output, Y, so that Y/K—the ratio of the firm's output to its existing capital stock—also increases. The result is higher profit, which leads the firm to undertake more investment. Over time, the higher level of investment leads to a higher capital stock, K, so that Y/K decreases back to normal. Profit per unit of capital returns to normal, and so does investment. Thus, in response to a permanent increase in sales, investment may increase a lot initially and then return to normal over time.

How much more volatile is investment than consumption? The answer is given in Figure 8-4, which plots yearly rates of change in Canadian consumption and investment since 1960. To make the figure easier to interpret, both rates of change are plotted as deviations from the average rate of change, so that they are on average equal to zero.

The figure has two characteristics. The first is that consumption and investment usually move together; recessions, for example, are associated with decreases in *both* investment and consumption. Given our discussion, which has emphasized that consumption and investment depend largely on the same determinants, this should not come as a surprise.

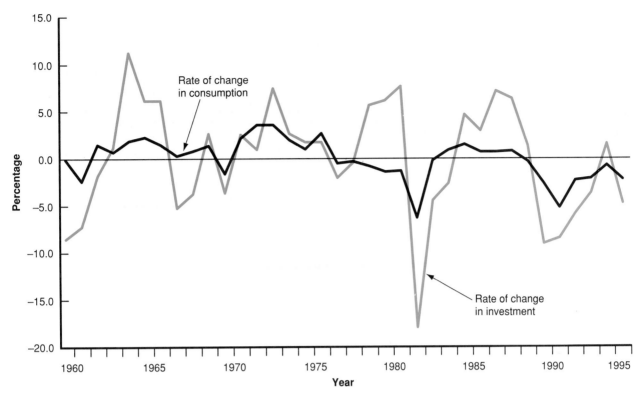

FIGURE 8-4

Rates of Change of Consumption and Investment in Canada, 1960–1996
Relative movements in investment are much larger than relative movements
in consumption.

Source: Statistics Canada, CANSIM Series D14443 and D14456.

The second makes the point we just discussed: Investment is much more
volatile than consumption. Relative movements in investment range from minus to
plus 15%, while relative movements in consumption range only from minus to plus
4%. Another way of stating the same fact is that whereas investment is much
smaller on average than consumption (remember from Chapter 3 that investment
accounts for 15% of spending, versus 58% for consumption), the range of *absolute*
movements in investment (that is, the range of changes in the level of investment
from one year to the next) is roughly the same as the range of *absolute* movements
in consumption. Both components contribute roughly equally to fluctuations in output over time.

Summary

....

■ Consumption depends on both current income and
wealth. Wealth is the sum of nonhuman wealth (financial
and housing wealth) and human wealth (the present
value of after-tax labour income).

■ The response of consumption to changes in income depends on whether consumers perceive these changes as
transitory or as permanent.

■ Consumption is likely to respond less than one for one
to movements in current income, and consumption may
move even if current income does not change.

■ Investment depends on both current profit and the
present value of expected future profits.

■ Under the simplifying assumption that firms expect future profits and interest rates to be the same in the future as they are today, we can think of investment as depending on the ratio of profit to the user cost of capital, where the user cost is the sum of the real interest rate and the depreciation rate.

■ Movements in profit are closely related to movements in output. Thus, we can think of investment as depending indirectly on current and future expected output movements. Firms that anticipate a long output expansion, and thus a long sequence of high profits, will invest. Movements in output that are not expected to last will have less of an effect on investment.

■ Investment is much more volatile than consumption. While investment accounts only for 15% of GDP and consumption accounts for 58%, movements in investment and consumption are of roughly equal magnitude.

KEY TERMS

- permanent income theory of consumption, 141
- life cycle theory of consumption, 141
- panel data sets, 142
- human wealth, 143
- nonhuman wealth, 143
- total wealth, 143

- natural experiment, 146
- depreciation rate, 149
- user cost or rental cost of capital, 151
- profitability, 153
- cash flow, 153

QUESTIONS AND PROBLEMS

1. Think about your own spending behaviour. Does your current consumption level depend in part on your *current* income rather than simply on your expected lifetime income? The text lists three reasons why current consumption is often dependent on current income. Which reasons apply to you? Be specific.

2. Suppose that it suddenly became possible for anyone to borrow money from a bank using his or her expected future earnings to guarantee the loan. What impact do you think this would have on the level of current consumption? What impact would it have on the marginal propensity to consume out of current income? Why?

3. "Most young, university-educated Canadians are millionaires." Justify this statement, using the economist's concept of *total wealth*.

4. A pretzel manufacturer is considering buying another pretzel-making machine that costs $50 000. The machine will depreciate by 10% per year. It will generate real profits equal to $10 000 this year, $10 000 (1 – 0.1) next year (that is, the same real profits, but adjusted for depreciation), $10 000 (1 – 0.1)2 two years from now, and so on. Determine whether the manufacturer should buy the machine if the real interest rate is assumed to remain constant at
   a. 5%
   b. 10%
   c. 15%

5. A consumer with nonhuman wealth of $100 000 will earn $50 000 this year, and expects her salary to rise by 5% in real terms each year for the following two years. She will then retire. The real interest rate is equal to zero and is expected to remain equal to zero in the future. Labour income is taxed at the rate of 40%.
   a. What is this consumer's human wealth?
   b. What is her total wealth?
   c. If this consumer expects to live for another ten years and wants her consumption to remain the same every year, how much should she consume this year?
   d. If this consumer were given a bonus of $20 000 in the current year only, with all future salary payments remaining as stated earlier, by how much would her current consumption rise?

6. An investor can sell a bottle of wine today for $7000, or put it in storage and sell it in 30 years for a real (constant dollar) price of $20 000. If the real interest rate is expected to remain constant at 4%, what should the investor do?

7. A worker signs a contract that freezes her salary at $40 000 for the next three years. The real interest rate is expected to remain constant at 3%, and the inflation rate to remain constant at 5%. What is the present discounted value of her three-year salary?

APPENDIX

DERIVATION OF THE EXPECTED PRESENT VALUE OF PROFITS WHEN FUTURE PROFITS AND INTEREST RATES ARE EXPECTED TO BE THE SAME AS TODAY

In general, the expected present value of profits is given by equation (8.3) in the text:

$$V(\Pi_t^e) = \frac{1}{1+r_t} \Pi_{t+1}^e + \frac{1}{(1+r_t)(1+r_{t+1}^e)}(1-\delta)\Pi_{t+2}^e + \cdots \quad (8.3)$$

If firms expect both future profits (per unit of capital) and future interest rates to remain at the same level as today, so that $\Pi_{t+1}^e = \Pi_{t+2}^e = \cdots = \Pi_t$, and $r_{t+1}^e = r_{t+2}^e = \cdots = r_t$, the equation becomes

$$V(\Pi_t^e) = \frac{1}{1+r_t} \Pi_t + \frac{1}{(1+r_t)^2}(1-\delta)\Pi_t + \cdots$$
$$= \frac{1}{1+r_t} \Pi_t \left(1 + \frac{1-\delta}{1+r_t} + \cdots\right)$$

where the second line follows by factoring out $[1/(1+r_t)]\Pi_t$.

The final term in parentheses in this equation is a geometric series, a series of the form $1 + x + x^2 + \cdots$, where x here is equal to $(1-\delta)/(1+r_t)$. Thus, its sum is given by $1/(1-x) = (1+r_t)/(r_t+\delta)$. Replacing in the previous equation gives

$$V(\Pi_t^e) = \left(\frac{1}{1+r_t}\right)\left(\frac{1+r_t}{r_t+\delta}\right)\Pi_t$$

Simplifying gives the equation we use in the text:

$$V(\Pi_t^e) = \frac{\Pi_t}{r_t+\delta} \quad (8.5)$$

FINANCIAL MARKETS AND EXPECTATIONS

In our first look at financial markets in Chapter 5, we assumed that there were only two assets, money and just one type of bond. The purpose was to focus on the choice between money and all other assets. But it is now time to relax this assumption and look at the choice among nonmoney assets—between short-term and long-term bonds, between bonds and stocks, and so on.

    The purpose of this chapter is to look at these choices and draw the implications for the behaviour of bond and stock prices. These prices play a central role in the interaction between goods markets and financial markets. For example, movements in stock prices both reflect and affect movements in economic activity. This interaction, and the role of expectations, serve as a major theme of this and the next chapter.

    We start the chapter by looking at the determination of bond prices and the yield curve. We then look at the determination of stock prices and the interpretation of movements in the stock market.

9-1 BOND PRICES AND THE YIELD CURVE

Bonds differ in two basic dimensions. The first is their **default risk,** the risk that the issuer of the bond will not pay back the full amount promised by the bond. The second dimension is their maturity. A bond's **maturity** is the length of time over which it promises to make payments to the holder. A bond that promises to make one payment of $1000 in six months has a maturity of six months; a bond that promises $100 per year for the next 20 years and a final payment of $1000 at the end of those 20 years has a maturity of 20 years. This second dimension is more important for our purposes and is the dimension on which we shall focus here.

Bonds of different maturities each have a price and an associated interest rate called the *yield to maturity,* or simply the *yield.* By looking on any given day at the yields on bonds of different maturities, we can trace the relation between yields and maturity. This relation is called the **yield curve,** or the **term structure of interest rates** (the word *term* is synonymous with maturity).

Figure 9-1 gives the average term structure on Canadian government bonds for the months of June 1990 and January 1997. (If you want to find out what the term structure looks like currently, visit the Royal Bank's web page and look at

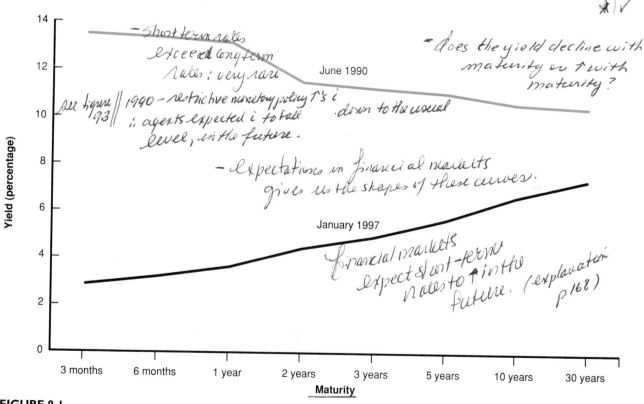

FIGURE 9-1

Canadian Yield Curves: June 1990 and January 1997

Source: Statistics Canada, CANSIM Matrix 2560.

their "Currency and Credit Market Report.") The choice of these two dates is not accidental; why we chose them will be made clear later in the chapter.

The yield curve in 1990 was both high and declining with maturity. The three-month rate (the interest rate on a three-month T-bill) was about 13.5%, while the rate on long-maturity bonds (the interest rate on bonds with a maturity of 20 to 30 years) was only 10.5%. Episodes where short-term interest rates exceed long-term interest rates are rather unusual (such a case is called an *inverted yield curve*). The shape of the January 1997 yield curve is much more common. Note how the entire yield curve had shifted down dramatically. The three-month rate was only 2.9%, while the long-maturity rate was 7.4%.

Why was the yield curve both high and downward-sloping in mid-1990, and why was it low but steeply upward-sloping in early 1997? What does the slope of the yield curve tell us about expectations in financial markets?

To answer these questions, we proceed in two steps. First, we look at the relation between the *prices of bonds* of different maturities. Second, we show the relation between *yields of bonds* of different maturities, and examine the determinants of the shape of the yield curve. (A brief introduction to the words used in bond markets is given in the Focus box entitled "The Vocabulary of Bond Markets.")

One reason why financial markets often seem so mysterious is that they have their own vocabulary, full of strange and complicated words. Here is a basic vocabulary review.

■ Bonds are issued by the government to finance its deficit, or by firms to finance their investment. If issued by the government or government agencies the bonds are called **government bonds.** If issued by firms, they are called **corporate bonds.**

■ In Canada, bonds are rated for their default risk by several private firms. The list includes Dominion Bond Rating Service(DBRS) and Canadian Bond Rating Service (CBRS), as well as the two large American firms, Standard & Poor's Corporation (S&P) and Moody's Investor Service. Moody's **bond ratings** range from AAA for bonds with practically no risk of default, such as Government of Canada bonds, to C for bonds where the default risk is high. A lower rating typically implies that the bond has to pay a higher interest rate. The difference between the interest rate paid on a given bond and the interest rate paid on a bond that promises the same sequence of payments but with the highest rating is called the **risk premium.**

Bonds with high default risk are known as **junk bonds.** Because they promised a very high interest rate, they became very popular with financial investors in the 1980s. After a few celebrated defaults, they have become less popular today.

■ Bonds that promise a single payment at maturity are called **discount bonds.** The single payment is called the **face value** of the bond.

■ Bonds that promise multiple payments before maturity and one payment at maturity are called **coupon bonds.** The payments before maturity are called **coupon payments.** The final payment is again called the face value of the bond. The ratio of coupon payments to the face value is called the **coupon rate.** The **current yield** is the ratio of the coupon payment to the price of the bond. A bond with coupon payments of $5, a face value of $100, and a price of $80 has a coupon rate of 5% and a current yield of 6.25%. From an economic viewpoint, neither the coupon rate nor the current yield are interesting measures. The correct measure of the interest rate on a bond is its *yield to maturity,* or simply *yield;* you can think of it as roughly the average interest rate paid by the bond over its life (we shall define it more precisely later in this chapter).

■ **Short-term, medium-term,** and **long-term bonds** typically refer to bonds with a maturity of 1 year or less, 1 to 10 years, and 10 years or more respectively.

■ Canadian government bonds range in maturity from a few days to 30 years. Bonds with a maturity of up to a year when they are issued are called **Treasury bills** or **T-bills.** They are discount bonds, making only one payment at maturity. Bonds with a maturity of 1 to 30 years when they are issued, that pay interest semiannually, are called **Canada bonds.**

■ In Canada, most bonds are **nominal bonds:** They promise a sequence of fixed nominal payments—pay-

ments in terms of dollars. However, other types of bonds exist. Among them are **indexed bonds,** bonds that promise not fixed nominal payments but rather payments adjusted for inflation, such as the federal government's Real Return Bonds. Instead of promising to pay, say, $100 in a year, a one-year indexed bond promises to pay $100(1 + \pi)$ dollars, where π is the rate of inflation that will take place over the coming year. Real Return Bonds have been growing in importance, but they remain a relatively small part of the government's financing. For example, in March 1996 there were $166 billion in Treasury bills outstanding, but less than $6 billion in Real Return Bonds.

BOND PRICES AS PRESENT VALUES

Consider two bonds. The first is a one-year bond that promises one payment of $100 in one year. The second is a two-year bond that promises one payment of $100 in two years.[1] Let their prices today be $\$P_{1t}$ and $\$P_{2t}$ respectively. How will these two prices be determined?

Take the one-year bond first. Let the current one-year nominal interest rate be i_{1t}. Note that we now denote the one-year interest rate in year t by i_{1t} rather than simply by i_t as we did in earlier chapters. This is to make it easier to remember that it is the *one-year* interest rate.

The price of the one-year bond today is the present value of $100 next year. Thus:

$$\$P_{1t} = \frac{\$100}{1 + i_{1t}} \tag{9.1}$$

The price of a one-year bond varies inversely with the current one-year nominal interest rate. We saw this relation in Chapter 5. Indeed, we saw that what is actually determined in the bond market is the price of one-year bonds, and the one-year interest rate is then inferred from the price according to equation (9.1). Reorganizing this equation, it follows that, if the price of one-year bonds is $\$P_{1t}$, then the current one-year interest rate is equal to $(\$100 - \$P_{1t})/\$P_{1t}$.

Turn now to the two-year bond. Its price must be equal to the present value of $100 in two years, so that

$$\$P_{2t} = \frac{\$100}{(1 + i_{1t})(1 + i^e_{1t+1})} \tag{9.2}$$

where i_{1t} denotes the one-year interest rate this year and i^e_{1t+1} denotes the one-year rate expected by financial markets for next year. The price of a two-year bond depends on both the current one-year rate and the one-year rate expected for next

---

[1]*Note that we take both bonds to be* discount bonds *(see the Focus box).*

year. In the same way we could write the price of an *n*-year bond—a bond that promises to pay, say, $100 in *n* years—as depending on the sequence of one-year rates expected by financial markets over the next *n* years.

Before exploring further the implications of equations (9.1) and (9.2), let's look at an alternative derivation of equation (9.2), based on the important notion of *arbitrage*. This way of thinking about the determination of prices in financial markets will prove very useful at many points in the rest of the book.

ARBITRAGE AND BOND PRICES

Suppose that you have decided to put some of your financial wealth in bonds. You have the choice between holding one- or two-year bonds. You care about how much you will have one year from now. Which bonds should you hold?

Suppose that you decide to hold one-year bonds. Then, for every dollar you put in one-year bonds, you will get $(1 + i_{1t})$ dollars next year. This relation is represented in the first line of Figure 9-2.

Suppose that you decide instead to hold two-year bonds. As the price of two-year bonds is $\$P_{2t}$, every dollar you put in two-year bonds buys you $1/\$P_{2t}$ bonds today. When next year comes, the bond has only one more year before maturity, and thus has become a one-year bond. Therefore, the price at which you can expect to sell it next year is $\$P^e_{1t+1}$, the expected price of a one-year bond next year. So for every dollar you put in two-year bonds, you can expect to receive $(\$P^e_{1t+1}/\$P_{2t})$ next year. This is represented in the second line of Figure 9-2.

Which bonds should you hold? Suppose that you, and other financial investors, care *only* about the expected return and thus will choose to hold only the bond with the higher expected return. This assumption is a strong one. It ignores differences in risk between the two bonds: The return from holding one-year bonds for one year is known with certainty, while the return from holding two-year bonds for one year depends on the price of one-year bonds next year and is therefore uncertain. Nevertheless, the assumption turns out to be a good approximation to reality, and it is the assumption we shall make here.

Under this assumption, and if there are positive amounts of one-year and two-year bonds in the economy, it follows that the two bonds will offer the same expected return. Suppose this were not true, and that the expected return on one-year bonds were lower than that on two-year bonds. Nobody would want to hold the existing supply of one-year bonds. The market for one-year bonds would not be in equilibrium. Only if the expected return is the same will financial investors be willing to hold both one-year bonds and two-year bonds.

If the two bonds offer the same expected return, it follows from Figure 9-2 that

$$1 + i_{1t} = \frac{\$P^e_{1t+1}}{\$P_{2t}} \qquad \text{(9.3)}$$

FIGURE 9-2
Returns from Holding One- and Two-Year Bonds for One Year

	Year *t*	Year *t* + 1
One-year bonds:	$1	$(1 + i_{1t})
Two-year bonds:	$1	$\dfrac{\$P^e_{1t+1}}{\$P_{2t}}$

The left-hand side of the equation gives the return per dollar from holding a one-year bond for one year; the right-hand side gives the expected return per dollar from holding a two-year bond for one year. We shall call equations such as (9.3)—equations which state that the expected returns on two assets have to be equal—**arbitrage** relations.[2]

Rewrite equation (9.3) as

$$\$P_{2t} = \frac{\$P^e_{1t+1}}{1 + i_{1t}} \qquad (9.4)$$

Arbitrage implies that the price of a two-year bond today is the present value of the expected price of the bond next year. This only raises the next question: What does ✳ the expected price of a one-year bond next year ($\$P^e_{1t+1}$) depend on?

The answer is straightforward. Just as the price of a one-year bond this year depends on this year's one-year interest rate, the price of a one-year bond next year will depend on the one-year interest rate next year. Writing equation (9.1) for the next year (year $t + 1$) and denoting expectations in the usual way gives

$$\$P^e_{1t+1} = \frac{\$100}{1 + i^e_{1t+1}} \quad \text{depends on} \; expectations \; of \; next \; year's \; one-year \; rate.$$

You expect the price of the bond next year to be equal to the final payment, $100, discounted by the one-year rate you expect for next year.

Replacing in equation (9.4) gives

$$\$P_{2t} = \frac{\$100}{(1 + i_{1t})(1 + i^e_{1t+1})} \quad \} \; P^e_{1t+1} \qquad \text{← this is how 9.2 is derived} \qquad (9.5)$$

This expression is the same as equation (9.2). What we have shown therefore is that *arbitrage* between one- and two-year bonds implies that the price of two-year bonds is the *present value* of the payment in two years, namely $100, discounted using current and next year's expected one-year rates. We could have used the same approach to derive the price of three-year bonds, and so on. Make sure that you can do it. The relation between arbitrage and present value is important; we shall use it repeatedly in this book.

FROM BOND PRICES TO BOND YIELDS (find 'i')

We have derived bond prices. We must now go from bond prices to bond yields.

Let's now give a precise definition of the yield to maturity. The **yield to maturity** on an *n*-year bond, or equivalently the ***n*-year interest rate,** is defined as the constant interest rate that makes the bond price today equal to the present value of future payments on the bond.

This definition is simpler than it sounds. For example, take the two-year bond we introduced earlier. Denote its yield by i_{2t}, where the subscript 2 reminds us that this is the yield to maturity on a two-year bond, or equivalently the two-year rate.

---

[2]*We shall use* arbitrage *to denote the proposition that the expected returns on two assets have to be the same. Many economists reserve the word arbitrage for the narrower proposition that riskless profit opportunities do not go unexploited.*

This yield is thus defined as the interest rate that, if it were constant over this year and next year, would make the present value of $100 in two years equal to the price of the bond today:

$$\$P_{2t} = \frac{\$100}{(1 + i_{2t})^2} \tag{9.6}$$

Suppose that the bond sells for $90 today. Then, the two-year rate is given by $\sqrt{100/90} - 1$, or 5.4%. In other words, holding the bond for two years—that is, until maturity—yields an interest rate of 5.4% per year.

What is the relation of the two-year rate to the current and expected one-year rates? All we need to do to answer this question is compare equation (9.6) with equation (9.5). Eliminating $\$P_{2t}$ between the two gives

$$\frac{\$100}{(1 + i_{2t})^2} = \frac{\$100}{(1 + i_{1t})(1 + i^e_{1t+1})}$$

or, rearranging,

$$(1 + i_{2t})^2 = (1 + i_{1t})(1 + i^e_{1t+1})$$

This gives the exact relation between the two-year rate and the current and expected one-year rates. A convenient approximation to this relation is given by[3]

$$i_{2t} \approx \frac{1}{2}\left(i_{1t} + i^e_{1t+1}\right) \tag{9.7}$$

Equation (9.7) is intuitive and important. It says that the two-year rate is (approximately) an average of the current one-year rate and next year's expected one-year rate. The relation extends to interest rates on bonds of higher maturity. *The n-year rate is (approximately) equal to the average of current and expected one-year rates over this and the next (n – 1) years:*

$$i_{nt} \approx \frac{1}{n}\left(i_{1t} + i^e_{1t+1} + \cdots i^e_{1t+n-1}\right)$$

These relations give us the key we need to interpret the yield curve. *An upward-sloping yield curve tells us that financial markets expect short-term interest rates to increase in the future. A downward-sloping yield curve tells us that financial markets expect short-term interest rates to decrease in the future.*

An example will make this clear. Return to the January 1997 yield curve in Figure 9-1. We can infer from it what financial markets expected the one-year interest rate to be one year hence. To do so, multiply both sides of equation (9.7) by 2, and reorganize to get

$$i^e_{1t+1} = 2i_{2t} - i_{1t} \tag{9.8}$$

In January 1997, i_{1t} (the one-year rate, the interest rate for 1997) was equal to 3.6%. The two-year rate was equal to 4.4%. The expected one-year rate for January 1998

---

[3]*We have seen a similar approximation when looking at the relation between nominal and real interest rates in Chapter 7. The approximation is derived in Proposition 3 in Appendix 3.*

Using 9.6.

$\frac{\sqrt{100}}{1+2 i_{2t}} = \sqrt{90}$

$\frac{\sqrt{100}}{\sqrt{90}} = 1 + i_{2t}$

$\sqrt{\frac{100}{90}} - 1 \longrightarrow$ solve for i_{2t}

was thus equal to $(2 \times 4.4\%) - 3.6\% = 5.2\%$, thus 1.6% above the January 1997 one-year rate.[4]

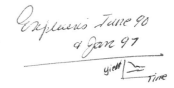

THE YIELD CURVE AND ECONOMIC ACTIVITY

Why was the yield curve so steep in early 1997? Why did financial markets expect short-term interest rates to increase in the future? The short answer is: because the Canadian economy was growing out of a recession. To see why, let's use the *IS-LM* diagram we developed in Chapter 6.[5]

To concentrate on the difference between interest rates of different maturities, let's leave aside the distinction between interest rates we introduced in Chapter 7, the distinction between real and nominal rates. More specifically, let's assume that expected inflation is equal to zero, so that the real and nominal rates are the same. Figure 9-3 draws the *IS-LM*, with the (nominal) interest rate on the vertical axis and output on the horizontal axis.

In June 1990, the Canadian economy was slipping into a recession. The Bank of Canada, concerned about inflation, had for several months made a concerted effort to constrain the money supply. In terms of Figure 9-3, monetary policy had shifted the *LM* curve from its usual position to *LM'*. Interest rates were driven up to i' in early 1990, but market participants expected the recession would lead the Bank to return to a less-restrictive policy stance and future interest rates would be lower. In mid-1990, the U.S. economy slipped into recession. What had been a modest slowdown in growth in Canada turned into a much deeper and prolonged recession than most had foreseen. For a variety of reasons, the recovery that began in 1992

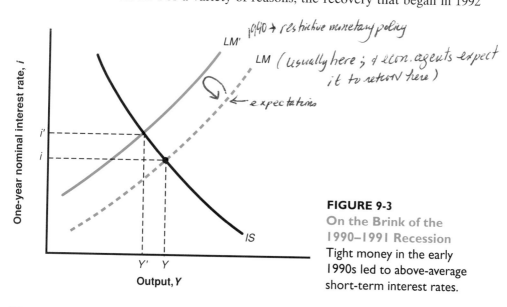

FIGURE 9-3

On the Brink of the 1990–1991 Recession

Tight money in the early 1990s led to above-average short-term interest rates.

---

[4]***Digging deeper.*** *A caveat is in order here. Recall that in deriving the arbitrage equation we ignored risk considerations. In fact, bonds of higher maturity are more risky to hold because, if they are sold before maturity, variations in their price typically lead to capital gains or losses. In practice, bond markets require a risk premium for bonds of higher maturity. Thus, a mildly upward-sloping yield curve is more likely to reflect a risk premium that increases with maturity rather than expectations of higher short-term rates in the future. The computation in the text does not take this risk premium into account.*

[5]*In Chapter 10 we extend the IS-LM model to take into account what we have learned about the effects of expectations on decisions. For the moment, the basic IS-LM will do.*

was extremely weak, both by historical standards and by comparison with the robust recovery enjoyed by our southern neighbours. Our governments worked to improve their finances by cutting expenditures and raising revenues. The labour market underwent a good deal of churning as both governments and many large firms were forced to "downsize" in order to adapt to changes in the economic environment. Increases in interest rates around the world and political uncertainty about the Quebec referendum made borrowing in Canada more expensive. The recovery stalled in 1995 and the economy grew well below trend in 1996.

By early 1997, however, it looked as though the Canadian economy was finally on the verge of some rapid growth. The U.S. economy was booming. The Bank of Canada was pleased with the low and stable rate of inflation and did its best to jump-start the economy. Look at Figure 9-4. The various shocks to the Canadian economy since 1990 had left us with a leftward shift of the *IS* curve to *IS''*. The expansionary monetary policy of the Bank had shifted the *LM* curve to *LM''*. Market participants realized that these shifts were transitory and that over time these curves would shift back to their usual levels and interest rates would return to *i*. The interest rate at *i''* looked lower than its future value.

Our discussion suggests a more general proposition: When short-term interest rates move, whether down (as in the 1990–1991 recession) or up, long-term interest rates are likely to move in the same direction, but by less. This is because financial markets are likely to assume that part of the movement in short-term interest rates will not last. Figure 9-5 shows how well this proposition indeed characterizes movements in Canadian short- and long-term interest rates.

Figure 9-5 plots monthly changes in the three-month interest rate on the horizontal axis and in the long-term interest rate (here the average yield on bonds of 10 or more years' maturity) on the vertical axis since 1960. The figure has three main features:

- First, monthly changes in the short-term interest rate range from −3.1% to 3.4%, while monthly changes in the long-term interest rate range only from −2.3% to 2.0%. Long-term rates move less than short-term rates.

- Second, most of the points lie in the northeast or the southwest quadrants of

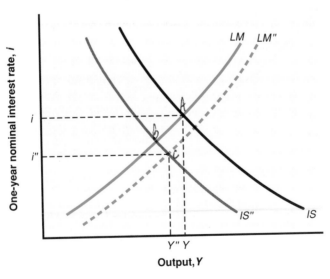

FIGURE 9-4

The Expected Path of Recovery, as of Early 1997

The market expected the recovery to bring higher interest rates.

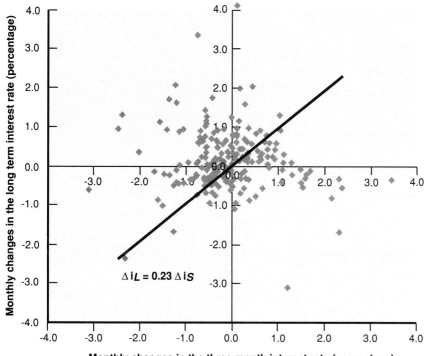

FIGURE 9-5

Monthly Changes in the Canadian Three-Month and Long-Term Interest
Rates, 1960–1996

Movements in the short-term interest rate are typically associated with smaller movements in the long-term interest rate in the same direction.

Source: Statistics Canada, CANSIM Series B14001 and B14013.

the diagram: Most of the time, short- and long-term interest rates move in the
same direction.

■ Third, the typical response of movements in the long-term interest rate to
movements in the short-term interest rate is less than one for one. The diagram plots the regression line. The equation associated with this regression is

$$\Delta i_L = 0.00 + 0.23\Delta i_S$$

where i_L (L for long) denotes the long-term interest rate, i_S (S for short) denotes the three-month rate, and Δ denotes the change in a rate from one
month to the next. This equation tells us that an increase in the short-term interest rate of 1% is typically associated with an increase in the long-term interest rate of only 0.23%.

This section has covered a lot of ground. Let us briefly summarize the main
points:

■ Arbitrage between bonds of different maturities implies that the price of a
bond is the present value of the payments on the bond, discounted using
current and expected short-term interest rates. Thus, higher current or expected short-term interest rates lead to lower bond prices.

■ The yield to maturity on a bond with a maturity of *n* years (or equivalently, the *n*-year rate) is approximately equal to the average of current and expected future one-year interest rates.

■ The slope of the yield curve tells us what financial markets expect to happen to short-term interest rates in the future. A downward-sloping yield curve implies that the market expects a decrease in short-term rates in the future; an upward-sloping yield curve implies that the markets expect an increase in short-term rates in the future.[6]

9-2 THE STOCK MARKET AND MOVEMENTS IN STOCK PRICES

We have so far focussed on bonds. But, while the government indeed finances itself primarily by issuing bonds, the same is not true of firms. Firms that need funds raise them in two ways. The first is through **debt finance;** debt may take the form of bonds or loans. The second is through **equity finance;** firms issue **shares** or **stocks.** Instead of paying predetermined amounts as bonds do, stocks pay **dividends** in an amount decided by the firm. Dividends are paid from the firm's profits. They are typically less than profits, as firms retain some of their profits to finance their own investment. But dividends typically move with profits: When profits increase, so do dividends.

Our focus in this section is on the determination of stock prices. As a way of introducing the issues, Figure 9-6 shows the behaviour of the *Toronto Stock Exchange Composite Index* (or the *TSE-300 Index*, for short) since 1960. Movements in the TSE index measure movements in the average stock price of 300 large firms in Canada. Similar indexes exist for other countries as well. The *Standard & Poor's 500 Composite Index (S&P 500)* measures the movements in a porfolio of 500 large companies traded on the New York Stock Exchange. (Another and better known American index is the *Dow Jones Industrial Index.* "The Dow" is a portfolio of a small number of very large firms, sometimes referred to as *blue chip*, and is less representative of U.S. stocks than the S&P index). The *Nikkei Index* reflects movements in stock prices in Tokyo, and the *FT* and *CAC* indexes reflect stock price movements in London and Paris respectively.)

Figure 9-6 plots two lines. The line labelled "Nominal index" gives the evolution of the index as it was published in newspapers or flashed on the evening news. The index shows near constancy until 1977 and a rapid increase since. The index, which was only 2000 in 1980, stood at 5270 in 1996. In late 1997, the index stood at 7000 (the value given in Figure 9-6 for 1997 is the average value for the year, which is lower than the value in late October).

This index, however, gives the evolution of stock prices in terms of dollars. Of more interest to us is the evolution of the index in real terms (that is, adjusted for inflation). This evolution is given by the line labelled "Real index"in Figure 9-6,

---

[6]*Once again, we remind you that we are ignoring risk premiums, which typically give a gently increasing term structure.*

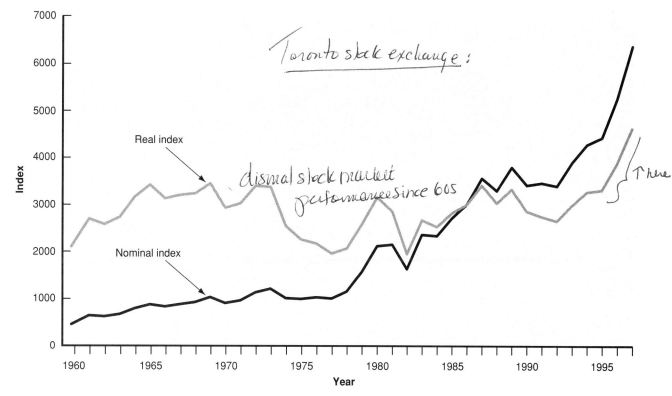

(handwritten annotations on figure: "Toronto stock exchange:"; "dismal stock market performance since 60s"; "There")

FIGURE 9-6

The TSE Index in Nominal and Real Terms, 1960–1997

While nominal stock prices went up more than tenfold from 1960 to 1997, real stock prices were relatively flat until the run-up of 1996–1997.

Source: Statistics Canada, CANSIM Series B4237 and P700000.

which plots the index divided by the consumer price index (CPI). The CPI is scaled to be equal to 1.0 in 1986, so that the nominal and real indexes are equal in 1986.

The "real" line gives a markedly different picture. It shows how dismal the stock market's performance was in the 1970s and early 1980s: Roughly constant nominal stock prices and a steadily increasing price level implied decreasing real stock prices. And while real stock prices have increased since about 1977, they took until 1996 to reach the level of the mid-1960s. In December 1996, the real index stood at 4333, compared with 3795 in May 1969 (the value given in Figure 9-6 for 1996 is the average value for the year, which is lower than the December value).

Why did the stock market do so badly for so long, and why has it rebounded since 1982? More generally, how do stock prices respond to changes in the economic environment? This is the issue we take up in the remainder of this section.

STOCK PRICES AS PRESENT VALUES

Take a stock that promises a sequence of dividends in the future. What will determine the price of that stock? By now, we are sure that the material in Chapter 7 has become second nature, and that you already know the answer: The stock price will be equal to the present value of expected dividends.

(handwritten: "Stock price = PV of future dividends")

Stock prices =
P.V of expected
future dividends

higher expected divs
lead to higher
stock prices;
Higher 'r'
leads to lower stock
prices.

Let $\$Q_t$ be the price of the stock. Let $\$D_t$ denote the dividend this year, so that $\$D_{t+1}^e$ denotes the expected dividend next year, $\$D_{t+2}^e$ denotes the expected dividend two years from now, and so on. Suppose that we look at the price of the stock just after the dividend has been paid this year—this price is known as the *ex-dividend price*—so that the first dividend to be paid after the purchase of the stock is next year's dividend. (This is just a matter of convenience; we could alternatively look at the price before this year's dividend has been paid.) The price of the stock is then given by

$$\$Q_t = \frac{\$D_{t+1}^e}{1 + i_{1t}} + \frac{\$D_{t+2}^e}{(1 + i_{1t})(1 + i_{1t+1}^e)} + \cdots \tag{9.9}$$

The price of the stock is equal to the present value of the dividend next year, discounted using the current one-year interest rate, plus the present value of the dividend two years from now, discounted using this year's one-year interest rate and the one-year interest rate expected for next year, and so on. As in the case of long-term bonds, the present value relation in equation (9.9) can be derived from arbitrage, from the assumption that the expected return per dollar from holding a stock for one year must be equal to the return from holding a one-year bond.[7] (The derivation is given in the appendix to this chapter. Going through it is good practice and will help improve your understanding of arbitrage and present values, but it can be skipped without harm.)

Equation (9.9) gives the stock price as the present value of *nominal* dividends, discounted by *nominal* interest rates. We know from Chapter 7 that we can rewrite it to get the *real* stock price as the present value of *real* dividends, discounted by *real* interest rates. Thus, we can rewrite the real stock price as

$$Q_t = \frac{D_{t+1}^e}{1 + r_{1t}} + \frac{D_{t+2}^e}{(1 + r_{1t})(1 + r_{1t+1}^e)} + \cdots \tag{9.10}$$

Q_t and D_t, now denoted without a dollar sign, are the real price and real dividends at time t. *The real stock price is the present value of expected future real dividends, discounted by the sequence of one-year expected real interest rates.*

This relation has two important implications. Higher expected dividends lead to a higher stock price. Higher current and expected one-year real interest rates lead to a lower stock price. Let's now see what light this relation sheds on movements in the stock market.

PREDICTING AND EXPLAINING THE STOCK MARKET

Figure 9-6 showed the large movements in stock prices over the last 35 years. It is not unusual for the index to go up or down by 15% within a year. In 1974, the stock market went down by 35% (in real terms); from June 1982 to June 1983 it went up by 70%. Daily movements of 2% or more are also not unusual. Where do these movements come from?

---

[7]**Digging deeper.** *That arbitrage implies that the price of a stock is the present value of dividends is true except in the presence of speculative bubbles or fads, which we discuss in Section 9-3.*

One point to be made is that these movements are for the most part unpredictable. The reason why is best understood by thinking in terms of arbitrage between stocks and bonds. If it were widely believed that the price of a stock was going to increase by 20% over the next year, holding the stock would be unusually attractive, much more attractive than holding short-term bonds. There would be a very large demand for the stock, and its price would increase *today* to the point where the expected return from holding the stock was back in line with the expected return on other assets. In other words, the expectation of a high stock price in the future would lead to a high stock price today.

There is a saying in economics that it is a sign of a well-functioning stock market that movements in stock prices are unpredictable. The saying is too strong: A few financial investors may indeed have better information or simply be better at reading the future. If there are only a few, they may not buy enough of the stock to bid its price all the way up today. Thus, they may get large expected returns. But the basic idea is nevertheless right. The financial market gurus who regularly predict large imminent movements in the stock market over the next few months are quacks. Major movements in stock prices cannot be predicted.

If movements in the stock market are unpredictable, if they are the result of the arrival of new information, where does this leave us? We can do two things. We can be Monday-morning quarterbacks, looking back and identifying the news to which the market reacted. (This is harder than it sounds. Economists are still puzzling about what caused the great crash of October 17, 1987.) We can also ask what role the variables that affect asset and money market equilibrium play in determining movements in the stock market.

The key is to look at the components of equation (9.9). Stock prices depend upon the path of expected future dividends and interest rates. Dividends tend to rise and fall with the business cycle, so, at least to a first approximation, we can replace the numerator in (9.9) with some fraction of expected output. How can we predict the path of output and interest rates? The basic *IS-LM* model gives us a good place to start. However, this model is only meant to describe short-run movements, so we need to make some brave assumptions about how agents forecast these variables in the distant future. Let's assume that everyone believes the economy will tend to return to the values i^* and y^*. (We'll look at how these long-run variables are determined in Chapters 22 to 24.) Also, to keep this simple, let's again leave aside the distinction between real and nominal interest rates, by assuming that expected inflation is equal to zero.

Many details will be refined as we extend the *IS-LM* model, but we can use it now to illustrate the basic links between developments in the goods and money markets and the response of the stock market.

A monetary expansion.[8] Suppose a more expansionary monetary policy shifts the *LM* curve down in Figure 9-7. Equilibrium output and interest rates shift from *A* to *A'*. The market realizes that this shift is temporary and that the economy will eventually return to *A*. How will the stock market react? The answer is: It depends. If the stock market had anticipated the expansionary policy, then it would not react at

---

[8]As we shall see in Chapter 12, there tends to be a close link between U.S. and Canadian monetary policy. We ignore these links for now.

Anticipated: although there is an ↑y ↓r market think that the shift is temporary & they realize it will return to A; ∴ no Δ in stock prices because no Δ in expected divs or expected future 'r';

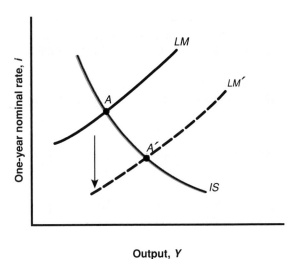

FIGURE 9-7

An Expansionary Monetary Policy and the Stock Market

A monetary expansion decreases the interest rate and increases output. What it does to the stock market depends on whether or not the stock market anticipated the monetary expansion.

Unanticipated —
Lower r, higher y
= higher stock prices.
(b/c of higher dividends)

all. Neither its expectations of true dividends nor its expectations of future interest rates are affected by a move it has already anticipated.

Now suppose that the move is at <u>least partly unexpected.</u> In that case, stock prices will increase. A more expansionary monetary policy implies lower current and future interest rates. It also implies higher output, higher profits, and in turn higher dividends. As equation (9.9) tells us, both lower interest rates and higher dividends, current and expected, lead to an increase in stock prices.

Finally, if the market had anticipated an even more expansionary policy, it would be disappointed. Stock prices will fall.

An increase in consumer spending and the stock market. Now consider an <u>unexpected shift of the *IS* curve to the right,</u> resulting, for example, from stronger-<u>than-expected consumer spending.</u> As a result of the shift, equilibrium output in Figure 9-8 increases from *A* to *B.* Will stock prices go up? One is tempted to say yes. A stronger economy means again higher profits and higher dividends. But this answer is incomplete, for at least two reasons.

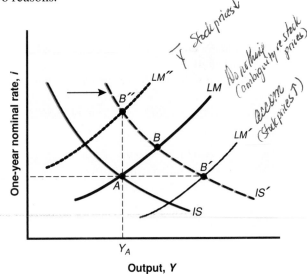

FIGURE 9-8

An Increase in Consumption Spending and the Stock Market

An increase in consumption spending leads to a higher interest rate and a higher level of output. What it does to the stock market depends on the slope of the <u>LM curve and on the central bank's behaviour.</u>

↑ Consumer confidence

First, it ignores the effect of higher activity on interest rates: The movement along the *LM* curve from *A* to *B* also implies an increase in interest rates. Higher interest rates decrease stock prices. Which effect, higher profits or higher interest rates, dominates? The answer depends on the slope of the *LM* curve. A very steep *LM* curve implies large increases in interest rates, small increases in output, and a fall in stock prices. A very flat *LM* curve leads to small increases in interest rates, large increases in output, and an increase in stock prices.

Second, it ignores the effect of the shift in the *IS* curve on the central bank's behaviour. In practice, this is the effect that financial investors often care the most about.

#1) LM effect
effects depend on shape of
LM curve (r vs y)

#2) but also central bank's behaviour

- Will the central bank "accommodate" the shift in the *IS* curve—that is, increase the money supply in line with money demand in order to avoid an increase in the interest rate? **Accommodation** corresponds to a downward shift of the *LM* curve, from *LM* to *LM'* in Figure 9-8. In this case the economy will go from point *A* to *B'*. The stock market will go up (that is, stock prices will increase): Output is expected to be higher, and interest rates are not expected to increase.

Output effect ∴ stock prices ↑ ; ī

- Will the central bank instead keep the same monetary policy, leaving the *LM* curve unchanged? In that case the economy will go from *A* to *B*. As we saw earlier, what happens to the stock market is ambiguous. The economy will have higher profits, but also a higher interest rate.

No action by central bank;
Leave LM unchanged;

- Or will the central bank worry that economic activity is becoming too strong, that an increase in output above Y_A may lead to overheating of the economy and inflation? This may be the case if the unemployment rate associated with Y_A is already quite low. The central bank may then decide to counteract the rightward shift of the *IS* curve with a monetary contraction, an upward shift of the *LM* curve from *LM* to *LM''*, so that output does not change. In that case the stock market will surely go down—that is, the prices of stocks will go down: There is no change in expected profits, but the interest rate is now likely to be higher for some time.

r↑ ȳ ∴
stock prices ↓

■ July 1993. Bad news on the economy leading to an increase in stock prices:

The Dow Jones Industrial Average climbed to within a point of a record as investors cast aside downbeat economic news and concentrated on a rally in bonds. . . . The rally came despite a series of disappointing developments yesterday morning. The Commerce Department estimated gross domestic product grew at an annual rate of 1.6% in the second quarter, short of the 2.2% pace analysts had largely forecast.

Cartoon by Wasserman, *The Boston Globe,* April 9, 1994; by permission of LA Times Syndicated.

In summary, changes in output may or may not be associated with changes in stock prices in the same direction. Whether they are depends on (1) what the market expected in the first place, (2) the source of the shocks, and (3) how the market expects the central bank to react to the output change. The Focus box gives you a number of quotes from *The Wall Street Journal* from May to July 1993. See whether you can make sense of them, using what you've just learned.

9-3 BUBBLES, FADS, AND STOCK PRICES

Do all movements in stock prices come from news about future dividends or interest rates? Many economists doubt it. They point to times such as Black October in 1929, when the U.S. stock market fell by 23% in two days, or to October 19, 1987, when the Dow Jones index fell by 22.6% in a single day. They point to the amazing rise of the Nikkei in the 1980s, followed by a sharp fall in the 1990s. The Nikkei increased from around 5000 in 1980 to around 35 000 in 1989, only to decline back to around 16 000 in 1992. In each case they point out the lack of obvious news, or at least of news important enough to justify such enormous movements.

What they argue is that stock prices are not always equal to their **fundamental value,** defined as the present value of expected dividends given in equation (9.10), and that stocks are sometimes underpriced or overpriced. Overpricing eventually comes to an end, sometimes with a crash, as in October 1987, or with a long slide, as in the case of the Nikkei index.

$$(9.10) \quad Q_t = \frac{\$D^e_{t+1}}{(1 + r_{1t})} + \frac{\$D^e_{t+2}}{(1 + r_{1t})(1 + r^e_{1t+1})} + \cdots$$

Under what conditions can such mispricing occur? The surprising answer is that it can occur even when investors are smart and when arbitrage holds. To see why, consider the case of a truly worthless stock (that is, a stock of a company that does not and never will pay dividends). Putting D_{t+1}^e, D_{t+2}^e, and so on equal to zero in equation (9.10) yields a simple and unsurprising answer: The fundamental value of the stock is equal to zero.

Might you nevertheless be willing to pay a positive price for the stock? The answer is yes. You might if you expect the price at which you can sell the stock next year to be higher than this year's price. And the same applies to a buyer next year: He may well be willing to buy at a high price if he expects to sell at an even higher price in the following year. This process suggests that stock prices may increase for a while just because investors expect them to. Such movements in stock prices are called **rational speculative bubbles.** The adjective "rational" emphasizes that financial investors may well be behaving rationally as the bubble inflates. Even those investors who hold the stock at the time of the crash, and therefore sustain a large loss, may also have been rational. They may have realized that there was a chance of a crash, but also a chance that the bubble would keep going, and that they could sell at an even higher price.

To make things simple, our example assumed the stock to be fundamentally worthless. But the argument is general and applies to stocks with a positive fundamental value as well. People might be willing to pay more than the fundamental value if they expect the price to further increase in the future. The same argument applies to other assets, such as housing, gold, and paintings. Two such bubbles, one in seventeenth-century Holland, the other in Russia in the 1990s, are described in the Global Macro box on page 178.

Are all deviations from fundamental values in financial markets rational bubbles? Probably not. Many financial investors may not be fully rational. An increase in stock prices in the past, due to a succession of good news, for example, may create overoptimism. If investors simply extrapolate from past returns to predict future returns, a stock may become "hot" (high-priced) for no reason other than the fact that its price has increased in the past. Such deviations of stock prices from their fundamental values are called **fads.** We are all aware of the existence of fads outside the stock market; there are good reasons to believe that they exist in the stock market as well.

How much of the movement in stock prices is due to movements in the fundamental value of stocks, and how much to fads and bubbles, is an active and difficult research topic that is far from settled. But it is an important question, not only for finance but also for macroeconomics. The stock market is more than just a sideshow. Stock prices affect firms' decisions by telling them at what price they can issue stocks and thus raise funds for investment. Stock prices also affect consumers' decisions by determining the value of their financial wealth. Stock market crashes can have large effects on activity; many believe that the U.S. stock market crash of 1929 was one of the sources of the Great Depression.[9] The long and large decline of the Nikkei after what was probably in large part a speculative bubble is one of the causes of the slump in Japan's economy since the early 1990s. These observations naturally take us to the topics of the next chapter, the interactions between financial markets and goods markets, and the role of expectations in those markets.

---

[9]*See Chapter 20.*

TULIPMANIA IN HOLLAND

In the seventeenth century, tulips became increasingly popular in Western European gardens. A market developed in Holland for both rare and more common forms of tulip bulbs.

The episode called the "tulip bubble" took place from 1634 to 1637. In 1634, the price of rare bulbs started increasing. The market went into a self-described frenzy, with speculators buying tulip bulbs in anticipation of even higher prices later. The price of a bulb called "Admiral Van de Eyck," for example, increased from 1500 guineas in 1634 to 7500 guineas in 1637, the equivalent of the price of a house at the time. There are stories about a sailor mistakenly eating bulbs, only to realize his costly mistake later. In early 1637, prices increased faster. Even the price of more common bulbs exploded, rising by a factor of up to 20 in January. But in February 1637, prices collapsed. A few years later, bulbs were trading for roughly 10% of their value at the peak of the bubble.*

THE MMM PYRAMID IN RUSSIA

In 1994 a Russian "financier," Sergei Mavrody, created a company called MMM and proceeded to sell shares. In issuing these shares, he promised shareholders a rate of return of at least 3000% per year!

*Source: This account is taken from Peter Garber, "Tulipmania," *Journal of Political Economy*, June 1989, 535–560. (Chicago, IL: University of Chicago Press.)

The company was an instant success. The share price increased from 1600 rubles (then $1) in February to 105 000 rubles ($51) in July. By July, according to the company claims, the number of shareholders had increased to 10 million people.

The trouble was that the company was not involved in any production and held no assets, except its 140 offices in Russia. The shares were intrinsically worthless. The company's initial success was based on a standard pyramid scheme, with MMM using the funds from the sale of new shares to pay the promised returns on the old shares. Despite repeated warnings by government officials, including Boris Yeltsin, that MMM was a scam and that the increase in the price of shares was a bubble, the promised returns were just too attractive to many Russian people, especially in the midst of a deep economic recession.

The scheme could work only as long as the number of new shareholders increased fast enough. By the end of July 1994, the company could no longer make good on its promises and the scheme collapsed. The company closed. Mavrody tried to blackmail the government into paying the shareholders, claiming that not doing so would trigger a revolution or a civil war. The government refused, leading many shareholders to be angry at the government rather than at Mavrody. Later in the year, Mavrody actually ran for Parliament, as a self-appointed defender of the shareholders who had lost their savings, and won.

SUMMARY

■ Arbitrage between bonds of different maturities implies that the price of a bond is the present value of the payments on the bond, discounted using current and expected short-term interest rates. Thus, higher current or expected short-term interest rates lead to lower bond prices.

■ The yield to maturity on a bond with a maturity of n years (or equivalently, the n-year rate) is approximately equal to the average of current and expected future one-year interest rates over this and the next $(n - 1)$ years.

■ The slope of the yield curve (equivalently, the term structure) tells us what financial markets expect to happen to short-term interest rates in the future. A downward-sloping yield curve implies that the market expects a decrease in short-term rates; an upward-sloping yield curve implies that the market expects an increase in short-term rates.

■ The fundamental value of a stock is the present value of expected future real dividends, discounted using current and future expected one-year real interest rates. In the absence of bubbles or fads, the price of a stock is equal to its fundamental value.

■ An increase in expected dividends leads to an increase in the fundamental value of stocks; an increase in current and expected one-year interest rates leads to a decrease in their fundamental value.

■ Changes in output may or may not be associated with changes in stock prices in the same direction. Whether they are depends on (1) what the market expected in the first place, (2) the source of the shocks, and (3) how the market expects the central bank to react to the output change.

■ Stock prices can be subject to bubbles or fads that lead a stock price to differ from its fundamental value. Bubbles are episodes where financial investors buy a stock for a price higher than its fundamental value in anticipation of reselling the stock at an even higher price. A fad is a general term for a time when, for reasons of fashion or overoptimism, financial investors are willing to pay more than the fundamental value of the stock.

KEY TERMS

- default risk, 161
- maturity, 161
- yield curve, or term structure of interest rates, 161
- government bonds, 162
- corporate bonds, 162
- bond ratings, 162
- risk premium, 162
- junk bonds, 162
- discount bonds, 162
- face value, 162
- coupon bonds, 162
- coupon payments, 162
- coupon rate, 162
- current yield, 162

- short-, medium-, and long-term bonds, 163
- Treasury bills, or T-bills, 163
- Canada bonds, 163
- indexed bonds, 163
- arbitrage, 165
- yield to maturity, or n-year interest rate, 165
- debt finance, 170
- equity finance, 170
- shares, or stocks, 170
- dividends, 170
- accommodation, 175
- fundamental value, 176
- rational speculative bubble, 177
- fads, 177

QUESTIONS AND PROBLEMS

1. Determine the yield to maturity of each of the following bonds. [*Hint:* The general formula relating a discount bond's price (P), face value (F), and yield (i) is $P = F/(1 + i)^n$.]
   a. A discount bond with a face value of $1000, a maturity of two years, and a price of $800
   b. A discount bond with a face value of $1000, a maturity of two years, and a price of $900
   c. A discount bond with a face value of $1000, a maturity of three years, and a price of $900
   d. Using your answers to parts **a** and **b**, what happens to a bond's yield when everything else remains the same but the price of the bond increases by $100?

2. If the interest rate for the current year is 10% and the interest rate expected for next year is 5%, determine the yield on a two-year bond using
   a. The approximation formula
   b. The exact formula

3. Suppose that the interest rate this year is 4% and financial markets expect the interest rate to increase by 1% each year thereafter. Determine the yield to maturity on a
   a. One-year bond
   b. Two-year bond
   c. Three-year bond

4. Suppose that a country's central bank unexpectedly pursues an expansionary monetary policy that financial markets believe will be reversed in the future. What effect will the expansionary monetary policy have on the *LM* curve? On the yield curve? Why?

5. A share will pay a dividend of $100 in one year and will be sold for an expected price of $1000 at that time. Determine the current price of the stock if the current one-year interest rate is
   a. 5%
   b. 10%

6. A share is expected to pay a real dividend of $100 next year, and thereafter the real value of dividend payments is expected to increase by 3% per year, forever. Determine the current price of the stock if the real interest rate is expected to remain constant at

   a. 5%

   b. 10%

7. Using the *IS-LM* model, determine the impact of each of the following on stock prices. If the effect is am-biguous, explain what additional information would be needed to reach a conclusion.

   a. An unexpected contractionary monetary policy, with no change in fiscal policy

   b. An unexpected contractionary fiscal policy, with no change in monetary policy

   c. An unexpected contractionary fiscal policy, with a monetary policy that keeps output unchanged

   d. An unexpected contractionary fiscal policy, with a monetary policy that keeps interest rates unchanged

FURTHER READING

A very good and fun book on the stock market and stock prices is Burton Malkiel, *A Random Walk Down Wall Street,* 6th ed. (New York: W. W. Norton, 1995).

   An account of historical bubbles is given by Peter Garber in "Famous First Bubbles," *Journal of Economic Perspectives,* Spring 1990, 35–54.

An interesting discussion of the causes of the 1987 stock market crash between Fischer Black, Kenneth French, and Robert Shiller, three financial economists with very different opinions, is given in the *NBER Macroeconomics Annual,* 1988, 269–297.

Examples in the Chapter:

Expectations re future 'r'

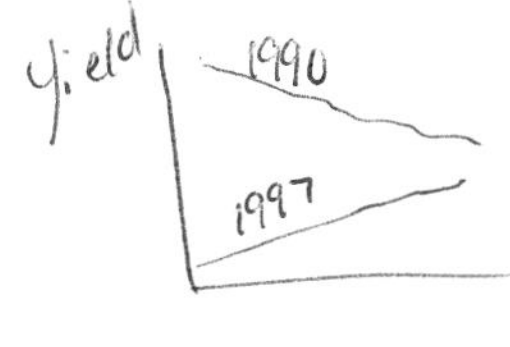

[handwritten graph: axes labeled "yield" and "t", with lines labeled 1990 and 1997]

#1 Cda 1990: Following contractionary monetary policy the fin. market expectation was that r was too high & the LM would eventually go back to its usual location (lower r, higher y ∴ Stock prices ↑

   Transitory Contractionary monetary policy. Shift in LM

#2 Cda 1997: Following significant shocks that shifted IS to left, fin markets expected Central bank to ↑ MS; shift LM to right (expansionary) ∴ expectation is that r will ↑ in future when LM goes back to its usual location (∴ Stock prices ↓ with ↑ r ↓ y)

   Transitory expansionary monetary policy.

#3 Bank decides to use more expansionary monetary policy. (effect on Stock Prices)

   Anticipated: Nothing happens; fin markets don't react; ∴ no expectations of future divs or interest rates

   Partly unexpected: Stock prices rise; fin markets: they expect higher output, lower 'r' in future;

#4 ↑ Consumer confidence: LM effect (steep or flat) (effect on Stock prices.) + 3 effects re: central bank

---

APPENDIX

ARBITRAGE AND STOCK PRICES

The purpose of this appendix is to show that, in the absence of rational speculative bubbles, arbitrage between stocks and bonds implies that the price of a stock is equal to the expected present value of dividends.

Suppose that you face the choice of investing either in one-year bonds or in stocks for a year. Which should you choose?

Suppose that you decide to hold one-year bonds. Then, for every dollar you put in one-year bonds, you will get $(1 + i_{1t})$ dollars next year. This payoff is represented in the first line of Figure 9A-1.

Suppose that you decide instead to hold stocks for a year. This implies buying a stock today, receiving a dividend next year, and then selling the stock. As the price of a stock is $\$Q_t$, every dollar you put in stocks buys you $\$1/\$Q_t$ stocks. And for each stock you buy, you expect to receive ($\$D_{t+1}^e + \$Q_{t+1}^e$), the sum of the expected dividend and the stock price next year. Thus, for every dollar you put in stocks, you expect to receive ($\$D_{t+1}^e + \$Q_{t+1}^e$)/$\$Q_t$. This payoff is represented in the second line of Figure 9A-1.

Let's use the same arbitrage argument we used for bonds in the text. If financial investors hold only the asset with the higher expected rate of return, then equilibrium requires that the expected rate of return from holding stocks for one year be the same as the rate of return on one-year bonds:

$$\frac{\$D_{t+1}^e + \$Q_{t+1}^e}{\$Q_t} = 1 + i_{1t}$$

Rewrite this equation as

$$\$Q_t = \frac{\$D_{t+1}^e}{1 + i_{1t}} + \frac{\$Q_{t+1}^e}{1 + i_{1t}} \qquad (9A.1)$$

Arbitrage implies that the price of the stock today must be equal to the present value of the sum of the expected dividend and price next year.

The next step is to think about what determines $\$Q_{t+1}^e$, the expected stock price next year. Next year, financial investors will again face the choice between stocks and one-year bonds. Thus, the same arbitrage

FIGURE 9A-1

Rates of Return on Holding One-Year Bonds or Stocks for One Year

relation will hold. Writing equation (9A.1), but now for time $t + 1$, and taking expectations into account gives

$$\$Q_{t+1}^e = \frac{\$D_{t+2}^e}{1 + i_{1t+1}^e} + \frac{\$Q_{t+2}^e}{1 + i_{1t+1}^e}$$

The expected price next year is simply the present value next year of the sum of the expected dividend and price two years from now. Replacing the expected price $\$Q_{t+1}^e$ in equation (9A.1) gives

$$\$Q_t = \frac{\$D_{t+1}^e}{1 + i_{1t}} + \frac{\$D_{t+2}^e}{(1 + i_{1t})(1 + i_{1t+1}^e)} + \frac{\$Q_{t+2}^e}{(1 + i_{1t})(1 + i_{1t+1}^e)}$$

The stock price is the present value of the expected dividend next year, plus the present value of the expected dividend two years from now, plus the expected price two years from now.

If we replace the expected price in two years by the present value of the expected price and dividends in three years, and so on for n years, we get

$$\$Q_t = \frac{\$D_{t+1}^e}{1 + i_{1t}} + \cdots + \frac{\$D_{t+n}^e}{(1 + i_{1t}) \ldots (1 + i_{1t+n-1}^e)}$$

$$+ \frac{\$Q_{t+n}^e}{(1 + i_{1t}) \ldots (1 + i_{1t+n-1}^e)}$$

Look at the last term, which is the present value of the expected price in n years. As long as people do not expect the stock price to explode in the future, then, as we keep replacing Q^e_{t+n} and n increases, this term will go to zero. To see why, suppose that the interest rate is constant and equal to i, and that people expect the price of the stock to converge to some value $\$\overline{Q}$ in the far future. Then, the last term becomes $\$\overline{Q}/(1+i)^n$. If the interest rate is positive, the term goes to zero as n becomes large.* The previous expression reduces to equation (9.9) in the text: The price is the present value of expected future dividends.

An extension of the present value formula. Suppose that because stocks are perceived as more risky than bonds, people require a *risk premium* to hold stocks rather than bonds. Let the premium be denoted by θ. If θ is equal, for example, to 5%, then people will hold stocks only if the expected rate of return on stocks exceeds the expected rate of return on short-term bonds by 5% a year.

In that case, the arbitrage equation between stocks and bonds becomes

$$\frac{\$D^e_{t+1} + \$Q^e_{t+1}}{\$Q_t} = 1 + i_{1t} + \theta$$

The only change is the presence of θ on the right-hand side of the equation. Going through the same steps as before, the stock price is then equal to

$$\$Q_t = \frac{\$D^e_{t+1}}{1 + i_{1t} + \theta} + \cdots + \frac{\$D^e_{t+n}}{(1 + i_{1t} + \theta) \ldots (1 + i^e_{1t+n-1} + \theta)} + \cdots$$

The stock price is still equal to the present value of expected future dividends. But the discount rate is now equal to 1 plus the interest rate plus the premium. Note that the higher the risk premium, the lower the stock price. The average risk premium on stocks over bonds has been positive on average. But (in contrast to the assumption we made above) it is not constant over time. Variations in the risk premium are yet another source of fluctuations in stock prices.

Digging deeper. *When prices are subject to rational bubbles as discussed in Section 9-3, the condition that the expected stock price does not explode is not satisfied. This is why, when there are bubbles, the stock price need not be equal to the present value of expected dividends.*

Expectations, Policy, and Output

In Chapters 8 and 9 we explored the role of expectations in consumption and investment decisions and in the determination of bond and stock prices. In this chapter we put the pieces together and take a second look at the effects of fiscal and monetary policy in a closed economy. (We took our first look in Chapter 6.)

10-1 EXPECTATIONS AND DECISIONS: TAKING STOCK

Let's start by reviewing what we have learned and discuss how we should modify the IS and LM relations we've been working with.

EXPECTATIONS AND THE *IS* RELATION

A major theme of Chapters 8 and 9 has been that consumption and investment decisions both depend very much on expectations of future income and interest rates. The various channels through which expectations affect spending are summarized in Table 10-1.

TABLE 10-1

SPENDING AND EXPECTATIONS: THE CHANNELS

	DEPENDS ON:	DEPENDS ON EXPECTATIONS OF:
Consumption		
	• Current after-tax labour income	
	• Human wealth ⟶	• Future after-tax labour income
		• Future real interest rates
	• Nonhuman wealth	
	• Stocks ⟶	• Future real dividends
		• Future real interest rates
	• Bonds ⟶	• Future nominal interest rates
Investment		
	• Current cash flow	
	• Present value of after-tax profits ⟶	• Future after-tax profits
		• Future real interest rates

A model that gave a detailed treatment of consumption and investment along the lines shown in Table 10-1 could be very complicated, and although this can be done—and indeed is done in the large empirical models that macroeconomists build to understand the economy and analyze policy—here is not the place to try. What we want to capture instead is the essence of what we've learned so far, namely the dependence of consumption and investment on expectations of the future.

The first step we can take is to divide time into two periods: (1) a *current* period, which you can think of as the current year, and (2) a *future* period, all future years lumped together.[1] This way we do not have to keep track of expectations about each future year.

How should we then write the *IS* relation for the current period? Let's first go back to the *IS* relation we wrote before thinking about the role of expectations:

$$Y = C(Y - T) + I(Y, r) + G$$

This is the equation we saw in Chapter 7, which takes into account the distinction between real and nominal interest rates we introduced in that chapter. Before we introduce expectations into this equation, it will prove convenient to rewrite it in more compact form but without changing its content. Define $A(Y, T, r) \equiv C(Y - T) + I(Y, r)$. Think of the letter A as standing for **aggregate private spending,** or simply **private spending.** With this notation we can rewrite the *IS* relation as

$$Y = A(Y, T, r) + G \qquad (10.1)$$
$$(+, -, -)$$

The properties of aggregate private spending, A, follow from the properties of consumption and investment that we laid down earlier. Aggregate private spending is an increasing function of current income Y: Higher income (equivalently, output)

---

[1] *This way of dividing time between "today" and "later" is the way many of us organize our own lives: We think of "things to do today" versus "things that can wait."*

increases consumption and investment. It is a decreasing function of taxes T: Higher taxes decrease consumption. It is a decreasing function of the real interest rate r: A higher interest rate decreases investment.

All we have done so far is simplify notation. We can now turn to the question at hand. How should we modify equation (10.1) to reflect the role of expectations? The natural extension is to allow spending to depend not only on current variables but also on their expected values in the future period, thus to write

$$Y = A(\overbrace{Y, T, r}^{shifters}, \overbrace{Y'^e, T'^e, r'^e}^{shifters}) + G \qquad (10.2)$$
$$(+, -, -, \quad +, \quad -, \quad -)$$

IS equilibrium or goods market equil.

where primes denote future values, so that Y'^e, T'^e, and r'^e denote future expected income, taxes, and real interest rate respectively. The notation is a bit heavy, but what it captures is straightforward:

Consumers feel wealthier + spend more.

- Increases in either current or expected future income increase private spending.

- Increases in either current or expected future taxes decrease private spending.

- Increases in either the current or expected future real interest rate decrease private spending.

With goods-market equilibrium now given by equation (10.2), Figure 10-1 draws the new *IS* curve. As usual, to draw the curve, we take all variables other than current output, Y, and the current real interest rate, r, as given. Thus, the *IS* curve is drawn for given values of current and future expected taxes, as well as for given values of expected future output and the expected future real interest rate.

The *IS* curve is still downward-sloping, and the reason is the same as before: A decrease in the current real interest rate leads to an increase in spending, which leads, through a multiplier effect, to an increase in output. We can say more, however. The *IS* curve is likely to be much steeper than it was earlier. Put another way, a decrease in the current interest rate is likely to have only a small effect on the level of output required to keep investment equal to savings.

- What moves output is not a ↓ in current 'r' but rather, IS shifters like expectations shifters r^e;

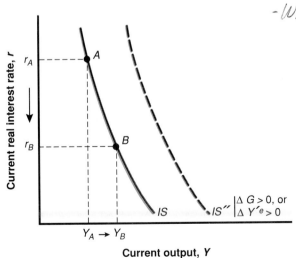

FIGURE 10-1

The New *IS* Curve

Given expectations, a decrease in the real interest rate leads to a (small) increase in output.

To see why this is so, take point A on the IS curve in Figure 10-1 and consider the effects of a decrease in the real interest rate. The effect of the decrease in the real interest rate on output depends on the strength of two effects. The first is the effect of the real interest rate on spending given income. The second is the multiplier. Let's examine each of these effects in turn.

A decrease in the current real interest rate, *given unchanged expectations of the future real interest rate,* does not have much effect on spending. We saw why in Chapters 8 and 9: A change in only the current real interest rate does not lead to large changes in present values, and thus does not lead to large changes in spending. For example, firms are not likely to change their investment plans very much in response to a decrease in the current one-period real interest rate if they do not expect future interest rates to be lower as well.

Second, the multiplier is likely to be quite small. The multiplier's size depends on the size of the effect of a change in current income (output) on spending. But a change in current income, *given unchanged expectations of future income,* is unlikely to have a large effect on spending. The reason is the same: Changes in income (output) that are not expected to last have only a limited effect on both consumption and investment. For example, consumers who expect their income to be higher for only a year will increase consumption, but by much less than the increase in income.

Putting things together, a large decrease in the current real interest rate—from r_A to r_B in Figure 10-1—requires only a small increase in output, from Y_A to Y_B, to clear the goods market. The IS curve, which goes through points A and B, is steeply downward-sloping.

As before, changes in current taxes (T) or in government spending (G) shift the IS curve. An increase in current government spending, for example, increases spending at a given interest rate and thus shifts the IS curve to the right, from IS to IS'' in Figure 10-1.[2]

Equation (10.2) makes it clear that changes in expected future variables also shift the IS curve. For example, an increase in expected future output, Y'^e, shifts the IS curve to the right: Higher expected future income leads consumers to feel wealthier and spend more. Higher expected future output implies higher expected profits, which leads firms to invest more. Higher spending leads, through the multiplier effect, to higher output. By a similar argument, an increase in expected taxes leads consumers to decrease current spending and shifts the IS curve to the left; so does an increase in the expected future real interest rate.

The *LM* Relation Revisited

The *LM* relation we wrote earlier was[3]

$$\frac{M}{P} = YL(i) \tag{10.3}$$

---

[2]*Because primes denote future values of the variables, to avoid confusion we shall use double primes (such as in IS″) or triple primes (IS‴) to denote shifts in curves in this chapter.*

[3]*See Chapter 5, in which we first derived the LM relation, and Chapter 7, in which we argued that it is the nominal interest rate (rather than the real interest rate) that affects the demand for money.*

where M/P was the supply of money and $YL(i)$ was the demand for money. The demand for money depended on real income and on the short-term nominal interest rate—the opportunity cost of holding money. We derived this demand for money, however, before thinking about expectations. The question we must now consider is whether, in light of our focus on the role of expectations, we should modify our *LM* equation. The answer—we are sure you will find this to be good news—is that we do not.

No adjustments to LM equation (expectations are not an issue)

Think of your own demand for money. How much money you want to hold today depends on your current <u>level of transactions</u>, not on the level of transactions you expect next year or the year after; there will be time to adjust your money balances to your transaction level if it changes in the future. <u>And the opportunity cost of holding money now depends on the current nominal interest rate,</u> not on the expected nominal interest rate next year or the year after. For example, if increases in short-term interest rates in the future increase the opportunity cost of holding money, the time to reduce your money balances will be then, not now.

So, in contrast with the consumption decision, the decision as to how much money to hold is quite myopic, depending primarily on current income and the current short-term nominal interest rate.

This discussion implies that we can still think of the demand for money as depending on the current level of output and the current nominal interest rate, and use equation (10.3) to describe the determination of the nominal interest rate in the current period.

✓ Y, i

10-2 A FIRST EXERCISE: MONETARY POLICY, EXPECTATIONS, AND OUTPUT

In the basic *IS-LM* model of Chapter 6, there was only one interest rate, i, which entered both the *IS* relation and the *LM* relation. When the central bank expanded the money supply, "the" interest rate went down, and spending increased. We now have to keep two distinctions in mind. The first is t<u>he distinction between the nominal interest rate and the real interest rate.</u> The second is the <u>distinction between current and expected future interest rates.</u> The rate that enters the *LM* relation, and thus the rate that the central bank affects directly, is the *current nominal interest rate*. In contrast, spending in the *IS* relation depends on both the *current real interest rate* and the *expected future real interest rate*. Economists sometimes state this distinction even more starkly by saying that, although the central bank controls the short-term nominal interest rate, what matters for spending and output is the long-term real interest rate.

✓ key.

Remember from Chapter 7 that the real interest rate is approximately equal to the nominal interest rate minus expected inflation:

$$r \approx i - \pi^e$$

Similarly, the expected future real interest rate is approximately equal to the expected future nominal interest rate minus expected future inflation:

$$r'^e \approx i'^e - \pi'^e$$

Thus, when the central bank adopts an expansionary monetary policy and decreases the current nominal interest rate i, the effect on current and expected future real interest rates depends on two factors. The <u>first</u> is whether the decrease in the current nominal interest rate, i, leads financial markets to revise their expectations of the future nominal interest rate, i'^e, as well. The <u>second</u> is whether the decrease in the current nominal interest rate leads financial markets to revise their expectations of current and future inflation, π^e and π'^e.[4] As the two equations above show, whether and how these inflation expectations change determines what happens to the real interest rate, given the nominal interest rate.

Let's explore these linkages further. For the time being we shall leave aside the second factor, the role of changing expectations of inflation, and focus on the first, the role of changing expectations of the future nominal interest rate.[5] Thus assume that expected current and future inflation are both equal to zero. In this case we need not distinguish between the nominal and the real interest rate, as they are equal, and we can use the same letter to denote both. Let r and r'^e denote the current and expected future real (and nominal) interest rates.

With this simplification we can rewrite the IS and LM relations in equations (10.2) and (10.3) as

$$IS: \qquad Y = A(Y, T, r, Y'^e, T'^e, r'^e) + G \qquad (10.4)$$

$$LM: \qquad \frac{M}{P} = Y\,L(r) \qquad (10.5)$$

The corresponding IS and LM curves are drawn in Figure 10-2. The vertical axis measures the current real interest rate r; the horizontal axis measures current output Y. The IS curve is downward-sloping, and steep. We saw the reason earlier: For given expectations, a change in the current interest rate has a limited effect on spending, and the multiplier is small. The LM curve is upward-sloping. An increase in income leads to an increase in the demand for money; given the supply of money, the result is an increase in the interest rate.

Consider an expansionary monetary policy, and assume for the moment that it does not change expectations of either the future interest rate or future output. The LM shifts down, from LM to LM''. The equilibrium moves from point A to point B, with higher output and a lower interest rate. The <u>steep IS curve,</u> however, implies that the increase in money has only a small effect on output. We saw why earlier: Changes in the current interest rate, unaccompanied by changes in expectations, have only a small effect on spending.

But is it reasonable to assume that expectations are unaffected by an expansionary monetary policy? Isn't it possible that, as the central bank decreases the current interest rate, financial markets anticipate a lower interest rate in the future as well, along with higher future output stimulated by this lower future interest rate? What happens if they do? At a given current interest rate, prospects of a lower future interest rate and higher future output both increase spending and output; they <u>shift the IS curve to the right, from IS to IS''.</u> The new equilibrium is

---

[4]*Think of "expected current inflation" as expected inflation over the course of this year.*

[5]*We return to the issue of changing expectations of inflation and their effects on interest rates in Chapter 19, after we have developed a theory of inflation.*

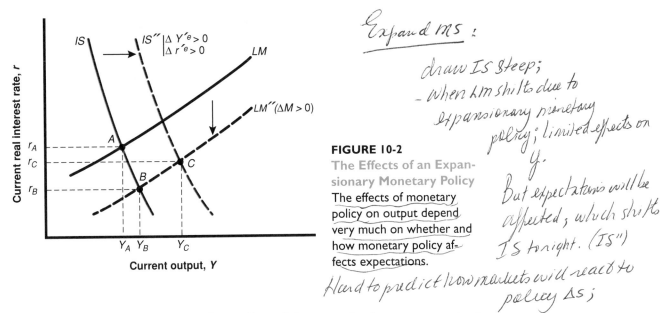

FIGURE 10-2

The Effects of an Expansionary Monetary Policy

The effects of monetary policy on output depend very much on whether and how monetary policy affects expectations.

Handwritten notes:

Expand MS :

draw IS steep;
— when LM shifts due to expansionary monetary policy; limited effects on y.

But expectations will be affected; which shifts IS tonight. (IS")

Hard to predict how markets will react to policy Δs;

given by point *C*. Thus, while the direct effect of the expansion in money on output is limited, the full effect, once changes in expectations are taken into account, is much larger.

There is an important lesson here. *The effects of monetary policy, indeed of any type of macroeconomic policy, are far from mechanical.* How much an increase in money affects output depends crucially on expectations. In Chapter 3 we constructed a simple model of the economy where governments were omnipotent and could (with the right choice of policy) achieve exactly the target level of output they wanted. We have now seen two reasons to qualify this optimistic view of policy. The first, explored in Chapter 4, is the presence of lags in the effects of policy on activity, making it harder for the government to act in time. The second, explored in this chapter, is the importance of expectations and the difficulty of predicting how they will react to changes in policy.

Nonetheless, saying that the effect of policy depends on its effect on expectations is not the same as saying that anything can happen. Expectations are not arbitrary. Financial investors, firms thinking about investment, and people thinking about retirement give a lot of thought to what may happen in the future. Indeed, we can think of them as forming expectations about the future by assessing the course of future expected policy and then working out the implications for future activity. If they do not do it themselves—and indeed, most of us do not spend our time solving macroeconomic models before making decisions—they do so indirectly by watching TV and reading newsletters and newspapers, which themselves rely on the forecasts of public and private forecasters. Economists refer to this forward-looking method of forming expectations, as opposed to simple extrapolations of the past, as **rational expectations.** The introduction of the assumption of rational expectations is probably the most important development in macroeconomics in the last 20 years, and is discussed further in the Focus box entitled "Rational Expectations."

To see how the assumption that people have rational expectations can be used, and what it implies, let's now study the effects of fiscal policy under the assumption of rational expectations.

Most macroeconomists today routinely solve their models under the assumption of rational expectations. But this was not always the case. Indeed, the last 20 years in macroeconomic research are often called the "rational expectations" revolution.

The importance of expectations is an old theme in macroeconomics. But, until the early 1970s, macroeconomists formalized the formation of expectations in one of two ways. One was as **animal spirits** (from an expression Keynes introduced in the *General Theory* to refer to movements in investment that could not be explained by movements in current variables): Shifts in expectations were taken as unexplained. The other was as backward-looking rules. For example, people were assumed to have **adaptive expectations,** to assume that if their income had grown fast in the past it would continue to do so in the future, to revise their expectations of future inflation upward if they had underpredicted in the past, and so on.

In the early 1970s, a group of macroeconomists led by Robert Lucas and Thomas Sargent argued that these assumptions did not do justice to the way people form expectations.* They argued that economists should assume that people have rational expectations, that people look to the future and do the best job they can in predicting it. This is not the same as assuming that people know the future, but rather that they use the information they have in the best possible way. Using the popular macroeconomic models of the time, Lucas and Sargent showed how replacing traditional assumptions about expectations formation with the assumption of rational expectations could fundamentally alter the results. We shall see their most dramatic example in Chapter 17 when we study the relation between inflation and unemployment known as the Phillips curve. Their research showed the need for a complete rethinking of macroeconomic models under the assumption of rational expectations, and this is what has happened since.

Most macroeconomists today use rational expectations as a working assumption in the models they build and in their analyses of policy. Surely there are times of overoptimism or overpessimism. But these are more the exception than the rule, and it is not clear that economists can say much about those times anyway. In thinking about the probable effects of a certain policy, the best assumption seems to be that financial markets, people, and firms will do the best they can to work out its implications. Designing a policy on the assumption that people will make systematic mistakes in responding to it is surely unwise.

So why did it take until the 1970s for rational expectations to become a standard assumption? Largely because of technical problems. Under rational expectations, what happens today depends on expectations of what will happen in the future. But what happens in the future depends in turn on what happens today. The success of Lucas and Sargent in convincing most macroeconomists to use rational expectations comes not only from the strength of their case, but also from showing how it could actually be done. Much progress has been made since in developing solution methods for larger and larger models. Today, a number of large macroeconometric models are solved under rational expectations.

*Robert Lucas received the Nobel Prize in 1995 for his work on expectations.

10-3 DEFICIT REDUCTION, EXPECTATIONS, AND ECONOMIC ACTIVITY

Prime Minister Chrétien and his Liberals won the election of 1993 and, in their first Budget, Finance Minister Paul Martin spelled out a bold plan for deficit reduction.

Although there was general agreement among both politicians and economists that deficit reduction was necessary, there was much less agreement as to the speed at which it should be implemented. With this question in mind, let's use the model we have developed in this chapter to look at the effects of anticipated deficit reductions on current activity.

Assume that the economy is described by equation (10.4) for the *IS* relation and equation (10.5) for the *LM* relation. Now suppose that the government announces a deficit-reduction program, with the reduction to be achieved through increases in current and future taxes, *T* and *T'*.[6] What are the effects likely to be on the interest rate and activity *this period*?

By now we know [from equation (10.4)] that the answer depends on whether and how this announcement leads to changes in expected future output Y'^e, and in the expected future interest rate r'^e. Thus, the first question we must take up is: If the announcement of future tax increases is credible—that is, if people trust that deficit reduction will indeed be implemented—how will they revise their expectations of future output and the future interest rate?

To answer this question we need a model of output and interest rate determination in the future. In effect, we already have such a model. Under our convenient simplification that there are only two periods, the current period and the future period, there is no period after the future period. Put in less-metaphysical terms, when we think about equilibrium in the future, we do not have to worry about expectations about the future further out.[7] Thus, we can think of the *IS* and *LM* relations in the future period as being given simply by

$$IS: \qquad Y' = A(Y', T', r') + G' \qquad (10.6)$$

$$LM: \qquad \frac{M'}{P'} = Y'L(r') \qquad (10.7)$$

Primes denote future-period variables. Thus, in the *IS* relation, future output depends on future private and government spending. And future private spending depends on future output, future taxes, and the future interest rate. The *LM* relation requires that the future supply of money be equal to the future demand for money, which in turn depends on the future interest rate and the future level of income. Remove the word "future" and all these effects are familiar ones from our basic *IS-LM* model. The only difference is that we are now looking at equilibrium in the future.

So, when the government announces an increase in future taxes, how should people adjust their expectations of future output and the future interest rate? Figure 10-3 shows the *IS* and the *LM* curves characterizing equilibrium in the future and corresponding to equations (10.6) and (10.7). As usual, the *IS* curve is downward-sloping and the *LM* curve is upward-sloping.

An increase in *T'* reduces disposable income and thus reduces private spending at a given interest rate. The *IS* curve shifts to the left, from *IS* to *IS''*. The equilibrium point moves from *A* to *B*. Thus, if people today anticipate no other policy change in the future than the increase in taxes, the announcement will lead them to expect lower output as well as a lower interest rate in the future. The effect of the deficit-reduction program will thus be to increase T'^e, decrease Y'^e, and decrease r'^e.

---

[6] *Although the Liberals also cut government spending, we shall focus on tax increases, because that was the major source of federal deficit reduction.*

[7] *This is why it is so convenient to think of just two periods. More-realistic models, which clearly cannot make this assumption, must deal with the complication that the equilibrium next period depends on expectations about the period after, and so on.*

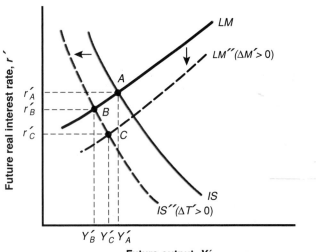

FIGURE 10-3

The Effects of a Future Increase in Taxes on Future Output and the Future Interest Rate

What happens to future output and the future interest rate depends on the central bank's reaction to the fiscal contraction.

Assumption 2: This argument assumes that the only change in policy that people anticipate is the increase in taxes. Suppose, however, that the Bank of Canada announces that it fully endorses the goal of deficit reduction, and that, if the government indeed embarks on a serious deficit-reduction program, the Bank will use monetary policy to moderate any adverse impact on spending and thus on output. In terms of Figure 10-3, the Bank announces that if the *IS* curve shifts to the left it will stand ready to shift the *LM* down, decreasing the interest rate so as to offset some of the contraction in output. In this case the new equilibrium will be at point *C,* with lower output and a lower interest rate. Therefore, if people anticipate a change in both future fiscal and future monetary policies, the effect of the announcement will be to increase $T^{\prime e}$, but decrease $Y^{\prime e}$ and $r^{\prime e}$.

We can now return to the question of what happens *this period* in response to the announcement and start of the deficit-reduction program. Figure 10-4 draws the current-period *IS-LM* model. In response to the policy change, there are now four factors shifting the *IS* curve:

net effects? ⊕ or ⊖? hard to say!

(1) ■ Current taxes go up, leading to a shift of the *IS* curve to the left.

(2) ■ Expected future taxes go up, also leading the *IS* curve to shift to the left. *Contractionary policy.*

(3) ■ Expected future income goes down, leading to yet another shift of the *IS* curve to the left. The more people expect the Bank to offset the adverse effects of fiscal contraction in the future, the weaker this effect is.

(4) ■ The expected future interest rate goes down, leading to a rightward shift of the *IS* curve as the lower interest rate stimulates spending. The more the Bank is expected to offset the fiscal contraction through monetary expansion, the stronger this effect is.[8] *expansionary*

(5) ───────────────────

[8]*There is actually a fifth effect, which we are not yet equipped to analyze but which we must mention. If lower deficits lead to higher capital accumulation—and this is indeed often the main motivation behind deficit-reduction plans—higher capital in the future means an increase in the level of output that the economy can produce in the future. This may lead to expectations of higher output in the future, leading to a shift of the* IS *curve to the right. A discussion of the long-term effects of deficits must wait until we have developed the supply side of the economy and looked at growth. It is given in Chapter 29.*

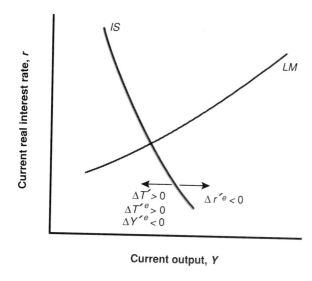

FIGURE 10-4

The Effects of a Deficit-Reduction Package on Current Output
When account is taken of its effect on expectations and of the central bank's reaction to the change in fiscal policy, a deficit reduction package need not lead to a decrease in output.

Without knowing a lot more about the specifics of the *IS* and *LM* relations and the details of the timing of tax increases, we cannot say whether the net effect of these four shifts is negative or positive. Increases in current and future expected taxes and decreases in future expected output are contractionary; decreases in the future expected interest rate are expansionary.

Our analysis thus implies that the net effect of the tax-increase package may lead to an *increase* in output this period, in sharp contrast to the conclusions of the basic *IS-LM* model. Under what conditions is this most likely to happen? Consider the following conditions:

Suppose that the tax package is very much _backloaded:_ Most of the increase in taxes is to take place in the future, so that there is only a small increase in current taxes, and thus a small direct effect on current spending through current taxes. Also suppose (as some of the evidence we saw in Chapter 8 suggests) that consumers do not react very much to expected future tax increases, so that the announced increase in future taxes also has little direct effect on their current spending. Finally, suppose that financial markets believe that the central bank will indeed offset the effects of contractionary fiscal policy in the future by increasing the money supply and thus lowering the interest rate, leaving future output and income completely unaffected. Given this monetary policy, anticipations of future output are unchanged, and the interest rate is expected to decrease.

Putting everything together, we can conclude that if these conditions are satisfied—a lower expected future interest rate, unchanged expectations of future output, and little increase in current taxes—the deficit-reduction package may well lead to a net increase in spending and thus an increase in output today.

The preceding paragraph should give you a good sense of the uncertainties associated with policy. It *may* indeed be the case that a cleverly designed deficit-reduction package does not lead to a contraction in activity. But all the "suppose" assumptions we made along the way can go wrong. For example, backloading tax increases implies that painful measures are left to later. Thus, deferring the increase in taxes may well decrease the program's **credibility**—the probability that the government will actually take these painful steps in the future. Indeed, financial markets may conclude that the missed opportunity makes it more rather than less likely that deficits will remain high, and this may lead to an increase, not a decrease, in the expected future interest rate. You can surely think of other scary possibilities, and so can markets.

10-4 ON TO THE OPEN ECONOMY

We have come to the end of the first major extension of our basic *IS-LM* model, the incorporation of expectations. We now turn to the second one: recognizing that economies are open and interdependent. This topic will occupy the next four chapters.

In Chapter 3, we warned you that policy makers would never again have it so good. We have seen how dynamics and expectations can make their job more difficult. Once again, their life is about to get harder.

SUMMARY

■ Spending in the goods market depends on both current and expected future output, and on the current and expected future real interest rate. Thus, changes in expected future output, or in the expected future real interest rate, lead to changes in spending today. Put another way, the effect of any policy on spending and output depends on whether and how policy affects expectations of future output and of the future real interest rate.

■ Changes in the money supply affect the short-term nominal interest rate. However, activity depends on current and expected future *real* interest rates. Thus, the effect of monetary policy on activity depends crucially on whether and how changes in the short-term nominal interest rate lead to changes in the current and expected real interest rates.

■ The assumption of rational expectations is the assumption that people, firms, and participants in financial markets form expectations of the future by assessing the course of future expected policy and then working out the implications for future output, interest rates, and so on. While it is clear that most people do not go through this exercise themselves, we can think of them as doing so indirectly by watching TV and reading newspapers, which themselves rely on the forecasts of public and private forecasters.

Although there may be cases where people, firms, or financial investors do not have rational expectations, the assumption seems to be the best benchmark to evaluate the potential effects of alternative policies. Designing a policy on the assumption that people will make systematic mistakes in responding to it is surely unwise.

■ When account is taken of its effect on expectations, a Budget-deficit reduction plan may lead to an increase rather than a decrease in output. The more the central bank is expected to offset the deficit reduction's adverse effect on spending, the more likely this is to happen.

KEY TERMS

■ aggregate private spending, or private spending, 184
■ rational expectations, 189
■ animal spirits, 190

■ adaptive expectations, 190
■ credibility, 193

QUESTIONS AND PROBLEMS

1. For each of the following, determine whether the IS curve, the LM curve, both, or neither will shift. In each case, assume that expected current and future inflation are equal to zero, and that no other exogenous variables are changing.
   a. An increase in expected future taxes
   b. An increase in the expected future interest rate
   c. An increase in the current interest rate
   d. An increase in current output
   e. An increase in the current money supply

2. "If a credible, phased-in deficit reduction is announced, and if the central bank is expected to keep the same interest rate in the future as was expected before the announcement, the IS curve must shift leftward and both output and the interest rate must fall." (By "phased in," we mean a program that is implemented piece by piece over the course of several years.)

   This statement seems at first glance to contradict the ambiguous effect of deficit reduction on the IS curve discussed in the text (Section 10-3), but it does not. What is the crucial difference between this statement and the discussion in the text? How can we know for certain that the statement is correct? [Hint: Assume that the statement is not correct—that the IS curve shifts rightward—and obtain a contradiction.]

3. A new prime minister—who promised during the campaign that she would decrease future income taxes—has just been elected. Assume that people trust that the new prime minister will keep her promise. Using the IS-LM model with only two periods (current and future) and zero expected current and future inflation, determine the impact on current output, the current interest rate, and current private spending, under each of the following assumptions:
   a. The central bank does nothing.
   b. The central bank will act to prevent any change in current and future output.
   c. The central bank will act to prevent any change in the current and future interest rate.

4. Does the rational-expectations assumption require everyone to be a brilliant economist? Why or why not? In what ways could your own expectations conform to this assumption, even though you have not yet completed your intermediate macroeconomics course?

5. What are your own predictions for the growth rate of output and the real interest rate in the coming year? (State them, even if you have to guess.) On what information, if any, are you basing your predictions? To what extent are your predictions "adaptive" or "rational"?

this Chapter:

FURTHER READINGS

#1)– Table 10-1 mem.
2) Steep IS → explain
3) Equations for aggr. private spending.

In "Can Severe Fiscal Contractions Be Expansionary? Tales of Two Small European Countries," *NBER Macroeconomics Annual*, 1990, 75–110, Francesco Giavazzi and Marco Pagano argue that large decreases in budget deficits in Denmark and Ireland in the 1980s were associated with increases, not decreases, in activity.

① EXAMPLE #1 : *Where expansionary monetary policy shifts IS curve, because of changing expectations; fin. markets expect $y^e ↑ r^e ↓$;*

② *Deficit reduction plan: factors affected T T^e y^e r^e*
   Role of Monetary policy to offset IS shift
   — effects on y^e r^e

   — If tax pkge is backloaded: perhaps few effects on output.

OPENNESS IN GOODS AND FINANCIAL MARKETS

We have assumed so far that the economy did not interact with the rest of the world. It is time to relax this assumption. Nearly all economies in the world are open and are very much affected by what happens in the rest of the world.

Openness covers three distinct notions:

(1) **Openness in goods markets:** the opportunity for consumers and firms to choose between domestic and foreign goods. In no country is this choice completely free of restrictions: Even the countries most committed to free trade have tariffs and quotas on at least some foreign goods. (**Tariffs** are taxes on imported goods; **quotas** are restrictions on the quantities of goods that can be imported.) At the same time, in most countries average tariffs are low and getting lower.

(2) **Openness in financial markets:** the opportunity for financial investors to choose between domestic and foreign financial assets. Until recently even some of the richest countries, such as France, Italy, and Japan, had **capital controls,** tight restrictions on the foreign assets their domestic residents could hold as well as on the domestic assets foreigners could hold. These restrictions are rapidly disappearing. As a result, world financial markets are becoming more and more closely integrated.

(3) Openness in factor markets: the opportunity for firms to choose where to locate production, and for workers to choose where to work and whether or not to migrate. Here again trends are clear. Firms are thinking harder about where to locate their new plants. Multinational companies operate plants in many countries and move their operations around the world to take advantage of low costs. Much of the debate about the **North American Free Trade Agreement (NAFTA)** signed in 1993 by the United States, Canada, and Mexico centred on its implications for the relocation of U.S. and Canadian firms to Mexico. And immigration from low-wage countries to higher-wage countries is a hot political issue in all rich countries.

This and the next three chapters focus on the implications of openness in both goods and financial markets. We shall leave aside for the moment the implications of openness in factor markets. This is because, to deal with those, we need to look at the supply side of the economy in more detail than we have done so far. For the moment, we shall keep our assumption—our last major oversimplifying assumption—that firms are willing to supply whatever is demanded at the existing price level.

[handwritten margin notes: direct foreign investment]

[handwritten margin notes:
In this Chapter : Intro
① Nominal xd rate $\boxed{E}$ eg: $\frac{1}{.7} = 1.428$ U.S dollars per cdn $ dollar.
real exchange rate
$\boxed{\varepsilon} = \dfrac{EP^*}{P}$]

11-1 OPENNESS IN GOODS MARKETS

Figure 11-1 plots the evolution of Canadian exports and imports as ratios of GDP since 1926. "Canadian exports" means exports *from* Canada; "Canadian imports" means imports *to* Canada. What is striking is how these ratios have increased in recent years. Exports and imports, which were equal to about 20% of GDP as recently as the 1960s, now stand at almost 40% of GDP. This means that Canada trades substantially more with the rest of the world than it did just 30 years ago.

A closer look at Figure 11-1 reveals two other interesting features as well. The first is the sharp decline in both exports and imports with the onset of the Great Depression. In the early 1930s, countries around the world responded to the decline in economic activity by trying to protect domestic industries. They sharply increased tariffs in order to keep out foreign goods. The result was a sharp reduction in world trade and a further decline in economic activity. Canadian exports recovered more quickly than imports. Throughout most of the Great Depression and World War II, Canada ran a large trade surplus with the rest of the world. (Recall from Chapter 3 that the trade balance is equal to the difference between exports and imports. A positive trade balance is a *trade surplus*; a negative trade balance is a *trade deficit*.)

[handwritten margin notes:
real appreciation of domestic currency $\varepsilon \downarrow$
real depn of domestic currency $\varepsilon \uparrow$
② Can relate it all to McCallum's Cube article.
③ p 210 choie between domestic & foreign assets
$i = i^* + \dfrac{E^e_{t+1} - E_t}{E_t}$]

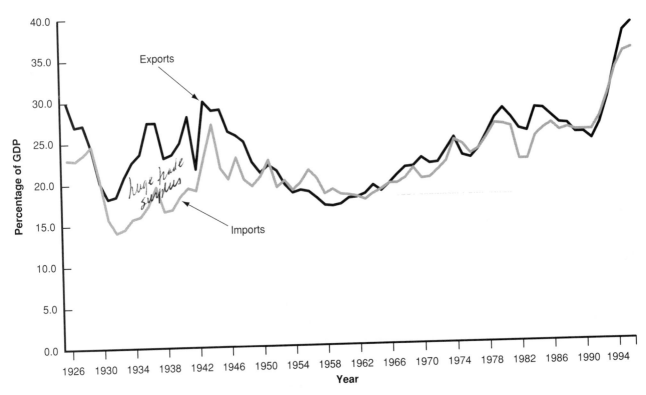

FIGURE 11-1

Canadian Exports and Imports as Ratios of GDP, 1926–1996

Canadian exports and imports have grown sharply with NAFTA.

Source: Statistics Canada, CANSIM Series D11011, D11030, and D11033.

Since World War II, imports and exports have moved more or less in tandem. On average, Canada has run a small trade surplus.

The second feature is that a large role for foreign trade is nothing new for the Canadian economy. Despite all the talk about increasing globalization and international competitiveness, it was 1994 before the share of exports to GDP exceeded the level in 1926. Historically, the two decades following World War II look like an aberration.

As large as 40% of GDP may sound to you, the volume of trade does not adequately capture the openness of the Canadian economy. Many sectors can be exposed to foreign competition without the effects of this competition showing up in increased imports. By being competitive and keeping their prices low enough, these sectors can retain their domestic market share and keep imports out. This suggests that a better index of openness than export or import ratios may be the proportion of aggregate output composed of **tradeable goods**—goods that compete with foreign goods in either domestic or foreign markets. In Canada today, tradeable goods may account for as much as 70% of GDP.

It remains true that, with exports around 40% of GDP, Canada has one of the larger ratios of exports to GDP among the rich countries of the world. Table 11-1 gives ratios for a number of OECD countries. The table shows that, with ratios around 10%, the United States and Japan are at the low end of the range of export

TABLE 11-1

RATIOS OF EXPORTS TO GDP FOR SELECTED OECD COUNTRIES, 1996

COUNTRY	EXPORT RATIO (%)	COUNTRY	EXPORT RATIO (%)
Japan	9.4	Canada	38.4
United States	11.3	Switzerland	36.3
Germany	23.3	Belgium	72.6
United Kingdom	29.2	Luxembourg	94.4

The number for Luxembourg is for 1994.

Source: International Financial Statistics, IMF, December 1997.

[handwritten margin note: depends on geography (distance from other markets or proximity to U.S. market) & size. (smaller economies are more dependent on exports/imports)]

ratios. The large European countries, such as Germany and the United Kingdom, have ratios that are comparable to Canada's. And the evidence from the smaller European countries is even more striking. Export ratios range from 36% in Switzerland, to 72% in Belgium, to 94% in Luxembourg! (Luxembourg's 94% ratio of exports to GDP raises an odd possibility. Could a country have exports larger than its GDP and thus an export ratio greater than 1? The answer is yes, and is developed in the Focus box entitled "Can Exports Exceed GDP?")

Do these numbers indicate that the United States has more trade barriers than, say, the United Kingdom or Luxembourg? No. The main factors behind these differences are <u>geography and size.</u> <u>Distance from other markets</u> explains a good part of the

■ CAN EXPORTS EXCEED GDP?

[handwritten note: Singapore: Exports/GDP ratio 140% in 1996.]

Can a country have exports larger than its GDP, and thus an export ratio greater than 1?

At first, the answer would seem to be that countries cannot export more than they produce, so that the export ratio must be less than 1. But this answer is incorrect. The trick is to realize that exports and imports may be exports and imports of <u>intermediate</u> <u>goods.</u>

An example will help here. Take a country that imports intermediate goods for $1 billion. Suppose that it transforms them into final goods using only labour. Say that total wages are equal to $200 million and there are no profits. The value of final goods is thus equal to $1200 million. Assume that $1 billion worth of final goods is exported and the rest is consumed in the country.

Exports and imports are therefore both equal to $1 billion. What is GDP in this economy? Remember that GDP is value added in the economy (see Chapter 2); it is thus equal to $200 million, so that the ratio of exports to GDP is equal to 5.

So, exports can exceed GDP. This is indeed the case for many small countries organized around a harbour and import-export activities, where exports and imports are indeed equal to many times GDP. This is even the case for small countries where manufacturing plays an important role. <u>Singapore is such a country. In 1996, its ratio of exports to GDP was 140%.</u>

Much of the recent increase in the ratios of exports and imports to GDP in Canada is due to a sharp increase in trade of intermediate goods, particularly in the auto sector.

In the text we shall typically think of exports and imports as exports and imports of final goods. But this is just for simplicity. Remember that exports and imports, as they are measured, refer to exports and imports of all goods and services, from raw materials to intermediate products to final goods.

low Japanese ratio, and proximity to the U.S. market helps explain the high Canadian ratio. Also, the smaller the country, the more it has to specialize in only a few products and thus have both high imports and high exports: Belgium's GDP is about 5% of U.S. GDP, and Luxembourg's GDP is less than 0.3% of U.S. GDP. It is clear that neither Belgium nor Luxembourg can afford to produce the range of goods produced by, say, the United States.

How does openness in goods markets force us to rethink the way we look at equilibrium in the goods market? When thinking about consumers' decisions in the goods market, we have thus far focussed on their decision to save or to consume. But when goods markets are open, domestic consumers face a second major decision: whether to buy domestic goods or to buy foreign goods. Other domestic buyers—firms, the government—and foreign buyers also face this decision. If they decide to buy more domestic goods, the demand for domestic goods increases, and so does domestic output. If they decide to buy more foreign goods, then it is foreign output that increases.

Central to consumers' and firms' decisions is the price of foreign goods in terms of domestic goods. We call this relative price the **real exchange rate.** The real exchange rate is not directly observable, and you will not find it by flipping through the pages of newspapers. What you will find there are *nominal exchange rates,* the relative prices of currencies. Let's start by looking at these, then see how we can use them to construct real exchange rates.

NOMINAL EXCHANGE RATES

Nominal exchange rates between currencies are quoted in two ways: (1) the number of units of foreign currency you can get for one unit of domestic currency or (2) the number of units of domestic currency you can get for one unit of foreign currency. In December 1997, for example, the nominal exchange rate between the Canadian dollar (C$) and the U.S. dollar (US$) was quoted as either US $0.70 for one Canadian dollar, or, equivalently, C$1.43 for one U.S. dollar.[1]

In this book, we shall define the **nominal exchange rate** as *the number of units of domestic currency you can get for one unit of foreign currency,* or equivalently, as *the price of foreign currency in terms of domestic currency,* and we shall denote it by E. For example, when looking at the exchange rate between Canada and the United States from the viewpoint of Canada (so that the Canadian dollar is the domestic currency), E will denote the number of Canadian dollars one can get for US $1—thus, as of December 1997, 1.43. We shall write it for short as the C$/US$ exchange rate.[2] To convert C$ into US$, simply divide by E. To convert US$ into C$, multiply by E.

Exchange rates between foreign currencies and the dollar change every day, indeed every minute during the day. These changes are called *nominal appreciations* or *nominal depreciations*—appreciations or depreciations for short. An **appreciation** of the domestic currency is an increase in the price of the domestic currency in terms of a foreign currency. Given our definition of the exchange rate as the price of

---

[1] *Because Canada and the U.S. (along with a dozen other countries around the world) call their currency the "dollar," it's easy to get confused. We will always refer to the American currency as the U.S. dollar (US$).*

[2] *A warning: The convention of defining exchange rates as the price of foreign currency in terms of domestic currency is quite common. However, some economists use the alternative definition, defining exchange rates as the price of domestic currency in terms of foreign currency (say, dollars per pound in the United Kingdom).*

Let's say Cdn $ is .70¢

$\frac{\$1\ US}{.70C} = 1.428 = E$

1 × 1.428 = .70 ¢

$\frac{1}{E}$ → Converts from U.S to Cdn.

If Cdn appreciates (↑ to .85)

$\frac{1}{.85} = 1.176 = E$ (lower xD rate)

2$US per 1 lb

the foreign currency in terms of domestic currency, an appreciation corresponds to a *decrease* in the exchange rate, E.

Similarly, a **depreciation** of the domestic currency means that the Canadian dollar is going down in terms of a foreign currency, and thus corresponds to an *increase* in E.[3]

That an appreciation corresponds to a decrease in the exchange rate, and a depreciation to an increase, will almost surely be confusing to you at first—it confuses many professional economists—but it will eventually become second nature as your understanding of open-economy macroeconomics deepens. Until then, you may find it useful to consult Figure 11-2, which summarizes the terminology.

With these preliminaries out of the way, Figure 11-3 plots the C$/US$ exchange rate since 1960. It has three important features:

1. *The trend increase in the exchange rate,* or the trend depreciation of the Canadian dollar vis-à-vis the U.S. dollar which has occurred over the last 40 years. Although there was no trend for the first 15 years, the C$ has depreciated noticeably since then. It was virtually on a par with the US$ in January 1977, but the exchange rate had risen to about 1.42 in late 1997.

2. For most of the 1960s, we had a *fixed exchange rate.* Although most of the world had been on fixed exchange rates since the end of World War II, the Canadian dollar was allowed to float during the 1950s and early 1960s. In 1962, Prime Minister John Diefenbaker's government decided to fix the Canadian dollar in a narrow band of 92.5 US cents (this value became known as the "Diefenbuck"). In our notation, this corresponds to E = 1.081. Canada abandoned fixed exchange rates in early 1970 and within a few years so had the most of the world.

Nominal exchange rate (E)   Price of US$ in Canadian dollars
equivalently:
Number of C$ per US$
(C$/US$)

Appreciation of the dollar: Price of US$ in Canadian dollars decreases
Value of dollar increases   equivalently:
Number of C$ per US$ decreases
$E\downarrow$

Depreciation of the dollar: Price of US$ in Canadian dollars increases
Value of dollar decreases   equivalently:
Number of C$ per US$ increases
$E\uparrow$

FIGURE 11-2
The Nominal Exchange Rate, Appreciation, and Depreciation: Canada and the United States (from the Viewpoint of Canada)

---

[3]*You may have encountered two other words for changes in exchange rates: "revaluations" and "devaluations." These terms are used when countries operate under **fixed exchange rates**—a system in which two or more countries maintain a fixed exchange rate between their currencies. Under such a system, decreases in the exchange rate, which are by definition infrequent events, are called **revaluations** (rather than appreciations). Increases in the exchange rate are called **devaluations** (rather than depreciations). We discuss fixed exchange rates in detail in Chapter 13.*

FIGURE 11-3

The C$/US$ Nominal Exchange Rate, 1960–1997

Source: Statistics Canada, CANSIM Series B3400.

[handwritten margin note: — the negative irrational exuberance that is from time to time characteristic of the market.]

3. *There are large swings in the exchange rate.* In the nine years from January 1977 to January 1986, the exchange rate increased approximately 50%. About half of this increase disappeared in the six years that followed, only to be restored by 1994. Many economists expected the Canadian dollar to appreciate during 1997, but in mid-December its value was close to an all-time low.

Figure 11-3 tells us only about swings in the relative price of the two currencies. However, to American tourists thinking of visiting Canada, the question is not only how many dollars they can get for 1 US$, but also how many goods their dollar will buy. It does them little good to get more Canadian dollars per US$ if the dollar prices of goods in Canada have increased proportionately. In the same way, the Canadian firm thinking of exporting to the United States needs to know not only the nominal exchange rate but also the price in US$ of American products with which it will have to compete. This takes us closer to where we want to go, to the construction of real exchange rates.

REAL EXCHANGE RATES

How do we construct the real exchange rate between Canada and the United States?

Suppose that the United States produced only one good, Cadillac Seville cars (this is one of those completely counterfactual "Suppose" statements, but we shall become more realistic below), and that Canada also produced only one good, say Chrysler Town & Country minivans.

Constructing the real exchange rate, the price of American goods in terms of Canadian goods, would be straightforward: The price of a Cadillac in the United States is US $45 000. The first step would be to convert this price in US$ to a price in C$. A US$ is worth C$1.43, so the price of a Cadillac in dollars is $45\,000 \times 1.43 =$ C$64 350. The second step would be to compute the ratio of the price of the Cadillac in Canadian dollars to the price of the minivan in Canadian dollars. The price of a minivan in Canada is C$30 000. Thus, the price of a Cadillac in terms of minivans—that is, the real exchange rate between Canada and the United States—would be C$64 350/C$30 000 = 2.145.

But the United States and Canada produce more than Cadillacs and minivans, and we want to construct a real exchange rate that reflects the relative price of *all* the goods produced in the United States in terms of *all* the goods produced in Canada. The computation in the preceding paragraph gives us a hint about how to proceed. Rather than use the US$ price of a Cadillac and the C$ price of a minivan, we must use a US$ price index for all goods produced in the United States and a C$ index for all goods produced in Canada. This is exactly what the GDP deflators we introduced in Chapter 2 do: They are by definition price indexes for the set of final goods and services produced in an economy.

Thus, let P be the GDP deflator for Canada, P^* be the GDP deflator for the United States (as a rule, we shall denote foreign variables by a star), and E be the nominal exchange rate between C$ and US$. Figure 11-4 shows the steps in the construction of the real exchange rate. The price of American goods in US$ is P^*. Multiplying it by the exchange rate, E, gives us the price of American goods in Canadian dollars, EP^*. The price of Canadian goods in C$ is P. The real exchange rate, the price of U.S. goods in terms of Canadian goods, which we shall call ε (the Greek lowercase epsilon), is thus given by

$$\varepsilon = \frac{EP^*}{P} \tag{11.1}$$

Note that, unlike the price of Cadillacs in terms of minivans, the real exchange rate is an index number and thus does not have a natural level. This is because the GDP deflators used in the construction of the real exchange rate are indexes and thus have no natural level; as we saw in Chapter 2, they are equal to 1 in whatever year is chosen as the base year. But while the *level* of the real exchange rate is arbitrary, its *movements* are not. They tell us whether foreign goods are becoming relatively more or less expensive than domestic goods.

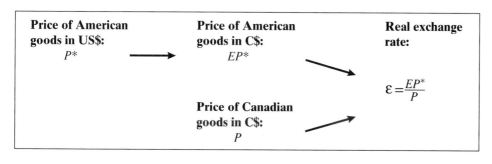

FIGURE 11-4
The Construction of the Real Exchange Rate

FIGURE 11-5

The Real Exchange
Rate, Real Appreciation,
and Real Depreciation:
Canada and the United
States (from the View-
point of Canada)

Real exchange rate(ε):	Price of U.S. goods in terms of Canadian goods
Real appreciation: Canadian goods are relatively more expensive	Price of U.S. goods in terms of Canadian goods decreases: $\varepsilon\downarrow$
Real depreciation: Canadian goods are relatively less expensive	Price of U.S. goods in terms of Canadian goods increases: $\varepsilon\uparrow$

An increase in the relative price of domestic goods in terms of foreign goods is called a **real appreciation;** a decrease is called a **real depreciation.** The word *real,* as opposed to *nominal,* indicates that we are now referring to changes in the relative price of *goods* rather than in the relative price of currencies. Given our definition of the real exchange rate as the price of foreign goods in terms of domestic goods, a real appreciation corresponds to a *decrease* in the real exchange rate, ε. Similarly, a real depreciation corresponds to an *increase* in ε. These relations are summarized in Figure 11-5, which corresponds to Figure 11-2, but this time for the real exchange rate.

Figure 11-6 plots the evolution of the real exchange rate between Canada and the United States from 1970 to 1996. For convenience, it also reproduces the evolution of the nominal exchange rate from Figure 11-3. The GDP deflators have been chosen to be equal to 1 in 1986, so that in that year the nominal and real exchange rates are equal by construction.

Figure 11-6 has two major features:

(1) The real depreciation since 1970 has been smaller than the nominal depreciation. While the US$ has increased in value by 0.9% a year on average with respect to the Canadian dollar, the price of U.S. goods in terms of Canadian goods has increased by only 0.6% a year on average.

To understand where the difference comes from, note that by definition the real exchange rate is equal to the nominal exchange rate (E) times the ratio of the American price level to the Canadian price level, $P*/P$. Since 1970, inflation in the U.S. has been lower than in Canada by 0.3% a year on average, so that the American price level has increased more slowly than the Canadian price level: $P*/P$ has decreased. Thus, while the nominal exchange rate has increased by 0.9% a year on average, the price ratio has decreased by 0.3% a year on average, leading to an increase in the real exchange rate of only 0.6% a year.

To make this explanation more intuitive, think of American tourists who last visited Canada in 1970. In preparing for their trip this time, they will find that a US$ buys more Canadian dollars (the nominal depreciation of the Canadian dollar), but that the cost of things in Canada has increased more than in the United States (due to the differential in inflation). Thus the trip is cheaper now than it was in 1970, but by less than the large nominal dollar depreciation might suggest.

(2) The large swing in the nominal exchange rate from 1986 to 1992 that we saw in Figure 11-3 also shows up in the real exchange rate. This is not very surprising. As inflation rates are not very different in the U.S. and Canada, the movements in the price ratio $P*/P$ are slow. Thus, from year to year, or even over a few years, move-

FIGURE 11-6

Real and Nominal Exchange Rates between Canada and the United States, 1970–1996

The real and nominal exchange rates have moved very much together since 1970.

Sources: Statistics Canada, CANSIM Series B3400 and D20556, and FRED[4]

Handwritten notes on figure:
① real depn has been smaller than nominal depn (Cdn price level has been higher)
②

ments in the real exchange rate ε are driven mostly by movements in the nominal exchange rate *E*.

We have one last step to take. We have so far concentrated on the nominal and real exchange rates between Canada and the United States. While the U.S. is our most important partner, Canada also trades with many other countries. Table 11-2 gives the geographic composition of Canadian trade for both exports and imports. The numbers refer only to **merchandise trade,** exports and imports of goods; they do not include exports and imports of services, such as travel services and tourism, for which this decomposition is not available.

The United States accounts for the bulk of Canadian merchandise trade: It absorbs over 80% of our exports and it is the source of more than two-thirds of our imports. In 1996, Canada ran a large merchandise trade surplus with the United States, and a small deficit with the rest of the world. Although Japan's large trade surplus with the rest of the world is often the source of major tension, note that in 1996 it had run a small deficit in its merchandise trade with Canada. Historically, Canada has run a trade surplus that is the combination of a large surplus in merchandise trade combined with a more modest deficit in services.

Handwritten note in margin: 80% of our exports: merchandise trade to U.S

---

[4]*For a description of FRED, see Appendix 1.*

TABLE 11-2

THE COUNTRY COMPOSITION OF CANADIAN MERCHANDISE TRADE, 1996

	EXPORTS TO		IMPORTS FROM	
Country	$ Billions	Percentage	$ Billions	Percentage
United States	223.5	81.0	157.5	67.6
Unitd Kingdom	4.0	1.4	5.9	2.5
Other E.E.C.	11.7	4.2	16.8	7.2
Japan	11.1	4.0	10.4	4.5
Other OECD	4.5	1.6	15.8	6.8
Other Countries	21.0	7.6	26.6	11.4
Total	275.9	100.0	233.1	100.0

Source: Statistics Canada, CANSIM Matrices 3618 and 3619.

How do we go from **bilateral exchange rates** (whether nominal or real), such as the exchange rate between Canada and the United States (the "bi" in bilateral means two, thus referring to the exchange rate between a pair of countries) to **multilateral exchange rates?** The answer is straightforward. If we want to measure the average Canadian exchange rate relative to all of its trading partners, we should use the weights of the Canadian share of trade with each country. Using export shares, we can construct an "export" exchange rate, and using import shares, we can construct an "import" exchange rate. To simplify life, economists typically use an exchange rate that takes an average of export and import shares as the weight. Other names for the multilateral exchange rate (whether real or nominal) are the **trade-weighted exchange rate**, or the **effective exchange rate**.

Figure 11-7 shows the paths of both the C$/US$ nominal exchange rate and the nominal multilateral exchange rate, from 1971 to 1996. The multilateral exchange rate is computed as an index of the Canadian dollar against the G10 currencies (the data on the index begin in 1971). The G10 index is scaled to equal 1 on average in 1981. Other than this difference in levels, the two series are very similar. This is exactly what we should expect on the basis of Table 11-2: for the Canadian economy, it is an excellent approximation to act as if the United States *is* the rest of the world. Because inflation rates in the advanced countries move closely together, the nominal and real multilateral exchange rates are extremely similar, so we do not report the latter.

11-2   OPENNESS IN FINANCIAL MARKETS

Openness in financial markets allows financial investors to hold both domestic and foreign assets, thus providing the opportunity of further diversifying their portfolios, and to speculate on movements in foreign versus domestic interest rates, exchange rates, and so on. And diversify and speculate they do. Given that buying or selling foreign assets implies, as part of the operation, buying or selling foreign currency (sometimes called **foreign exchange**), the size of transactions in foreign-exchange markets gives a sense of the importance of international financial

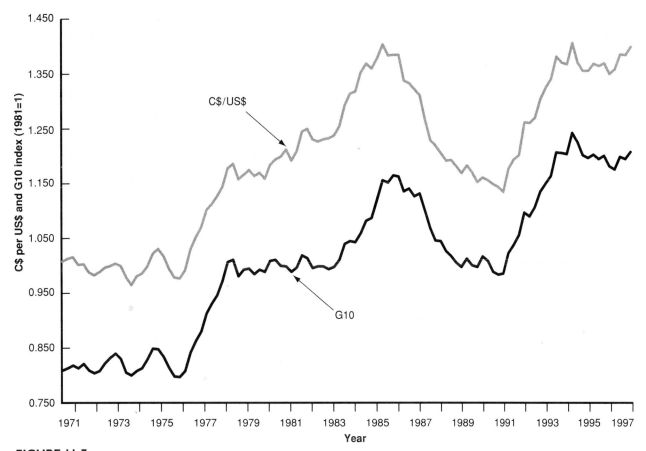

FIGURE 11-7

The C$/US$ Exchange Rate and the Canadian Multilateral Exchange Rate, 1971–1997

Because it dominates our trade, the C$/US$ exchange rate is an excellent proxy for the Canadian multilateral exchange rate.

Source: Statistics Canada, CANSIM Series B3400 and B3418.

transactions. In 1996 the *daily* volume of foreign-exchange transactions in the world was about $2 trillion.

To get a sense of the magnitude of these numbers, recall from Chapter 3 that the value of all goods and services produced by the Canadian economy in 1996 was roughly $800 billion. This number is for the whole year and thus corresponds to about $2 billion a day. Thus, transactions in foreign-currency markets each day equal about 1000 times the value of Canada's daily production or about 20 times the entire planet's production of goods and services. Most of the transactions are thus associated not with trade, but with purchases and sales of financial assets. The volume of transactions in foreign-exchange markets is not only high but also rapidly increasing. The volume of foreign-exchange transactions in New York is now more than 16 times what it was in 1980. Again, this activity reflects mostly an increase in financial transactions rather than an increase in trade over the last 15 years.

For a country as a whole, openness in financial markets has another important implication. It allows the country to run trade surpluses and trade deficits. Recall that a country running a trade deficit is buying more from the rest of the world than it is selling to rest of the world. Thus, it must borrow the difference. It does so by making it attractive for foreign financial investors to increase their holdings of domestic assets, in effect to lend to the country. Let's now look at the relation between trade and financial flows more closely.

THE BALANCE OF PAYMENTS

A country's transactions with the rest of the world are summarized by a set of accounts called the **balance of payments.** Table 11-3 presents the Canadian balance of payments for 1996. Note that it has two parts separated by a line. For this reason, transactions are referred to as either transactions **above the line** or transactions **below the line.**

The current account. Let's first look at the transactions above the line, all of which record payments to and from the rest of the world. These are called **current account** transactions.

The first two lines record exports and imports of goods and services. Exports lead to payments from the rest of the world, imports to payments to the rest of the world. In 1996, exports exceeded imports, leading to a Canadian trade surplus of $25.2 billion. (Note that the numbers for exports and imports are different from those in Table 11-2; this is because the numbers here include both goods *and* services.)

Exports and imports are not the only sources of payments to and from the rest of the world. Canadian residents receive **investment income** on their holdings of foreign assets, and foreign residents receive investment income on their holdings of Canadian assets. In 1996, investment income received from the rest of the world was $17.8 billion and investment income paid to foreigners was $45.8 billion, for a net balance of –$28 billion.

Finally, countries give and receive foreign aid, immigrants give and receive gifts and inheritances, and the Canadian government collects a withholding tax on

....
TABLE 11-3
THE CANADIAN BALANCE OF PAYMENTS, 1996

CURRENT ACCOUNT		
Exports	306.5	
Imports	281.4	
Trade balance (deficit = –) (1)		25.2
Investment income received	17.8	
Investment income paid	45.8	
Net investment income (2)		–28.0
Net transfers received (3)		1.2
Current account balance (deficit = –) (1) + (2) + (3)		–1.6
CAPITAL ACCOUNT		
Increase in foreign holdings of Canadian assets	35.8	
Increase in Canadian holdings of foreign assets	30.0	
Net capital inflows/capital account		
balance (deficit = –)		–5.8
Statistical discrepancy		7.4

Source: Statistics Canada, CANSIM Matrix 2333.

foreigners; the net value of these payments is recorded as **net transfers received.** These amounted in 1996 to $1.2 billion.

Adding all payments to and from the rest of the world, net payments were thus equal to $25.2 − $28 + $1.2 = −$1.6 billion. This total is the *current account balance.* In 1996, Canada ran a current account deficit of $1.6 billion, about 0.2% of its GDP. The year 1996 was exceptional, however. Historically, the current account deficit has been much larger.

The capital account.   The Canadian current account deficit in 1996 implied that Canada had to borrow $1.6 billion from the rest of the world—or, equivalently, that net foreign holdings of Canadian assets had to increase by $1.6 billion. The numbers below the line describe the way this result was achieved. Transactions below the line are called **capital account** transactions.

The increase in Canadian holdings of foreign assets in 1996 was $30 billion. But at the same time, the increase in foreign holdings of Canadian assets in 1996 was $35.8 billion. Thus, the capital account balance which measures the net increase in Canadian foreign indebtedness, also called **net capital flows** to Canada, was − $35.8 + $30 = − $5.8 billion.

Shouldn't the sum of net capital flows and the current account deficit be zero? The answer is: conceptually, yes, but in practice it is not. The numbers for current and capital account transactions come from different sources; while they should add to zero, they typically do not. In 1996, the number that needed to be added to get the sum of the balances to zero, the **statistical discrepancy,** was equal to $7.4 billion.[5]

Now that we have looked at the current account, we can return to an issue we touched on briefly in Chapter 2, the difference between GDP (the measure of output

---

[5]*Another statistical problem: It is clear that the sum of the current account deficits of all countries should be equal to zero. One country's deficit should show up as a surplus for the other countries taken as a whole. This is not the case. If we just add the published current account deficits of all the countries in the world, it would appear that the world is running a (measured) large current account deficit. Some economists jokingly speculate that the explanation is unrecorded trade with the Martians. Most others believe that mismeasurement is the explanation.*

GDP VERSUS GNP: THE EXAMPLE OF KUWAIT

GLOBAL MACRO

Should value added in an open economy be defined as the value added domestically (that is, within the country) or as the value added by domestically owned factors of production? These two need not be the same: Some domestic output may be produced by capital owned by foreigners, while some foreign output may be produced by capital owned by Canadian residents.

The answer is that either definition is fine, and economists use both. **Gross domestic product (GDP),** the measure we have used so far, corresponds to value added domestically. **Gross national product (GNP)** corresponds to the value added by domestically owned factors of production. GNP is thus equal to GDP plus net factor payments from the rest of the world. While GDP is now the measure most commonly mentioned, GNP was widely used until a few years ago, and you will still often encounter it in newspapers and academic publications.

For most countries, the difference between GNP and GDP is typically small, because factor payments to and from the rest of the world roughly cancel. For Canada in 1996, for example, GDP was $820 billion compared with $793 billion for GNP.

But there are a few exceptions. One is Kuwait.

When oil was discovered in Kuwait, Kuwait's government decided that a portion of oil revenues would be saved and invested abroad rather than spent, so as to provide future Kuwaiti generations with investment income when oil revenues came to an end. As a result, Kuwait accumulated large foreign assets and thus now receives substantial investment income from the rest of the world. Table I gives GDP, GNP, and net factor payments for Kuwait.

Note how much larger GNP is compared with GDP throughout the period. But note also how net factor payments have decreased since 1989. This is because Kuwait has had to pay its allies for part of the cost of the Gulf War and to pay for reconstruction after the war. As a result, Kuwait's foreign assets and factor payments decreased steadily until 1995.

TABLE 1

GDP, GNP, AND NET FACTOR PAYMENTS IN KUWAIT, 1989–1996

YEAR	GDP	GNP	NET FACTOR PAYMENTS
1989	7 143	9 562	2 419
1990	5 328	7 560	2 232
1991	3 131	4 699	1 538
1992	5 826	7 364	1 538
1993	7 231	8 318	1 083
1994	7 380	8 207	827
1995	7 942	9 337	1 395
1996	9 277	10 749	1 472

Source: IMF, *International Financial Statistics,* August 1997.

All numbers in millions of Kuwaiti dinars. 1 dinar = C\$4.55 (1996).

we have used so far) and GNP, another measure of aggregrate output. This is done in the Global Macro box entitled "GDP versus GNP: The Example of Kuwait."

THE CHOICE BETWEEN DOMESTIC AND FOREIGN ASSETS

depends on interest!

Openness in final markets implies that financial investors face the choice of holding domestic versus foreign assets.

It might appear that we have to think about at least *two* new decisions, the choice of holding domestic versus foreign *money,* and the choice of holding domestic versus foreign *interest-paying assets.* But remember why people hold money: to engage in transactions. For somebody who lives in Canada, whose transactions are thus mostly or fully in dollars, there is little point in holding foreign currency. It cannot be used for transactions, and, if the goal is to hold foreign assets, holding foreign currency is clearly less desirable than holding foreign bonds, which pay interest.[6] Thus, the only new choice we have to think about is that between domestic and foreign interest-paying assets. Let's think of them for the time being as domestic and foreign one-year bonds.

Unless hyperinflation!

Consider, for example, the choice between Canadian and U.S. one-year bonds. Suppose that you decide to hold Canadian bonds. Let i_t be the one-year Canadian nominal interest rate in year t. Then, as Figure 11-8 shows, for every dollar you put in Canadian bonds, you will get $(1 + i_t)$ dollars next year.

---

[6]*There are two qualifications to this statement. First, those involved in illegal activities often hold U.S. dollars, because these can be exchanged easily and cannot be traced. Also, in Canada many of us hold small amounts of U.S. currency because of the frequent trips we make down south. We shall ignore this aspect of reality here.*

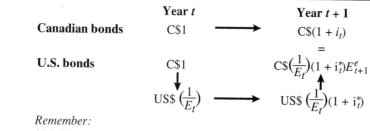

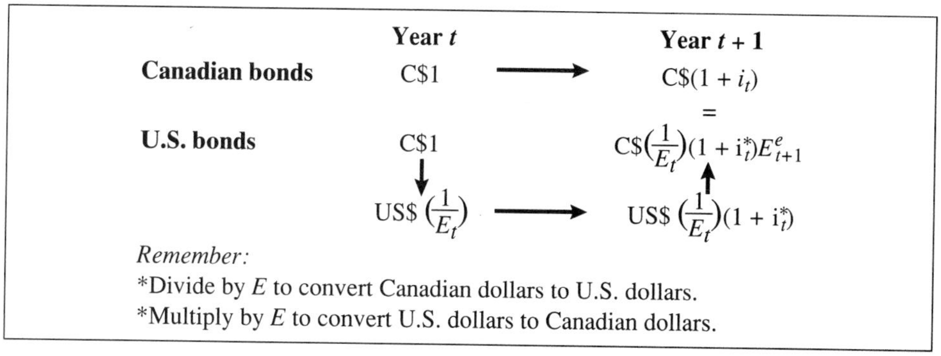

<figure>

	Year t		Year $t + 1$
Canadian bonds	C\$1	$\longrightarrow$	C\$$(1 + i_t)$
			$=$
U.S. bonds	C\$1		C\$$\left(\frac{1}{E_t}\right)(1 + i_t^*)E_{t+1}^e$
	US\$ $\left(\frac{1}{E_t}\right)$	$\longrightarrow$	US\$ $\left(\frac{1}{E_t}\right)(1 + i_t^*)$

Remember:
*Divide by E to convert Canadian dollars to U.S. dollars.
*Multiply by E to convert U.S. dollars to Canadian dollars.

</figure>

FIGURE 11-8
Expected Returns from Holding One-Year Canadian and U.S. Bonds

Suppose you decide instead to hold American bonds. To buy American bonds, you must first buy U.S. dollars. Let E_t be the nominal exchange rate between our dollar and the U.S. dollar. Thus for every dollar you get $(1/E_t)$ US\$. Let i_t^* denote the one-year nominal interest rate on American bonds (in US\$). When next year comes, you will have $(1/E_t)(1 + i_t^*)$ US\$. You will then have to convert your US\$ back into C\$. If you expect the nominal exchange rate next year to be E_{t+1}^e, you can expect to have $(1/E_t)(1 + i_t^*)E_{t+1}^e$ Canadian dollars next year for every Canadian dollar you invested. This set of steps is represented in the second part of Figure 11-8.

We shall look at this expression in more detail soon. But note already its basic implication. In assessing the attractiveness of American bonds, you cannot look just at the Canadian and U.S. interest rates; you must also assess what you think will happen to the C\$/US\$ exchange rate between this year and next.

Let's now make the same assumption we made in Chapter 9 when discussing the choice between short- and long-term bonds, or between bonds and stocks. Let's assume that you and other financial investors want to hold only the asset with the highest rate of return. In that case, if both Canadian and U.S. bonds are to be held, they must have the same expected rate of return, so that the following *arbitrage relation* must hold:

return from
holding Cdn Bonds

$$1 + i_t = \left(\frac{1}{E_t}\right)(1 + i_t^*)(E_{t+1}^e) \quad \text{Arbitrage relation} \quad (11.2)$$

return from US bonds

For rich countries of the world, this is a realistic approximation

Equation (11.2) is called the **uncovered interest parity relation,** or simply the **interest parity condition.**[7]

The assumption that financial investors will hold only the bonds with the highest expected rate of return is obviously too strong, for two reasons. First, it ignores transaction costs: Going in and out of American bonds requires three separate transactions, each with a transaction cost. Second, it ignores risk: The exchange rate a year from now is uncertain; thus, holding American bonds is more risky, for a Canadian investor, than holding Canadian bonds. But as a characterization of capital movements among the major world financial markets (New York, Frankfurt, London, Tokyo, and

---

[7] *Digging deeper. The word* uncovered *is to distinguish it from another relation called the* covered interest parity condition. *That condition is derived by looking at the following choice. The first option is to hold Canadian bonds for one year. The second is to buy US\$ today, buy one-year American bonds with the proceeds, and agree to sell the US\$ for C\$ a year ahead at a predetermined price, called the* forward exchange rate. *The rate of return to these two alternatives, which can both be realized at no risk today, must be the same. The covered interest parity condition is a riskless arbitrage condition.*

Toronto), it is not far off. Small changes in interest rates and rumours of impending appreciation or depreciation can lead to movements of billions of dollars within minutes. For the rich countries of the world, the arbitrage assumption in equation (11.2) is a good approximation of reality. Other countries whose capital markets are smaller and less developed, or that have various forms of capital controls, have more leeway in choosing their domestic interest rate than is implied by equation (11.2).

To get a better sense of what arbitrage implies, rewrite equation (11.2) as

Subtract $\frac{E_t}{E_t}$ from each side

$$1 + i_t = (1 + i_t^*)\left(1 + \frac{E_{t+1}^e - E_t}{E_t}\right) \tag{11.3}$$

This gives a relation between the domestic nominal interest rate, the foreign nominal interest rate, and the expected rate of depreciation. Remember that an increase in E is a depreciation, so that $(E_{t+1}^e - E_t)/E_t$ is the expected rate of depreciation of the domestic currency. (If the domestic currency is expected to appreciate, then this term is negative.) As long as interest rates or the expected rate of depreciation are not too large—say, below 20% a year—a good approximation to this equation is given by[8]

higher XR rate means that our currency has depreciated.

of domestic currency

$$i_t \approx i_t^* + \frac{E_{t+1}^e - E_t}{E_t}$$ *} expected dep^n of domestic currency.   if term is ⊖ then expected apprec* $\tag{11.4}$

This is the relation you must remember: Arbitrage implies that *the domestic interest rate must be (approximately) equal to the foreign interest rate plus the expected depreciation rate of the domestic currency.*

5.2 ≈ 5.5 – .3 expected appreciation of Cdn currency. (when you convert your $ back to Cdn $ it'll be worth less so you expect a higher U.S 'r;

Let's apply this equation to U.S. versus Canadian bonds. Suppose that the U.S. and Canadian one-year nominal interest rates are equal to 5.5% and 5.2% respectively (these were the values of these two interest rates in December 1997). Should you hold Canadian or U.S. bonds? It depends on whether you expect the Canadian dollar to appreciate vis-à-vis the US$ by more or less than 5.5% – 5.2% = 0.3% over the coming year. If you expect our dollar to appreciate by more than 0.3%, then despite the fact that the interest rate is lower in Canada than in the United States, investing in Canadian bonds is more attractive than investing in U.S. bonds. By holding U.S. bonds, you will get more US$ a year from now, but US$ will also be worth less in terms of C$ a year from now, making investing in Canadian bonds more attractive than investing in U.S. bonds. However, if you expect the C$ to appreciate by less than 0.3% or even to depreciate, the reverse holds and U.S. bonds are more attractive than Canadian bonds. Arbitrage tells us that financial markets must be expecting on average an appreciation of the C$ with respect to the US$ of about 0.3% over the coming year, and this is why financial investors are willing to hold Canadian bonds despite their lower interest rate. (Another example is provided in the Global Macro box entitled "Should You Buy Brazilian Bonds?")

This ex. is assuming there is just a change in future E^e_t (expected dep^n) – nothing is happening to E_t yet ✓ |

The arbitrage relation between interest rates and exchange rates in equation (11.4) will play a central role in the following chapters. It suggests that, unless participants in financial and foreign exchange markets expect large depreciations or appreciations, domestic and foreign interest rates are likely to move very much together. Take the extreme case of two countries that commit to maintaining their bilateral exchange rate at a fixed value. If markets have faith in this commitment, they will expect the exchange rate to remain constant, and the expected depreciation will

---

[8]*This follows from Proposition 3 in Appendix 3.*

Put yourself back in September 1993. (We choose this date because the very high interest rate in Brazil at the time makes the point we want to make very dramatically.) Brazilian bonds are paying a monthly interest rate of 36.9%. This seems very attractive compared to an *annual* rate of 5.5% on Canadian. bonds. Shouldn't you buy Brazilian bonds?

The discussion in the text tells you that, to decide, you need one more crucial element, the expected rate of appreciation of the dollar vis-à-vis the cruzeiro (the name of the Brazilian currency at the time; the currency is now called the real). You need this information because (as Figure 11-8 makes clear) the return in dollars from investing in Brazilian bonds for a month is

$$(1 + i_t^*)\frac{E_{t+1}^e}{E_t} = (1.369)\frac{E_{t+1}^e}{E_t}$$

What rate of cruzeiro depreciation should you expect over the coming month? Assume that the rate of depreciation next month will be equal to the rate of depreciation last month. You know that 100 000 cruzeiros, worth $1.29 at the end of July 1993, were worth only $0.98 at the end of August 1993. Thus, if depreciation continues at the same rate, the return from investing in Brazilian bonds for a month is

$$(1 + i_t^*)\frac{E_{t+1}^e}{E_t} = (1.369)\left(\frac{0.98}{1.29}\right) = 1.04$$

If their currency is depr. then your domest. currency is appreciating (expected) (apprse)

The expected rate of return in dollars from holding Brazilian bonds is thus only (1.04 − 1) = 4% per month, not the 36.9% per month that looked so attractive. Note that 4% per month is still much higher than the monthly rate of return on Canadian bonds. But think of the risk and the transaction costs—all the elements we ignored when we wrote the arbitrage condition. When these are taken into account, you may well decide to keep your funds out of Brazil.

---

be equal to zero. In that case, the arbitrage condition implies that interest rates in the two countries will have to move together exactly. Most of the time, as we shall see, governments do not make such absolute commitments to maintain the exchange rate, but they often do try to avoid large movements in the rate. This in turn puts sharp limits on how much they can allow their interest rate to deviate from interest rates elsewhere in the world.

How much do nominal interest rates actually move together between major countries? Figure 11-9 plots nominal interest rates in the United States and Canada for each year since 1970. The impression is indeed one of very closely related but not identical movements. Interest rates were high (by historical standards) in both countries in the mid-1970s, and very high in the early 1980s. They have been generally decreasing since then in both countries. At the same time, differences between the two are sometimes large. In 1990, the Canadian interest rate was nearly 5 percentage points above the U.S. interest rate. In 1997, by contrast, the Canadian interest rate was nearly 2 percentage points below the U.S. interest rate.

eg: 1990

$$i = i^* + \frac{E_{t+1}^e - E_t}{E_t}$$

14% = 9% + 5%

expected depr of domestic currency.

11-3 CONCLUSIONS AND A LOOK AHEAD

We have now set the stage for the open economy. Openness in goods markets allows for a choice between domestic and foreign goods. This choice depends primarily on the *real exchange rate*, the relative price of foreign goods in terms of domestic

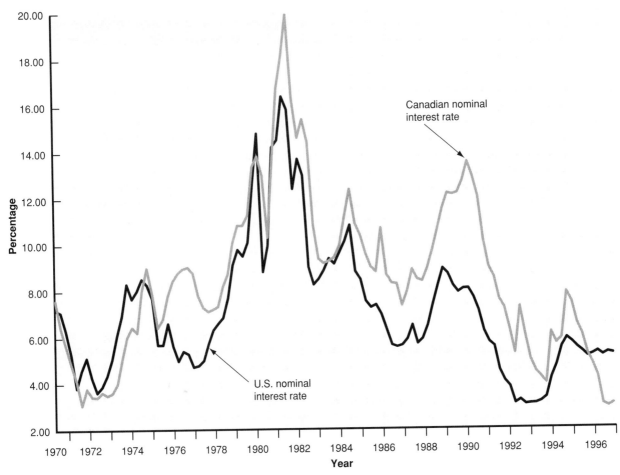

FIGURE 11-9

Canadian and U.S. Three-Month Treasury Bill Rates, 1970–1997

Canadian and U.S. nominal interest rates have moved closely together since 1970.

Source: Statistics Canada, CANSIM Series B14001 and B54409.

goods. Openness in financial markets allows for a choice between domestic and foreign assets. This choice depends primarily on their relative rate of return, which depends in turn on domestic and foreign interest rates, and on the expected rate of depreciation of the domestic currency.

In the next chapter we look at the implications of openness in goods markets. Chapter 13 then brings in openness in financial markets. In Chapter 14 we reintroduce the role of expectations and thus put together the two major extensions of the *IS-LM* model we have seen so far, the role of expectations and the openness of the economy.

SUMMARY

■ Openness in goods markets allows people and firms to choose between domestic and foreign goods. Openness in financial markets allows financial investors to hold domestic or foreign financial assets.

■ The nominal exchange rate is the price of one unit of foreign currency in terms of domestic currency. Thus, from the viewpoint of Canada, the nominal exchange rate between Canada and the States is the number of C$ per US$.

■ A nominal appreciation (or appreciation, for short) is an increase in the relative price of the domestic currency in terms of foreign currencies. Given the definition of the exchange rate, it corresponds to a decrease in the exchange rate. A nominal depreciation (or depreciation, for short) is a decrease in the relative price of the domestic currency in terms of foreign currencies; it corresponds to an increase in the exchange rate.

■ The real exchange rate is the relative price of foreign goods in terms of domestic goods. It is equal to the nominal exchange rate times the foreign price level divided by the domestic price level.

■ A real appreciation is an increase in the relative price of domestic goods in terms of foreign goods. It corresponds to a decrease in the real exchange rate. A real depreciation is a decrease in the relative price of domestic goods. It corresponds to an increase in the real exchange rate.

■ The multilateral real exchange rate, or real exchange rate for short, is a weighted average of bilateral real exchange rates, with weights equal to trade shares.

■ The balance of payments records a country's transactions with the rest of the world. The current account balance is equal to the sum of the trade balance, net investment income, and net transfers received from the rest of the world. The capital account balance is equal to capital flows from the rest of the world, minus capital flows to the rest of the world.

■ The current account and the capital account are mirror images of each other. A current account deficit is financed by net capital flows from the rest of the world, thus by a capital account surplus. Similarly, a current account surplus corresponds to a capital account deficit.

■ Uncovered interest parity, or interest parity for short, is an arbitrage condition stating that the expected rates of return in terms of domestic currency on domestic and foreign bonds must be equal. Interest parity implies that the domestic interest rate is approximately equal to the foreign interest rate plus the expected depreciation rate of the domestic currency.

KEY TERMS

- openness in goods markets, 196
- tariffs, 196
- quotas, 196
- openness in financial markets, 196
- capital controls, 196
- openness in factor markets, 197
- North American Free Trade Agreement (NAFTA), 197
- tradeable goods, 198
- real exchange rate, 200
- nominal exchange rate, 200
- appreciation (nominal, real), 200, 204
- depreciation (nominal, real), 201, 204
- fixed exchange rates, 201
- revaluation, 201
- devaluation, 201
- merchandise trade, 205
- bilateral exchange rates, 206

- multilateral exchange rates, 206
- trade-weighted exchange rate, or effective exchange rate, 206
- foreign exchange, 206
- balance of payments, 208
- above the line, below the line, 208
- current account, 208
- investment income, 208
- net transfers received, 209
- capital account, 209
- net capital flows, 209
- statistical discrepancy, 209
- gross domestic product (GDP) versus gross national product (GNP), 209
- uncovered interest parity relation, or interest parity condition, 211

QUESTIONS AND PROBLEMS

1. In what ways do imports affect your life? (Think of imported goods or services that you have used recently.)

2. Which of the following goods and services are *tradeable,* and which are not? In each case, explain your answer briefly.
   a. Automobiles
   b. Computers
   c. Haircuts
   d. Restaurant meals

3. Assume the following:
   (i) The only goods in the world are Canadian wheat and French wine.
   (ii) In Canada, the price of a bushel of wheat is $5. Determine the real exchange rate (the price of foreign goods in terms of domestic goods) between Canada and France when
   a. A French franc is worth $0.20, and the price of a bottle of wine in France is 25 francs.

b. A French franc is worth \$0.20, and the price of a bottle of wine in France is 30 francs.

c. A French franc is worth \$0.25, and the price of a bottle of wine in France is 30 francs.

4. Brian Mulroney was prime minister of Canada from 1984 to 1993. Using Figure 11-7, would you have been better off taking a trip around the world at the beginning, middle, or end of his term? Why?

5. Suppose that it takes five French francs to buy a dollar, the price level in France is 1.2, and the price level in Canada is 1.5.

a. What is the *real* exchange rate between Canada and France (the price of French goods in terms of Canadian goods)? [*Hint:* First, calculate the nominal exchange rate as the price of a franc in dollars.]

b. What would happen to the real exchange rate if the dollar rose to eight French francs? (Give a numerical answer.)

c. Comparing your answers in parts **a** and **b**, is this a real *appreciation* or a real *depreciation* of the dollar? Of what percentage?

6. Suppose the following:

(i) The interest rate in Canada is 6%.

(ii) The interest rate in Japan is 1%.

(iii) The current nominal exchange rate (dollar price of a yen) is 0.01.

(iv) The expected nominal exchange rate next year is 0.011.

a. How many dollars would a Canadian resident expect to earn for each dollar invested in Japanese bonds for one year?

b. Ignoring risk and transaction costs, should a Canadian resident prefer to invest in Canadian or Japanese bonds?

c. How many yen would a resident of Japan expect to earn for each yen invested in Canadian bonds for one year?

d. Ignoring risk and transaction costs, should a resident of Japan prefer to invest in Canadian or Japanese bonds?

e. What is the expected rate of appreciation or depreciation of the dollar? (State which.)

f. Show that the data in (i)–(iv) are not consistent with uncovered interest parity. [Use the approximation formula (11.4) in the text.]

FURTHER READING

If you want to learn more about international trade and international economics, a very good textbook is Paul Krugman and Maurice Obstfeld, *International Econom-* *ics, Theory and Policy,* 4th ed. (New York: HarperCollins, 1996).

APPENDIX

AN EXTENSION OF THE INTEREST PARITY CONDITION

The relationship between domestic and foreign interest rates is more complex than we have described in the text. A better model than the interest parity condition of (11.3) allows for the possibility that investors will demand (or pay) a premium to hold Canadian bonds. The modified interest parity condition is

$$1 + i_t = (1 + i_t^*)\left(1 + \frac{E_{t+1}^e - E_t}{E_t} + \theta_t\right)$$

where θ_t denotes the premium. (We first encountered such a premium in the appendix to Chapter 9. Although the two premiums are conceptually quite distinct, we will use the same symbol, θ_t.) With this modification, the approximation given in (11.4) becomes

$$i_t \approx i_t^* + \frac{E_{t+1}^e - E_t}{E_t} + \theta_t$$

There is some controversy over whether to think of θ_t as a premium for *risk* (what looks risky for a Canadian investor is very different from what looks risky to a foreign investor). The premium would be zero if the Bank of Canada was believed to be committed to achieving a fixed appreciation or depreciation. Among advanced economies, the premiums tend to be small and highly persistent but are sometimes very volatile. The premium required for holding Canadian rather that U.S. bonds historically has been positive on average, and, although usually it is fairly small, it became quite large in 1995 in response to a serious threat of Quebec separation. Because it is small and we don't really understand how it is determined, we shall ignore the premium in our discussion in the text.

THE GOODS MARKET IN AN OPEN ECONOMY

When, after three years of little or no growth, the U.S. recovery became stronger in 1993, Canada (and other countries around the world) cheered. This was not out of love for the United States, but because we saw higher U.S. output as implying higher demand not only for U.S. goods but for our goods as well. This higher demand meant higher exports to the United States, an improvement in our trade position, an increase in our output, and thus a chance to grow more quickly out of our own recession.

This story raises a series of questions. Can a foreign expansion really lift another country out of a recession? If there are such strong interactions between countries, shouldn't macroeconomic policies be coordinated between countries? If so, why does it seem so difficult to achieve such coordination? These are some of the questions we take up in this chapter.

12-1 THE *IS* RELATION IN THE OPEN ECONOMY

When we assumed in Chapters 3 to 10 that the economy was closed to trade, there was no need to distinguish between the domestic demand for goods and the demand for domestic goods: They were clearly the same. Now, the distinction is important. Some domestic demand falls on foreign goods, and some of the demand for domestic goods comes from foreigners. Let's look at this distinction more closely.

THE DEMAND FOR DOMESTIC GOODS

In an open economy, the **demand for domestic goods** is given by

$$Z \equiv C + I + G - \varepsilon Q + X \tag{12.1}$$

The sum of the first three terms—consumption (C), investment (I), and government spending (G)—constitutes the **domestic demand for goods.** If the economy were closed, $C + I + G$ would also be the demand for domestic goods. This is why, until now, we looked only at $C + I + G$. But now we have to make two adjustments.

First, we must subtract imports, that part of domestic demand that falls on foreign goods. We must be careful here. Foreign goods are different from domestic goods, so we cannot just subtract the quantity of imports, Q; if we were to do so, we would be subtracting oranges (foreign goods) from apples (domestic goods). We must first express the value of imports in terms of domestic goods. This is what εQ in equation (12.1) stands for. As we saw in Chapter 11, ε is the real exchange rate, the relative price of foreign goods in terms of domestic goods. Thus εQ is the value of imports in terms of domestic goods.[1]

Second, we must add exports, the demand for domestic goods that comes from abroad. This is captured by the term X in equation (12.1).

THE DETERMINANTS OF THE DEMAND FOR DOMESTIC GOODS

Having listed the components of demand, our next task is to specify their determinants. Let's start with the first three: C, I, and G.

The determinants of C, I, and G.   Now that we are assuming that the economy is open, how should we modify our earlier descriptions of consumption, investment, and government spending? The answer is: not very much, if at all. How much consumers decide to spend still depends on their income and their wealth. While the real exchange rate surely affects the "composition" of consumption spending between domestic and foreign goods, there is no obvious reason why it should affect the overall *level* of consumption. The same is true of investment. The real exchange rate may well affect whether firms buy domestic or foreign machines, but it should not affect total investment.

---

[1]*In Chapter 3, we ignored this point and just subtracted Q. This was wrong; our excuse is that we did not want to have to talk about the real exchange rate—and thus complicate matters—too early in the book.*

This is good news because it implies that we can use the descriptions of consumption, investment, and government spending that we developed earlier. We do so and assume that domestic demand, $C + I + G$, is given by

$$C(Y - T) + I(Y, r) + G$$
$$(\ +\ )\quad (+, -)$$

We assume that consumption depends on disposable income $(Y - T)$ and that investment depends on production (Y) and the real interest rate (r).[2] We take government spending (G) as given. Note that we leave aside for the moment the many refinements we introduced in earlier chapters, such as dynamics and expectations. The point is to take things one step at a time to understand the effects of opening the economy; we shall reintroduce some of those refinements later.

The determinants of imports. What does the quantity of imports, Q, depend on? Primarily on the overall level of domestic demand: The higher domestic demand, the higher the demand for all goods, both domestic and foreign. But Q also clearly depends on the real exchange rate: The higher the price of foreign goods relative to domestic goods, the lower the relative demand for foreign goods, and the lower the quantity of imports.

Thus, we write imports as

$$Q = Q(Y, \varepsilon)$$
$$(+, -)$$

(12.2)

[handwritten: → increase ε means that our $ has depreciated ∴ Imports fall]

[handwritten right margin: ↓ε $\frac{1}{ε.8} = 1.25$ ↑ε $\frac{1}{ε.7} = 1.42$ Cdn $ depn.]

Imports depend on income, Y: Higher income leads to higher imports.[3] Imports also depend on the real exchange rate. Recall that the real exchange rate, ε, is defined as the relative price of foreign goods in terms of domestic goods. A higher real exchange rate makes foreign goods relatively more expensive and thus leads to a decrease in the quantity of imports, Q.[4] This negative effect of the real exchange rate on imports is captured by the negative sign under ε in the import equation.

The determinants of exports. The export of one country is, by definition, the import of another. Thus, in thinking about what determines Canadian exports, we can ask, equivalently, what determines foreign imports. In doing so, we can draw on our discussion of the determinants of imports in the preceding paragraph. Foreign imports are likely to depend on foreign activity and on the relative price of foreign goods. Thus, we write

$$X = X(Y^*, \varepsilon)$$
$$(+, +)$$

(12.3)

[handwritten: ↑$\frac{1}{\varepsilon}$ means Cdn $ has depreciation (see above say from .8 to .7) ∴ we export more because of our lower ε]

---

[2] *In an interdependent world, domestic investment will also depend on foreign income, Y^*. We shall ignore this possibility.*

[3] ***Digging deeper.*** *We cheat a bit here. Our discussion suggests that we should be using domestic demand, $C + I + G$, instead of income, Y. You might even dispute the assumption that imports depend on the sum of domestic demand and not on its composition: It may well be that the proportion of imports in investment differs from that of imports in consumption. For example, many poor countries import most of their capital equipment but consume mostly domestic goods. In that case, the composition of demand would matter for imports. We leave these complications aside, but you may want to explore them on your own.*

[4] *Note that as ε goes up while Q goes down, what happens to the value of imports in terms of domestic goods, εQ, is ambiguous. We return to this point later in the chapter.*

$Y*$ is output in the rest of the world, or simply *foreign output* (recall that asterisks refer to foreign variables). An increase in foreign output leads to an increase in the foreign demand for all goods, some of which falls on Canadian goods, thus leading to higher Canadian exports. An increase in ε—an increase in the relative price of foreign goods in terms of Canadian goods—makes Canadian goods relatively more attractive and thus leads to an increase in exports.

We can summarize what we have learned so far in Figure 12-1(a), which plots the various components of demand against output, keeping constant all other variables that affect demand (the interest rate, taxes, government spending, foreign output, and the real exchange rate).

The line DD plots domestic demand, $C + I + G$, as a function of output. This relation between demand and output is familiar from Chapters 3 and 6. Under our standard assumptions, its slope is positive but less than 1: An increase in output— equivalently, in income, as the two are still the same in the open economy—in- creases demand but less than one for one. In the absence of good reasons to the contrary, we draw this and other relations in this chapter as lines rather than curves.

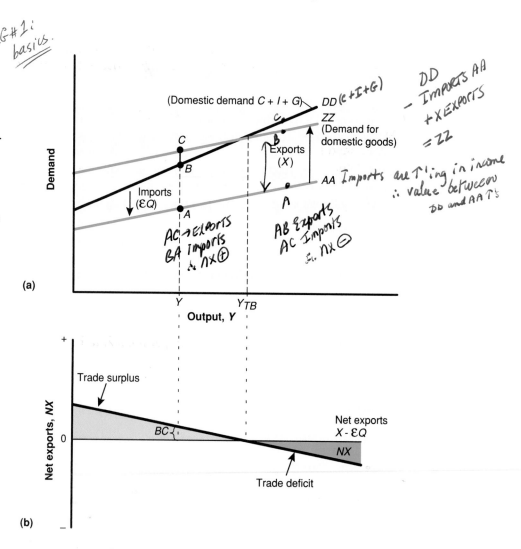

FIGURE 12-1

The Demand for Domestic Goods and Net Exports

The domestic demand for goods is an increasing function of income. The demand for domestic goods is obtained by subtracting the value of imports from domestic demand and then adding exports.

This is purely for aesthetic convenience, and none of the discussions that follow depend on that assumption.

To arrive at the demand for domestic goods, we must first subtract imports. This gives us the line AA: The distance between DD and AA is equal to the value of imports, εQ. Because the quantity of imports increases with income, the distance between the two lines increases with income. We can establish two facts about line AA, which will be useful later in the chapter. First, AA is flatter than DD. As income increases, some of the additional domestic demand falls on foreign goods rather than on domestic goods. Thus, as income increases, the domestic demand for domestic goods increases less than total domestic demand. Second, as long as some of the additional demand falls on domestic goods, AA has a positive slope: An increase in income leads to some increase in the demand for domestic goods.

Finally, we must add exports. This gives us the line ZZ, which is above AA. The distance between ZZ and AA is equal to exports. Because exports do not depend on domestic output, the distance between ZZ and AA is constant, so that the two lines are parallel. Because AA is flatter than DD, ZZ is flatter than DD as well.

From Figure 12-1(a) we can also characterize the behaviour of "net" exports—the difference between exports and imports $(X - \varepsilon Q)$—as a function of output. (Recall that *net exports* is synonymous with trade balance, so that positive net exports correspond to a trade surplus, negative net exports to a trade deficit.) At output level Y, for example, exports are given by the distance AC and imports by the distance AB, so that net exports are given by the distance BC.

The relation between net exports and output is given by the line denoted NX (for "Net eXports") in Figure 12-1(b). Net exports are a decreasing function of output: As output increases, imports increase and exports are unaffected, leading to lower net exports. Call Y_{TB} (TB for "trade balance") the level of output at which the value of imports is just equal to exports, so that net exports are equal to zero. Levels of output above Y_{TB} lead to higher imports, and thus to a trade deficit. Levels of output below Y_{TB} lead to lower imports, and thus to a trade surplus.

12-2 EQUILIBRIUM OUTPUT AND THE TRADE BALANCE

The goods market is in equilibrium when domestic output is equal to the demand for domestic goods, thus when

$$Y = Z$$

Collecting the relations we derived for the components of Z gives

$$Y = C(Y - T) + I(Y, r) + G - \varepsilon Q(Y, \varepsilon) + X(Y^*, \varepsilon) \qquad (12.4)$$

This equilibrium condition determines output as a function of all the variables we take as given, from taxes to the real exchange rate to foreign output. This is not a particularly simple relation, and Figure 12-2 gives a more user-friendly graphical characterization. Demand is measured on the vertical axis, output (equivalently production

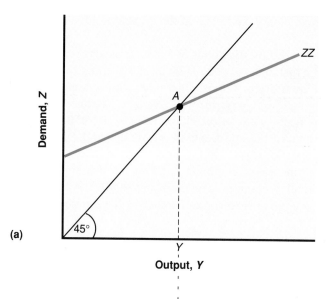

FIGURE 12-2

Equilibrium Output and Net Exports

The goods market is in equilibrium when production is equal to the demand for domestic goods. At the equilibrium level of output, the trade balance may show a deficit or a surplus.

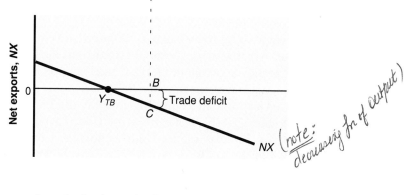

or income) on the horizontal axis. The line ZZ plots demand as a function of output; this line just replicates the line ZZ in Figure 12-1(a). As we saw in that figure, ZZ is upward-sloping, but with slope less than 1.

Equilibrium output is at the point where demand is equal to output, thus at the intersection of the line ZZ and the 45-degree line—point A in the figure, with associated output level Y.

Figure 12-2(b) replicates Figure 12-1(b), drawing <u>net exports as a decreasing function of output</u>. There is no reason for the equilibrium level of output, Y, to be the same as the level of output at which trade is balanced, Y_{TB}. As we have drawn the figure, equilibrium output is associated with a trade deficit, equal to the distance BC.

We now have the tools needed to answer the questions we asked at the beginning of this chapter.

12-3 INCREASES IN DEMAND, DOMESTIC OR FOREIGN

Let's start with a variation of what is by now an old favourite, an increase in government spending, and then turn to the effects of an increase in foreign activity.

INCREASES IN GOVERNMENT SPENDING

Suppose that the economy is in recession and the government is thinking about an increase in government spending. What will be the effects on output and on the trade balance?

The answer is given in Figure 12-3. Demand is initially given by ZZ, and the equilibrium is at point A, where output is equal to Y. Let's assume—though, as we saw, there is no reason why this should be true in general—that trade is initially balanced, so that Y and Y_{TB} are the same.

What happens if the government increases spending by ΔG? At any level of output, demand is higher by ΔG, so that the demand line shifts up by ΔG, from ZZ to ZZ'. The equilibrium point moves from A to A', and output increases from Y to Y'. The increase in output is clearly larger than the increase in government spending: There is a multiplier effect.

So far, the story sounds very similar to what happened in the closed economy earlier in the book. However, let's look more closely.

There is now an effect on the trade balance. Because government spending enters neither the exports relation nor the imports relation directly, the line representing net

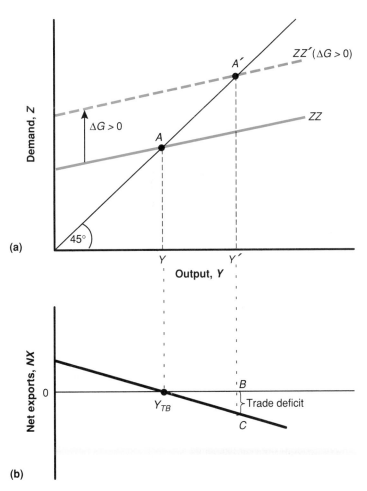

FIGURE 12-3

The Effects of Higher Government Spending

An increase in government spending leads to an increase in output and a trade deficit.

exports as a function of output in Figure 12-3(b) does not shift. The increase in output from Y to Y' thus leads to a trade deficit equal to BC.

Not only does government spending now generate a trade deficit, but its effect on output is smaller than in the closed economy. Recall from Chapter 3, Section 3-3, that the smaller the slope of the demand relation, the smaller the multiplier. For example, if ZZ is flat, the multiplier is just equal to 1. And recall from Figure 12-1 that the demand relation ZZ is flatter than the demand relation in the closed economy, DD. Thus, the multiplier is smaller in the open economy.

The trade deficit and the smaller multiplier arise from the same cause: An increase in demand now falls not only on domestic goods, but also on foreign goods. Thus, when income increases, the effect on the demand for domestic goods is smaller than it would be in a closed economy, leading to a smaller multiplier. And, because some of the increase in demand falls on imports—and exports are unchanged—the result is a trade deficit.

These two implications are important. In an open economy, an increase in domestic demand has a smaller effect on output than in a closed economy, as well as an adverse effect on the trade balance. Indeed, the more open the economy, the smaller the effect on output and the larger the adverse effect on the trade balance. For example, in Canada the ratio of imports to GDP is about 40%. This number implies that, when demand increases in Canada, roughly 40% of this increased demand goes to higher imports and only 60% to an increase in the demand for domestic goods. The effect of an increase in government spending is thus likely to be a large increase in the trade deficit and a small increase in output, making domestic demand expansion a rather unattractive policy. Even for the United States, which has an import ratio of only about 10%, an increase in demand will be associated with some deterioration in the trade position. (This conclusion is discussed at greater length in the Focus box entitled "Multipliers: Canada versus the United States.")

MULTIPLIERS: CANADA VERSUS THE UNITED STATES

If we assume that the various relations in equation (12.4) are linear, we can compute the effects of government spending, foreign output, and so forth on both output and the trade balance. Here we will focus on the differences between the effects of government spending in a large country such as the United States and in a small country such as Canada.

Assume that consumption increases with disposable income, and that investment increases with output and decreases with the real interest rate:

$$C = c_0 + c_1(Y - T)$$
$$I = d_0 + d_1Y - d_2r$$

For simplicity, ignore movements in the real exchange rate and assume that the real exchange rate is equal to 1. Also assume that imports are proportional to domestic output and exports are proportional to foreign output:

$$Q = q_1Y$$
$$X = x_1Y^*$$

In the same way as we referred to c_1 as the marginal propensity to consume, q_1 is the **marginal propensity to import**.

The equilibrium condition is that output be equal to the demand for domestic goods:

$$Y = C + I + G - Q + X$$

(Recall that we are assuming that ε is equal to 1.) Replacing the components by the expressions from above gives

$$Y = [c_0 + c_1(Y - T)] + (d_0 + d_1Y - d_2r) + G - q_1Y + x_1Y^*$$

Regrouping terms gives

$$Y = (c_1 + d_1 - q_1)Y + (c_0 + d_0 - c_1T - d_2r + G + x_1Y^*)$$

Bringing the terms in output together, and solving, gives

$$Y = \frac{1}{1 - (c_1 + d_1 - q_1)}(c_0 + d_0 - c_1T - d_2r + G + x_1Y^*)$$

MPC tax MPI

Output is equal to the multiplier times the final term in parentheses, which captures the effect of all the variables we take as given in explaining output.

Consider the multiplier. More specifically, consider the term $(c_1 + d_1 - q_1)$ in the denominator. As in the closed economy, $(c_1 + d_1)$ gives the effects of an increase in output on consumption and investment demand; $(-q_1)$ captures the fact that some of the increased demand falls not on domestic goods but on foreign goods. In the extreme case where all the additional demand falls on foreign goods—that is, when $q_1 = c_1 + d_1$—an increase in output has no effect back on the demand for domestic goods; in that case, the multiplier is equal to 1. In general, q_1 is less than $(c_1 + d_1)$, so that the multiplier is larger than 1. But it is smaller than it would be in a closed economy.

Using this equation, we can easily characterize the effects of an increase in government spending of ΔG. The increase in output is equal to the multiplier times the change in government spending, thus

$$\Delta Y = \frac{1}{1 - (c_1 + d_1 - q_1)}\Delta G$$

The increase in imports that follows from the increase in output implies a change in net exports of

$$\Delta NX = -q_1\Delta Y$$

$$= -\frac{q_1}{1 - (c_1 + d_1 - q_1)}\Delta G$$

Let's see what these formulas imply by choosing numerical values for the parameters. Take $c_1 + d_1$ equal to 0.6. What value should we choose for q_1? We saw in Chapter 11 that, in general, the larger the country, the more self-sufficient it is, and the less it imports from abroad. So let's choose two values of q_1—a small value, say 0.1, for a large country such as the United States, and a larger one, say 0.3, for a small country such as Canada. Note that we can think of $q_1/(c_1 + d_1)$ as the proportion of an increase in demand that falls on imports, so that an equivalent way of stating our choices of q_1 is that, in the large country, 1/6 of demand falls on imports, versus 1/2 in the small country.

For the large country, the effects on output and the trade balance are given by

$$\Delta Y = \frac{1}{1 - (0.6 - 0.1)}\Delta G = 2.0\Delta G$$

and

$$\Delta NX = -0.1\Delta Y = \frac{-0.1}{1 - (0.6 - 0.1)}\Delta G = -0.2\Delta G$$

For the small country, the effects are given by

$$\Delta Y = \frac{1}{1 - (0.6 - 0.3)}\Delta G = 1.43\Delta G$$

and

$$\Delta NX = -0.3\Delta Y = \frac{-0.3}{1 - (0.6 - 0.3)}\Delta G = -0.43\Delta G$$

These computations show how different the tradeoffs faced by the two countries are. In the large country, the effect of an increase in G on output is large and the effect on the trade balance is small. In the small country, the effect on output is small, and the deterioration of the trade balance is equal to one-third of the increase in government spending.

This example makes clear how drastically openness binds the hands of policy makers in small countries. We shall see more examples of this proposition as we go along.

Eq #4

INCREASES IN FOREIGN DEMAND

Consider now an increase in foreign activity, an increase in Y^*. This could be due to an increase in foreign government spending, G^*—the same policy that we just analyzed, but now taking place abroad. But we do not need to know where it comes from to analyze the effects on the Canadian economy.

Figure 12-4 shows the effects of an increase in Y^* on domestic output and the trade balance. The initial demand for domestic goods is given by ZZ in Figure 12-4(a). The equilibrium is at point A, with output level Y. Let's assume that trade is balanced, so that in Figure 12-4(b) net exports associated with Y are equal to zero.

It will be convenient to draw in Figure 12-4(a) the line DD corresponding to domestic demand for goods $C + I + G$ as a function of income. As we showed in Figure 12-1, this line is steeper than ZZ. The difference between ZZ and DD is equal to net exports, so that if trade is balanced at point A, then ZZ and DD intersect at point A.

FIGURE 12-4

The Effects of Higher Foreign Demand

An increase in foreign demand leads to an increase in output and to a trade surplus.

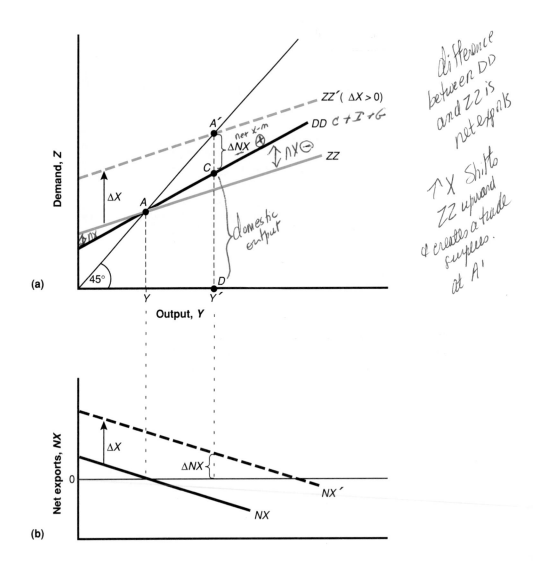

difference between DD and ZZ is net exports

↑X shifts ZZ upward & creates a trade surplus at A'

226   OPENNESS

Now consider the effects of an increase in foreign output, ΔY^*. The direct effect is to increase Canadian exports by some amount, call it ΔX. For a given level of output, this increase in exports leads to an increase in the demand for Canadian goods by ΔX, so that the line giving the demand for domestic goods as a function of output shifts up by ΔX, from ZZ to ZZ'. As exports increase by ΔX at a given level of output, the line giving net exports as a function of output in Figure 12-4(b) also shifts up by ΔX, from NX to NX'.

The new equilibrium is at point A', with output level Y'. The increase in foreign output leads to an increase in domestic output. The channel is clear: Higher foreign output leads to higher exports of domestic goods, which in turn increases domestic output and the domestic demand for goods through the multiplier.

What happens to the trade balance? We know that exports go up. But could it be that the increase in domestic output leads to such a large increase in imports that the trade balance actually deteriorates? The answer is no. The trade balance must improve.

To see why, return to Figure 12-4(a). When foreign demand increases, the demand for domestic goods shifts up from ZZ to ZZ'. But the line DD, which gives the domestic demand for goods as a function of output, does not shift. At the new equilibrium level of output Y', domestic demand is given by the distance DC, and the demand for domestic goods is given by DA'. Net exports are thus given by the distance CA'—which, because DD is necessarily below ZZ', is necessarily positive. Thus, while imports increase, they do not offset the increase in exports, and the trade balance improves.

GAMES THAT COUNTRIES PLAY

We have derived two basic results so far:

- An increase in domestic demand leads to an increase in output, but also to a trade deficit. (We looked at an increase in government spending, but the results would have been the same for a decrease in taxes, an increase in consumer spending, and so on.)

- An increase in foreign demand (which could come from the same types of changes taking place abroad) leads to an increase in domestic output and a trade surplus.

Governments do not like trade deficits, and for good reasons. The main one is this: A country that consistently runs a trade deficit accumulates debt vis-à-vis the rest of the world and thus has to pay steadily higher interest payments to the rest of the world. Thus, it is no wonder that countries prefer increases in foreign demand (which lead to an improvement in the trade balance) to increases in domestic demand (which lead to a deterioration in the trade balance).

These preferences may have disastrous implications. Consider a group of countries, all doing a large amount of trade with each other, so that an increase in demand in any one country falls largely on the goods produced in the other countries. Suppose that all these countries are in recession and that each of them has roughly balanced trade. Each country may be very reluctant to take measures to increase domestic demand. Were it to do so, the result might be a small increase in output but also a larger trade deficit. Thus each country may just wait for others to

In May 1981, the Socialist party won the elections in France. Faced with an economy suffering from more than 7% unemployment, the Socialists created a program aimed at increasing demand through more generous social policies and subsidies to job creation. Welfare benefits and pensions were increased. Public jobs were created, as were new training programs for the young and the unemployed. Table I summarizes the macroeconomic results of the policy.

The fiscal expansion is quite visible in the data. The budget, which was balanced in 1980, was in deficit by 2.8% of GDP in 1982. The effects on growth are equally visible. Average growth in 1981–1982 was equal to 1.85%—not an impressive growth rate, but still much above the EU's dismal 0.45% average growth rate over the same two years.

Nevertheless, the Socialists abandoned their policy in March 1983. The last line of Table I tells us why. As France was expanding faster than its trading partners, it experienced a <u>sharp increase in its trade deficit</u>. While the government may have tolerated those trade deficits, financial markets—which were very nervous

about the Socialists in the first place—forced three <u>devaluations of the franc in 18</u> months. (Recall from Chapter 11 that when countries try to maintain a fixed exchange rate—as was the case for France at the time—depreciations are called devaluations. We shall see the mechanisms that lead to such devaluations in the next two chapters.) The first was in October 1981, by 8.5% against the DM; the second in June 1982, by 10% against the DM; and the third in March 1983, by 8% against the DM. In March 1983, unwilling to face further attacks on the franc and worried about the trade deficit, the French government gave up its attempt to use demand policies to decrease unemployment and shifted to a new <u>policy of "austerity"</u>—a policy aimed at <u>achieving low inflation, budget and trade balance, and no further devaluations</u>. This policy has been maintained by the various French governments, from both the left and the right, to this day.

Reference

For more on French economic policy in the 1980s, read *Reflation and Austerity: Economic Policy under Mitterand,* by Pierre-Alain Muet and Alain Fonteneau (New York: Berg, 1990).

TABLE I

MACROECONOMIC AGGREGATES, FRANCE: 1980–1983

	1980	1981	1982	1983
GDP growth (%)	1.6	1.2	2.5	0.7
EU growth (%)	1.4	0.2	0.7	1.6
Budget surplus	0.0	−1.9	−2.8	−3.2
Current account surplus	−0.6	−0.8	−2.2	−0.9

Budget and *current account surplus* are measured as ratios to GDP, in percentages. A minus sign indicates a deficit.
EU growth refers to the average growth rate for the countries of the European Union.

Source: OECD Economic Outlook, December 1993.

When a country runs a trade deficit, it has to let the currency depreciate in order to bring things back into balance

increase their own demand. But if they all wait, nothing happens and the recession may endure.

Is there a way out of this situation? Indeed there is, at least in theory. If all countries coordinate their macroeconomic policies so as to increase domestic demand simultaneously, they can all expand without increasing their trade deficit (vis-à-vis each other; their trade deficit with respect to the rest of the world will still increase). The reason is clear: The coordinated increase in demand leads to increases in both exports and imports in each country. It is still true that domestic demand expansion leads to larger imports; but this increase in imports is offset by the increase in exports, which comes from the foreign demand expansions.

Coordination is indeed a word that governments often invoke. The seven major economic powers of the world, the so-called **G7** (the United States, Japan, France, Germany, the United Kingdom, Italy, and Canada), meet regularly to discuss their economic situations; the final communiqué rarely fails to mention coordination. But the evidence is that there is much less than full macro-coordination among countries. Economists have identified some good reasons why this is so.

First, coordination may imply that some countries have to do more than others. Suppose, in our example, that only some countries are in recession. Those that are not will be reluctant to increase their own demand; but, if they do, those countries that expand will run a trade deficit vis-à-vis those that do not. Or suppose that some countries are already running a large budget deficit. These countries will not want to cut taxes or increase spending further, and will ask other countries to take on more of the adjustment. Those other countries may well be reluctant to do so.

The second problem is that countries have a strong incentive to promise to coordinate, and then not to deliver on that promise. Once all countries have agreed, say, to an increase in spending, each country has an incentive not to deliver, thus benefiting from the increase in demand elsewhere and incurring an improvement in its trade position. But if each country cheats, or at least does not do everything it promised, then there will be insufficient demand expansion to get out of the recession.

These are far from abstract concerns. Countries in the European Union, which are indeed highly integrated with one another, have in the past 20 years often suffered from such coordination problems. In the late 1970s, a bungled attempt at coordination left most countries weary of trying again. In the early 1980s, an attempt by the French socialists to go at it alone led to a large French trade deficit, and eventually to a change in policy. (This episode is described in the Global Macro box entitled "The French Socialist Expansion: 1981–1983.") Thereafter, most countries decided that it was better to wait for an increase in foreign demand than to increase their own demand. Some economists attribute part of Europe's poor macro performance in the 1980s precisely to this waiting game.

12-4 DEPRECIATION, THE TRADE BALANCE, AND OUTPUT

Suppose that the Canadian government takes policy measures that lead to a depreciation of the dollar. (We shall see what these policies may be in Chapter 13; for the moment we shall assume that the government can simply choose the exchange rate.)

Recall that the real exchange rate is given by

$$\varepsilon \equiv \frac{EP^*}{P}$$

The real exchange rate, ε (the relative price of foreign goods in terms of domestic goods) is equal to the nominal exchange rate, E (the relative price of foreign currency in terms of domestic currency) times the foreign price level, P^*, divided by the domestic price level, P. Under our maintained assumption that price levels are given, a nominal depreciation is thus reflected one for one in a real depreciation. More concretely, if the dollar depreciates vis-à-vis the yen by 5% (a nominal depreciation), and if the price levels in Japan and Canada do not change, Canadian goods will be 5% cheaper compared with Japanese goods (a real depreciation).[5]

Let's now ask what the effects of this real depreciation will be on the Canadian trade balance and on Canadian output.

DEPRECIATION AND THE TRADE BALANCE: THE MARSHALL-LERNER CONDITION

Recall the definition of net exports:

$$NX \equiv X - \varepsilon Q$$

Replacing X and Q by their expressions from equations (12.2) and (12.3) gives

$$NX = X(Y^*, \varepsilon) - \varepsilon Q(Y, \varepsilon)$$

As the real exchange rate ε enters in three places, this equation makes it clear that the real depreciation—an increase in ε—affects the trade balance through three channels.

1. *X increases.* The depreciation, which makes Canadian goods relatively cheaper abroad, leads to an increase in foreign demand, and thus to an increase in Canadian exports.

2. *Q decreases.* The depreciation, which makes foreign goods relatively more expensive in Canada, leads to a shift in domestic demand toward domestic goods, leading to a decrease in the quantity of imports.

3. *The relative price of imports, ε, increases.* This tends to increase the import bill, εQ. The same quantity of imports now costs more to buy.

Thus, for the trade balance to improve following a depreciation, exports must increase enough and imports must decrease enough to compensate for the increase in the price of imports. The condition under which a real depreciation leads to an increase in net exports is known as the **Marshall-Lerner condition.**[6] It is derived more formally in the appendix to this chapter. It turns out—with a caveat we shall state when we introduce dynamics later in this chapter—that this condition is satisfied in reality. Thus, for the rest of this book, we shall assume that an increase in ε, a real depreciation, leads to an increase in net exports.

THE EFFECTS OF A DEPRECIATION

We have looked so far at the direct effects of a depreciation on the trade balance—that is, the effects *given Canadian and foreign output.* But the effects do not end

---

[5]*Once we have discussed the supply side, we shall return (in Chapter 19) to the effects of a nominal depreciation and allow for price-level adjustments.*

[6]*The name comes from the two economists who stated it first, Alfred Marshall and Abba Lerner.*

there. The change in net exports in turn changes domestic output, which affects net exports further.

Because the effects of a real depreciation are very much like those of an increase in foreign output, we can use the same figure that we used to show the effects of an increase in foreign output earlier, in Figure 12-4.

Just like an increase in foreign output, a depreciation leads, at any level of output, to an increase in net exports (assuming, as we do, that the Marshall-Lerner condition holds). Thus, both the demand relation [ZZ in Figure 12-4(a)] and the net exports relation [NX in Figure 12-4(b)] shift up. The equilibrium moves from A to A'; output increases from Y to Y'. By the same argument that we used earlier, the trade balance improves: The increase in imports induced by the increase in output is less than the direct improvement in the trade balance induced by the depreciation. In summary, *the depreciation leads to a shift in demand, both foreign and domestic, towards domestic goods. This leads in turn both to an increase in domestic output and to an improvement in the trade position.*

While a depreciation and an increase in foreign output have the same effect on domestic output and the trade balance, there is a subtle but important difference between the two. A depreciation works by making foreign goods relatively more expensive. But this means that, given their income, people—who now have to pay more to buy foreign goods—are worse off. This mechanism is strongly felt in countries that go through a major depreciation. Governments that try to achieve a major depreciation often find themselves with strikes and riots in the streets, as people react to the much higher prices on imported goods. A recent example is Indonesia, where the large depreciation of the rupiah in 1997–1998—from 3400 rupiahs per US$ in June 1997 to 14 000 rupiahs per US$ in January 1998—led to a large decline in workers' standard of living and strong social tensions.[7]

COMBINING EXCHANGE-RATE AND FISCAL POLICIES

Suppose that a government wants to reduce the trade deficit without changing the level of output. A depreciation will not do; it would reduce the trade deficit, but it would also increase output. A fiscal contraction will not work either; it would reduce the trade deficit, but it would decrease output. What should the government do? The answer is straightforward: Use the right combination of depreciation and fiscal contraction. Figure 12-5 on the next page shows what this combination should be.

The initial equilibrium in Figure 12-5(a) is at point A, associated with output Y, and the trade deficit is given by the distance BC in Figure 12-5(b). If the government wants to eliminate the trade deficit without changing output, it must do two things.

First, it must achieve a depreciation sufficient to eliminate the trade deficit at the initial level of output. Thus, the depreciation must be such as to shift the net exports relation from NX to NX' in Figure 12-5(b).

---

[7]***Digging deeper.*** *There is an alternative to strikes and riots: namely, asking for and obtaining an increase in wages. But if wages increase, presumably the prices of domestic goods will increase as well, leading to a smaller real depreciation. To discuss this mechanism, we need to look at the supply side in more detail than we have done so far. We return to the dynamics of depreciation, wage movements, and price movements in Chapter 19.*

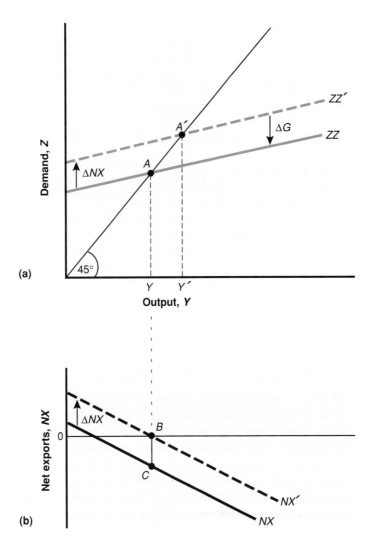

FIGURE 12-5

Reducing the Trade Deficit without Changing Output

To reduce the trade deficit without changing output, the government must both achieve a depreciation and decrease government spending.

The problem is that this depreciation, and the associated increase in net exports, also shifts the demand relation in Figure 12-5(a) from ZZ to ZZ'. In the absence of other measures, the equilibrium would move from A to A', and output would increase from Y to Y'. To avoid this effect, the government must reduce government spending so as to shift ZZ' back to ZZ. This combination of a depreciation and a fiscal contraction leads to the same level of output and an improved trade balance.

There is a general point behind this example. To the extent that governments care about *both* the level of output and the trade balance, they have to use *both* fiscal and exchange-rate policies together. We just saw one such combination. Table 12-1 shows others, depending on the initial output and trade positions. Take, for example, the entry in the top right corner of the table. Initial output is too low (put another way, unemployment is too high), and the economy has a trade deficit. A depreciation will help on both the trade and the output fronts: It reduces the trade deficit and increases output. But there is no reason for the depreciation to achieve

TABLE 12-1

EXCHANGE-RATE AND FISCAL POLICY COMBINATIONS

INITIAL CONDITIONS	TRADE SURPLUS	TRADE DEFICIT
Low output	$\varepsilon? G\uparrow$	$\varepsilon\uparrow G?$
High output	$\varepsilon\downarrow G?$	$\varepsilon? G\downarrow$

both the right increase in output and the elimination of the trade deficit. Depending on the initial situation and the relative effects of the depreciation on output and the trade balance, the government may need to complement the depreciation with either an increase or a decrease in government spending. This ambiguity is captured by the question mark in the entry. Make sure that you understand the logic behind each of the other three entries.

12-5 LOOKING AT DYNAMICS: THE J-CURVE

We have ignored dynamics so far in this chapter. It is time to reintroduce them. The dynamics of consumption, investment, sales, and production we saw earlier are as relevant to the open economy as they are to the closed economy. But there are additional dynamic effects which come from the dynamics of exports and imports. These are the ones we want to focus on here.

Let's return to the effects of the exchange rate on the trade balance. We argued earlier that a depreciation leads to an increase in exports and to a decrease in imports. But these effects surely do not happen overnight. Think of the dynamic effects of, say, a 10% dollar depreciation. In the first few months following the depreciation, the effect of the depreciation is likely to be reflected much more in prices than in quantities. The price of imports in Canada goes up, the price of Canadian exports abroad goes down.[8] But the quantity of imports and exports is likely to adjust slowly. It takes a while for consumers to realize that relative prices have changed, it takes a while for firms to shift to cheaper suppliers, and so on. Thus a depreciation may well lead to an initial deterioration of the trade balance; ε increases, but neither X nor Q adjusts very much initially, leading to a decline in net exports, $X - \varepsilon Q$.

As time passes, the effects of the changes in the relative prices of both exports and imports become stronger. Exports increase, imports decrease. If the Marshall-Lerner condition eventually holds—and we have argued that it does—the response of exports and imports eventually becomes stronger than the adverse price effect, and the eventual effect of the depreciation is to improve the trade balance.

---

[8]***Digging deeper.*** *The price of imported goods may not go up by 10%, however. The price would go up 10% if importers adjusted their dollar price fully for the dollar depreciation. To keep their market share, or because they are committed under previous contracts to deliver at a given dollar price, importers may decide instead to pass along only part of the dollar depreciation and take a reduction in their profit margins. This is indeed what we observe in practice: While import prices respond to a depreciation, they respond less than one for one. The same logic applies to the prices of exports. We abstract from these complications here.*

FIGURE 12-6

The J-Curve

A real depreciation leads initially to a deterioration, then to an improvement, in the trade balance.

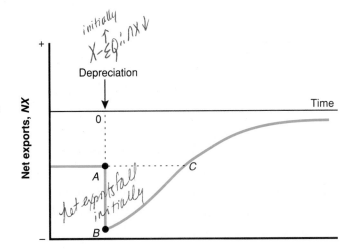

Figure 12-6 captures this adjustment by plotting the evolution of the trade balance against time in response to a real depreciation. The pre-depreciation trade deficit is equal to OA. The depreciation initially increases the trade deficit to OB: ε goes up, but neither Q nor X changes right away. Over time, exports increase and imports decrease, reducing the trade deficit. Eventually (if the Marshall-Lerner condition is satisfied), the trade balance improves beyond its initial level; this is what happens from point C on in the figure. Economists refer to this adjustment process as the **J-curve** because—admittedly with a bit of imagination—the curve in the figure resembles a "J": first down, then up.

Figure 12-7 shows the relation between the trade balance and the real Canada/U.S. exchange rate since 1970. One of the features that stands out is the large appreciation of about 25% from 1985 to 1990 that was then reversed in the next five years. Figure 12-7 suggests some interesting behaviour:

1. Although the trends are slightly different, for most of the sample, the broad swings in the trade balance (measured as the ratio of net exports to GDP) closely parallel the movements in the exchange rate. The appreciation of 1985 to 1990 was accompanied by a large deterioration in the trade balance, and the later depreciation was associated with a large improvement in the trade balance.

2. The timing between the swings is complicated. At times, the exchange rate appears to be responding to movements in the trade balance. In the 1990s, however, the depreciation in the Canadian dollar appeared to lead by many quarters the improvement in the trade balance.

In general, the econometric evidence on the dynamic relation between exports, imports, and the real exchange rate suggests that in all OECD countries a real depreciation eventually leads to a trade balance improvement. But it also suggests that this process takes some time, typically between six months and a year. These lags have implications not only for the effects of a depreciation on the trade balance but also for the effects of a depreciation on output. If a depreciation initially decreases net exports, it also initially exerts a contractionary effect on output. Thus, if a government relies on a depreciation to both improve the trade balance and expand domestic output, the effects will go the "wrong" way for a while.

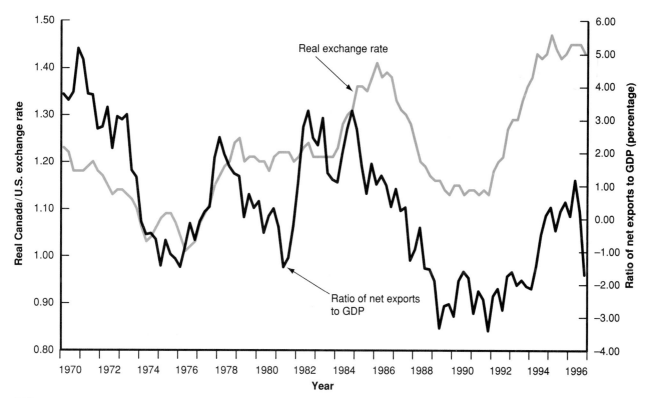

FIGURE 12-7

The Real Canada/U.S. Exchange Rate and the Ratio of Net Exports to GDP:
Canada, 1970–1996

The real exchange rate and the trade balance display similar broad swings.

Sources: Statistics Canada, CANSIM Series D20463, D20476, and D20480, and FRED.

12-6 SAVING, INVESTMENT, AND TRADE DEFICITS

In Chapter 3 we saw how we could rewrite the condition for equilibrium in the goods market as the condition that investment equals saving, private and public. We now derive the corresponding condition for the open economy. Start from our equilibrium condition:

$$Y = C + I + G - \varepsilon Q + X$$

Subtract $C + T$ from both sides, and use the fact that private saving is given by $S = Y - C - T$ to get

$$S = I + G - T - \varepsilon Q + X$$

Using the definition of net exports $NX \equiv X - \varepsilon Q$, and reorganizing, gives

$$NX = S + (T - G) - I \quad \text{(Same as Partinn)} \quad (12.5)$$

See Marilu's text p

NFI, S, Trade surplus, Trade def., E, S,I

$$NX = (S-I) + (NT-G)$$

S1, S0, I

(12.1) $Z = C + I + G + X - \varepsilon Q$
⇕ equivalent
(12.5) $NX = S + (T-G) - I$
Private svg, Public svg

McCallum

This condition says that in equilibrium, the trade balance (*NX*) must be equal to saving—private (*S*) and public (*T − G*)—minus investment (*I*). Thus, a trade surplus corresponds to an excess of saving over investment. A trade deficit corresponds to an excess of investment over saving.

One way of getting more intuition for this relation is to recall from the discussion of the current and capital accounts in Chapter 11 that a trade surplus implies net lending to the rest of the world, and a trade deficit implies net borrowing from the rest of the world. Thus, take a country that invests more than it saves, so that $S + (T − G) − I$ is negative. That country must be borrowing the difference from the rest of the world; it must therefore be running a trade deficit.

Note some of the things that equation (12.5) says:

■ It implies that an increase in investment must be reflected in either an increase in private or public saving or in a trade deficit.

■ It implies that an increase in the budget deficit must be reflected in an increase in private saving, a decrease in investment, or a trade deficit.

■ It implies that a country with a high saving rate, private and public, must have either a high investment rate or a large trade surplus.

Note also what equation (12.5) *does not say*. It does not say, for example, whether and under what conditions a budget deficit will be reflected in a trade deficit, in an increase in private saving, or in a decrease in investment. To answer that question, we must explicitly solve for what happens to output and its components using the assumptions that we have made about consumption, investment, exports, and imports. We can do so using either equation (12.1)—as we have done throughout this chapter—or equation (12.5), as the two are equivalent. However, let us strongly recommend that you use equation (12.1). Using (12.5) can, if one is not careful, be very misleading. Consider, for example, the following argument (which is so common that you may well have read it in some form in newspapers):

> It is clear that a country cannot reduce its trade deficit through a depreciation. Look at equation (12.5). It shows that the trade deficit is equal to investment minus saving, private and public. Why should a depreciation affect either saving or investment? Thus, how can a depreciation affect the trade deficit?

The argument may sound convincing, but we know that it is wrong. We showed earlier that a depreciation leads (if the Marshall-Lerner condition holds) to an increase in output and an improvement in the trade position. So what is wrong with the argument above? A depreciation actually affects saving and investment, by affecting the demand for domestic goods and thus by increasing output. Higher output leads to an increase in saving over investment, or equivalently to a decrease in the trade deficit.

A good way of making sure that you understand the material in this chapter is to go back and look at the various cases we have considered, from changes in government spending, to changes in foreign output, to combinations of depreciation and fiscal contraction, and so on. Trace what happens in each case to each of the four components of equation (12.5): private saving, public saving (equivalently, the budget surplus), investment, and the trade balance. Make sure, as always, that you can tell the story in words. If you can, you are ready to go on to Chapter 13.

SUMMARY

■ In an open economy, the demand for domestic goods is equal to the domestic demand for goods (consumption, plus investment, plus government spending) minus imports, plus exports.

■ An increase in domestic demand leads to a smaller increase in output in an open economy than in a closed economy, because some of the additional demand goes to imports. It also leads to a deterioration of the trade balance.

■ An increase in foreign demand leads to an increase in domestic output (as a result of increased exports) and an improvement in the trade balance.

■ Because increases in foreign demand improve the trade balance and increases in domestic demand worsen it, a government may be tempted to wait for increases in foreign demand to move its country out of a recession.

When a group of countries is in recession, coordination can help them get out of it.

■ If the Marshall-Lerner condition is satisfied—and econometric evidence suggests that it is—a real depreciation leads to an increase in net exports and thus an improvement in the trade position.

■ The typical response of the trade balance to a real depreciation is first a deterioration, and then an improvement. This adjustment process is known as the J-curve.

■ The condition for equilibrium in the goods market can be rewritten as the condition that saving (public and private) minus investment must be equal to the trade balance. A trade surplus corresponds to an excess of saving over investment, a trade deficit to an excess of investment over saving.

KEY TERMS

- demand for domestic goods, 218
- domestic demand for goods, 218
- marginal propensity to import, 224
- coordination, 229
- G7, 229
- Marshall-Lerner condition, 230
- J-curve, 234

QUESTIONS AND PROBLEMS

1. In this chapter we draw a careful distinction between the "demand for domestic goods" and the "domestic demand for goods." Conceptually, what is the difference between them? If you know the value of the domestic demand for goods, how would you obtain the value of the demand for domestic goods?

2. It is often stated that budget deficits lead to trade deficits. Using what you have learned in this chapter, explain why this might be true.

3. During a recession, should small countries or large countries be more interested in policy coordination? Why?

4. Suppose that the goods market in an open economy is characterized by the following behavioural equations:

$$C = 400 + 0.5Y_D$$
$$I = 700 - 4{,}000i + 0.2Y$$
$$G = 200$$
$$T = 200$$
$$X = 100 + 0.1Y^* + 100\varepsilon$$
$$Q = 0.1Y - 50\varepsilon$$
$$\varepsilon = 2.0$$
$$Y^* = 1000$$

where X is the quantity of exports, Q the quantity of imports, ε the real exchange rate, and Y^* foreign output.

a. Assuming that the interest rate is 10% (.10), find equilibrium GDP. [*Hint:* Use the equilibrium condition for the goods market in an open economy, equation (12.4) in the text.]

b. Determine the values of C, I, G, and net exports, and verify that the demand for domestic goods is equal to the value you found in part a.

c. Suppose that government spending increases from 200 to 400.
   (i) Solve again for equilibrium GDP.
   (ii) Solve for C, I, G, and net exports, and verify the equality of the demand for domestic goods and GDP.
   (iii) What has happened to net exports as a result of the increase in G? Explain.

d. Go back to the original assumptions of this problem. Continue to assume that the interest rate is 10%. Now, suppose that foreign output increases from 1000 to 1200.
   (i) Solve again for equilibrium GDP.

(ii) Solve for *C, I, G,* and net exports, and verify that the demand for domestic goods equals GDP.

(iii) What has happened to net exports? Explain why the answer is different from the one you obtained when government spending increased by the same amount.

5. We have pointed out that a major depreciation may lead to strikes and riots in the streets, since it raises the prices of imported goods and thus the cost of living. Why, then, wouldn't a government want to pursue the opposite policy: a major *appreciation?*

6. What combination of fiscal and exchange-rate changes should a government pursue if it wants to:

   a. Increase output while improving the trade balance?

   b. Decrease output while improving the trade balance?

7. Suppose that the government wants to eliminate a trade deficit but does not want the level of output to change. What measures should it take? Explain what happens to private spending, public spending, and investment.

8. Suppose that investment in Canada depends upon the level of U.S. income, in addition to domestic income and interest rates. How would this affect Figure 12-4?

9. In the early 1990s, the Ontario economy entered a very deep recession. The NDP government of the day, headed by Premier Bob Rae, chose to fight the recession with an expansionary fiscal policy. Other than generating an enormous fiscal deficit, the efforts had very little impact. Why is this not surprising?

FURTHER READING

A good discussion of the relation among trade deficits, budget deficits, private saving, and investment is given in

Barry Bosworth's *Saving and Investment in a Global Economy* (Washington, DC: Brookings Institution, 1993).

APPENDIX

DERIVATION OF THE MARSHALL-LERNER CONDITION

Start from the definition of net exports, $NX \equiv X - \varepsilon Q$, and assume trade to be initially balanced, so that $X = \varepsilon Q$. The Marshall-Lerner condition is the condition under which a real depreciation, an increase in ε, leads to an increase in net exports.

To derive this condition, consider an increase in the real exchange rate of $\Delta \varepsilon$. The change in the trade balance is thus given by

$$\Delta NX = \Delta X - \varepsilon \Delta Q - Q \Delta \varepsilon$$

The first term on the right (ΔX) gives the change in exports, the second ($\varepsilon \Delta Q$) the real exchange rate times the change in the quantity of imports, and the third ($Q \Delta \varepsilon$) the quantity of imports times the change in the real exchange rate.

Divide both sides of the equation by X to get

$$\frac{\Delta NX}{X} = \frac{\Delta X}{X} - \varepsilon \frac{\Delta Q}{X} - \frac{Q \Delta \varepsilon}{X}$$

Use the fact that $\varepsilon Q = X$ to replace ε / X by $1/Q$ in

the second term on the right, and to replace Q/X by $1/\varepsilon$ in the third term on the right. These substitutions give

$$\frac{\Delta NX}{X} = \frac{\Delta X}{X} - \frac{\Delta Q}{Q} - \frac{\Delta \varepsilon}{\varepsilon}$$

This equation says that the change in the trade balance in response to a real depreciation, normalized by exports, is equal to the sum of three terms. The first is the proportional change in exports, $\Delta X/X$, induced by the real depreciation. The second term is equal to minus the proportional change in imports, $-\Delta Q/Q$, induced by the real depreciation. The third term is equal to minus the proportional change in the real exchange rate, $-\Delta \varepsilon/\varepsilon$, or, equivalently, minus the rate of real depreciation.

The Marshall–Lerner condition is the condition that the sum of these three terms be positive. If it is satisfied, a real depreciation leads to an improvement in the trade balance.

A numerical example will help here. Suppose

that a 1% depreciation leads to a relative increase in exports of 0.9% and to a relative decrease in imports of 0.8%. (Econometric evidence on the relation of exports and imports to the real exchange rates suggests that these are indeed reasonable numbers.) In that case, the right-hand side of the equation is equal to 0.9% − (−0.8%) − 1% = 0.7%. Thus the trade balance improves: The Marshall-Lerner condition is satisfied.

#1) determinants of imports (equation $Q = Q(Y, \varepsilon)$
    (+) (−)

#2) determinants of exports (equation $X = X(Y^*, \varepsilon)$
    (+) (+)

3) Basic diagram, ZZ, AA, DD & NX

4) ↑G → effect on trade balance (trade deficit ↑'s)   Diagram

5) ↑ Foreign demand for our exports (due to ↑ Y^*)   Diagram
    (improves trade balance)

6) Discussion of multiplier: MPM in U.S (smaller)
                                     vs Cda.

7) France: Socialist Party's dilemma in 1980s;
        Why countries are reluctant to stimulate domestic
        demand (↑G & T) during a recession ⇒ linked to trade deficit

8) Define & explain Marshall Lerner conditions.

9) Combining X∆ rate policy with fiscal policy in order to maintain Y.
        (diagram & explain)

10) J Curve: explain

11) In each case above trace what happens to private sug; public sug,
                                Investment
        through the eqns. $S = I + G - T - \varepsilon Q + X$
                            $NX = S + (T - G) - I$
                                        (p235)
                            $S = (Y - T - C) + (T - G)$

13

OUTPUT, THE INTEREST RATE, AND THE EXCHANGE RATE

In Chapter 12 we treated the exchange rate as one of the policy instruments available to the government. But the exchange rate is not a policy instrument. Rather, it is determined in the foreign-exchange market—a market in which there is an enormous amount of trading, as we saw in Chapter 11. This fact raises two obvious questions: What determines the exchange rate? How can the government affect it?

These are the questions that motivate this chapter. More generally, we examine the implications of simultaneous equilibrium in both the goods market and financial markets, including the foreign-exchange market. This allows us to characterize the joint movements of output, the interest rate, and the exchange rate in an open economy. The model we develop is an extension to the open economy of the *IS-LM* model we saw in Chapter 6. It is known as the **Mundell-Fleming model,** after the two economists, Robert Mundell[1] and Marcus Fleming, who first put it together in the 1960s. (The model presented here keeps the spirit but differs in its details from the original Mundell-Fleming model.)

---

[1] *Robert Mundell is one of Canada's most famous economists. He wrote pathbreaking papers on monetary policy in an open economy, and his ideas have proven both insightful and durable. In the 1960s, Mundell held an appointment at the University of Chicago. He and fellow Canadian Harry Johnson (also at Chicago at the time) were almost as famous for their hard living as for their work. They shared a secretary who often showed up to work in the morning to find a dustbin filled with empty rye bottles and two papers neatly stacked on her desk waiting to be typed.*

13-1 EQUILIBRIUM IN THE GOODS MARKET

Equilibrium in the goods market was the focus of Chapter 12, where we derived the following equilibrium condition:

$$Y = C(Y - T) + I(Y, r) + G + NX(Y, Y^*, \varepsilon) \qquad (13.1)$$
$$(\ +\ )\quad(+,-)\qquad\qquad(-,\ +,+)$$

[handwritten annotation: ↑ a real deprec. of domestic currency.]

For the goods market to be in equilibrium, output (the left-hand side of the equation) must be equal to the demand for domestic goods (the right-hand side of the equation). This demand is equal to consumption plus investment plus government spending plus exports minus imports. For simplicity, we have regrouped the last two terms under net exports, defined as exports minus imports, $X - \varepsilon Q$.

Consumption depends positively on disposable income. Investment depends positively on output and negatively on the real interest rate. Government spending is taken as given.

Net exports depend on domestic output, foreign output, and the real exchange rate. An increase in domestic output increases imports and thus decreases net exports. An increase in foreign output increases exports and thus increases net exports. An increase in ε—a real depreciation—leads (under the Marshall-Lerner condition, which we shall assume to hold throughout this chapter) to an increase in net exports.

The important implication of equation (13.1) for our purposes is the dependence of demand, and thus of output, on both the real interest rate and the real exchange rate:

[handwritten annotation: ✓✓]

- An increase in the real interest rate leads to a decrease in investment spending, and thus to a decrease in the demand for domestic goods. This leads, through the multiplier, to a decrease in output.

- An increase in the real exchange rate—a real depreciation—leads to a shift in demand toward domestic goods, and thus an increase in net exports. The increase in net exports increases demand and output.

For the remainder of the chapter, we shall make two simplifications to equation (13.1). First, given our focus on the short run, we have assumed in our previous treatment of the *IS-LM* model that the (domestic) price level was given. We shall extend this assumption to the foreign price level as well. Under this assumption, the real exchange rate ($\varepsilon \equiv EP^*/P$) and the nominal exchange rate (E) move together. A nominal depreciation leads, one for one, to a real depreciation. If, for notational convenience, we choose P and P^* so that $P^*/P = 1$ (and we can do so because they are index numbers), then $\varepsilon = E$ and we can replace ε by E in equation (13.1).

[handwritten annotation: Assumptions: Short run (very); P̄ P̄; ε = E P̄*/P̄ ; No inflation]*

Second, since we take the domestic price level as given, there is no inflation, actual or expected. Thus, the nominal and the real interest rates are the same, and we can replace the real interest rate, r, in equation (13.1) by the nominal interest rate, i.

With these simplifications, equation (13.1) becomes

$$Y = C(Y - T) + I(Y, i) + G + NX(Y, Y^*, E) \qquad (13.2)$$
$$(\ +\ )\quad(+,-)\qquad\qquad(-,\ +,+)$$

Output depends on both the nominal interest rate and the nominal exchange rate. These two simplifications will make it easier to think about the interactions between the goods market and financial markets.

13-2 EQUILIBRIUM IN FINANCIAL MARKETS

When we looked at financial markets in a closed economy, we simplified our task by assuming that people chose between only two financial assets, money and bonds. Now that we look at a financially open economy, we must allow for a second choice, that between domestic bonds and foreign bonds. Let's consider each choice in turn.

MONEY VERSUS BONDS

When looking at a closed economy in Chapter 6, we wrote the condition that the supply of money be equal to the demand for money as

$$\frac{M}{P} = YL(i) \tag{13.3}$$

We took the real supply of money [the left-hand side of equation (13.3)] as given. We assumed that the real demand for money [the right-hand side of equation (13.3)] depended on the level of transactions in the economy, measured by real output (Y), and on the opportunity cost of holding money rather than bonds (the nominal interest rate on bonds, i).

How should we change this characterization now that the economy is open? The answer is not very much, if at all.

In an open economy, the demand for domestic money is still mostly a demand by domestic residents. There is not much reason for, say, Germans to hold Canadian currency or demand deposits. They cannot use it for transactions in Germany, which require payment in German money. If they want to hold dollar-denominated assets, they are better off holding Canadian bonds, which at least pay a positive interest rate. And the demand for money by domestic residents still depends on the same factors as before: their level of transactions that we proxy by domestic real output, and the opportunity cost of holding money, the nominal interest rate on bonds.[2]

Thus, we can still use equation (13.3) to characterize the determination of the nominal interest rate in an open economy. The interest rate must be such that the supply and the demand for money are equal. *An increase in the money supply leads to a decrease in the interest rate. An increase in money demand, say as a result of an increase in output, leads to an increase in the interest rate.*

---

[2]***Digging deeper***. *Given that domestic residents can now hold both domestic and foreign bonds, the demand for money should depend on the expected rates of return on both domestic and foreign bonds. But our next assumption—interest parity—implies that these two expected rates of return are equal, so that we can write the demand for money directly as we did in equation (13.3).*

DOMESTIC BONDS VERSUS FOREIGN BONDS

In looking at the choice between domestic bonds and foreign bonds, we rely on the assumption we introduced in Chapter 11: Financial investors, domestic or foreign, go for the highest expected rate of return. In equilibrium, both domestic bonds and foreign bonds must be held. Both must therefore have the same expected rate of return.

As we saw in Chapter 11, this conclusion implies that the following arbitrage relation must hold:

$$i_t = i_t^* + \frac{E_{t+1}^e - E_t}{E_t}$$

Assumes perfect capital mobility which is ok in advanced economies with sophisticated fin. markets.

This equation, (11.4), which is known as the *interest parity condition*, says that the domestic interest rate i_t, must be equal to the foreign interest rate i_t^* plus the expected rate of depreciation of the domestic currency $(E_{t+1}^e - E_t)/E_t$.[3]

For now, we shall take the expected future exchange rate as given and denote it as $\overline{E}^e$ (we shall relax this assumption in Chapter 14).[4] Under this assumption, and dropping time indexes, the interest parity condition becomes

Key relation:
$$i = i^* + \frac{\overline{E}^e - E}{E} \qquad (13.4)$$

Guarantees the Forex market is in equilibrium. NX = NCF

$i - i^* = \frac{E^e}{E} - \frac{E}{E} / 1$

Bringing the terms in the current exchange rate, E, to the left-hand side and dividing both sides by $(1 + i - i^*)$ gives us an expression for the current exchange rate as a function of the expected future exchange rate and the domestic and foreign interest rates:

$i - i^* + 1 = \frac{E^e}{E}$

Key relation:
$$E = \frac{\overline{E}^e}{1 + i - i^*} \qquad (13.5)$$

$E = \frac{E^e}{1 + i - i^*}$

Equation (13.5) implies a negative relation between the domestic interest rate and the exchange rate. Given the expected future exchange rate and the foreign interest rate, *an increase in the domestic interest rate leads to a decrease in the exchange rate, thus to an appreciation. Symmetrically, a decrease in the domestic interest rate leads to an increase in the exchange rate, thus to a depreciation.*

Note i and domestic apprec/depn move together.

↑i ↓E which implies domestic appreciation

This relation between the exchange rate and the domestic interest rate will play a central role in the rest of this chapter. The best way to understand it is to think about the sequence of events that takes place in financial and foreign-exchange markets after an increase in the Canadian interest rate to above, say, the U.S. interest rate. Start from a situation in which the Canadian and U.S. interest rates are equal, so that $i = i^*$. This implies, from equation (13.5), that the current C$/US$ exchange rate is equal to the expected future exchange rate.

---

[3] *For notational convenience, we shall replace the earlier approximation symbol (≈) by an equal sign (=) in what follows.*

[4] ***Digging deeper.*** *All that is needed for our purposes is that the future expected exchange rate responds less than one for one to movements in the current exchange rate. Check, for example, how the results of this chapter are modified if you assume that the expected future exchange rate is given by $E^e = \lambda E + (1 - \lambda)\overline{E}$, with λ between zero and 1. The assumption we use in the text corresponds to the case where λ is equal to zero. But all the qualitative results we derive still hold when λ is between zero and 1. However, the reader should be warned: The data suggest λ is indeed very close to 1. In the limiting case with $\lambda = 1$, equation (13.5) becomes $i = i^*$: The domestic interest rate must equal the foreign interest rate.*

(handwritten top margin) ↑ higher 'r' Forex (us $)

(handwritten top margin) as investors want to purchase cdn bonds they need cdn $, ∴ excess supply of foreign $US, ↓ XS rate ↓, cdn $ appreciates

(handwritten near graph) ↑ appreciating Cdn $

(handwritten left margin) Immediately due to higher r + excess demand for cdn $, the Cdn $ appreciates. But then according to the eqn, there is a future expected depn.

Suppose that, as a result of a Canadian monetary contraction, the Canadian interest rate increases. At an unchanged exchange rate, Canadian bonds become more attractive, so that financial investors want to shift out of U.S. bonds and into Canadian bonds. To do so, they must sell American bonds for US$, then sell US$ for C$, then use the proceeds to buy Canadian bonds. But as investors sell U.S. dollars for Canadian dollars, the Canadian dollar appreciates.

The intuition for why an increase in the Canadian interest rate leads to an appreciation of the dollar is straightforward: An increased demand for dollars leads to an increase in the price of dollars. What is less obvious is by how much the dollar must appreciate. The important point here is the following: <u>If financial investors do not change their expectation of the future exchange rate, then *the more the dollar appreciates today,* the more investors expect it to *depreciate in the future.*</u> Other things being equal, this expectation makes U.S. bonds more attractive: When the dollar is expected to depreciate, a given rate of return in US$ means a higher rate of return in C$. The initial dollar appreciation must therefore be such that the expected future depreciation compensates for the increase in the domestic interest rate. When this is the case, investors are again indifferent and equilibrium prevails.

A numerical example will help here. Assume that one-year Canadian and U.S. interest rates are both equal to 4%. Suppose that the Canadian interest rate now increases to 10%. The Canadian dollar will then appreciate by 6% today. To see why, note that if the Canadian dollar appreciates by 6% today and investors do not change their expectation of the exchange rate one year ahead, the C$ is now expected to depreciate by 6% over the coming year. Put another way, the US$ is expected to appreciate by 6% over the C$ over the coming year, so that holding U.S. bonds yields an expected rate of return of 4%, the rate of return in US$, plus 6%, the expected appreciation of the US$. Holding Canadian bonds or holding U.S. bonds both yield an expected rate of return of 10% in Canadian dollars. Financial investors are willing to hold either one, so there is equilibrium in the foreign-exchange market.

In terms of equation (13.4),

(handwritten) ↑↓EV: ↓E = Ēe

(handwritten) H i - i* (larger den) ↑

(handwritten) / cdn currency appreciates immediately

$$ i = i^* + \frac{\bar{E}^e - E}{E} $$

(handwritten) expected depn, not actual in the immediate sense

$$ 10\% = 4\% + 6\% $$

(handwritten) but in terms of future:

(handwritten) ↑ i Ēe : move opposite directions these's an expected depn of cdn currency.

The rate of return from holding Canadian bonds (the left-hand side) is equal to 10%. The expected rate of return from holding U.S. bonds, expressed in C$ (the right-hand side), is equal to the U.S. interest rate, 4%, plus the expected depreciation of the Canadian dollar, 6%.

Figure 13-1 plots the relation between the interest rate and the exchange rate implied by equation (13.5)—the *interest parity relation.* It is drawn for a given expected future exchange rate, $\bar{E}^e$, and a given foreign interest rate, i^*. The lower the interest rate, the higher the exchange rate: The relation is thus drawn as a downward-sloping curve. Equation (13.5) also implies that when the domestic interest rate is equal to the foreign interest rate, the exchange rate is equal to the expected future exchange rate. When $i = i^*$, then $E = \bar{E}^e$. This point is denoted by A in the figure.

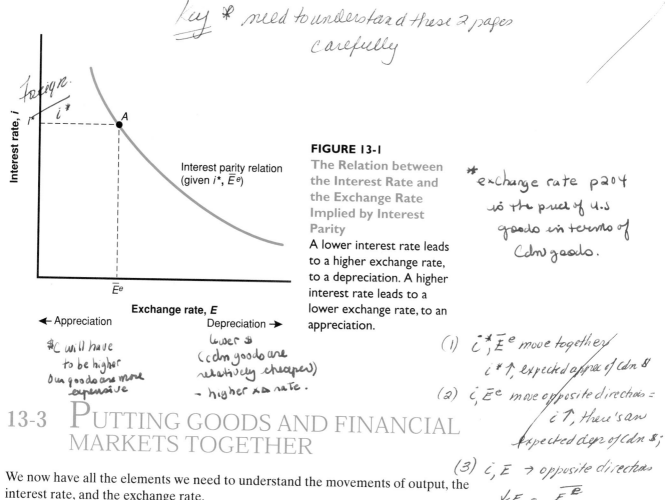

Foreign

Interest rate, i

i^*

A

Interest parity relation
(given i^*, $\bar{E}^e$)

$\bar{E}^e$

Exchange rate, E

← Appreciation   Depreciation →

FIGURE 13-1

The Relation between the Interest Rate and the Exchange Rate Implied by Interest Parity

A lower interest rate leads to a higher exchange rate, to a depreciation. A higher interest rate leads to a lower exchange rate, to an appreciation.

*exchange rate p204
is the price of U.S
goods in terms of
Cdn goods.

$C will have
to be higher
our goods are more
expensive

lower $
(cdn goods are
relatively cheaper)
- higher xR rate.

(1) i^*, $\bar{E}^e$ move together
$i^* \uparrow$, expected apprec of Cdn $

(2) i, $\bar{E}^e$ move opposite directions =
$i \uparrow$, there's an
expected depn of Cdn $;

(3) i, E → opposite directions
$\downarrow E = \dfrac{\bar{E}^e}{1+i-i^*}$
$\uparrow$

(4) i^*, E move together:
$\uparrow E = \dfrac{\bar{E}^e}{1+i-i^* \uparrow}$ } smaller denominator results

13-3 PUTTING GOODS AND FINANCIAL MARKETS TOGETHER

We now have all the elements we need to understand the movements of output, the interest rate, and the exchange rate.

Goods-market equilibrium implies that output depends on, among other factors, the interest rate and the exchange rate:

$$Y = C(Y - T) + I(Y, i) + G + NX(Y, Y^*, E)$$

The interest rate, in turn, is determined by the equality of money supply and money demand:

$$\frac{M}{P} = YL(i)$$

The interest parity condition implies a negative relation between the domestic interest rate and the exchange rate:

$$E = \frac{\bar{E}^e}{1 + i - i^*}$$

Together, these three relations determine output, the interest rate, and the exchange rate. Working with three relations is not very easy. But we can easily reduce them to two by using the interest parity condition to eliminate the exchange rate in the goods-market equilibrium relation. Doing this gives us the following two equations, the open-economy versions of our old *IS* and *LM* relations:

EG p212 $i = i^* + \dfrac{E^e - E}{E}$

$5.2 = 5.5 - .3$ (exp apprec)

Immediately there's a depn of
Cdn $; $i \downarrow E \uparrow$;
but in terms of future E^e,
$i \downarrow$, expected future apprec
of Cdn $

$$\text{IS:} \quad Y = C(Y - T) + I(Y, i) + G + NX\left(Y, Y^*, \frac{\overline{E}^e}{1 + i - i^*}\right)$$

$$\text{LM:} \quad \frac{M}{P} = YL(i)$$

movement along IS

Take the *IS* relation first and consider the effects of an <u>increase in the interest rate</u> on output. An increase in the interest rate now has two effects.

■ The first, which was already present in a closed economy, is the direct effect on investment. A higher interest rate leads to a decrease in investment, and thus to a decrease in the demand for domestic goods.

■ The second, which is new, is the effect through the exchange rate. As we saw earlier, an increase in the domestic interest rate leads to an appreciation. The appreciation, which makes domestic goods relatively more expensive, leads in turn to a decrease in net exports and thus to a decrease in the demand for domestic goods.

Both effects work in the same direction: An increase in the interest rate leads to a decrease in demand and—through the multiplier—to a decrease in output. The multiplier is smaller than in the closed economy. This is because part of demand falls on foreign goods rather than on domestic goods.

The *IS* relation between the interest rate and output is drawn in Figure 13-2(a) for given values of all the other variables in the relation, namely T, G, Y^*, i^*, and $\overline{E}^e$. The *IS* curve is downward-sloping, as an increase in the interest rate leads

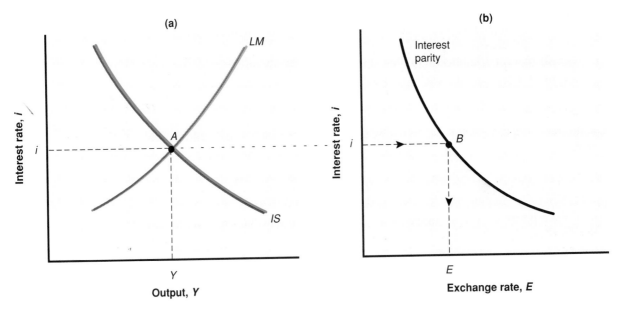

FIGURE 13-2

The *IS-LM* Model in the Open Economy

An increase in the interest rate reduces output both directly and indirectly, through the exchange rate: The *IS* curve is downward-sloping. Given the real money stock, an increase in income increases the interest rate: The *LM* curve is upward-sloping.

to a decrease in output. It looks very much the same as in the closed economy, but it hides a more complex relation than before: The interest rate affects output not only directly, but also indirectly through the exchange rate.

The *LM* relation, by contrast, is exactly the same as it was in the closed economy. The *LM* curve is upward-sloping. For a given value of the real money stock *(M/P)*, an increase in output leads to an increase in the demand for money, and thus to an increase in the equilibrium interest rate.

Equilibrium in the goods and financial markets is attained at point *A* in Figure 13-2(a), with output level *Y* and interest rate *i*. The equilibrium value of the exchange rate cannot be read directly from the graph. But it is easily obtained from Figure 13-2(b), which replicates Figure 13-1 and gives the exchange rate associated with a given interest rate. The exchange rate associated with the equilibrium interest rate *i* is equal to *E*.

13-4 THE EFFECTS OF POLICY IN AN OPEN ECONOMY

Having derived the *IS-LM* model for the open economy, we can now put it to good use and look at the effects of policy.

THE EFFECTS OF FISCAL POLICY IN AN OPEN ECONOMY

Let's look, once again, at a change in government spending. Suppose that, starting from budget balance, the government decides to increase health spending and thus to run a budget deficit. What happens to the level of output and its composition, to the interest rate, and to the exchange rate?

The answer is given in Figure 13-3(a). The economy is initially at point *A*. An increase in government spending from *G* to *G'* increases output at a given interest rate and thus shifts the *IS* curve to the right, from *IS* to *IS'*. Because government spending does not enter the *LM* relation, the *LM* curve does not shift. The new equilibrium is at point *A'*, with a higher level of output and a higher interest rate. As shown in Figure 13-3(b), the higher interest rate leads to a decrease in the exchange rate—an appreciation. Thus, *an increase in government spending leads to an increase in output, an increase in the interest rate, and an appreciation.*

How do we tell this story in words? An increase in government spending leads to an increase in demand, and thus to an increase in output. As output increases, so does the demand for money, leading to upward pressure on the interest rate. The increase in the interest rate, which makes domestic bonds more attractive, also leads to an appreciation of the domestic currency. The higher interest rate and the appreciation both decrease the domestic demand for goods, offsetting some of the effect of government spending on demand and output.

Can we tell what happens to the various components of demand? Consumption and government spending clearly both go up, consumption because of the increase in income, government spending by assumption. But what happens to investment is ambiguous. Recall that investment depends on both output and the interest rate: *I = I(Y, i)*. On the one hand, output goes up, leading to an increase in

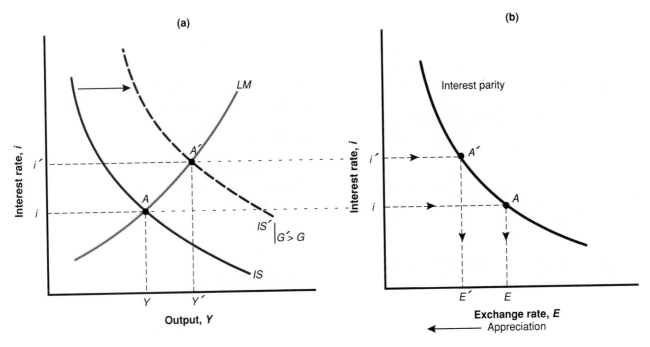

FIGURE 13-3

The Effects of an Increase in Government Spending

An increase in government spending leads to an increase in output, an increase in the interest rate, and an appreciation.

ambiguous effect of G on investment in closed + open economy ✓

investment. But, on the other, the interest rate also goes up, leading to a decrease in investment. Depending on which of these two effects dominates, investment can go up or down. The effect of government spending on investment was ambiguous in the closed economy; it remains ambiguous in the open economy.

Turn next to net exports. Recall that net exports depend on foreign output, domestic output, and the exchange rate: $NX = NX(Y, Y^*, E)$. Thus, both the appreciation and the increase in output combine to decrease net exports: The appreciation decreases exports and increases imports, and the increase in output increases imports further. The budget deficit leads to a deterioration of the trade balance. If trade was balanced to start, then the budget deficit leads to a trade deficit.

THE EFFECTS OF MONETARY POLICY IN AN OPEN ECONOMY

EG #2:

The effects of our other favourite policy experiment, a monetary contraction, are shown in Figure 13-4. At a given level of output, a decrease in the money stock, from M/P to M'/P, leads to an increase in the interest rate: The LM curve therefore shifts up, from LM to LM'. Because money does not directly enter the IS relation, the IS curve does not shift. The equilibrium moves from point A to point A' in Fig-

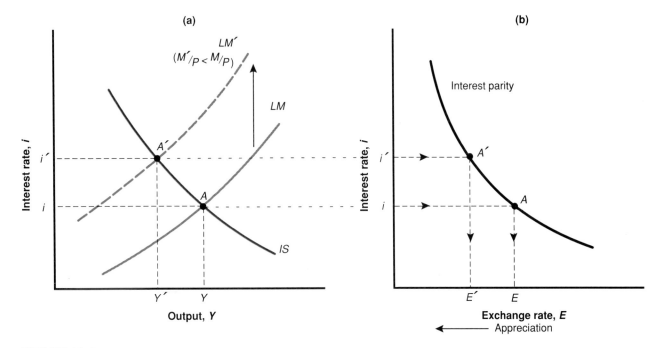

FIGURE 13-4
The Effects of a Monetary Contraction
A monetary contraction leads to a decrease in output, an increase in the interest rate, and an appreciation.

ure 13-4(a). As shown in Figure 13-4(b), the increase in the interest rate leads to an appreciation of the domestic currency.

Thus, *a monetary <u>contraction</u> leads to a decrease in output, an increase in the interest rate, and an appreciation.* The story is easy to tell. A monetary contraction leads to an increase in the interest rate, making domestic bonds more attractive and triggering an appreciation. The higher interest rate and the appreciation both decrease demand and output. As output decreases, the induced decrease in the demand for money decreases the interest rate, offsetting some of the initial increase in the interest rate and some of the initial appreciation.

How well do the implications of this model fit the facts? To answer this question, one could hardly design better experiments than the sharp monetary and fiscal policy changes the U.S. economy went through in the late 1970s and early 1980s.[5] This is the topic taken up in the In Depth box entitled "Monetary Contraction and Fiscal Expansion: The United States in the Early 1980s." The Mundell-Fleming model and its predictions pass with flying colours.

---

[5] *That these experiments were instructive for economists does not imply that they were necessarily good for the U.S. economy (or their neighbours).*

In 1979, the Chairman of the Fed, Paul Volcker, concluded that U.S. inflation was too high and had to be reduced. The result was a drastic shift in monetary policy and, for much of the following three years, a sharp monetary contraction. As we have not yet developed a theory of inflation, we defer a detailed examination of the "Volcker disinflation" to later in this book (Chapter 18). Suffice it to say that the effects of that monetary contraction match well the implications of the model that we have just developed. From 1980 to 1982, interest rates increased sharply, the dollar appreciated, and output contracted significantly. This is the context in which the other major policy shift—the change in fiscal policy—took place, from 1981 on.

In 1980, Ronald Reagan had been elected on the promise of more conservative policies, namely a scaling down of taxation and the government's role in economic activity. This commitment was the inspiration for the *Economic Recovery Act* of August 1981. Personal income taxes were to be cut by a total of 23%, in three installments over the period 1981 to 1983. Corporate taxes (that is, taxes on corporations) were also reduced. Tax packages are not written in stone, and what Congress does, Congress can undo. But in this case, Congress made only minor modifications later. The main one, the 1982 *Tax Equity and Fiscal Responsibility Act,* removed some of the tax cuts on corporations. In 1986 a major tax reform was put in place, with a reduction in the highest marginal income tax rates and the elimination of many loopholes; but this takes us beyond the period on which we want to focus.

The 1981 to 1983 tax cuts were not accompanied by corresponding decreases in government spending, and the result was a steady increase in budget deficits, which reached a peak in 1983 at 5.6% of GDP. Table 1 gives spending and revenue numbers for 1980 to 1984.

What were the Reagan administration's motivations for cutting taxes without implementing corresponding cuts in spending? These are still being debated today, but there is agreement that there were two main motivations.

One was based on the beliefs of a fringe, but influential, group of economists called the **supply siders,** who argued that a cut in tax rates would lead people and firms to work much harder and more productively, and that the resulting increase in activity would lead to an increase, not a decrease, in tax revenues. Whatever the merits of the argument appeared to be then, it proved wrong after the fact. Even if some people did work harder and more productively after the tax cuts, tax revenues decreased and the fiscal deficit increased.

The other was the hope that the cut in taxes, and the resulting increase in deficits, would scare Congress into cutting spending, or at the very least into not increasing spending further. This motivation turned out to be partly right; Congress indeed found itself under enormous pressure not to increase spending, and the growth of spending in the 1980s was surely lower than it would have been otherwise. Nonetheless, this decrease in spending was not enough to offset the shortfall in taxes and avoid the rapid increase in deficits.

· · · · · · · · · · · ·
TABLE 1

THE EMERGENCE OF LARGE U.S. BUDGET DEFICITS, 1980–1984

	1980	1981	1982	1983	1984
Spending	22.0	22.8	24.0	25.0	23.7
Revenues	20.2	20.8	20.5	19.4	19.2
Personal taxes	9.4	9.6	9.9	8.8	8.2
Corporate taxes	2.6	2.3	1.6	1.6	2.0
Budget surplus (−: deficit)	−1.8	−2.0	−3.5	−5.6	−4.5

Numbers are for fiscal years, which start in October of the preceding calendar year. They are expressed as a percentage of GDP.

Source: Office of Management and Budget, Budget of the United Sates Government (1998), *Historical Tables.*

Whatever the reason for the deficits, the effects of the monetary contraction and fiscal expansion were very much in line with what the Mundell-Fleming model predicts. Table 2 gives the evolution of the main macroeconomic variables from 1980 to 1984.

From 1980 to 1982, the evolution of the economy was dominated by the effects of the monetary contraction. Interest rates, both nominal and real, increased sharply, leading both to a large dollar appreciation (a decrease in the real exchange rate) and a recession. The goal of lowering inflation was achieved, although not right away; by 1982, inflation was down to about 4%. Lower output and dollar appreciation had opposing effects on the trade balance (lower output leading to lower imports and an improvement in the trade balance, the appreciation leading to a deterioration), resulting in little change in the trade deficit before 1982.

From 1982 on, the evolution of the economy was dominated by the effects of the fiscal expansion. As our model predicts, these effects were strong output growth, high interest rates, and further dollar appreciation. The effects on the trade balance now went the same way: High output growth and dollar appreciation both led to a trade deficit of 2.7% by 1984. By the mid-1980s, the main U.S. macroeconomic policy issue had become that of the **twin deficits,** the budget and the trade deficits.

....

TABLE 2

MAJOR U.S. MACROECONOMIC VARIABLES, 1980–1984

	1980	1981	1982	1983	1984
GDP growth (%)	−0.5	1.8	−2.2	3.9	6.2
Unemployment rate (%)	7.1	7.6	9.7	9.6	7.5
Inflation (CPI) (%)	12.5	8.9	3.8	3.8	3.9
Interest rate (%):					
Nominal	11.5	14.0	10.6	8.6	9.6
Real	2.5	4.9	6.0	5.1	5.9
Real exchange rate	117	99	89	85	77
Trade surplus (%) (−: deficit)	−0.5	−0.4	−0.6	−1.5	−2.7

[handwritten margin notes: fiscal expansion ↑Y; ↓fiscal expansion ∴ i↑'s; apprec of U.S currency; → doesn't really correlate with i because there was also ↓Y which improved trade balance. (↓Imports); fiscal exp.]

Inflation: rate of change of the CPI. The *interest rate* is the three-month T-bill rate. The *real interest rate* is equal to the nominal rate minus the forecast of inflation by DRI, a private forecasting firm. The *real exchange rate* is the trade-weighted real exchange rate, with 1973 = 100. The *trade surplus* is expressed as a ratio to GDP.

Sources: Economic Report of the President, 1995; Data Resources Incorporated.

*13-5 FIXED EXCHANGE RATES

We have assumed so far that the central bank chose the money supply and let the exchange rate adjust in whatever manner was implied by equilibrium in the foreign-exchange market. In most countries, however, this assumption does not reflect reality. Central banks act under implicit or explicit exchange-rate targets and use monetary policy to achieve those targets.[5] The targets are sometimes implicit, sometimes explicit; they are sometimes specific values, sometimes bands or ranges. Let's briefly survey the arrangements across countries.

---

[5]*The Bank of Canada's Monetary Conditions Index (MCI) is a weighted average of the nominal interest rate and the nominal exchange rate. We will discuss it further in Chapter 28.*

PEGS, CRAWLING PEGS, BANDS, AND THE EMS

At one end of the spectrum stand countries such as the United States and Japan. These countries have no explicit exchange-rate targets. And, while their central banks surely do not ignore movements in the exchange rate, they have shown themselves quite willing to let their exchange rates fluctuate considerably. There were large swings in the U.S. dollar in the 1980s. Until recently, the main story in foreign-exchange markets was the steady appreciation of the yen.

At the other end of the spectrum stand countries that operate under **fixed exchange rates.** These countries maintain a fixed exchange rate in terms of some foreign currency. Some **peg** their currency to the dollar: The list ranges from the Bahamas to Hong Kong. Others peg their currency to the French franc; most of these are former French colonies in Africa. Others peg to a basket of currencies, with the weights reflecting the composition of their trade. The label "fixed" is misleading, however. It is not the case that the exchange rate in countries with fixed exchange rates never changes. But, typically, the changes are rare. An extreme case is that of the African countries pegged to the French franc. When their exchange rates were readjusted in January 1994, this adjustment was the first one in 45 years. Because these changes are rare, economists use specific words to distinguish them from the daily changes that occur under flexible exchange rates. They refer to an increase in the exchange rate under a system of fixed exchange rates as a **devaluation** rather than a depreciation, and to a decrease in the exchange rate under a system of fixed exchange rates as a **revaluation** rather than an appreciation.

Between these extremes stand countries with various degrees of commitment to an exchange-rate target. Although Canada abandoned fixed exchange rates in 1970, since then the Bank of Canada has displayed a good deal of concern in maintaining a stable C$/US$ exchange rate. The Bank has no formal policy but appears to operate under the following practice: It tries to smooth out short-term fluctuations (swings as large as half a penny per day are extremely uncommon), but it tries not to stand in the way of relatively slow but long-lived appreciations and depreciations.

Other nations have more formal ties to the U.S. dollar. For example, many countries operate under a **crawling peg.** The name describes it well. These countries typically have rates of inflation that exceed that of the United States. If they were to peg their nominal exchange rate against the dollar, the more rapid increase in the domestic price level over the U.S. price level would lead to a steady real appreciation and rapidly make their goods noncompetitive. To avoid this effect, these countries choose a predetermined rate of depreciation against the dollar. They choose to "crawl" vis-à-vis the dollar.

Yet another arrangement is for a group of countries to maintain their bilateral exchange rates (that is, the exchange rate between each pair of countries) within some bands. A prominent example of such a system was the **European Monetary System (EMS),** introduced in 1978, and gradually adopted by most of the countries of the European Union. Under the rules of their **exchange-rate mechanism (ERM),** member countries agree to maintain their exchange rate vis-à-vis the other currencies in the system within narrow limits or **bands** around a **central parity.** Changes in the central parity and devaluations or revaluations of specific currencies can occur, but only by common agreement among member countries. In practice, adjustments in parity have been infrequent, but not exceptional. For example, Italy—which consistently had a higher rate of inflation than the average of its EMS partners—went through seven devaluations between 1978 and 1994. Germany, which consistently had a lower rate of inflation than its partners, went through seven revaluations during the same time period. In 1992, the EMS began to fall apart: Two of its major members, the United Kingdom and Italy, suspended their participation, and the bands for the remaining members were widened. We examine the origins of this crisis in the Global Macro on German unification on pages 254 and 255, as well as in Chapter 14. But first we must understand how, and then why, countries operate under such arrangements.

Pegging the Exchange Rate, and Monetary Control

How does a country peg its exchange rate? It cannot just announce an exchange rate and stand there. Rather, it must take measures so that its chosen exchange rate will prevail in the foreign-exchange market. Let's look at the implications and mechanics of pegging.

Pegging or no pegging, under the assumption of perfect capital mobility, the exchange rate and the nominal interest rate must satisfy the interest parity condition:

$$i_t = i_t^* + \frac{E_{t+1}^e - E_t}{E_t}$$

Now suppose that the country pegs the exchange rate at some level, call it $\bar{E}$, so that $E_t = \bar{E}$. If financial and foreign-exchange markets believe that the exchange rate will indeed remain fixed, then their expectation of the future exchange rate is also equal to $\bar{E}$, and the interest parity relation becomes

$$i_t = i_t^* + \frac{\bar{E} - \bar{E}}{\bar{E}} = i_t^* \qquad \text{\textit{Expectation equals the fixed xa rate}}$$

If financial investors expect the exchange rate to remain unchanged, they will require the same nominal interest rate in both countries. Thus, *under a fixed exchange rate and perfect capital mobility, the domestic interest rate must be equal to the foreign interest rate.* $\quad i = i^*$

This condition has one further important implication. Return to the equilibrium condition that the supply of and demand for money be equal. Now that $i = i^*$, this condition becomes

$$\frac{M}{P} = YL(i^*) \qquad (13.6)$$

EG #3. ↑ IS under fixed xa rate system.

Suppose that an increase in domestic output increases the demand for money. In a closed economy, the central bank could leave the money stock unchanged, leading to an increase in the equilibrium interest rate. The same is true in an open economy under flexible rates. In that case, the result will be an increase in the interest rate and an appreciation. But now, because of its commitment to maintain the exchange rate, the central bank can no longer keep the money stock unchanged. To prevent an increase in the domestic interest rate above the foreign interest rate, it must increase the supply of money in line with the increase in demand so that the equilibrium interest rate does not change. Given the price level, P, nominal money M must adjust so that equation (13.6) holds.

Put more starkly: *Under fixed exchange rates, the central bank gives up monetary policy as a policy instrument.* A fixed exchange rate implies a domestic interest rate equal to the foreign rate. And the money supply must adjust so as to maintain that interest rate. (Note that these results depend very much on the assumption of perfect capital mobility. The case of fixed exchange rates with imperfect capital mobility, which is more relevant for middle-income countries, is treated in the appendix to this chapter.)

FISCAL POLICY UNDER FIXED EXCHANGE RATES

If monetary policy can no longer be used under fixed exchange rates, what about fiscal policy? To answer this question, we use Figure 13-5.

Figure 13-5 starts by replicating Figure 13-3(a), which we used earlier to analyze the effects of fiscal policy under flexible exchange rates. In that case, we saw that a fiscal expansion shifted the IS curve to the right. If the money stock were to remain unchanged, this would lead, as we saw in Figure 13-3, to a movement in the equilibrium from point A to point B, leading to an increase in output from Y_A to Y_B, an increase in the interest rate, and a decrease in the exchange rate—an appreciation. Under fixed exchange rates, however, the central bank cannot let the exchange rate decrease. As the increase in output leads to an increase in the demand for money, the central bank must accommodate this increased demand for money by increasing the money supply. In terms of Figure 13-5, the central bank must shift the LM curve down as the IS curve shifts to the right, so that the interest rate, and thus the exchange rate, do not change. The equilibrium therefore moves from point A to point C, with higher output Y_C and unchanged interest and exchange rates. Thus,

FIGURE 13-5

The Effects of a Fiscal Expansion under Fixed Exchange Rates
Under flexible exchange rates, a fiscal expansion increases output from Y_A to Y_B. Under fixed exchange rates, output increases from Y_A to Y_C.

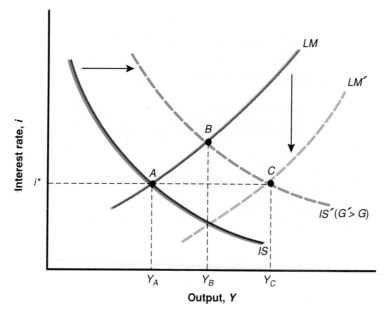

under fixed exchange rates, fiscal policy is more powerful than it is under flexible exchange rates. This is so because fiscal policy triggers monetary accommodation.

As this chapter comes to an end, a question should have started to form in your mind. Why would a country—particularly a small country—choose to fix its exchange rate? By doing so, it gives up a powerful tool for correcting trade imbalances or changing the level of economic activity. By committing to a particular exchange rate, it also gives up control of its interest rate. Furthermore, it has to match movements in the foreign interest rate, at the risk of unwanted effects on its own activity. (This is the topic of the Global Macro box entitled "German Unification, Interest Rates, and the EMS.") And while the country retains control of fiscal policy, one policy instrument is not enough. As we saw in Chapter 12, for example, a fiscal expansion can help the economy get out of a recession, but only at the cost of a larger trade deficit. And a country that wants to decrease its budget deficit cannot use monetary policy to offset the contractionary effect of its fiscal policy on output.

So, why do countries (and provinces) do it? This is a question to which we shall return many times in this book. Here is a preview of the arguments:

- Fixed exchange rates make business easier and less costly for firms. In making production and investment decisions, firms do not have to worry about changes in the exchange rate. This is one of the main reasons why many European countries have chosen to adopt a common currency.

- Another reason, which we explore in Chapter 14, is that, while fixed exchange rate systems may indeed have shortcomings, so do flexible exchange rate systems. Many economists believe that, like changes in stock prices, some movements in exchange rates are not related to fundamentals (see Chapter 9) and that, under flexible exchange rates, exchange rates often move too much. These exchange-rate movements, in turn, lead to unwelcome movements in imports, exports, and output. Fixing the exchange rate eliminates these problems.

- Yet another argument is the fact that if the real exchange rate needs to be adjusted—say, to eliminate a trade deficit—this can be done even under fixed exchange rates. What is fixed under fixed exchange rates is the *nominal* exchange rate, not the *real* exchange rate. Thus, in principle, adjustments in the domestic and foreign price levels—which so far we have taken as given—can achieve the same result as a change in the nominal exchange rate. We explore this argument in Chapter 19, after we have developed a theory of price movements.

- A final argument is that, by removing the scope for independent monetary policy, fixed exchange rates eliminate the risk that the central bank will misbehave. This is part of a larger argument that tight constraints on policy, whether fiscal or monetary, can actually make things better rather than worse. We explore it in Chapter 27.

The arguments are sufficiently complex that the case for or against fixed exchange rates will remain controversial. Indeed, it is likely to remain one of the main policy issues in many countries for a long time to come.

GERMAN UNIFICATION, INTEREST RATES, AND THE EMS

GLOBAL MACRO

Under a system of fixed exchange rates such as the EMS (let's ignore here the degree of flexibility afforded by the bands), no individual country can change its interest rate if the others do not change theirs as well. So, how do interest rates actually change? Two arrangements are possible. One is for the member countries to coordinate all changes in their interest rates. Another is for one of the countries to take the lead and for the other countries to follow. This is indeed what happened in the EMS, with Germany in the role of the leader.

During the 1980s, most European central banks shared similar goals and were happy to let the Bundesbank (the German central bank) take the lead. But in 1990 German unification led to a sharp divergence in goals between the Bundesbank and the other EMS nations' central banks. Recall the macroeconomic implications of unification from Chapter 6: The need for large transfers to Eastern Germany and an investment boom both led to a large increase in demand in Germany. The Bundesbank's fear that this shift would generate too strong an increase in activity led it to adopt a restrictive monetary policy. The result was, as we saw, strong growth in Germany together with a large increase in interest rates.

This may well have been the right policy mix for Germany. But for other countries the effects of German unification were less appealing. The other countries had not experienced the same increase in demand, but to stay in the EMS they had to match German interest rates. The results were sharp decreases in demand and in output in the other countries. These results are presented in Table 1, which gives nominal and real interest rates, inflation rates, and GDP growth from 1990 to 1992 for Germany and for two of its EMS partners, France and Belgium.

Note first how the high German nominal interest rates were matched by both France and Belgium. Nominal interest rates were actually higher in France than in Germany in all three years! This is because France needed higher interest rates than Germany to maintain the DM/franc parity; the reason is that financial markets were not sure that France would actually keep the parity of the franc vis-à-vis the DM. Thus, worried about a possible devaluation of the franc, they asked for a higher interest rate on French bonds than on German bonds.

While they had to match—or, as we have just seen, more than match—German nominal rates, France and Belgium had less inflation than Germany. The result was very high real interest rates, indeed higher than in Germany. In both France and Belgium, average real interest rates from 1990 to 1992 were close to 7%. And, in both countries, the period 1990–1992 was characterized by slow growth and rising unemployment. Unemployment in France in 1992 was 10.4%, up from 8.9% in 1990. The corresponding numbers for Belgium were 12.1% and 8.7%.

While we have looked at only two of Germany's EMS partners, a similar story was unfolding in the other EMS countries. Average unemployment in the European Union, which stood at 8.7% in 1990, had increased to 10.3% in 1992. The effects of high real interest rates on spending were not the only source of this slowdown, but they were the main one.

By 1992, an increasing number of countries were wondering whether to keep defending their EMS parity or to give up and lower their interest rates. Worried by the risk of devaluations, financial markets started to ask for higher interest rates in those countries where they thought devaluation was more likely. The result was two major exchange rate crises, one in the fall of 1992 and the other in the summer of 1993. By the end of these two crises, two countries, Italy and the United Kingdom, had left the EMS. We look at these crises, their origins and their implications, in Chapter 14.

Summary: Early 90s → German unification leads to ↑I, ↑Y boom in economy; high growth rates; 1990 → 5.1% 1991 → 4.5%; Central bank wants to dampen the boom ∴ ↓MS ↑i

Other European economies: not the right policy mix ∴ dilemma they're forced to match German "i" ∴ ↓Y (France .7 in 91 and 1.4 in 92, Belgium also).

— Turns out that France has to keep i higher than Germany to maintain DM/Franc parity; fin. markets are worried about potential devaluation of franc ∴ want higher i) (an other countries also)

Inflation: Germany → extra economic activity ⟹ higher π ∴ lower 'r'; not the case in France/Belgium = both much higher 'r' than Germany ∴ ↓Y ↑Un etc

....

TABLE 1

GERMAN UNIFICATION, INTEREST RATES, AND OUTPUT GROWTH: GERMANY, FRANCE, AND BELGIUM, 1990–1992

	NOMINAL INTEREST RATES			INFLATION		
	1990	1991	1992	1990	1991	1992
Germany	8.5	9.2	9.5	2.7	3.7	4.7
France	10.3	9.6	10.3	2.9	3.0	2.4
Belgium	9.6	9.4	9.4	2.9	2.7	2.4
	REAL INTEREST RATES			GDP GROWTH		
	1990	1991	1992	1990	1991	1992
Germany	5.7	5.5	4.8	5.7	4.5	2.1
France	7.4	6.6	7.9	2.5	0.7	1.4
Belgium	6.7	6.7	7.0	3.3	2.1	0.8

(handwritten annotations: "Fin markets" near France 1990–1992 nominal rates; "less than Germany" near France/Belgium 1992 real interest rates; "more Un." near France/Belgium 1992 GDP growth)

The *nominal interest rate* is the short-term nominal interest rate. The *real interest rate* is the ex post real interest rate over the year—that is, the nominal interest rate minus actual inflation over the year.

Source: OECD Economic Outlook, December 1993.

(handwritten notes:)
This chapter: 2 key relations: $i = i^* + \dfrac{E^e - E}{E}$ and $E = \dfrac{\overline{E}}{1 + i - i^*}$

P244-5 key to explain
Eg 1: Fiscal Policy
Eg 2: Monetary contraction

SUMMARY

....

■ In an open economy, the demand for goods depends on both the interest rate and the exchange rate. A decrease in the interest rate increases the demand for goods. An increase in the exchange rate—a depreciation—also increases the demand for goods.

■ The interest rate is determined by the equality of money demand and money supply. The exchange rate is determined by the interest parity condition, which states that the domestic interest rate must equal the foreign interest rate plus the expected rate of depreciation. Given the expected exchange rate and the foreign interest rate, increases in the domestic interest rate lead to a decrease in the exchange rate (an appreciation) and decreases in the domestic interest rate lead to an increase in the exchange rate (a depreciation).

■ Under flexible exchange rates, an expansionary fiscal policy leads to an increase in output, an increase in the interest rate, and an appreciation. A contractionary monetary policy leads to a decrease in output, an increase in the interest rate, and an appreciation.

■ There are many types of exchange-rate arrangements. They range from fully flexible exchange rates to crawling pegs, to pegs, to fixed exchange rates. Under fixed exchange rates, a country maintains a fixed exchange rate in terms of a foreign currency or a basket of currencies.

■ Under fixed exchange rates and perfect capital mobility, a country must maintain an interest rate equal to the foreign interest rate. Thus, the central bank loses the use of monetary policy as a policy instrument. Fiscal policy becomes more powerful, however, because fiscal policy triggers monetary accommodation and thus does not lead to offsetting changes in the interest rate and exchange rate.

(handwritten notes:)
- Eg of U.S economy in 80s: Combo of fiscal + monetary policy + twin deficits
- Fixed X∆ rates = Eg 3: ↑ IS under Fixed X∆ rate
    - reasons for & against Fixed X∆rate
    - Summary of European economies

CHAPTER 13: OUTPUT, THE INTEREST RATE, AND THE EXCHANGE RATE 257

KEY TERMS

- Mundell-Fleming model, 240
- supply siders, 250
- twin deficits, 251
- fixed exchange rates, 252
- peg, 252
- devaluation, 252

- revaluation, 252
- crawling peg, 252
- European Monetary System (EMS), 252
- exchange-rate mechanism (ERM), 252
- bands, 252
- central parity, 252

QUESTIONS AND PROBLEMS

1. Suppose that an investor regularly invests in foreign bonds and cares only about expected return, not risk. If interest parity holds, would she prefer a fixed exchange rate, freely fluctuating exchange rates, or neither? Explain.

2. Suppose that the interest parity condition holds and the expected exchange rate between the French franc and the Canadian dollar in one year is .2 (20 cents per franc). Determine the current exchange rate for the following pairs of annual interest rates:
   a. France, 7%; Canada, 5%
   b. France, 7%; Canada, 7%
   c. France, 7%; Canada, 9%

3. Suppose that, initially, the Canadian and German interest rates are equal. Equation (13.5) in the text suggests that a rise in the Canadian interest rate will lead to an *appreciation* of the dollar. But then the Canadian interest rate will be higher than the German rate, and this is only possible when people expect a *depreciation* of the dollar. Is this a contradiction? Explain.

4. "The United States cannot eliminate its trade deficit until it eliminates its budget deficit." Comment.

5. Suppose that a formerly closed economy becomes open. What will happen to the effectiveness of fiscal and monetary policy in affecting GDP if the economy operates under flexible exchange rates? (Answer separately for fiscal and monetary policy.)

6. Under fixed exchange rates and perfect capital mobility, there is no such thing as a "pure fiscal policy" (that is, a fiscal policy *not* accompanied by monetary policy). Explain.

7. Under fixed exchange rates, a fear of devaluation can lead to a recession. Explain.

8. Suppose that the German interest rate is 6%, the Canadian interest rate is 2%, and the exchange rate is 0.8 dollars per DM. If interest parity holds:
   a. What is the exchange rate expected to be one year from today?
   b. Is the dollar expected to appreciate or depreciate? By what percentage?

9. In this chapter, we have assumed that the foreign interest rate is fixed. But when a very large country like the United States changes its interest rate the foreign rate will change as well. In this case, a better assumption might be that the foreign interest rate (i^*) adjusts as follows:

$$i^* = (1 - a)\bar{i}^* + ai$$

where

   i is the domestic (U.S.) interest rate
   $\bar{i}^*$ is a target foreign interest rate
   a is an adjustment parameter, between zero and one

   a. What can we say about the determination of the foreign interest rate if $a = 0$? If $a = 1$? If $a = 0.5$?
   b. Suppose that a is equal to one-half, and the target foreign interest rate $\bar{i}^*$ is equal to 6%. Suppose that the U.S. interest rate rises from 4% to 5%. What will happen to the current foreign interest rate?

   Suppose that the expected exchange rate (the price of foreign currency in terms of dollars) is equal to 10. If interest rate parity holds, what happens to the current exchange rate when the U.S. interest rate rises from 4% to 5%?
   c. "The closer the parameter a is to 1, the more monetary policy works as in a closed economy." Explain.

FURTHER READING

A fascinating account of the politics behind fiscal policy in the Reagan administration is given by David Stockman—who was then the director of the Office of Management and Budget (OMB)—in *The Triumph of Politics: Why the Reagan Revolution Failed* (New York: Harper & Row, 1986).

APPENDIX
FIXED EXCHANGE RATES, INTEREST RATES, AND CAPITAL MOBILITY

The assumption of perfect capital mobility is a good approximation to what happens in countries with highly developed financial markets and few capital controls, such as the United States, the United Kingdom, and Canada. But the assumption is more questionable in countries that have less developed financial markets or have a battery of capital controls in place. There, domestic financial investors may have neither the savvy nor the right to move easily into foreign bonds when domestic interest rates are low. Thus, the central bank may be able to both decrease interest rates and maintain a given exchange rate.

To look at these issues, let us start with the balance sheet of the central bank. In Chapter 6 we assumed that the only asset held by the central bank was domestic bonds. In an open economy, the central bank actually holds two types of assets: (1) domestic bonds and (2) **foreign-exchange reserves,** which we shall think of as foreign currency, although they also take the form of foreign bonds or foreign interest-paying assets. Think of the balance sheet of the central bank as represented in Figure 13A-1. On the asset side are bonds and foreign currency reserves, and on the liability side is the monetary base. There are now two ways in which the central bank can change the money supply: by purchases or sales of bonds in the bond market or by purchases or sales of foreign currency in the foreign-exchange market.

Perfect capital mobility, fixed exchange rates. Consider first the effects of an open-market operation under the joint assumptions of perfect capital mobility and fixed exchange rates (the assumptions we made in the last section of the chapter).

Assume that the domestic and foreign nominal interest rates are initially equal, so that $i = i^*$. Suppose that the central bank embarks on an expansionary open-market operation, buying bonds in the bond market in amount ΔB, and creating money in exchange. This purchase of bonds leads to a decrease in the domestic interest rate, i.

This is, however, only the beginning of the story. Now that the domestic interest rate is lower than the foreign interest rate, financial investors prefer to hold foreign bonds. To buy foreign bonds, they must first buy foreign currency. Thus, they go to the foreign-exchange market and sell domestic currency for foreign currency. If the central bank did not intervene, the price of domestic currency would fall, and the result would be a depreciation. But under its commitment to a fixed exchange rate, the central bank must intervene in the foreign-exchange market and sell foreign currency for domestic currency. As it does so, and buys domestic money, the money supply decreases.

How much foreign currency must the central bank sell? The answer is that it must keep selling until the money supply is back to its pre–open-market operation level, so that the domestic interest rate is again equal to the foreign interest rate. Only then are financial investors willing to hold domestic bonds. How long do all these steps take? Under perfect capital mobility, all this may happen within minutes or so of the original open-market operation.

After these steps, the balance sheet of the central bank looks as in Figure 13A-2. Bond holdings are up by ΔB, reserves of foreign currency are down by ΔB,

Assets	Liabilities
Bonds	Money
Foreign currency reserves	

FIGURE 13A-1
Balance Sheet of the Central Bank

Assets	Liabilities
Bonds: ΔB	Money: $\Delta B - \Delta B$
Reserves: $-\Delta B$	$= 0$

FIGURE 13A-2
Balance Sheet of the Central Bank After an Open Market Operation, and the Induced Intervention in the Foreign-Exchange Market

and the stock of central bank money is unchanged, having gone up by ΔB in the open market operation and down by ΔB as a result of the sale of foreign currency in the foreign-exchange market. Thus, under fixed exchange rates and perfect capital mobility, the only effect of the open-market operation is to change the *composition* of the central bank's balance sheet but not the money supply. This is the result we relied on in Section 13-5.

Imperfect capital mobility, fixed exchange rates. Let's now move away from the assumption of perfect capital mobility. Suppose that it takes some time for financial investors to shift between domestic and foreign bonds. Now an expansionary open market operation can initially bring the domestic interest rate below the foreign interest rate. But, over time, investors shift to foreign bonds, leading to an increase in the demand for foreign currency in the foreign-exchange market. To avoid a depreciation, the bank must again stand ready to sell foreign currency and buy domestic currency. Eventually, the central bank buys enough domestic currency to offset the effects of the initial open market operation. The money stock is back to normal, and so is the interest rate. The central bank holds more bonds and smaller reserves of foreign currency.

The difference between this case and the preceding one is that by accepting a loss in foreign-exchange reserves, the central bank is now able to decrease interest rates for some time. If it takes just a few days for bondholders to adjust, the tradeoff is rather unattractive—as many countries have discovered. But, if the central bank can affect the domestic interest rate for a few weeks or months, it may, in some circumstances, be willing to do so.

Now let's deviate further from perfect capital mobility. Suppose that in response to a decrease in the domestic interest rate, financial investors are either unwilling or unable to move much of their portfolio into foreign bonds. After an expansionary open-market operation, the domestic interest rate decreases, making domestic bonds less attractive. Some domestic investors move into foreign bonds, selling domestic currency for foreign currency. To maintain the exchange rate, the central bank must buy domestic currency and supply foreign currency. However, the foreign-exchange intervention may now be small compared with the initial open-market operation. And, if capital controls truly prevent investors from moving into foreign bonds at all, there may be no need at all for such an intervention.

Even leaving this extreme case aside, the net effect of an open-market operation is likely to be an increase in the money supply, a decrease in the domestic interest rate, an increase in the central bank's bond holdings, and some—but smaller—loss in reserves of foreign currency. Thus, with imperfect capital mobility, a country has some freedom to move the domestic interest rate while maintaining its exchange rate. This case is the most relevant one for most middle-income countries, from Latin America to the countries of Eastern Europe.

KEY TERMS

- foreign-exchange reserves, 259

QUESTIONS AND PROBLEMS

1. Suppose that Belgium operates under perfect capital mobility and fixed exchange rates, and that the central bank of Belgium conducts a contractionary open-market operation, selling $1 billion of Belgian government bonds.

   a. Explain step by step what will happen, and the ultimate effect on the Belgian money supply.
   b. Illustrate the changes in the balance sheet of the Belgian central bank after all adjustments have taken place.

EXPECTATIONS, EXCHANGE-RATE MOVEMENTS, AND EXCHANGE-RATE CRISES

Y, i, E connected

How can we explain fluctuations in the dollar? We took a first stab at answering this question in Chapter 13. But we did so under a convenient but restrictive assumption: namely, that whatever happened today did not affect expectations of the future exchange rate. It is time to relax this assumption and to look at the role of expectations in the determination of exchange rates. In doing so, we integrate two major themes of this book: the importance of expectations and the openness of the economy.

We start the chapter by showing that we can think of the real exchange rate as determined by two sets of factors: (1) factors related to trade and (2) differences between domestic and foreign long-term real interest rates. Next, we return to the effects of monetary policy under flexible exchange rates. Finally, turning to fixed exchange rates, we show how the approach can be used to think about exchange-rate crises, such as the crises in the European Monetary System (EMS) in the 1990s.

14-1 THE DETERMINANTS OF THE REAL EXCHANGE RATE

We saw in Chapter 11 how the interest parity condition [equation (11.2)] led to a relation between the *one-year* domestic and foreign *nominal* interest rates on the one hand, and the current and expected future *nominal* exchange rates on the other. The purpose of this section is to show that the same condition can be used to derive a relation between *long-term* domestic and foreign *real* interest rates on the one hand, and current and expected future *real* exchange rates on the other. If this sounds like a mouthful, do not worry: It is simpler than it sounds. This relation will provide us with a way of thinking about the determination of the exchange rate.

REAL INTEREST RATES AND THE REAL EXCHANGE RATE

Consider, as in Chapter 11, the choice between one-year Canadian and U.S. bonds. But instead of expressing the two rates of return in dollars as we did there, let's express both of them in terms of Canadian goods.

Suppose that you decide to invest the equivalent of one Canadian good—to "invest one Canadian good" for short in what follows. Also suppose that you decide to hold Canadian bonds. Let r_t be the one-year Canadian real interest rate, the interest rate on one-year Canadian bonds in terms of Canadian goods. By the definition of the real interest rate, you will get $(1 + r_t)$ Canadian goods next year. This is represented by the first line in Figure 14-1.

Suppose that you decide instead to hold American bonds. This involves exchanging C\$ for US\$, holding American bonds for a year, and selling US\$ for C\$ in a year. Let ε_t be the real exchange rate, the relative price of U.S. goods in terms of Canadian goods. A real exchange rate of ε_t means that you get $1/\varepsilon_t$ U.S. goods for every Canadian good you invest. Let r_t^* be the one-year U.S. real interest rate, the interest rate on one-year U.S. bonds in terms of U.S. goods. Let the expected real exchange rate a year from now be ε_{t+1}^e. Then, for every Canadian good you invest in one-year U.S. bonds, you can expect to get $(1/\varepsilon_t)(1 + r_t^*)\varepsilon_{t+1}^e$ Canadian goods next year. The three steps involved in the transaction are represented in the second part of Figure 14-1.

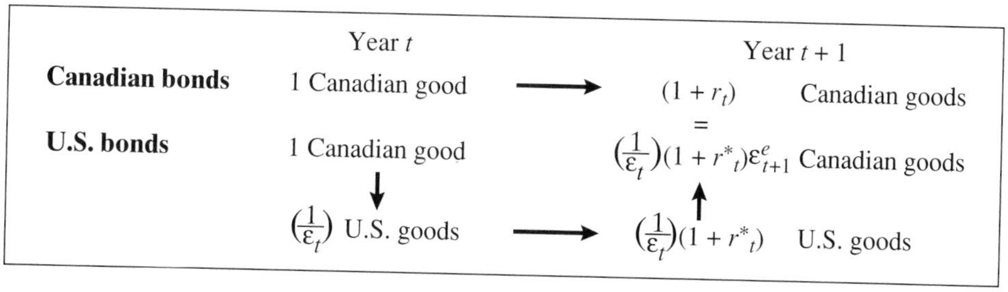

	Year t	Year $t+1$
Canadian bonds	1 Canadian good $\longrightarrow$	$(1 + r_t)$ Canadian goods
		$=$
U.S. bonds	1 Canadian good	$(\frac{1}{\varepsilon_t})(1 + r^*_t)\varepsilon^e_{t+1}$ Canadian goods
	$\downarrow$	$\uparrow$
	$(\frac{1}{\varepsilon_t})$ U.S. goods $\longrightarrow$	$(\frac{1}{\varepsilon_t})(1 + r^*_t)$ U.S. goods

FIGURE 14-1

Expected Returns, in Terms of Canadian Goods, from Holding Canadian or U.S. Bonds

If we assume, as we did earlier, that expected returns expressed in the same units (here Canadian goods) must be equal in equilibrium, the following condition must hold:

$$1 + r_t = \frac{1}{\varepsilon_t}(1 + r^*_t)\varepsilon^e_{t+1} \qquad (14.1)$$

— For every Cdn good you invest in 1 year U.S bonds you can expect to get $(\frac{1}{\varepsilon_t})(1 + r^*_c)\varepsilon^e_{t+1}$ cdn goods next year

This equation gives us a relation between the domestic and the foreign real interest rates on the one hand, and the current and expected future real exchange rates on the other. You may wonder whether this is a different condition from the condition derived in terms of nominal interest rates and nominal exchange rates in Chapter 11, equation (11.2). The answer is no. The two conditions are equivalent. Indeed, the appendix to this chapter shows how to go from one to the other. (The derivation is not much fun, but it is good practice and a useful way of brushing up on the relation between nominal and real interest rates, and nominal and real exchange rates.) The reason why they are equivalent is simple. The interest parity condition states that the expected returns *when expressed in common units*—whatever these units are, as long as they are common—must be equal. In Chapter 11, the common unit was the dollar. Here, the common unit is a Canadian good.

As in Chapter 11, it is more convenient to work with an approximation to equation (14.1). As long as the domestic and foreign real interest rates are not too different, a good approximation to equation (14.1) is given by

$$r_t \approx r^*_t + \left(\frac{\varepsilon^e_{t+1} - \varepsilon_t}{\varepsilon_t}\right) \qquad (14.2)$$

In words: *The domestic real interest rate must be (approximately) equal to the foreign real interest rate plus the expected rate of real depreciation.* If the domestic real interest rate is higher than the foreign real interest rate, then financial investors must be expecting a real depreciation.

Let's apply this equation to U.S. versus Canadian bonds. In December 1997, the one-year nominal interest rate was 5.2% in Canada. Expected inflation, as reflected by commercial forecasts, was around 2.1%, so that the Canadian real interest rate was 5.2% − 2.1% = 3.1 percent. The one-year nominal interest rate in the United States was 5.5%; expected inflation was around 2.6%, so that the real interest rate in the United States was 5.5% − 2.6% = 2.9%. Equation (14.2) tells us that financial investors were expecting a real depreciation of the C$ vis-à-vis the US$ of 3.1% − 2.9% = 0.2% over the following year.

3.1   2.9
5.2 = 5.5

3.1 = 2.9 + .2

LONG-TERM REAL INTEREST RATES AND THE REAL EXCHANGE RATE

We have looked so far at the choice of holding domestic versus foreign bonds *for one year*. But we can apply the same logic to the choice of holding domestic versus foreign bonds *for many years*. Let's now do that.

Suppose that you decide to invest the equivalent of one Canadian good— to invest one Canadian good for short—for *n* years in either *n*-year Canadian bonds or *n*-year U.S. bonds (think of *n* as, say, 10 years).

Suppose that you decide to hold *n*-year Canadian bonds. Let r_{nt} be the *n*-year Canadian real interest rate. Recall from Chapter 9 that, by the definition of an *n*-year interest rate, r_{nt} is the average yearly real interest rate you can expect to get if you hold the *n*-year bond for *n* years. Then, by the definition of the *n*-year real interest rate, you can expect to get $(1 + r_{nt})^n$ goods in *n* years. This is represented in the first line of Figure 14-2.

Now suppose that you decide to hold *n*-year U.S. bonds instead. Let ε_t be the real exchange rate. Let r_{nt}^* be the *n*-year U.S. real interest rate. Let the expected real exchange rate *n* years from now be ε_{t+n}^e. Then, for every Canadian good you invest in *n*-year U.S. bonds, you can expect to get $(1/\varepsilon_t)(1 + r_{nt}^*)^n \varepsilon_{t+n}^e$ Canadian goods in *n* years. This set of steps is represented in the second part of Figure 14-2.

If we assume again that expected returns have to be the same, then the following condition must hold:

$$(1 + r_{nt})^n = \left(\frac{1}{\varepsilon_t}\right)(1 + r_{nt}^*)^n \varepsilon_{t+n}^e$$

Again, a good approximation (which is derived as an exercise in Appendix 3 at the end of the book) is given by

$$n r_{nt} = n r_{nt}^* + \frac{\varepsilon_{t+n}^e - \varepsilon_t}{\varepsilon_t} \tag{14.3}$$

Let's again take a numerical example. Suppose that the 10-year domestic nominal interest rate is 10%, and that yearly inflation is expected to be 6% on average for the next 10 years. Thus, the 10-year domestic real interest rate, r_{nt}, is equal to $10\% - 6\% = 4\%$. Suppose that the 10-year foreign interest rate is 6%, with average expected inflation of 3%, so that the 10-year foreign real interest rate is equal to $6\% - 3\% = 3\%$.

Then, if you hold domestic bonds, you can expect to receive 4% a year in terms of domestic goods; if you hold foreign bonds, you can expect to receive 3% a

FIGURE 14-2

Expected Returns, in Terms of Canadian Goods, from Holding *n*-year Canadian or U.S. Bonds for *n* Years

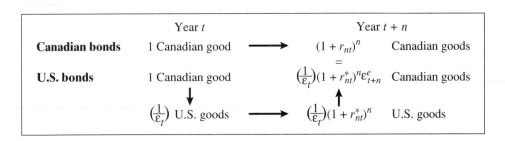

year in terms of foreign goods. When will you be indifferent between holding domestic bonds and holding foreign bonds? Only when you expect the relative price of foreign goods to increase by 10% over the next 10 years—that is, only when you expect a real depreciation of 10% over the next 10 years. In that case, the two expected returns, in terms of domestic goods, are equal. Or, in terms of equation (14.3),

$$nr_{nt} \quad = \quad nr^*_{nt} \quad + \quad \frac{\varepsilon^e_{t+n} - \varepsilon_t}{\varepsilon_t}$$

(10) 4% = (10) 3% + 10%

As in Chapter 13, it will be more convenient to rewrite equation (14.3) so that the current exchange rate is on the left. This gives

$$\varepsilon_t = \frac{\varepsilon^e_{t+n}}{1 + n(r_{nt} - r^*_{nt})} \qquad (14.4)$$

This equation gives the real exchange rate today as a function of the expected future real exchange rate n years from now, and of the differential between n-year domestic and foreign real interest rates. Let's now look at this relation more closely.

THE REAL EXCHANGE RATE, TRADE, AND INTEREST RATE DIFFERENTIALS

The first determinant of the current real exchange rate in equation (14.4) is the expected future real exchange rate, ε^e_{t+n}. If we take the number of years, n, to be large (say, 10 years or more), we can think of ε^e_{t+n} as the real exchange rate that financial market participants expect to prevail in the long run. Indeed, for simplicity, we shall now call it the *long-run real exchange rate*. ε^e_{t+n}

How should we think of the long-run real exchange rate? In the long run, we can reasonably assume that trade will be roughly balanced. No country can run trade deficits forever, and no country wants to run trade surpluses forever.[1] If trade is balanced in the long run, the long-run exchange rate must be such as to ensure trade balance. Thus, we can think of the expected future real exchange rate as that exchange rate consistent with trade balance in the long run.

The second determinant of the current real exchange rate is the difference between domestic and foreign long-term real interest rates. *An increase in domestic over foreign long-term real interest rates leads to a decrease in the real exchange rate—a real appreciation.* We focussed on this mechanism when discussing a similar relation between the interest rate and the exchange rate in Chapter 13; let's go through its logic again.

Suppose that the long-term domestic real interest rate goes up, making domestic bonds more attractive than foreign bonds. As investors try to shift out of foreign bonds into domestic bonds, they sell foreign currency and buy domestic

---

[1] **Digging deeper.** *This statement is not quite right. What must be balanced in the long run is the current account. (For a refresher on the difference between the trade balance and the current account, see Chapter 11.) Thus, a country that is a net debtor vis-à-vis the rest of the world can, even in the long run, run a trade surplus to finance interest payments on its net foreign liabilities. We ignore the difference between trade and current account balance here.*

currency, so that the domestic currency appreciates. Because the exchange rate is expected to return eventually to its long-run value, the more the domestic currency appreciates today, the more it is expected to depreciate in the future. Thus, the domestic currency appreciates today to the point where the expected future depreciation exactly offsets the fact that the long-term domestic real interest rate is higher than the long-term foreign real interest rate. At that point, financial investors are again indifferent between holding domestic or foreign bonds.

We now have a way of thinking about exchange-rate movements: *The real exchange rate today depends first on the long-run real exchange rate—the exchange rate that ensures trade balance in the long run—and, second, on the difference between long-term domestic and foreign real interest rates.*

We illustrate both effects by looking at the fluctuations of the U.S. dollar against the mark and the yen.

14-2 THE DANCE OF THE U.S. DOLLAR IN THE 1980s

The US\$/DM exchange rate in the 1980s saw a sharp real appreciation during the first half of the decade, and a sharp real depreciation during the second half. In light of the theory we just developed, we can ask: Were these movements due more to movements in long-term real interest rates in the United States and in Germany, or more to movements in the long-run U.S. real exchange rate?

Note from equation (14.4) that, if the long-run real exchange rate were constant, there would be an exact negative relation between the domestic minus the foreign long-term real interest rate ($r_{nt} - r_{nt}^*$), and the real exchange rate, ε_t. Equivalently—this way of stating it will be more convenient to set up the graph that follows—there would be an exact positive relation between the foreign minus the domestic long-term real interest rate ($r_{nt}^* - r_{nt}$), and the real exchange rate, ε_t.

This statement suggests the following approach to interpreting movements in the real exchange rate: Construct for each year the difference between the long-term foreign and domestic real interest rates. Then, plot the real exchange rate against this difference. If the two series move closely together, differences in long-term real interest rates are the dominant factor in explaining movements in the real exchange rate. If they do not, changes in the long-run real exchange rate must play an important role.

Figure 14-3 implements this strategy. It focusses on the bilateral real exchange rate between the United States and Germany from 1980 to 1990.

The real exchange rate is defined and constructed in the same way as in Figure 11-6 in Chapter 11. It is given by $E_t P_t^*/P_t$, where E_t is now the US\$/DM exchange rate, and P_t and P_t^* are the GDP deflators in the United States and Germany, respectively. The real exchange rate is normalized to 1 in 1987, and is measured on the scale at the left of the figure.

The long-term U.S. real interest rate for each year is constructed as the nominal interest rate on a 10-year U.S. bond minus average expected inflation over the following 10 years. As a proxy for expected inflation in each year, we use commercial forecasts (from DRI, a firm specializing in economic forecasting) of future inflation over the following 10 years. For example, for 1985 we use forecasts as of

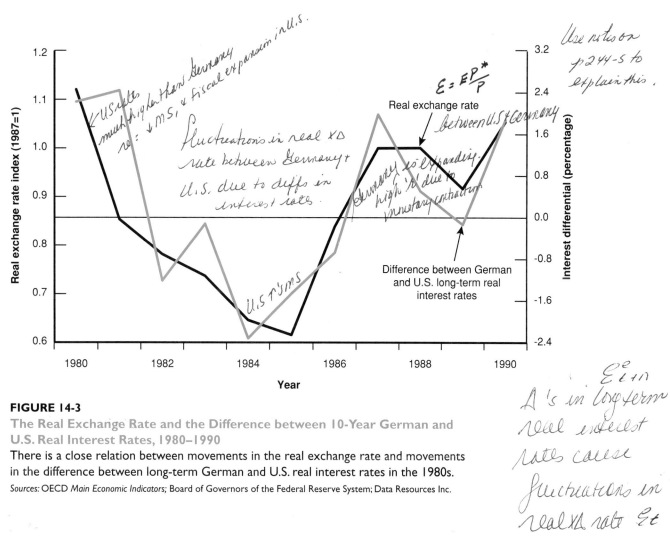

FIGURE 14-3

The Real Exchange Rate and the Difference between 10-Year German and U.S. Real Interest Rates, 1980–1990

There is a close relation between movements in the real exchange rate and movements in the difference between long-term German and U.S. real interest rates in the 1980s.

Sources: OECD *Main Economic Indicators;* Board of Governors of the Federal Reserve System; Data Resources Inc.

December 1984 of inflation for 1985 to 1994, and construct a measure of average expected inflation over 1985 to 1994, which we then subtract from the nominal interest rate. The long-term German real interest rate is constructed in similar fashion. The second line in the figure then plots the difference between the German and the U.S. real interest rates. This difference is measured as a percentage on the scale at the right of the figure.

The fit between the two series in Figure 14-3 is strikingly good. The figure thus yields a simple conclusion: *Most of the movements in the US$/DM exchange rate in the 1980s can be traced to large movements in the difference between U.S. and German real interest rates.* The main reason why the US$ was so high in the mid-1980s was that long-term U.S. real interest rates were very attractive in the mid-1980s. The main reason why the dollar went down was that U.S. real interest rates became less attractive.

Where, in turn, did these movements in long-term real interest rates come from? As discussed in Chapter 13, a monetary contraction/fiscal expansion of the early 1980s led to high real interest rates in the United States. In Chapter 6 we saw how German unification led to a mix of fiscal expansion and tight money in Germany from 1990 on. Let's now put these elements together and tell the complete story.

The real appreciation of U.S. goods in the early 1980s was due to a steady increase in U.S. long-term real interest rates over German rates; the causes were primarily monetary contraction and fiscal expansion in the United States. From 1985 on, U.S. long-term real rates decreased relative to German rates, leading to a real depreciation. In the second half of the 1980s, the main cause of this depreciation was a less restrictive monetary policy in the United States, which led to a decline in U.S. long-term interest rates. This evolution was reinforced first by an expansion in Germany in the late 1980s, and then by German unification in 1990, both leading the Bundesbank to increase German interest rates to avoid overheating of the German economy.[2]

14-3 THE APPRECIATION OF THE YEN IN THE 1990s

Are movements in real exchange rates always caused by movements in long-term real interest rates? The answer is no. Movements in real exchange rates often come from movements in the long-run real exchange rate. From 1990 to 1995, the yen appreciated in real terms by nearly 40% vis-à-vis the US$; in other words, Japan's real exchange rate vis-à-vis the United States decreased by nearly 40%. This phenomenon cannot be explained by an increase in relative Japanese long-term real interest rates. Thus, it must come from a change in the long-run real exchange rate.

The relevant facts are given in Table 14-1. Column (1) gives the evolution of the yen/U.S. dollar exchange rate. In 1990, a U.S. dollar was worth 144 yen. In 1994 it was worth only 101 yen. In December 1995, after a further decline and then a recovery during 1995, it was back at 101 yen, implying an average annual rate of yen appreciation of about 7% since 1990.[3] Although we concentrate on the yen/US$ rate, it is representative of what has happened to the Japanese multilateral real exchange rate: The yen has appreciated with respect to most currencies.

Columns (2) and (3) give inflation rates in Japan and the United States. Both inflation rates were low. The Japanese inflation rate was 1% to 2% lower than the U.S. inflation rate. Lower inflation in Japan implied that the yen's real appreciation was somewhat smaller than its nominal appreciation but still about 5% per year.[4]

Columns (4) and (5) give long-term *nominal* interest rates. The rates in both countries declined after 1990, less so in the United States than in Japan. What matters for our purposes, however, is what happened not to nominal interest rates but to *real* interest rates. Constructing them would again require constructing expected inflation, as we did to draw Figure 14-3. We shall not do so formally here, but the evidence from commercial forecasts is that expected future inflation decreased by

---

[2]*If you had no problem understanding this paragraph, realize how much progress you have made already. The paragraph relies on many mechanisms and many interactions that we have built one by one over the past 14 chapters.*

[3]*The value of 92 for the 1995 yen/US$ exchange rate in Table 14-1 refers to the average for 1995. In December 1995, the exchange rate had increased back to 101. By December 1997, it had climbed to 130.*

[4]*In going through this section, remember that we look at things from the point of view of Japan. P is the Japanese price level, P* is the U.S. price level. The yen is the domestic currency, the U.S. dollar the foreign one; r_{nt} stands for the long-term real interest rate in Japan, r_{nt}^* for the long-term real interest rate in the United States.*

[Handwritten marginalia:]

real appreciation leads to an expected depreciation

$$r_t^\varepsilon \approx r_t^* + \frac{(\varepsilon_{t+1}^e - \varepsilon_t)}{\varepsilon_t}$$

U.S ↑ due to monetary contraction & fiscal expansion

$$\varepsilon_{t+n}^e$$

$$\varepsilon = \frac{EP^*}{P}$$

THE YEN APPRECIATION, INTEREST RATES, AND THE JAPANESE TRADE SURPLUS, 1990–1995

YEAR	(1) YEN/US$ EXCHANGE RATE	(2) INFLATION RATE: JAPAN	(3) INFLATION RATE: U.S.	(4) LONG-TERM NOMINAL INTEREST RATE: JAPAN	(5) LONG-TERM NOMINAL INTEREST RATE: U.S.	(6) JAPAN: TRADE BALANCE
1990	144	2.2	4.3	7.4	8.7	1.8
1991	134	1.9	3.8	6.4	8.2	2.3
1992	126	1.6	2.7	5.1	7.5	2.9
1993	110	0.8	2.2	4.0	6.5	2.8
1994	101	0.1	2.0	4.2	7.4	2.7

The *exchange rate* is the average over the year. *Inflation* is the change in the GDP deflator from year to year. The *trade balance* is expressed as a ratio to GDP; a positive number indicates a surplus. Numbers for 1995 are forecasts for the year, as of June 1995.

Source: OECD Economic Outlook, June 1995.

roughly the same amount in both countries. Together these facts imply that, if anything, the long-term U.S. real interest rate declined by less than the long-term Japanese real interest rate. Put another way, Japan's long-term real interest rate was lower relative to that of the United States in 1995 than it was in 1990. From equation (14.4) this should have led to a real depreciation of the yen, not to the observed real appreciation. The relative evolution of long-term real interest rates cannot therefore explain the large real yen appreciation.

So, what accounts for the steady decreases in the exchange rate—the steady appreciation of the yen? Our theory leaves us only one alternative: that *the long-run real exchange rate steadily decreased after 1990*. The question, then, is why. Column 6 of Table 14-1 gives us a hint. Despite the appreciation of the yen since 1990, which made Japanese goods relatively more expensive year after year, the Japanese trade surplus remained very large, increasing from 1.8% in 1990 to 2.9% of GDP in 1992, and remaining at 2.6% in 1995.

Why did Japan's trade surplus grow so large? There are three potential reasons, and each has played its part:

(1) The first reason comes from the dynamic effects of what we called the J-curve in Chapter 12. As we saw there, as long as the Marshall-Lerner condition holds, an appreciation eventually reduces the trade surplus but may initially increase it. (We actually saw the converse proposition: that a depreciation eventually reduces the trade deficit but may initially increase it. These propositions are clearly equivalent.) Until the quantity of exports and imports starts responding to the real exchange rate, the positive effect may dominate. In Japan, which experienced five years of steady appreciation, the J-curve effects were still at work.

(2) The second reason comes from the effects of activity on the trade balance. We have seen that a country in recession typically sees an improvement in its trade position: Its imports (which depend on domestic activity) go down, while its exports, which depend on activity in the rest of the world, are unaffected. After 1992, Japan

grew more slowly than its trading partners. GDP growth in Japan from 1992 to 1995 averaged 0.7% per year, compared to 3% for the OECD as a whole.

(3) The two factors we have just analyzed should not have led financial markets to change their expectations of the *long-run* real exchange rate. But the assessment of financial markets was that, even after these factors disappeared, it would take a lower real exchange rate for Japan to achieve trade balance. In other words, the steady appreciation of the yen reflected financial markets' belief that this is what it would take to achieve an eventual return to trade balance in Japan. Was the financial markets' assessment correct? The deterioration in the yen since 1995 suggests that the financial markets were mistaken.

14-4 EXCHANGE RATES, MONETARY POLICY, AND NEWS

We have seen how exchange-rate movements reflect both changes in the expected long-run exchange rate as well as differences between domestic and foreign long-term interest rates. With our newly acquired knowledge, let us return to how monetary policy works in an open economy with flexible exchange rates.

MONETARY POLICY, INTEREST RATES, AND EXCHANGE RATES

Let us start with a blatantly unrealistic case. Because it is unrealistic, it is easier to analyze. Yet it provides a good base on which to build a more realistic discussion.

Assume that there is no inflation, here or abroad, current or expected, so that we do not need to distinguish between nominal and real interest rates, or between nominal and real exchange rates. Suppose further that, initially, domestic and foreign interest rates are expected to be constant and equal to each other.

Now suppose that the central bank unexpectedly announces that, to increase activity, it has decided to decrease interest rates. The central bank announces that one-year interest rates will be 2% lower for the next five years, after which they will return to normal. Financial markets fully believe this announcement.

What happens at the time of the announcement? Short-term interest rates go down by 2%, and so do rates on bonds with maturities less than or equal to five years. Yields on bonds with longer maturity go down, but by less.[5]

What is the effect on the exchange rate today? To answer this question, work backward in time. Start five years in the future. The exchange rate five years from now depends on what is expected to happen to interest rates thereafter, as well as on the long-run exchange rate. Because the announcement does not change expectations of interest rates beyond the first five years, and presumably does not change the long-run exchange rate either, there is no change in the expected real exchange rate five years from now. What happens between today and five years hence? Think about it this way. The interest parity condition tells us that for each of the next five

---

[5]*Can you show by how much the 10-year interest rate goes down, using the term structure relations we derived in Chapter 9? (The answer is 1%.)*

years, there must be an expected appreciation of the domestic currency of 2% per year, so that the expected rates of return on holding domestic and foreign bonds are equal. Thus, there must be an expected cumulative appreciation of $5 \times 2\% = 10\%$ over the next five years. As the expected exchange rate five years hence is unchanged, there must therefore be a depreciation today of 10%, so as to generate the expected appreciation of 10% over the next five years. In other words, if the domestic currency *depreciates by 10% today* and is then expected to *appreciate by 2% a year for the next five years,* financial investors will be willing to hold domestic bonds, although the domestic interest rate is 2% lower than the foreign interest rate.

The expected paths of the one-year interest rate and of the exchange rate are shown in Figure 14-4. Before the announcement, the domestic interest rate was expected to be the same as the foreign interest rate forever. After the announcement, the domestic interest rate is expected to be 2% lower than the foreign interest rate for five years. The effect of the announcement is to lead to an increase in the exchange rate of 10% at the time of the announcement (a 10% depreciation), followed by a decrease by 2% a year (an expected appreciation of 2% a year) over the next five years. Note how much the exchange rate initially moves and overshoots its long-run value—increasing first, only to decrease back to its initial value five years later. For that reason, this exchange-rate adjustment is often referred to as **overshooting.**

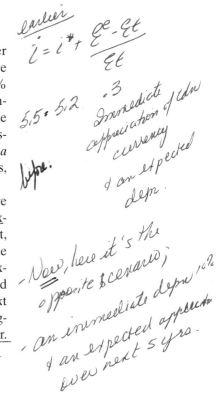

(handwritten margin notes)
earlier
$i = i^* + \dfrac{\varepsilon^e - \varepsilon t}{\varepsilon t}$

$5.5 = 5.2$ .3
Immediate appreciation of cdn currency & an expected dep.

- Now, here it's the opposite scenario;
- an immediate depn 10%
& an expected apprec over next 5 yrs.

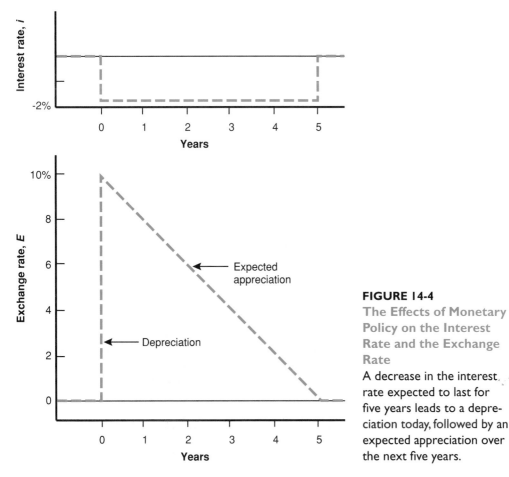

FIGURE 14-4
The Effects of Monetary Policy on the Interest Rate and the Exchange Rate
A decrease in the interest rate expected to last for five years leads to a depreciation today, followed by an expected appreciation over the next five years.

(handwritten margin notes)
depn today but expected apprec for 5yr.
$i_t \approx i_t^* + \dfrac{\varepsilon^e_{t+1} - \varepsilon t}{\varepsilon t}$

This result is an important one. When countries moved in the early 1970s from fixed to flexible exchange rates, the large fluctuations in exchange rates that followed came as a surprise to most economists. For a long time, the fluctuations were thought to be the result of irrational speculation in foreign-exchange markets. It was not until the mid-1970s that economists realized that these large movements could be explained, as we have done here, by the rational reaction of financial markets to differences in expected future interest rates.[6]

We can now build on what we learned in previous chapters to tell the full story of the effects of monetary policy in an open economy. After the announcement, the decrease in interest rates and the depreciation both lead to an increase in domestic demand and net exports, and thus to an increase in output. As time passes and the exchange rate returns to normal, the effect on net exports disappears. And as the interest rate returns to normal, so does domestic demand. If the purpose of the expansionary monetary policy was to strengthen demand for a while (for example, to strengthen a recovery), then this is exactly what the policy does. The effect is strongest at the beginning, and its effects disappear over time.

POLICY AND EXPECTATIONS

In the example we just looked at, we dismissed many complications. We assumed in particular that the central bank announced what it was going to do over the next five years and that financial markets fully believed the announcement. In practice, this is not the way things happen.

When the central bank cuts interest rates, financial markets have to assess whether this action signals a major shift in monetary policy and is just the first of many such cuts, or whether this cut is just temporary. Announcements by the central bank itself may not be very useful; the central bank itself may not even know what it will do in the future. Typically, it will be reacting to early signals, which may be reversed later. Financial markets also have to assess how foreign central banks will react: whether they will stay put, or follow suit and cut their interest rates. As we saw in Chapter 13, what happens to the exchange rate depends very much on the reactions of other central banks.

Thus, how changes in short-term interest rates affect both long-term interest rates and the exchange rate depends on expectations. Sometimes a small decrease in short-term interest rates may convince markets that monetary policy has changed substantially, resulting in a large decrease in long-term interest rates and a large depreciation. But if markets anticipated a large cut and the central bank announces a smaller cut than was anticipated, the effect may actually be an appreciation, not a depreciation. This happened in 1994 and 1995 in response to decreases in German interest rates by the Bundesbank. Because markets were expecting larger decreases in interest rates than were actually implemented, the markets' reaction to many of those small decreases was an *appreciation* of the mark. The reaction of foreign exchange and financial markets to policy news is the topic of the Focus box entitled "News and Movements in Foreign-Exchange, Stock, and Bond Markets."

---

[6]*This result was first derived by Rudiger Dornbusch from MIT in "Expectations and Exchange Rate Dynamics,"* Journal of Political Economy, 1976, *1161–1176.*

Remember the game we played in Chapter 9: "Why did the stock market go up/down today?" Now we can play the advanced version of the game. Why did the domestic/foreign stock/bond markets go up/down today, and why did the dollar also go up/down? The game is a good way for you to test your understanding of the various mechanisms at work. It can be played every day. Listen to the news and try to predict what happened to the various markets, and why. (This is not an easy game to play. If you are like most economists, you will be wrong about as often as you are right. But it gives you a sense of the difficulties policy makers face when they try to predict the effects of a change in policy.)

Here is one example: In late 1997, the Canadian economy was showing signs of strength and the Canadian dollar was close to an all-time low against the U.S. dollar. In anticipation of inflationary pressures emerging in 1998, the Bank of Canada on October 1 increased its *Bank rate* (an overnight lending rate) by 25 basis points (or one-quarter of one percent) to 3.75%. In response the Canadian dollar appreciated by about one-third of a U.S. cent, long-term interest rates fell by 10 basis points as the market appeared to agree that the Bank's actions would lead to lower inflation, and the TSE index improved modestly relative to the DJIA. Within a few weeks, however, our dollar again came

[handwritten: ↑ in appreciation]

under pressure. On November 25, the Bank rate was increased to 4%. Although the dollar rallied briefly, the markets were not impressed. On December 12, the rate was increased once again, this time by 50 basis points, to 4.5%.

Before you read what actually happened, try to predict what happened to Canadian bond yields, the dollar, and stock prices. *[handwritten: Apprec. $; ↓ Stock prices]*

What actually happened is the following: Financial markets concluded that a financial crisis in Asia required even higher interest rates to support our dollar. The crisis was predicted to lead to lower prices for Canada's commodity exports (a higher real exchange rate) and a shift in investor sentiment towards U.S. bonds as a "safe haven" (a higher premium required to hold C$-denominated assets). Interest rates on bonds with less than three years' maturity rose with the increase in the Bank rate, but longer-term interest rates fell. The yield on 30-year bonds fell by about 10 basis points initially, but within a few days the entire term structure of interest rates shifted up so that the yield on long-term bonds returned to their previous level. The Canadian dollar rallied briefly by about one-third of a U.S. cent, but within a few days it fell again below 70 U.S. cents. The Toronto Stock Exchange index appeared to be largely unaffected by the increase in the Bank rate.

14-5 FIXED EXCHANGE RATES AND EXCHANGE-RATE CRISES

[handwritten: How this approach can be used to think about XD rate crisis: ie (EMS) in 90s.]

As we saw in Chapter 13, many countries operate under fixed exchange rates, announcing a given exchange rate vis-à-vis some foreign currency (or basket of currencies) and using monetary policy to sustain the announced exchange rate, or to stay within tight bands around the announced parity. Exchange rates are typically not fixed forever and countries reserve the right to change them as circumstances dictate.

Fixed exchange-rate systems often face crises during which the government is forced to readjust the exchange rate. One of the reasons for such crises lies, paradoxically, in the governments' option to change the parity if they feel it appropriate. In short, the belief by financial markets that a government is about to change the exchange rate may trigger an exchange-rate crisis and lead the government to adjust earlier than it wanted to, or to adjust even if it did not intend to.

To see why this is so, we can start from our basic interest parity condition:

$$i_t = i_t^* + \frac{E_{t+1}^e - E_t}{E_t} \tag{14.5}$$

In Chapter 13 we interpreted this equation as a relation between the *one-year* domestic and foreign nominal interest rates, the current exchange rate, and the expected exchange rate a year hence. But the choice of one year as the time period was arbitrary. The relation holds over a day, a week, a month. For example, if financial markets expect the exchange rate to be 2% higher in a month, they will hold domestic bonds only if the monthly domestic interest rate exceeds the monthly foreign interest rate by 2% (or, if we express interest rates at an annual rate, if the domestic interest rate exceeds the foreign interest rate by about $2\% \times 12 = 24\%$).

Under fixed exchange rates, the current exchange rate E_t is fixed at some level, say $\bar{E}$. If the markets expect that the parity will be maintained over the period, then $E_{t+1}^e = \bar{E}$, and the interest parity condition simply states that the domestic interest rate and the foreign interest rate must be equal to maintain the existing parity.

But now suppose that financial markets believe that there may soon be an exchange-rate adjustment, say a devaluation. Why might they believe that?

(1) They may believe that the domestic currency is overvalued.[7] At the real exchange rate implied by the current parity, there is a large trade deficit, and eliminating it will require, sooner or later, a devaluation. Overvaluation often happens in countries that fix the nominal exchange rate while experiencing a higher inflation rate than the country to which they are pegging. Higher relative inflation implies a steadily increasing relative price of domestic goods compared with foreign goods and a steady worsening of the trade position.

(2) Financial markets may believe that internal conditions require a decrease in domestic interest rates, and thus a devaluation. This may be the case if the country suffers from high unemployment and financial markets believe that political realities will soon force the government to lower interest rates to increase demand and reduce unemployment.

Suppose that, for whatever reason, financial markets start anticipating a devaluation—an increase in the exchange rate. Suppose that they believe that, over the coming month, there is a 50% chance that the parity will be maintained and a 50% chance that there will be a 10% devaluation. Thus, the term $(E_{t+1}^e - E_t)/E_t$ in the interest parity equation (14.5), which we assumed equal to zero earlier, is now equal to $(0.5 \times 0.0) + (0.5 \times 10\%)$—a 50% chance of no change plus a 50% chance of a devaluation of 10%—which equals 5%.

This implies in turn that, to maintain the existing parity, the central bank must now offer a monthly interest rate 5% higher, thus $12 \times 5\% = 60\%$ higher at an annual rate! Sixty percent is the interest differential needed to convince investors to hold domestic bonds in view of the risk of a devaluation.

[7]*Expectations matter. It is enough that the market thinks the currency will be overvalued. The crisis in Southeast Asia that emerged in 1997 was of this sort. As the crisis spread, a wider circle of currencies came under attack because it was believed that foreign investors would shy away from the region and competition from sharply devalued currencies elsewhere would hurt exports in places like Korea. Thus, even the anticipation of a large current account deficit could lead to immediate attacks on the currency.*

[handwritten marginalia:]

$\varepsilon = \frac{EP^*}{P}$

real appreciation in domestic currency (overvaluation) — worsens the trade position↓

this should rise with ↑P, implying a ↓ dep'n of the domestic currency ↓ improvement in trade balance (but $\bar{E}$ remains constant)

$E\uparrow$ dep'n of currency

5% higher ↑ expected dep'n of 5%

$i_t = i_t^* + \frac{E_{t+1}^e - E_t}{E_t}$

What, then, are the choices confronting the government and the central bank?

(1) The government and the central bank can try to convince markets that they have no intention of devaluing. This is indeed always the first line of defence: Communiqués are issued, and prime ministers go on TV to reiterate their absolute commitment to the existing parity. But words are cheap, and they rarely convince financial markets.

(2) The central bank can increase the interest rate, but by less than is needed to satisfy equation (14.5)—in our example, by less than 60% annually. Thus, although domestic interest rates are high, they are not high enough to compensate fully for the perceived risk of devaluation. This action typically leads to a large capital outflow, as financial investors decide that foreign bonds are more attractive. To maintain the parity, the central bank must buy domestic currency and sell foreign currency in the foreign-exchange market. In doing so, it often loses most of its reserves of foreign currency. (The mechanics of central bank intervention were described in the appendix to Chapter 13.)

(3) Eventually—and this may be after a few hours or a few months—the choice for the central bank becomes one of accepting the very high interest rates or of validating the markets' expectations and devaluing. Setting very high short-term domestic interest rates can have a devastating effect on demand and on output. This course of action makes sense only if (a) the perceived probability of a devaluation is small, so that the interest rate does not have to be too high, and (b) the government believes that markets will eventually become convinced that no devaluation is coming. Otherwise, the only option is to devalue.

In summary: The belief that a devaluation is coming in the future can trigger a devaluation much earlier. The devaluation may happen even if the belief was initially groundless. Even if the government initially had no intention of devaluing, it may be forced to do so if financial markets believe that it will. The cost of not devaluing may be a long period of very high interest rates needed to maintain the parity.

CRISES IN THE EUROPEAN MONETARY SYSTEM

All the mechanisms we've just discussed were very much in evidence after 1992 in the EMS. From 1987 to 1992, there had been only two **realignments**—adjustments of parities—among member countries. There was increasing talk about narrowing the bands further and even moving to the next stage, a common currency. But in 1992 financial markets became increasingly convinced that more realignments were soon to come.

The reason was one we have analyzed already, the implications of German reunification. Because of the pressures coming from reunification, the Bundesbank was maintaining high interest rates.[8] While Germany's trading partners needed

---

[8]See the Global Macro box in Chapter 13 on the implications of German unification.

$$i = i^* + \frac{E^e_{t+1} - E_t}{E_t}$$

then this must rise

if there's an expected depn (ie devaluation)

* Otherwise huge capital outflows

lower interest rates to reduce a growing unemployment problem, they had to match the German interest rates to maintain their EMS parities. To financial markets, the position of Germany's partners looked increasingly untenable. Lower interest rates outside Germany, and thus devaluations of many currencies vis-à-vis the DM, appeared increasingly likely.

Throughout 1992, the perceived probability of a devaluation forced a number of Germany's trading partners to maintain higher nominal interest rates than Germany. But the first major crisis did not come until September 1992.

The day-by-day story is told in the Global Macro box entitled "Anatomy of a Crisis: The September 1992 EMS Crisis." The belief that a number of countries were soon going to devalue led in early September to speculative attacks on a number of currencies, with financial investors selling in anticipation of the oncoming devaluation. All the lines of defence described earlier were used. First, solemn communiqués were issued, but with no discernible effect. Then, interest rates were increased, up to 500% for the **overnight interest rate** (the rate for lending and borrowing overnight) in Sweden (expressed at an annual rate). But they were not increased enough to prevent capital outflows and large losses of foreign exchange reserves by the central banks under pressure. The next step was different courses of action in different countries: Spain devalued its exchange rate, Italy and the United Kingdom suspended their participation in the EMS, and France decided to tough it out through higher interest rates until the storm was over.

By the end of September, financial markets believed that no further devaluations were imminent. Some countries were no longer in the EMS, others had devalued but remained in the EMS, and those that had maintained their parity had shown their determination to stay in the EMS, even at a very high cost in terms of interest rates. But the underlying problem—namely, the high German interest rates—was still present, and it was only a matter of time until the next crisis. In November 1992, further speculation forced a devaluation of the Spanish peseta, the Portuguese escudo, and the Swedish krona. The peseta and the escudo were further devalued in May 1993. In July 1993, after yet another large speculative attack, EMS countries decided to adopt large bands of fluctuations (plus or minus 15%) around central parities, in effect moving to a system that allows for very large exchange-rate fluctuations.

Some economists believe that the best way to avoid such crises is for the EU to move as quickly as possible to a *common currency*. Leaving aside its symbolic significance—a very important element in the debate, in a Europe that still remembers World War II—a common currency has all the advantages of a fixed exchange-rate system and more. As we discussed briefly in Chapter 13, it makes for easier transactions and a more efficient organization of production within the common currency area. And, under a common currency, exchange rates disappear, and so do exchange-rate crises.

advantages of common currency

Others believe that while a common currency will indeed eliminate exchange-rate crises, it will not eliminate the underlying reason for such crises—namely, the fact that different interest rates, as well as adjustments in exchange rates, are needed to lower the unemployment levels in the different countries of the European Union. For example, Spain, with its 20% unemployment rate, may well need lower interest rates and a depreciation of the peseta compared with its European partners, in order to return to a single-digit unemployment rate. In other words, they believe that the constraints imposed by fixed exchange rates (or a common currency) on national macroeconomic policies exceed the benefits, and that, at least for the time being, Europe should return to more flexible exchange rates.

At this point, the plan is to begin moving to a European common currency, called the "Euro," sometime in 1998 or 1999, with the complete conversion to a common currency to be completed three years later. But which countries will adopt the Euro in 1998 is unclear. Two of the conditions set by the **Maastricht treaty** (the treaty signed in 1991 that defines the steps of the transition to a common currency) for participation in the common currency area are a budget deficit-to-GDP ratio smaller than 3% and a debt-to-GDP ratio smaller than 60%. At this point, few countries are in a position to meet these criteria. But there appears to be some flexibility in the criteria, and in early 1998 as many as eleven countries appeared ready to adopt the Euro.

① higher r
② large losses in Forex Reserves

ANATOMY OF A CRISIS: THE SEPTEMBER 1992 EMS CRISIS

GLOBAL MACRO

■ **September 5–6.** The Ministers of Finance of the European Union meet in Bath, England. The official communiqué at the end of the meeting reaffirms their commitment to maintaining existing parities within the exchange-rate mechanism (ERM) of the European Monetary System (EMS).

■ **September 8: The first attack.** The attack comes not against one of the currencies in the EMS, but rather against the currencies of Scandinavian countries, which are also pegged to the DM. The Finnish authorities give in and decide to let their currency, the markka, **float**—that is, be determined in the foreign-exchange market without central bank intervention. The markka depreciates by 13% vis-à-vis the DM. Sweden decides to maintain its parity and increases its overnight interest rate to 24% (at an annual rate). Two days later, it increases it further to 75%.

■ **September 10–11: The second attack.** The Bank of Italy intervenes heavily to maintain the parity of the lira, leading the bank to sustain large losses of foreign exchange reserves. But on September 13 the lira is devalued by 7% vis-à-vis the DM.

■ **September 16–17: The third and major attack.** Speculation starts against the English pound, leading to large losses in reserves by the Bank of England. The Bank of England increases its overnight rate from 10% to 15%. However, speculation continues against both the pound and (despite the previous devaluation) against the lira. Both England and Italy announce that they are temporarily suspending their participation in

the ERM. Over the following weeks, both currencies depreciate by roughly 15% vis-à-vis the DM.

■ **September 16–17.** With the pound and the lira out of the ERM, the attack turns against the other currencies. To maintain its parity, Sweden increases its overnight rate to 500%! Ireland increases its overnight rate to 300%. Spain decides to stay in the ERM but to devalue by 5%.

■ **September 20.** French voters narrowly approve the Maastricht treaty in a referendum. A negative vote would surely have amplified the crisis. The narrow but positive vote is seen as the sign that the worst may be over, and that the treaty will eventually be accepted by all EU members.

■ **September 23–28.** Speculation against the franc forces the Banque de France to increase its short-term interest rate by 2.5%. To defend their parity without having to resort to very high short-term interest rates, both Ireland and Spain reintroduce capital controls.

■ **End of September.** The crisis ends. Two countries, the United Kingdom and Italy, have left the ERM and let their currencies depreciate. Spain remains within the ERM, but only after a devaluation. The other countries have maintained their parity, but, for some of them, at the cost of large foreign-exchange reserve losses.

Source: World Economic Outlook, October 1993.

To prevent capital outflow

Summary

■ The real exchange rate depends both on the long-run real exchange rate and on the difference between long-term domestic and foreign real interest rates. An increase in the long-term domestic real interest rate over the foreign interest rate leads to a real appreciation, a decrease to a real depreciation.

■ Movements in the difference between U.S. and foreign interest rates explain most of the movements in the U.S. real exchange rate ("the dance of the dollar") in the 1980s. The appreciation of the yen from 1990 to 1995 appears instead due primarily to a decrease in Japan's long-run real exchange rate, and thus to trade factors.

■ The interest parity condition implies that changes in monetary policy may lead to large variations in the exchange rate. An increase in interest rates leads to a large initial appreciation, followed by a slow depreciation over time. The large initial movement of the exchange rate is known as overshooting.

■ The interest parity condition also implies that, under fixed exchange rates, the perception by financial markets that a currency may be devalued will lead to very high interest rates in the country under attack. These high interest rates have adverse macroeconomic effects. They may even force a devaluation, even if there were no plans for such a devaluation in the first place.

■ Several exchange-rate crises affected the European Monetary System in the 1990s. These led some economists to advocate a more rapid move to a common European currency, which would eliminate the potential for such crises. They led others to recommend a return to more flexible exchange rates.

Key Terms

■ overshooting, 271
■ realignment, 275
■ overnight interest rate, 276

■ Maastricht treaty, 277
■ float, 277

Questions and Problems

1. Suppose that a French citizen decides to invest in one-year British bonds, which offer a real interest rate of 5%. The real exchange rate ε (the relative price of British goods in terms of French goods) is 1.25, but is expected to be 1.4 in one year.
   a. For each "French good" invested, how many *British* goods will the French citizen expect to earn in one year?
   b. For each "French good" invested, how many *French* goods will the French citizen expect to earn in one year?
   c. What real rate of return can the French citizen expect from investing in British bonds?

2. State whether each of the following statements is true or false, and explain your answer briefly.
   a. If interest parity holds, then foreign and domestic real interest rates must be equal.
   b. If interest parity holds and markets expect a real appreciation, then the domestic real interest rate must exceed the foreign real interest rate.
   c. If interest parity holds and the real exchange rate is expected to increase by 10%, then the foreign real interest rate must be 10 percentage points higher than the domestic real interest rate.

3. A five-year British bond pays a nominal interest rate of 10%, while a five-year Canadian bond pays a nominal interest rate of 7%. Inflation in both countries is expected to average 4% per year for the next five years. If interest parity holds, by how much is the dollar expected to appreciate or depreciate vis-à-vis the pound over the next five years?

4. Country A has had lower inflation than country B, and country A's currency has experienced a *real* appreciation relative to country B's currency. Can we say whether country A's currency has also had a *nominal* appreciation relative to country B's currency? Why or why not?

5. Statistics Canada announces preliminary estimates for GDP during the last quarter. These estimates show that GDP during the last quarter was much higher than had been anticipated. This announcement leads to (1) a large increase in long-term interest rates, (2) an appreciation of the dollar, and (3) a drop in stock market prices. Provide a possible explanation for the

reaction of financial and foreign-exchange markets.

6. Why is it sometimes said that a fear of devaluation can become a "self-fulfilling prophecy"?

7. Suppose that the DM-French franc exchange rate is fixed, and that financial markets suddenly believe that the franc may be devalued some time during the next month. More specifically, they believe that there is a 50% chance that the franc will be devalued by 5% within the month, and a 50% chance that the exchange rate will remain unchanged. By how much must the French central bank increase the one-month interest rate (expressed at an annual rate) to maintain the exchange rate today?

8. Many of the emerging economies in Southeast Asia have had fixed exchange rates against the U.S. dollar. Many, however, have strong trade links with Japan but compete in export markets against Japan (and against each other). What implications would the rise and fall of the yen against the US$ in the 1990s have for the long-run exchange rates of these emerging economies? What about their long-term real interest rates relative to those in the United States?

FURTHER READING

The classic reference on the pros and cons of a common currency in Europe is a report by the European Commission of the European Union called "One Market, One Money" (*European Economy*, 1990).

A discussion of the macroeconomic implications of the European Monetary Union is given in David Begg et al., *European Monetary Union: The Macro Issues* (London: Centre for Economic Policy Research, 1991).

The crisis in Asia took many observers by surprise in 1997. It began as a fairly typical speculative attack on the Thai baht, but spread beyond initial expectations. It is still too soon to comment on this crisis, but the *World Economic Outlook: Interim Assessment* (I.M.F., December 1997) gives a good synopsis of events to late 1997.

APPENDIX

FROM NOMINAL TO REAL INTEREST PARITY

The purpose of this appendix is to show how one can go from the interest parity condition in terms of nominal interest and exchange rates derived in Chapter 11 to equation (14.1) in this chapter, the interest parity condition in terms of the real interest and exchange rates.

Let us start from the nominal interest parity condition, equation (11.2) in Chapter 11:

$$1 + i_t = (1 + i_t^*)\frac{E_{t+1}^e}{E_t}$$

Recall the definition of the real interest rate from Chapter 7:

$$1 + r_t \equiv \frac{1 + i_t}{1 + \pi_t^e}$$

where $\pi_t^e \equiv (P_{t+1}^e - P_t)/P_t$ is the expected rate of inflation. Similarly, the foreign interest rate is given by

$$1 + r_t^* = \frac{1 + i_t^*}{1 + \pi_t^{*e}}$$

where $\pi_t^{*e} \equiv (P_{t+1}^{*e} - P_t^*)/P_t^*$ is the expected foreign rate of inflation.

Use these two relations to eliminate nominal interest rates in the interest parity condition, so that

$$1 + r_t = (1 + r_t^*)\left[\frac{E_{t+1}^e(1 + \pi_t^{*e})}{E_t(1 + \pi_t^e)}\right]$$

Note from the definition of inflation that $1 + \pi_t^e = P_{t+1}^e/P_t$ and, similarly, $1 + \pi_t^{*e} = P_{t+1}^{*e}/P_t^*$. Using these two relations in the term in brackets gives

$$\frac{E_{t+1}^e(1 + \pi_t^{*e})}{E_t(1 + \pi_t^e)} = \frac{E_{t+1}^e\, P_{t+1}^{*e}/P_t^*}{E_t\, P_{t+1}^e/P_t} = \frac{E_{t+1}^e\, P_{t+1}^{*e}/P_{t+1}^e}{E_t\, P_t^*/P_t} = \frac{\varepsilon_{t+1}^e}{\varepsilon_t}$$

where the second equality follows from reorganizing terms, and the third follows from the definition of the real exchange rate.

Replacing in the preceding equation gives

$$1 + r_t = (1 + r_t^*)\frac{\varepsilon_{t+1}^e}{\varepsilon_t}$$

which is equation (14.1) in the text.

QUESTIONS AND PROBLEMS

1. Assume the following information for Canada and Germany:

	NOMINAL INTEREST RATE (%)	EXPECTED INFLATION RATE (%)	INITIAL PRICE LEVEL
Canada	5	3	1.0
Germany	8	4	1.0

The current nominal exchange rate (dollars per mark) is 0.8.

a. If interest parity holds, what is the expected nominal exchange rate in one year?

b. What is the current real exchange rate?

c. Use your answers above to show that, if nominal interest parity holds, then real interest parity must hold as well. [*Hint:* First calculate the real interest rate and next year's (expected) price level in both countries.]

15

THE LABOUR MARKET

Think about the sequence of events that takes place when firms respond to an increase in demand by stepping up production. Higher production requires an increase in employment. Higher employment leads to lower unemployment. The tighter labour market leads to higher wages. And higher wages increase production costs, forcing firms in turn to increase prices.

Until now, we have ignored this mechanism and assumed that firms were able and willing to supply any amount of output at a given price level. It is time to relax this assumption. This will be our task in this and the next four chapters. (Remember the three major simplifying assumptions that we made in laying down the basic *IS-LM* model earlier: no treatment of expectations, no allowance for openness of the economy, and a fixed price level. We relaxed the first two earlier. We now relax the third.)

At the centre of the adjustment process described in the first paragraph is the labour market, the market in which wages are determined. This chapter thus starts with an overview of the labour market and takes a first pass at the characterization of equilibrium in the labour market. It derives the central notion of the *natural rate of unemployment*. The next four chapters then combine our earlier treatment of goods and financial markets with our newly acquired knowledge of the labour market. These four chapters are truly exercises in *general-equilibrium* macroeconomics: They examine the behaviour of output and the other main macroeconomic aggregates, taking into account equilibrium conditions in all markets—goods, financial, and labour.

15-1 A TOUR OF THE LABOUR MARKET

The total Canadian population in 1996 was 29.9 million. The relevant population available for civilian employment is called the **civilian noninstitutionalized population** of working age (15 years of age or older). Specifically excluded are full-time members of the Canadian Armed Forces and inmates of institutions (penal institutions or long-term care facilities). Also excluded, for the most part because they are difficult to sample, are residents of the two Territories and persons living on native reserves. Together, the excluded groups represent approximately 2% of the working age population. In 1996, the civilian noninstitutionalized population in Canada was 23.3 million (Table 15-1).

The **labour force**—the sum of those either working or looking for work—was, however, only 15.1 million. The other 6.6 million people were **not in the labour force,** neither working in the marketplace nor looking for work. (Work in the home, such as cooking or raising children, is not classified as work in official statistics. The reason is simply the difficulty of measuring these activities, not a value judgement as to what is work or not work.) The **participation rate,** defined as the ratio of the labour force to the noninstitutionalized population, was thus equal to 15.1/23.3, or 64.9%. This participation rate has steadily increased over time; it stood at only 61.5% in 1976. This increase reflects the steadily increasing participation rate of women.

Of those in the labour force, 13.7 million were employed, and 1.47 million were unemployed. The **unemployment rate,** defined as the ratio of the unemployed to the labour force, was thus equal to 1.47/13.7 = 9.7%. A convenient approximation to keep in mind is that a 1% increase in the Canadian unemployment rate corresponds roughly to 150 000 more people unemployed.

These numbers tell us where people were (whether they were employed, unemployed, or not in the labour force) at one point in time—in this case, 1996. But they do not tell us what typically happens to them *over time.* Do those out of the labour force stay out all the time, or do they go back and forth between participation and nonparticipation? How long do the unemployed remain unemployed? How long do the employed remain employed? To answer these questions, we must turn to the evidence on flows rather than stocks.

....

TABLE 15-1

POPULATION, LABOUR FORCE, EMPLOYMENT, AND UNEMPLOYMENT (MILLIONS) IN CANADA, 1996

Total population	29.9
Civilian noninstitutionalized population	23.3
Labour force	15.1
Employed	13.7
Unemployed	1.47

Source: Statistics Canada, CANSIM Matrix 3472.

THE LARGE FLOWS OF WORKERS

A slowly changing aggregate level of employment may reflect two very different realities. It may reflect an active labour market, with many **separations** (workers leaving or losing their jobs) and many **hires;** or it may reflect a stagnant labour market, with few separations and few hires. Finding out which scenario best reflects reality requires data on movements of workers. Such data are available in Canada from a monthly survey called the Labour Force Survey (LFS). Average monthly flows computed from the LFS for Canada from February 1976 to October 1991 are reported in Figure 15-1.[1] (For more on the ins and outs of the LFS, see the Focus box entitled "The Labour Force Survey.")

Figure 15-1 has three striking characteristics:

(1) *The size of the flows into and out of employment.* The average monthly flow into employment is almost half a million: 0.24 million from unemployment plus 0.25 million from not in the labour force. The average monthly flow out of employment is 0.37 million: 0.19 million to unemployment plus 0.28 million to not in the labour force. Put another way, over this sample period, hires by firms and separations from firms equalled, respectively, 4.4% and 3.3% of employment *each month.*[2]

Why are these flows so large? About half of separations (the flows from employment) are **quits,** workers leaving their jobs in search of a better alternative. The other half are **layoffs.** These come mostly from changes in employment levels across firms: The slowly changing aggregate employment numbers hide a reality of continual job destruction and job creation across firms. At any time, some firms are experiencing decreases in demand and decreasing their employment; others are experiencing increases in demand and increasing employment.

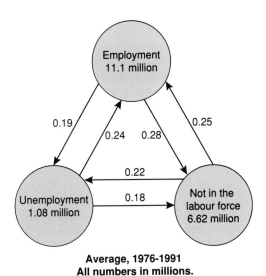

**Average, 1976-1991
All numbers in millions.**

FIGURE 15-1

Average Monthly Flows between Employment, Unemployment, and Nonparticipation in Canada, Febraury 1976 to October 1991.

(1) The flows of workers into and out of employment are large. (2) The flows into and out of unemployment are large in relation to the number of unemployed. (3) There are also large flows into and out of the labour force, much of it directly to and from employment.

Source: See footnote 1.

---

[1] *The data used in Figure 15-1 are taken from Stephen R.G. Jones, "Cyclical and Seasonal Properties of Canadian Gross Flows of Labour," Canadian Public Policy, XIX:1 (1993).*

[2] *This is actually an underestimate, because it excludes movements of workers directly from one job to another.*

(2) *The size of the flows into and out of unemployment in relation to the total number of unemployed.* The average monthly flow out of unemployment each month is 0.42 million: 0.24 million to employment plus 0.18 million to not in the labour force. Put another way, the proportion of unemployed leaving unemployment is equal to 0.42/1.08, or 39% each month. Put yet another way, the average **duration of unemployment** is about three months.[3]

This fact has an important implication. You should not think of unemployment in Canada as a stagnant pool of workers waiting indefinitely for jobs. For most of the unemployed, being unemployed is more a way station between jobs: For the period 1976–1991, the proportion of unemployed getting a job was just over 20% (0.24/1.08) each month. In this respect, Canada resembles the U.S. more closely than Europe. Evidence from Western Europe indicates lower flows into and out of unemployment compared to the stock of unemployed, thus a longer average duration of unemployment. As we shall see in Chapter 26, evidence from Eastern Europe since the beginning of the transition indicates even lower flows and a very long duration of unemployment.

---

[3]*For those who know statistics: If p is the probability of finding a job each month, the expected duration of unemployment is equal to 1/p. Here p is equal to 39%, so the expected time—equivalently, the duration of unemployment—is 1/0.39, or 2.6 months.*

THE LABOUR FORCE SURVEY

FOCUS

Since its inception in 1945, the Labour Force Survey (LFS) is the main source of statistics on the labour force, employment, and participation in Canada.

The LFS has been a monthly survey since 1952. Its coverage has expanded over the years and there were major redesigns of the survey content in 1976 and again in 1997. The most recent changes provide a number of improvements including data on wages and union status, and more detailed information on job status and hours worked. Since July 1995, a total of 52 350 households across Canada are surveyed each month; this amounts to over 100 000 respondents. The sample is not random, but it is chosen to obtain reasonably accurate estimates for different demographic groups (determined by age and gender) and at various geographic levels: national, provincial, census metropolitan areas (large cities), and employment insurance regions. The LFS follows a rotating panel design; each household stays in the survey for six months before it is replaced, so there is a five-sixths month-to-month overlap. The initial interview is done through a personal visit and a large amount of sociodemographic information for each person in the household is collected. Subsequent interviews are conducted by tele-phone. Since 1994, the data has been entered directly into a laptop computer to reduce processing time and transcription errors.

Although the survey is conducted monthly, labour force status is determined by looking at what happened during a single week, called the **reference week**, for that month. Persons are classified as employed if they do any work at all at a job or business during the reference week (including unpaid work for a family business), or if they had a job but were away because of illness, vacation, labour dispute, or similar reasons. The concept of being unemployed is a bit fuzzier. Anyone available for work is called unemployed who (a) is on a temporary layoff, (b) is without work and has been actively looking for work in the past four weeks, or (c) is waiting to start a job at a future date. Full-time students looking for full-time work are not included in this list. Anyone not considered employed or unemployed is deemed to be not in the labour force. The notion of "actively looking for work" is a bit loose. In the United Sates, the criteria require the individual to actually contact potential employers; In Canada, it is enough to check the newspaper.

(3) *The size of the flows into and out of the labour force.* One might have expected these flows to be small, composed on one side of those finishing school and entering the labour force for the first time, and on the other side of workers going into retirement. But these groups actually represent a small fraction of the total flows. The fact is that many of those classified as not in the labour force are in fact willing to work, and move back and forth between participation and nonparticipation. The flow from not in the labour force to employment is larger than the flow from unemployment to employment.

This fact also has an important implication. The sharp focus on the unemployment rate by economists, policy makers, and newspapers is partly misdirected. Some of those classified as not in the labour force are in fact very much like the unemployed; they are in effect **discouraged workers,** and, although they are not actively looking for a job, they will take it if they find it. This is why economists sometimes focus on the **employment rate,** the ratio of employment to population, rather than the unemployment rate. We shall follow tradition and focus on the unemployment rate, but keep in mind that the unemployment rate typically underestimates the number of people available for work.

DIFFERENCES ACROSS WORKERS

The aggregate picture is one of large flows between employment, unemployment, and nonparticipation. However, this picture conceals important differences across groups of workers. Table 15-2, based on evidence from the LFS, shows some of the differences in separation rates by age and sex, again for the period 1976–1991.

On average, 8.7% of workers aged 15 to 24 leave their jobs each month. By contrast, the rate is only 2.9% among female workers aged 25 and over. Slightly more women than men leave their jobs every month, 4.9% versus 3.7%; the difference is largely due to a higher rate of going from employment to not in the labour force. These different separation rates are reflected in different unemployment rates for the different groups. In November 1997, when the aggregate unemployment rate was 9.0%, the rate of unemployment among young males aged 15 to 24 was 16.8%.

Where do these differences come from? Young people often hold low-paying jobs, the proverbial McDonald's jobs, which are often only marginally more attractive than unemployment. Young workers have little seniority and are often the first

....

TABLE 15-2

MONTHLY SEPARATION RATES FOR DIFFERENT GROUPS, 1976–1991

CATEGORY	MONTHLY SEPARATION RATE (%) (QUITS AND LAYOFFS)
Young (15–24 years old)	8.7
Older (25+)	2.9
Male	3.7
Female	4.9

The separation rate is defined as the monthly flow out of employment divided by the initial level of employment. The number is the average separation rate for the period 1976–1991.

Source: Statistics Canada; also see footnote 1.

to be laid off when a firm needs to downsize its work force. Thus, young workers frequently move between jobs, unemployment, and nonparticipation. Middle-aged males, in contrast, tend to keep the jobs they have. This is because the jobs are typically better, and because family responsibilities make it much more difficult to give up the jobs they have for a chance at a better one.

The variations in labour-market experiences reflect in part life-cycle considerations, with the young going from one job to the next until they eventually find one they like and settle down. But they also reflect permanent differences among workers, including education level, skill, and race. Unskilled workers, whatever their age, typically have higher unemployment rates. So do workers in Newfoundland. In November 1997, the unemployment rate in Newfoundland was 17.4%, nearly twice the national average.

These differences across jobs and workers have sometimes led macroeconomists and labour economists to model the labour market as a **dual labour market** that includes a **primary labour market** where jobs are good, wages are high, and turnover is low, and a **secondary labour market** where jobs are poor, wages are low, and turnover is high. This is not a distinction that we shall pursue in this book.

MOVEMENTS IN UNEMPLOYMENT

How does unemployment move over time, and how do these movements affect individual workers? Figure 15-2 shows the average value of the Canadian unemployment rate for each year from 1948 to 1996.

Figure 15-2 has two main characteristics. First, there is evidence of a noticeable upward trend. Unemployment seems to be fluctuating around a slowly increasing average rate, 4% to 5% in the 1960s, perhaps 9% in the 1980s and 1990s.[4] Second, there are large fluctuations in the unemployment rate. The fluctuations are closely associated with recessions and expansions. Note, for example, the last two peaks in unemployment. The most recent, in which unemployment peaked at 11.3%, was associated with the recession of 1990–1991. The one before, in which unemployment peaked at 11.9% (a postwar high) was during the recession of 1980–1983.[5]

How do fluctuations in the *aggregate unemployment rate* affect *individual workers?* This is an important question. The answer determines both the welfare cost of unemployment—the effect of unemployment on the welfare of the unemployed—and the effect of unemployment on wages.

Think about how firms can decrease their employment in response to a decrease in demand for the goods or services they produce. They can hire fewer new workers and/or they can lay off existing workers. Typically, firms prefer to first slow or stop the hiring of new workers, relying on quits and retirements to achieve a decrease in employment. But if the decrease in demand is large this may not be enough, and firms then turn to layoffs.

Now think about the implications for workers, employed or unemployed. If most of the adjustment takes place through decreased hires, the effect is to decrease

---

[4]*In this respect, the experience of Canada has been very similar to that of Western Europe. The unemployment rate in Europe has increased steadily, and the average unemployment rate in the European Union, which stood at 2% to 3% in the 1960s, is now above 10%. We discuss the European unemployment experience in Chapter 20.*

[5]*11.9% is the average unemployment rate for 1983. The unemployment rate actually peaked at 12.8% in December 1982.*

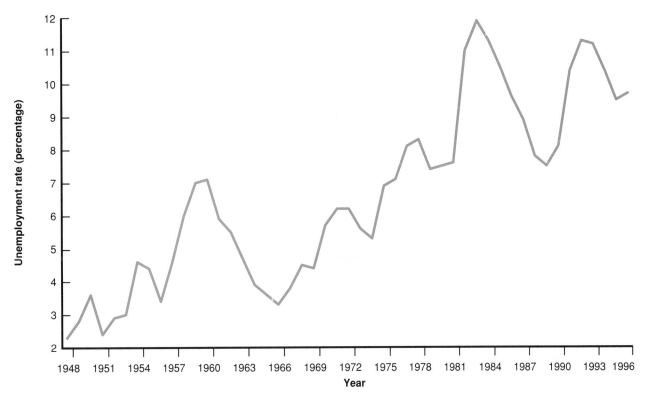

FIGURE 15-2

Movements in the Canadian Unemployment Rate, 1948–1996

Since 1948, the average yearly unemployment rate in Canada has fluctuated from 2% to 12%, with a noticeable upward trend.

Source: Statistics Canada, CANSIM Series D44950 and *Historical Statistics of Canada.*

the chance that an unemployed worker will find a job. Fewer hires means fewer job openings; higher unemployment means more applicants. Fewer openings and more applicants both lead to an increase in the duration of unemployment. If the adjustment takes place through fewer hires *and* more layoffs, higher unemployment also implies a higher chance that employed workers will lose their jobs. Higher unemployment thus means both a lower chance of finding a job if unemployed and a higher chance of losing it if employed.

Movements in the proportion of unemployed finding work and in the separation rate again hide important differences across workers. Typically, the risk of losing a job increases much more in a recession for those who are younger, have fewer skills, or are less educated. Consequently, increases in the average unemployment rate are reflected in much larger increases in the unemployment rates of the young and the unskilled. In this respect, the recession of 1990–1991 was unusual. The proportion of skilled, prime-age workers who lost their jobs was higher than in a standard recession, although their unemployment rate was still much lower than the average. Many economists see this as a sign that the labour market is changing, that firms are being forced to reduce their commitment to job security—even for older, more skilled workers.

15-2 WAGE DETERMINATION

Having looked at the characteristics of unemployment, let's turn to wage determination and the relation of wages to unemployment.

Wages are set in many ways. In Canada, collective bargaining plays only a limited role, especially outside manufacturing and the government sector. The percentage of workers covered by collective bargaining agreements stands at about 30% today. For the rest, wages are either simply set by employers or by bilateral bargaining between the employer and the employee. The higher the skills involved, the more typical is bilateral bargaining. Wages offered for entry-level jobs at McDonald's are on a take-it-or-leave-it basis. New college graduates can typically negotiate a few aspects of their contract. CEOs and sports stars can negotiate a lot more.

There are also large differences across countries. Collective bargaining is slightly less important in the United States, but it plays a very important role in Japan and in most European countries. Negotiations may take place at the firm level, at the industry level, or at the national level. Contract agreements sometimes apply only to those firms that have signed the agreement; sometimes they are automatically extended to all firms and all workers in the sector or the economy.

Countries (and provinces) also differ widely in the relative level and importance of the minimum wage. In the United States, the minimum wage is low relative to the average wage. In 1994, it stood at 38% of the average wage. By contrast, the minimum wage in France in the same year stood at 50% of the average wage.

Given these large institutional differences across countries and time, can we hope for anything like a general theory of wage determination? The answer is yes. Although institutional differences indeed play a role, common forces are at work. Two sets of facts stand out:

1. Workers are typically paid a wage that exceeds their **reservation wage,** the wage that would make them indifferent to working or becoming unemployed. In other words, most workers very much prefer to be employed than unemployed.

2. Wages typically depend on labour-market conditions. Put informally, the lower the unemployment rate, the higher are wages. (We shall make this statement more precise shortly.)

To think about these facts, economists have focussed on two broad lines of explanation. The first is that, even in the absence of collective bargaining, workers have some bargaining power, which they can and do use to extract higher wages from firms. The second is that firms themselves may, for a number of reasons, want to pay wages higher than the reservation wage. Let's look at each of these explanations in turn.

BARGAINING

Many employed workers have some **bargaining power.** This power comes from the fact that, were a worker to leave, the firm would have to find another worker. Finding another worker who fits the job description may be time consuming. The current worker may have learned how to do the job right; training a new worker may take time and money.

How much bargaining power a worker has clearly depends on the nature of his job. Replacing a worker at McDonald's may not be very costly; the required skills can be taught quickly, and typically a large number of willing applicants have already filled out job application forms. In this situation the worker is unlikely to have much bargaining power. If he asks for a higher wage, the firm can fire him at minimum cost. In contrast, a highly skilled worker who has proven unusually good at her job may be very difficult to replace. This gives her more bargaining power. If she asks for a higher wage, the firm may well decide to give it.

Labour-market conditions also affect workers' bargaining power. If the unemployment rate is low, the firm will find it more difficult to find an acceptable replacement. This in turn will increase the bargaining power of employed workers, who will then be able to obtain higher wages. In a market with high unemployment, finding good replacements (even for highly specialized workers) is much easier. In this situation employed workers have less bargaining power, and they may be forced into accepting wage cuts.

EFFICIENCY WAGES

Leaving aside bargaining power on the part of workers, firms themselves may want to pay more than the reservation wage. Firms want their workers to be productive, and the wage can help them achieve that goal. Consider two examples:

- If it takes a while for a worker to learn how to do a job correctly, firms will want workers to stay. If workers were paid just their reservation wage, they would be indifferent between staying or leaving, and turnover would be high. Paying a wage higher than the reservation wage, and thus making it financially attractive for workers to stay, decreases the quit rate and increases productivity.

- Suppose that firms are worried about shirking on the job. Systematically monitoring workers may be costly or even impossible. Paying a wage higher than the reservation wage makes it costly for workers to lose their jobs if they are found shirking. This leads them to shirk less, thus increasing productivity.

These are just two examples. Many economists think that there is a more general mechanism at work. Most firms want their workers to feel good about their jobs, and thus to feel good about the firm. Feeling good promotes good work, and this in turn leads to higher productivity. Paying a high wage is one instrument the firm can use to achieve these goals. (See the Focus box on Henry Ford.) Economists lump all these arguments under the common heading **efficiency wages** to capture the fact that wages have an effect on workers' efficiency (*efficiency* here is a synonym for *productivity*).

Efficiency-wage considerations suggest that wages again depend on the nature of the job and on labour-market conditions. First, the more responsibilities a worker has, and/or the harder it is to monitor his performance, the higher the wage the firm will pay to make sure that the worker does not shirk. Workers who operate expensive machinery—for example, a crane operator or a train engineer—may have to be paid more than other workers. Firms that see employees' morale and commitment as essential to the quality of their work—for example, high-tech firms—will also pay more than firms in sectors where activity is more routine.

In 1914, Henry Ford—the builder of the most popular car in the world, the model T—made a striking announcement. His company would pay all qualified employees a minimum of $5 a day for an eight-hour day. This was a very large salary increase for the majority of employees, who had previously been earning $2.34 for nine-hour days. Although profits were substantial, this increase in pay was far from negligible; it represented about half of the company's profits at the time.

What Ford's motivations were is not entirely clear. Ford himself gave too many reasons for us to know exactly which ones he actually believed. The reason was not that the company had a hard time finding workers at the previous wage. But the company clearly had a hard time retaining workers. There was a very high turnover rate, as well as high dissatisfaction among workers.

Whatever the reasons behind Ford's decision, the results of the wage increase were dramatic. They are shown in Table 1.

The annual turnover rate (the ratio of separations to employment) plunged from a high of 370% in 1913 to a low of 16% in 1915.* The layoff rate collapsed from 62% to nearly 0%. Other measures point in the same

Workers put the finishing touches on Ford Model T's, circa 1915.

....

TABLE 1
ANNUAL TURNOVER AND LAYOFF RATES (%) AT FORD, 1913–1915

	1913	1914	1915
Turnover rate	370	54	16
Layoff rate	62	7	0.1

Source: See source note for this box.

An annual turnover rate of 370% means that on average 31% of the company's workers left each month, so that over the year the ratio of separations to employment was 31% × 12 ≈ 370%.

direction. The average rate of absenteeism, which ran at 10% in 1913, was down to 2.5% a year later. There is little question but that higher wages were the main source of these changes.

Did productivity at the Ford plant increase enough to offset the cost of increased wages? The answer to this question is less clear. Productivity was much higher in 1914 than in 1913; estimates of productivity increases range from 30% to 50%. Despite higher wages, profits were also higher in 1914 than in 1913. But how much of these increases was due to changes in behaviour and how much was due to the increasing success of model-T cars is harder to establish.

Thus, while the effects support efficiency-wage theories, it may well be that the increase in wages to $5 a day was excessive, at least from the point of view of profit maximization. But Henry Ford probably had other objectives as well, such as keeping the unions out—which he did—to generating publicity for himself and the company—which he did as well.

Source: Dan Raff and Lawrence Summers, "Did Henry Ford Pay Efficiency Wages?," *NBER Working Paper*, 2101, December 1986.

Second, labour-market conditions will affect the wage. A tight labour market makes it more attractive for employed workers to quit: Even if they experience a brief period of unemployment, the chance of finding other jobs quickly is high. Thus, a firm that wants to avoid an increase in quits will increase wages as the labour market tightens. Or consider the case of a firm that uses high wages to deter shirking. The potential cost of shirking to a worker is to be found out and fired. The cost of being fired depends on labour-market conditions. If the labour market is tight, the worker may be able to find another job quickly. Thus, a tighter labour market will force the firm to pay a higher wage to avoid shirking.

WAGES AND UNEMPLOYMENT

Our discussion of wage determination suggests a wage equation of the form

$$W = P^e F(u, z) \qquad\qquad (15.1)$$
$$(-, +)$$

This equation says that the aggregate nominal wage, W, depends on three factors. The first is the expected price level, P^e. The second is the unemployment rate, u. The third is a catchall variable, z, that stands for all other variables that affect the outcome of wage setting. Let's look at these three factors in turn.

The expected price level.   Leave aside first the difference between the expected and the actual price level and ask: Why does the price level affect wages? Quite simply because workers and firms care about *real wages,* not nominal wages. Workers care not about how many dollars they receive but rather about how many goods they can buy with their wages. In other words, they care about their wage in terms of goods, about W/P. In the same way, firms care not about the nominal wages they pay workers but about the nominal wages they pay in terms of the price of the output they sell, thus again about W/P. Thus, if both sides knew that the price level was going to double, they would agree to doubling the nominal wage as well. This relation is what is captured in equation (15.1). A doubling in the (expected) price level leads to a doubling of the nominal wage chosen in wage setting.

Let's now turn to the issue we left aside in the preceding paragraph. Why do wages depend on the *expected price level, P^e,* rather than the *actual price level, P?* The answer is, because wages are set in nominal terms, and when they are set what the relevant price level will be is not yet known. For example, in most union contracts, nominal wages are largely predetermined for two years. In other words, unions and firms have to decide on what nominal wages will be over the following two years based on what they expect the price level to be over those two years. Even when wages are set unilaterally by firms, nominal wages are typically set for a year. If the price level goes up unexpectedly during the year, nominal wages are typically not readjusted. How workers and firms form expectations of the price level will occupy us for much of the next three chapters; we defer a discussion until then.

The unemployment rate.   The second factor affecting the wage in equation (15.1) is the unemployment rate. As indicated by the minus sign under u, an increase in the unemployment rate decreases wages. This is one of the main implications of our earlier discussion about wage determination. Higher unemployment weakens workers' bargaining power, forcing them to accept lower wages.

The other factors.   The last variable affecting the wage in equation (15.1), z, is a catchall variable that stands for all the factors that affect wages, given the expected price level and the unemployment rate. By convention, z is defined in such a way that an increase in z leads to an increase in the wage. Our earlier discussion suggests a long list of such factors. For instance:

- Unemployment insurance offers workers protection from complete loss of income if and when they become unemployed. There are good reasons why society should provide at least partial insurance to workers who lose a job and find it difficult to find another. But there is little question that, by making the prospects of unemployment less distressing, more generous unemployment benefits do increase wages. To take an extreme example, suppose employment insurance did not exist. Workers would then be willing to accept very low wages to avoid being unemployed. But employment insurance does exist, and it allows unemployed workers to hold out for higher wages. Thus, we can think of z as standing for the level of unemployment benefits.

- Suppose that the economy goes through a period of structural change, so that the rates of job creation and job destruction both increase, leading to larger flows into and out of unemployment. Thus, at a given level of unemployment, the chance of getting a job while unemployed increases. If it is easier to get a job while unemployed, then unemployment is less of a threat to workers. At a given level of unemployment, workers are in a stronger bargaining position and the wage increases. Thus, we can think of an increase in z as standing for an increase in the rate of structural change in the economy.

It is easy to think of other examples, from changes in minimum-wage legislation to changes in restrictions on firing and hiring, and so on. We shall explore the implications of some of these as we go along.

15-3 PRICE DETERMINATION

Having looked at the determination of wages given (expected) prices, let's now look at the determination of prices given wages.

Prices depend on costs. Costs in turn depend on the nature of the **production function**—that is, the relation between the inputs used in production and the quantity of output produced. We start with a simple assumption about production and will extend it in later chapters. We assume that firms produce goods using labour as the only factor of production and according to the following production function:

$$Y = AN$$

where Y is output, N is employment, and A is labour productivity. We assume that **labour productivity**—the ratio of output to the number of workers—is constant and equal to A.

It is useful to flag the simplifications implicit in this equation. In reality, firms use factors of production other than labour, from raw materials to capital. In reality, labour productivity is not constant, but instead increases steadily over time. We shall introduce raw materials in Chapter 16 when we discuss the oil crises of the 1970s, and we shall focus on the role of capital and productivity in Chapters 22 to 26. For the moment, the simple relation between output and employment will serve our purposes.

Given our assumption that labour productivity, *A,* is constant, we can make one further notational simplification. We can choose the units for output so that one worker produces one unit of output—so that $A = 1$:

$$Y = N \qquad (15.2)$$

This way, we do not need to carry the parameter capturing the level of productivity *A* explicitly, and this will make for simpler notation.

The production function $Y = N$ implies that the cost of producing an additional unit of output is the cost of employing one more worker, and is thus equal to the wage, *W.* Using the terminology introduced in your microeconomics course: The marginal cost of production is equal to *W.* If there were perfect competition in the goods market, the price of a unit of output would be equal to the marginal cost: *P* would be equal to *W.* But many goods markets are not competitive, and firms charge a price higher than their marginal cost. A simple way of capturing this fact is to assume that firms set their price according to

$$P = (1 + \mu)W \qquad (15.3)$$

where μ (the Greek letter mu) is the markup of price over cost. If goods markets were perfectly competitive, the price would simply be equal to the cost and μ would equal zero. To the extent that they are not, and to the extent that firms have market power, the price will be higher than the cost, and μ will be positive.

15-4 THE NATURAL RATE OF UNEMPLOYMENT

determined by the model of wage setting / price setting :

Let's now look at the implications of wage and price determination. Let's do so under the assumption that in wage determination nominal wages depend on the actual price level, *P,* rather than on the expected price level, P^e. (Why we make this assumption will become clear soon.)

In this case, as we shall now see, wage setting and price setting determine both the equilibrium real wage and the equilibrium rate of unemployment. In effect, the equilibrium rate of unemployment must be such that the real wage implied by wage-setting decisions is consistent with the real wage implied by firms' price-setting decisions. Let's now go through this argument step by step.

THE WAGE-SETTING RELATION

Given the assumption that nominal wages depend on the actual price level (*P*) rather than on the expected price level (P^e), equation (15.1), which characterizes wage determination, becomes

$$W = PF(u, z)$$

Dividing both sides by the price level, we obtain

$$\frac{W}{P} = F(u, z) \qquad (15.4)$$

FIGURE 15-3

The Wage- and Price-Setting Relations

The real wage chosen in wage setting is a decreasing function of the unemployment rate. The real wage implied by price setting is constant, independent of the unemployment rate.

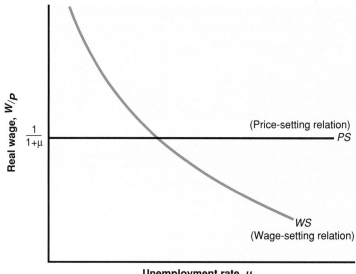

Wage determination implies a negative relation between the real wage, W/P, and the unemployment rate, u: *The higher the unemployment rate, the lower will be the real wage chosen by wage setters.*[6] The intuition is simple: The higher the unemployment rate, the weaker are workers in bargaining, and thus the lower the real wage.

Let's call this relation between the real wage and the rate of unemployment the *wage-setting relation,* and draw it as shown in Figure 15-3. The real wage is measured on the vertical axis. The unemployment rate is measured on the horizontal axis. The wage-setting relation is drawn as the downward-sloping curve WS (for wage setting): the higher the unemployment rate, the lower the real wage.

THE PRICE-SETTING RELATION

Turn now to the implications of price determination. If we divide both sides of the price-determination equation, (15.3), by the nominal wage, we get

$$\frac{P}{W} = 1 + \mu$$

The ratio of the price level to the wage implied by the price-setting behaviour of firms is equal to 1 plus the markup. Now invert both sides of this equation to get the implied real wage:

$$\frac{W}{P} = \frac{1}{1 + \mu} \tag{15.5}$$

Price-setting decisions determine the real wage paid by firms. An increase in the markup leads firms to increase prices given wages; it thus leads to a decrease in the real wage.

---

[6]*"Wage setters" stands for unions and firms if wages are set by collective bargaining, for individual workers and firms if wages are set in bilateral bargaining, or for firms if wages are set on a take-it-or-leave-it basis.*

The *price-setting relation* in equation (15.5) is drawn as a horizontal line PS (for price setting) in Figure 15-3. The real wage implied by price setting is constant, equal to $1/(1 + \mu)$, and is thus independent of the unemployment rate.

EQUILIBRIUM REAL WAGES, EMPLOYMENT, AND UNEMPLOYMENT

Equilibrium in the labour market requires that the real wage implied by wage setting be equal to the real wage implied by price setting. (This way of stating equilibrium may sound a bit strange to those who learned to think of labour supply and labour demand in their microeconomics course. The relation between the wage- and price-setting relations on the one hand and labour supply and labour demand on the other is closer than it looks at first and is discussed further in the appendix to this chapter.)

Substituting equation (15.4) into equation (15.5) gives

$$F(u, z) = \frac{1}{1 + \mu} \tag{15.6}$$

The unemployment rate must be such that the real wage implied by wage setting (the left-hand side of the equation) is equal to the real wage implied by price setting (the right-hand side of the equation). Let's denote this unemployment rate by u_n. It is characterized graphically in Figure 15-4. The equilibrium is at point A, with associated real wage $W/P = 1/(1 + \mu)$ and unemployment rate u_n.

This unemployment rate—the unemployment rate such that the price and wage decisions are consistent—is called the **natural rate of unemployment** (which is why we have used the subscript n to denote it).

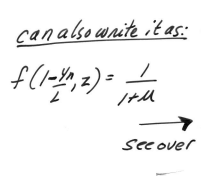

can also write it as:

$$f\left(1 - \frac{Y_n}{L}, z\right) = \frac{1}{1 + \mu}$$

$\longrightarrow$

see over

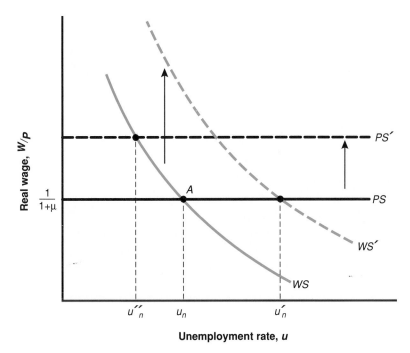

FIGURE 15-4

The Natural Rate of Unemployment

The natural rate of unemployment is such that the real wage chosen in wage setting is equal to the real wage implied by price setting. An increase in unemployment benefits leads to an increase in the natural rate of unemployment. A decrease in the markup leads to a decrease in the natural rate of unemployment.

The terminology has become standard, and we shall adopt it, but the word *natural* is a misnomer. It suggests a constant of nature, one that is unaffected by institutions and policy.[7] As its derivation makes clear, the "natural" rate of unemployment is anything but natural. The positions of the wage-setting and price-setting curves in Figure 15-4, and thus the position of the equilibrium point, depend on both z and μ. Take, for example, an increase in unemployment benefits that increases the real wage chosen in wage setting at a given rate of unemployment, and thus shifts the wage-setting relation from WS to WS', leading to an increase in the natural rate of unemployment from u_n to u'_n. Or consider a more stringent competition policy, which decreases the markup and thus shifts the price-setting relation PS up to PS', leading to a decrease in the natural rate of unemployment from u_n to u''_n. Factors such as the generosity of unemployment benefits or competition policy can hardly be thought of as the result of nature. Rather, they reflect various structural characteristics of the economy. For that reason, a better name for the equilibrium rate of unemployment might be the **structural rate of unemployment**,[8] but so far the name has not caught on.

From Unemployment to Output

Associated with the natural rate of unemployment is a **natural level of employment,** the level of employment that prevails when unemployment is equal to its natural rate. The relation between the two is given by

$$u \equiv \frac{U}{L} = \frac{L - N}{L} = 1 - \frac{N}{L}$$

where the first step follows from the definition of the unemployment rate (u), the second follows from the fact that the level of unemployment (U) is equal to the labour force (L) minus employment (N), and the third follows from simplifying the fraction: The unemployment rate is equal to 1 minus the employment rate.

Rearranging to get employment as a function of the labour force and the unemployment rate gives

$$N = L(1 - u)$$

If the natural rate of unemployment is u_n, the natural level of employment N_n is given by $N_n = L(1 - u_n)$. For example, if the labour force is 10 million and the natural rate of unemployment is 6%, then the natural level of employment is equal to 9.4 million.

Finally, associated with the natural level of employment is a **natural level of output,** the level of production when employment is equal to the natural level of employment. Given the production function we have been using in this chapter ($Y = N$), the relation is particularly simple here: The natural level of output, Y_n, is simply equal to N_n. For reference in Chapter 16, note that, using equation (15.6) and the relations between the unemployment rate, employment, and output we just derived, the natural level of output is defined implicitly by

---

[7]*A common definition of* natural *is: "in a state provided by nature, without human intervention."*
[8]*See Edmund Phelps,* Structural Slumps *(Cambridge, MA: Harvard University Press, 1994).*

$$F\left(1 - \frac{Y_n}{L}, z\right) = \frac{1}{1 + \mu} \qquad (15.7)$$

The natural level of output is that level of output such that the associated rate of unemployment is equal to the natural rate—the unemployment rate such that the real wage chosen by wage setters is equal to the real wage implied by firms' pricing decisions.

15-5 WHERE WE GO FROM HERE

At this point, you should be asking the following question: We have just seen how equilibrium in the labour market determines the natural rate of unemployment and, in turn, the natural level of output, so what have we been doing in the first 14 chapters of this book? If our primary goal was to understand the determination of aggregate output, why did we spend so much time looking at the goods and financial markets? What about our earlier conclusions that the level of aggregate output was determined by factors such as monetary policy, fiscal policy, consumer confidence, and so on—all factors that do not enter equation (15.7) and thus do not affect the natural level of output?

The key to the answer is both simple and important. We have derived the natural rate of unemployment, and the associated levels of employment and output, under two assumptions. The first was equilibrium in the labour market. The second was that the price level was equal to the expected price level.

There is no reason for the second assumption to be true in the short run. The price level may well turn out to be different from what was expected by wage setters when nominal wages were set. Thus, in the short run, there is no reason for unemployment to be equal to the natural rate, or for output to be equal to its natural level. As we shall see in Chapter 16, the factors that determine movements in output *in the short run* are indeed those we focussed on in the first 14 chapters: monetary policy, fiscal policy, and so on. Your time was not wasted.

But expectations of the price level are unlikely to be systematically wrong (say, always too high, or always too low) forever. Thus, in the long run, unemployment returns to the natural rate, and output returns to its natural level. *In the long run,* the factors that determine unemployment and output are indeed the factors that appear in equations (15.6) and (15.7).

Developing this answer in detail will be our task in the next four chapters.

SUMMARY

■ The labour force is composed of those who are working (employed) or looking for work (unemployed). The unemployment rate is equal to the ratio of the number of unemployed to the labour force. The participation rate is equal to the ratio of the labour force to the noninstitutionalized civilian population of working age.

■ The Canadian labour market is characterized by large flows between employment, unemployment, and not in the labour force. Each month, on average, more than one-third of the unemployed move out of unemployment, either to take a job or to drop out of the labour force.

■ Many people who are not actively searching for jobs and are thus not counted as unemployed are in fact willing to work if they find a job. The unemployment rate is thus an imperfect measure of the number of people not working but willing to work.

■ There are important differences across groups of workers in terms of their average unemployment rate and in terms of their average duration of unemployment. Average unemployment rates, and flows into and out of employment, are typically higher for the young, for the low-skilled, and for minorities.

■ Unemployment is high in recessions, low in expansions. During periods of high unemployment, the probability of losing one's job increases, and the probability of finding a job if unemployed decreases.

■ Wages depend negatively on the unemployment rate. A higher unemployment rate leads to lower wages. Wages depend positively on expected prices. The reason why wages depend on expected rather than actual prices is that wages are set in nominal terms, and for some period of time. During that time, even if prices turn out to be different from what was expected, wages are typically not readjusted.

■ Prices set by firms depend on wages and on the markup of prices over wages. Higher wages lead to higher costs, which lead in turn to higher prices.

■ Equilibrium in the labour market requires that the real wage implied by wage setting be consistent with the real wage implied by price setting. Under the additional assumption that the actual price level is equal to the expected price level, equilibrium in the labour market determines the unemployment rate. This unemployment rate is known as the natural rate of unemployment.

■ In general, the actual price level may turn out to be different from what was expected by wage setters, and therefore the unemployment rate need not be equal to the natural rate of unemployment. The coming chapters will show that in the short run unemployment and output are determined by the factors we focussed on in the first 14 chapters (monetary policy, fiscal policy, and so on), but that in the long run unemployment tends to return to the natural rate.

KEY TERMS

- civilian noninstitutionalized population, 282
- labour force, 282
- not in the labour force, 282
- participation rate, 282
- unemployment rate, 282
- separations, 283
- hires, 283
- quits, 283
- layoffs, 283
- duration of unemployment, 284
- reference week, 284
- discouraged workers, 285
- employment rate, 285

- dual labour market, 286
- primary labour market, 286
- secondary labour market, 286
- reservation wage, 288
- bargaining power, 288
- efficiency wages, 289
- production function, 292
- labour productivity, 292
- natural rate of unemployment, 295
- structural rate of unemployment, 296
- natural level of employment, 296
- natural level of output, 296

QUESTIONS AND PROBLEMS

1. Suppose that, during a given month, the civilian non-institutionalized population is 23 million, of which 13 million are employed and 1 million are unemployed. Calculate:
   a. The labour force participation rate
   b. The unemployment rate
   c. The employment rate
2. Think about jobs you may have held in the past. In which of them were you paid your reservation wage? In which were you paid more than your reservation wage? How was your on-the-job behaviour different

in each case? Do your answers help to shed light on efficiency wage theory? Explain.
3. Respond to each of the following statements.
   a. "In any given month, almost all of those entering the labour force are new entrants, and almost all of those leaving are retiring."
   b. "If the unemployment rate remains a constant 7% for one year, it means that 7% of the labour force has remained permanently unemployed for the entire year."
   c. "Since all firms are profit maximizers, they will al-

ways pay the lowest wage necessary to attract the number of workers they need."

4. In any given month, could there be an increase in the number of people who lose their jobs and, at the same time, a decrease in the unemployment rate? Explain.

5. Suppose that the firm's markup over costs is 10%, and that the wage-determination equation is $W = P(1 - u)$, where u is the unemployment rate.

a. What is the real wage as determined by the price-setting equation?

b. What is the natural rate of unemployment?

c. Suppose that the markup over costs increases to 20%. What will happen to the natural rate of unemployment? Explain the logic behind your result.

6. Explain how and why more vigorous competition policy enforcement may affect the natural rate of unemployment.

FURTHER READING

An in-depth discussion of unemployment along the lines of this chapter is given by Richard Layard, Stephen Nickell, and Richard Jackman in *The Unemployment Crisis* (Oxford: Oxford University Press, 1994).

APPENDIX

WAGE- AND PRICE-SETTING RELATIONS VERSUS LABOUR SUPPLY AND LABOUR DEMAND

You may have seen in your microeconomics course a representation of equilibrium in a competitive labour market in terms of labour supply and labour demand. How does the representation in terms of wage setting and price setting relate to that of your microeconomics course?

In an important sense, the two representations are similar. To see why, let's redraw Figure 15-3, but in terms of the real wage and the level of employment (rather than the unemployment rate). We do so in Figure 15-A1 on the next page. Employment (N) is measured on the horizontal axis. The level of employment must be somewhere between zero and L, the labour force. The higher the level of employment, the lower will be the unemployment rate, and thus the higher the real wage implied by wage setting. Thus, the

FIGURE 15-A1

Wage and Price Setting and the Natural Level of Employment

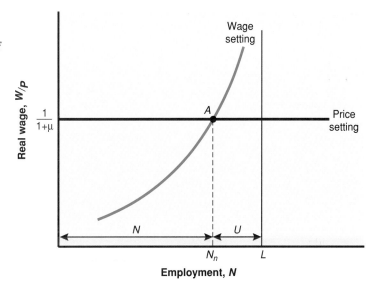

wage-setting relation is now upward-sloping. The price-setting relation is still a horizontal line at $W/P = 1/(1 + \mu)$. The equilibrium is given by point A, where the employment level is the natural level of employment—and the implied unemployment rate is equal to the natural rate of unemployment.

Note that in this figure the wage-setting relation looks like a standard labour-supply relation. As the level of employment increases, the real wage paid to workers increases as well. For that reason, the wage-setting relation is sometimes called the "labour supply" relation (in quotes).

What we have called the price-setting relation looks like a flat labour-demand relation. The reason it is flat rather than downward-sloping has to do with our simplifying assumption of constant returns to labour in production. Had we assumed, more conventionally, that there were decreasing returns to labour in production, our price-setting curve would, like the standard labour-demand curve, be downward-sloping. Because of increasing marginal cost, as employment increased, firms would have to increase their price given the nominal wage; in other words, the real wage implied by price setting would decrease as employment increased.

But in another, equally important sense, the two approaches are very different.

The standard labour-supply relation gives the wage at which a given number of workers are willing to work. In contrast, the wage corresponding to a given level of employment in the wage-setting relation is the result of a complex process of bargaining between workers and firms, or unilateral wage setting by firms. Factors such as the structure of collective bargaining or the use of wages to deter quits affect the wage-setting relation; they have no place in the standard labour-supply relation.

The standard labour-demand relation gives the level of employment chosen by firms at a given real wage. It is derived under the assumption that firms operate in competitive goods and labour markets and therefore take wages and prices—and by implication the real wage—as given. The price-setting relation takes into account the fact that in most markets firms actually set prices. Factors such as the degree of competition in the goods market affect the price-setting relation by affecting the markup; these factors have no place in the standard labour-demand relation.

PUTTING ALL MARKETS TOGETHER

We are now ready to think about **general equilibrium**—that is, the determination of output taking into account equilibrium in *all* markets (goods, financial, and labour) at the same time.

We shall keep this task manageable by reducing the equilibrium to two relations. The first, which we shall call *aggregate supply,* captures equilibrium in the labour market; it builds on what we saw in Chapter 15. The second, which we shall call *aggregate demand,* characterizes equilibrium in both the goods and financial markets; it builds on what we saw in earlier chapters, especially Chapter 6. Together, the two relations determine the general equilibrium level of output and prices.

In this chapter we develop a basic model. When confronted with macroeconomic questions, it is a good place from which to begin to organize your thoughts. For some questions, however, it must be refined and extended. This is what we do in the next three chapters. In Chapters 17 and 18 we extend it to look at movements in inflation, changes in unemployment, and the role of expectations in labour markets. In Chapter 19 we reintroduce the two major themes that we developed earlier: the role of expectations in goods and financial markets, and the implications of openness.

16-1 AGGREGATE SUPPLY

The *aggregate supply relation* captures the effects of <u>output on the price level</u>. It is derived from equilibrium in the labour market.

THE DERIVATION OF THE AGGREGATE SUPPLY RELATION

Recall our characterization of wage and price determination in Chapter 15:

$$W = P^e F(u, z)$$
$$P = (1 + \mu)W$$

The nominal wage (W) depends on the expected price level (P^e), on the unemployment rate (u), and on the catchall variable z that stands for all the institutional factors that affect wage determination, from unemployment benefits to the form of collective bargaining.

The price level (P) is equal in turn to the nominal wage (W) times 1 plus the markup (μ).

Combine these two equations by replacing the wage in the second equation by its expression from the first, to get

$$P = P^e(1 + \mu)F(u, z)$$

This equation gives the price level as a function of the expected price level and the unemployment rate. It will be more convenient in this chapter to express the price level as a function of the level of output rather than the unemployment rate. To do this, recall from Chapter 15 the relation between the unemployment rate, employment, and output:

$$u \equiv \frac{U}{L} = 1 - \frac{N}{L} = 1 - \frac{Y}{L}$$

The first step follows from the definition of the unemployment rate, the second from the definition of unemployment ($U \equiv L - N$). The last follows from our specification of the production function, which says that one unit of output requires one worker, so that $Y = N$.

Replacing u in the preceding equation gives the **aggregate supply relation** between the price level, the expected price level, and output (the name comes from the fact that the relation gives the price level at which firms are willing to *supply* a given level of output):

$$P = P^e(1 + \mu) \, F\left(1 - \frac{Y}{L}, z\right) \tag{16.1}$$

This aggregate supply relation has two main characteristics:

1. *A higher expected price level leads, one for one, to a higher actual price level.* For example, if the expected price level doubles, then the price level will also double. This effect <u>works through wage setting</u>. If wage setters expect higher prices, they set higher nominal wages. This, in turn, leads firms to set higher prices.

Note:
AS: expectations of price level 1: P^e have a huge effect on P^a

① $\uparrow \uparrow P^e \Rightarrow \uparrow \uparrow P^a$
Works through wage setting.

2. *An increase in output leads to an increase in the price level.* This effect is the result of four underlying steps:

■ An increase in output leads to an increase in employment. This result follows from the production function linking output and employment.

■ The increase in employment leads to a decrease in unemployment, and thus a decrease in the unemployment rate.

■ The lower unemployment rate leads to an increase in nominal wages. This result follows from wage determination.

■ The increase in nominal wages leads to an increase in costs, which leads firms to increase prices. This result follows from price determination.

The aggregate supply relation between output and the price level is represented by the *aggregate supply curve AS* in Figure 16-1. It is upward-sloping, as an increase in output leads to an increase in the price level. The curve is drawn for a given value of the expected price level, P^e.

THE AGGREGATE SUPPLY RELATION AND THE NATURAL LEVEL OF OUTPUT

How does this aggregate supply relation relate to our derivation of the natural level of output in Chapter 15? Along the aggregate supply curve, which is drawn for a given value of the expected price level (P^e), the price level P increases with the level of output. Thus, for low enough values of output, P is lower than P^e; for high enough values of output, P is higher than P^e. Consider the value of output such that the price level turns out exactly as wage setters expected, so that $P = P^e$. We know from Chapter 15 that if $P = P^e$ then output is equal to its natural level, Y_n (this is how we derived the natural level of output). This reasoning implies that the aggregate supply curve goes through point A in Figure 16-1: When output is equal to its natural level (when $Y = Y_n$), then the price level turns out to be equal to the price level expected by wage setters ($P = P^e$).

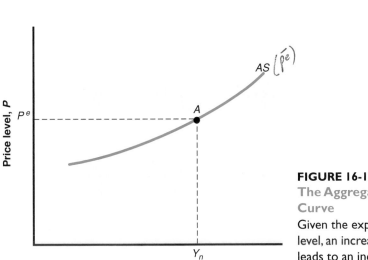

FIGURE 16-1
The Aggregate Supply Curve
Given the expected price level, an increase in output leads to an increase in the price level.

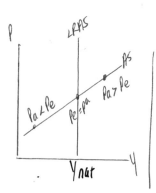

This will prove to be a very useful result. Let us state it in words. When output is equal to its natural level—equivalently, when the unemployment rate is equal to the natural rate—the price level is equal to what wage setters expected. When output is above its natural level—when the unemployment rate is below the natural rate—the price level turns out to be higher than wage setters expected. In effect, the tight labour market leads wage setters, given their expectations of prices, to set high nominal wages, which lead in turn to high prices—prices higher than were expected. When output is below its natural level—when the unemployment rate is above the natural rate—the price level turns out to be lower than expected. The depressed labour market leads to low nominal wages, and in turn to prices lower than expected.

16-2 AGGREGATE DEMAND

The *aggregate demand relation* captures the effect of the price level on output. It is derived from equilibrium in the goods markets and financial markets.

While we have spent several chapters developing progressively more sophisticated and realistic models of equilibrium in those two markets, we shall for the moment use the simplest one, ignoring altogether expectations, open-economy implications, and the distinction between real and nominal interest rates (we shall reintroduce these in Chapter 19).

Borrowing from Chapter 6, the two equations we use to characterize equilibrium in goods and financial markets are:

$$IS: \qquad Y = C(Y - T) + I(Y, i) + G$$

$$LM: \qquad \frac{M}{P} = YL(i)$$

Equilibrium in the goods market requires that the supply of goods be equal to the demand for goods: The sum of consumption, investment, and government spending. This is the *IS* relation.

Equilibrium in financial markets requires that the supply of money must be equal to the demand for money. This is the *LM* relation. Note that what appears on the left-hand side of the *LM* equation is the real money supply, *M/P.* We have focussed so far on changes in the real money supply that came from changes in nominal money, *M*—monetary contractions or expansions implemented by the central bank. But it is clear that changes in *M/P* can also come from changes in the price level. A 10% increase in the price level has the same effect on *M/P* as a 10% decrease in the stock of nominal money. Both lead to a 10% decrease in the real money stock.

Figure 16-2 derives the relation between the price level and output implied by the *IS-LM* model. Figure 16-2(a) draws the *IS* and *LM* curves. The *IS* curve is downward-sloping: An increase in the interest rate leads to a decrease in demand and in output. The *LM* curve is upward-sloping: An increase in output increases the demand for money, and the interest rate must increase so as to maintain equality of money demand and the—unchanged—money supply. The initial equilibrium is at point *A*.

Now consider an increase in the price level from *P* to *P'*. Given the stock of nominal money, *M,* the increase in the price level decreases the real money supply,

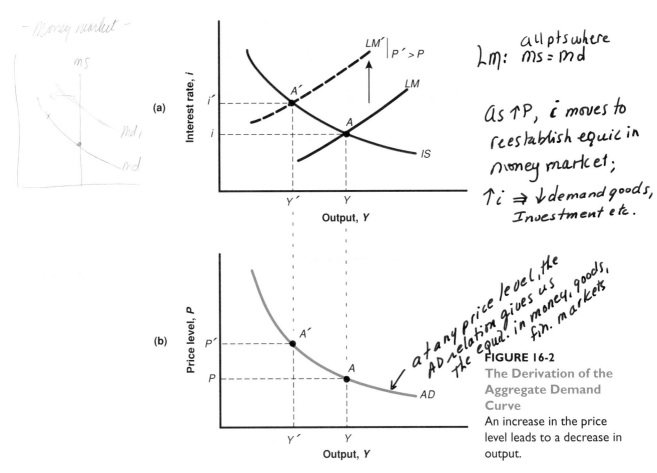

Handwritten annotations (top and right margins):

if P↑'s : this describes what happens :

— Money market —

ms

md₁

md

Lm: all pts where
ms = md

as ↑P, i moves to
reestablish equic in
money market;
↑i ⇒ ↓demand goods,
Investment etc.

at any price level, the
AD relation gives us
the equil. in money, goods,
fin. markets

FIGURE 16-2
The Derivation of the
Aggregate Demand
Curve
An increase in the price
level leads to a decrease in
output.

M/P, and the LM curve shifts up. The equilibrium moves from A to A'; the interest rate increases from i to i', and output decreases from Y to Y'. Thus, an increase in the price level leads to a decrease in output.

In words: As the price level increases, the demand for *nominal* money increases. Put less formally, when the price of goods increases in dollars, people want to hold more money for transactions. Because the supply of nominal money is fixed, the interest rate must increase so as to decrease the demand for money and reestablish equilibrium. The increase in the interest rate leads in turn to a decrease in the demand for goods and a decrease in output.

✓ Keynes interest rate effect.

The implied negative relation between output and the price level is drawn as the downward-sloping curve AD in Figure 16-2(b). Points A and A' in Figure 16-2(b) correspond to points A and A' in Figure 16-2(a). An increase in the price level from P to P' leads to a decrease in output from Y to Y'. We call this curve the *aggregate demand curve,* and we call the underlying negative relation between output and the price level the **aggregate demand relation.** Its name comes from the fact that, at any price level, it gives the demand for output consistent with equilibrium in both goods and financial markets.

│ AD defined

Any variable other than the price level that shifts either the IS curve or the LM curve in Figure 16-2(a) also shifts the aggregate demand relation in Figure 16-2(b). Take, for example, an increase in consumer confidence, which shifts the IS curve to the right and thus leads to higher output. At the same price level, output is higher; the aggregate demand curve thus shifts to the right. Or take a contractionary open-market operation, which shifts the LM curve up and decreases output. At the same price level, output is lower; the aggregate demand curve thus shifts to the left.

We summarize the aggregate demand relation by

$$Y = Y\left(\frac{M}{P}, G, T\right) \qquad (16.2)$$
$$(+, \ +, \ -)$$

Output is an increasing function of the real money stock, an increasing function of government spending, and a decreasing function of taxes. Other factors, such as consumer confidence, could be introduced in this equation; we omit them for simplicity. Given monetary and fiscal policy—that is, given M, G, and T—an increase in the price level leads to a decrease in the real money stock, M/P, and thus to a decrease in output. This is the relation captured by the AD curve in Figure 16-2(b).

16-3 MOVEMENTS IN OUTPUT AND PRICES

At the end of Chapter 15, we argued informally that output tends to return over time to its natural level. Let's now look at that argument more closely.

Because the expected price level has such a strong effect on the actual price level in the aggregate supply relation, equation (16.1), the dynamics of output and prices depend very much on how wage setters form their expectations. In this chapter we shall use the simple assumption that wage setters expect the price level this year to be the same as last year, so that—reintroducing time indexes, which will be needed from now on—$P_t^e = P_{t-1}$. We shall spend the next two chapters discussing and relaxing this assumption. But the assumption will do here to show the basic dynamics at work.

Introducing time indexes and assuming that $P_t^e = P_{t-1}$, the aggregate supply and aggregate demand relations become

$$AS: \qquad P_t = P_{t-1}(1 + \mu)\, F\left(1 - \frac{Y_t}{L}, z\right) \qquad (16.3)$$

$$AD: \qquad Y_t = Y\left(\frac{M}{P_t}, G, T\right) \qquad (16.4)$$

We can now ask the following question: If the economy is left to itself, will it return to its natural level of output, and, if so, how? Or, more formally: Suppose that all the parameters and exogenous variables (μ, z, and L in the aggregate supply relation, M, G, and T in the aggregate demand relation) remain constant. What will happen to output over time?

Start with what happens *this year*, year t. Let the aggregate supply curve and the aggregate demand curve be given by AS and AD in Figure 16-3. The equilibrium is given by the intersection of aggregate supply and aggregate demand, point A. In other words, at point A, the goods, financial, and labour markets are all in equilibrium. Output and the price level are given by Y_t and P_t.

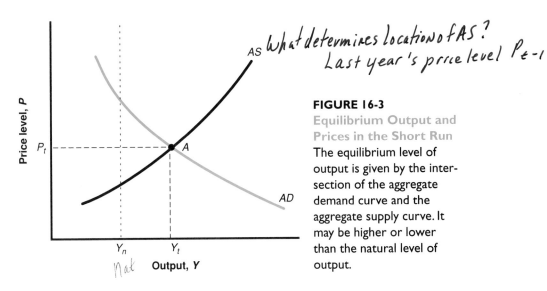

(handwritten) What determines location of AS?
Last year's price level P_{t-1}

FIGURE 16-3
Equilibrium Output and Prices in the Short Run
The equilibrium level of output is given by the intersection of the aggregate demand curve and the aggregate supply curve. It may be higher or lower than the natural level of output.

(axis labels) Price level, P — P_t — A — AS — AD — Y_n — Y_t — *(handwritten)* Nat — Output, Y

Because its position depends on last year's price level, which could have been anything, the aggregate supply curve can be anywhere in Figure 16-3. Thus, the intersection of the aggregate supply and aggregate demand curves can occur at any level of output. Put another way, there is no reason why equilibrium output this year, Y_t, should be equal to Y_n, the natural level of output. As we have drawn the aggregate supply curve in Figure 16-3, Y_t is larger than Y_n: The economy in year t is operating above its natural level. Equivalently, thinking in terms of unemployment, the unemployment rate is below the natural rate.

Because output exceeds its natural level, the price level P_t is higher than was expected, thus higher than P_{t-1}. Recall the argument from Section 16-1: When output is above its natural level, the unemployment rate is below its natural rate; the tight labour market leads to higher wages; and higher wages in turn lead to higher prices, higher than were expected.

Now turn to what happens *next year,* year $t + 1$. As equation (16.3) makes clear, the position of the aggregate supply curve depends on the price level in the preceding year. As this year's price level (P_t) is higher than last year's price level (P_{t-1}), aggregate supply next year (whose position depends on P_t) is higher than aggregate supply this year (whose position depends on P_{t-1}). Draw next year's aggregate supply curve as AS' in Figure 16-4. The equilibrium next year is thus at point A'. The price level is higher than in year t, and output is lower than in year t, thus closer to its natural level. The mechanism works like this: As long as the economy is operating above its natural level, prices are increasing. Higher prices lead to lower real money balances and, hence, to a higher interest rate. The higher interest rate leads to lower demand and lower output.[1]

(handwritten in right margin) Dynamics of return to y_n

$P_t > P_{t-1}$

What happens in the following years is now easy to describe. As long as output is higher than the natural level, prices keep increasing and the aggregate

---

[1] **Digging deeper.** *We can be more specific about the position of the aggregate supply curve in year* $t + 1$. *Remember that, in any year, the aggregate supply curve is such that if output is equal to its natural level,* Y_n, *the price level is equal to the expected price level—here the price level in the preceding year. Thus, the aggregate supply curve in year* $t + 1$ *goes through point B in Figure 16-4, where* $Y = Y_n$ *and the price level is equal to the expected price level for year* $t + 1$, *namely* P_t.

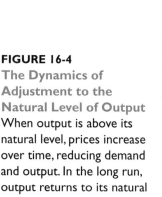

FIGURE 16-4

The Dynamics of Adjustment to the Natural Level of Output
When output is above its natural level, prices increase over time, reducing demand and output. In the long run, output returns to its natural level.

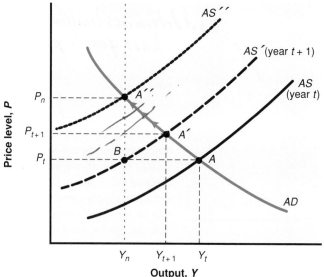

supply curve keeps shifting up. Output keeps decreasing. The economy moves up along the *AD* curve in Figure 16-4, until it eventually reaches point *A″*. At point *A″*, the *AS* curve is given by *AS″* and output is equal to its natural level. There is no more pressure on prices, and the economy settles at Y_n, with associated price level P_n.

This is the basic mechanism through which the economy returns to its natural level. We shall use it in the next section to understand the dynamic effects of various shocks and changes in policy. But we can already draw two important lessons:

■ In the short run, output can be above or below its natural level. Changes in any of the variables that enter either the aggregate supply or the aggregate demand relation lead to changes in output and prices.

■ In the long run, however, output eventually returns to its natural level. The adjustment process works through prices. Output in excess of the natural level leads to higher prices. Higher prices decrease demand and output. Symmetrically, output below the natural level leads to lower prices, which in turn increase demand and output.

We can now use the framework we have developed to look at the dynamic effects of changes in policy or in the economic environment. We shall focus on three such changes. The first two are old favourites: an open-market operation, which changes the stock of nominal money, and a reduction in the budget deficit. The third, which we could not examine until we had developed a theory of wage and price determination, is an increase in the price of oil.

Each of these shocks is interesting in its own right. Monetary policy was responsible for the recession experienced by many O.E.C.D. countries in 1980–1982. Budget deficit reduction made headlines throughout the 1990s. And increases in the price of oil were the main cause of the 1973–1975 recession.

EXAMPLE #1 in the chapter.

Good eg of money neutrality explanation

16-4 THE EFFECTS OF A MONETARY EXPANSION

What are the short- and long-run effects of an expansionary monetary policy, say an increase in nominal money from M to M'?[2]

THE DYNAMICS OF ADJUSTMENT

Assume that, before the change in nominal money, output was at its natural level. In Figure 16-5, aggregate demand and aggregate supply cross at point A, and the level of output at A is equal to Y_n.

Recall the specification of aggregate demand from equation (16.4):

$$Y_t = Y\left(\frac{M}{P_t}, G, T\right) \quad \text{* key}$$

Thus, for a given price level P_t, the increase in money leads to an increase in M/P_t, and thus to an increase in output. The aggregate demand curve shifts to the right, from AD to AD'. The equilibrium moves from point A to A'. Output is higher, and so is the price level.

Over time, the adjustment of expectations comes into play. As long as output exceeds its natural level, wages rise and the price level increases, shifting the aggregate supply curve up. The economy moves up along the aggregate demand curve AD'. The adjustment process stops when output has returned to its natural level. In the long run, the economy goes to point A'': Output is back to its natural level, and the price level is higher. Indeed, we can pin down exactly the size of the increase in

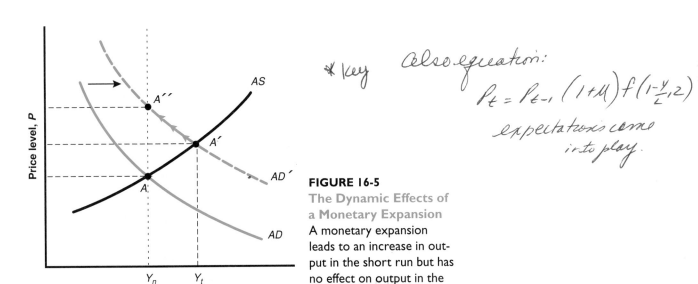

* key

Also equation:

$$P_t = P_{t-1}(1+\mu)f\left(1-\frac{Y}{L}, z\right)$$

expectations come into play.

FIGURE 16-5
The Dynamic Effects of a Monetary Expansion
A monetary expansion leads to an increase in output in the short run but has no effect on output in the long run.

---

[2]We leave the more difficult question of the effects of changes in the growth rate of money—rather than changes in the level of money—to the next two chapters.

the price level. If output is back to its natural level, the real money supply must also be back to its initial value. In other words, the proportional increase in prices must be equal to the proportional increase in the nominal money stock: If the initial increase in nominal money is equal to 10%, then the price level ends up 10% higher as well.

LOOKING BEHIND THE SCENES

It is useful to look behind the scenes at what happens in terms of the underlying *IS-LM* model. We do this in Figure 16-6.

The initial equilibrium is at point *A*, which corresponds to point *A* in Figure 16-5. Output is equal to its natural level, Y_n, and the interest rate is given by *i*. The short-run effect of the monetary expansion is to shift down the *LM* curve from *LM* to *LM'*, moving the equilibrium from point *A* to point *A'*, which corresponds to point *A'* in Figure 16-5. The interest rate is lower, output is higher. There are two effects at work behind the shift in the *LM* curve:

1. The increase in nominal money shifts the *LM* curve down to *LM''*. If the price level did not change—as was our assumption in Chapter 6—the economy would move to point *B*.

2. But, even in the short run, the price level increases with output as the economy moves along the aggregate supply curve. This increase in the price level shifts the *LM* curve upward from *LM''* to *LM'*, partially offsetting the effect of the increase in nominal money.[3]

Over time, the price level increases, reducing the real money supply and shifting the *LM* back up. The economy thus moves along the *IS* curve: The interest rate

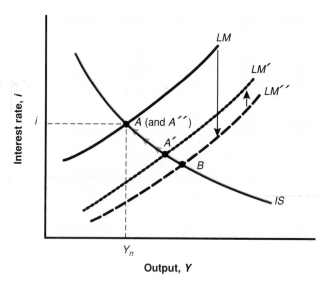

FIGURE 16-6
The Dynamic Effects of the Monetary Expansion on Output and the Interest Rate
The increase in nominal money initially shifts the *LM* curve down, decreasing the interest rate and increasing output. Over time, the price level increases, shifting the *LM* curve back up until output is back at its natural level.

[3]*Why only partially? Suppose that the increase in the price level cancelled the increase in nominal money and thus left the real money supply unchanged. If the real money supply were unchanged, output would remain unchanged as well. But if output were unchanged, the price level would not increase, in contradiction to our premise.*

310   THE SUPPLY SIDE

increases and output declines. Eventually, the *LM* curve returns to where it was before the increase in nominal money. The economy ends up at point *A,* which corresponds to point *A″* in Figure 16-5. The increase in nominal money is exactly offset by a proportional increase in the price level, which leaves the real money supply unchanged. With the real money supply unchanged, output is also back to its initial value, Y_n, and the interest rate also returns to its initial value, *i.*

THE NEUTRALITY OF MONEY

Let's summarize our findings in words. In the short run, a monetary expansion leads to an increase in output, a decrease in the interest rate, and an increase in the price level. Because of the increase in the price level, the increase in the real money supply is smaller than the increase in nominal money. Thus, the effects on output and the interest rate are smaller than they would be if prices were constant.

How much of the initial effect falls on output and how much falls on prices depends on the slope of the aggregate supply curve. In previous chapters we assumed implicitly that the aggregate supply curve was flat, so that prices did not increase at all in response to an increase in output. This was a simplification, but empirical evidence shows that the initial effect of changes in output on prices is indeed quite small.

Over time, prices increase, and the effect on output and the interest rate disappears. *In the long run, the increase in nominal money is reflected entirely in a proportional increase in the price level; it has no effect on output and the interest rate.* (The length of the "long run" is the topic of the In Depth box "How Long-Lasting Are the Real Effects of Money?") Economists refer to the absence of long-run effects of money on output and the interest rate by saying that money is neutral in the long run. The **neutrality of money** does not mean that monetary policy cannot or should not be used. An expansionary monetary policy can, for example, help the economy move out of a recession and return faster to its natural level. But it is a warning that monetary policy cannot sustain higher output forever.

▶ **HOW LONG-LASTING ARE THE REAL EFFECTS OF MONEY?**

IN DEPTH

How long-lasting are the effects of an increase in money on output? Put another way, how much time does it take for the real effects of money to disappear, for money to become neutral?

One way to answer this question is to turn to large macroeconometric models. These models are large-scale versions of the aggregate supply and aggregate demand model presented in the text. One of the most recent has been built by John Taylor of Stanford University. It embodies most of the developments we have studied in this book. Recognizing the importance

of interactions across countries, it is a multicountry model, solving for equilibrium in all G7 countries (the United States, Japan, the United Kingdom, Germany, France, Italy, and Canada) simultaneously. It emphasizes the forward-looking nature of financial markets and solves for the equilibrium under the assumption of rational expectations in financial and foreign exchange markets. It formalizes wage determination as the result of wage setting in overlapping labour contracts, in which wage setters care about real wages as well as wages relative to other workers. Because models with rational expectations are harder to solve, it is relatively small by the standard of macroeconometric models. It includes 98 estimated equations. Some models include up to 1000 equations.

Figure 1 shows the effects in that model of a 3% permanent increase in U.S. nominal money, starting in 1975:1 (this initial date is arbitrary). The increase in nominal money is implemented over four quarters: 0.1% in the first quarter, 0.7% after the second, 1.9% after the third, and 3.0% after the fourth. Nominal money is assumed unchanged in the other G7 countries.

Note how the effects of money on output (as measured by real GNP) increase initially, reaching a maximum after three quarters. By then, output is 1.8% higher than it would have been without the increase in nominal money. In the simplified model in the text, the effect of money is largest at the very beginning. But the Taylor model has a more sophisticated formalization

of aggregate demand than our simple model. An increase in money decreases the interest rate. The effects of a lower interest rate on consumption and investment spending take some time, leading to a slower increase in demand at the beginning.

Over time, prices increase and output returns to its natural level. Four years later, in 1979:1, prices (as measured by the GNP deflator) are up by 2.5%, while output is up by only 0.3%. Therefore, the Taylor model suggests, it takes roughly four years for money to be neutral.

Many economists are skeptical of the results of simulations from large models. In building a model, one must make decisions about which equations to include, which variables to include in each equation, and which ones to leave out. Some decisions are bound to be wrong. Because the models are so large, it is difficult to know how each of these decisions affects the outcome

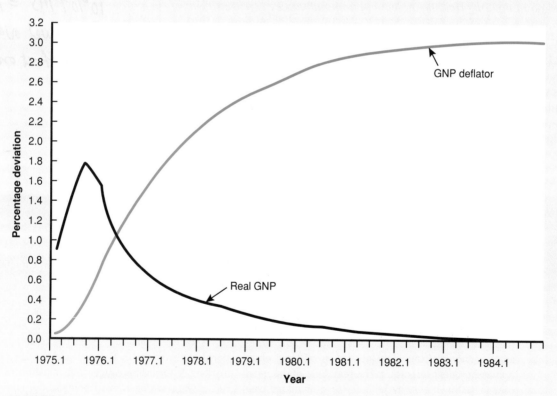

FIGURE 1

The Effects of an Expansion in Nominal Money in the Taylor Model of a particular simulation. Thus, whenever possible, they argue, one should try to use simpler methods.

One such method is simply to trace out, using econometrics, the effects of a change in money on output. This method is not without its pitfalls. A strong relation between money and output may not come from an effect of money on output but rather from an effect of output on the conduct of monetary policy and thus on nominal money (this is an example of the problem of simultaneity that we discussed in Section 4-2). But, keeping in mind such caveats, the method can provide a useful first pass. The results of such a study by Frederic Mishkin, building on earlier work by Robert Barro, are summarized in Table 1.

Following Barro, Mishkin first decomposes movements in nominal money into those movements that could have been predicted based on the information available up to that time (a component he calls **anticipated money**) and those that could not (a component he calls **unanticipated money**). The motivation for this distinction should be clear from this chapter. If wage setters anticipate increases in money, they may expect the price level to be higher and ask for higher nominal wages. Thus, to the extent that they are anticipated, changes in money may have a

larger effect on prices and a smaller effect on output.

The results in Table 1 confirm that changes in money indeed have stronger effects when they are unanticipated. Whether anticipated or unanticipated, the effects on output peak after about two quarters. The effects are substantially larger than in the Taylor model (which looked at a 3% increase in nominal money; Table 1 looks at the effects of a 1% increase). As in the Taylor model, the effects disappear after three to four years (12 to 16 quarters).

Thus, while results using the two approaches are not identical, they have the same general features. Money has a strong effect on output in the short run. But the effect is largely gone after three or four years. By then, the effect of higher nominal money is largely reflected in higher prices, not higher output.

Sources:

Figure 1 is reproduced from John Taylor, *Macroeconomic Policy in a World Economy* (New York: W. W. Norton, 1993), Figure 5-1A, p. 138.

Table 1 is from Frederic Mishkin, *A Rational Expectations Approach to Macroeconometrics* (Chicago: NBER and University of Chicago, 1983), Table 6.5, p. 122.

The study by Mishkin builds in turn on Robert Barro, "Unanticipated Money Growth in the United States," *American Economic Review,* March 1977, pp. 101–115.

....

TABLE 1

THE EFFECTS OF A 1% INCREASE IN NOMINAL MONEY, ANTICIPATED AND UNANTICIPATED

	EFFECT ON OUTPUT					
Quarters	0	2	4	6	12	16
Anticipated	1.3	1.9	1.8	1.3	0.7	−0.6
Unanticipated	2.0	2.3	2.2	2.0	0.5	−0.4

16-5 A DECREASE IN THE BUDGET DEFICIT

Example #2 :
reallocation from
G to I
ultimately

The policy we just looked at—a monetary expansion—led to a shift in aggregate demand coming from a shift in the *LM* curve. Let's now look at the effects on aggregate demand of a shift in the *IS* curve. Let's consider a decrease in the budget deficit coming, say, from a decrease in government spending, *G*.

Suppose that the government, which was running a budget deficit, decides to eliminate it by decreasing health care spending while leaving taxes unchanged. How will this affect the economy in the short, the medium, and the long run?

Assume that output is initially at its natural level, so that the economy is at point A in Figure 16.7. Output is equal to Y_n. The decrease in government spending shifts the aggregate demand curve to the left, from AD to AD': At a given price level, the demand for output is lower. The economy therefore moves from A to A', leading to lower output and lower prices. The initial effect of deficit reduction is thus to trigger a recession. We saw this in Chapter 3, and it remains true in the short run (subject to the qualifications we saw in Section 10-3 when the effect of deficit reduction on expectations is taken into account).

What happens over time? With output now below its natural level, this year's prices are lower than last year's, and the aggregate supply curve next year is lower than this year. As long as output is below its natural level, the aggregate supply curve keeps shifting down, and the economy moves down along the aggregate demand curve AD' until the aggregate supply curve is given by AS'' and the economy reaches point A''. By then, the initial recession is over, and output is back at Y_n.

So, just as with an increase in nominal money, the effects on output of a decrease in the budget deficit do not last forever. Eventually, output returns to its natural level; unemployment returns to the natural rate. But there is an important difference between the effects of a change in money and the effects of a change in the deficit. At point A'', not everything is the same as before. Output is back to its natural level, but the price level and the interest rate are now lower than before the shift. The best way to see why is to look at the adjustment in terms of the underlying IS-LM model.

THE BUDGET DEFICIT, OUTPUT, AND THE INTEREST RATE

The IS and LM curves underlying the aggregate demand relation are drawn in Figure 16-8. The initial equilibrium is at point A, which corresponds to point A in Figure 16-7; output is equal to its natural level, Y_n, and the interest rate is equal to i.

As the government reduces the budget deficit, the IS curve shifts to the left, to IS'. If prices were fixed, the economy would move from point A to point B. But, because prices decline in response to the decrease in output, the real money stock increases, leading to a partly offsetting shift of the LM curve downward, to LM'. The initial effect of deficit reduction is thus to move the economy from A to A'; point A' corresponds to point A' in Figure 16-7. Both output and the interest rate are lower than before. As we discussed in earlier chapters, what happens to investment is ambiguous: Lower output decreases investment but the lower interest rate increases investment.

Over time, output below the natural level—equivalently, unemployment above the natural rate—leads to a further decrease in prices. As long as output is below its natural level, prices decrease, and the LM curve shifts down. The economy moves down from point A' along IS', and eventually reaches A'' (which corresponds to A'' in Figure 16-7). At A'', the LM curve is given by LM''. Output is back at its natural level. But the interest rate is now equal to i'', lower than it was before

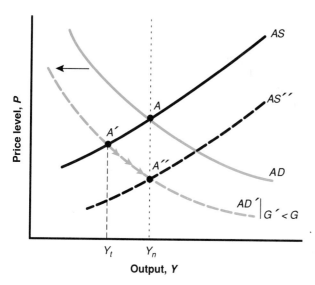

FIGURE 16-7

The Dynamic Effects of a Decrease in the Budget Deficit

A decrease in the budget deficit leads initially to a decrease in output. Over time, output returns to its natural level.

deficit reduction. Indeed, the composition of output is now different. To see how and why, let us rewrite the *IS* relation, taking into account that at both *A* and *A″* output is at its natural level, so that $Y = Y_n$:

note the reallocation from G to I

$$Y_n = C(\bar{Y}_n - \bar{T}) + I(\bar{Y}_n, i) + G$$

Because income and taxes have not changed, consumption is the same as before deficit reduction. By assumption, government spending, *G*, is lower than before; thus investment, *I*, must be higher than before deficit reduction. Indeed, investment must be higher by an amount exactly equal to the decrease in the budget deficit. Put another way, in the long run, a reduction in the budget deficit unambiguously leads to a decrease in the interest rate and an increase in investment.

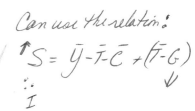

Can use the relation:

$$\uparrow S = \bar{Y} - \bar{T} - \bar{C} + (\bar{T} - G)$$

$$\therefore \quad \downarrow$$

$$I$$

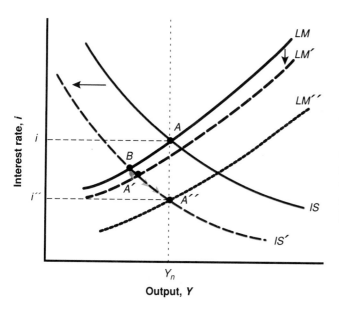

FIGURE 16-8

The Dynamic Effects of a Decrease in the Budget Deficit on Output and the Interest Rate

Deficit reduction leads in the short run to a decrease in output and in the interest rate. In the long run, output returns to its natural level, while the interest rate declines further to a lower long-run value.

BUDGET DEFICITS, OUTPUT, AND INVESTMENT

Our model delivers two conclusions: *p190*

I.(4)

1. In the short run, a budget deficit reduction leads to a decrease in output and may lead to a decrease in investment. The richer treatment of expectations we explored in Section 10-3 (but did not introduce in this chapter) suggests that this conclusion must be qualified. To the extent that deficit reduction leads people and firms to expect lower interest rates in the future, it may lead to an increase in private spending that offsets some of the decrease in public spending. It is even possible for a reduction in the budget deficit to lead to a net increase in total spending, and thus an increase in output in the short run. But such a case must be seen as the exception rather than the rule. The basic message remains that, in the short run, deficit reduction is likely to lead to a decrease in output.

2. The long-run effects of a budget deficit reduction are very different, however. Output returns to its natural level, the interest rate is lower, and investment is higher. We have not taken into account yet the effects of investment on capital accumulation and the effects of capital on production (we shall do so from Chapter 22 on). But it is easy to see how our conclusions would be modified if we did. In the long run, a lower budget deficit leads to higher investment, thus higher capital accumulation and higher output.

We have looked so far in this section at the effects of changes in public saving. (Recall that a deficit reduction is a decrease in public dissaving.) The results would have been similar had we looked at changes in private saving instead. Consider, for example, the effects of a decrease in consumer confidence, which leads consumers to save more at any level of income, thus to increase their marginal propensity to save. *leads to a paradox of svg*

(Paradox p56)

■ In the short run, this increase in the marginal propensity to save is likely to lead to a decrease in output and a decrease in investment, just as a budget deficit reduction would. This is the paradox of saving, which we saw in Chapter 3: An attempt to save more (dissave less in the case of deficit reduction) may actually lead to a decrease in output and a decrease in saving and investment.

■ In the long run, however, the paradox goes away. An increase in the marginal propensity to save or a reduction in the budget deficit eventually leads to an increase in investment, and, through capital accumulation, to an increase in output.

Disagreements among economists about the effects of measures aimed at increasing either private saving or public saving often come from differences in time frames. Those concerned with short-run effects worry that such measures may create a recession and decrease saving and investment for some time. Those who look further out see the eventual increase in saving and investment, and emphasize the long-run increase in output. We discuss these issues at more length in Chapter 29.

Example #3
AS (OPEC)

In the 1970s, the price of oil increased dramatically. This was the result of the formation of OPEC (the Organization of Petroleum Exporting Countries), a cartel of oil producers. Behaving as a monopolist would, OPEC reduced the supply of oil and in doing so increased its price. Figure 16-9, which plots the ratio of the price of crude petroleum to the producer price index since 1960, shows the effects of the formation of OPEC. The relative price of petroleum, which had remained roughly constant throughout the 1960s, almost tripled between 1970 and 1982. This was the result of two particularly sharp increases in the price, the first in 1973–1975 and the second in 1979–1981.

From 1982 on, however, the cartel became unable to enforce the production quotas it had set for its members. Some member countries started to produce more than their quotas, and the supply of oil steadily increased, leading to a large decline in the price. As Figure 16-9 shows, the breakdown of OPEC led to a steady decline

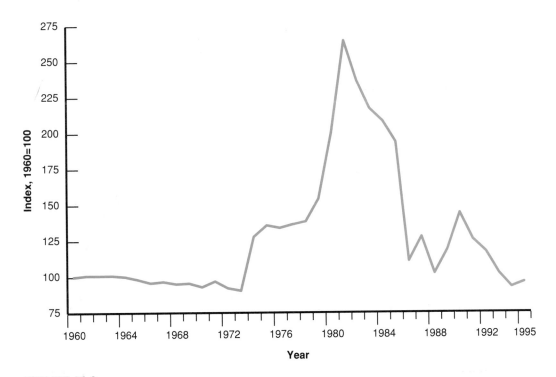

FIGURE 16-9

The Price of Crude Petroleum, 1960–1995

There were two sharp increases in the relative price of oil in the 1970s, followed by a decrease in the 1980s.

Source: U.S. Department of Labour, Bureau of Labour Statistics.

in the relative price of crude petroleum. From a high of 264 in 1981, it has fallen back to the values seen in the 1960s.

In thinking about the effects of an increase in the price of oil, we face an obvious problem. The price of oil appears neither in our aggregate supply relation nor in our aggregate demand relation! The reason is that we have assumed thus far that output was produced using only labour. One way of proceeding would be to relax this assumption, recognize explicitly that output is produced using labour and other inputs (including energy), and derive the implications for the relation of prices to wages and the price of oil. We shall instead use a shortcut and capture the increase in the price of oil by an increase in μ, the markup of prices over wages. The justification is straightforward: Given wages, an increase in the price of oil increases non-labour costs and thus forces firms to increase prices.

We can then track the dynamic effects of an *increase in the markup* on output and prices. It is best to work backward in time, to start by asking what happens in the long run, and then work out the dynamics of adjustment.

EFFECTS ON THE NATURAL RATE OF UNEMPLOYMENT

Let's first ask what happens to the natural rate of unemployment as a result of the increase in the price of oil. Figure 16-10 reproduces the characterization of labour-market equilibrium from Chapter 15. The wage-setting curve is downward-sloping. The price-setting relation is represented by the horizontal line at $W/P = 1/(1 + \mu)$. The initial equilibrium is at point A, and the initial natural rate of unemployment is u_n.

An increase in the markup leads to a downward shift of the price-setting line, from PS to PS': The higher the markup, the lower is the real wage implied by price setting. The equilibrium moves from A to A'. The real wage is lower. The natural unemployment rate is higher; getting workers to accept the lower real wage requires an increase in unemployment.

FIGURE 16-10

The Effects of an Increase in the Price of Oil on the Natural Rate of Unemployment
An increase in the price of oil leads to a lower real wage and a higher natural rate of unemployment.

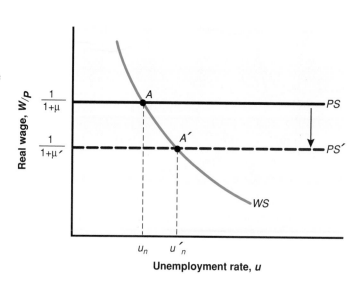

The increase in the natural rate of unemployment implies a decrease in the natural level of employment. If we assume that the relation between employment and output is unchanged—that is, that each unit of output still requires one worker, in addition to the energy input—then the decrease in the natural level of employment leads to an identical decrease in the natural level of output. In short, an increase in the price of oil leads to a decrease in the natural level of output.

THE DYNAMICS OF ADJUSTMENT

Let's now turn to dynamics. Suppose that before the increase in the price of oil the economy is at point A in Figure 16-11, with output at its natural level, Y_n, and a constant price level (so that $P_t = P_{t-1}$). We have just established that the increase in the price of oil decreases the natural level of output from Y_n to, say, Y_n'. We now want to know what happens in the short run and how the economy moves from Y_n to Y_n'.

Recall that the aggregate supply relation is given by

$$P_t = P_{t-1}(1+\mu)\, F\!\left(1 - \frac{Y_t}{L}, z\right)$$

An increase in the markup leads to an increase in the price level at a given level of output. Thus, in the short run, the aggregate supply curve shifts up.

We can be more specific about the size of the shift, and this will be useful in what follows. We know from Section 16-1 that the aggregate supply curve always goes through the point at which output is equal to its natural level and the price level is equal to the price level expected by wage setters. Thus, before the increase in the price of oil, the aggregate supply curve goes through point A, where output is equal to Y_n and the price level is equal to P_{t-1} (as we are assuming that expectations of the price level are such that $P_t^e = P_{t-1}$). After the increase in the price of oil, the new aggregate supply curve goes through point B, where output is equal to the new lower natural level Y_n' and the price level is equal to P_{t-1}. Thus, the aggregate supply curve shifts from AS to AS'.

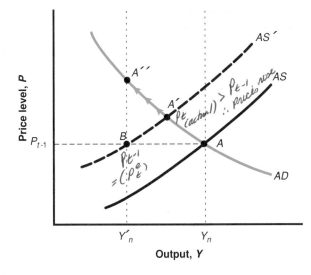

FIGURE 16-11

The Dynamic Effects of an Increase in the Price of Oil

An increase in the price of oil leads, in the short run, to a decrease in output and an increase in the price level. Over time, output decreases further and the price level increases further.

Does the aggregate demand curve shift as a result of the increase in the price of oil? There are many channels through which demand might be affected at a given price level. The higher price of oil may lead firms to change their investment plans, cancelling some investment projects and/or shifting to less energy-intensive equipment. The increase in the price of oil also redistributes income from oil buyers to oil producers. This was indeed the case in Canada in the 1970s, with income falling in central Canada while booming in Alberta. We take the easy way out here: Because some of the effects shift the aggregate demand curve to the right and others shift the aggregate demand curve to the left, we shall simply assume that the effects cancel each other out and that aggregate demand does not shift.

In the short run, the economy therefore moves from A to A'. The increase in the price of oil leads firms to increase prices; the increase in prices in turn decreases demand and output. Note the different effects of adverse demand and supply shocks: Adverse demand shocks (of which we saw an example in Section 16-5) lead to lower output and lower prices. Adverse supply shocks (in this case, an increase in the price of oil) lead to lower output and *higher* prices.

What happens over time? While output has decreased, the natural level of output has decreased even more. At point A' the economy is still above the new natural level of output Y'_n. This leads to a further increase in prices. The economy therefore moves over time from A' to A''. At point A'' output is equal to its new natural level, and prices are higher than before the oil shock. Thus, shifts in aggregate supply affect output not only in the short run but in the long run as well.

How does our analysis compare with what actually happened after the first oil shock? Table 16-1 gives the basic macroeconomic facts.

From 1973 to 1975, the price of imported crude petroleum quadrupled. The price paid by Canadian consumers and businesses, although it jumped dramatically, went up by much less because the federal government imposed price controls. Canada was not a big importer of petroleum at the time; while eastern Canada imported oil, western Canada was an exporter, so much of the impact of the oil price increase was really a redistribution of income within Canada. Nonetheless, the results of the oil price increase were pretty much what our model predicts: a recession and large increases in prices. GDP growth dropped sharply in 1974 and by 1975 the unemployment rate started to climb. Inflation (as measured by the GDP deflator) jumped in 1974, then moderated slightly in 1975. In the United States and Europe, which relied much more on imported petroleum, the effects were even larger. The combination of stagnation and inflation—which was baptized **stagflation**—came as a surprise to most economists. It was the trigger for a large

TABLE 16-1

THE EFFECTS OF THE INCREASE IN THE PRICE OF OIL, 1973–1975

	1973	1974	1975
Rate of change of the price of imported petroleum (%)	14.0	321.2	27.8
Rate of change of the GDP deflator (%)	8.8	11.4	9.9
Rate of change of GDP growth (%)	7.7	4.4	2.6
Unemployment rate (%)	5.6	5.3	6.9

Source: CANSIM series B1352, D10000, D14442 and D44950.

amount of research on the effects of supply shocks for the rest of the decade. By the time of the second oil shock in the late 1970s, macroeconomists were better equipped to understand it.

Many questions still remain, however. One of the most intriguing is whether the effects of changes in oil or other raw material prices are symmetric, whether increases and decreases have symmetrical effects on output. The motivation for the question is the fact that the positive effects on output of the large *decrease* in oil prices since 1982 appear to have been weaker than the negative effects of the increases in oil prices of the 1970s.

16-7 CONCLUSIONS

This has been a long but important chapter. Let us repeat and develop some of its conclusions.

THE SHORT RUN VERSUS THE LONG RUN

One basic message of the chapter is that changes in policy, or, more generally, changes in the economic environment—from changes in consumer confidence to changes in the price of oil—typically have different short-run and long-run effects. We looked at the effects of a monetary expansion, of a deficit reduction, and of an increase in the price of oil. The main results are summarized in Table 16-2. A monetary expansion, for example, affects output in the short run but not in the long run. A decrease in the budget deficit decreases output and the interest rate in the short run, and may therefore decrease investment. But in the long run the interest rate decreases and output returns to the natural rate, so that investment necessarily increases. An increase in the price of oil decreases output, not only in the short run but also in the long run. And so on.

This difference between the short- and long-run effects of policies is one of the main reasons why economists disagree in their policy recommendations. Some economists believe that the economy adjusts quickly to its long-run equilibrium, and thus

TABLE 16-2

SHORT- AND LONG-RUN EFFECTS OF A MONETARY EXPANSION, A BUDGET DEFICIT REDUCTION, AND AN INCREASE IN THE PRICE OF OIL ON OUTPUT, THE INTEREST RATE, AND THE PRICE LEVEL *(3 models summarized.)*

	SHORT RUN			LONG RUN		
	Output Level	Interest Rate	Price Level	Output Level	Interest Rate	Price Level
Monetary expansion	increase	decrease	increase (small)	no change	no change	increase
Budget deficit reduction	decrease	decrease	decrease (small)	no change	decrease	decrease
Increase in the price of oil	decrease	increase	increase	decrease	increase	increase

emphasize long-run implications of policy. Others believe that the adjustment mechanism through which output returns to its natural level is a slow one at best, and put more emphasis on the short-run effects of policy. They are more willing to use active monetary policy or budget deficits to get out of a recession, even if money is neutral in the long run, and budget deficits have adverse long-run implications.[4]

SHOCKS AND PROPAGATION MECHANISMS

This chapter also gives you a general way of thinking about **output fluctuations** (sometimes called **business cycles**)—movements in output around its trend (a trend that we have ignored so far, but on which we shall focus in Chapters 22 to 25).

The economy is constantly buffeted by **shocks** to aggregate supply, or to aggregate demand, or to both. These shocks may be shifts in consumption coming from changes in consumer confidence, shifts in investment, shifts in foreign demand for domestic goods, shifts in portfolio behaviour, and so on. Or they may come from changes in policy—from the introduction of a new tax law, to a new program of infrastructure investment, to the central bank's decision to fight inflation through tight money.[5]

Each shock has dynamic effects on output and its components. These dynamic effects are called the shock's **propagation mechanism.** Propagation mechanisms are different for different shocks. The effects on activity may be largest at the beginning and then decrease over time. Or the effects may build up for a while, and then decrease and disappear. We saw, for example, that, in the United States, the effects of an increase in money on output peak after six to nine months and then slowly decline afterward, as prices eventually increase in proportion to the increase in money. Some shocks have effects even in the long run. This is the case for any shock that has a permanent effect on aggregate supply, such as a permanent change in the price of oil.

Fluctuations in output come from the constant appearance of new shocks, each with its own propagation mechanism. At times, some shocks are sufficiently bad, or come in sufficiently bad combinations, that they create a recession. The two recessions of the 1970s were due largely to increases in the price of oil; the recession of the early 1980s was due to a sharp change in monetary policy. At other times, the shocks combine instead to produce long expansions; as we saw in the In Depth box in Chapter 13 (page 250), the long expansion of the 1980s was due in part to the effects of the recovery from the earlier recession and in part to the effects of large budget deficits. What we call economic fluctuations are the result of these shocks and their dynamic effects on output.

---

[4]*We shall return to these issues many more times in the book. See in particular Chapter 20, which focusses on periods of sustained high unemployment such as the Great Depression, and Chapters 27 to 29, which look at macroeconomic policy in more detail.*

[5]*How to define* shocks *raises quasi-philosophical issues. Suppose that a failed economic program in a foreign country leads to the fall of democracy in that country, which leads to an increase in the risk of nuclear war, which leads to a fall in domestic consumer confidence in our country, which leads to an increase in saving given current income. What is the "shock"—the failed program, the fall of democracy, the increased risk of nuclear war, or the decrease in consumer confidence? In practice, we have to cut the chain of causation somewhere. Thus, we may refer to the drop in consumer confidence as "the shock," ignoring its underlying causes.*

OUTPUT, UNEMPLOYMENT, AND INFLATION

In developing the model of this chapter, we made the assumption that the nominal money supply was constant. That is, although we considered the effects of a change in the level of nominal money (in Section 16-4), we did not allow for sustained nominal money growth. One implication of that assumption was that the price level was constant in the long run, that there was no inflation. We must now relax this assumption and allow for nominal money growth. Only by doing so can we explain why inflation is typically positive, and think about the relation between economic activity and inflation. Movements in unemployment, output, and inflation are the topics of the next two chapters.

SUMMARY

■ The model of aggregate supply and aggregate demand describes the movements in output and prices when account is taken of equilibrium in the goods, financial, and labour markets.

■ The aggregate supply relation captures the effects of output on the price level. It is a relation between the price level, the expected price level, and the level of output, implied by equilibrium in the labour market. A higher expected price level leads, one for one, to a higher actual price level. An increase in output decreases unemployment, increasing wages and, in turn, the price level.

■ The aggregate demand relation captures the effects of the price level on output. It is derived from equilibrium in the goods and financial markets. An increase in the price level decreases the real money supply, increasing interest rates and decreasing output.

■ In the short run, movements in output come from shifts in either aggregate demand or aggregate supply. In the long run, output returns to its natural level, which is determined by equilibrium in the labour market.

■ An expansionary monetary policy leads in the short run to an increase in the real money supply, a decrease in the interest rate, and an increase in output. Over time, the price level increases, leading to a decrease in the real money supply until output has returned to its natural level. In the long run, money is neutral: it does not affect output, and changes in money are reflected in proportional increases in the price level.

■ A decrease in the budget deficit leads in the short run to a decrease in the demand for goods and thus a decrease in output. Over time, the price level decreases, leading to an increase in the real money supply and a decrease in the interest rate. In the long run, output is back to its natural level, but the interest rate is lower and investment is higher.

■ An increase in the price of oil leads, in both the short run and the long run, to a decrease in output. In the short run, it leads to an increase in prices, which decreases the real money supply and leads to a contraction of demand and output. In the long run, it decreases the real wage paid by firms, increases the natural rate of unemployment, which in turn decreases the natural level of output.

■ The difference between short- and long-run effects of policies is one of the main reasons why economists disagree in their policy recommendations. Some economists believe that the economy adjusts quickly to its long-run equilibrium, and thus emphasize long-run implications of policy. Others believe that the adjustment mechanism through which output returns to its natural level is a slow one at best, and put more emphasis on short-run effects.

■ Economic fluctuations are the result of a constant stream of shocks to aggregate supply or to aggregate demand, and of the dynamic effects of each of these shocks on output. Sometimes the shocks are sufficiently adverse, alone or in combination, that they lead to a recession. Sometimes they lead instead to a sustained expansion.

KEY TERMS

■ general equilibrium, 301
■ aggregate supply relation, 302
■ aggregate demand relation, 305
■ neutrality of money, 311
■ anticipated versus unanticipated money, 313

■ stagflation, 320
■ output fluctuations, or business cycles, 322
■ shocks, 322
■ propagation mechanism, 322

QUESTIONS AND PROBLEMS

1. For each of the following changes, state which curve or curves are affected initially (*IS, LM, AS,* and *AD*) and in which direction they will initially shift.
   a. An increase in government spending
   b An increase in the nominal money supply
   c. An increase in the price of oil
   d. A decrease in consumer confidence

2. Respond to each of the following statements.
   a. "Since money is neutral, there is no point at all in trying to use monetary policy to affect output."
   b. "Since fiscal policy cannot change the natural level of output, we can say that government spending, too, is neutral."
   c. "All recessions over the past 25 years have been caused by government policy changes."

3. Suppose that unemployment benefits are increased permanently.
   a. What will happen in the long run? Illustrate with
      (i) the *AD* and *AS* diagram
      (ii) the wage-setting and price-setting diagram
   b. What will happen in the short run? Illustrate with
      (i) the *AD* and *AS* diagram
      (ii) the *IS-LM* diagram

4. Suppose that the *IS* curve were vertical.
   a. What would this imply about economic behaviour?
   b. What would the aggregate demand curve look like?
   *c. What would happen over time if the level of output implied by aggregate demand were different from the natural level of output?

5. What kinds of changes in the economy would probably shorten the time needed for long-run money neutrality to occur?

6. Some economists argue that the key to a low interest rate is a more expansionary monetary policy. Other economists argue that the key is larger private saving and/or the reduction of budget deficits. Who is right?

this Chapter: relation between p^a p^e and Un

FURTHER READING

An in-depth discussion of the effects of supply shocks is given in Michael Bruno and Jeffrey Sachs, *Economics of* *Worldwide Stagflation* (Cambridge, MA: Harvard University Press, 1985).

(1) Show AD/AS relation algebraically (define them)

(2) 2 main characteristics of AS: (a) $\uparrow p^e \Rightarrow \uparrow Pa$ (works thru wage setting)
(b) $\uparrow y \Rightarrow \uparrow Pa$ (also thru wage setting)

(3) Diagram of AS relation & relation to $p^e = P$ $\quad pa < pe \quad pa = pe$
$\quad pa > pe$

(4) Relation between AD and IS/LM (diagram)

(5) Assume $\uparrow$ Price level; Draw diagram of relation between money market; IS on /AD model

(6) Assume AD AS > Yn; will economy return to Yn? Show how thru expectations.

(7) What determines AS's location along AD?

(8) Show effects of $\uparrow$ MS on AS/AD diagram; Start at Yn; Short run effect on i?

(9) Define/discuss concept of money neutrality (how long it takes for $ to be neutral)

(10) $\downarrow$ Budg deficit; effects on I = f(y,i); long run effects on I and i
$\quad$ Public vs Private sug

(11) Explain Paradox of Sug = SR vs longrun
$\quad$ (start w/ a $\downarrow$ in consumer confidence)

(12) How oil shocks affect Yn, AS curve/wage-setting relation in 70s
$\quad$ Any effects on AD?

(13) Summary of LR/SR effects $\uparrow$MS, $\downarrow$ deficit $\uparrow$ oil stocks: prices

(14) Propagation mechanism (explain)

THE PHILLIPS CURVE

In 1958, A. W. Phillips drew a diagram that was to become famous. He plotted the rate of inflation against the rate of unemployment in the United Kingdom for each year from 1861 to 1957. He found clear evidence of a negative relation between inflation and unemployment: When unemployment was low, inflation was high, and when unemployment was high, inflation was low, often even negative.[1]

Two years later, Paul Samuelson and Robert Solow replicated Phillips' exercise for the United States, using data from 1900 to 1960. Figure 17-1 shows the relationship between inflation and unemployment in Canada from 1927 to 1960.[2] Apart from the period of very high unemployment during the 1930s (the years from 1931 to 1939 are denoted by coloured diamonds and are clearly to the right of the other points in the figure), there appeared also to be a stable negative relation between inflation and unemployment in Canada.

This relation, which Samuelson and Solow baptized the **Phillips curve,** rapidly became central to macroeconomic thinking and policy. It appeared to imply that, leaving aside such episodes as the Great Depression, countries could choose between different combinations of unemployment and inflation. They could achieve low unemployment if they were ready to tolerate higher inflation, or they could achieve

---

[1]When no confusion can arise, we shall refer from now on to the inflation rate simply as "inflation," and to the unemployment rate simply as "unemployment."

[2]Figure 17-1 uses the rate of change of the GDP deflator as the measure of inflation. Samuelson and Solow used the rate of change of nominal wages, as did Phillips.

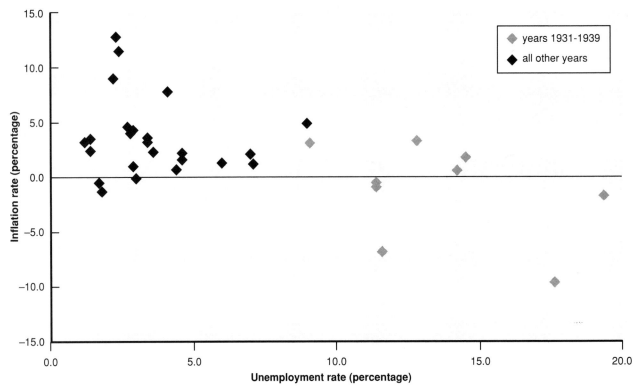

FIGURE 17-1

Inflation and Unemployment in Canada, 1927–1960

From 1927 to 1960, there appears to have been a negative relationship between inflation and unemployment in Canada. The years 1931 to 1939 are plotted with blue diamonds.

Source: Statistics Canada, CANSIM Series D11000, D14442, and D44950, and *Historical Statistics of Canada.*

price level stability—zero inflation—if they were ready to tolerate higher unemployment. Macroeconomic policy became focussed on choosing the preferred point on the Phillips curve.

By the 1970s, however, the relation broke down. In both Canada and most OECD countries, there was both high inflation *and* high unemployment, a clear contradiction to the original Phillips curve. When a relation reappeared, it was now a relation between the unemployment rate and the *change* in the inflation rate. Today high unemployment appears to lead not to low inflation, but rather to a decrease in inflation.

The purpose of this chapter is to explore the mutations of the Phillips curve and, more generally, to understand the relation between inflation and unemployment.

17-1 INFLATION, EXPECTED INFLATION, AND UNEMPLOYMENT

We derived in Chapter 16 the following aggregate supply relation between the *price level*, the *expected price level*, and the *unemployment rate:*[3]

$$P_t = P_t^e (1 + \mu) F(u_t, z)$$

This relation can be rewritten as a relation between the *inflation rate*, the *expected inflation rate*, and the *unemployment rate:*

$$\pi_t = \pi_t^e + (\mu + z) - \alpha u_t \qquad (17.1)$$

where π_t denotes the rate of inflation, defined as the rate of change in prices from last year to this year, and π_t^e denotes the corresponding expected rate of inflation. The parameter α (the Greek lowercase letter alpha) reflects the effect of unemployment on inflation; the larger α, the stronger the (negative) effect of unemployment on inflation. In short, equation (17.1) tells us that *inflation depends positively on expected inflation and negatively on unemployment.*

The derivation of equation (17.1) from the aggregate supply relation is not difficult, but it is tedious and is better left to an appendix at the end of this chapter. You can skip the appendix, but you should make sure that you understand each of the effects at work in equation (17.1):

■ *Higher expected inflation leads to higher inflation.* We saw in Chapter 16 how higher expected prices led to higher nominal wages and in turn to higher prices. But note that given last year's prices, higher prices this year imply higher inflation this year; similarly, higher expected prices imply higher expected inflation. So we can restate our earlier result as: Higher expected inflation leads to higher actual inflation.

■ *Given expected inflation, the higher the markup chosen by firms, μ, or the higher the factors that affect wage determination, z, the higher will be inflation.* We saw in Chapter 16 how a higher markup led to higher prices given wages, and thus to higher prices given expected prices. We can restate this proposition as follows: A higher markup leads to higher inflation given expected inflation. The same argument applies to increases in any of the factors that affect wage determination.

■ *Given expected inflation, the higher unemployment, the lower will be inflation.* We saw in Chapter 16 that, given expected prices, a higher unemployment rate led to lower wages, and thus to lower prices. We can restate this proposition as follows: Given expected inflation, a higher unemployment rate leads to lower actual inflation.

We now have the relation we need to understand the tribulations of the Phillips curve.

---

[3] *After deriving this relation in Chapter 16, we replaced the unemployment rate by its expression in terms of output to obtain the aggregate supply relation between the price level, the expected price level, and output. It will be more convenient in this chapter not to take this additional step, and to stay with the relation in terms of unemployment rather than output.*

17-2 THE PHILLIPS CURVE

Let's start with the relation between unemployment and inflation as it was first discovered by Phillips, Solow, and Samuelson, circa 1960.

THE EARLY INCARNATION

Think of an economy where inflation is positive in some years, negative in others, and is equal on average to zero. This is not the way things are in Canada today: The last year during which inflation was negative—the last year during which there was **deflation**—was 1953, when inflation was −0.1%. But average inflation *was* close to zero during much of the period that Phillips, Samuelson, and Solow were examining.

Think of wage setters choosing nominal wages for the coming year and thus having to assess what inflation will be over the year. With the average inflation rate equal to zero in the past, it is reasonable for them to expect that inflation will be equal to zero over the next year as well. Assuming that $\pi_t^e = 0$ in equation (17.1) gives the following modified relation between unemployment and inflation:

$$\pi_t = (\mu + z) - \alpha u_t \qquad (17.2)$$

This is precisely the negative relation between unemployment and inflation that Phillips, Solow, and Samuelson found for the United Kingdom and for the United States. The story behind it is simple: Given expected prices, which workers simply take to be last year's prices, lower unemployment leads to higher nominal wages. Higher nominal wages lead in turn to higher prices. Putting the steps together, lower unemployment leads to higher prices this year compared with last year—that is, to higher inflation.

This mechanism has sometimes been called the **wage-price spiral,** and this phrase captures well the basic mechanism at work: Low unemployment leads to higher nominal wages. In response to higher nominal wages, firms increase their prices and the price level increases. The higher price level leads workers to ask for higher nominal wages next year, prompting yet another increase in prices set by firms, and so on, leading to steady inflation.

MUTATIONS

The combination of an apparently reliable empirical relation, together with a plausible story to explain it, led to the instant acceptance of the Phillips curve by macroeconomists and policy makers alike. Macroeconomic policy in the 1960s was aimed at maintaining unemployment in the range that appeared consistent with moderate inflation. And, through most of the 1960s, the negative relation provided a reliable guide to the relation between unemployment and inflation. Figure 17-2 plots the combinations of inflation and unemployment in Canada for each year from 1948 to 1969. Note how well the relation held during the long expansion of the 1960s. (The years 1961 to 1968 are denoted by blue diamonds in the figure.) From 1961 to 1966, the unemployment rate declined steadily from 5.9% to 3.3%; the inflation rate steadily increased, from 0.5% to 4.8%.

However, in 1967 and 1968 we begin to see signs that something is not right. As the unemployment rate increased, inflation fell but not by as much as previous

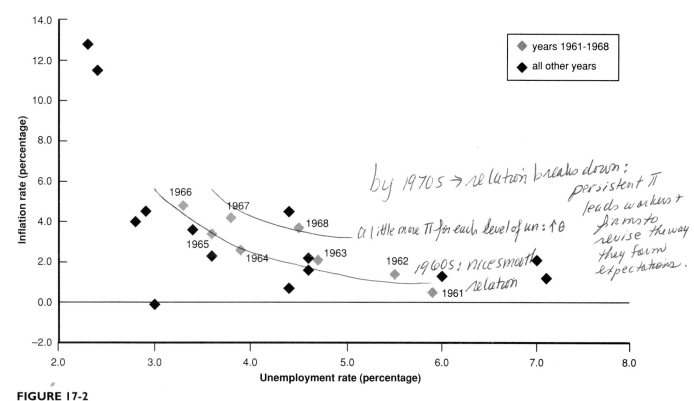

Handwritten annotations on figure:
by 1970s → relation breaks down; persistent π leads workers + firms to revise the way they form expectations.
a little more π for each level of un: ↑θ
1960s: nice smooth relation

FIGURE 17-2

Inflation and Unemployment in Canada, 1948–1969

The steady decline in Canadian unemployment from 1961–1966 was associated with a steady increase in inflation.

Source: Statistics Canada, CANSIM Series D11000, D14442, and D44950, and *Historical Statistics of Canada.*

episodes would have led us to expect. By the 1970s, the relationship had broken down. Figures 17-3 gives the combination of inflation and unemployment rates in Canada for every year since 1948. The points are scattered in a roughly symmetric cloud: There is no longer any relation between the unemployment rate and the inflation rate.

Why did the original Phillips curve vanish? The main reason is that firms and workers changed the way they formed expectations. This change came from a change in the process of inflation itself. Look at Figure 17-4, which plots the Canadian inflation rate for each year since 1927. Starting at around 1960 (at the vertical bar), there was a clear change in the inflation process. Rather than going from positive to negative, as it had for the first part of the century, inflation became both more persistent and consistently positive. High inflation in one year became more likely to imply high inflation the next year as well.

The persistence of inflation led workers and firms to revise the way they formed expectations. When inflation is consistently positive, expecting that prices this year will be the same as last year becomes systematically incorrect, indeed foolish. People do not like to make the same mistake repeatedly. Thus, as inflation became consistently positive and more persistent, expectations started to incorporate the presence of inflation. This change in expectation formation changed the nature of the relation between unemployment and inflation.

Handwritten annotation in right margin:
When π>0, expecting that this year's prices will be same as last year → incorrect. People don't make these kinds of mistakes.

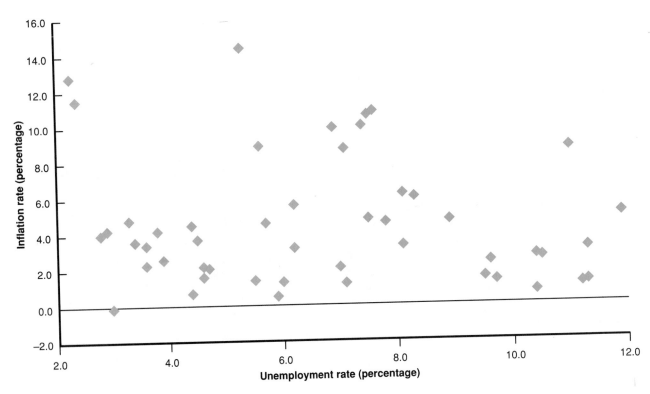

FIGURE 17-3

Inflation and Unemployment in Canada, 1948–1996

When we add the data since the 1970s, the negative relationship between inflation and unemployment disappears.

Source: Statistics Canada, CANSIM Series D11000, D14442, and D44950, and *Historical Statistics of Canada.*

To understand what happened, suppose that expectations are formed according to the relation

$$\pi_t^e = \theta \pi_{t-1} \tag{17.3}$$

The value of the parameter θ (the Greek lowercase letter theta) captures the effect of last year's inflation rate on this year's expected inflation rate. The higher the value of θ, the more last year's inflation leads workers and firms to revise their expectations of what inflation will be this year, and thus the higher expected inflation.

We can then think of what happened from 1967 on as an increase in the value of θ over time. As long as inflation was low and not very persistent, it was reasonable for workers and firms to ignore past inflation and to assume that this year's price level would be roughly the same as last year's. Thus, for the period that Phillips, Samuelson, and Solow had looked at, θ was close to zero, expectations were roughly given by $\pi_t^e = 0$, and the relation between the inflation and unemployment rates was given by equation (17.2).

But, as inflation became more persistent, workers and firms started changing the way they formed expectations. They started assuming that if inflation had been high last year inflation was likely to be high this year as well. The parameter θ, the effect of last year's inflation rate on this year's expected inflation rate, increased

FIGURE 17-4

Canadian Inflation, 1927–1996

With the exception of during World War II, before 1960 inflation rates were volatile and often switched signs. Since 1960, they have been more persistent and have not been negative.

Source: Source: Statistics Canada, CANSIM Series D11000 and D14442.

steadily. By the 1970s, the evidence is that people formed expectations by expecting this year's inflation rate to be the same as last year's—in other words, that θ was now equal to 1.

To see the implications of different values of θ for the relation between inflation and unemployment, replace π_t^e in equation (17.1) with the right-hand side of equation (17.3). Doing so gives

$$\pi_t = \overbrace{\theta\pi_{t-1}}^{\pi_t^e} + (\mu + z) - \alpha u_t$$

When θ is equal to zero, we get the original Phillips curve, a relation between the inflation rate and the unemployment rate. When θ is positive, the inflation rate depends not only on the unemployment rate but also on last year's inflation rate. When θ is equal to 1, the aggregate supply relation becomes (moving last year's inflation rate to the left-hand side of the equation)

$$\pi_t - \pi_{t-1} = (\mu + z) - \alpha u_t \tag{17.4}$$

Δ in infl rate [handwritten]

Modified P.C. or expectations-augmented P.C. [handwritten]

Thus, when $\theta = 1$, the unemployment rate affects not the inflation rate, but rather the *change* in the inflation rate. High unemployment leads to decreasing inflation; low unemployment leads to increasing inflation.

To distinguish it from the original Phillips curve, equation (17.2), equation (17.4) is often called the **modified Phillips curve,** or the **expectations-augmented Phillips curve** (to indicate that the term π_{t-1} stands for expected inflation), or the **accelerationist Phillips curve** (to indicate that a low unemployment rate leads to an increase in the inflation rate and thus an *acceleration* of the price level). We shall simply call it the Phillips curve, and refer to the earlier incarnation, equation (17.2), as the *original* Phillips curve.

This discussion gives the key to what happened from 1970 on. As θ increased from zero to 1, the simple relation between unemployment and inflation disappeared. This is what we saw in Figure 17-3. But equation (17.4) tells us what to look for, namely a relation between unemployment and the *change* in inflation. This relation is shown in Figure 17-5, which plots the change in the inflation rate versus the unemployment rate for each year since 1970. It shows indeed a clear negative relation between unemployment and the change in inflation. Using econometrics, we can find the line that best fits the scatter of points for the period 1970–1996. It is given by

$$\pi_t - \pi_{t-1} = 4.3 - 0.52u_t \tag{17.5}$$

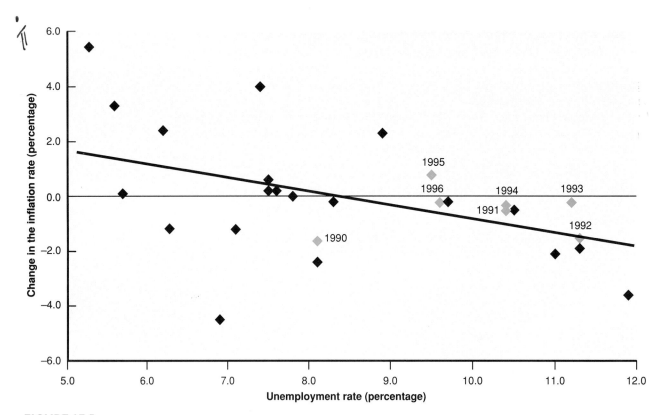

FIGURE 17-5

Change in Inflation versus Unemployment in Canada, 1970–1996

Since 1970, there has been a negative relation between the change in the inflation rate and the level of unemployment in Canada. This relationship appears to be breaking down again in the 1990s.

Source: Statistics Canada, CANSIM Series D11000, D14442, and D44950, and *Historical Statistics of Canada.*

The estimated relation implies that an increase in the unemployment rate of 1% for a year leads on average to a *decrease* of 0.52% in inflation. The regression line is drawn in Figure 17-5. The coloured diamonds denote the years 1991 to 1996; they are fairly close to the regression line, but on closer inspection there appears to be little relationship between the two variables in the 1990s.

BACK TO THE NATURAL RATE OF UNEMPLOYMENT

The history of the Phillips curve is closely related to the discovery of the concept of the natural rate of unemployment that we developed in Chapter 15.

The major implication of the original Phillips curve was that there was *no* natural rate of unemployment. If they were willing to tolerate a higher inflation rate, policy makers could maintain a lower unemployment rate forever.

In the late 1960s, and even while the original Phillips curve still gave a good description of the data, two economists—Milton Friedman and Edmund Phelps—questioned the existence of such a tradeoff between unemployment and inflation. They questioned it on logical grounds. They argued that such a tradeoff could exist only if wage setters systematically underpredicted inflation, and that they were unlikely to do so forever. They also argued that, if the government attempted to sustain lower unemployment by accepting higher inflation, the tradeoff would ultimately disappear and the unemployment rate could not be sustained below a certain level that they called the "natural rate of unemployment." Events proved them right, and the tradeoff between the unemployment rate and the inflation rate indeed disappeared. (See the Focus box entitled "Theory Ahead of the Facts: Milton Friedman and Edmund Phelps.") Today, most economists accept the notion of a *natural rate of unemployment,* subject to the many caveats that we state in the next section.

Let's make the connection between the Phillips curve and the natural rate of unemployment explicit. By definition (see Chapter 15), the natural rate of unemployment is that unemployment rate at which the actual price level turns out to be equal to the expected price level. Equivalently, and more conveniently here, the natural rate of unemployment is the unemployment rate at which the actual inflation rate is equal to the expected inflation rate. Denote the natural unemployment rate by u_n. Then, imposing the condition that actual and expected inflation be the same ($\pi_t = \pi_t^e$) in equation (17.1) gives

$$0 = (\mu + z) - \alpha u_n$$

Solving for the natural rate yields

$$u_n = \frac{\mu + z}{\alpha} \tag{17.6}$$

Thus, the higher the markup, μ, or the higher the factors that affect wage setting, z, the higher is the natural rate. The stronger the effect of unemployment on inflation given expected inflation—in other words, the higher α—the lower is the natural rate.

From equation (17.6), $\alpha u_n = \mu + z$. Replacing ($\mu + z$) by αu_n in equation (17.1) and rearranging gives

$$\pi_t = \pi_t^e - \alpha(u_t - u_n) \tag{17.7}$$

If—as appears to be the case in Canada today—the expected rate of inflation (π_t^e) is well approximated by last year's inflation rate (π_{t-1}), the relation finally becomes

$$\pi_t - \pi_{t-1} = -\alpha(u_t - u_n) \tag{17.8}$$

This relation links the actual rates of unemployment, the natural rate of unemployment, and the change in inflation. *The change in inflation depends on the difference between the actual and the natural rates of unemployment. When the actual unemployment rate exceeds the natural rate, inflation decreases; when actual unemployment is less than the natural rate, inflation increases.* Equation (17.8) gives another way of thinking about the natural rate of unemployment: It is the rate of unemployment required to keep inflation constant. This is why the natural rate is also called the **nonaccelerating inflation rate of unemployment,** or **NAIRU.**

What is the natural rate of unemployment in Canada today? To answer that question, we can return to the estimated equation (17.5) and ask: What is the unemployment rate at which inflation is constant? Putting the change in inflation equal to zero in equation (17.5) implies a value for the natural unemployment rate of

THEORY AHEAD OF THE FACTS: MILTON FRIEDMAN AND EDMUND PHELPS

FOCUS

Economists are usually not very good at predicting major changes before they happen, and most of their insights are derived after the fact. But here is an exception to the rule.

In the late 1960s—precisely as the original Phillips curve relation was working like a charm—two economists, Milton Friedman and Edmund Phelps, argued that the appearance of a tradeoff between inflation and unemployment was an illusion.

Here are a few quotes from Milton Friedman. Talking about the Phillips curve, he said:

Implicitly, Phillips wrote his article for a world in which everyone anticipated that nominal prices would be stable and in which this anticipation remained unshaken and immutable whatever happened to actual prices and wages. Suppose, by contrast, that everyone anticipates that prices will rise at a rate of more than 75% a year—as, for example, Brazilians did a few years ago. Then, wages must rise at that rate simply to keep real wages unchanged. An excess supply of labor will be reflected in a less rapid rise in nominal wages than in anticipated prices, not in an absolute decline in wages.

He went on to say:

To state [my] conclusion differently, there is always a temporary tradeoff between inflation and unemployment; there is no permanent tradeoff. The temporary tradeoff comes not from inflation per se, but from a rising rate of inflation.

He then tried to guess how much longer the apparent tradeoff between inflation and unemployment would last in the United States:

But how long, you will say, is "temporary"? . . . I can at most venture a personal judgment, based on some examination of the historical evidence, that the initial effect of a higher and unanticipated rate of inflation lasts for something like two to five years; that this initial effect then begins to be reversed; and that a full adjustment to the new rate of inflation takes as long for employment as for interest rates, say, a couple of decades.

Friedman could not have been more right. A few years later, the original Phillips curve started to disappear, in exactly the way that Friedman had predicted.

Source: Milton Friedman, "The Role of Monetary Policy," March 1968, *American Economic Review* 58-1, 1–17. (The article by Phelps, "Money-Wage Dynamics and Labor-Market Equilibrium," *Journal of Political Economy,* August 1968, part 2, 678–711, made the same points more formally.)

4.3%/0.52 = 8.3%. This estimate is a bit higher than the estimates from more sophisticated studies; most estimates of the natural rate today in Canada are around 7.5%. In other words, the evidence suggests that in Canada today the rate of unemployment required to keep inflation constant is around 7.5%. However, there is enormous uncertainty around this estimate. The best that can be said is that it is now somewhere in the range of 5% to 8.5% and that it has drifted up by 3% or 4% since the 1960s.

17-3 A SUMMARY AND MANY WARNINGS

Let us summarize what we have learned so far. The aggregate supply relation can be proxied in Canada today by the Phillips curve, which is a relation between the change in the inflation rate and the deviation of the unemployment rate from the natural rate [equation (17.8)]. The natural unemployment rate appears today to be around 7.5%, but the estimate is not very precise. When the unemployment rate exceeds the natural rate, inflation decreases. When the unemployment rate is below the natural rate, inflation increases.

This relation has held reasonably well over the last 20 years. But its earlier history points out he need for a number of warnings. All of them are on the same theme: The relation can shift, and indeed often has.

THE INFLATION PROCESS AND THE PHILLIPS CURVE

Recall how the Canadian Phillips curve changed as inflation became more persistent and the formation of expectations changed as a result. The lesson is a general one: The relation between unemployment and inflation is likely to change with the inflation process. Evidence from countries with high inflation confirms this lesson. Not only does the process through which workers and firms form expectations change, but institutional arrangements change as well.

When the inflation rate becomes high, inflation also tends to become more variable. Workers and firms become more reluctant to enter into labour contracts that predetermine nominal wages for a long period of time. If inflation turns out to be higher than expected, real wages may plunge and workers may suffer a large cut in their standard of living. If inflation turns out to be lower than expected, real wages may explode and firms may go bankrupt.

For this reason, the structure of wage agreements changes with the level of inflation. Nominal wages are set for shorter periods of time, down from a year to a month or even less. **Wage indexation,** a rule that increases wages automatically in line with inflation, becomes more prevalent.

These changes lead in turn to a stronger response of inflation to unemployment. To see this, an example based on wage indexation will help. Think of an economy that has two types of labour contracts. A proportion λ (the Greek lowercase letter lambda) of labour contracts is indexed: Nominal wages in those contracts move one for one with variations in the actual price level.[4] A proportion $1 - \lambda$ of labour contracts is not indexed; nominal wages are set on the basis of expected inflation. Expected inflation is equal to last year's inflation.

Under this assumption, equation (17.7) becomes

$$\pi_t = \lambda \pi_t + (1 - \lambda)\pi_{t-1} - \alpha(u_t - u_n)$$

When $\lambda = 0$, all wages are set on the basis of expected inflation—which we have assumed to be equal to last year's inflation, π_{t-1}—and the equation reduces to equation (17.8). When λ is positive, however, a proportion λ of wages is set on the basis of actual rather than expected inflation.

Reorganizing the equation gives

$$\pi_t - \pi_{t-1} = -\frac{\alpha}{1 - \lambda}(u_t - u_n) \tag{17.9}$$

Indexation increases the effect of unemployment on inflation. More formally, the higher the proportion of indexed contracts—the higher λ—the larger will be the effect of the unemployment rate on the change in inflation—the higher the coefficient $\alpha/(1 - \lambda)$.

The intuition is as follows. Without indexation, lower unemployment increases wages, which in turn increases prices. But, because wages do not respond to prices right away, there is no further effect within the year. With wage indexation, however, an increase in prices leads to an increase in wages within the year, which in turn increases prices, and so on, so that the effect of unemployment on inflation within the year is higher.

If and when λ gets close to 1—when most labour contracts allow for wage indexation—small movements in unemployment can lead to very large changes in inflation. Conversely, there can be large changes in inflation with little movement in unemployment. This is indeed what happens in countries where inflation is very high: The relation between inflation and unemployment becomes more and more tenuous and eventually disappears altogether.[5]

DIFFERENCES IN THE NATURAL RATE ACROSS COUNTRIES

Recall from equation (17.6) that the natural rate of unemployment depends on all the factors that affect wage setting, summarized by the catchall variable z; on the markup set by firms μ; and on the response of inflation to unemployment, as summarized by the parameter α. To the extent that these factors differ across countries, there is no reason to expect different countries to have the same natural rate.

Indeed, natural rates differ across countries, sometimes considerably. Compare Canada and the United States. Until 1981, unemployment rates in the two countries moved very closely together, but since then the two rates have diverged. In late 1997, the Canadian rate stood at 9.1%, but the U.S. rate was just over half

---

[4]*This assumption is actually too strong. Indexation clauses typically adjust wages not for current inflation (which is not yet known), but for inflation in the recent past, so there remains a short lag between inflation and wage adjustments. We ignore this lag here.*

[5]*High inflation and its implications are the topic of Chapter 21.*

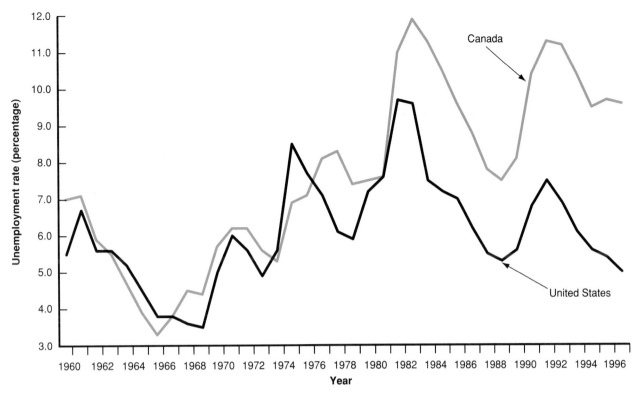

FIGURE 17-6

Unemployment Rates in Canada and the United States, 1960–1997

Unemployment rates in Canada and the United States move closely together, but a gap has opened since 1982.

Source: OECD Economic Outlook, December 1997 and previous issues..

that amount at 4.6%. Natural rates are not directly observable, but under the assumption that the economy gravitates around them, a simple strategy to get an estimate of the natural rate is to look at the unemployment rate over a decade or so. In the 1960s, the average unemployment rates in Canada and the United States were very similiar at 5.0% and 4.8 % respectively; these numbers climbed in the 1970s by virtually the same amount to 6.7% and 6.2% respectively. In the 1980s, the unemployment rates began to diverge. The average Canadian unemployment rate climbed to 9.4%. The U.S. rate also climbed, but by much less, to 7.3%. The gap widened in the 1990s. From 1990 to 1997, the Canadian unemployment rate averaged 10%, but the U.S. rate fell back down to 6.1%. Thus there is no question but that the Canadian natural rate now looks much larger than the U.S. natural rate.

Why have the natural rates in Canada and the United States diverged since 1982? From Figure 17-6, we can see that the gap between the two rates can be traced to two episodes. The unemployment rates in both countries increased in the 1980–1981 and 1990–1991 recessions. In both episodes, however, the increase in Canada was about 2% greater than that experienced in the United States. This gap

[handwritten top: A in composition of jobs.]

The explanation for the increase in the U.S. natural rate of unemployment over the last two or three decades given by Chinhui Juhn, Kevin Murphy, and Robert Topel—three economists from the University of Chicago—is that it is the result of a shift in demand away from unskilled workers and towards skilled workers.

Their argument is as follows: Over the last 20 years, the United States has seen a steady decrease in the demand for unskilled workers relative to skilled workers. This change has led to a decrease in the relative wage of unskilled workers vis-à-vis skilled workers, and often to a decrease in the absolute real wage of unskilled workers (more on this in Chapter 25). Lower-paying jobs have led unskilled workers to care less about employment, to spend more time either unemployed or out of the labour force. While skilled workers have seen an increase in their relative wage, their unemployment rate was low to start with and could not decrease much. Thus, on net, the shift in relative demand for workers has led to an increase in unemployment.

Juhn, Murphy, and Topel base their conclusions on an examination of CPS data from 1967 to 1989 on wages, employment, and participation of adult men with one to 30 years in the labour market. They look at changes in unemployment and participation from the period 1967–1969 to the period 1987–1989. Both periods come at the end of long expansions. The overall unemployment rate for adult men during 1967–1969 was 3.7%. It was 5.6% in 1987–1989.

Table 1 summarizes their findings. The way to read Table 1 is as follows. In 1987–1989, the unemployment rate of workers in the bottom 10% of the wage distribution was 7.1 percentage points higher than in 1967–1969; their participation rate was 9.2 percentage points lower. By contrast, the unemployment rate of those in the top 40% of the distribution was only 0.4% higher, and their participation rate was actually 0.3% higher than in 1967–1969. Thus, it appears indeed that much of the increase in the natural rate can be traced to a higher unemployment rate among the unskilled, which appears in turn to be due to lower relative wages.

Does this explanation account for the full increase in the natural rate in the United States? The answer is no. If the change in relative wages were the only factor behind the increase in unemployment, the unemployment rate of skilled workers should have decreased, not (slightly) increased as it did. So other factors are probably at work. But the authors may well have accounted for the main factor behind the increase in the natural rate of unemployment.

Source: Chinhui Juhn, Kevin M. Murphy, and Robert H. Topel, "Why Has the Natural Rate of Unemployment Increased over Time?" *Brookings Papers on Economic Activity*, 1991:2, 75–142.

[handwritten: Note: skilled workers: ↑ W Un low to begin with.]

[handwritten left margin: ↑ U.S. nat rate Un; due to ↓ Ld unskilled workers ↓ wage ∴ less attachment to labour force: Un or drop out of L. force ∴ ↑ nat rate due to higher Un among unskilled workers due to lower relative wages.]

TABLE 1

CHANGES IN U.S. ADULT MALE UNEMPLOYMENT AND PARTICIPATION BY WAGE PERCENTILE, BETWEEN 1967–1969 AND 1987–1989

WAGE PERCENTILE	PERCENTAGE CHANGE IN UNEMPLOYMENT	PERCENTAGE CHANGE IN PARTICIPATION
1–10 (lowest)	7.1	−9.2
11–20	5.6	−6.6
21–40	3.1	−3.5
41–60	1.5	−0.9
61–100 (highest)	0.4	0.3

Source: See source note for this box.

was not closed in the subsequent recovery. As a result, the average unemployment rate was 2% higher in Canada during the 1980s and 4% higher during the 1990s. There is a good deal of controversy over why these gaps opened up and why they were not closed. The usual suspects have been rounded up, but so far none have been found guilty beyond a reasonable doubt. A closer look at the sources of unemployment show that the incidence of unemployment (the probability of becoming unemployed) was lower in Canada than in the United States until 1990, and about the same since then. But the duration of unemployment (the average length of an unemployment spell) in Canada has climbed from 1.2 times the U.S. level in 1981 to 1.5 times the U.S. level in 1996. Why are unemployment spells in Canada so much longer? Many economists point to Canada's relatively generous employment insurance programs. Although we are sympathetic to this hypothesis, it must be noted that reforms initiated in 1995 have made it harder to get employment insurance (formerly called "unemployment insurance") and have reduced the number of weeks of eligibility, but have had no discernible impact on the unemployment rate differential. Most economists expect the gap to narrow in 1998. The Canadian unemployment rate should fall if the expansion in Canada continues, but the U.S. rate seems unsustainably low. Time will tell.

VARIATIONS IN THE NATURAL RATE OVER TIME

In estimating equation (17.6), we implicitly treated $\mu + z$ as a constant. But there is in fact no reason to believe that μ and z are constant over time. The composition of the labour force, the structure of wage bargaining, the system of unemployment benefits, and so on, are likely to change over time, leading to changes in the natural rate of unemployment.

Movements in the natural rate over time are hard to measure. The reason is again that we observe the actual rate of unemployment, not the natural rate. But broad evolutions can again be established by comparing average unemployment rates across decades. We have seen that the average unemployment rates over the past four decades have varied significantly in both Canada and the United States. The estimates suggest that the natural rate has increased over time. More sophisticated methods suggest that the Canadian natural rate has increased by about 3% since the 1960s and that the U.S. natural rate is about 1.5% higher, although in both cases the exact amount is hard to pin down.

Why have the natural rates increased over time? In Canada, many point to changes in our employment insurance programs. There must be some truth to this, but other factors are also clearly at work. The United States has also experienced increases in the natural rate, so perhaps by looking there we can gain some insight into a common explanation of what is going on in both countries. For the United States, one of the most convincing explanations lies in the change in the composition of jobs and is presented in the Focus box entitled "Low-Wage Jobs and the U.S. Natural Rate." In short, the hypothesis is that a decrease in the wage paid to unskilled workers has led them to become less attached to their jobs, and to spend more time either in unemployment or out of the labour force.

The Limits of Our Understanding

The theory of the natural rate gives macroeconomists directions as to where to look for differences in natural rates across countries or for variations in the natural rate over time in a given country. But the truth is that macroeconomists' understanding of exactly which factors determine the natural rate of unemployment is still very limited. In particular, there is considerable uncertainty about the exact list of factors behind z, and about the dynamic effects of each one on the natural rate.

Let's return, for example, to an increase in the price of oil. When we examined the effects of such an increase in Chapter 16—capturing it by an increase in the markup, μ—we concluded that the result would be an increase in the natural rate of unemployment. The same conclusion holds in equation (17.6): An increase in μ increases the natural rate.

But there is in fact more uncertainty about the effects of an increase in the price of oil than our analysis let appear. Some of the bargaining models that underlie the wage-determination equation imply that workers may eventually accept a wage cut without a need for an increase in the unemployment rate. In terms of our equation, these models suggest that when μ increases z eventually decreases, so that $\mu + z$, and thus the natural rate of unemployment $u_n = (\mu + z)/\alpha$, remains unchanged. Other models imply that, while z may decrease in response to an increase in μ, the offset is only partial, so that an increase in the price of oil has a permanent effect on the natural rate.[6]

The fact that different models lead to different conclusions is not unusual; the way to decide which is most appropriate is to see which fits the data best. In this case, the data do not speak very clearly and have so far been unable to give us precise answers. Thus, whether changes in the price of oil have a permanent effect on the natural rate of unemployment is still very much an open question.

The limits of our understanding are particularly clear and painful in times of high unemployment, at precisely those times when it is most crucial to understand what is happening. This is indeed the case in Europe today. You may recall our discussion of the evolution of European unemployment in Chapter 1. The European unemployment rate, which until the early 1970s had been relatively low, has steadily increased since. In 1997, the unemployment rate in the European Union stood at 11.3%, compared with 9.7% for Canada and 5.0% for the United States.

This high unemployment rate could be due either to a large deviation of the actual unemployment rate from the natural rate, or to the fact that the natural rate itself is high. As EU countries have roughly stable inflation, it seems safe to conclude that the actual unemployment rate is not far from the natural rate. This point is made in Figure 17-7, which plots the change in the EU inflation rate against the unemployment rate for each year since 1971. The line gives the estimated relation between the change in inflation and the unemployment rate using data since

---

[6]*The two recent and important books on unemployment we mentioned in Chapter 15 indeed reach different conclusions on this point. The book by Layard et al. (see Suggested Reading in Chapter 15) argues that factors such as the price of oil, indirect taxes, the real exchange rate, and the real interest rate have no permanent effect on the natural rate. In contrast, the book by Edmund Phelps (see Chapter 15, footnote 8) argues that these factors have permanent effects on the natural rate and indeed explain a good part of the movements in unemployment in OECD countries over the last 20 years.*

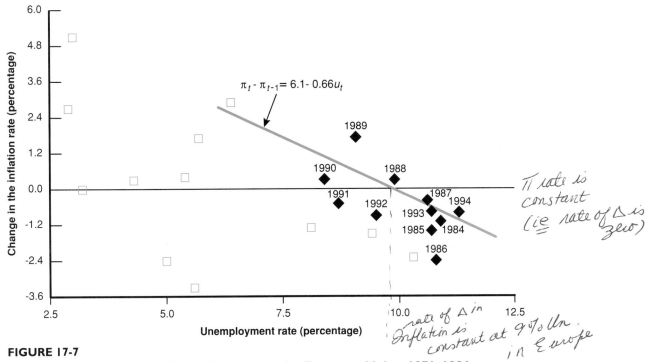

The figure contains the equation $\pi_t - \pi_{t-1} = 6.1 - 0.66u_t$

Handwritten annotations:
π rate is constant (ie rate of Δ is zero)
rate of Δ in Inflation is constant at 9% Un. in Europe

FIGURE 17-7

Change in Inflation versus Unemployment in the European Union, 1971–1996
The natural rate of unemployment in the European Union is now around 9%, much higher than it was two or three decades ago.

Source: OECD Economics Outlook, December 1997 and previous issues.

1984 (the years since 1984 are denoted by black diamonds in the figure). The line suggests that the natural rate—the unemployment rate at which inflation remains constant—is now around 9% and is thus much higher than it was two or three decades ago. (The average EU unemployment rate in the 1960s was about 3 percent. If we take this average as an estimate of what the natural rate was then, the natural rate is therefore about 6% higher now than it was in the 1960s.)

Why was the natural rate so much higher in the 1930s in the United States, and why is it so high today in Canada and Europe? The question is sufficiently important that we shall spend all of Chapter 20 trying to answer it. But be warned: Although this is one of the major economic questions of our time, there is little agreement as to the answer.

SUMMARY

■ The aggregate supply relation can be expressed as a relation between inflation, expected inflation, and unemployment. The higher expected inflation, the higher is inflation. The higher unemployment, the lower inflation.

■ When inflation is not very persistent, expected inflation does not depend very much on past inflation. Thus, the aggregate supply relation becomes a relation between inflation and unemployment. This is what Phillips in the United Kingdom, and Solow and Samuelson in the United States, discovered when they looked in the early 1960s at the joint behaviour of unemployment and inflation.

■ As inflation became more persistent in the 1970s and 1980s, expected inflation became increasingly dependent on past inflation. In Canada today, the aggregate supply relation takes the form of a relation between unemployment and the *change* in inflation. High unemployment leads to decreasing inflation; low unemployment leads to increasing inflation.

■ The natural unemployment rate is the unemployment rate required to keep inflation constant. When the actual unemployment rate exceeds the natural rate, inflation decreases; when the actual unemployment rate is less than the natural rate, inflation increases. The Canadian natural unemployment rate is currently around 7.5%.

■ Changes in the inflation process change the way wage setters form expectations and how much they use wage indexation. When wage indexation is widespread, small movements in unemployment can lead to very large changes in inflation. At high rates of inflation, the relation between inflation and unemployment disappears altogether.

■ The natural rate of unemployment depends on many factors that differ across countries and can change over time. Thus, the natural rate varies across countries. It is much lower in the United States than in Canada. Also, the natural rate varies over time. In Canada the natural rate appears to have increased by 3% to 4% since the 1960s. In Europe, it appears to have increased by more.

KEY TERMS

- Phillips curve, 325
- deflation, 328
- wage-price spiral, 328
- modified, or expectations-augmented, or accelerationist Phillips curve, 332
- nonaccelerating inflation rate of unemployment (NAIRU), 334
- wage indexation, 335

QUESTIONS AND PROBLEMS

1. "If the government wants to reduce the natural rate of unemployment, all it needs to do is increase the demand for goods and services. With greater demand, there will be more production and fewer people unemployed." Comment.

2. Suppose that the Phillips curve in an economy is given by the equation $\pi_t - \pi_t^e = 0.18 - 3u_t$, where $\pi_t^e = \theta \pi_{t-1}$. Further, suppose that in period $t-1$ the unemployment rate is equal to the natural rate, and the inflation rate is 0%.

   a. What is the natural rate of unemployment in this economy?

   b. Suppose that, beginning in period t, the authorities bring the unemployment rate down to 5% and keep it there indefinitely. Determine the inflation rate in periods, t, $t+1$, $t+2$, and $t+3$ when $\theta = 0$. Then do the same for $\theta = 1$.

   c. For which of the two values of θ does $u_t < u_n$ imply an *acceleration* of the price level (a continually increasing rate of inflation)?

   d. Suppose that the authorities do not know the natural rate of unemployment. Can they find out what it is? How?

3. Suppose that the Phillips curve is as specified in problem 2 (with $\theta = 1$). In period $t-1$, unemployment is at its natural rate, and the inflation rate is zero. Beginning in period t, two things happen:

   (1) The authorities move the actual unemployment rate to 5% and keep it there.

   (2) Half of all workers sign indexed labour contracts.

   a. Give the *new* equation for the Phillips curve.

   b. Calculate the inflation rates in period t, $t+1$, and $t+2$.

   c. Comparing your answers in problems **2b** and **3b**, what does indexing imply about the impact of maintaining the unemployment rate below the natural rate?

4. If Canada experienced three or four years of low (under 1%) inflation, would it return to the original Phillips curve of the 1950s and 1960s? Why or why not?

5. During the 1980s and early 1990s, the price of oil *decreased.* What has been the probable effect of this on the natural rate of unemployment?

6. Add the most recent years' data to the scatter plot in Figure 17-5. Are these data points consistent with the Phillips curve line drawn in the figure? If not, can you account for the deviation?

FURTHER READING

Canadian Public Policy "The Canada-U.S. Unemployment Rate Gap," in Vol. 2A Supplement, February 1998.

APPENDIX

FROM THE AGGREGATE SUPPLY RELATION TO THE PHILLIPS CURVE

The purpose of this appendix is to derive equation (17.1) in the text, the relation between inflation, expected inflation, and unemployment.

The starting point is the aggregate supply relation between the price level, the expected price level, and the unemployment rate that we derived in Chapter 16:

$$P_t = P_t^e(1 + \mu)F(u_t, z)$$

Assume a specific form for the function F (which captures the effects of the unemployment rate, u_t, and of the other factors that affect wage setting, as summarized by the catchall variable z):

$$F(u_t, z) = 1 - \alpha u_t + z$$

This captures the notion that the higher the unemployment rate, the lower will be the wage; and the higher z, the higher is the wage. Replacing in the aggregate supply equation gives

$$P_t = P_t^e(1 + \mu)(1 - \alpha u_t + z)$$

Dividing both sides by last year's price level, P_{t-1}, gives

$$\frac{P_t}{P_{t-1}} = \frac{P_t^e}{P_{t-1}}(1 + \mu)(1 - \alpha u_t + z) \qquad (17.A1)$$

Rewrite the fraction P_t/P_{t-1} as

$$\frac{P_t}{P_{t-1}} = 1 + \frac{P_t - P_{t-1}}{P_{t-1}} = 1 + \pi_t$$

where the first equality follows from rearranging terms, and the second from the definition of the inflation rate. Do the same for the fraction P_t^e/P_{t-1} on the right-hand side, so that

$$\frac{P_t^e}{P_{t-1}} = 1 + \frac{P_t^e - P_{t-1}}{P_{t-1}} = 1 + \pi_t^e$$

where π_t^e is expected inflation in year t. Replacing both fractions in equation (17A.1) gives

$$1 + \pi_t = (1 + \pi_t^e)(1 + \mu)(1 - \alpha u_t + z)$$

Divide both sides by $(1 + \pi_t^e)(1 + \mu)$, so that

$$\frac{1 + \pi_t}{(1 + \pi_t^e)(1 + \mu)} = 1 - \alpha u_t + z$$

As long as inflation, expected inflation, and the markup are not too large, a good approximation to this equation is given by (see Propositions 3 and 6 in Appendix 3 at the end of the book):

$$1 + \pi_t - \pi_t^e - \mu = 1 - \alpha u_t + z$$

Rearranging gives

$$\pi_t = \pi_t^e + (\mu + z) - \alpha u_t$$

This is equation (17.1) in the text. The inflation rate depends on the expected inflation rate, the markup μ and other factors z, and the unemployment rate u_t.

① Describe relation between π, π^e, u_t (equation) & explain the relation between the variables & actual π

② Early P.C. (no nat. rate of un) $\pi_t = (\mu + z) - \alpha u_t$ $\pi_t^e = 0$
  – illustrate on diagram: breakdown in 70s
  – define/explain wage spiral
  – discussion of $\pi_t^e = \theta \pi_{t-1}$ θ?

③ expectations-augmented P.C: $\pi_t - \pi_{t-1} = (\mu + z) - \alpha u_t$
  Δ Infl. rate is affected by unemployment;
  $\equiv$ Cda 1% ↑Un, Δ Infl. rate by .5% (½ of 1%)

④ equation for nat rate of Un $= \dfrac{\mu + z}{\alpha}$

⑤ eqn for P. curve with π_t^e & explain
  $$\pi_t = \pi_t^e - \alpha (u_t - u_n)$$
  or $\pi_t - \pi_{t-1} = -\alpha (u_t - u_n)$
  When $u_t = u_n \Rightarrow$ no inflation

⑥ Effects of Inflation on other institutional relationships ie: wage indexation of labour contracts
  $$\pi_t - \pi_{t-1} = \frac{\alpha}{1-\lambda}(u_t - u_n)$$

  Effects of Un on π
  What if λ is close to 1?

18

INFLATION, DISINFLATION, AND UNEMPLOYMENT

In late 1979, inflation in Canada was running at close to 12% a year. Five years later, and after a deep recession, inflation was down to 4% a year.

This chapter focusses on the relation between money growth, inflation, and economic activity. The first two sections develop a model linking inflation, output, and unemployment. The third section discusses alternative disinflation strategies, very much as they were discussed in the late 1970s. We now, however, have one advantage that macroeconomists at the time did not have: hindsight. We can actually look at what policy was and what the effects turned out to be. This is what we do in the last section of the chapter.

18-1 INFLATION, OUTPUT, AND UNEMPLOYMENT

In thinking about the interactions between inflation, output, and unemployment, one must keep in mind three relations:

- The first, which we developed in Chapter 17, is the Phillips curve, the relation between *unemployment* and the *change in inflation.*
- The second, Okun's law, is the relation between *output growth* and the *change in unemployment.*

■ The third, which reflects the behaviour of aggregate demand, is the relation between *output growth, money growth,* and *inflation.*

In this section we look at each relation on its own. In Section 18-2 we look at their joint implications.

THE PHILLIPS CURVE: UNEMPLOYMENT AND THE CHANGE IN INFLATION

Higher unemployment leads to lower inflation. More specifically, we saw in Chapter 17 that we can write the relation between inflation, expected inflation, and unemployment as follows [equation (17.7)]:

$$\pi_t = \pi_t^e - \alpha(u_t - u_n) \tag{18.1}$$

Inflation depends on expected inflation and on the deviation of unemployment from the natural rate.

We also saw that in Canada today expected inflation appears to be well approximated by last year's inflation, so that we can replace π_t^e by π_{t-1}. Thus the relation between inflation and unemployment takes the form

$$\pi_t - \pi_{t-1} = -\alpha(u_t - u_n) \tag{18.2}$$

Unemployment above the natural rate leads to a decrease in inflation; unemployment below the natural rate leads to an increase in inflation. The parameter α captures the effect of unemployment on the change in inflation. In Canada, since 1970, α is roughly equal to 0.5. This means that an unemployment rate of 1% above the natural rate for one year leads to a decrease in the inflation rate of about 0.5%. We shall refer to this relation as the Phillips curve.[1]

OKUN'S LAW: OUTPUT GROWTH AND CHANGES IN UNEMPLOYMENT

When we wrote the relation between unemployment and output in Chapter 15, we made two convenient but very restrictive assumptions. We assumed that output and employment moved together, that Y was simply equal to N. We assumed also that the labour force L was constant, so that movements in employment were reflected one for one in opposite movements in unemployment.

To see why we must now move beyond these assumptions, think about what they imply for the relation between the rate of output growth and the unemployment rate. As output and employment move together, a 1% increase in output leads to a 1% increase in employment. And because movements in employment are reflected in opposite movements in unemployment, a 1% increase in employment leads to a decrease of 1% in the unemployment rate.[2] Let g_{yt} denote the growth rate of output. Then, under our two assumptions, the following relation should hold:

---

[1] *It would be more logical to call it the "Phillips relation," and reserve "Phillips curve" for the curve that represents the relation. But the tradition is to use "Phillips curve" to denote equation (18.2). We shall follow tradition.*

[2] **Digging deeper.** *This last step is only approximately correct. Remember the definition of the unemployment rate $u \equiv 1 - N/L$. Thus, if the labour force is fixed, $\Delta u = -\Delta N/L = -(\Delta N/N)(N/L)$, where the last equality follows from multiplying and dividing by N. If N/L is equal to, say, 0.95, a 1% increase in employment leads to a decrease of 0.95% in the unemployment rate. The result in the text is based on approximating N/L by 1, so that a 1% increase in employment leads to a decrease of 1% in the unemployment rate.*

$$u_t - u_{t-1} = -g_{yt} \quad \overleftarrow{\text{growth rate of output}} \qquad (18.3)$$

The change in the unemployment rate should be equal to the negative of the growth rate of output. If output growth is, say, 4%, then the unemployment rate should decline by 4%.

Contrast this with the *actual* relation between output growth and the change in the unemployment rate, the relation known as **Okun's law.** Figure 18-1 plots the change in the unemployment rate against the rate of output growth for each year since 1975. It also plots the regression line that best fits the scatter of points. The relation corresponding to the line is given by

$$u_t - u_{t-1} = -0.4(g_{yt} - 3\%) \quad \longrightarrow \text{L.f growth + productivity growth} \text{ (called "Normal growth rate")} \qquad (18.4)$$

Equation (18.4) differs in two ways from equation (18.3).

First, annual output growth has to be at least 3.0% to prevent the unemployment rate from rising. Why is this so? Because of two factors we have neglected so far, labour-force growth and labour-productivity growth.

Suppose that the labour force is growing at 1% a year. To maintain a constant unemployment rate, employment must grow at the same rate as the labour force, thus at 1% a year. Suppose now that labour productivity—output per worker—is

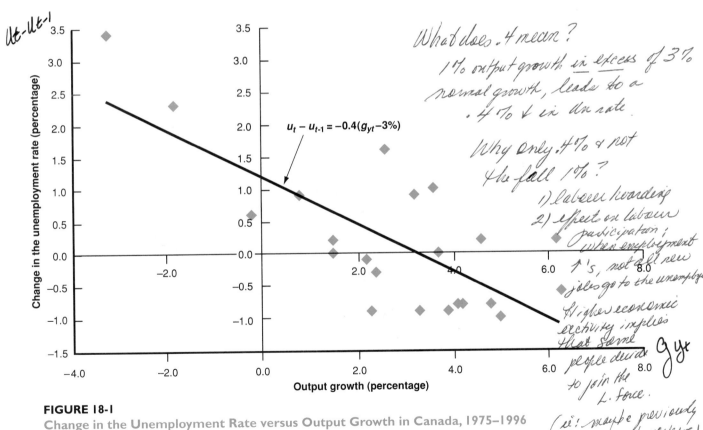

(Handwritten notes in margin:)
What does .4 mean?
1% output growth in excess of 3% normal growth, leads to a .4% ↓ in un rate.

Why only .4% ↓ not the full 1%?
1) labour hoarding
2) effect on labour participation; when employment ↑'s, not all new jobs go to the unemployed. Higher economic activity implies that some people decide to join the L. force. (ie: maybe previously discouraged workers)

FIGURE 18-1

Change in the Unemployment Rate versus Output Growth in Canada, 1975–1996
High output growth is associated with a reduction in the unemployment rate; low output growth is associated with an increase in the unemployment rate.

Sources: Statistics Canada, CANSIM Series D14442 and D44950.

The coefficient β in Okun's law gives the effect on the unemployment rate of deviations of output growth from normal. A value of β of 0.4, for example, tells us that output growth 1% above normal for one year decreases the unemployment rate by 0.4 percentage points.

The coefficient β depends in part on firms' decisions regarding how to adjust employment in response to temporary deviations in output. These decisions depend in turn on such factors as firms' internal organization and the legal and social constraints on hiring and firing. One would thus expect the coefficient to be different across countries, and indeed it is. Table 1 gives the estimated coefficient β for a number of OECD countries. The first column gives estimates of

β based on data from 1960 to 1980. The United States has the highest coefficient, followed by Germany, Canada, the United Kingdom, and Japan.

The ranking in the first column fits well what we know about the behaviour of firms and the structure of firing/hiring regulations across countries. The coefficient β is smallest in Japan. Japanese firms offer a high degree of job security to their workers, so variations in output have little effect on employment and thus on unemployment. The coefficient β is largest in the United States, where there are few social and legal constraints on firms' adjustment of employment. And legal restrictions on firing—from severance pay to obtaining legal permission from the state to terminate employment—explain why the coefficients estimated for Canada and the two European countries are in between those of Japan and the United States.

The second column gives estimates based on data from 1981 to 1994. The coefficient is uniformly larger. This again fits with what we know about firms and regulations. Increased competition in goods markets since the early 1980s has led firms in most countries to reconsider and reduce their commitment to job security. And, at the urging of firms, legal restrictions on hiring and firing have been considerably weakened. Both factors have led to a larger response of employment to fluctuations in output, thus to a larger value of β.

··················

TABLE 1

OKUN'S LAW COEFFICIENTS ACROSS COUNTRIES AND TIME

COUNTRY	1960–1980	1981–1994
Canada	0.17	0.44
United States	0.40	0.47
Germany	0.27	0.42
United Kingdom	0.17	0.49
Japan	0.15	0.23

Source: Authors' estimates.

growing at 2% a year. Thus an employment growth rate of 1% implies an annual growth rate of output of 1% + 2% = 3%. In other words, to keep unemployment constant, output growth must equal 3%.

In Canada, the sum of the rate of labour-force growth and labour-productivity growth has indeed been close to 3% on average since 1975, and this is why the number 3% appears on the right-hand side of equation (18.4). In what follows we shall call the rate of output growth needed to maintain a constant unemployment rate the **normal growth rate**.

Second, the coefficient on the deviation of output growth from normal is equal to −0.4 in equation (18.4) rather than −1.0 in equation (18.3). Put another way, output growth of 1% in excess of normal growth leads to only a 0.4% reduction in the unemployment rate rather than a 1% reduction. There are two reasons for this:

1. Firms adjust employment less than one for one in response to deviations of output growth from normal. More specifically, a 1% increase in output above

normal leads to only a 0.45% increase in the employment rate. One reason is that some workers are needed no matter what the level of output is. The accounting department of a firm, for example, needs roughly the same number of employees whether the firm is selling more or less than normal. Another reason is that training new employees is costly. Thus, many firms prefer to keep current workers around rather than lay them off when demand is low, and ask them to work overtime rather than hire new employees when demand is high. Thus, in bad times, firms in effect hoard labour; this effect is called **labour hoarding.**

2. An increase in the employment rate does not lead to a one-for-one decrease in the unemployment rate. More specifically, a 0.45% increase in the employment rate leads to only a 0.4% decrease in the unemployment rate. The reason is that labour participation increases. When employment increases, not all the new jobs are filled by the unemployed. Some of the jobs go to those who were classified as *not in the labour force,* who were not officially looking for a job. And, as labour-market prospects improve for the unemployed, some discouraged workers—who were previously classified as not in the labour force—regain courage: They now start looking actively for a job and thus become classified as unemployed.

Using letters rather than numbers, we write the relation between output growth and changes in the unemployment rate as

$$u_t - u_{t-1} = -\beta(g_{yt} - \bar{g}_y) \quad (18.5)$$

where $\bar{g}_y$ is the normal growth rate of the economy (about 3% for Canada), and the parameter β (the Greek lowercase letter beta) tells us how growth in excess of normal growth translates into decreases in the unemployment rate. In Canada, β is equal to 0.4. (The evidence for other countries is given in the Global Macro box entitled "Okun's Law across Countries.")

THE AGGREGATE DEMAND RELATION: OUTPUT GROWTH, MONEY GROWTH, AND INFLATION

We have focussed so far on the aggregate supply side, on how unemployment affects inflation, and how movements in output in turn affect unemployment. We now ask what determines these movements in output, and in particular how they depend in turn on inflation. Doing so requires us to look at the aggregate demand side of the economy.

The central link here is the effect of the real money supply on output. One of the recurring themes of this book has been that changes in the real money supply affect demand, which in turn affects output. The channels vary depending on whether the economy is open or closed. In a closed economy, an increase in the real money supply leads to a decrease in the interest rate and an increase in demand and output. In an open economy with flexible exchange rates, an increase in the real money supply leads to both a decrease in interest rates and a depreciation, both leading in turn to an increase in demand and output.

In Chapter 16 we wrote the aggregate demand relation as a relation between output and the real money supply, government spending, and taxes [equation (16.2)]. To focus on the relation between the real money supply and output, we shall

[Handwritten margin notes:]
explains β

new participants in l.f. when ↑ y.
(not all new jobs go to the unemployed workers)

Same idea as Mankiw p417
Δ cyclical un rate =
−½ × (% Δ GDP − 4% Nat)
Suppose Real GDP falls 1%
−½ × (−1% − 4%)
= 2.5% rise in cyclical un rate.

4 Cdn
3% Cda normal growth.

* domestic bonds less attractive

ignore movements in all factors other than real money here, and write the aggregate demand relation simply as

$$Y_t = \gamma \frac{M_t}{P_t}$$

where γ (the Greek lowercase letter gamma) is a positive parameter. This equation assumes that the demand for goods, and thus output, is proportional to the real money supply. This simplification makes our life easier and captures what we want to capture here—namely, that increases in the real money supply increase demand and output, and that decreases in the real money supply decrease demand and output.

It will be more convenient to work with this relation in terms of growth rates. As before, let g_{yt} be the growth rate of output. Let g_{mt} be the growth rate of nominal money, and let π_t be the growth rate of prices—the rate of inflation. Then, it follows from the equation above that

$$g_{yt} = g_{mt} - \pi_t \tag{18.6}$$

The growth rate of output is equal to the growth rate of nominal money minus the rate of inflation.[3] When inflation is equal to nominal money growth, the real money supply is constant and so is demand and, in turn, output (remember that we keep all factors other than real money constant): Output growth is equal to zero. Suppose, however, that inflation far exceeds nominal money growth. Then, the real money supply steadily decreases, leading to a steady decrease in the demand for goods, and thus a steady decrease in output: Output growth is negative.

In short, the aggregate demand relation implies that, given money growth, the higher the rate of inflation, the lower will be the rate of output growth.

18-2 THE LONG RUN

Let's now collect the three relations between inflation, unemployment, and output growth that we derived in Section 18-1.

The first is the Phillips curve, which relates the change in inflation to the deviation of the unemployment rate from its natural rate [equation (18.2)]:

$$\pi_t - \pi_{t-1} = -\alpha(u_t - u_n)$$

The second is Okun's law, which relates the change in the unemployment rate to the deviation of output growth from normal [equation (18.5)]:

$$u_t - u_{t-1} = -\beta(g_{yt} - \bar{g}_y)$$

The third is the aggregate demand relation, which relates output growth to the difference between nominal money growth and inflation [equation (18.6)]:

$$g_{yt} = g_{mt} - \pi_t$$

---

[3] *Appendix 3 gives a refresher on how to go from levels to growth rates. Equation (18.6) is derived as an application of Proposition 8 in that appendix.*

Our task for the rest of this chapter will be to draw the implications of these three relations for the behaviour of inflation, output, and unemployment. It is easiest to work backward in time, to start by looking at the long run and then turn to dynamics. Let's focus on the long run in this section. Let's ask: What determines inflation, output, and unemployment in the long run? How do these three variables depend on the rate of money growth? Suppose, for example, that the central bank decides to reduce its annual rate of money growth from 10% to 3%. What will be the long-run effects of this decrease in money growth on inflation, output growth, and unemployment?

Assume that in the long run the central bank maintains a constant growth rate of nominal money, call it $\bar{g}_m$. We can then proceed in three steps.

(1) In the long run, the unemployment rate must return to some constant value: It cannot increase or decrease forever. Thus, from Okun's law, and the condition that the change in the unemployment rate is equal to zero, it follows that *in the long run output must grow at its normal growth rate,* $\bar{g}_y$. Only when $g_{yt} = \bar{g}_y$ does $u_t = u_{t-1}$.

(2) If output grows at its normal rate, the aggregate demand relation implies that

$$\bar{g}_y = \bar{g}_m - \pi$$

or equivalently,

$$\pi = \bar{g}_m - \bar{g}_y$$

In the long run, *inflation is equal to nominal money growth minus normal output growth.* It will be convenient to call nominal money growth minus normal output growth **adjusted nominal money growth,** so that this result can be stated as: In the long run, inflation must be equal to adjusted nominal money growth. Suppose, for example, that nominal money growth is equal to 10%, and normal output growth is 3%. Then inflation will be equal to 10% − 3% = 7%.

One way to think about this result is as follows. A higher level of output implies a higher level of transactions and thus a higher demand for real money balances. If output is growing at 3% per year, then to meet this increased demand the real money supply must also grow at 3% per year. If the nominal money supply grows at a rate different from 3% per year, the difference must show up in inflation (or deflation). For example, if nominal money growth is 10%, then inflation must be equal to 7%.

It follows that, in the long run, inflation moves one for one with money growth. If money growth increases by 1%, adjusted money growth also increases by 1%, and thus inflation also increases by 1%. Or, as Milton Friedman once put it, (in the long run) *inflation is always and everywhere a monetary phenomenon.* Put another way, unless they lead to higher nominal money growth, factors such as the monopoly power of firms, strong unions, strikes, fiscal deficits, the price of oil, and so on, have no effect on inflation *in the long run.*[4]

(3) Turn finally to the implications of the Phillips curve. We have just seen that if money growth is constant, inflation is also constant. But as the Phillips curve tells

---

[4]*The "unless" qualification is important. When we study episodes of high inflation in Chapter 21, we shall see that fiscal deficits often lead to money creation, and thus to higher nominal money growth.*

FIGURE 18-2

Inflation and Unemploy-
ment in the Long Run
In the long run, unemployment is equal to the natural
rate, and inflation is equal
to adjusted money growth.

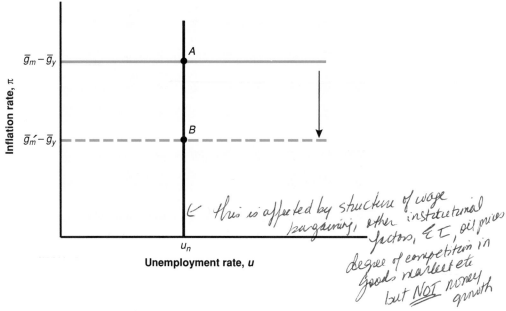

this is affected by structure of wage bargaining, other institutional factors, E.T., oil prices degree of competition in goods market etc but NOT money growth

us, for inflation to be constant the unemployment rate must be equal to its natural rate. This is our third result: *In the long run, the unemployment rate is equal to the natural rate.* As we saw in Chapter 15, this natural rate depends on the structure of wage bargaining, the degree of competition in goods markets, the system of unemployment benefits, the price of oil, and so on. But it does not depend on money growth.

Let's summarize our results with the help of a figure. Figure 18-2 plots the unemployment rate on the horizontal axis and the inflation rate on the vertical axis. In the long run, the unemployment rate is equal to the natural rate. The economy must thus be somewhere on the vertical line at $u = u_n$. And inflation is equal to adjusted money growth—the rate of nominal money growth minus the normal rate of output growth. This is represented by the horizontal line at $\pi = \bar{g}_m - \bar{g}_y$.

A decrease in nominal money growth from $\bar{g}_m$ to $\bar{g}_m'$ shifts the horizontal line downward, moving the equilibrium from point A to point B. The inflation rate decreases by the same amount as the decrease in nominal money growth. There is no change in either the unemployment rate or output growth. In Chapter 16 we saw that changes in the *level* of nominal money were *neutral;* they had no effect on output in the long run, but led one for one to changes in the price level. The result extends here to changes in the *growth rate* of nominal money. In the long run, changes in the growth rate of nominal money are also neutral; they affect neither output growth nor the unemployment rate, but lead one for one to changes in the inflation rate.

Having looked at the long run, we can now turn to the dynamics of adjustment. The next two sections focus on the adjustment of inflation, output, and unemployment in response to a decrease in money growth.

18-3 DISINFLATION: THE ISSUES

Suppose that the economy is in long-run equilibrium: Unemployment is at its natural rate, and output growth is sufficient to maintain stable unemployment. But the inflation rate is high, and there is a growing consensus that it should be reduced.

The inflation that accompanied the first oil price shock of 1974 proved to be stubbornly persistent. Double-digit rates of inflation were common over the next few years, and inflation was actually accelerating by the end of the decade. Everyone agreed that inflation needed to be reduced. Some observers claimed that all the old rules were no longer valid and that high inflation was now inevitable. Economists knew that to reduce the rate of inflation the Bank of Canada had to rein in the rate of growth of the money supply. Although the Bank announced as early as 1975 that it intended to target inflation more aggressively, the Bank was unwilling to engineer a rapid and large appreciation of the Canadian dollar, and so Canada essentially adopted its monetary policy from the U.S. Federal Reserve (recall the discussion in Section 13-5, page 251). By late 1979, the Fed had seen enough. Unemployment was close to its natural rate and growth was fairly good. The unemployment rate was 5.8% and GDP growth was 2.5%. But the U.S. rate of inflation, as measured by the CPI, was a high 13.8%.[5] The question was no longer whether to reduce inflation, but how fast to reduce it. This is the question that we take up here.

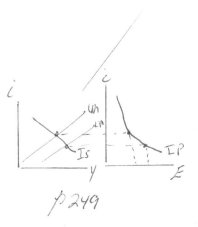

p 249

A FIRST PASS

Let's start with the following assumption: After the change in monetary policy, wage setters continue to form expectations in the same way as before, so the relation between the inflation rate and the unemployment rate remains as given by equation (18.2):

$$PC\ relation\quad \pi_t - \pi_{t-1} = -\alpha(u_t - u_n) \quad (18.2)$$

When $u_t > u_n$, RHS ⊖ ∴ π rate falls.

The equation makes it clear that **disinflation**—that is, a decrease in inflation—can be obtained only at the cost of higher unemployment. For the left-hand side of the equation to be negative—that is, for inflation to decrease—the term $(u_t - u_n)$ must be positive: The unemployment rate must exceed the natural rate.

The equation actually has a stronger and quite startling implication. Define a **point-year of excess unemployment** as a difference between the actual and the natural unemployment rate of one percentage point for one year. For example, if the natural rate is equal to 6%, an actual unemployment rate of 9% four years in a row corresponds to 4 times $(9 - 6) = 12$ point-years of excess unemployment.

Suppose now that the central bank wants to reduce inflation by x percentage points. To make things simpler, let's use specific numbers. Assume that the central bank wants to reduce inflation from 14% to 4%, so that x is equal to 10. Let's also assume that α is equal to 1.

Then equation (18.2) tells us that the central bank can achieve its goal by having, for example, one year of unemployment at 10% above the natural rate. In this case, the right-hand side of the equation is equal to –10%, and thus the inflation

How many point-years of excess un is required to reduce inflation?

↓π = ↑ut above un

So if ↓π = 10% & un = 4%, then

You need 1 yr of un at 14% (ut = 14%)

or it can be spread out over a # of years.

but the total # of years of excess un = 10 pts

---

[5]*This raises the question of where this opposition to inflation comes from, if growth is proceeding at a normal rate and unemployment is at its natural rate. We need to go through a few more steps before we can fully discuss the costs of inflation, and why policy makers take steps to prevent or decrease inflation. We shall do so in Chapter 28.*

rate decreases by 10% within a year. Or it can achieve its goal by having two years of unemployment at 5% above the natural rate. During each of the two years, the right-hand side of the equation is equal to −5%, so that the inflation rate decreases by 5% each year, thus by 10% over two years. Or, by the same reasoning, it can achieve its goal by having five years of unemployment at 2% above the natural rate, or by having 10 years of unemployment at 1% above the natural rate.

Note that in each case the number of point-years of excess unemployment is the same, namely 10: 1 times 10 in the first scenario, 2 times 5 in the second, 10 times 1 in the last. The implication is straightforward: The central bank may be able to choose the distribution of excess unemployment over time, but it cannot change the total number of point-years of excess unemployment. We can state this conclusion another way. Define the **sacrifice ratio** as the number of point-years of excess unemployment needed to achieve a decrease in inflation of 1%. Then, equation (18.2) implies, this ratio is independent of policy and simply equal to $1/\alpha$. If α is roughly equal to 0.5, as the estimated Phillips curve for Canada suggests, then the sacrifice ratio is roughly equal to 2 as well. In the United States, α is closer to 1.

The fact that the number of point-years required to decrease the inflation rate is fixed does not imply, however, that the speed of disinflation is unimportant. Suppose that the natural rate of unemployment is 6% and α is equal to 1. If the goal is a decrease in the inflation rate of 10%, achieving the full 10% decrease in one year implies that unemployment must increase to 16% for a year:

$$\pi_t - \pi_{t-1} = -\alpha(u_t - u_n)$$
$$-10\% = -1(u_t - 6\%)$$
$$16\% = u_t$$

Given Okun's law [equation (18.5)], and taking a value of 0.4 for β and a normal output growth rate of 3% implies in turn that output growth must be equal to −22% for a year:

$$u_t - u_{t-1} = -\beta(g_{yt} - \bar{g}_y)$$
$$16\% - 6\% = -0.4(g_{yt} - 3\%)$$
$$10\% = -0.4(g_{yt} - 3\%)$$
$$-22\% = g_{yt}$$

It is fair to say that macroeconomists do not know with great confidence what would happen if monetary policy were aimed at inducing such an extreme decrease in output. But most would surely be unwilling to try. The increase in the overall unemployment rate would lead, as we saw in Chapter 15, to extremely high unemployment rates for some groups—specifically the young and the unskilled. Not only would the welfare costs for these groups be large, but such high unemployment might leave permanent scars.[6] The sharp drop in output would most likely also lead to a large number of bankruptcies, with long-lasting effects on economic activity.

Thus, a more reasonable approach is to aim at achieving disinflation over a number of years. How should the central bank engineer this disinflation? Clearly, it does not control either inflation or unemployment directly. What it controls is

---

[6]***Digging deeper.*** *This suggests that very high unemployment may have long-lasting effects on the natural rate. This will indeed be one of the questions we take up when looking at European unemployment in Chapter 20.*

money growth.[7] Let's now solve for the required path of money growth that will achieve the desired disinflation.

Let's assume that the central bank decides to decrease inflation from 14% to 4% in five years. Let's make the same numerical assumptions as before. Normal output growth is 3%. The natural rate of unemployment is 6%. The parameter α in the Phillips curve is equal to 1; the parameter β in Okun's law is equal to 0.4. Table 18-1 then shows how to derive the path of money growth needed to achieve this disinflation.

In year 0, before disinflation, output growth is proceeding at its normal rate of 3%, and unemployment is at the natural rate, 6%. Inflation is running at 14%. Nominal money growth is equal to 17%. Thus, real money growth is equal to 17% − 14% = 3%, equal to output growth. The decision is then made to reduce inflation to 4% over five years, starting in year 1.

The easiest way to solve for the path of money growth that will achieve this disinflation is to start from the desired path of inflation, find the required path of unemployment and the required path of output growth, and finally derive the required path of money growth.

The first line of Table 18-1 gives the *target path of inflation*. Inflation starts at 14% before the change in monetary policy, decreases by 2% a year from year 1 to year 5, and then remains at its lower level of 4% thereafter.

The second line gives the required *path of unemployment* implied by the Phillips curve. If inflation is to decrease at 2% a year and $\alpha = 1$, the economy must accept five years of unemployment at 2% above the natural rate ($5 \times 2\% = 10\%$, the required decrease in inflation). Thus, from year 1 to year 5, the unemployment rate must be equal to 6% + 2% = 8%.

The third line gives the required *path of output growth*. From Okun's law we know that the initial increase in unemployment requires lower output growth. Under the assumption that β is equal to 0.4, an initial increase in unemployment of 2% requires output growth to be lower than normal by 2%/0.4 = 5%. Given a normal growth rate of 3%, the economy must therefore have a growth rate of 3% − 5% = −2% in year 1. From years 2 to 5, growth must proceed at a rate sufficient to maintain the unemployment rate constant at 8%. Thus, output must grow at its normal rate, 3%. In other words, from years 2 to 5, the economy grows at a normal rate but has an unemployment rate that exceeds the natural rate by 2%. Then, once disinflation is achieved, a burst of growth in year 6 is needed to return unemployment to normal. To decrease the unemployment rate by 2% in a year, growth must exceed normal growth by 2%/0.4, thus by 5%. The economy must therefore grow at 3% + 5% = 8% for one year.

TABLE 18-1

ENGINEERING DISINFLATION

	Before	Disinflation					After		
	0	1	2	3	4	5	6	7	8
Inflation (%)	14	12	10	8	6	4	4	4	4
Unemployment rate (%)	6	8	8	8	8	8	6	6	6
Output growth (%)	3	−2	3	3	3	3	8	3	3
Nominal money growth (%)	17	10	13	11	9	7	12	7	7

[7]*Even there, as we saw in Chapter 5, what it controls is not the money supply but rather the monetary base. We ignore this complication here.*

Handwritten margin annotations:

Year 1: Just use the equations.

$\pi_t - \pi_{t-1} = -(u_t - u_n)$

$-2\% = (u_t - 6\%)$ ← assumed

(by assumption this is the required ↓ in inflation)

$-2 + u_t = 6\%$

$u_t = 8\%$ (see below)

(for 5 years it's the same eqn)

Okun's relation:

$8\% - 6\% = -.4(g_{yt} - 3\%)$

$2\% = -.4 g_{yt} + 1.2$

$.8 = -.4 g_{yt}$

$-2\% = g_{yt}$ ✓ (see below)

after Yr 1 it is: (Yr 1)

$3\% - 8\% = -.4(g_{yt} - 3\%) = 3\%$

Nominal money growth:

Use:

$g_{yt} = g_{mt} - \pi_t$

$-2\% = g_{mt} - 12\%$

$12 - 2\% = g_{mt}$

$g_{mt} = 10\%$

See below ↓

Year 2:

$3 = g_{mt} - 10$

$g_{mt} = 13\%$

Table annotations: π_t, u_t, g_{yt} labels; −2% changes between inflation values; ✓ marks on year 1 column.

The fourth line gives the implied path of *nominal money growth.* From the aggregate demand relation [equation (18.6)] we know that output growth is equal to nominal money growth minus inflation, or, equivalently, that nominal money growth is equal to output growth plus inflation. Thus, adding the numbers for inflation in the first line and for output growth in the third gives us the required path for the rate of nominal money growth. The path looks surprising at first: Money growth goes down sharply in year 1, then up again, then slowly down for three years, then up again in the year following disinflation. But the explanation is easy to give.

To start disinflation, the central bank must induce an increase in unemployment. This requires a sharp contraction in money growth in year 1. The decrease in nominal money growth—from 17% to 10%—is much sharper than the decrease in inflation—from 14% to 12%. The result is thus a sharp decrease in real money growth, thus a decrease in demand and output, and in turn an increase in the unemployment rate.

For the next four years, monetary policy is aimed at maintaining unemployment at 8%, not at increasing unemployment further. Thus nominal money growth is aimed at allowing demand and thus output to grow at the normal growth rate. Put another way, nominal money growth is set equal to inflation plus the nominal growth rate of 3%. And as inflation decreases—because of high unemployment—so does nominal money growth.

At the end of the disinflation the central bank must allow unemployment to go back to its natural rate. It thus provides in year 6 a one-time increase in money growth before returning, from year 7 on, to the now-lower rate of money growth.

Figure 18-3 shows the path of unemployment and inflation implied by this disinflation path. In year 0, the economy is at point *A:* The unemployment rate is 6% and the inflation rate is 14%. Years 1 to 5 are years of disinflation, during which the economy moves from *A* to *B.* Unemployment is higher than the natural rate, leading to a steady decline in inflation. Inflation decreases until it reaches 4%. From year 6 on, the economy remains at point *C,* with unemployment back down to its natural rate and an inflation rate of 4%. *In the long run, money growth and inflation are lower, and the unemployment rate and output growth are back to normal;* this is the neutrality result we obtained in Section 18-2. *But the transition to lower money growth and lower inflation is associated with a period of higher unemployment.*

The disinflation path drawn in Figure 18-3 is one of many possible paths. We could have looked instead at a path that front-loaded the increase in the unemployment rate and allowed it to return slowly to its natural rate, thus avoiding the awkward increase in money growth that takes place at the end of our scenario (year 6 in Table 18-1). Or we could have looked at a path where the central bank decreased the rate of money growth from 14% to 4% at once, letting inflation and unemployment adjust over time.[8]

But all paths share one characteristic: The unemployment cost—that is, the number of point-years of excess unemployment—is the same under all scenarios. Put another way, *unemployment must remain above its natural rate by a large enough amount, and/or for enough years, to achieve disinflation.*

---

[8] ***Digging deeper.*** *If you trace the effects of an instant decrease in the rate of money growth from 14% to 4%, you will find that it leads to a complicated path of inflation and unemployment, with inflation actually undershooting its new lower long-run value for some time.*

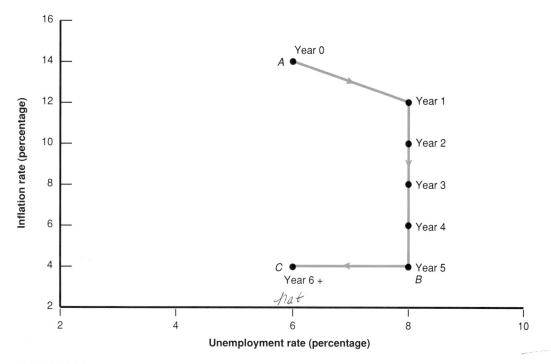

FIGURE 18-3

Disinflation Path

Five years of unemployment above the natural rate lead to a permanent decrease in inflation.

The analysis we have just developed is very much the type of analysis economists at central banks were conducting in the late 1970s. The econometric models they used, as well as most econometric models in use at the time, shared our simple model's property that policy could change the timing but not the number of point-years of excess unemployment. We shall call this the *traditional approach* in what follows. But the traditional approach was challenged by two groups of academic economists. We now look at their arguments.

EXPECTATIONS AND CREDIBILITY: THE LUCAS CRITIQUE

The conclusions of the first group were based on the work of Robert Lucas and Thomas Sargent, from the University of Chicago. In what has become known as the **Lucas critique,** Lucas pointed out, that when trying to predict the effects of a major change in policy, it could be very misleading to take as given the relations estimated from past data.

In the case of the Phillips curve, taking equation (18.2) as given was equivalent to assuming that wage setters would keep expecting inflation in the future to be the same as in the past, that expectation formation would not change in response to the change in policy. This was an unwarranted assumption, Lucas argued: Why shouldn't wage setters take policy changes into account? If wage setters believed that the central bank was committed to lower inflation, they might well expect inflation to be lower than in the past. In turn, if they lowered their expectations of

$$\pi_t - \pi_{t-1} = -\alpha(u_t - u_n)$$

inflation, then actual inflation would decline without the need for a protracted recession.

The logic of Lucas's argument can be seen by returning to equation (18.1):

$$\pi_t = \pi_t^e - \alpha(u_t - u_n)$$

If wage setters kept forming expectations of inflation by looking at last year's inflation (if $\pi_t^e = \pi_{t-1}$), then the only way to decrease inflation would be to accept higher unemployment for some time; this is the mechanism we explored earlier. But, if wage setters could be convinced that inflation was indeed going to be lower than in the past, they would decrease their expectations of inflation. This would in turn reduce actual inflation, without necessarily any change in the unemployment rate. For example, if wage setters were convinced that inflation, which had been running at 14% in the past, would be only 4% in the future, and if they formed expectations accordingly, then inflation would decrease to 4% *even if unemployment remained at the natural rate:*

→ Recall this is $(\mu+z)$ p333

$$\pi_t = \pi_t^e - \alpha(u_t - u_n)$$
$$4\% = 4\% - 0\%$$

Put another way, money growth, inflation, and expected inflation could all be reduced without the need for a recession.

Lucas and Sargent did not believe that disinflation could really happen without some increase in unemployment. But Sargent, looking at the historical evidence of the unemployment cost associated with ending several very high inflations, concluded that the increase in unemployment could indeed be small. The essential ingredient of successful disinflation, he argued, was **credibility** of monetary policy— the belief by wage setters that the central bank was truly committed to reducing inflation. Only credibility would lead wage setters to change the way they formed expectations. Furthermore, he argued, a clear and quick disinflation program was much more likely to be credible than a protracted one that offered plenty of opportunities for reversal and political infighting along the way.

NOMINAL RIGIDITIES AND CONTRACTS

A contrary view was taken by Stanley Fischer, from MIT, and John Taylor, then at Columbia University. Both emphasized the presence of **nominal rigidities**—the fact that, in modern economies, many wages and prices are set in nominal terms for some time and are typically not readjusted when there is a change in policy.

Fischer argued that, even with credibility, too rapid a decrease in money growth would lead to higher unemployment. Even if the central bank fully convinced workers and firms that money growth was going to be lower, the wages set before the change in policy would reflect expectations of inflation prior to the change in policy. In effect, inflation would already be built into existing wage agreements and could not be reduced costlessly and instantaneously. At the very least, Fischer said, a policy of disinflation should be announced sufficiently in advance of its actual implementation to allow wage setters to take it into account when setting wages.

Taylor's argument went one step further. An important characteristic of wage contracts, he argued, is that they are not all signed at the same time; instead, they

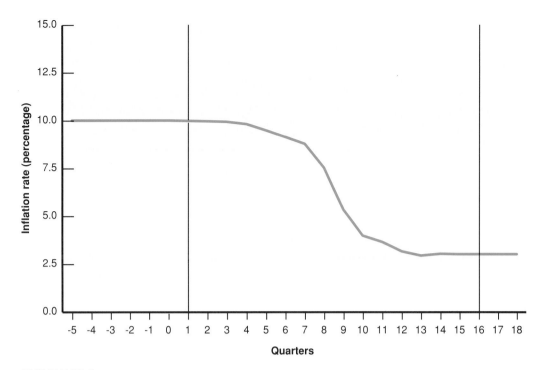

FIGURE 18-4
Disinflation without Unemployment in the Taylor Model
With staggering of wage decisions, disinflation must be phased in slowly to avoid an increase in unemployment.

Source: John Taylor, "Union Wage Settlements during a Disinflation," *American Economic Review,* December 1983, 981–993.

—) wage cntracts

prevents a fast disinflation mechanism .: a recession is triggered; Tun

are staggered over time. He showed that this **staggering of wage decisions** imposed strong limits on how fast disinflation could proceed without triggering higher unemployment, even if the central bank's commitment to inflation were fully credible. Why? If workers cared about relative wages—that is, cared about their wages relative to the wages of other workers—each wage contract would choose a wage not very different from wages in the other contracts in force at the time. Too rapid a decrease in nominal money growth would not lead to a proportional decrease in inflation. Rather, the real money supply would decrease, triggering a recession and an increase in the unemployment rate.

Taking into account the time pattern of wage contracts in the United States, Taylor then showed that under full credibility of monetary policy, there *was* a path of disinflation consistent with no increase in unemployment. This path is shown in Figure 18-4.

Disinflation starts in quarter 1 and lasts for 16 quarters. Once it is achieved, the inflation rate, which started at 10%, stands at 3%. The most striking feature is how slowly disinflation proceeds at the beginning. One year (four quarters) after the announcement of the change in policy, inflation still stands at 9.9%. But then disinflation becomes faster. By the end of the third year inflation is down to 4%, and by the end of the fourth year the desired disinflation is achieved.

The reason for the slow decrease in inflation at the beginning—and behind the scene, for the slow decrease in nominal money growth— is straightforward. Wages in force at the time of the policy change are the result of decisions made before the policy change, so that the path of inflation in the near future is largely predetermined. If nominal money growth were to decrease sharply, inflation could not decrease very much right away, and the result would be a decrease in real money and a recession. Thus, the best policy is for the central bank to proceed slowly at the beginning while announcing that it will proceed faster in the future. This announcement leads new wage settlements to take the new policy into account. When most wage decisions in the economy come from decisions made after the change in policy, disinflation can proceed much faster. This is what happens in the third year following the policy change.

Like Lucas and Sargent, Taylor did not believe that disinflation could really be implemented without increasing unemployment. For one thing, he realized that the path of disinflation drawn in Figure 18-4 might not be credible. The announcement this year that money growth will be decreased two years from now is likely to run into a serious credibility problem. Wage setters are likely to ask themselves: If the decision has been made to disinflate, why should the central bank wait two years? Without credibility, inflation expectations might not change, defeating the hope of disinflation without an increase in the unemployment rate. But Taylor's analysis had two clear messages. First, expectations are potentially important. Second, a slow but credible disinflation might have a cost lower than that implied by the traditional approach.

With this discussion in mind, let us look at what happened in Canada and the United States from 1979 to 1985.

18-4 THE U.S. AND CANADIAN DISINFLATION, 1979–1985

In August 1979, President Carter appointed Paul Volcker Chairman of the Federal Reserve Board. Volcker, who had served in the Nixon administration, was considered an extremely qualified man who would and could lead the fight against inflation.

In October 1979, the Fed announced a number of changes in its operating procedures. In particular, it indicated that it would shift from targeting a given level of the short-term interest rate to targeting the growth rate of money.

This change would hardly seem to be the stuff of history books. The Fed made no announcement of a battle against inflation, of a targeted path of disinflation, or various other ambitious-sounding plans. Nevertheless, financial markets widely interpreted this technical change as a sign of a major change in monetary policy. In particular, the change was interpreted as indicating that the Fed was now committed to reducing inflation and, if necessary, ready to let interest rates increase, perhaps to very high levels.

The Bank of Canada made no such dramatic announcements. But by keeping its eye on the C\$/US\$ exchange rate, it decided in effect to let Mr. Volcker choose the path for disinflation in Canada too.

Beginning in October 1979, the Fed aggressively raised interest rates to slow the growth of the U.S. money supply. The federal funds rate (the rate at which U.S.

banks can borrow or lend reserves overnight) increased by more than 6%, from 11.4% in September 1979 to 17.6% in April 1980. But then there was a halt, followed by a rapid reversal. By July 1980, the rate was back down to 9%, an 8.6% drop in four months. The rollercoaster movement of the federal funds rate is shown in Figure 18-5. Also shown are the rate of change of the CPI over the previous 12 months, for both the United States and Canada, and the value of the C$/US$ exchange rate from January 1979 to December 1985. Figure 18-5 shows that the Bank of Canada decided to accept the U.S. shift in monetary policy; although interest rates gyrated wildly in the United States, our bilateral exchange rate was kept very steady.

The decrease in the interest rate in mid-1980 was due to the accumulation of signs that the U.S. economy was in a sharp recession. In March 1980, believing that high consumer spending was one of the causes of inflation, the Carter administration imposed controls on consumer credit—limits on how much consumers could borrow to buy some durable goods. The effect of these controls was much larger than the Carter administration had anticipated. The combination of the fear of a sharp recession and the political pressure coming from the proximity of presidential elections was enough to lead the Fed to decrease interest rates sharply.

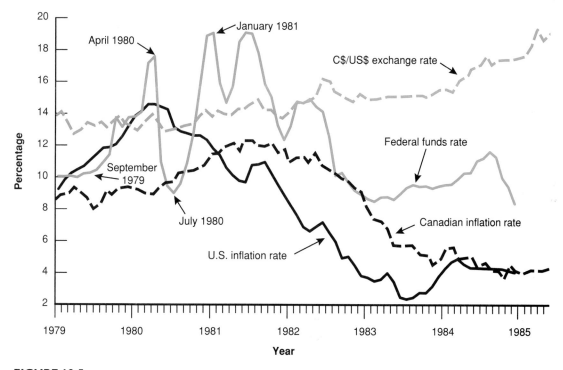

FIGURE 18-5

Disinflation in Canada and the United States, 1979–1985

A sharp increase in the federal funds rate from September 1979 to April 1980 was followed by a sharp decline in mid-1980, and then a second and sustained increase from January 1981 on, lasting for most of 1981 and 1982.

Sources: U.S. Department of Labour, Bureau of Labor Statistics; Board of Governors of the Federal Reserve System; and Statistics Canada, CANSIM Series B3400 and B700000.

By the end of 1980, with the U.S. economy apparently in recovery, the federal funds rate increased sharply again. By January 1981, the rate was back up to 19%.

By the end of 1981, signs had accumulated that the very high real interest rates had triggered a second recession. The Fed decided not to repeat its mistake of 1980, the abandonment of its disinflation target in the face of a recession. Thus, in contrast with its actions in 1980, it kept interest rates high. The federal funds rate decreased to 12.4% in December 1981, but then increased back to 14.9% in April 1982.

We have gone through the events of 1979–1982 in some detail to show the practical difficulties of establishing "credibility." Paul Volcker had credibility when he came to office. However, the credibility of the Fed's disinflation stance was surely eroded by the Fed's behaviour in 1980. Credibility was progressively reestablished in 1981 and 1982, especially when—despite clear indications that the economy was in recession—the Fed increased the federal funds rate in the spring of 1982.

Did this credibility—to the extent that it was indeed present—lead to a more favourable tradeoff between unemployment and disinflation than implied by the traditional approach? Table 18-2 gives the relevant numbers. The upper half of the table makes it clear that there was no expectation miracle: Disinflation was associated with substantial unemployment. The average unemployment rate in the U.S. was above 9% in both 1982 and 1983, peaking at 10.8% in December 1982. As we have seen, Canadian rates went even higher.

The answer to whether the unemployment cost was lower than implied by the traditional approach is given in the bottom half of the table. Under the traditional approach, each point of disinflation in the United States is predicted to require about one point-year of excess unemployment. Thus, line 4 computes the cumulative number of point-years of excess unemployment from 1980 on, assuming a natural rate of 6.0%. Line 5 computes cumulative disinflation—the decrease in inflation starting from its 1979 level. Line 6 gives the sacrifice ratio, the ratio of the cumulative point-years of unemployment above the natural rate to cumulative disinflation.

The table shows that there were no obvious "credibility gains." By 1982, the sacrifice ratio looked quite attractive: The cumulative decrease in inflation since 1979 was nearly 9.5%, at a cost of 6.4 point-years of unemployment. But in 1982

TABLE 18-2

U.S. INFLATION AND UNEMPLOYMENT, 1979–1985

	1979	1980	1981	1982	1983	1984	1985
1. GDP growth (%)	2.5	−0.5	1.8	−2.2	3.9	6.2	3.2
2. Unemployment rate (%)	5.8	7.1	7.6	9.7	9.6	7.5	7.2
3. CPI inflation (%)	13.3	12.5	8.9	3.8	3.8	3.9	3.8
4. Cumulative unemployment		1.1	2.7	6.4	10.0	11.5	12.7
5. Cumulative disinflation		0.8	4.4	9.5	9.5	9.4	9.5
6. Sacrifice ratio		1.37	0.61	0.67	1.05	1.22	1.33

Cumulative unemployment is the sum of point-years of excess unemployment from 1980 on, assuming a natural rate of 6.0%.
Cumulative disinflation is the difference between inflation in a given year and inflation in 1979.
The *sacrifice ratio* is the ratio of cumulative unemployment to cumulative disinflation.

Source for lines 1–3: Economic Report of the President.

unemployment was still very high. By 1983, with unemployment not yet back to its natural rate of 6.0%, the sacrifice ratio was nearly exactly equal to 1. By 1985 it had reached 1.33. A 10% disinflation had been achieved with close to 13 point-years of excess unemployment, an outcome a bit less favourable than the prediction made by the traditional approach (10 point-years).

In short, the U.S. disinflation of the early 1980s was associated with a substantial increase in unemployment. Canada's experience was even worse. The Phillips curve relation between the change in inflation and the deviation of the unemployment rate from the natural rate proved more robust than many economists anticipated. Was this close relation due to a lack of credibility of the change in monetary policy, or to the fact that credibility is not enough to reduce substantially the cost of disinflation? One way of learning more is to look at other disinflation episodes. This is the approach followed in a recent paper by Laurence Ball, from Johns Hopkins. Ball has estimated sacrifice ratios for 65 disinflation episodes in 19 OECD countries over the last 30 years.[9] He reaches three main conclusions. First, disinflations typically lead to higher unemployment for some time. Put another way, even if it is neutral in the long run, a decrease in money growth leads to an increase in unemployment in the short and the medium run. Second, faster disinflations are associated with smaller sacrifice ratios. This conclusion provides some evidence to support the expectation and credibility effects emphasized by Lucas and Sargent. Third, sacrifice ratios are smaller in countries that have shorter wage contracts. This provides some evidence to support Fischer and Taylor's emphasis on the importance of the structure of wage settlements.

---

[9]Laurence Ball, "What Determines the Sacrifice Ratio?" in N. Gregory Mankiw, ed., Monetary Policy (NBER and the University of Chicago, 1994), 155–194.

SUMMARY

■ There are three important relations linking inflation, output, and unemployment.

The first is the Phillips curve, which relates the change in inflation to the deviation of unemployment from the natural rate. In Canada since 1970, an unemployment rate 1% above the natural rate for a year leads to a decrease in inflation of about 0.5%.

The second is Okun's law, which relates the change in unemployment to the deviation of output growth from normal. In Canada since 1975, output growth of 1% above normal for a year leads to a decrease in the unemployment rate of about 0.4%.

The third is the aggregate demand relation, which relates output growth to the rate of growth of real money balances. The growth rate of output is equal to the growth rate of nominal money minus the rate of inflation. Given nominal money growth, higher inflation leads to a decrease in output growth.

■ In the long run, unemployment is equal to the natural rate, and output grows at its normal growth rate. Money growth determines the inflation rate: A 1% increase in money growth leads to a 1% increase in inflation. In the long run inflation is always and everywhere a monetary phenomenon.

■ In the short run, however, a decrease in money growth leads to a slowdown in growth and an increase in unemployment for some time. Thus, disinflation (a decrease in the inflation rate) can be achieved only at the cost of some unemployment. How much unemployment is required is a controversial issue.

■ The traditional approach to disinflation assumes that people do not change the way they form expectations when monetary policy changes, so that the relation between inflation and unemployment is unaffected by the change in policy. This approach implies that disinflation can be achieved by a short but large increase in unemployment, or by a longer and smaller increase in unemployment. But it also predicts that policy cannot affect the total number of point-years of excess unemployment.

■ An alternative view of disinflation is as follows. If the change in monetary policy is credible, expectation formation may change, leading to a smaller increase in unemployment than predicted by the traditional approach. In its extreme form, this alternative view implies that, if policy is fully credible, it can achieve disinflation at no cost in un-

employment. A less extreme form recognizes that while expectation formation may change, the presence of nominal rigidities is likely to imply some increase in unemployment, although less than implied by the traditional answer.

■ The U.S. disinflation of the early 1980s, during which inflation decreased by approximately 10%, was associated with a large recession, and its unemployment cost was close to the predictions of the traditional approach.

KEY TERMS

■ Okun's law, 347
■ normal growth rate, 348
■ labour hoarding, 349
■ adjusted nominal money growth, 351
■ disinflation, 353
■ point-year of excess unemployment, 353

■ sacrifice ratio, 354
■ Lucas critique, 357
■ credibility, 358
■ nominal rigidities, 358
■ staggering of wage decisions, 359

QUESTIONS AND PROBLEMS

1. Find the change in the inflation rate and the deviation of the average unemployment rate from the natural rate over the most recent 12-month period. (Use the CPI to compute inflation, and assume a natural unemployment rate of 7.5%.) Are the data consistent with the Phillips curve equation (18.2), assuming $\alpha = 0.5$? If not, what special events over these 12 months may account for the difference?

2. Use equation (18.4) to answer the following questions.
   a. What growth rate of output will reduce the unemployment rate by one percentage point in one year?
   b. If the unemployment rate is currently 10%, what annual growth rate of output will bring it down to 6% in two years? (Assume that u decreases by 2% in each of the two years.)
   c. Would your answers above be different if labour productivity began growing faster? Explain.

3. Suppose that the economy can be described by the following three equations:

$$u_t - u_{t-1} = -0.4(g_{yt} - 0.03) \quad \text{(Okun's law)}$$
$$\pi_t - \pi_{t-1} = -0.5(u_t - 0.075) \quad \text{(Phillips curve)}$$
$$g_{yt} = g_{mt} - \pi_t \quad \text{(Aggregate demand relation)}$$

   a. What is the natural rate of unemployment for this economy?

   b. Suppose that inflation is running at 10% each year and the economy is operating at the natural rate of unemployment. To *keep* unemployment at its natural rate, what must be
      (i) The growth rate of output?
      (ii) The growth rate of the money supply?
   c. Suppose that conditions are as in part **b**, and then, in year t, the authorities use monetary policy to reduce the inflation rate to 5% and keep it there. What will happen to the unemployment rate and output growth in years t, $t + 1$, and $t + 2$? What money growth rate in years t, $t + 1$, and $t + 2$ will accomplish this goal?

4. Suppose you were advising a government that wants to reduce its annual inflation rate from 20% to 10%. The policy choices are (1) an immediate (one-year) reduction to 10% inflation or (2) a gradual reduction to 10% inflation over several years. Neither choice is pleasant. What are the chief arguments against choosing option 1? Against choosing option 2?

5. "From early 1979, the Fed pursued a consistently credible policy of reducing inflation, yet the markets did not believe it. This shows that inflationary expectations are not influenced by credible policy changes." Comment.

6. Suppose an economy is described by the following three equations:

$$u_t - u_{t-1} = -0.4(g_{yt} - 0.03) \quad \text{(Okun's law)}$$
$$\pi_t - \pi_{t-1} = -0.5(u_t - 0.075) \quad \text{(Phillips curve)}$$
$$g_{yt} = g_{mt} - \pi_t \qquad \text{(Aggregate demand relation)}$$

a. Reduce the three equations to two by substituting the aggregate demand relation into Okun's law.

b. Assume initially that $u_t = u_{t-1} = 0.075$, $g_{mt} = 0.10$, and $\pi_t = 0.07$, and that this year money growth is permanently reduced from 10% to 0%.

(i) Determine the impact on unemployment and inflation *this year.*

(ii) Determine the impact on unemployment and inflation *next year.*

(iii) Determine the *long-run* impact on unemployment and inflation.

FURTHER READING

The Lucas critique was first presented by Robert Lucas in "Econometric Policy Evaluation: A Critique," in *The Phillips Curve and Labor Markets,* Carnegie Rochester Conference, Volume 1, 1976, 19–46.

The article by Stanley Fischer arguing that credibility would not be enough to achieve costless disinflation is "Long-Term Contracts, Rational Expectations, and the Optimal Money Supply Rule," *Journal of Political Economy,* 85, 1977, 163–190.

The article that derived the path of disinflation reproduced in Figure 18-4 is by John Taylor, "Union Wage Settlements during a Disinflation," *American Economic Review,* December 1983, 981–993.

(All three articles above are relatively technical.)

A description of U.S. monetary policy in the 1980s is given by Michael Mussa in Chapter 2 of Martin Feldstein, ed., *American Economic Policy in the 1980s* (Chicago: University of Chicago Press and NBER, 1994), 81–164. One of the comments on the chapter is by Paul Volcker, who was Chairman of the Fed from 1979 to 1987.

INFLATION, INTEREST RATES, AND EXCHANGE RATES

Our treatment of aggregate demand in the preceding four chapters was minimalist. This was justified because what was new, and what we wanted to concentrate on, was aggregate supply. But the time has come to extend our treatment of the demand side, building on what we saw in earlier chapters. We examine two sets of issues in this chapter. First, we examine the relation between money growth, inflation, and interest rates—a perennial issue in macroeconomics. Second, we look at the role of the exchange rate in how an open economy adjusts to shocks, then reexamine the effects and the role of devaluations under fixed exchange rates.

19-1 MONEY GROWTH, INFLATION, AND INTEREST RATES

The Bank of Canada's decision to allow for higher money growth is the main factor behind this year's decline in interest rates.

　　　　　　　　　　　　　　　　　—circa November 1996

The Bank of Canada's reluctance to curtail the sharp growth in money has led financial analysts to worry about higher inflation and has led to upward pressure on long-term interest rates.

　　　　　　　　　　　　　　　　　—circa May 1997[1]

Which of these two statements is correct? Does higher money growth lead to lower interest rates or to higher interest rates? The answer is: both. The key is the distinction between the short run and the long run. Higher money growth leads to lower interest rates in the short run but to higher interest rates in the long run. The purpose of this section is to develop this answer and draw its implications.

REAL AND NOMINAL INTEREST RATES

Our derivation of the aggregate demand relation in Chapter 16 was based on a version of the *IS-LM* model that did not distinguish between nominal and real interest rates. But the distinction will be important here. Borrowing from what we saw in Section 7-3, let us write the *IS-LM* model as:

IS:　　　　　　　　$Y = C(Y - T) + I(Y, r) + G$

LM:　　　　　　　　$\dfrac{M}{P} = YL(i)$

Real interest rate:　　$r = i - \pi^e$

　　Spending decisions depend on the *real interest rate*—the interest rate in terms of goods. Thus, it is the real interest rate that enters the *IS* relation. But it is the nominal interest rate that enters the *LM* relation: The opportunity cost of holding money rather than bonds is the *nominal interest rate*—the interest rate in terms of dollars. The two interest rates are related in a simple way: The real interest rate is equal to the nominal interest rate minus expected inflation, π^e.

THE SHORT RUN

In Chapter 16 we looked at the effects of a one-time increase in nominal money. We are looking here at the effects of an increase in the *growth rate* of money. But in the short run, the two have the same basic effect: They both lead to a higher nominal money supply than would have been the case without the change in policy.

　　In Chapter 16 also, we ignored the distinction between nominal and real interest rates that we have just reintroduced. But in the short run, expected inflation is

---

[1] *These two quotes are made up, but they are composites of what was written at the time.*

unlikely to change very much.[2] In other words, in the short run, the nominal and the real interest rates are likely to move largely together.

Thus, in thinking about the short run, we can use the conclusions of Section 16-4. Let us restate them as follows. In the short run, an increase in nominal money growth leads to an increase in the real money supply; the price level increases, but by less than the increase in nominal money. Higher real money leads to lower nominal and real interest rates, and to a higher level of output.

THE LONG RUN

Let's now turn to the opposite end of the time scale, to the long-run effects of money growth on interest rates.

Recall first the main result of Chapter 18: In the long run, higher nominal money growth affects neither unemployment nor output, but translates one for one into higher inflation. Although we derived this result in a model with a simpler aggregate demand relation than the *IS-LM* model we are using here, it holds here as well. The fact that money growth does not affect unemployment follows directly from the propositions that in the long run unemployment returns to the natural rate, and that the natural rate of unemployment does not depend on money growth. And if money growth does not affect unemployment, then it does not affect employment, nor does it affect output.

Money growth and the real interest rate. Assume for simplicity that the normal rate of output growth is equal to zero, so that in the long run output returns to its natural level, Y_n, which does not increase over time.[3] One way of thinking about the *IS* relation is that it tells us what real interest rate is needed to sustain a given level of demand, and thus a given level of output. So let r_n be the real interest rate associated with the natural level of output Y_n—that is, the real interest rate that generates a level of demand and output equal to Y_n:

$$Y_n = C(Y_n - T) + I(Y_n, r_n) + G$$

If, as we have just established, output returns to Y_n in the long run, it follows that the real interest rate returns to r_n in the long run. This conclusion extends the money neutrality result of Chapter 18 to read: *In the long run, money growth affects neither unemployment, nor output, nor the real interest rate.*

Money growth and the nominal interest rate. From our *IS-LM* model, the nominal interest rate is given by

$$i = r + \pi^e$$

We have just established that in the long run higher money growth does not affect the real interest rate, which remains equal to r_n. Let's look at the second term, π^e. In the long run, it is reasonable to assume that expected inflation is equal to actual inflation: $\pi^e = \pi$. In turn, actual inflation moves one for one with nominal money

---

[2] *This statement should probably be qualified. If the central bank announced that it had decided to increase the rate of money growth from 0% to 50% a year, then expectations of inflation would probably be revised quickly and drastically.*

[3] *If we allowed for output growth, we would have to take into account that the natural level of output increases over time. This would complicate the argument but not change any of the basic conclusions.*

growth. Indeed, under our simplifying assumption that the normal growth rate of output is equal to zero, in the long run inflation is simply equal to nominal money growth: $\pi = g_m$. Putting everything together yields

$$i = r_n + \pi$$
$$i = r_n + g_m$$

In the long run, the nominal interest rate moves one for one with inflation, and thus one for one with nominal money growth. Put another way, in the long run *nominal money growth is reflected one for one in a higher nominal interest rate.* A permanent increase in nominal money growth of, say, 10% is eventually reflected in a 10% increase in the inflation rate and a 10% increase in the nominal interest rate, leaving the real interest rate unchanged. This result is known as the **Fisher effect,** or the **Fisher hypothesis,** after Irving Fisher, who first stated it and its logic at the beginning of the twentieth century.[4]

Money growth and real money balances. Higher nominal money growth leads to a higher nominal interest rate, and thus to a higher opportunity cost of holding money rather than bonds. This effect in turn leads people to reduce their real demand for money. In terms of our *LM* relation, a higher value of the nominal interest rate, i, leads to a lower real demand for money, $YL(i)$.

In equilibrium, real money demand must equal real money supply: $M/P = YL(i)$. Thus, the real money supply (M/P) must also be lower. This implies that the price level must increase relative to the nominal money supply, so that M/P goes down. *In the long run, higher nominal money growth therefore leads to a decrease in real money balances.*

Let's now put our long-run results together. Higher money growth has no effect on output or on the real interest rate in the long run. Rather, higher money growth is reflected one for one in higher inflation and in a higher nominal interest rate.

DYNAMICS

We now have a reconciliation of the two quotes that started this section. In the short run, higher nominal money growth leads to a lower nominal interest rate. In the long run, higher nominal money growth increases the nominal interest rate. What happens between the short run and the long run? A complete characterization of the movements of real and nominal interest rates over time would take us beyond what we can do here. But the basic features of the adjustment process are easy to establish.

In the short run, real and nominal interest rates go down. Why don't they remain lower forever? In short, because as long as the real interest rate remains below its initial value, inflation increases. The increase in inflation forces the real interest rate to return to its initial value; correspondingly, the nominal interest rate must converge to a new higher value.

Let's look at this mechanism more closely. As long as the real interest rate is below its initial value (the value corresponding to the natural level of output),

---

[4]*Irving Fisher,* The Rate of Interest *(New York: Macmillan, 1906). After stating the result, however, Fisher proceeded to qualify it heavily. We shall return to some of the issues he raised later in this chapter.*

output is higher than the natural level. Equivalently, unemployment is below the natural rate. From the Phillips curve relation, we know that as long as unemployment is below the natural rate, inflation increases.

Given nominal money growth, higher inflation means lower real money growth. As inflation increases, it eventually becomes higher than nominal money growth, leading to negative real money growth. When real money growth turns negative, the nominal interest rate starts increasing. And, given expected inflation, so does the real interest rate.

Eventually, the real interest rate increases back to its initial value. Output is then back to its natural level, unemployment is back to its natural rate, and inflation is no longer changing. As the real rate converges back to its initial value, the nominal interest rate converges to a new higher value, equal to the real interest rate plus the new, higher rate of nominal money growth.

Figure 19-1 summarizes our results, showing the adjustment over time of the real and the nominal interest rates to an increase in nominal money growth from 5% to 10%, starting at time t. Before time t, both interest rates are constant. The real interest rate is equal to r_n. The nominal interest rate exceeds the real interest rate by an amount equal to the rate of money growth, thus by 5%.

At time t, the rate of money growth increases from 5% to 10%. The increase in the rate of nominal money growth leads, for some time, to an increase in real money and to a decrease in the nominal interest rate. To the extent that expected inflation increases, the real interest rate decreases more than the nominal interest rate.

Eventually, the nominal and real interest rates start increasing. The real interest rate returns over time to its initial value r_n. Inflation and expected inflation converge to the new rate of money growth, thus 10%. The nominal interest rate converges to a value equal to the real interest rate plus 10%.

Let's end our analysis by going back to the second quote that opened this section. Can long-term interest rates go *up* when a central bank increases money

---

FIGURE 19-1

An increase in nominal money growth leads initially to a decrease in both the real and the nominal interest rate. Over time, the real interest rate returns to its initial value. The nominal interest rate converges to a new higher value. This new higher value is equal to the initial value plus the increase in money growth.

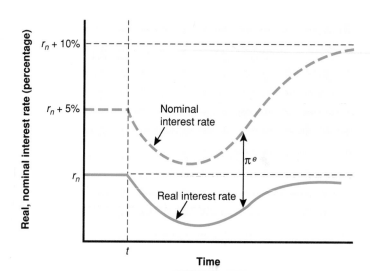

growth? The answer is yes. Recall from Chapter 9 that long-term interest rates are weighted averages of current and future expected short-term interest rates. We have just seen that higher nominal money growth leads to lower nominal interest rates in the short run but to higher nominal interest rates in the long run. Thus, to the extent that financial markets anticipate these higher nominal interest rates in the future, long-term nominal interest rates may indeed go up when the central bank increases money growth.

EVIDENCE ON THE FISHER HYPOTHESIS

There is plenty of evidence that a monetary expansion initially decreases nominal interest rates and, symmetrically, that a monetary contraction initially increases nominal interest rates, as we saw in Figure 6-13 (in Section 6-5). But how much evidence is there for the Fisher hypothesis, the proposition that in the long run increases in inflation lead to one-for-one increases in nominal interest rates?

Economists have tried to answer this question by looking at two types of evidence. The first is the relation between nominal interest rates and inflation *across countries*. Because the relation holds only in the long run, we should not expect inflation and nominal interest rates to be close to each other in any one country at any given time, but the relation should hold on average. This line of inquiry is explored further in the Global Macro box entitled "Nominal Interest Rates and Inflation across Latin America," which looks at Latin American countries in the early 1990s and finds substantial support for the Fisher hypothesis.

The second type of evidence is the relation between the nominal interest rate and inflation over time for one country. Again, the Fisher hypothesis does not suggest that the two should move together from year to year. But it does suggest that the long-run movements in inflation should eventually be reflected in similar movements in the nominal interest rate. To capture these long-run movements, it is thus important to look at as long a period of time as we can. Figure 19-2 looks at the nominal interest rate and inflation in Canada since 1946. The nominal interest rate

GLOBAL MACRO

NOMINAL INTEREST RATES AND INFLATION ACROSS LATIN AMERICA

Figure 1 on page 372 plots nominal interest rate–inflation pairs for eight Latin American countries (Argentina, Bolivia, Chile, Ecuador, Mexico, Peru, Uruguay, and Venezuela) for both 1992 and 1993. Because the Brazilian numbers would dwarf those from other countries, they are not included in the figure. (In 1992, Brazil's inflation rate was 1008% and its nominal interest rate was 1560%. In 1993, inflation was 2140% and the nominal interest rate was 3240%.) The numbers for inflation refer to the rate of change of the consumer price index. The numbers for nominal interest rates refer to the "lending rate." The exact definition of this term varies with each country, but you can think of it as corresponding to the prime rate in Canada—the rate charged to borrowers with the best credit rating.

Note the wide range of inflation rates, from 10% to about 100%. This is precisely why we have chosen to present numbers from Latin America in the early 1990s. With this much variation in inflation, we can learn a lot about the relation between nominal interest rates and inflation. And the figure indeed shows a clear relation between inflation and nominal interest rates. The line drawn in the figure plots what the nominal interest rate should be under the Fisher hypothesis, assuming an underlying real interest rate of 10%, so that $i = 10\% + \pi$. The slope of the line is 1: Under the Fisher hypothesis, a 1% increase in inflation should be reflected in a 1% increase in the nominal interest rate.

As the eye can see, the line fits well; roughly half of the points are above the line, the other half below. The Fisher hypothesis appears roughly consistent with the cross-country evidence from Latin America in the early 1990s.

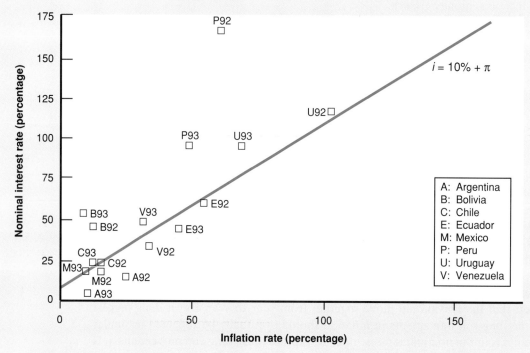

FIGURE 1

Nominal Interest Rates and Inflation: Latin America, 1992 and 1993

Source: International Financial Statistics, International Monetary Fund.

is the three-month Treasury-bill rate, and inflation is the rate of change of the GDP deflator.

Figure 19-2 has a number of interesting features.

1. The increase in inflation from the early 1960s to the early 1980s was associated with a roughly parallel increase in the nominal interest rate. The decrease in inflation since the mid-1980s has come with a decrease in the nominal interest rate. These evolutions support the Fisher hypothesis.

2. Evidence of the short-run effects that we discussed earlier is easy to see. The nominal interest rate lagged behind the increase in inflation in the 1970s, while the disinflation of the early 1980s was associated with an initial *increase* in the nominal rate and a much slower decline in the nominal interest rate than in inflation.

3. The other episode of inflation, during and after World War II, underlines the importance of the "long-run" qualifier in the Fisher hypothesis. During that period, inflation was high but short-lived. And it was gone before it had time

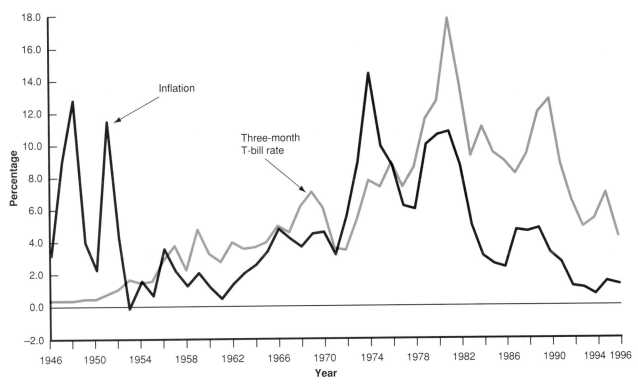

FIGURE 19-2

The Three-Month Treasury Bill Rate and Inflation, 1946–1996

The increase in inflation from the early 1960s to the early 1980s was associated with an increase in the nominal interest rate. The decrease in inflation since the mid-1980s has been associated with a decrease in the nominal interest rate.

Source: Statistics Canada, CANSIM Series B14001, D11000, and D14442.

to be reflected in a higher nominal rate. The nominal interest rate remained very low throughout the 1940s.

More sophisticated studies confirm our basic conclusion. The <u>Fisher hypothesis</u> that in the long run increases in inflation are reflected in a higher nominal interest rate appears to fit the data quite well. But this conclusion comes with two caveats.

First, <u>the long run is indeed very long.</u> The data confirm the speculation by Milton Friedman, which we quoted in a Focus box in Chapter 17 (page 334), that it takes a "couple of decades" for nominal interest rates to reflect the higher inflation rate.

Second, although it is clear that inflation leads to a higher nominal interest rate, it is <u>not clear that the effect is one for</u> one. One potential explanation, which was advanced by Irving Fisher himself and has been explored by other economists since, is **money illusion,** which Fisher defined as "the failure to perceive that the dollar, or any other unit of money, expands or shrinks in value over time." A formal argument on how money illusion can lead to the failure of the Fisher effect is presented in the Focus box entitled "Money Illusion and the Failure of the Fisher Effect." Money illusion as a behavioural trait has been documented at length since Fisher. Its quantitative implications for macroeconomics are still hotly debated, with little agreement in sight.

MONEY ILLUSION AND THE FAILURE OF THE FISHER EFFECT

There is a lot of anecdotal evidence that many people fail to adjust properly for inflation in financial computations. Recently, economists and psychologists have looked at money illusion more closely. In a recent study, two psychologists, Eldar Shafir from Princeton and Amos Tversky from Stanford, and one economist, Peter Diamond from MIT, designed a survey aimed at finding the presence and the determinants of money illusion.[*] Among the many questions they asked of people in various groups (people at Newark International Airport, people at two New Jersey shopping malls, and a group of Princeton undergraduates) is the following:

> Suppose that Adam, Ben, and Carl each received an inheritance of $200 000 and each used it immediately to purchase a house. Suppose that each of them sold the house one year after buying it. Economic conditions were, however, different in each case:
>
> ■ During the time that Adam owned the house, there was a 25% deflation—the prices of all goods and services decreased by approximately 25%. A year after Adam bought the house, he sold it for $154 000 (23% less than he had paid).
> ■ During the time that Ben owned the house, there was no inflation or deflation—the prices of all goods and services did not change significantly during the year. A year after Ben bought the house, he sold it for $198 000 (1% less than he had paid).
> ■ During the time that Carl owned the house, there was a 25% inflation—the prices of all goods and services increased by approximately 25%. A year after Carl bought the house, he sold it for $246 000 (23% more than he had paid).
>
> Please rank Adam, Ben, and Carl in terms of the success of their house transactions. Assign "1" to the person who made the best deal, and "3" to the person who made the worst deal.

It is clear that in *nominal* terms, Carl made the best deal, followed by Ben, followed by Adam. But in *real* terms—that is, adjusting for inflation—the ranking is reversed: Adam, with a 2% real gain, made the best deal, followed by Ben (with a 1% loss), followed by Carl.

The answers to the question were the following:

RANK	ADAM	BEN	CARL
First	37%	16%	48%
Second	10%	73%	16%
Third	53%	10%	36%

Thus, Carl was ranked first by 48% of the respondents, and Adam was ranked third by 53% of the respondents. These answers are very suggestive of money illusion.

If money illusion is indeed prevalent, the Fisher hypothesis will fail. To take an extreme but transparent case, suppose that firms base their investment decisions on the nominal interest rate rather than on the more appropriate real rate, and therefore investment demand is given by $I(Y, i)$ rather than by $I(Y, r)$ as we have assumed so far.

In the long run, output still returns to its natural level Y_n, which is independent of inflation. Because spending now depends on the nominal interest rate, the nominal interest rate must also return to a value that is independent of inflation. (Contrast this with our result earlier in the text that, when investment depends on the real interest rate, the real interest rate returns to a value that is independent of inflation.) But if in the long run the nominal interest rate does not vary with inflation, then the real rate varies negatively with inflation. For example, if the nominal interest rate remains equal to 6% when inflation goes from, say, 0 to 5%, it follows that the real interest rate decreases from 6% to 1% as inflation increases. The Fisher hypothesis fails badly.

This case is admittedly extreme. Money illusion is likely to be only partial; firms are likely to correct at least partly for inflation. And if inflation is high and persistent, money illusion is likely to fade away eventually. Thus, the Fisher effect may still hold in the long run, but with the long run being very long indeed.

[*]"On Money Illusion," mimeo, MIT, October 1995.

FIXED EXCHANGE RATES AND DEVALUATIONS

Sooner or later, countries operating under a fixed exchange rate face the following question: Should they devalue? Those economists in favour of a devaluation typically argue that a devaluation can help reestablish the country's competitiveness, or help the economy out of a recession, or both. Those against argue that a devaluation undermines the very principle of a fixed-exchange-rate regime and is likely to prove ineffective or even counterproductive. As we saw in Section 14-5, whether or not to devalue was a recurring issue in many countries of Western Europe in the early 1990s. And, as we shall see, many economists trace the Mexican crisis of 1994 in part to Mexico's decision not to devalue earlier.

Whether and when a country should devalue is the topic of this section. To discuss it, we develop a model of aggregate demand and aggregate supply for the open economy. Once this is done, we look at the adjustment process with and without a devaluation. The section ends with a general discussion of the pros and cons of devaluations.

AGGREGATE DEMAND UNDER FIXED EXCHANGE RATES

To derive the aggregate demand relation between output and the price level in an open economy under fixed exchange rates, let us first review what we learned in Chapter 13.

First, and by the definition of a fixed-exchange-rate régime, the *nominal* exchange rate is fixed. Let us denote this fixed rate by $\bar{E}$. Recall the basic terminology here. The nominal exchange rate is defined as the price of foreign currency in terms of domestic currency. Under fixed exchange rates, changes in the parity are called revaluations or devaluations. A *devaluation*, which makes the domestic currency cheaper, corresponds to an increase in the nominal exchange rate. A *revaluation*, which makes the domestic currency more expensive, corresponds to a decrease in the nominal exchange rate.

A fixed exchange rate and perfect capital mobility imply that the nominal interest rate must be equal to the foreign interest rate. To see why, recall the interest parity condition between domestic and foreign bonds:

$$i = i^* + \left(\frac{E^e - E}{E}\right) = 0$$

The domestic interest rate must be equal to the foreign interest rate plus the expected rate of change of the exchange rate (E^e is the expected exchange rate next period). Under fixed exchange rates, the expected exchange rate is equal to today's exchange rate: $E^e = E = \bar{E}$, so the second term on the right in the interest parity condition is equal to zero.[5] It follows that the domestic interest rate must be equal to the foreign interest rate. This condition implies in turn that the central bank is no longer free to choose the money supply: The money supply must be such that it will maintain an interest rate equal to the foreign interest rate.

---

[5] *If, however, financial markets start expecting a devaluation, then the first equality ($E^e = E$) no longer holds. This was one of the topics we discussed in Section 14-4, and we shall return to it later in this chapter.*

Equilibrium in the goods market, the *IS* relation, can thus be written as

$$Y = C(Y - T) + I(Y, i^*) + G + NX\left(Y, Y^*, \frac{\overline{E}P^*}{P}\right)$$

where we have used the fact that the nominal exchange rate is fixed, and the implication that the interest rate is equal to the foreign interest rate, so that i^* rather than i appears in the investment function.[6]

The important link here is the link between the real exchange rate, $\overline{E}P^*/P$, and output. It implies a negative relation between the price level and output: A higher price level leads to a *real appreciation*—a decrease in the real exchange rate. It makes domestic goods relatively more expensive, decreasing net exports, thus decreasing demand and output.

Thus, as in a closed economy, an increase in the price level leads to a decrease in output. But the channel is different. Rather than working through the real money supply, the price level affects output through the real exchange rate. The price level does not affect output through the real money supply because the interest rate is fixed at i^*.

AGGREGATE DEMAND AND AGGREGATE SUPPLY

We are now ready to put aggregate demand and aggregate supply together.

Let's summarize what we have just seen by writing the following *aggregate demand relation:*

$$Y_t = Y\left(\frac{\overline{E}P^*}{P_t}, G, T\right) \tag{19.1}$$
$$= (\ +\ ,\ +,\ -)$$

Output is an increasing function of the real exchange rate, an increasing function of government spending, and a decreasing function of taxes. The other factors that enter the open-economy *IS* relation, and thus affect the demand for goods—i^* and Y^* in particular—could be introduced in equation (19.1); They are omitted for simplicity. Note that, in anticipation of the study of dynamics, we have introduced time indexes for both output and the price level. The nominal exchange rate is fixed and thus does not have a time index. The same applies to the foreign price level, to government spending, and to taxes.

An increase in the price level leads to a decrease in the real exchange rate—a real appreciation—which leads in turn to a decrease in net exports and a decrease in output. This relation is drawn as the *AD* curve in Figure 19-3. As always, the relation is drawn for given values of all other variables, in particular for a given value of the nominal exchange rate.

Let's now turn to aggregate supply and the determination of the price level. We shall rely here on the version of the aggregate supply relation we derived in Chapter 16 [equation (16.3)]:[7]

---

[6]*Note, however, that we have ignored the distinction between nominal and real interest rates we focussed on earlier in this chapter. One can keep only so many balls in the air at one time!*

[7]*We derived this equation in Chapter 16 under a number of simplifying assumptions, from the assumption that expected prices are equal to last year's prices, to the assumptions that the labour force is fixed and output is equal to employment. These simplifications are not essential for our purposes here.*

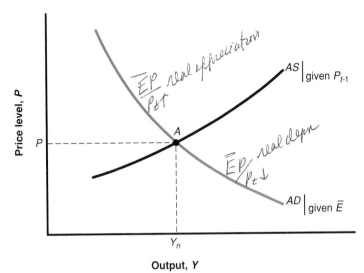

FIGURE 19-3

Aggregate Demand and Aggregate Supply in an Open Economy under Fixed Exchange Rates

An increase in the price level leads to a real appreciation and a decrease in output: The aggregate demand curve is downward-sloping. An increase in output leads to an increase in the price level: The aggregate supply curve is upward-sloping.

$$P_t = P_{t-1}(1+\mu) F\left(1 - \frac{Y_t}{L}, z\right) \tag{19.2}$$

The price level depends on the price level last year and on the current level of output. Last year's price level matters because it affects nominal wages this year, which in turn affect prices this year. Higher output matters because it leads to lower unemployment, which leads to higher wages, which lead in turn to higher prices. The aggregate supply curve is drawn as *AS* in Figure 19-3 for a given value of last year's price level. It is upward-sloping: Higher output leads to higher prices.[8]

The short-run equilibrium. The short-run equilibrium is given by the intersection of the aggregate demand curve and the aggregate supply curve, point *A* in Figure 19-3. Assume that at point *A* output happens to be equal to Y_n, the natural level of output. When output is equal to its natural level, the aggregate supply curve does not shift. And in the absence of shocks or changes in policy, the aggregate demand curve does not shift either. Thus, in the absence of shocks or changes in policy, the economy would stay at point *A*.

---

[8]***Digging deeper.*** *There is an important shortcut involved in using equation (19.2) as the aggregate supply relation for an open economy. Note that the price level on the left-hand side of equation (19.2) is the price set by firms, thus the price of domestic goods. The (lagged) price level on the right-hand side stands for the (lagged) consumer price index. It is there because workers care about the (expected) price of the goods they buy, thus about the consumer price index. In the closed economy, we assumed that the price of domestic goods and the consumer price index were the same. In the open economy, we cannot make this assumption; because consumers consume both domestic and foreign goods, the consumer price index depends on both the price of domestic goods and the price of foreign goods.*

To see what this implies, consider the effects of a devaluation that increases the price of foreign goods and thus increases the consumer price index. The next time wages are set, workers—who face an increase in the consumer price index—are likely to ask for an increase in nominal wages. This increase in nominal wages leads firms to increase the price of domestic goods. Thus, a devaluation is likely to lead to an increase in the price of domestic goods, even given the level of output. It would take us too far afield to keep track of the two price indexes. Thus equation (19.2) ignores the difference between them. As a result, it does not allow for a direct effect of devaluations on the price level, but the effect is relevant in the real world, especially in countries where a large proportion of consumption is composed of imports.

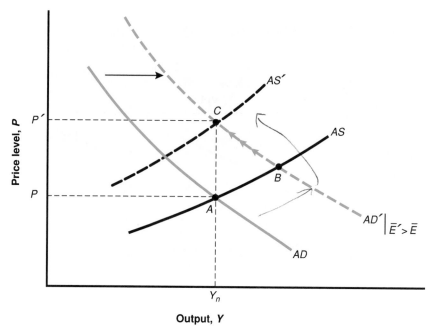

FIGURE 19-4

Dynamic Effects of a Devaluation

A devaluation leads, in the short run, to a real depreciation and an increase in output. Over time, the price level increases, undoing the effects of the devaluation, and output returns to its natural level.

(19.1)
$$Y_t = Y\left[\frac{\bar{E}P^*}{P_t}, G, T\right]$$
$$\qquad\qquad + \quad + \quad -$$

(19.2)
$$P_t = P_{t-1}(1+\mu)F\left(1-\frac{Y_t}{L}, z\right)$$

Now suppose that, starting from point A, the government devalues, so that the exchange rate increases from $\bar{E}$ to $\bar{E}'$ (recall that a devaluation is an increase in the exchange rate). At a given price level, the devaluation translates into a real depreciation—at a given P, the increase in $\bar{E}$ leads to an increase in $\bar{E}P^*/P$, an increase in the real exchange rate (the relative price of foreign goods in terms of domestic goods). Because domestic goods are now cheaper, net exports increase, leading to an increase in the demand for domestic goods and an increase in output. In Figure 19-4, the aggregate demand curve shifts to the right, say from AD to AD', and the equilibrium moves from point A to point B.

Thus, in the short run, a devaluation leads to an increase in output. It also leads to an increase in the price level, but the increase in prices does not cancel the effect of the devaluation on the real exchange rate. That is, P does not increase proportionately as much as $\bar{E}$, so $\bar{E}P^*/P$ increases.[9]

The dynamics. What happens over time? With output now above its natural level, the aggregate supply curve shifts up over time. As the price level increases, the effects of the devaluation on the real exchange rate steadily diminish. The aggregate supply curve keeps shifting up until it reaches AS' and the economy reaches point C. At that point, output is back to its natural level Y_n. And the price level is higher by an amount exactly proportional to the initial devaluation. To see why, look at the aggregate demand equation (19.1). Aggregate demand depends on the real exchange rate. Thus, if output is back to Y_n, the real exchange rate must be back to its initial value as well.

---

[9]*Digging deeper. The easiest way to see this is by contradiction. Suppose that P increased proportionately more than $\bar{E}$. The net result would be a real appreciation. Net exports would decrease, and so would output. But if output decreased, the price level would decrease, in contradiction to our initial assumption.*

In words: A devaluation leads initially to a real depreciation, an increase in net exports, and an increase in output. Over time, however, the increase in prices steadily erodes the effects of the devaluation. Eventually, the increase in prices cancels the effects of the devaluation. The real exchange rate is back to its original value, and output is back at its natural level.

We have therefore another important neutrality result, this time with respect to the nominal exchange rate. Just as with the level of nominal money in a closed economy, *the level of the nominal exchange rate is neutral in the long run. Prices and wages eventually adjust so that the real variables are unaffected in the long run.*

WHAT DEVALUATIONS CAN DO

Does the neutrality result we just saw imply that there is no potential role for nominal exchange rate adjustments? The answer is clearly no. A devaluation can affect output for some time; the neutrality result serves as a warning that policy makers should not count on these effects remaining forever. Let's look at two examples.

Overvaluation and recession. First, take a country with an *overvalued real exchange rate*. Because domestic goods are too expensive relative to foreign goods, the country is running a large trade deficit. And because of low demand for domestic goods, the country is also in a recession—say, at point A in Figure 19-5. Getting back to the natural level of output clearly requires a real depreciation, a decrease in the relative price of domestic goods. The government has two options.

First, it can maintain the existing nominal exchange rate. Over time, the economy will go from point A to point B. Because output is initially below its natural level, the aggregate supply curve will shift down. Prices will decline until the economy is at point B. Domestic goods will steadily become more competitive. At point

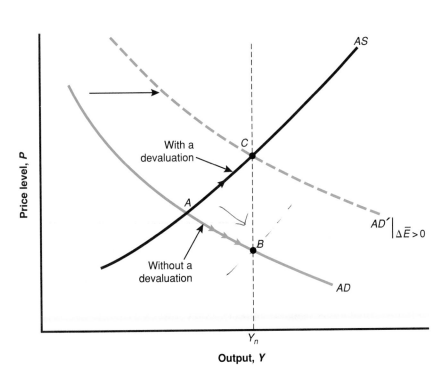

FIGURE 19-5

The Adjustment to an Overvaluation

A country can adjust to an overvaluation either by letting the price level adjust or by adjusting the nominal exchange rate.

B, output will be back at its natural level, and the required real depreciation will have been achieved by a decline in prices.

Second, it can choose to achieve the real depreciation by adjusting the nominal exchange rate. In theory, a devaluation of the right size can shift aggregate demand to the right so as to take the economy from point *A* exactly to point *C,* where output is at its natural level. The real depreciation is the same as in the first option, but has been achieved instead by a devaluation rather than a decline in prices. In reality, the complicated dynamic effects of a devaluation on net exports, and of net exports on output, are such that the devaluation is unlikely to exactly achieve that outcome.[10] But there is no doubt that a devaluation can hasten the return of the economy to its natural output level.

Eliminating a trade deficit without changing output. Now take an economy that is at its natural level of output but has an overvalued real exchange rate and is running a trade deficit. Suppose that the government desires to reduce the trade deficit but has no desire to trigger a recession or a boom. What should it do?

A devaluation *by itself* is surely not the right instrument. Why? The initial real depreciation will improve net exports, but in so doing it will increase the demand for domestic goods and increase output beyond its natural level. The boom in output will lead to an increase in prices over time. As prices increase, the real exchange rate will return to its initial value. As it does, output will return to its initial level, but so will the trade deficit.

Thus, the government must take measures that improve the trade balance *at the natural level of output.* Remember from Section 12-6 that net exports are equal to the excess of private saving plus public saving (which is positive if the government is running a surplus, negative if it is running a deficit) over investment. Thus, to improve the trade balance, the government must take measures that *at the natural level of output* either increase private saving, decrease investment, or increase public saving. Only such changes will lead to a long-run improvement in the trade balance.

Taken alone, however, such measures imply an initial recession. Take, for example, a decrease in government spending aimed at reducing the budget deficit. This decrease will initially shift the aggregate demand curve to the left, leading to a recession. Over time, as prices decline, the real exchange rate will steadily increase until the economy is back at its natural level. When output is back to its natural level, the budget deficit is smaller. And so is the trade deficit, because of the real depreciation. Looking at it another way, government spending is smaller, but the effect on the demand for goods is offset by an increase in net exports, due to the real depreciation.

The government can, however, accomplish the same long-run goal while avoiding the initial recession. If it combines the decrease in government spending with a devaluation, the adverse effects of the decrease in spending on demand can be partly offset by the increase in net exports brought about by a higher exchange rate and thus an improvement in the trade balance. In theory, just the right mix of spending cuts and devaluation can leave the total demand for domestic goods unchanged and achieve an improvement in the trade balance without affecting output. In reality, such a feat is again unlikely. The complex dynamic effects of the devalua-

---

[10]*Remember the J-curve dynamics we saw in Chapter 12.*

$$Y\left[\frac{EP^*}{P}, G, T\right]$$

Trade balance improves.

$$NX = (Y-T-C) + (T-G)$$
P svg   Public svg

Eg 1   ↓G recession → shift in AS back to natural level → ↓Price level → real depn → improved trade balance

Eg 2:

tion are unlikely to exactly match the adverse effects of the decrease in spending. But the basic point remains: The combination of a decrease in the budget deficit and a devaluation can lead to a long-run improvement in the trade balance while reducing the short-run adverse effects on activity.

WHETHER AND WHEN TO DEVALUE

We have seen that, when properly used, a devaluation can help an economy get out of a recession faster. In combination with other measures—in our example, tighter fiscal policy—it can also help improve the trade balance without a recession. This leads to a question that has been much debated in economics: Under what circumstances should governments devalue, even if they are committed to the principle of a fixed exchange rate? Let's look at the arguments on both sides.

The arguments in favour of devaluation. Those who argue in favour of a devaluation make the arguments we have just presented. If a country suffers from overvaluation, why rely on a recession rather than on a much less painful adjustment of the nominal exchange rate? Perhaps the most forceful presentation of this view was made more than 70 years ago by Keynes, who argued against Winston Churchill's decision to return the English pound in 1923 to its pre–World War I parity. His arguments are presented in the Global Macro box entitled "The Return of Britain to the Gold Standard: Keynes versus Churchill." Most economic historians believe that history proved Keynes right, and that overvaluation of the pound was indeed one of the main reasons for Britain's poor economic performance after World War I.

Those in favour of devaluation also argue that not devaluing may not only be costly in terms of unemployment, but may prove ultimately impossible. Their argument is based on the mechanism behind exchange rate crises we discussed in Section 14-5. When a country suffers from overvaluation, foreign exchange markets start anticipating that the government might devalue. As a result, to maintain the parity, the government must increase the domestic interest rate enough to compensate for the risk of devaluation. The result is that the country that resists devaluation faces two costs: The first is the overvaluation itself; the second is very high interest rates. This second cost may be so high that the government has no choice other than to devalue, thus validating financial markets' expectations.

As we saw in Section 14-5, this is exactly what happened in Europe in 1992 and 1993. Financial markets' assessment that many countries would soon be forced to devalue led to very high interest rates in the countries under suspicion and to a series of devaluations. Some countries, most notably France, maintained their parity. Others, including the United Kingdom and Italy, dropped out of the EMS and experienced large depreciations; so far the evidence is that they have indeed done better, in terms of net exports and output, than their partners have.

This scenario also describes what happened in Mexico in 1994—see the Global Macro box entitled "The Mexican Crisis of 1994." Financial markets' belief that the peso had become overvalued led to a major exchange rate crisis.

The arguments against devaluation. Those who oppose devaluations make three main arguments.

(1) They point out that, even without a devaluation, the economy will adjust, through the adjustment of prices rather than the nominal exchange rate.

THE RETURN OF BRITAIN TO THE GOLD STANDARD: KEYNES VERSUS CHURCHILL

In 1925, Britain decided to return to the **gold standard.** The gold standard was a system in which each country fixed the price of its currency in terms of gold and stood ready to exchange gold for currency at the stated parity. This system in turn implied fixed nominal exchange rates between countries.

The gold standard had been in place from 1870 until World War I. Because of the need to finance the war, and to do so in part by money creation, Britain suspended the gold standard in 1914. In 1925, Winston Churchill, then Britain's Chancellor of the Exchequer (the English equivalent of the Minister of Finance in Canada), decided to return to the gold standard, and to do so at the prewar parity—that is, at the prewar value of the pound in terms of gold. But because prices had increased faster in Britain than in many of its trading partners, returning to the prewar parity implied a large real appreciation: At the same nominal exchange rate as before the war, British goods were now relatively more expensive than foreign goods.

Keynes severely criticized the decision to return to the prewar parity. In *The Economic Consequences of Mr. Churchill,* a book he published in 1925, Keynes argued as follows. If Britain was going to return to the gold standard, it should have done so at a higher price of gold in terms of currency, at a nominal exchange rate higher than the prewar nominal exchange rate. In a newspaper article, he articulated his views as follows:

> There remains, however, the objection to which I have never ceased to attach importance, against the return to gold in actual present conditions, in view of the possible consequences on the state of trade and employ-

ment. I believe that our price level is too high, if it is converted to gold at the par of exchange, in relation to gold prices elsewhere; and if we consider the prices of those articles only which are not the subject of international trade, and of services, i.e. wages, we shall find that these are materially too high—not less than 5 per cent, and probably 10 per cent. Thus, unless the situation is saved by a rise of prices elsewhere, the Chancellor is committing us to a policy of forcing down money wages by perhaps 2 shillings in the Pound.

I do not believe that this can be achieved without the gravest danger to industrial profits and industrial peace. I would much rather leave the gold value of our currency where it was some months ago than embark on a struggle with every trade union in the country to reduce money wages. It seems wiser and simpler and saner to leave the currency to find its own level for some time longer rather than force a situation where employers are faced with the alternative of closing down or of lowering wages, cost what the struggle may.

For this reason, I remain of the opinion that the Chancellor of the Exchequer has done an ill-judged thing—ill judged because we are running the risk for no adequate reward if all goes well.[*]

Keynes's prediction turned out to be right. While other countries were growing, Britain was in recession for the rest of the decade. Most economic historians attribute a good part of the blame to the initial overvaluation.

[*]From The Nation, *May 2, 1925.*

(2) They argue that, by their nature, devaluations defeat the purpose of fixed exchange rates. One of the main arguments for fixed exchange rates is that they allow firms to plan production and sales without having to worry about changes in the nominal exchange rate. Devaluations reintroduce that risk.

(3) They argue that, while a devaluation may help in a particular instance, it may hurt in the long run. They point out that if a government develops the reputation of being willing to devalue, then, whenever there are signs of overvaluation, financial markets will expect a devaluation and require higher interest rates. In contrast, if a government is known to oppose devaluations, interest rates will increase less, and

In the second half of the 1980s, Mexico embarked on both macroeconomic stabilization and economic reform. One of the elements of the program was the reduction of inflation. After a successful reduction of the inflation rate from 159% in 1987 to about 20% in 1991, the Mexican government decided to maintain a roughly constant exchange rate of the peso vis-à-vis the U.S. dollar. This decision proved to be one of the causes of the peso crisis of December 1994.

While the nominal exchange rate vis-à-vis the U.S. dollar was approximately constant from 1990 on, inflation in Mexico remained substantially higher than in the United States. The result was a substantial real appreciation. This is shown in Table 1, which gives the nominal and the real exchange rates between Mexico and the United States for 1990–1994. By 1994, Mexican goods were 22% more expensive relative to U.S. goods than they had been five years earlier. The effect of this real appreciation was, unsurprisingly, a large trade deficit. As Table 1 shows, by 1994 the Mexican trade deficit was equal to 7.2% of GDP.

A middle-income country such as Mexico would typically have problems financing such a large trade deficit. The assumption of perfect capital mobility—which implies that as long as the domestic interest rate is equal to the foreign interest rate, capital flows will be large enough to finance any trade deficit—is never completely right, even for rich countries. It is even less so for middle-income countries, where attracting capital flows sufficient to finance a large trade deficit may require offering very high interest rates.[*]

But attracting capital flows was not a problem for Mexico, at least until 1994. The reason is that Mexican economic reform was widely considered a success, the Mexican stock market was booming, and foreign investors were eager to invest in Mexico. Thus, there was no strong pressure from the foreign exchange market to reduce the trade deficit.

During 1994, however, it became increasingly clear to many economists and to foreign investors that peso overvaluation was becoming a serious issue, and that a devaluation could probably not be avoided. By December, fear of a devaluation led to large capital outflows. Mexico tried to maintain the parity by offering high interest rates. But it was too late. The peso had to be devalued by 50% in December 1994. One year later, in December 1995, the peso stood at 7.75 pesos to the U.S. dollar, compared with 3.45 pesos in November 1994. The reason why the depreciation was so large is that many foreign investors decided to get out of Mexico altogether. Those who stayed required very high interest rates. Short-term nominal interest rates from January to December 1995 averaged 50%; despite rising inflation, these implied high real interest rates as well. It is estimated that Mexico's GDP declined in 1995 by nearly 10%, with the effects of high interest rates dominating the effects of the peso devaluation.

Fortunately, decisive action by the Mexican policy makers led to a quick recovery, and within two years the Mexican economy was back on track.

[*]See the appendix on imperfect capital mobility in Chapter 13.

TABLE 1

NOMINAL AND REAL EXCHANGE RATES BETWEEN MEXICO AND THE UNITED STATES, 1990–1994

	1990	1991	1992	1993	1994
Pesos per dollar, E	2.81	3.01	3.09	3.11	3.37
U.S. price level, P^*	100	100.2	100.8	102.3	103.6
Mexican price level, P	100	120.5	136.7	148.8	158.9
Real exchange rate, EP^*/P	100	89	81	76	78

The U.S. and Mexican price levels are producer price indexes, equal to 100 in 1990. The real exchange rate is also normalized to 100 in 1990. A minus sign in the last line denotes a trade deficit.

Source: International Financial Statistics, International Monetary Fund.

the cost of not devaluing will be accordingly smaller.[11] This argument extends beyond foreign exchange markets. Under fixed exchange rates, unions that ask for too high a wage, or firms that choose too high a price, know that they will become uncompetitive. Union members may find themselves unemployed, and firms may lose markets. Knowing these implications, unions may be more careful in their wage demands, and firms may be more restrained in their price setting. These discipline effects may be lost if unions and firms expect the government to bail them out through a devaluation.

In short, economists who oppose devaluations typically do so for the same reason that they favour fixed exchange rates in the first place. If the government is known to be willing to rely on devaluations, they argue, most of the advantages of fixed exchange rates will be lost.

Summary. There are two main lessons to be drawn from this discussion.

First, when there are strong advantages for a group of countries to operate under fixed exchange rates, it may be better for these countries to adopt a common currency. True, a common currency eliminates the possibility of using a devaluation to eliminate an overvaluation. But, by eliminating the possibility of a devaluation, it also eliminates the danger that overvaluation may lead to high interest rates and an exchange rate crisis. Most economists agree, for example, that the common currency area composed of the 50 states of the United States is indeed justified. The advantages of all 50 states using the same currency are obvious. And these advantages probably more than compensate for the fact that states that suffer from an adverse shift in demand cannot rely on changes in the exchange rate to return to equilibrium.[12] A similar argument is made by the proponents of a common currency in Europe, although the case is less strong than for the states of the United States.[13] As we noted earlier, Quebec nationalists plan to keep using the Canadian dollar in the event of separation.

Second, when a common currency is not an option, trying to correct a large overvaluation through price adjustment rather than through a devaluation is likely to be unwise. The reason is as much the unemployment cost of the adjustment as its feasibility: Trying to maintain the parity in the face of a large overvaluation may require prohibitively high interest rates. What a "large overvaluation" means, however, is likely to remain a matter of disagreement. Thus, when and whether devaluations should be used is not a question with a simple solution.

---

[11] *This argument is a special case of a more general one, which states that policies that appear to be best in the short run may still be undesirable because of their long-term implications. We examine this argument at length in Chapter 27.*

[12] *The evidence is that much of the adjustment takes place through the movement of workers. U.S. states that suffer from a decrease in the demand for their goods typically lose workers to other states until unemployment has returned to normal. For more evidence, see Olivier Blanchard and Lawrence Katz, "Regional Evolutions,"* Brookings Papers on Economic Activity, *1992:1, 1–75.*

[13] *For more discussion, see the Further Reading listed at the end of Chapter 14.*

SUMMARY

Money Growth, Inflation, and Interest Rates

■ An increase in nominal money growth decreases nominal interest rates in the short run, but increases them in the long run.

■ The Fisher hypothesis is the proposition that an increase in inflation leads in the long run to an equal increase in the nominal interest rate, so that the real interest rate remains constant.

■ Under the Fisher hypothesis, an increase in nominal money growth—which is reflected one for one in an increase in inflation—has no effect on the real interest rate in the long run and increases the nominal interest rate by an amount equal to the increase in money growth.

■ Evidence across countries and across long periods of time suggests that the Fisher hypothesis may be a good approximation to reality, but that the long run may be very long indeed.

Fixed Exchange Rates and Devaluations

■ A devaluation leads to a real depreciation in the short run but has no effect on the real exchange rate in the long run.

■ Starting from a situation where the currency is overvalued, a devaluation may help eliminate the overvaluation at a smaller unemployment cost than the cost implied by simply waiting for the price level to adjust. Starting from a trade deficit, a reduction in the budget deficit combined with a devaluation can achieve a permanent decrease in the trade deficit while avoiding a recession.

■ The main argument for devaluations is that they can help an economy adjust faster and at a lower unemployment cost. The main argument against devaluations is that they defeat the purpose of adopting a fixed exchange regime in the first place.

■ Governments may in fact suffer from having the option to devalue. The anticipation that a devaluation might occur may lead financial markets to require higher interest rates, compounding the effects of an overvaluation with high interest rates and possibly forcing a devaluation, even if the government initially had no intention of devaluing.

KEY TERMS

■ Fisher effect, or Fisher hypothesis, 369
■ money illusion, 373

■ gold standard, 382

QUESTIONS AND PROBLEMS

1. *Exam* "Macroeconomists are confused about the impact of reducing money growth on interest rates. Sometimes, they say it will raise interest rates; sometimes, they say it will lower them; and sometimes they say that interest rates won't change at all." Comment.

2. In this chapter we assumed that the normal growth rate of output is zero, and found that the nominal interest rate should equal the real interest rate plus the growth rate of the money supply.

   a. How does this relationship change if we assume a more realistic normal growth rate of output of 3% per year?

   b. Does this more realistic assumption change the nature of the Fisher effect—that is, do we still expect a 1% rise in the money growth rate to cause a 1% rise in the nominal interest rate in the long run? Why or why not?

3. Look at Figure 1 in the Global Macro box on page 372 and notice that the line drawn through the scatter of points does *not* go through the origin. Does the "Fisher effect" suggest that it *should* go through the origin? Explain.

4. Suppose that starting from an initial equilibrium at the natural level of output, a country *revalues* its currency (that is, it *decreases* its nominal exchange rate and makes its currency *more* expensive in terms of foreign currency).

   a. Draw an *AD-AS* diagram illustrating the short-run impact of this policy.

   b. In the short run, what happens to the real exchange rate, net exports, and output?

   c. After all adjustments have taken place, indicate on your diagram the final long-run equilibrium of the economy.

d. In the long run, what happens to the real exchange rate, net exports, and output (compared with the initial equilibrium)?

5. Suppose that a country switches from floating to fixed exchange rates. If everyone believes that the government is committed to maintaining the exchange rate, would you expect the natural rate of unemployment to rise or fall? Why?

6. Sketch a graph showing the effects explained under the heading "Eliminating a trade deficit without changing output" on page 380.

7. The chapter discusses the short-run impact of an increase in money growth and *no* change in expected inflation. How would the analysis change if an increase in money growth increases the expected inflation rate in the short run?

8. Suppose the central bank announced an immediate and permanent increase in nominal money growth. What would be the likely impact on the yield curve?

FURTHER READING

More on the gold standard and the history of exchange rate arrangements

Paul Krugman and Maurice Obstfeld, *International Economics, Theory and Policy*, 4th ed. (New York: HarperCollins, 1996).

More on Mexico

For a general description of Mexican economic policy since the mid-1980s, read Pedro Aspe, *Economic Transformation the Mexican Way* (Cambridge, MA: MIT Press, 1993).

For a study of the macroeconomic problems of Mexico leading to the 1994 crisis, read Rudiger Dornbusch and Alejandro Werner, "Mexico: Stabilization, Reform and No Growth," *Brookings Papers on Economic Activity,* 1994:1, 253–315.

For an assessment of the Mexican crisis, read Lawrence Summers, "Ten Lessons to Learn," *The Economist,* December 23, 1995, 46–48.

For an (early) comparison of the Mexican crisis of 1994 with the Asian crisis of 1997, see *World Economic Outlook: Interim Assessment* (Washington, DC: IMF, December 1997), especially Section III.

20

PATHOLOGIES I: HIGH UNEMPLOYMENT

Most of the time, economies go through relatively mild economic fluctuations. But once in a while things go very wrong. Unemployment remains very high for a very long time. Inflation increases to high levels, sometimes extraordinarily high levels, until drastic policy measures lower it.

These episodes raise an obvious question. Are they the result of unusually large shocks? Or are they the result of unusual and perverse propagation mechanisms that enormously amplify the effects of the initial shocks?[1] This question is the topic of this chapter and the next. In this chapter we look at two episodes of high unemployment: (1) the Great Depression and (2) high unemployment in Western Europe today. In Chapter 21, we examine episodes of high inflation.

20-1 THE GREAT DEPRESSION

No matter how often we look at it, the **Great Depression** of the 1930s inspires both awe and fear. From 1929 to 1933, real GDP in Canada fell by almost 30%, and output per capita did not recover until well past the onset of World War II. Over the same period, corporate profits fell from $400 million to *negative* $100 million. The

---

[1]*See Section 16-7 for a discussion of shocks and propagation mechanisms.*

decade saw double-digit unemployment rates in all but perhaps the last years, peaking at over 20% in 1933. The Great Depression was worldwide. (While there is no agreed-upon precise definition, economists use the word **depression** to describe a deep and long-lasting recession.) The average unemployment rate from 1930 to 1938 was 13.3% in Canada, 16.5% in the United States, 15.4% in the United Kingdom, 10.2% in France, and 21.2% in Germany.[2] Being unemployed is never pleasant, but the lot of those unemployed in the 1930s appears to have been exceptionally mean.

Such a cataclysmic event should inspire great study.[3] You may be surprised to learn that the flow of research on the Great Depression in Canada is a very modest trickle. One thing seems certain, however. Canada did not create the Great Depression. We imported it. The 1930s stands as a potent reminder that to understand economic developments in Canada we often must look elsewhere.

Soon after World War I, the United States replaced the United Kingdom as both our largest export market and our largest source of investment finance. While the United Kingdom still played an important role in Canadian economic life at the time, if we want to understand the Great Depression in Canada we must turn our gaze south. Table 20-1 gives the evolution of the unemployment rate, the growth rate of output, the price level, and the money supply, for both Canada and the United States, from 1929 to 1942. Although there are some interesting differences,

TABLE 20-1

CANADIAN AND U.S. UNEMPLOYMENT, OUTPUT GROWTH, PRICES, AND MONEY, 1929–1942

YEAR	UNEMPLOYMENT RATE (%) CANADA	U.S.	OUTPUT GROWTH RATE (%) CANADA	U.S.	PRICE LEVEL CANADA	U.S.	NOMINAL MONEY SUPPLY CANADA C$ millions	U.S. US$ billions
1929	2.9	3.2	0.9	−9.8	100.0	100.0	787.8	26.4
1930	9.1	8.7	−3.3	−7.6	96.7	97.4	722.0	25.4
1931	11.6	15.9	−11.2	−14.7	90.1	88.8	683.6	23.6
1932	17.6	23.6	−9.3	−1.8	81.8	79.7	605.8	19.4
1933	19.3	24.9	−7.2	9.1	80.2	75.6	603.6	21.5
1934	14.5	21.7	10.4	9.9	81.8	78.1	633.3	25.5
1935	14.2	20.1	7.2	13.9	82.6	80.1	704.2	29.2
1936	12.8	16.9	4.6	5.3	85.1	80.9	757.3	30.3
1937	9.1	14.3	8.8	−5.0	87.6	83.8	851.1	30.0
1938	11.4	19.0	1.4	8.6	86.8	82.2	857.0	30.0
1939	11.4	17.2	7.5	8.5	86.8	81.0	929.6	33.6
1940	9.2	14.6	13.3	16.1	90.9	81.8	1146.6	39.6
1941	4.4	9.9	13.3	12.9	97.5	85.9	1453.6	46.5
1942	3.0	4.7	17.6	13.2	102.5	95.1	1844.9	55.3

Sources:
For Canada: Unemployment rate: Statistics Canada, CANSIM Series D31253 and D31254; output growth, GDP growth: Statistics Canada, CANSIM Series D14442; price level, GDP deflator (1929=100): Statistics Canada, CANSIM Series D14476; money supply, M1: C. Metcalf, A. Redish, R. Shearer, "New Estimates of the Canadian Money Stock, 1871–1967," *Canadian Journal of Economics*, 1998.
For U.S.: Unemployment rate: Historical Statistics of the United States, U.S. Department of Commerce, Series D85-86; output growth, GNP growth: Series F31; price level, CPI (1929=100): Series E135; money stock, M1: Series X414; *Historical Statistics of the United States*, U.S. Department of Commerce

[2]*A warning: The quality of the unemployment statistics is much lower for the pre-World War II period than it is for post-World War II numbers. Comparisons across countries are particularly perilous.*

[3]*Indeed, many of the most influential postwar economists were products of that horrible decade who were inspired to understand its causes and to avoid its repetition.*

the most striking impression that emerges from Table 20-1 is the similarity in the experiences of the two countries. Output fell dramatically from the peak, and by 1933 the rate of unemployment had increased sevenfold. Both economies grew quickly after 1933, but unemployment did not return to 1929 levels until well after each of the two countries had entered World War II. There is no contradiction here, just an application of Okun's law: A long period of high growth was needed to steadily decrease the high unemployment rate.[4]

In what follows, we shall first examine the main sources of the Great Depression in the United States. We shall see that the episode is well explained by a sharp decline in aggregate demand coming first from a sharp decline in spending that was then aggravated by a steep fall in the money supply. We will then contrast the U.S. experience with that of Canada and discuss how the downturn in the United States was transmitted to Canada and the rest of the world.

THE FALL IN SPENDING IN THE UNITED STATES

Popular accounts often say that the Great Depression was caused by the stock market crash of 1929. This is an overstatement. A recession had started before the crash in the United States, and other factors played an important role later in the Depression.

Nevertheless, the crash was important. The stock market had boomed from 1921 to 1929. Stock market prices had increased much faster than the dividends paid by firms, and as a result the dividend/price ratio had decreased from 6.5% in 1921 to 3.5% in 1929. On October 28, 1929, the stock market price index dropped from 298 to 260. The next day, it dropped further to 230. This was a fall of 23% in just two days, and a drop of 40% from the peak of early September. By November the index was down to 198. A stock market recovery in early 1930 was followed by a steady decline as the size of the depression became increasingly clear.

Was the October crash caused by the sudden realization that a depression was coming? The answer is no. There is no evidence of major news in October. The source of the crash was almost surely the end of a speculative bubble. Stockholders who had purchased stocks at high prices in the anticipation of further increases in prices got scared and attempted to sell their stocks, leading to further large price declines.[5]

The effects of the crash were to decrease consumers' wealth, to increase uncertainty, and to decrease demand. Unsettled by the crash and now uncertain about the future, consumers and firms decided to see how things evolved and to postpone purchases of durable goods and investment goods. For example, there was a large decrease in car sales—the type of purchase that can easily be deferred—in the months just following the crash.[6] In terms of the *IS-LM* model in Figure 20-1, the effect of the crash was to shift the *IS* curve to the left, from *IS* to *IS'*, leading to a sharp decrease in output from Y to Y'. (Note that we have the real interest rate r rather than the nominal interest rate i on the vertical axis in Figure 20-1. We shall discuss why a bit later in this chapter.) Industrial production, which had declined by

---

[4]*For a review of Okun's law, see Section 18-1.*

[5]*See the discussion of bubbles and crashes in Section 9-3.*

[6]*The argument that increased uncertainty was important is developed by Christina Romer in "The Great Crash and the Onset of the Great Depression,"* Quarterly Journal of Economics, *August 1990, 105, 597–624.*

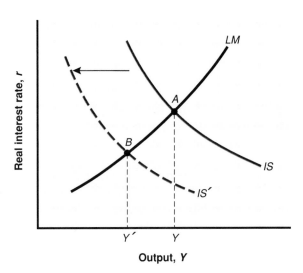

FIGURE 20-1

The Great Depression and the *IS-LM: I*
The effect of the stock market crash was to decrease wealth and increase uncertainty, leading the *IS* curve to shift to the left.

1.8% from August to October 1929, declined by 9.8% from October to December, and by another 24% from December 1929 to December 1930.

THE CONTRACTION IN THE U.S. NOMINAL MONEY

When we described the effects of a decrease in aggregate demand in Chapter 16, we showed how the economy would, after the initial recession, eventually return to the natural level of output. The mechanism was the following: Low output led to high unemployment. High unemployment led to lower wages. Lower wages led to lower prices. Lower prices led to an increase in the real money supply, steadily shifting the *LM* curve down until output returned to its natural level.

You can see from Table 20-1 that low output did indeed lead to lower prices in the years following 1929. The consumer price index decreased from 100.0 in 1929 to 75.6 in 1933, a 24% decrease in the price level in four years. But the rest of the mechanism failed, for two reasons. First, there was a large decrease in nominal money, leaving real money roughly unchanged. Second, the deflation itself had perverse effects, leading to a deeper decline in output.

Let's focus on the decrease in nominal money first. Table 20-2 gives the evolution of nominal money—both the money stock (measured by *M1*) and the monetary base (*H*)—from 1929 to 1933. It also shows the evolution of the money multiplier (*M1/H*) and of the real money stock (*M1/P*), where *P* is the consumer price index. From 1929 to 1933, nominal money, *M1*, *decreased* by 27%. To understand what happened, let's recall the basic mechanics of money creation.[7]

Nominal money, *M1* (the sum of currency in circulation and chequable deposits), is equal to the monetary base, *H* (currency plus banks' reserves), which is under the central bank's control, times the money multiplier:

$$M1 = H \times \text{money multiplier}$$

The money multiplier in turn depends both on how much reserves banks keep in proportion to their deposits and what proportion of money people keep in the form of currency as opposed to demand deposits.

---

[7]*See Section 5-3 for a review.*

U.S. MONEY SUPPLY, NOMINAL AND REAL, 1929–1933

YEAR	M1	H	$\dfrac{M1}{H}$	$\dfrac{M1}{P}$
1929	26.4	7.1	3.7	26.4
1930	25.4	6.9	3.7	26.0
1931	23.6	7.3	3.2	26.5
1932	20.6	7.8	2.6	25.8
1933	19.4	8.2	2.4	25.6

Source: M1: Series X414; H: Series X422 plus series X423; P: Series E135. *Historical Statistics of the United States,* U.S. Department of Commerce.

Note that from 1929 to 1933 the monetary base, *H,* increased from $7.1 to $8.2 billion. Thus the decrease in *M*1 did not come from a decrease in the monetary base. Rather, it came from a decrease in the money multiplier, *M*1/*H,* which went from 3.7 in 1929 to 2.4 in 1933.

This decrease in the multiplier was the result of bank failures.[8] With the large decline in output, more and more borrowers found themselves unable to repay their loans to banks, and more and more banks became insolvent and closed down. Bank failures increased steadily from 1929 until 1933, when the number of failures reached a peak of 4000, out of about 20 000 banks in operation at the time.

Bank failures had a direct effect on the money supply; chequable deposits at the failed banks became worthless. But the major effect on the money supply was an indirect one. Worried that their banks might also fail, many people shifted from deposits to currency. The increase in the ratio of currency to deposits led to a decrease in the money multiplier, and thus to a decrease in the money supply.

Think of the mechanism this way. If people had liquidated *all* their deposits and asked banks for currency in exchange, the multiplier would have decreased to one, and *M*1 would have been just equal to the monetary base. The shift was less dramatic than that. Nevertheless, the multiplier went down from 3.7 in 1929 to 2.4 in 1933, leading to a decrease in the money supply despite an increase in the monetary base. (Some economists have argued that the shift from chequable deposits to currency had implications that went beyond the effect on the money multiplier. Their argument is presented in the Focus box titled "Money versus Bank Credit.")

With a decrease in the nominal money supply from 1929 to 1933 roughly proportional to the decrease in prices, the real money supply remained roughly constant, eliminating one of the mechanisms that could have led to a recovery. In other words, the *LM* curve remained roughly unchanged—it did not shift down as it would have if the real money supply had increased. Indeed, Milton Friedman and Anna Schwartz have argued that the Fed was responsible for the depth of the Great Depression, that it should have expanded the monetary base even more than it did to offset the decrease in the money multiplier.

---

[8]*The classic treatment here is by Milton Friedman and Anna Schwartz,* A Monetary History of the United States, 1867–1960 *(Princeton, NJ: Princeton University Press, 1963).*

We have focussed in this text on the effects of the shift from chequable deposits to currency on the money multiplier. A number of economists have argued that this shift had implications beyond those for the money multiplier. Faced with a decrease in deposits, banks had to call existing loans. Those who had borrowed from the banks were unable to find other sources of borrowing, and this was a further source of output contraction.

The argument has been developed by Ben Bernanke, from Princeton University.[*] He starts from the observation that banks play a special role in credit markets. Banks make loans to borrowers who are typically too small, or not known well enough, to be able to issue bonds. Before making a loan to a firm, a bank acquires knowledge about the firm, and once the loan has been made, monitors the firm's decisions closely. If, for whatever reason, a bank decides to decrease its volume of loans, those who lose their loans cannot borrow elsewhere. Other banks do not have the first bank's specialized knowledge, and neither do bond markets. Thus, the borrower may be forced to cancel investment plans, curtail production, or close altogether. [Note that, in writing the investment equation as $I = I(i, Y)$ so far in this book, we have implicitly assumed that firms could borrow as much as they wanted at the given interest rate i; we have therefore implicitly excluded the effect we are looking at now.]

By putting together many pieces of evidence, Bernanke builds a strong case that the credit channel was indeed important in first deepening, and then prolonging, the Great Depression. One citation from a large survey of firms, carried out in 1934–1935, puts it clearly: "[We find that there is] a genuine unsatisfied demand for credit by solvent borrowers, many of whom could make economically sound use of capital. The total amount of this unsatisfied demand for credit is a significant factor, among many others, in retarding business recovery."

Reference

For more on the role of banks and the credit channel of monetary policy, read Anil Kashyap and Jeremy Stein, "Monetary Policy and Bank Lending," in *Monetary Policy*, N. Gregory Mankiw, ed. (Chicago: University of Chicago Press and NBER, 1994), 221–262.

[*]*Ben Bernanke, "Nonmonetary Effects of the Financial Crisis in the Propagation of the Great Depression,"* American Economic Review, *1983, 257–276.*

THE ADVERSE EFFECTS OF DEFLATION

In addition to the decrease in nominal money, deflation (the decrease in the price level) itself was a further source of decline in output during the Great Depression. Building on Section 7-3, we can use the *IS-LM* diagram to see why, keeping in mind the distinction between the real interest rate and the nominal interest rate.

In Figure 20-2, think of the pre-crash equilibrium as given by point *A*. We argued earlier that the effect of the crash was, through its effects on wealth and uncertainty, to shift the *IS* curve to *IS'*, taking the economy from point *A* to point *B*. We then argued that, from 1929 to 1933, the combined effects of the decrease in prices and the decrease in nominal money left the *LM* curve roughly unchanged.

Now consider the effects of deflation on the difference between nominal and real interest rates. Recall that the real interest rate is equal to the nominal interest rate minus the expected rate of inflation (equivalently, plus the expected rate of deflation). By 1931, the rate of deflation exceeded 10% a year, and by then the evidence is that people expected deflation to continue. This expectation implied that even low nominal rates—in 1931, the U.S. three-month Treasury-bill rate stood at

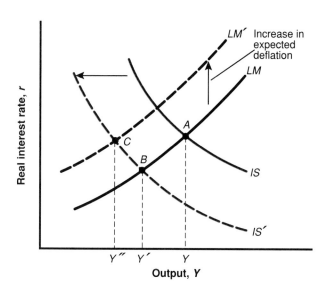

FIGURE 20-2

The Great Depression and the *IS-LM: II*

The effect of expected deflation was to shift the *LM* curve up, leading to a further decline in output.

only 1.4%—implied high real interest rates. When expected deflation is 10%, even a nominal rate of 0% implies a real interest rate of 0% − (−10%) = 10%!

In terms of Figure 20-2, the effect of the increase in expected deflation was to shift the *LM* curve up by an amount equal to the increase in expected deflation. At a given level of income, the nominal interest rate consistent with financial markets was unchanged; thus the real interest rate was higher by an amount equal to the increase in expected deflation. The result of this shift from *LM* to *LM′* was to move the economy from point *B* to point *C,* and to decrease output further and deepen the depression.

THE U.S. RECOVERY

The recovery started in 1933. Except for another sharp decrease in the growth rate of output in 1937 (see Table 20-1), growth was consistently high, running at an average of 7.7% for the period from 1933 to 1941. Macroeconomists and economic historians have studied the recovery much less than they studied the initial decline. And many questions remain.

One of the factors that contributed to the recovery is clear. Following the election of Franklin Roosevelt in 1932, there was a dramatic increase in nominal money growth. From 1933 to 1936, nominal money increased by 50%, and the real money stock increased by 40%. From 1933 to 1941, nominal money increased by 140%, real money by 100%. These increases were due to increases in the monetary base, not in the money multiplier. In a controversial recent article, Christina Romer has argued that if monetary policy had been unchanged from 1933 on, output would have been 25% lower than it actually was in 1937, and 50% lower than it was in 1942.[9] These are very large numbers indeed. Even if one believes that these numbers overestimate the effect of monetary policy, the conclusion that money played an important role in the recovery is still surely warranted.

---

[9]*Christina Romer, "What Ended the Great Depression?"* Journal of Economic History, *December 1992, 757–784.*

The role of other factors, from budget deficits to the **New Deal**—the set of programs put in place by the Roosevelt administration to get the U.S. economy out of the Great Depression—is much less clear.

One New Deal program was aimed at improving the functioning of banks by creating the *Federal Deposit Insurance Corporation* (*FDIC*) to insure demand deposits, and thus to avoid bank runs and bank failures. And, indeed, there were few bank failures after 1933.

THE GREAT DEPRESSION IN CANADA

Canada entered the Great Depression at almost the same time as the United States. How was the downturn in the United States transmitted so quickly? What were the differences and similarities in monetary and fiscal responses to the downturn in the two countries?

The western parts of Canada and the United States suffered a drought in 1929. The effects were felt in both countries, but played a larger role in explaining the Canadian downturn. Wheat production peaked at 567 million bushels in 1928 and then, largely because of the drought, fell to 302 million bushels in 1929. Even so, wheat and wheat flour still managed to account for 35% of Canada's exports in 1929. Prices in 1929 did not reflect the downturn, but by 1930 wheat prices had fallen in half and stayed low until the end of the decade. What had been a *quantity* shock in 1929 became an adverse *terms of trade* shock in 1930 as the Great Depression spread to Europe. Not only wheat was affected: About 80% of Canada's exports in 1929 were based on natural resources or agricultural products. The prices of these primary commodities were particularly hard hit by the worldwide slump.

In 1928, 38% of Canada's exports went to the United States. It is not surprising that the sharp drop in income in the United States was transmitted to Canada. The decline in exports could account for a leftward shift of the *IS* curve in Canada, just as shown in Figure 20-1. But other factors were also at work in reducing aggregate demand. Canada had a large external debt in 1928, 70% of which was held by U.S. residents. Direct foreign ownership of Canadian firms was extremely high. The same forces that made American entrepreneurs reluctant to invest in their U.S. enterprises after the stock market crash of 1929 also reduced investment in Canada. Nor could Canadian entrepreneurs remain unaffected by what was going on around them. From 1929 to 1933, gross domestic investment fell from 22.1% to 8.6% of GDP, while real GDP itself was falling. Consumers in Canada reacted much like their American cousins. Automobile manufacturing had been one of the important sources of growth in the 1920s, but this changed with the onset of the Depression. Automobiles were viewed by many as a luxury and were either kept longer or not replaced. By 1932, the number of automobiles registered in Canada actually declined.

The decline in the money supply in Canada was not as dramatic as in the United States. Canada had early on adopted a branch banking system, so there were only 31 active chartered banks in 1900. Canada's banking system underwent a number of crises; failures and mergers reduced that number to 11 by 1929, with the largest three controlling over 70% of chartered bank assets. Although there was further consolidation, bank failures played no role in reducing the money supply.

The last chartered bank to fail in Canada was the Home Bank that closed its doors in 1923. Not a single bank failed in Canada during the Great Depression.[10] Another important difference to keep in mind is that Canada had no central bank until 1935. The Department of Finance had a monopoly on notes in amounts under five dollars, but private banks also issued their own notes. Rather than keep deposits at a central bank as reserves, the Canadian banks kept deposits in large New York banks. So, in Canada, not only was the multiplier determined by the private sector, so was the monetary base!

You might ask: What pinned down the level of the money supply? From the late nineteenth century until World War I, Canada along with the United States and most of the countries of Europe was on the gold standard. Among other things, the gold standard was a fixed-exchange-rate system. Each country fixed the value of its currency in terms of gold—that is, it stood ready to exchange its currency for gold on demand. By arbitrage, this determined the relative values of any two currencies. The gold standard also provided a limit to the growth of the domestic money supply. Banks in Canada were allowed to issue notes, but these had to be redeemable in gold or other legal tender such as Dominion of Canada notes (the ability to redeem bank notes in gold was called *convertibility*). Canada had returned to the gold standard in 1926 but convertibility was suspended in 1928. Nonetheless, many market participants anticipated a return to convertibility, and this appears to have restrained private note issue. This expectation changed rather abruptly when the United Kingdom formally abandoned the gold standard in 1931 and Canada quickly followed suit. For the next four years, the chartered banks were literally licensed to print money. But they did not.[11]

The gold standard is widely viewed as the mechanism by which the Great Depression in the United States was spread throughout the world. The decline in the U.S. money supply and the falling prices that ensued put pressure on other countries to deflate as well in order to maintain convertibility. The abandonment of the gold standard allowed currencies to float and to insulate their economies from the U.S. downturn. The Canadian dollar depreciated vis-à-vis the American dollar after 1931 and this helped to speed up our recovery.

Canada's fiscal response to the Great Depression was also very passive. Although towns and the provinces greatly increased their relief payments, these levels of government were constrained by their constitutional restriction to raise revenue only through direct taxes. The federal government had access to larger sources of revenue, but for the most part tried to balance its budget and allow market forces to restore output and employment. By 1935, Prime Minister Bennett seemed open to some radical changes and proposed his own "New Deal" for Canada. But he lost the election to Mackenzie King. Just as Canada had imported the Great Depression, it waited for foreign forces to pull it out.

---

[10]*The introduction of deposit insurance in the 1960s, and changes in the rules governing trust companies, brought in a number of new trust companies in the 1970s. A few of these companies did go bankrupt.*

[11]*One of the more interesting stories of the Great Depression in Canada was the Social Credit Party. Among other things, they proposed simply printing money and giving it to private citizens as a way out of the Depression.*

20-2 UNEMPLOYMENT IN EUROPE

Although far smaller than it was during the Great Depression, European unemployment is nevertheless very high. In 1997, the average unemployment rate in the European Union (EU) stood at about 11%.

Until 1970, the unemployment rate had been much lower in Europe than in Canada or the United States. But Figure 20-3, which gives the average unemployment rates in these three regions since 1970, shows that European unemployment began increasing steadily in the 1970s.[12] By 1981, EU and North American unemployment rates were roughly similar, standing at around 7%. In all three regions there was a further large increase in the early 1980s. But from 1982 on, as the U.S. rate decreased steadily, the Canadian and European rates remained very high. A decrease in European unemployment in the late 1980s was reversed by the reces-

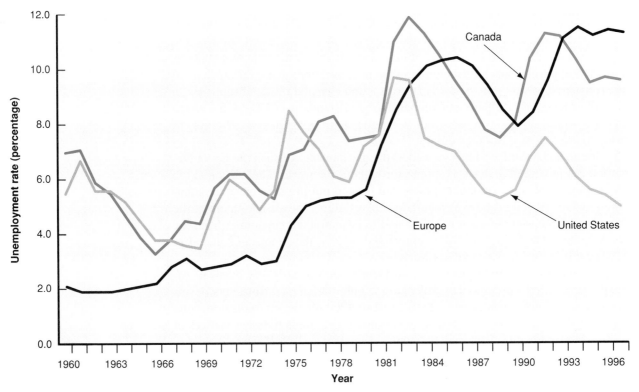

FIGURE 20-3

Unemployment Rates in Canada, Europe, and the United States, 1960–1997
The European unemployment rate, which was lower than the U.S. and Canadian rates until the 1970s, is now higher.

Sources: Statistics Canada, CANSIM Series D44950, and Historical Statistics of Canada; U.S. Department of Labor, Bureau of Labor Statistics; and OECD Economic Outlook, various issues.

---

[12]*The membership of the European Union has expanded through time, from the six original countries in 1957 to 15 countries in 1995. Whenever we refer to the European Union unemployment rate, we mean the unemployment rate, now or in the past, in the group of 15 countries that compose the European Union today. See Table 20-3 (page 398) for a list of the members of the European Union.*

FIGURE 20-4

EU Unemployment and Inflation, 1970–1997

While unemployment is high today in Europe, inflation is no longer decreasing.

Sources: OECD Economic Outlook, various issues.

sion of the early 1990s, and in 1994 the European unemployment rate was higher than it had ever been since the Great Depression. While European rates have stayed high, Canadian rates have started to inch back down towards U.S. levels.

Turn now to the joint behaviour of inflation and unemployment in the European Union. Figure 20-4 plots the EU unemployment and inflation rates since 1970. Three points come out:

1. The *increase* in unemployment in the 1970s was associated with an *increase* in inflation. This is suggestive of aggregate supply shocks. Recall from Chapter 16 that aggregate demand shocks move unemployment and inflation in opposite directions, but that aggregate supply shocks move them in the same direction. (Remember that a negative aggregate supply shock decreases output and thus increases unemployment.) There are indeed strong suspects here, namely the two large increases in OPEC oil prices in the mid and late 1970s.[13]

---

[13]*See Section 16-6 for a discussion of the effects of an increase in the price of oil on activity and the price level.*

TABLE 20-3

UNEMPLOYMENT RATES ACROSS COUNTRIES, 1997

European Union:		Portugal	6.8
Austria	6.1	Spain	21.0
Belgium	12.7	Sweden	8.1
Denmark	7.9	United Kingdom	6.9
Finland	14.6		
France	12.4		
Germany	11.4	**Other countries, non-European:**	
Greece	10.5	Australia	8.7
Ireland	10.3	Canada	9.2
Italy	12.3	New Zealand	6.7
Luxembourg	3.7	United States	5.0
Netherlands	5.8		

Source: *OECD Economic Outlook,* Annex Table 21, December 1997.

2. The further *increase* in European unemployment in the early 1980s was associated with a large *decrease* in inflation. Building on our study of disinflation in Chapter 18, this suggests that the increase in unemployment in the 1980s was due in large part to a shift in monetary policy aimed at decreasing inflation by lowering money growth.

3. Inflation started increasing again in 1987 while the unemployment rate was around 10%. In 1997, with unemployment around 11%, inflation was declining but only modestly. Building on our discussion of the natural rate of unemployment in Chapters 15 and 17, this suggests that the current unemployment rate in the European Union is not far above the natural rate. (Recall from Chapter 18 that we can think of the natural unemployment rate as that unemployment rate at which inflation remains constant.)

Does it make sense to talk about *the* high European unemployment rate, or are there substantial differences across countries? Table 20-3 gives the unemployment numbers for each of the countries of the European Union, as well as for a few non-European countries. The table suggests two conclusions:

1. Unemployment rates are indeed high in most European countries. But there are exceptions. Look at Austria, which had a rate of 6.1% in 1995, or Luxembourg with a rate of 3.7%. Also look at Portugal, whose economy in many ways resembles that of Spain but had an unemployment rate of only 6.8%, compared with 21% in Spain. (Unemployment in Spain is the subject of the In Depth box at the end of this chapter titled "Spanish Unemployment: Diagnosis and Policy Options.")

2. High unemployment is not limited to Europe. Canada has high unemployment. On the other side of the world, so does Australia.

So one can speak of *a* European unemployment problem. But it is important, especially when testing different explanations, to keep in mind the differences across countries and the fact that the problem is not limited to Europe.

LABOUR MARKET RIGIDITIES

The dominant view in Europe today is that high European unemployment is the result of **labour market rigidities.** These put too many restrictions on firms, prevent them from adjusting to changes in the economic environment, make the cost of doing business too high, and, the argument goes, lead to high unemployment. The word **Eurosclerosis** has been coined to denote this view. (*Sclerosis* means hardening of the tissues; the argument is that the many rigidities imposed on business are leading to the sclerosis of the economic structure.)

Here is a list of what are seen as the main labour-market rigidities in Europe today:

■ Wages represent only a fraction of the cost of labour. To these must be added employers' contributions to social insurance, pensions, and so on. Employers' contributions are typically much larger in Europe than in the United States. Canada's employer costs are closer to those in Europe.

■ Firms that want to lay off workers face large firing costs. These costs include large **severance payments** (payments that must be paid to laid-off workers) and/or a complex and lengthy legal process to obtain the authorization to lay them off in the first place. These large costs not only make it hard to lay off unneeded workers, increasing the cost of doing business, but they also make firms think twice about hiring workers in the first place.

■ Unions are much more powerful in Europe than they are in the United States. They work to secure higher wages and, by imposing restrictions on the organization of work within firms, limit firms' flexibility to adapt to changes, again increasing costs. Again, Canada lies between these two extremes.

■ Unemployment benefits are more generous in Europe than in the United States. They are often larger as a proportion of wages. They are also easier to qualify for and last longer, providing limited incentives for the unemployed to find work. Canada's unemployment benefits were cut substantially in 1997, but they are still more generous than those in the United States.

■ In many European countries, minimum wages are high in proportion to the average wage. Together with high nonwage labour costs, they often make it unprofitable to hire low-skill workers. Thus, low-skill workers remain unemployed. They also lose the opportunity for on-the-job training and thus the opportunity to become more skilled. Minimum wages in Canada and the United States are roughly similar.

How can these factors lead to a high natural rate of unemployment? Can they account for why Canada's unemployment rate lies between those of Europe and the United States? To answer the question, recall our discussion of the determinants of the natural rate in Chapter 15. We can think of the natural rate as determined by two relations.[14]

The first is the wage-setting relation:

$$\frac{W}{P} = F(u, z)$$

---

[14]*See Section 15-4.*

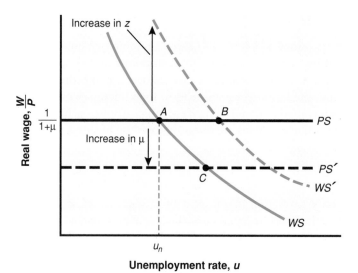

FIGURE 20-5
The Determinants of
the Natural Rate of
Unemployment
Increases in z and
increases in μ both
increase the natural rate
of unemployment.

This relation follows from our description of wage setting, together with the assumption that the expected price level is equal to the actual price level. The real wage is a decreasing function of the unemployment rate, *u*, and an increasing function of all the other factors that affect wage setting, captured by *z*. It is represented as the downward-sloping curve *WS* (for wage setting) in Figure 20-5.

The second relation is the price-setting relation, which implies that

$$\frac{W}{P} = \frac{1}{1+\mu}$$

where μ is the markup of prices over wages. It is represented by the horizontal line *PS* (for price setting) in Figure 20-5. Equilibrium is given by point *A* and the natural rate of unemployment is equal to u_n.

Increases in *z*, which increase the real wage at a given unemployment rate, shift the *WS* curve up, moving the equilibrium from *A* to *B* and leading to an increase in the natural rate of unemployment. Increases in μ, which shift the *PS* curve down, lead the equilibrium to move from *A* to *C* and thus also lead to an increase in the natural rate of unemployment.

The list of factors we just presented traces high unemployment in Europe to factors that increase either *z* or μ. High indirect or hidden labour costs lead to higher costs, and thus to a higher μ, a higher markup of prices over wages. Unions decrease firms' flexibility, leading to higher costs and thus to a higher μ. By increasing workers' bargaining power, unions also lead to higher wages, increasing *z* and increasing the wage at a given unemployment rate. Unemployment benefits and minimum wages have the same effect.

Evaluating the Eurosclerosis view. How convincing is the Eurosclerosis view of European unemployment?

While it is true that European labour markets indeed offer workers more job security than do their U.S. counterparts, this security is in no way a new development. Many of the rules were in place in the 1960s, when European unemployment was very low. And, if anything, the movement since the early 1980s has been in the direction of making labour markets more flexible. Many of these "rigidities" are less important today than they were a decade ago.

For example, the power of unions is clearly declining. A number of countries, most notably the United Kingdom under Margaret Thatcher, have passed legislation limiting the role of unions. **Union density,** the proportion of the work force that is unionized, has decreased in most European countries since the early 1980s.

Many countries have also passed legislation making it easier for firms to rely on part-time employment or to offer limited-time employment contracts, thus avoiding the need to pay severance payments at the end of the contract. As a result, part-time employment has increased substantially in Europe.

Thus, to validate the argument that labour market rigidities are responsible for the increase in European unemployment one has to argue as follows. While European labour markets have become more flexible, substantial changes in the economic environment require even more flexibility than has actually been introduced. Growth has slowed down, structural change is more rapid, and competition between firms has become more intense. In this environment, rigidities matter more. To take a simple example, to a firm that faces stable and growing demand and thus never needs to lay off workers, restrictions on firing are irrelevant. But in an environment where demand becomes more variable, and where the firm must adapt faster to survive, such restrictions become much more relevant indeed.

But is there evidence that European economies are indeed going through more rapid structural change than in earlier decades? Given the talk of increased international competition and the rapid development of new high-tech service sectors, you may be surprised to learn that economists have found little evidence that the pace of European structural change is higher now than it was in earlier decades.

One measure of "structural change" that economists have constructed is the dispersion of rates of change in employment across sectors. If all sectors grow at roughly the same rate, the dispersion will be small, indicating that the economy is undergoing little structural change. If some sectors are growing rapidly while others are shrinking, the dispersion will be large, indicating large underlying structural change. Dispersion measures constructed for each European country and each year show no consistent movement across countries and time. They are typically no higher today than they were in the 1970s or the 1960s. According to this measure, therefore, structural change in Europe is no higher now than it was in previous decades.

However, change in the sectoral composition of employment is only one of the dimensions of structural change. There is at least one dimension in which the last 15 or so years appear different from earlier decades. In Europe as well as in Canada and the United States, the demand for unskilled workers appears to have steadily declined compared with the demand for skilled workers.[15] And, as Table 20-4 shows, the response of wages and employment has been different in North America and in Europe.

In Canada and the United States, the decrease in demand for unskilled workers has led to a decrease in their real wage. In Europe, it has not. The relative unemployment rate of unskilled workers has remained stable in North America, while it has increased in Europe.

---

[15]*Whether the shifts have been similar in North America and Europe is the topic of much research. We leave the issue aside here but take it up in Chapter 25.*

Europe recently:
↓ rigidities
↑ limitations of unions (legislation)
↑ part time work / contracts
↓ prop. of L.F that is unionized

Economic Δ's

TABLE 20-4

RELATIVE WAGE GROWTH AND UNEMPLOYMENT OF UNSKILLED WORKERS

	REAL WAGE GROWTH, UNSKILLED WORKERS (%)	RELATIVE UNEMPLOYMENT RATE, UNSKILLED VERSUS SKILLED WORKERS (%)	
Country		Late 1970s	Late 1980s
Canada	−0.9	2.6	2.6
United States	−1.2	3.5	3.4
France	0.4	1.5	2.7
Germany	2.5	2.2	2.6
Italy	1.5	0.5	1.0
United Kingdom	0.8	3.2	5.7

First column: Annual real wage growth of male workers in the bottom 10th percentile of the overall earnings distribution. *Dates:* Canada, 1981–1990; United States, 1980–1989; France, 1980–1987; Germany, 1983–1990; Italy, 1980–1987; United Kingdom, 1980–1992. *Second and third columns:* Adult male unemployment rate for the bottom quartile of the labour force, ranked by educational qualifications, divided by the rate for the top quartile.

Source: OECD Jobs Study, 1994, Charts 14 and 15.

One must be careful, however, not to attribute all of the large European increase in relative unemployment to too high a wage for unskilled workers. When overall unemployment increases, as it has in Europe, the unemployment rate of unskilled workers always goes up more than the unemployment rate of skilled workers. Skilled workers often work in jobs that have to be performed even when activity is lower. Also, firms get rid of their less skilled workers first and, when they need to hire, choose the more skilled workers from among the unemployed. How much of the increase in the unskilled workers' unemployment rate in Europe is due to the lack of decline of their relative wage is still uncertain. It is reasonable to think, though, that part of the difference between European and North American unskilled workers' unemployment rates comes from the difference in the adjustment of relative wages.

Hysteresis

The weaknesses of the Eurosclerosis case have led a number of macroeconomists to put forth an alternative line of argument known as **hysteresis.** The argument goes as follows.

In the late 1960s and then again in the 1970s, European countries were hit by a series of aggregate supply shocks: labour unrest in the late 1960s and increases in oil prices in the mid and late 1970s. The result was stagflation—that is, increasing unemployment and increasing inflation.

In the early 1980s, European countries made the decision to reduce inflation via monetary contraction. The United Kingdom under Margaret Thatcher was first to act, followed a few years later by most other countries. As in North America, the result was disinflation and a sharp increase in unemployment.

In the United States, monetary contraction was followed by a fiscal expansion and the very large deficits of the Reagan administration. The result was a rapid output expansion from 1982 on. There was no such expansion in Europe. Thus, while the U.S. unemployment rate decreased rapidly in the mid-1980s, there was no such decrease in European unemployment. And, because European monetary and fiscal policies have remained tight to this day, unemployment has remained much higher there than in the United States.

The analysis so far raises an obvious question, however. If unemployment in Europe is the result of an adverse shift in aggregate demand that has increased the unemployment rate far above the natural rate, we should be observing a rapid decrease in inflation. But as Figure 20-4 clearly shows, this is not the case. Inflation is low in Europe, but it is not decreasing rapidly.

This is where *hysteresis* (which means the dependence of a variable on its whole history) comes in. The natural unemployment rate, the argument goes, is not, as we have assumed until now, independent of actual unemployment. Instead, the "natural rate" itself depends on the history of actual unemployment. In particular, a long period of high unemployment leads to an increase in the natural rate. Thus, persistently high unemployment is likely to be associated with less and less downward pressure on inflation. This is why inflation is no longer decreasing much in Europe.[16] To return to our discussion of the Great Depression, hysteresis may also explain why, from 1933 on, deflation stopped in the United States despite very high unemployment.

How does actual unemployment affect the natural rate over time? Research has identified a number of potential channels.

Increased unemployment benefits. When confronted with persistently high unemployment, society adjusts in many ways to make unemployment less painful. Although these adjustments do alleviate some of the burden of unemployment, many also increase the natural rate.

Table 20-5 gives the evolution of an index of the generosity of unemployment benefits in a number of countries since the 1960s. Unemployment benefits have many dimensions, ranging from eligibility conditions to how long benefits last; the index shown in Table 20-5 takes these different dimensions into account by computing the ratio of unemployment benefits to the wage for workers with different initial wages and different durations of unemployment, and then computing the average ratio over all these cases.

TABLE 20-5
THE GENEROSITY OF UNEMPLOYMENT BENEFITS (AS A PERCENTAGE OF THE WAGE)

COUNTRY	1960s	1970s	1980s	1991
Canada	22	26	27	27
France	25	24	34	37
Germany	32	30	29	28
Spain	15	18	32	34
United Kingdom	28	26	24	20
United States	10	13	15	12

Source: OECD Jobs Study, 1994, Chart 16.

---

[16]See Olivier Blanchard and Lawrence Summers, "Hysteresis and European Unemployment," NBER Macroeconomics Annual, *1986, 14–89.*

The table shows that in both France and Spain (two of the countries with the highest unemployment rates) unemployment benefits actually became more generous in the 1980s. In those two countries, the reason for the increased generosity was the large increase in unemployment in the first place. Faced with high unemployment, the governments of those countries felt that they had no choice other than to provide enough benefits for the unemployed to survive. One of the unintended results of more generous unemployment benefits may, however, have been a shift upward of the wage-setting relation in Figure 20-5 (an increase in z), leading to an increase in the natural rate of unemployment.

The emergence of long-term unemployment.   High persistent unemployment comes with an increase in long-term unemployment. In 1995, the proportion of unemployed workers who had been unemployed for more than a year exceeded 30% in most European countries. The number was close to 60% in Ireland, Belgium, and Italy. There are two good reasons to worry about long-term unemployment.

The first is simply the human cost. It is one thing to be unemployed for a few months; it is another to be unemployed for a year or more. The evidence is that many of those who become long-term unemployed end up losing skills and work habits—or not acquiring them in the case of youth unemployment, which is also very high in many European countries. The result is a vicious circle in which employers become reluctant to hire the long-term unemployed, who in turn give up searching for jobs. The final result can be the permanent loss of employment, the loss of self-esteem, and psychological depression.

The second reason is macroeconomic in nature. The emergence of long-term unemployment leads to an increase in the natural rate of unemployment. Take the extreme case in which the long-term unemployed become simply unemployable. They then become completely irrelevant to the process of wage determination. Employers cannot credibly threaten to hire the long-term unemployed to extract wage concessions from their current workers. Currently employed workers, were they to find themselves unemployed, do not have to worry about competing with the long-term unemployed.

The irrelevance of the long-term unemployed in wage determination has a simple implication: The higher the proportion of long-term unemployed, the less pressure a given unemployment rate exerts on wages, thus the higher is the wage set in wage setting. In terms of Figure 20-6, an increase in the proportion of long-term unemployed shifts the wage-setting relation upward, from WS to WS'. This shift in turn results in a higher natural unemployment rate, an increase from u_n to u'_n.

This increase in the natural rate can explain why an economy that has had high unemployment for a long time, and thus has a high proportion of long-term unemployed, may have both high unemployment and stable inflation (that is, the same inflation rate every year). Take, for example, an economy in which the unemployment rate is 15% but where 12% of the labour force are long-term unemployed. Let's maintain our extreme assumption that these long-term unemployed workers are unemployable and thus irrelevant to the process of wage determination. Then, the unemployment rate relevant for wage determination (the unemployment rate taking into account only those workers who are employable) is only 15% − 12% = 3%. The labour market is in effect very tight, and there may well be wage pressure and increasing inflation.

The case we have just looked at is too extreme. Many of the long-term unemployed are employable, and thus long-term unemployment exerts some pressure on wages in wage determination. But the basic lesson is general. Prolonged high unemployment leads to an increase in the proportion of long-term unemployed. As the long-term unemployed become less employable, the same rate of unemployment exerts less downward pressure on wages, and thus less downward pressure on inflation.

Implications of hysteresis.   The theory of hysteresis has two important implications, one specific and one general.

In the case of Europe today, it implies that there may be room for unemployment to decrease without fundamental changes in the organization of the labour market, regardless of the desirability of such reforms on other grounds. How fast and how much unemployment can decrease depends on how fast the hysteresis mechanisms can work in reverse, as unemployment decreases rather than increases. For example, can the long-term unemployed be "re-enfranchised," and, if so, how quickly? Will those who have given up searching for jobs start searching again if labour-market conditions improve? Will those who have lost their skills be able to reacquire them quickly, or should specific training programs be put in place?

More generally, hysteresis implies that disinflation may be more costly than we concluded in Chapter 18. The increase in unemployment needed to decrease inflation may lead to an increase in the natural rate, and thus to long-lasting unemployment costs.

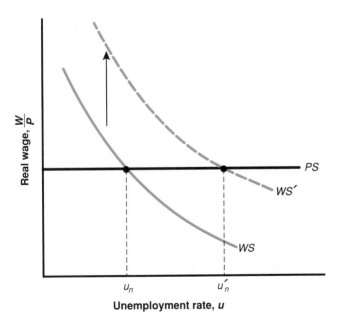

FIGURE 20-6
The Effects of an Increase in the Proportion of Long-Term Unemployed on the Natural Unemployment Rate
An increase in the proportion of long-term unemployed increases the natural rate of unemployment.

In 1975, Generalisimo Francisco Franco died, ending nearly 40 years of dictatorship in Spain. Labour relations, which had been tightly controlled by the state, broke down. In an attempt to establish a constituency in the new democratic regime, socialist and communist unions competed in their demands for higher wages. The result was a rapid increase in wages, leading to a rapid increase in prices, leading to further increases in wages, and so on. With money growth accommodating these increases in prices and wages, inflation reached 25% in Spain in 1977. Figure 1, opposite, shows the behaviour of inflation and unemployment in Spain since 1971.

Macroeconomic developments during the 10 years following 1997 (these years are shaded in Figure 1) were dominated by the fight against inflation. From 1978 on, the Bank of Spain maintained a policy of tight money; real interest rates, which had been negative in the mid-1970s (with expected inflation far exceeding nominal interest rates) turned positive, averaging 6% in the first half of the 1980s. By 1987, inflation was down to 5%. But the unemployment rate had increased to 20%, an increase of 15 percentage points in 10 years.

There was then a brief decrease in the unemployment rate at the end of the 1980s, reflecting a general economic expansion in Europe. But even at the peak of the expansion in Spain the unemployment rate was still 17%. Since then, the unemployment rate has increased again. In 1995, it stood at 23%, its second-highest level ever. (The highest level, 24.2%, was reached in 1994.)

(1) Are the unemployment numbers accurate?

Economists' first reaction, when confronted with such high unemployment numbers, is to question whether the numbers are right. In the case of Spain, a natural question is whether high measured unemployment actually hides a large underground economy. A survey commissioned by the Spanish government to measure unreported activity in 1985 in Spain concluded that between 10% and 15% of employment was irregular—that is, not properly registered with the social insurance system. But it also concluded that most of those jobs were held by people already employed, so that adjustment for the underground economy would decrease the unemployment rate by 3.5 percentage points at most. There is no reason to believe that this

situation has changed since 1985. Thus, most of measured unemployment is indeed genuine.

(2) What are the causes of unemployment?

Does high unemployment in Spain reflect a high natural rate, or a large deviation from the natural rate? Recall (from Chapter 17) that, if high unemployment reflects a large deviation from the natural rate, it should be associated with a large decrease in inflation. But inflation was roughly the same in 1994 as it was in 1987. This suggests that the average unemployment rate during that time, 19.4%, is not far from the natural rate.

Where does this high natural rate come from? Here, as in the rest of Europe, there are two conflicting diagnoses.

■ The first argues that Spain suffers from very high labour market rigidities, primarily high firing costs. One of the legacies of the Franco period was a high degree of job protection. Workers could be dismissed only under limited conditions, and dismissals sometimes involved substantial severance payments. The OECD Jobs Study cited in the source note to Table 20-4 has computed an index of employment protection. Using that index, Spain has the second-highest value of all OECD countries. Since 1984, however, Spanish firms have been allowed to offer fixed-term contracts, contracts that allow firms to lay off workers at the expiration of the contract at no cost. As a result, the proportion of employment under fixed-term contracts increased from 10% in 1984 to 30% in 1994.

The problem with the "labour market rigidities" diagnosis is the same for Spain as for other European countries. Firing restrictions have substantially decreased since the early 1980s, and so have the power of unions and most other labour market rigidities. So why is it that the natural rate has increased so much during that time?

■ The alternative diagnosis relies on hysteresis. It argues that, initially, the increase in unemployment was due to tight money and the fight against inflation. But the high actual unemployment rate has led over time to a high natural rate. This increase, in turn, is due to two main factors:

Firing restrictions protect employed workers from the risk of unemployment, and thus decrease the effect

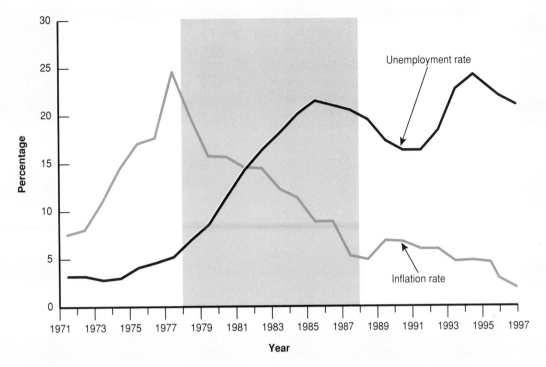

FIGURE I
Unemployment and Inflation in Spain, 1971–1997
Source: OECD Economic Outlook, various issues.

of unemployment on wage determination. Indeed, the movement toward fixed-term employment contracts may have had the perverse effect of further decreasing the effects of labour-market conditions on wages. Now that firms can adjust employment through the use of workers under fixed-term contracts, workers with permanent contracts are even more protected from the risk of unemployment than they were before. If these workers, who are typically more senior, dominate wage negotiations, this may explain why the movement toward fixed-term employment has led to less rather than more downward pressure of unemployment on wages, and thus to an increase in the natural rate of unemployment.

High unemployment leads to societal changes that decrease its human cost but also decrease its effect on wage determination. We referred earlier to the increase in unemployment benefits in Spain since the 1980s. But other mechanisms are at work that may well be more important. For example, high unemployment has increasingly fallen on the very young, who

typically can survive the loss of income more easily by staying home and relying on family income. (The unemployment rate of Spanish people age 16 to 19 was an amazing 50% in 1994.) More generally, most of the unemployed live in a household where somebody is working. At the end of 1993, while close to 25% of workers were unemployed, only 10% of households had no one employed in the household, and only 5% had no one employed and no one receiving unemployment benefits. This is probably why high unemployment has not led to riots. But it is also why the downward pressure of unemployment on wage determination is weaker than one would expect.

(3) **What should be done to reduce Spanish unemployment?** What should be done depends in part on the diagnosis that one makes of the sources of Spanish unemployment. Those who believe that labour market rigidities are responsible argue that giving firms more flexibility and limiting unemployment benefits will eventually decrease unemployment.

IN DEPTH

Those who believe that hysteresis plays an important role argue for a combination of aggregate demand and aggregate supply policies: demand policies aimed at increasing output and employment, supply policies such as a decrease in firing costs and a pact among unions, firms, and the government to avoid wage pressure and higher inflation. So far, however, forecasts are for no dramatic policy change, and for high unemployment for at least the rest of the decade.

Reference

For more on the facts and the policy options, read "Spanish Unemployment: Is There a Solution?" *Report of the CEPR Group on Spanish Unemployment*, CEPR, 1995. Also see Ramon Marimon, "Reconsidering Spanish Unemployment," *CREI*, 1997.

20-3 CONCLUSIONS

To return to the title of this chapter, it is not clear that there is a unique "high-unemployment pathology." The two major episodes of high unemployment in this century appear quite different, both in their causes and in their symptoms.

The Great Depression was a case of a sharp decrease in activity followed by a long period of recovery. The decline in activity clearly had its origin on the demand side. Once the economy reached a trough, recovery was strong, largely because of an increase in the money supply and the resulting demand expansion.

High unemployment in Europe has more the feeling of a long illness, with prospects for only a slow recovery. European unemployment crept up in the 1970s, before increasing faster in the early 1980s. There is no agreement as to its sources. Some blame problems on the demand side, mistakes in policy, and hysteresis. Others point to the supply side and to the sclerosis of the economic system. As of 1997, the strong growth rates that accompanied the recovery from the Great Depression are nowhere to be seen in Europe today.

SUMMARY

On the Great Depression in Canada

■ The unemployment rate increased from 2.9% in 1929 to 19.3% in 1933.

■ The origins of the Great Depression in Canada can be linked to poor weather, a decline in the relative prices of our exports, and a large adverse shift in demand. The Prairie provinces suffered a severe drought in 1929 that sharply reduced Canada's wheat exports. After 1929, the world price of wheat and other commodities that Canada exported dropped sharply. The United States, our largest export market and an important source of investment finance, slumped because of the stock market crash of 1929 and the resulting uncertainty. Canadian entrepreneurs and consumers behaved similarly to their American counterparts.

■ Both the United States and Canada experienced deflation from 1929 to 1933. The favourable effects of deflation on real money balances were offset, however, by roughly equal decreases in nominal money stocks. In the United States, the decline in the money stock can be linked to bank failures. But no banks failed in Canada.

■ The Canadian recovery began in 1933. The abandonment of the gold standard allowed the Canadian dollar to depreciate and helped our exports. Fiscal policy did not play an active role until the government sharply increased its spending as Canada entered World War II.

On European Unemployment

■ The European unemployment rate was much lower than the U.S. rate until the early 1970s. Both increased in the 1970s and the early 1980s. Since then, however, the U.S. rate has declined while the European rate has remained very high. In 1995, the unemployment rate in the European Union was about 11%.

■ While the increase in European unemployment in the 1980s was associated with a large decline in inflation, inflation is now roughly constant in Europe, suggesting that the current unemployment rate is close to the natural rate.

■ The first line of explanation for high unemployment in Europe holds that high unemployment reflects rigidities in European labour markets, from excessive employment protection to a too-generous system of unemployment benefits, to high minimum wages that price low-skilled workers out of the labour market. This line, however, does not explain well why unemployment has increased during a period when most of these rigidities have actually decreased.

■ The alternative line of explanation, called *hysteresis,* holds that, initially, high unemployment was due to disinflation policies, but that the high actual rate of unemployment has led to a high natural rate of unemployment. It argues in particular that high unemployment leads to a high proportion of long-term unemployed, and that the long-term unemployed have little effect on wage determination.

KEY TERMS

■ Great Depression, 387
■ depression, 388
■ New Deal, 394
■ labour market rigidities, 399

■ Eurosclerosis, 399
■ severance payments, 399
■ union density, 401
■ hysteresis, 402

QUESTIONS AND PROBLEMS

1. "Once the stock market crash of October 1929 caused the Great Depression, there was little the authorities could do." Comment.

2. Suppose that the Federal Deposit Insurance Corporation (FDIC), which insures deposits against bank failures in the United States, had existed in 1929. Which of the columns in Table 20-2 would have been affected? How?

3. Might the natural rate of unemployment have increased during the Great Depression? Explain.

4. One implication of hysteresis in labour markets is that using a recession to bring down inflation might cause the natural rate of unemployment to rise. Can you think of other "social costs" of recession that last beyond the recession itself? Explain.

5. Many believe that the relative demand for unskilled labour will decrease even more rapidly over the next few decades. If this does indeed happen, what are the implications for the natural rate of unemployment? Does your answer depend on government policy changes that might or might not occur during that time? Explain.

* 6. According to the hysteresis hypothesis, the natural rate of unemployment might be affected by the proportion of the jobless who are less "employable" than others because they have been unemployed so long.

To see how this works, suppose that price-setting is given by the equation

$$\frac{W}{P} = \frac{1}{(1 + .1)}$$

and wage determination is given by the equation

$$\frac{W}{P} = 1 - (u_S + .5u_L),$$

where

u_S = the short-term unemployed as a fraction of the labour force
u_L = the long-term unemployed as a fraction of the labour force

Suppose further that the fraction of the jobless who are long-term-unemployed is β, so that $u_L = \beta u$, and $u_S = (1 - \beta)u$.

a. According to this model, which type of unemployment has a greater impact on wages—long-term or short-term? Explain.

b. Obtain an expression for the natural rate of unemployment. (*Hint:* Substitute $u_L = \beta u$ and $u_S = (1 - \beta)u$ into the wage-determination equation;

then use the price-setting equation to solve for u. Your answer will depend on the parameter β.)

  c. Calculate the natural rate of unemployment when β is equal to

    (i) 0

    (ii) 0.5

    (iii) 0.75

Provide some economic intuition for the differences in your answers.

FURTHER READING

More on the Great Depression

Kenneth, Norrie, and Douglas, *a History of the Canadian Economy* (Toronto: Harcourt Brace, 1996), Chapter 17, gives the basic facts.

Ed Safarian's book, *The Canadian Economy in the Great Depression* (Toronto: University of Toronto Press, 1959) is the classic reference.

For the United States, Peter Temin, *Did Monetary Forces Cause the Great Depression?* (New York: W.W. Norton, 1976) looks more specifically at the macroeconomic issues. So do the articles in a symposium on the Great Depression in the *Journal of Economic Perspectives,* Spring 1993.

The view that high European unemployment is due primarily to labour market rigidities is presented in the *OECD Jobs Study,* published in 1994. The report comes in three parts: a short report, and two longer volumes, Parts I and II, which give a detailed and useful description of labour markets in each OECD country. A 1995 update, called "Implementing the Strategy," describes recent developments in policy and outcomes.

Balanced discussions of European unemployment are given by Charles Bean of the London School of Economics in "European Unemployment: A Survey," *Journal of Economic Literature,* June 1994, 573–619, and in *World Employment 1995,* a report from the ILO, the International Labour Organization based in Geneva.

PATHOLOGIES II: HIGH INFLATION

In 1913, the value of all currency and coin circulating in Germany was 6 billion marks. Ten years later, in October 1923, 6 billion marks was barely enough to buy a one-kilo loaf of rye bread in Berlin. A month later, the price of the loaf had increased to 428 billion marks.[1]

The German hyperinflation of the early 1920s is probably the most famous episode of hyperinflation. (**Hyperinflation** simply means very high inflation.) But it is not the only one. Table 21-1 summarizes the seven major hyperinflations that followed World Wars I and II. They share a number of features. They were all short (lasting for a year or so) but intense, with monthly inflation running at about 50% or more. In all cases, the increase in the price level was staggering. As you can see, the largest price increase was actually not reached during the German hyperinflation but in Hungary after World War II. What cost one Hungarian pengö in July 1946 cost 3800 trillions of trillions of pengös less than a year later.

Such rates of inflation have not been seen since. The closest case in the recent past is that of Bolivia. From January 1984 to September 1985, Bolivian inflation ran at an average of 40% per month, implying a roughly 1000-fold increase in the price level over 21 months.[2] But many countries, especially in Latin America, have been struggling with bouts of high inflation, often in excess of 15% to 20% per month, over many years.[3]

---

[1]From Steven Webb, Hyperinflation and Stabilization in the Weimar Republic *(New York: Oxford University Press, 1989).*

[2]*Inflation at 40% per month implies that the price level at the end of 21 months is equal to $(1 + 0.4)^{21} = 1171$ times the price level at the beginning.*

[3]*Transition in Eastern Europe in the 1990s has also often come with high inflation. See Chapter 26.*

TABLE 21-1
SEVEN HYPERINFLATIONS OF THE 1920s AND 1940s

COUNTRY	BEGINNING	END	P_T/P_0	AVERAGE MONTHLY INFLATION RATE (%)	AVERAGE MONTHLY MONEY GROWTH (%)
Austria	Oct. 1921	Aug. 1922	70	47	31
Germany	Aug. 1922	Nov. 1923	1.0×10^{10}	322	314
Greece	Nov. 1943	Nov. 1944	4.7×10^{6}	365	220
Hungary I	Mar. 1923	Feb. 1924	44	46	33
Hungary II	Aug. 1945	Jul. 1946	3.8×10^{27}	19 800	12 200
Poland	Jan. 1923	Jan. 1924	699	82	72
Russia	Dec. 1921	Jan. 1924	1.2×10^{5}	57	49

P_T/P_0: Price level in the last month of hyperinflation divided by the price level in the first month.

Source: Philip Cagan, "The Monetary Dynamics of Hyperinflation," in Milton Friedman, ed., *Studies in the Quantity Theory of Money* (Chicago: University of Chicago Press, 1956), Table 1.

Table 21-2 gives average monthly inflation rates for a number of Latin American countries since the mid-1970s. Note how Argentina, Brazil, Nicaragua, and Peru went through high inflation in the late 1980s but have since reduced their inflation rates.

What causes hyperinflations? We saw in Chapter 18 that *inflation ultimately comes from money growth.* This relation between money growth and inflation is confirmed by the last two columns of Table 21-1; in each country, high inflation is associated with correspondingly high nominal money growth. But this relation raises another question: *Why* was money growth so high? The answer turns out to be common to all hyperinflations: Money growth is high because the budget deficit is high. In turn, the deficit is high because the economy is affected by major shocks that make it difficult or impossible for the government to finance its expenditures in any way other than money creation. In the remainder of this chapter we look at this answer in more detail, relying on examples from various hyperinflations.

TABLE 21-2
HIGH INFLATION IN LATIN AMERICA SINCE 1975 (AVERAGE MONTHLY INFLATION RATE, %)

	1975–1980	1981–1985	1986–1989	1990	1991	1992	1994	1996
Argentina	9.3	12.7	17.4	30.3	8.6	1.8	0.3	0.3
Bolivia	1.3	17.7	1.1	1.2	1.6	1.2	0.6	1.0
Brazil	3.4	7.9	17.7	32.9	15.0	22.1	24.1	1.2
Nicaragua	1.4	3.6	33.6	43.4	32.1	1.5	0.6	0.9
Peru	3.4	6.0	18.8	43.4	14.5	4.7	1.3	0.9

Source: International Financial Statistics, IMF, various issues.

21-1 BUDGET DEFICITS AND MONEY CREATION

A government can finance its deficit in one of two ways.

It can do it in the same way that you would, namely by borrowing. Governments borrow by issuing bonds. But it can also do something that you cannot do. It can, in effect, finance the deficit by creating money. The reason for using the phrase "in effect" is that, as you will remember from Chapter 5, governments do not create money; the central bank does. But with the central bank's cooperation, the government can in effect finance itself by money creation. It can issue bonds and ask the central bank to buy them. The central bank then pays the government with money it creates, and the government in turn uses that money to finance its deficit. This process is called **debt monetization.**

Most of the time and in most countries, deficits are financed primarily through borrowing rather than through money creation. For example, less than 10% of the large Canadian budget deficits during the 1980s was financed by money creation. But, at the start of hyperinflations, two changes usually take place.

The first is a budget crisis. The source is typically a major social or economic upheaval. This may be a civil war or a revolution that destroys the state's ability to collect taxes, as in Nicaragua in the 1980s. Or it may come, as in the case of the hyperinflations in Table 21-1, from the aftermath of a war, which leaves the government with both smaller tax revenues and the large expenditures needed for reconstruction. Burdened in 1922 and 1923 with the war reparations it had to pay to Allied forces, Germany had a budget deficit equal to more than two-thirds of its expenditures. The budget crisis may also come from a large economic shock—for example, a large decline in the price of a raw material that is both the country's major export and the government's main source of revenues. As we shall see in the In Depth box later in this chapter, the decline in the price of tin, Bolivia's principal export, was one of the causes of the Bolivian hyperinflation in the 1980s.

The second is the government's increasing unwillingness or inability to borrow from the public or from abroad to finance its deficit. The reason is the size of the deficit itself. Worried that the government may not be able to repay the debt in the future, potential lenders start asking the government for higher and higher interest rates. Sometimes, foreign lenders decide to stop lending to the government altogether. As a result, the government increasingly turns to the other source of finance available—namely, money creation. Eventually, most of the deficit is financed by money creation.

How large is the rate of money growth needed to finance a given deficit? To answer this question, let's assume that the deficit is financed entirely by money creation, so that

$$\Delta M = \$ \text{ deficit}$$

This equation simply tells us that the government (through the central bank) must create enough new money to cover the nominal deficit. M is the nominal money supply, measured, say, at the end of each month.[4] (In the case of a hyperinflation, variables change quickly enough that it is useful to divide time into months, rather than quarters or years as we did in earlier chapters.) ΔM is the change in the nominal money supply from the end of last month to the end of this month; equivalently, it is equal to nominal money creation this month; "$ deficit" is the budget deficit, measured in nominal terms.

If we divide both sides of the equation by the price level during the month, P, and denote the real deficit by "deficit" without a dollar sign, we get

$$\frac{\Delta M}{P} = \text{deficit} \tag{21.1}$$

The revenues from money creation, $\Delta M/P$, are called **seigniorage.** The word is revealing: The right to issue money was indeed a precious source of revenues for the "seigneurs" of the past. Equation (21.1) says that the government must create enough money so that seigniorage is enough to finance the real deficit.

By multiplying top and bottom of $\Delta M/P$ by M, we can rewrite seigniorage as

$$
\underbrace{\frac{\text{seigniorage}}{\dfrac{\Delta M}{P}}} \quad = \quad \underbrace{\frac{\text{money growth}}{\dfrac{\Delta M}{M}}} \quad \times \quad \underbrace{\frac{\text{real money balances}}{\dfrac{M}{P}}} \tag{21.2}
$$

Seigniorage is the product of money growth ($\Delta M/M$) times real money balances (M/P). The larger the real money balances held in the economy, the larger the amount of seigniorage corresponding to a given rate of money growth.

To think about relevant magnitudes, it is convenient to divide both sides of equation (21.2) by real income, Y (measured at a monthly rate):

$$\frac{\Delta M/P}{Y} = \frac{\Delta M}{M}\left(\frac{M/P}{Y}\right) \tag{21.3}$$

This equation says that the ratio of seigniorage to real income (the term on the left) is equal to the rate of money growth (the first term on the right) times the ratio of real money balances to real income (the second term on the right). Suppose that the government is running a budget deficit equal to 10% of real income. If it finances the deficit through money creation, then seigniorage must be equal to 10% of real income as well. Suppose that people hold real balances equal to two months of income, so that $(M/P)/Y = 2$. Then, equation (21.3) tells us, the monthly growth rate of money must be equal to $10\%/2 = 5\%$.

Does this imply that the government can finance a deficit equal to 20% of real income through a money growth rate of 10%, a deficit of 40% of real income through money growth of 20%, and so on? No. As money growth increases, so does inflation. And as inflation increases, the opportunity cost of keeping money increases, leading people to reduce their real money balances. In terms of equation (21.2), an increase in $\Delta M/M$ leads to a decrease in M/P, so that an increase in money growth does not generate a proportional increase in seigniorage. What is crucial

---

[4]*We are taking a shortcut here. What should be on the left-hand side of the equation is* H, *the monetary base—the money created by the central bank—rather than* M, *the money supply. We ignore the distinction between the two in this chapter, as it does not play an important role in the arguments that follow.*

here is how much people adjust their real money balances in response to inflation, and it is to this topic that we now turn.

21-2 INFLATION AND REAL MONEY BALANCES

What determines the amount of real money balances that people hold?

Recall the *LM* relation we have used so far:

$$\frac{M}{P} = YL(i)$$

Higher real income leads people to hold larger real money balances. A higher nominal interest rate increases the opportunity cost of holding money rather than bonds and leads people to reduce their real money balances.

This characterization holds in both stable economic times and times of hyperinflation. But in times of hyperinflation, we can simplify it a bit further. Here's how. First, rewrite the *LM* relation using the relation between the nominal and the real interest rate, $i = r + \pi^e$:[5]

$$\frac{M}{P} = YL(r + \pi^e)$$

Real money balances depend on real income (Y), on the real interest rate (r), and on expected inflation (π^e). All three variables move during a hyperinflation, but expected inflation moves much more than the other two variables. During a typical hyperinflation, actual inflation—and thus presumably expected inflation—may move from close to 0% to 50% a month or more. Thus, it is not a bad approximation to simplify by assuming that both income and the real interest rate are constant, and focus just on the movements in expected inflation. So, let's write

$$\Big\downarrow \frac{M}{P} = \overline{Y} L(\overline{r} + \overset{\uparrow}{\pi^e}) \tag{21.4}$$

where the bars on Y and r mean that we now take both as constant. In times of hyperinflation, equation (21.4) tells us, we can think of real money balances as depending primarily on expected inflation. As expected inflation increases and it becomes more and more costly to hold money, people will reduce their real money balances.

And, indeed, during a hyperinflation people find many ways of reducing their real money balances. When the monthly rate of inflation is 100%, for example, keeping currency for a month implies losing half of its real value (because things cost twice as much a month later). Thus **barter,** the exchange of goods for other goods rather than for money, increases. Payments for wages become much more frequent (often twice weekly). Once they are paid, people rush to stores to buy goods. While the government often makes it illegal to use other currencies than the one it is printing, people shift to foreign currencies as stores of value. In describing

---

[5]*See Chapter 7 for a review.*

the Austrian hyperinflation of the 1920s, Keynes noted: "In Vienna, during the period of collapse, mushroom exchange banks sprang up at every street corner, where you could change your krone into Zurich francs within a few minutes of receiving them, and so avoid the risk of loss during the time it would take you to reach your usual bank."[6] And, even if illegal, an increasing proportion of transactions takes place in foreign currency. During the Latin American hyperinflations of the 1980s, people shifted not to Swiss francs but to U.S. dollars. The shift to dollars has become so widespread in the world that it has a name: **dollarization** (the use of U.S. dollars in domestic transactions).

By how much do real money balances actually decrease as inflation increases? Figure 21-1 examines the evidence from the Hungarian hyperinflation of the 1920s and provides some insights.

Figure 21-1(a) plots real money balances and the monthly inflation rate from November 1922 to February 1924. Note how movements in inflation are reflected in opposite movements in real money balances. The short-lived decline in Hungarian inflation from July to October 1923 is reflected in an equally short-lived increase in real money balances. At the end of the hyperinflation in February 1924, real money balances stood at a little more than half their level at the beginning.

Figure 21-1(b) presents the same numbers, but in the form of a scatter diagram. It plots monthly real money balances on the horizontal axis against inflation on the vertical axis. (We do not observe expected inflation, which is clearly the variable we would like to plot.) Note how the points nicely describe a downward-sloping demand for money: As inflation—actual and, presumably, expected—increases, the demand for money strongly decreases.[7]

We have looked at the numbers from the first Hungarian hyperinflation. But the conclusion is general: Increases in inflation lead people to economize on the use of money, and lead to a decrease in real money balances.

21-3 DEFICITS, SEIGNIORAGE, AND INFLATION

We have looked at the relation between deficits and money creation, and then at the relation between real money balances and inflation. By putting the two together, we can now show how the need to finance a budget deficit can lead not only to *high inflation,* but also, as is the case during hyperinflations, to *high and increasing inflation.*

THE CASE OF CONSTANT MONEY GROWTH

Suppose first that the government chooses a *constant* rate of money growth and maintains that rate forever. (This is clearly not what happens during hyperinflations, where the rate of money growth typically increases over the course of the hyperinflation; we shall be more realistic in the second part of this section.) How much

---

[6]*J. M. Keynes,* Tract on Monetary Reform *(New York: Harcourt Brace and Company, 1924), 51.*

[7]*Note that this decrease in real money balances explains why, in Table 21-1, average inflation is higher than average money growth in each of the seven postwar hyperinflations. The fact that real money balances, M/P, decrease during a hyperinflation implies that prices, P, must increase more than M. In other words, average inflation must be higher than average money growth.*

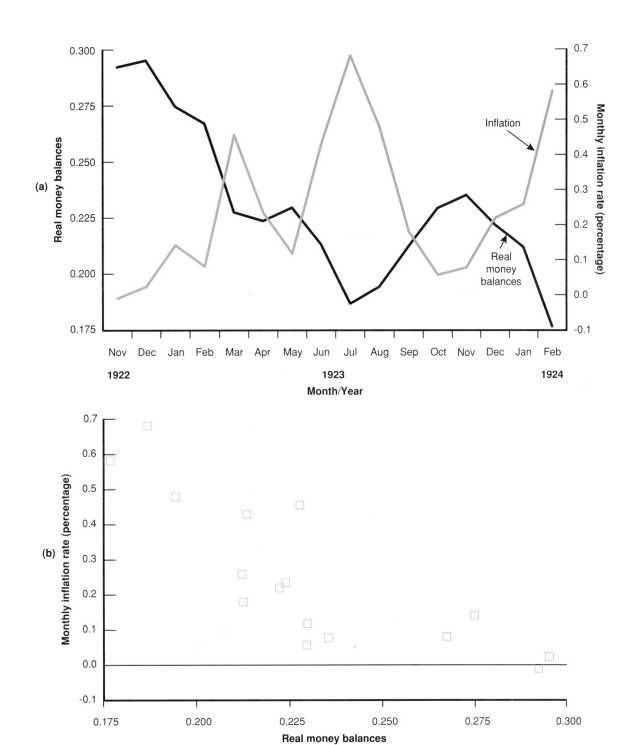

FIGURE 21-1

Inflation and Real Money Balances in Hungary, November 1922–February 1924

At the end of the Hungarian hyperinflation, real money balances stood at roughly half their pre-hyperinflation level.

Source: Philip Cagan, "The Monetary Dynamics of Hyperinflation," in Milton Friedman, ed., *Studies in the Quantity Theory of Money* (Chicago: University of Chicago Press, 1956), Table 1.

seigniorage will this constant rate of money growth generate? Put another way, how large a deficit will the government be able to finance?

Combining what we have learned, we can write

$$\text{seigniorage} = \frac{\Delta M}{M}\left(\frac{M}{P}\right)$$

$$= \frac{\Delta M}{M}\left[\overline{Y}L(\overline{r} + \pi^e)\right] \qquad (21.5)$$

The first line repeats equation (21.2): seigniorage is equal to the rate of money growth times real money balances. The second line incorporates what we learned in equation (21.4): that <u>real money balances depend negatively on expected inflation</u>.

If money growth is constant forever, then inflation and expected inflation must eventually be constant as well. For simplicity, assume that output growth is zero. Then, actual and expected inflation must be equal to money growth:

$$\pi^e = \pi = \frac{\Delta M}{M}$$

Replacing π^e by $\Delta M/M$ in equation (21.5) gives

$$\text{seigniorage} = \frac{\Delta M}{M}\left[\overline{Y}L\left(\overline{r} + \frac{\Delta M}{M}\right)\right] \qquad (21.6)$$

As money growth—the first term on the right in equation (21.6)—increases, real money balances—the term in square brackets on the right—decrease. Thus, what happens to the product of the two terms, to seigniorage, is ambiguous. The empirical evidence is that the relation between seigniorage and money growth looks as in Figure 21-2.

The relation is hump-shaped. At low rates of money growth, such as those we observe in the rich countries today, an increase in money growth leads to a small reduction in real money balances. Thus, it leads to <u>an increase in seigniorage</u>. When money growth (and therefore inflation) becomes very high, however, the reduction

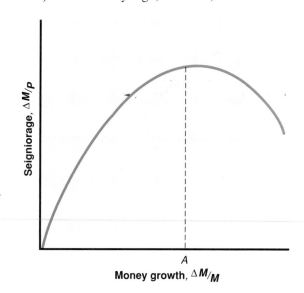

FIGURE 21-2
Seigniorage and Money Growth
Seigniorage is first an increasing function, then a decreasing function of money growth.

in real money balances induced by higher money growth becomes larger and larger. Eventually, there is a rate of money growth—shown by point A in Figure 21-2—beyond which further increases in money growth actually *decrease* seigniorage.

The shape of the relation in Figure 21-2 may look familiar to those of you who have studied the economics of taxation. Think of income taxes. Income tax revenues are equal to the income *tax rate* times income—the *tax base.* At low tax rates, the tax rate has little influence on how much people work, and tax revenues increase with the tax rate. But as tax rates increase further, some people start working less or stop declaring part of their income, and the tax base decreases. As the income tax reaches very high levels, increases in the tax rate lead to a decline in tax revenues. Obviously, tax rates of 100% lead to no tax revenue at all: Why work if the government takes all your income? This relation between tax revenues and the tax rate is often called the **Laffer curve,** named after the economist Arthur Laffer, who argued in the early 1980s that a cut in U.S. tax rates would lead to more tax revenues. He was clearly wrong about where the United States was on the curve; the effect of the decrease in tax rates was to lower tax revenues, not increase them. But the general point still stands: When tax rates are high enough, a further increase in the tax rate can indeed lead to a decrease in tax revenues.

There is more than a simple analogy here. Inflation can indeed be thought of as a tax on money balances. The tax rate is the rate of inflation, π, which reduces the real value of money holdings. The tax base is real money balances, M/P. The product of these two variables, $\pi(M/P)$, is called the **inflation tax.** There is a subtle difference between this and other forms of taxation. What the government receives from money creation at any point in time is not the inflation tax, but rather seigniorage: $(\Delta M/M)$ times (M/P). However, the two are closely related. When money growth is constant, inflation must eventually be equal to money growth, so that

$$\text{inflation tax} = \pi\left(\frac{M}{P}\right)$$

$$= \frac{\Delta M}{M}\left(\frac{M}{P}\right)$$

$$= \text{seigniorage}$$

What rate of money growth leads to the *most seigniorage,* and how much seigniorage does it generate? These are the questions that Philip Cagan asked in a classic paper on hyperinflations written in 1956.[8] In one of the earliest uses of econometrics, Cagan estimated the relation between the demand for money and expected inflation [equation (21.4)] during each of the hyperinflations in Table 21-1. Then, using equation (21.6), he computed the rate of money growth that maximized seigniorage, and the associated amount of seigniorage. The answers he obtained are given in the first two columns of Table 21-3. The third column repeats the actual money growth numbers from Table 21-1.

This table has a very interesting feature. In all seven hyperinflations, actual average money growth (given in column 3) far exceeded the rate of money growth

---

[8]Philip Cagan, "The Monetary Dynamics of Hyperinflation," in *Studies in the Quantity Theory of Money,* Milton Friedman, ed. (Chicago: University of Chicago Press, 1956).

TABLE 21-3

THE RATE OF MONEY GROWTH THAT MAXIMIZES SEIGNIORAGE, AND THE CORRESPONDING AMOUNT OF SEIGNIORAGE (MONTHLY RATES)

	RATE OF MONEY GROWTH MAXIMIZING SEIGNIORAGE (%)	IMPLIED SEIGNIORAGE (% OF OUTPUT)	ACTUAL RATE OF MONEY GROWTH (%)
Austria	12	13	31
Germany	20	14	314
Greece	28	11	220
Hungary 1	12	19	33
Hungary 2	32	6	12 200
Poland	54	4.6	72
Russia	39	0.5	49

Source: Philip Cagan, "The Monetary Dynamics of Hyperinflation," in *Studies in the Quantity Theory of Money,* Milton Friedman, ed. (Chicago: University of Chicago Press, 1956), Table 10.

that maximizes seigniorage (given in column 1). Compare the actual money growth in Hungary after World War II, 12 200%, to the rate of money growth that maximizes seigniorage, 32%. This would seem to be a serious problem for the story we have developed so far. If the reason for money creation was to finance the budget deficit, why was the rate of money growth so much higher than the number that maximizes seigniorage? There is a simple answer, which lies in the dynamics of the economy's adjustment to high money growth. We now turn to it.

DYNAMICS AND INCREASING INFLATION

Return to the argument we just developed. If maintained forever, a higher rate of money growth will eventually lead to a proportional increase in both actual and expected inflation, and lead to a decrease in real money balances. If money growth is higher than the amount that maximizes seigniorage, the increase in money growth will lead to a decrease in seigniorage.

The crucial words in the argument are "if maintained forever" and "eventually." Consider a government that needs to finance a suddenly much larger deficit and decides to do so by creating money. As the rate of money growth increases, it may take a while for inflation and expected inflation to adjust. Even as expected inflation increases, it will take a while longer for people to fully adjust their real money balances: Creating barter arrangements takes time, the use of foreign currencies develops slowly, and so on.

Let's state this conclusion more formally. Recall our equation for seigniorage:

$$\text{seigniorage} = \frac{\Delta M}{M}\left(\frac{M}{P}\right)$$

In the short run, an increase in the rate of money growth, $\Delta M/M$, leads to little change in real money balances, M/P. Put another way, if it is willing to increase money growth sufficiently, a government will be able to generate nearly any amount of seigniorage that it wants in the short run, far in excess of the numbers in the second column of Table 21-3. But over time, as prices adjust and real money balances decrease, this government will find that the same rate of money growth yields less

[handwritten margin note: Seigniorage is a good s/r policy but if you consider medium run → trouble!]

and less seigniorage. Thus, if the government keeps trying to finance a deficit larger than that shown in the second column of Table 21-3 (for example, if Austria tries to finance a deficit that is more than 13% of its GDP), it will find that it cannot do so in the long run with a constant rate of money growth. The only way it will succeed is by continually *increasing* the rate of money growth. This is why actual money growth exceeds the numbers in the first column, and why hyperinflations are nearly always characterized by increasing money growth and inflation.

There is also another effect at work. As inflation becomes very high, the budget deficit typically becomes worse. Part of the reason has to do with lags in tax collection. This effect is known as the **Tanzi-Olivera effect,** for Vito Tanzi and Julio Olivera, two economists who have emphasized its importance. As taxes are collected on past nominal income, their real value goes down with inflation. For example, if income taxes are computed on the basis of last year's income, the tax rate is 20%, and this year's price level is 100 times higher than last year's, the real value of income taxes collected this year is equal to only 20%/100 = 0.2% of income today. Thus, high inflation typically decreases real revenues, making the deficit problem worse. The problem is often compounded by other effects on the expenditure side: Governments often try to slow inflation by prohibiting firms under state control from increasing their prices, although their costs are increasing with inflation. The direct effect on inflation is small at best, but these firms' deficits then have to be financed by the government, further increasing the budget deficit. As the budget deficit increases, so does the need for more seigniorage, and thus for even higher money growth.

HYPERINFLATIONS AND ECONOMIC ACTIVITY

We have focussed so far on movements in money growth and inflation, which clearly dominate the scene during a hyperinflation. But hyperinflations affect the economy in many other ways.

Initially, higher money growth leads to an *increase* in output. The reason is that it takes some time for increases in money growth to be reflected in inflation, and during that time the effects of higher money growth are expansionary. As we saw in Chapter 19, the initial effects of an increase in nominal money growth are actually to *decrease* nominal and real interest rates, leading to an increase in demand and an increase in output.

But, as inflation becomes very high, the adverse effects of hyperinflation dominate. As inflation increases and people reduce their real money balances, the exchange system becomes less and less efficient. One famous example of inefficient exchange is the story of people using wheelbarrows to carry the currency needed for transactions at the end of the German hyperinflation. But many other decisions become distorted also. A joke told during the high inflation in Israel in the 1980s makes the point: "Why is it cheaper to take the taxi rather than the bus? Because in the bus you have to pay the fare at the beginning of the ride. In the taxi, you pay only at the end."

As inflation increases, price signals become less and less useful. Because prices change so often, it is difficult for consumers and producers to assess the relative prices of goods and to make informed decisions. Indeed, the evidence shows

During the German hyperinflation of the early 1920s, people used wheelbarrows to carry the currency they needed for transactions. Pictured here is a group of armed guards picking up their company's payroll in 1923 Berlin.

that the higher the rate of inflation, the higher the variation in the relative prices of different goods. Thus the price system, which is crucial to the functioning of a market economy, also becomes less and less efficient.

Swings in the inflation rate also become larger as inflation increases. It becomes harder to predict what inflation will be in the near future—whether it will be, say, 1000% or 1500% over the next year. Borrowing at a given nominal interest rate becomes more and more of a gamble. Borrowing and lending typically come to a stop in the last months of hyperinflation, leading to a large decline in investment.

Thus, as inflation increases and its costs become larger, there is an increasing consensus that it should be stopped.[9] This takes us to the last section of this chapter, how hyperinflations actually end.

21-4 HOW DO HYPERINFLATIONS END?

Hyperinflations do not die a natural death. Rather, they have to be stopped through what is known as a **stabilization program.**

THE ELEMENTS OF A STABILIZATION PROGRAM

What needs to be done to end a hyperinflation follows from our analysis of the causes of hyperinflation.

---

[9]*We have discussed here the cost of high inflation. We shall return to the costs of moderate inflation, say 5% to 10% a year, in Chapter 28.*

(1) There must be a fiscal reform and a credible reduction of the budget deficit. This reform must take place on both the expenditure side and the revenue side.

On the ~~expenditure side~~, reform typically implies reducing the subsidies that have often mushroomed during the hyperinflation. Obtaining a temporary suspension of interest payments on foreign debt also helps decrease expenditures. An important component of stabilization in Germany in 1923 was the reduction in reparation payments—precisely those payments that had triggered the hyperinflation in the first place.

On the revenue side, what is required is not so much an increase in overall taxation but rather a change in the composition of taxation. This is an important point. During a hyperinflation, people are in effect paying a tax, the inflation tax. Stabilization implies replacing the inflation tax with other taxes. The challenge is to put in place and collect these other taxes. This cannot be done overnight, but it is essential that people become convinced that it will be done and that the budget deficit will be reduced.

(2) The central bank must make a credible commitment that it will no longer automatically monetize the government debt. This credibility may be achieved in a number of ways. The central bank can be prohibited, by decree, from buying any government debt, so that no monetization of the debt is possible. Or the central bank can peg the exchange rate to the currency of a country with low inflation. An even more drastic step is to adopt dollarization officially, to make a foreign currency the country's official currency. This step is drastic because it implies giving up seigniorage altogether, and is often perceived as a decrease in the country's independence. An Israeli finance minister was fired in the 1980s for proposing such a measure as part of a stabilization program.[10]

(3) Are other measures needed as well? Some economists believe that incomes policies—that is, wage and/or price guidelines or controls—should be used, in addition to fiscal and monetary measures, to help the economy reach a new lower rate of inflation. **Incomes policies**, they argue, help coordinate expectations around a new lower rate of inflation. If firms know that wages will not increase, they will not increase prices. If workers know that prices will not increase, they will not ask for wage increases, and inflation will be eliminated more easily.

Others believe that credible deficit reduction and central bank independence are all that is required. They argue that the appropriate policy changes, if credible, can lead to drastic changes in expectations and thus to the elimination of expected and actual inflation nearly overnight. They point to the potential dangers of wage and price controls. Governments may end up relying on the controls and may not take the painful but needed fiscal and policy measures, leading ultimately to failure. Also, if the structure of relative prices is distorted to start with, price controls run the risk of maintaining these distortions.[11]

---

[10]*For more on the implications of money-based versus exchange-rate-based stabilization, see Rudiger Dornbusch, Federico Sturzenegger, and Holger Wolf, "Extreme Inflation: Dynamics and Stabilization,"* Brookings Papers on Economic Activity, *1990:2, 1–84.*

[11]*This last argument was particularly relevant in the case of stabilizations in Eastern Europe in the early 1990s, where, because of central planning, the initial structure of relative prices was very different from that in a market economy. See Chapter 26.*

Stabilization programs that do not include incomes policies are called **orthodox;** those that do are called **heterodox** (because they rely on both monetary-fiscal changes and income policies). The hyperinflations of Table 21-1 were all ended through orthodox programs. Many of the more recent Latin American stabilizations have relied instead on heterodox programs.

CAN STABILIZATION PROGRAMS FAIL?

Can stabilization programs fail? The answer is yes, and they often do. Argentina went through five stabilization plans from 1984 to 1989 before succeeding in 1990. Brazil went through six such plans from 1989 to 1995; the last appears to have been successful.

Sometimes failure comes from a botched or half-hearted effort at stabilization. A government puts wage controls in place but does not take the measures needed to reduce the deficit and money growth. Wage controls cannot work if money growth continues, and the stabilization program eventually fails.

Sometimes failure comes from political opposition. If social conflict was at the root of the hyperinflation, it may still be present and just as hard to resolve at the time of stabilization. Those who lose from the fiscal reform needed to decrease the deficit will oppose the stabilization program and may force the government to retreat. Often, workers who perceive an increase in the price of public services or an increase in taxation, but who do not fully perceive the decrease in the inflation tax, go on strike or even riot, leading to failure of the stabilization plan.

Failure can also come from the anticipation of failure. Suppose, for example, that the exchange rate is fixed to the U.S. dollar as part of the stabilization program.[12] Also suppose that participants in financial markets anticipate that the government will soon be forced to devalue. To compensate for the risk of devaluation, they require very high interest rates to hold domestic rather than U.S. bonds. These very high interest rates cause a large recession. The recession in turn forces the government to devalue, validating the markets' initial fears. If, instead, markets had believed that the government would maintain the exchange rate, the risk of devaluation would have been lower, interest rates would have been lower, and the government might have been able to proceed with stabilization. To many economists, the successes and failures of stabilization plans appear to have an element of self-fulfilling prophecies. Even well-conceived plans work only if they are expected to work. In other words, luck and good public relations play a role.

THE COSTS OF STABILIZATION

We saw in Section 18-4 how the U.S. and Canadian disinflations of the early 1980s were associated with a large recession and a substantial increase in unemployment. One might therefore expect the much larger disinflations associated with the end of a hyperinflation to be associated with very large recessions. This is not always the case. To understand why, recall our discussion of disinflation in Section 18-3. We saw three reasons why inflation may not decrease in line with a decrease in money growth, leading to a decrease in real money balances and a recession. The first was

---

[12]*This is a variation on the theme of exchange rate crises developed in Chapters 14 and 19.*

the fact that wages are typically set in nominal terms for some period of time, and as a result many of them are already determined when the decision to disinflate is made. The second was that wage contracts are typically staggered, making it difficult to implement a slowdown in all wages at the same time. The third was credibility.

Hyperinflation eliminates the first two problems. During hyperinflation, wages and prices are changed so often that the nominal rigidities, and the staggering of wage decisions, become nearly irrelevant.

But the issue of credibility remains. The fact that even coherent programs may not succeed implies that no program is fully credible from the start. If, for example, the government decides to fix the exchange rate, a high interest rate may be needed initially to maintain the parity. Those programs that turn out to be successful are those in which the government maintains the program and where increased credibility leads to a lower interest rate over time. But even in that case, the initial high interest rate often leads to a recession. Overall, the evidence is that most, but not all, stabilizations involve some cost in output. Much of the current research focusses on how stabilization packages should be designed to reduce this cost: orthodox versus heterodox, restrictions on money growth or fixing of the exchange rate, and so on.

▶ THE BOLIVIAN HYPERINFLATION OF THE 1980s

IN DEPTH

In the 1970s, Bolivia achieved strong output growth, in large part because of high world prices for its exports: tin, silver, coca, oil, and natural gas. But by the end of the decade the economic situation started deteriorating. The price of tin declined. Foreign lending, which had financed a large part of Bolivian spending in the 1970s, was sharply curtailed as foreign lenders started worrying about repayment prospects. Partly as a result, and partly because of long-running sociopolitical conflicts, political chaos ensued. From 1979 to 1982, the country had 12 presidents, nine from the military and three civilians.

When the first freely elected president in 18 years came to power in 1982, he faced a nearly impossible task. U.S. commercial banks and other foreign lenders were running scared. Not only did they not want to make new loans to Bolivia, they also wanted previous loans to be repaid. Net private (medium- and long-term) foreign lending to the Bolivian government had decreased from 3.5% of GDP in 1980 to −0.3% in 1982, and to −1% in 1983. Because the government felt it had no other choice, it turned to money creation to finance the budget deficit.

INFLATION AND BUDGET DEFICITS

The next three years were characterized by the interaction of steadily higher inflation and budget deficits.

Table 1 gives the budget numbers for the period 1981–1986. Because of the lags in tax collection, the effect of rising inflation was to reduce the real value of taxes dramatically. Also, the government's attempt to maintain low prices for public services was the source of large deficits for state-run firms. The result was subsidies to those firms and a further increase in the budget deficit. In 1984, the budget deficit reached a staggering 30.6% of GDP.

The result of higher budget deficits and the need for higher seigniorage was to increase inflation. (Again, remember the relation between money growth and inflation.) Inflation, which had run at an average 2.5% a month in 1981, increased to 7% in 1982 and to 11% in 1983. As shown in Figure 1, which gives Bolivia's monthly inflation rate from January 1984 to April 1986 (the vertical line indicates the beginning of stabilization, which we discuss below), inflation kept increasing in 1984 and 1985, reaching 183% in February 1985.

....

TABLE I

REVENUES, EXPENDITURES, AND THE DEFICIT, AS A PERCENTAGE OF BOLIVIAN GNP

	1981	1982	1983	1984	1985	1986
Revenues	9.4	4.6	2.6	2.6	1.3	10.3
Expenditures	15.1	26.9	20.1	33.2	6.1	7.7
Budget surplus (−: deficit)	−5.7	−22.3	−17.5	−30.6	−4.8	2.6

Revenues and expenditures of the central government.

Source: Jeffrey Sachs, "The Bolivian Hyperinflation and Stabilization," National Bureau of Economic Research, *Working Paper 2073,* November 1986, Table 3.

STABILIZATION

There were many attempts at stabilization along the way. Stabilization programs were launched in November 1982, November 1983, April 1984, August 1984, and February 1985. The April 1984 package was an orthodox program involving a large devaluation, an announcement of tax reform, and an increase in public-sector prices. But the opposition of trade unions was too strong, and the program was abandoned.

After the election of a new president, yet another attempt at stabilization was made in September 1985. This one proved successful. It was an orthodox stabi-

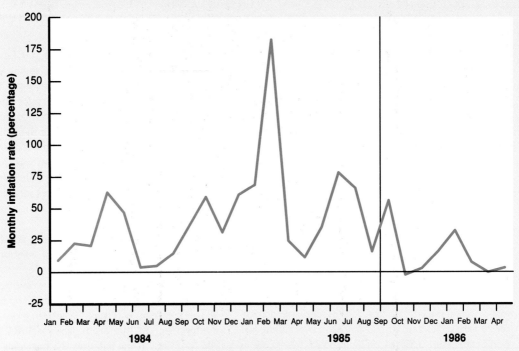

FIGURE I

Bolivian Monthly Inflation Rate, January 1984–April 1986

Source: Jeffrey Sachs, "The Bolivian Hyperinflation," NBER working paper, 1986.

lization plan, organized around the elimination of the budget deficit. Its main features were:

■ *Fiscal policy:* Public-sector prices were increased; food and energy prices were increased; public-sector wages were frozen; and a tax reform, aimed at reestablishing and broadening the tax base, was announced.

■ *Monetary policy:* The official exchange rate of the peso was adjusted to what the black market rate had been prestabilization. The exchange rate was thus set at 1.1 million pesos to the dollar, up from 67 000 pesos to the dollar the month before (a 1600% devaluation). The exchange rate was then left to float, within limits.

■ To reestablish international creditworthiness, negotiations with international organizations and commercial banks were announced. An agreement with foreign creditors and the IMF was reached nine months later, in June 1986.

As in the previous attempt at stabilization, the unions called a general strike. In response the government declared a state of siege, and the strike was quickly disbanded. After hyperinflation and so many failed attempts to control it, public opinion was clearly in favour of stabilization.

The effects on inflation were dramatic. By the second week of September, the inflation rate was actually negative! Inflation did not remain negative for very long, but the average monthly rate of inflation was below 2% during 1986–1989. As Table 1 shows, the budget deficit was drastically reduced in 1986, and the average deficit was below 5% of GNP for the rest of the decade.

Did stabilization have a negative effect on output? It probably did. Real interest rates remained very high for more than a year after stabilization. The full effect of these high real interest rates on output is hard to establish, however. At the same time that stabilization was implemented, Bolivia was hit with further large declines in the price of tin and natural gas. In addition, a major campaign against narcotics had the effect of disrupting coca production. How much of the Bolivian recession of 1986 was due to stabilization, and how much was due to these other factors, is difficult to assess.

References

The material in this box draws largely from Jeffrey Sachs, "The Bolivian Hyperinflation and Stabilization," NBER working paper, 1986. Sachs was one of the architects of the stabilization program.

See also Juan Antonio Morales, "The Transition from Stabilization to Sustained Growth in Bolivia," in *Lessons of Economic Stabilization and Its Aftermath,* Michael Bruno et al., eds. (Cambridge, MA: MIT Press, 1991).

21-5 CONCLUSIONS

This ends our two chapters on pathologies.

We argued in Chapter 20 that there is no unique high-unemployment pathology. We characterized the Great Depression as a sharp decrease in activity followed by a long and strong recovery, and high European current unemployment as a long and more insidious illness, with prospects for only a slow recovery.

By contrast, there is a clear hyperinflation pathology. High inflation has the feel of an intense but short-lived illness. Its causes are largely common across episodes: governments' inability to control their budget in the face of major shocks, economic or political. In addition, their symptoms are largely common across episodes: accelerating inflation and progressively larger real costs until stabilization is attempted and eventually achieved.

Summary

■ Hyperinflations are periods of high inflation. The most extreme ones took place after World War I and World War II in Europe. But Latin America has had episodes of high inflation as recently as the early 1990s.

■ High inflation comes from high money growth. High money growth comes in turn from the combination of large budget deficits and the inability to finance them through borrowing, either from the public or from abroad.

■ The revenues from money creation are called seigniorage. Seigniorage is equal to the product of money growth and real money balances. Thus, the smaller real money balances, the higher the required rate of money growth, and thus in turn the higher the rate of inflation required, to generate a given amount of seigniorage.

■ Hyperinflations are typically characterized by increasing inflation. There are two reasons why this is so. The first is that higher money growth leads to higher inflation,

inducing people to reduce real money balances, requiring even higher money growth (and thus leading to even higher inflation) to finance the same real deficit. The second is that higher inflation often decreases tax revenues and increases the deficit, which in turn requires higher money growth, and thus even higher inflation.

■ Hyperinflations are ended through stabilization programs. To be successful, stabilization programs must include fiscal measures aimed at reducing the deficit and monetary measures aimed at reducing or eliminating money creation as a source of financing the deficit. Some stabilization plans also include wage and price guidelines or controls.

■ A stabilization program that imposes wage and price controls without changes in fiscal and monetary policy will fail. But even coherent and well-conceived programs do not always succeed. Anticipations of failure may lead to failure of even a coherent plan.

Key Terms

- hyperinflation, 411
- debt monetization, 413
- seigniorage, 414
- barter, 415
- dollarization, 416
- Laffer curve, 419

- inflation tax, 419
- Tanzi-Olivera effect, 421
- stabilization program, 422
- incomes policies, 423
- orthodox stabilization program, 424
- heterodox stabilization program, 424

Questions and Problems

1. To see why small differences in inflation matter:
   a. Using 1.0 as the price index in the base year, calculate the price level after five years and the percentage change in the overall price level if the *monthly* rate of inflation is
      (i) 1%
      (ii) 2%
      (iii) 5%
      (iv) 10%
   b. For each of the monthly inflation rates in parts (i)–(iv) above, what would be the price of a 50-cent cup of coffee in five years?

2. Why might a government prefer an inflation rate of 1% per month to an inflation rate of 2% per month? Why might it choose a rate of 2% per month?

3. Suppose that the demand for real money balances is given by $Y[1 - (r + \pi^e)]$, where $Y = 1000$ and $r = 0.1$.
   a. Assume that, in the short run, π^e remains constant

at 0.25 or 25%. Calculate the amount of seigniorage in the short run when the money growth rate is
      (i) 25%
      (ii) 50%
      (iii) 75%
   b. In the long run, π^e is equal to $\Delta M/M$. Calculate the amount of seigniorage in the long run for each of the three money growth rates above.
   c. Are your answers consistent with the short-run and long-run behaviour of seigniorage as described in this chapter? Explain briefly.

4. In the text we ignored the distinction between money (M) and high-powered money (H). Suppose that the money multiplier is 2, so that $M = 2H$.
   a. Does it still make sense to define the amount of seigniorage (the government revenues from money creation) as $\Delta M/P$? If so, explain why. If not, give the new definition of seigniorage.

b. Will equation (21.2) still hold? If so, explain why. If not, give the new equation.

c. Does seigniorage still depend positively on both money growth and real money balances?

5. "There are only two ways for the government to finance its budget deficit: tax or borrow." Comment.

6. Would you expect the adjustment to lower real money balances to take longer when money growth increases from 10% to 200%, or when it increases from 200% to 400%? Why?

7. How would each of the following change the Tanzi-Olivera effect?

a. Requiring monthly instead of yearly tax payments by households

b. Assessing greater penalties for under-withholding of taxes from weekly paycheques

c. Calculating each month's income in end-of-year dollars (or domestic currency) and assessing taxes on the total

8. Consider the following variation on Milton Friedman's dictum (given in Chapter 18): "Inflation is always a fiscal phenomenon." Comment, in the light of what you have learned in this chapter.

FURTHER READING

Two good reviews of what economists know and don't know about hyperinflation are:

Rudiger Dornbusch, Federico Sturzenegger, and Holger Wolf, "Extreme Inflation: Dynamics and Stabilization," *Brookings Papers on Economic Activity,* 1990:2, 1–84.

Pierre Richard Agenor and Peter Montiel, *Development Macroeconomics* (Princeton, NJ: Princeton University Press, 1995), Chapters 8 to 11. Chapter 8 is easy reading; the other chapters are more difficult.

The experience of Israel, which went through high inflation and stabilization in the 1980s, is described in Michael Bruno, *Crisis, Stabilization and Economic Reform* (New York: Oxford University Press, 1993), especially Chapters 2 to 5. Michael Bruno was the head of Israel's central bank for most of that period.

Much recent research has focussed in particular on how to end hyperinflations:

One of the classic articles is "The Ends of Four Big Inflations," by Thomas Sargent, in Robert Hall, ed., *Inflation: Causes and Effects* (Chicago: NBER and the University of Chicago, 1982), 41–97. In that article, Sargent argues that a credible program can lead to stabilization at little or no cost in terms of activity.

Rudiger Dornbusch and Stanley Fischer, "Stopping Hyperinflations, Past and Present," in *Weltwirtschaftliches Archiv,* 1986:1, 1–47, gives a very readable description of the end of hyperinflations in Germany, Austria, Poland, and Italy in 1947, Israel in 1985, and Argentina in 1985.

THE FACTS OF GROWTH

Our perceptions of how the economy is doing tend to be dominated by year-to-year fluctuations in activity. A recession leads to gloom, an expansion to optimism. But if one steps back and looks at activity over longer periods of time—say, over the course of many decades—the picture changes markedly. Fluctuations fade in importance, and **growth**—the steady increase in aggregate output over time—dominates the picture.

Figure 22-1 shows the evolution of Canadian GDP (in 1986 dollars) since 1926. We have shaded both the years from 1929 to 1933—corresponding to the large decrease in output during the Great Depression—and the years 1980 to 1982, corresponding to the largest postwar recession. Even these episodes appear small compared to the steady increase in output over the last 70 years.[1]

---

[1]*The scale used to measure GDP on the vertical axis is called a **logarithmic scale.** It differs from the standard, linear, scale in the following way. Take a variable that grows over time at, say, 3% per year. Then, the larger the variable, the larger will be its increase from one year to the next. When GDP was $44 billion (in 1986 dollars) in 1926, a 3% increase was equal to $1.3 billion; in 1996, with GDP at $618 billion, a 3% increase was equal to $18.5 billion. If we were to plot GDP using a linear scale, the increments would become larger and larger over time. A logarithmic scale is one in which the same proportional increase represents the same distance on the scale. Thus, an increase of 3% is always represented by the same distance on the scale. Put another way, the behaviour of a variable that grows at a constant rate will be represented by an exploding curve when a linear scale is used, but by a straight line when a logarithmic scale is used. The slope of the line will be equal to the rate of growth: If a variable grows at 3% per year, the slope of the line will be 0.03. Even when a variable has a growth rate that varies from year to year, as is indeed the case for GDP, the slope at any point in time still gives the variable's growth rate at that point in time. This is the reason for using a logarithmic scale to plot variables that grow over time: By looking at the slope, we can easily see what is happening to the growth rate.*

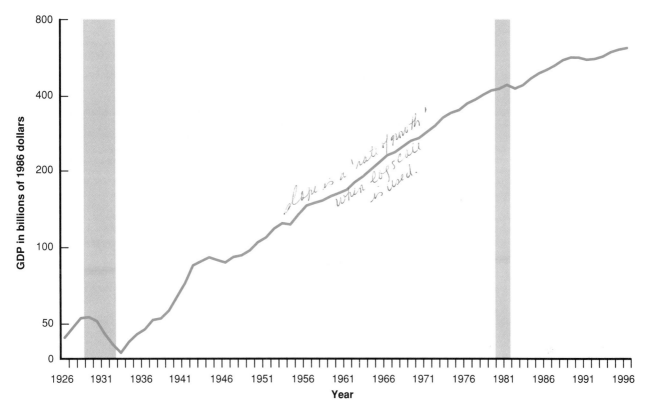

FIGURE 22-1

Canadian GDP, 1926–1996

Aggregate Canadian output has increased by a factor of 14 from 1926 to 1996.

Source: Statistics Canada, CANSIM Series D14442.

Our focus so far in this book has been on fluctuations. In this and the next four chapters, we focus on growth instead. This chapter presents the facts of growth, from Canada and elsewhere, and from the recent as well as the more distant past. It then articulates a framework for thinking about the determinants of growth. This framework is developed in the next two chapters. Chapter 23 focusses on capital accumulation; Chapter 24 focusses on technological progress.

Building on our understanding of both fluctuations and growth, Chapters 25 and 26 then focus on two current issues. Chapter 25 looks at the relation between technological progress, unemployment, and wages. It examines whether technological progress leads to unemployment, an old argument that is still heard in many countries today. It also looks at whether technological progress is at the root of the increase in wage inequality that has occurred in Canada and the United States during the last 20 years. Chapter 26 looks at the transition from centrally planned to market economies in Eastern Europe and examines why growth slowed down under the central planning system, how the transition has proceeded so far, and what the prospects for growth are in the future.

22-1 GROWTH IN RICH COUNTRIES SINCE 1950

Table 22-1 gives the evolution of **output per capita** (GDP divided by population) for the G7 countries—Canada, France, Germany, Italy, Japan, the United Kingdom, and the United States—since 1950. We have chosen these countries, not only because they are the world's major economic powers today, but also because their experience is broadly representative of that of advanced countries (the countries that are OECD members) over the last half century or so.[2]

Note that the table gives numbers for output *per capita* rather than for output. There are two reasons for focussing on the former rather than the latter. The evolution of a country's output per capita gives a better sense of that country's improvement in the standard of living than does the evolution of total output. And, when comparing countries with different populations, output numbers must be normal-

....
TABLE 22-1
THE EVOLUTION OF OUTPUT PER CAPITA IN G7 COUNTRIES SINCE 1950

	ANNUAL GROWTH RATE OUTPUT PER CAPITA (%)		REAL OUTPUT PER CAPITA (1985 U.S. DOLLARS)			
	1950–1973	1973–1992	1950	1973	1992	Ratio of real output per capita, 1992/1950
Canada	2.7	1.7	6 380	11 854	16 362	2.6
France	4.2	1.6	4 045	10 316	13 918	3.4
Germany	4.9	1.9	3 421	10 315	14 709	4.3
Italy	4.9	2.3	2 743	8 275	12 721	4.6
Japan	8.1	3.0	1 430	8 539	15 105	10.6
United Kingdom	2.5	1.5	5 395	9 609	12 724	2.4
United States	2.2	1.2	8 772	14 379	17 945	2.0
Average	4.2	1.9	4 598	10 470	14 783	3.2

Average is unweighted. Germany is West Germany only.

Source: Robert Summers and Alan Heston, *Penn World Tables*, 1995.

---
[2]*See the Focus box in Chapter 1 (page 18) for a list of OECD member countries.*

ized to take into account these differences in population size. This is exactly what output per capita does.

Before discussing the table, we must discuss the way the output numbers are constructed. So far, in constructing output numbers in dollars for countries other than Canada, we have used the straightforward method of taking that country's GDP expressed in that country's currency, then multiplying it by the current exchange rate to express it in terms of dollars (see Chapter 1). But this simple computation will not do here, for two reasons.

First, as we saw in Chapters 13 and 14, under flexible exchange rates there can be large fluctuations in exchange rates. (For example, this is the case when monetary policies lead to large interest rate differentials.) The dollar depreciated and then appreciated in the 1980s by roughly 25% vis-à-vis the currencies of its trading partners. But surely Canadian output per capita—considered an index of the Canadian standard of living—did not decrease and then increase by 25% compared with those of its trading partners in the 1980s. Yet this is the conclusion we would reach if we compared GDPs using current exchange rates.

The second reason goes beyond fluctuations in exchange rates. GDP per capita in India, using the current exchange rate, is equal to $458, compared with $26 241 in Canada. Surely nobody could live on $458 a year in Canada, but people live on it— admittedly not very well—in India. The reason is that the prices of basic goods, those needed for subsistence, are much lower in India than in Canada. Thus the average consumer in India, who consumes mostly necessities, is not in fact 57 times worse off than his or her Canadian counterpart. This pattern applies to other countries besides Canada and India: In general, the lower a country's income, the lower the prices of food and basic services in that country.

Thus, when our focus is on comparing standards of living, either across time or across countries (as it is here), we get more meaningful comparisons by correcting for the effects we have just discussed. This is what the numbers in Table 22-1 do. Further discussion and the details of computation are given in the Focus box entitled "The Construction of PPP Numbers." But the principle of construction is simple, that of using a common set of prices for similar goods and services produced in each economy. Such adjusted real GDP numbers, which you can think of as measures of **purchasing power** across time or across countries, are called **purchasing power parity (PPP)** numbers.

The differences between PPP numbers and the numbers based on current exchange rates can be substantial. Take our comparison between India and Canada. Using PPP numbers,[3] GDP per capita in Canada is equal to 13 times GDP per capita in India. This is still a large difference, but less than the 57 times we derived using current exchange rates. Table 22-2 on page 435 compares GDP per capita for some OECD countries, using both exchange rates and PPP. Although the differences are not as large as in our comparison between Canada and India, they are still substantial. In 1996, using 1996 exchange rates, GDP per capita in Canada was one of the lowest in the OECD and well below the average in the European Union. But using PPP numbers, Canada was one of the richest countries in the world, with a GDP per capita that was about 11% above the average in the EU.

---

[3] These data are from 1995, because they are the latest available that allow us to compare Canada and India.

Consider two countries, which for the sake of concreteness we shall call the United States and Russia, but without attempting to fit the facts of these two countries very closely.

In the United States, consumption per capita equals $20 000. Individuals buy two goods. They buy a new car every year, for $10 000, and spend the rest on food. The price of a yearly bundle of food is $10 000.

In Russia, consumption per capita equals 6 million rubles. People keep their cars for 20 years. The price of a car is 40 million rubles, so that individuals spend on average a twentieth of that—2 million rubles—a year on cars. They buy the same bundle of food as their U.S. counterparts, at a price of 4 million rubles.

Russian and U.S. cars are of identical quality, and so are Russian and U.S. food. (You may dispute the realism of these assumptions. Whether a car in country X is the same as a car in country Y is very much the type of problem confronting economists constructing PPP measures.) The exchange rate is such that one dollar is equal to 4000 rubles. What is relative consumption per capita in Russia compared with that in the United States?

We can attempt to answer this question by taking consumption per capita in Russia and converting it into dollars using the exchange rate. Using that method, Russian consumption in dollars is 6 000 000/4000 = $1500, thus 7.5% of U.S. consumption.

Does this answer make sense? True, Russians are poorer, but food is relatively much cheaper in Russia. A U.S. consumer spending all on food would buy ($20 000/$10 000) = 2 bundles of food. A Russian consumer spending all on food would buy (6 000 000 rubles/4 000 000 rubles) = 1.5 bundles of food. In terms of food bundles, the difference between U.S. and Russian income looks much smaller. And given that two-thirds of consumption in Russia goes to spending on food, this seems like a relevant computation.

So how can we improve on our initial answer? One way is to use the same set of prices for both countries and then measure the quantities of each good consumed in each country using this common set of prices. Suppose that we use U.S. prices. In terms of U.S. prices, consumption per capita in the United States is obviously still $20 000. What is it in Russia? The average Russian buys 0.05 cars a year and one bundle of food. Using U.S. prices—specifically, $10 000 for a car and $10 000 for a bundle of food—gives Russian consumption per capita as [(0.05 × $10 000) + (1 × $10 000)] = $10 500. This puts Russian consumption per capita at $10 500/$20 000 = 52.5% of U.S. consumption per capita, a better estimate of relative standards of living than we obtained using our first method (which gave 7.5%).

This type of computation underlies the PPP estimates we use in the text, which are the results of an ambitious project known as the "Penn World Tables." Led by three economists (Irving Kravis, Robert Summers, and Alan Heston) over the course of more than 15 years, this project has constructed PPP series not only for consumption (as we just did in our example), but more generally for GDP and its components, since 1950, for most countries in the world. The latest available numbers are for 1992.

Source: Robert Summers and Alan Heston, "The Penn World Table Mark 5: An Expanded Set of International Comparisons, 1950–1988," *Quarterly Journal of Economics*, 1991:2, 327–368.

Let us now return to the numbers in Table 22-1. You should draw three main conclusions from the table:

(1) First and foremost is how strong growth has been in all five countries, how much the standard of living has improved over the last 40 years. Growth since 1950 has increased real output per capita by a factor of 2.6 in Canada, by a factor of 4.3 in Germany, and by a factor of 10.6 in Japan.

These numbers show what is sometimes called the *force of compounding*. You have probably heard how saving even a little while you are young will build to a large amount by the time you retire. For example, if the interest rate is 6% a year, an investment of one dollar, with the proceeds reinvested every year, leads to about

TABLE 22-2

GDP PER CAPITA USING EXCHANGE RATES AND PPP FOR SOME OECD COUNTRIES, 1996.

	BASED ON EXCHANGE RATES	BASED ON PURCHASING POWER PARITIES
Canada	19 330	21 529
United States	27 821	27 821
Japan	36 509	23 235
Belgium	26 409	21 856
France	26 323	20 533
Germany	28 738	21 200
Greece	11 684	12 743
Italy	21 127	19 974
Luxembourg	40 791	32 416
Norway	36 020	24 364
Sweden	28 283	19 258
Switzerland	41 411	25 402
United Kingdom	19 621	18 636
European Union	23 042	19 333

Figures are expressed in U.S. dollars.

Source: OECD National Accounts Main Aggregates January 1998.

twelve dollars 42 years later [$(1 + 6\%)^{42} = 11.55$ dollars]. The same logic applies to the Japanese growth rate over the 1950–1992 period. The average annual growth rate in Japan over that 42-year period was equal to 5.8% {[(8.1% × 23 years) + (3.0% × 19 years)]/42}, leading to a more than tenfold increase in real output per capita. Clearly, a better understanding of growth, if it leads to the design of policies that stimulate growth, can have a very large effect on the standard of living. Policy measures that increased the growth rate from, say, 2% to 3% would lead to a standard of living 100% higher after 40 years than it would have been without the policy.

(2) Growth has slowed down since the mid-1970s.

The first two columns of Table 22-1 show growth rates for both pre- and post-1973. Pinpointing the exact date of the slowdown is difficult; 1973, the date used to split the sample in the table, is as good as any date in the mid-1970s.

Growth has slowed down in all the G7 countries. The slowdown has been stronger, however, in the countries that were growing fast pre-1973, such as France, Germany, and especially Japan, with the result that the differences in growth rates across countries are smaller post-1973 than they were pre-1973.

If it continues, this decline in growth will have profound implications for the evolution of income per capita in the future. At a growth rate of 4.2% per year (the unweighted average growth rate across our seven countries from 1950 to 1973) it takes 17 years for the standard of living to double. At a growth rate of 1.9% per year—the average from 1973 to 1992—it takes 37 years. Expectations of fast growth in individual income that developed in the 1950s and 1960s have had to confront the reality of slow growth since 1973. Indeed, for some socioeconomic groups, slow growth of income per capita for the economy as a whole, together with a decline in their income relative to the average, has led to an absolute decline in their real income. Typically, the most adversely affected by the growth slowdown have been the least skilled workers. We explore this topic in detail in Chapter 25.

(3) Levels of output per capita across the seven countries have converged over time. Put another way, those countries that were behind have grown faster, reducing the gap between them and the United States.

In 1950, output per capita in the United States was around twice that of the United Kingdom, Germany, and France, and more than six times that of Japan. In Japan and Europe, the United States was seen as the land of plenty, where everything was bigger and better. Today these perceptions have faded, and the numbers explain why. Using PPP numbers, U.S. output per capita is still the highest, but in 1992 it stood only 25% above German and French output, and only 20% above Japanese output.

This **convergence** of levels of output per capita across countries extends to the set of OECD countries. OECD convergence is shown in Figure 22-2, which plots the average annual growth rate of output per capita from 1950 to 1992 against the initial level of output per capita in 1950 for the set of countries that are members of the OECD today. There is a clear negative relation between the initial level of output per capita and the growth rate since 1950. The relation is not perfect: Turkey, which had roughly the same low level of output per capita as Japan in 1950, has had a growth rate equal to only about half that of Japan. But the relation is clearly there. Countries that were behind in 1950 have typically grown faster. ✓ *Catching up effect*

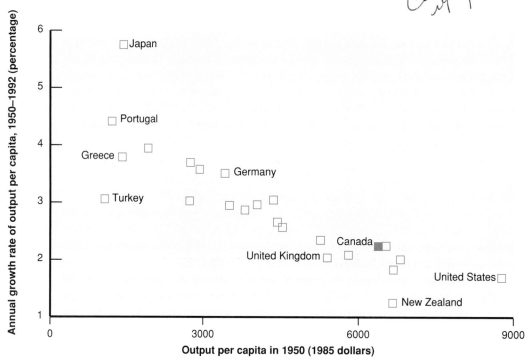

FIGURE 22-2

Growth Rate of GDP per Capita since 1950 versus GDP per Capita in 1950, OECD Countries

Countries that had a lower level of output per capita in 1950 have typically grown faster.

Source: Robert Summers and Alan Heston, "The Penn World Table Mark 5: An Expanded Set of International Comparisons, 1950–1988," *Quarterly Journal of Economics,* 1991:2, 327–368.

The issue of convergence of output per capita has been a hot topic of macroeconomic research over the past decade. Some have pointed to a potential flaw in graphs like Figure 22-2. By looking at the set of countries that are members of the OECD today, what we have done in effect is look at a club of economic winners. OECD membership is not officially based on economic success, but economic success is surely an important determinant of membership. But when one looks at a club whose membership is based on economic success, one will find that those who came from behind had the fastest growth; this is precisely why they made it into the club. Thus, the finding of convergence could come in part from the way we selected the countries in the first place.

A better way of looking at convergence is thus to define the set of countries not on the basis of where they are today—as we did in Figure 22-2—but on the basis of where they were in, say, 1950. For example, if we are interested in convergence among relatively rich countries, we can look at all countries that had an output per capita of, say, more than $2000 in 1950 (in 1985 U.S. dollars), then look for convergence within that group. It turns out that most of the countries in that group have indeed converged, and therefore convergence is not solely an OECD phenomenon. However, a few countries—Uruguay, Argentina, and Venezuela among them—have not converged. Perhaps the most striking case is that of Argentina. Output per capita in Argentina, which was equal to $4032 in 1950—similar to France then—stood at only $4706 in 1990, a meager 17% increase in 40 years, and far below the 1990 French level of $13 904.

22-2 A BROADER LOOK ACROSS TIME AND SPACE

The large increase in the standard of living since 1950, the slowdown in growth since the mid-1970s, and convergence among rich countries are the three basic facts we shall keep in mind and try to explain as we go along. But before we do so, we must have a broader look at the evidence, looking both across a longer time span and over a wider set of countries. Both extensions are needed to place the rich countries' recent experience in context.

LOOKING ACROSS TWO MILLENNIA

Have the currently rich economies always grown at growth rates similar to those in Table 22-1? Even without looking at history, logic implies that the answer must be no. For example, suppose that the annual growth rate in the seven countries of Table 22-1 had been as small as 0.5% per year since year 0 of the Christian calendar (clearly an arbitrary date here). Working backward, this implies that output per capita in year 0 was 72 cents per year (in 1985 U.S. dollars), clearly an absurdly small number.[4]

An actual look at history confirms this conclusion. Estimates of growth are clearly harder to construct as one goes further back in time. But there is

---

[4] *To check that this is indeed the solution, note that 72 cents $\times (1.005)^{1992}$ is indeed approximately equal to $14 783 (the average output per capita in Table 22-1 for 1992).*

THE REALITY OF GROWTH: A U.S. "WORKINGMAN'S BUDGET" IN 1851

Data on GDP per capita do not fully convey the reality of growth and the accompanying increase in the standard of living. An examination of a "workingman's budget" in 1851 Philadelphia gives a much better sense of the improvement.

Note how much a worker consumed on food, 41% of expenditures. Today's corresponding share—as reflected in the composition of the consumption basket

used to compute the U.S. Consumer Price Index—is only 14.4%. And food at home—as opposed to food in restaurants—accounts for only 8.9% of total consumption today. But perhaps more revealing is the composition of food consumption. Compare the food in the table with the richness and diversity of the food we eat today.

YEARLY "WORKINGMAN'S BUDGET," PHILADELPHIA, 1851

ITEM OF EXPENDITURE	AMOUNT (DOLLARS)	PERCENTAGE OF TOTAL
Butcher's meat (2 pounds a day)	72.80	13.5
Flour ($6\frac{1}{2}$ pounds a year)	32.50	6.0
Butter (2 pounds a week)	32.50	6.0
Potatoes (2 pecks a week)	26.00	4.8
Sugar (4 pounds a week)	16.64	3.1
Coffee and tea	13.00	2.4
Milk	7.28	1.4
Salt, pepper, vinegar, starch, soap, yeast, cheese, eggs	20.80	3.9
Total expenditures for food	221.52	41.1
Rent	156.00	29.0
Coal (3 tons a year)	15.00	2.8
Charcoal, chips, matches	5.00	0.9
Candles and oil	7.28	1.4
Household articles (wear, tear, and breakage)	13.00	2.4
Bedclothes and bedding	10.40	1.9
Wearing apparel	104.00	19.3
Newspapers	6.24	1.2
Total expenditures other than food	316.92	58.9

Source: William Baumol et al., *Productivity and American Leadership* (Cambridge, MA: MIT Press, 1989), Chapter 3, Table 3.2. The composition of expenditures today comes from Table 708 (Average Annual Income and Expenditures of All Consumer Units, 1991) in the *Statistical Abstract of the United States,* 1993.

agreement among economic historians about the main evolutions over the last 2000 years.

From the end of the Roman Empire to roughly year 1500, there was essentially no growth of output per capita in Europe; most workers were employed in agriculture, in which there was little technological progress. Because agriculture's share of output was so large, inventions with applications outside agriculture could contribute little to overall production and output. From about 1500 to 1700, growth of output per capita turned positive but small, around 0.1%, increasing to 0.2% from 1700 to 1820. Even during the Industrial Revolution, growth rates were not high by current standards. The growth rate of output per capita from 1820 to 1950 in the United States was only equal to 1.5%. (For a better understanding of what such growth really means see the Focus box entitled 'The Reality of Growth: A U.S. "Workingman's Budget" in 1851.')

On the scale of human history, therefore, growth of output per capita is a recent phenomenon. In the light of the growth record of the last 200 years or so, what appears unusual is the high growth rate achieved in the 1950s and the 1960s rather than the lower growth rate since 1973.

History also puts into context the convergence of OECD countries to the level of U.S. output per capita since 1950. The United States was not always the world's economic leader. Rather, history looks more like a long-distance race in which one country assumes leadership for some time, only to pass it on to another and return to the pack or disappear from sight. For much of the first millennium and until the fifteenth century, China probably had the world's highest level of output per capita. For a couple of centuries, leadership moved to the cities of northern Italy. It was then assumed by the Netherlands until around 1820, and then by the United Kingdom from 1820 to around 1870. Since then, the United States has had the lead. Seen in this light, history looks more like **leapfrogging** (in which countries get close to the leader and then overtake it) than like convergence (in which the race becomes closer and closer). If history is any guide, the United States may not remain in the lead much longer.

LOOKING ACROSS COUNTRIES

We saw how OECD countries appear to converge over time, in terms of output per capita. But what about the world's other countries? Are the poorest countries also growing relatively faster? Are they converging toward the United States, even if they are still far behind?

The answer is given in Figure 22-3, which plots the annual growth rate of output per capita from 1960 to 1992 against output per capita in 1960, now for 97 countries. Figure 22-3 shows no clear pattern.[5] Over the last 30 years, convergence has by no means been the rule. Countries that were relatively poorer in 1960 have not in general grown faster.

But the cloud of points in fact hides a number of interesting subpatterns, which appear when we put countries into different groups. We do this in Figure 22-4, which shows growth rates for three groups of countries. The first, denoted by diamonds, is the set of OECD countries that we looked at earlier. The second, denoted by squares, is the set of African countries. The third, denoted by triangles, is a set of four Asian countries: Singapore, Taiwan, Hong Kong, and South Korea. Together, these three groups account for 59 countries. To avoid cluttering, Figure 22-4 leaves out all other countries; these show less obvious patterns. The figure yields three main conclusions:

(1) The picture for OECD countries is very much the same as in Figure 22-2, which looked at a slightly longer period of time (from 1950 on, rather than from 1960 on here). Nearly all start at relatively high levels of output per capita (say, at least one-third of the U.S. level in 1960), and there is clear evidence of convergence.

(2) Convergence is also evident in the case of the four Asian countries. While Japan (a member of the OECD, and thus represented by a diamond) was the first of the

---

[5]*The numbers for 1950 are missing for too many countries to use 1950 as the initial year, as we did in Figure 22-2. The figure includes all the countries for which PPP estimates of GDP per capita exist for both 1960 and 1992 (or, in some cases, 1990 or 1991). There are some notable absences, such as China and a number of Eastern European countries, for which the numbers for 1960 are not available.*

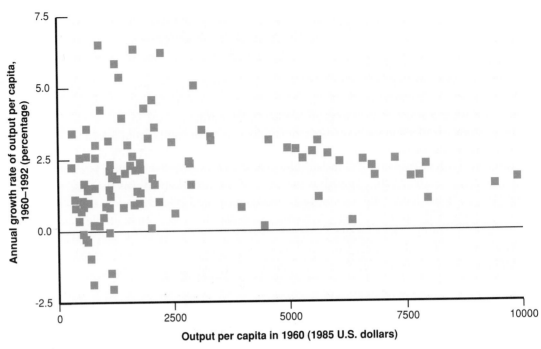

FIGURE 22-3

Growth Rate of GDP per Capita, 1960–1992, versus GDP per Capita in 1960: 97 Countries

There is no clear relation between the growth rate of output since 1960 and the level of output per capita in 1960.

Source: Robert Summers and Alan Heston, "The Penn World Table Mark 5: An Expanded Set of International Comparisons, 1950–1988," *Quarterly Journal of Economics,* 1991:2, 327–368.

Asian countries to grow rapidly and now has the highest level of output per capita in Asia, a number of other Asian countries are trailing it closely. Singapore, Taiwan, Hong Kong, and South Korea—sometimes called the **four tigers**—have had average annual growth rates of GDP per capita in excess of 6% over the last 30 years. In 1960, their average output per capita was about 16% of the U.S. number; by 1992, it had increased to 62% of U.S. output. A number of other Asian countries, among them Indonesia, Malaysia, China, and Thailand (which are not included in the figure because of the lack of data for 1960), have recently grown at similar rates and are also rapidly catching up.[6]

(3) The picture is very different for African countries. Convergence is certainly not the rule in Africa. Most African countries were very poor in 1960, and many have experienced negative growth of output per capita—an absolute decline in their standard of living—since then. Output per capita has declined at the rate of 1.3% annually in Chad and Madagascar since 1960; even in the absence of major wars, output in these two countries stands at 67% of its 1960 level. Why so many African countries are not growing, let alone converging, is one of the main questions facing development economists today.

---

[6]*The Asian crisis of 1997 has stalled growth in many of these countries, but it is too soon to judge whether the previous high growth rates in Asia will return.*

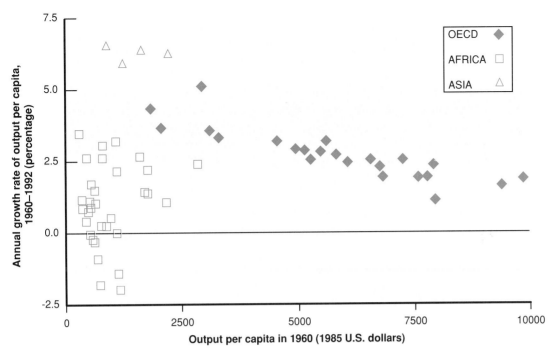

FIGURE 22-4
Growth Rate of GDP per Capita, 1960–1992, versus GDP per Capita in 1960: OECD, Africa, and Asia

Asian countries are converging to OECD levels. There is little evidence of convergence for African countries.

Source: See Figure 22-2.

We shall not take on the wider challenges raised by the facts presented in this section. Doing so would take us too far into economic history and development economics. But they put into perspective the three basic facts we discussed earlier for the OECD:

(1) Growth is not an historical necessity. There was little growth for most of human history, and in many countries today growth remains elusive. Thus theories that explain growth in the OECD today must also be able to explain its absence in the past and its absence in Africa today.

(2) Convergence of OECD countries to the United States may well be the prelude to leapfrogging, a stage when economic leadership passes from one country to another. Theories that explain convergence must therefore also allow for the possibility that it will be followed by leapfrogging and the appearance of a new leader.

(3) Finally, in a longer historical perspective, it is not so much the lower growth since 1973 in the OECD that is puzzling. More puzzling is the earlier period of exceptionally fast growth. Finding the explanation for a slower growth today may well come from understanding what factors contributed to fast growth post–World War II, and whether these factors have disappeared.

Convergence is not the rule in Africa today. Most African countries were very poor in 1960, and many have experienced negative output growth since then. Another problem is the civil wars that have plagued Africa in recent years; thousands of people turn to refugee camps for basic supplies like food and water.

22-3 THINKING ABOUT GROWTH: A PRIMER

How do we explain all these facts? What are the roles of capital accumulation and technological progress in growth? To think about and answer these questions, economists use a framework developed originally by Robert Solow, from MIT, in the late 1950s.[7] The framework has proven sturdy and useful, and we shall use it here. This section provides an introduction. Chapters 23 and 24 provide a more detailed analysis, first of the role of capital accumulation and then of technological progress.

THE AGGREGATE PRODUCTION FUNCTION

Think of aggregate output as being produced using two inputs, capital and labour:

$$Y = F(K, N) \tag{22.1}$$

As before, Y is aggregate output. K is capital—the sum of all the machines, plants, office buildings, and housing in the economy. N is labour—the number of workers in the economy. The function F, which tells us how much output is produced for given quantities of capital and labour, is called the **aggregate production function.**

This way of thinking about production is an improvement on our treatment in earlier chapters where we assumed that production required only labour. But it is still a drastic simplification. Surely, machines and office buildings play very different roles in the production of aggregate output and should be treated as separate inputs. Surely, workers with PhDs are different from high-school dropouts; yet, by constructing the labour input as simply the *number* of workers in the economy, we treat all workers as identical. We shall relax some of these simplifications later. For the time being, we shall use the production function in equation (22.1).

What does the aggregate production function F itself depend on? In other words, how much output can be produced for given quantities of capital and labour? The answer depends on the **state of technology.** A country with a more advanced technology will produce more output from the same quantities of capital and labour than will an economy with only a primitive technology.

What do we mean by technology? In a narrow sense we can think of the state of technology as the list of blueprints defining both the range of products that can be produced in the economy as well as the techniques available to produce them. We can also think of the state of technology in a broader sense. How much output is produced in an economy also depends on how well firms are run, on the organization and sophistication of markets, on the system of laws and their enforcement, on the political environment, and so on.[8] We shall think of the state of technology in the narrow sense for most of the next two chapters. We shall briefly return at the end of Chapter 24 to what we know about the role of the other factors, from the system of laws to the form of government.

---

[7]The article in which Solow presented this framework is called "A Contribution to the Theory of Economic Growth," Quarterly Journal of Economics, February 1956, 65–94. Solow received the Nobel Prize in 1987 for his work on growth.

[8]To take another example, we saw in Chapter 21 how hyperinflation disrupts production and the functioning of markets, and decreases output.

Returns to scale and returns to factors. What restrictions should we impose on the aggregate production function?

Consider a thought experiment in which we double both the number of workers and the amount of capital in the economy. It is reasonable to guess that output will roughly double as well. In effect, we have cloned the original economy, and the clone economy can produce output in the same way as the original economy. This property is called **constant returns to scale:** If the scale of operation is doubled—that is, if the quantities of capital and labour are doubled—then output will double:

$$2Y = F(2K, 2N)$$

Or, more generally, for any number λ,

$$\lambda Y = F(\lambda K, \lambda N) \tag{22.2}$$

Constant returns to scale refers to what happens to production when *both* capital and labour are increased. What should we assume when only *one* input—say capital—is increased?

It is surely reasonable to assume that output will increase as well. It is also reasonable to assume that the same increase in capital will lead to smaller and smaller increases in output as the level of capital increases. Why? Think, for example, of a secretarial pool, composed of a given number of secretaries. The introduction of just one computer will substantially increase the pool's production, as some of the more time-consuming tasks will now be done automatically by the computer. As the number of computers increases and more secretaries in the pool get their own PCs, production will further increase, although by less per additional computer than was the case when the first one was introduced. Once all the secretaries have their own PCs, increasing the number of computers further is unlikely to increase production very much, if at all. Additional computers may simply remain unused and left in their shipping boxes, and lead to no increase in output whatsoever.

We shall refer to the property that increases in capital lead to smaller and smaller increases in output as the level of capital increases as **decreasing returns to capital** (a property that will be familiar to those who have taken a course in microeconomics). A similar property holds for the other input, labour: Increases in labour, given capital, lead to smaller and smaller increases in output as the level of labour increases. There are **decreasing returns to labour** as well.

Output and capital per worker. The production function we have written and the two properties we have just assumed imply a simple relation between output per worker and capital per worker.

To see why, take λ equal to $1/N$ in equation (22.2), so that we get a relation between output and capital per worker:

$$\frac{Y}{N} = F\left(\frac{K}{N}, 1\right) \tag{22.3}$$

This equation says that the amount of output per worker depends on the amount of capital per worker. This relation between output per worker and capital per worker is drawn in Figure 22-5.

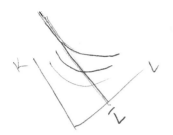

FIGURE 22-5

Output and Capital
per Worker

Increases in capital per
worker lead to smaller and
smaller increases in output
per worker.

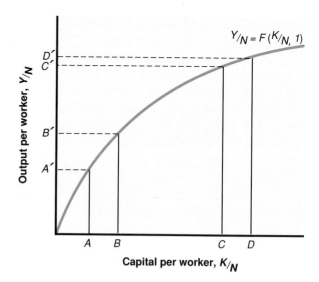

Output per worker (Y/N) is measured on the vertical axis, capital per worker (K/N) on the horizontal axis. The relation between the two is given by the upward-sloping curve. As capital per worker increases, so does output per worker. But, because of decreasing returns to capital, the curve is drawn so that increases in capital lead to smaller and smaller increases in output. At point *A,* where capital per worker is low, an increase in capital per worker equal to the distance *AB* leads to an increase in output per worker of *A'B'.* At point *C,* where capital per worker is larger, the same increase in capital per worker, *CD* (where the distance *CD* is equal to the distance *AB*) leads to a much smaller increase in output per worker, only *C'D'.* This is just as in our example of the secretarial pool, where additional computers led to less and less effect on total output.

THE SOURCES OF GROWTH

We are now ready to return to growth. Where does growth come from? Why does output per worker—or output per capita, if we assume the ratio of workers to the population as a whole remains roughly constant over time—go up over time? Equation (22.3) gives a simple answer:

- Increases in output per worker (Y/N) can come from increases in capital per worker (K/N). This is the relation we just looked at in Figure 22-5.
- Or they can come from improvements in the state of technology, which shift the production function, *F,* and lead to more output per worker *given* capital per worker.

Thus we can think of growth as coming from **capital accumulation** and/or from **technological progress**—the improvement in the state of technology. Let's look at each in turn.

Capital accumulation.   Can capital accumulation *by itself* sustain output growth forever? The answer is no. A formal argument will have to wait until Chapter 23.

But we can derive the basic intuition for this answer from Figure 22-5. Because of decreasing returns to capital, sustaining a steady increase in output per worker would require larger and larger increases in the level of capital per worker. At some stage, society will not be willing to save enough to further increase capital, and output per worker will stop growing.

savings (+ Investment)

Does this mean that an economy's **saving rate**—the proportion of income that is saved—is irrelevant? The answer is again no. True, a higher saving rate cannot permanently sustain *higher growth of output*. But it can sustain a higher *level* of output. To understand this important distinction, think about the following example.

Consider two countries, *A* and *B*. Assume that country *A* has a higher saving rate than country *B*: $s_A > s_B$. Assume also that both economies have the same rate of technological progress. Then, ignoring fluctuations, the evolution of output per capita in the two countries will look as in Figure 22-6 (which uses a logarithmic scale to measure output, so that an economy where output grows at a constant, or steady, rate is represented by a line with slope equal to the growth rate).[9]

Because both countries have the same rate of technological progress, output per capita grows at the same rate in both countries; this is reflected by the fact that the two lines are parallel. But, at any point in time, the country with a higher saving rate will have a higher level of output per capita, and thus be richer than the other; this is reflected by the fact that the line for country *A* is above the line for country *B*.

Country w/ higher svg rate shifts intercept & therefore per capita output is higher; but it won't result in a sustained higher level of output.

How and how much the saving rate affects the level of output, and whether a country such as Canada (which has a very low saving rate) should try to increase its saving rate, are the topics taken up in Chapter 23.

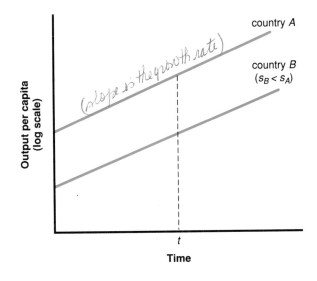

(slope is the growth rate)

country *A*

country *B*
$(s_B < s_A)$

Output per capita
(log scale)

t

Time

FIGURE 22-6

The Effects of Different Saving Rates on Growth
A higher saving rate leads to a higher level of output, but does not permanently affect the economy's growth rate.

[9] *See footnote (page 430).*

Technological progress. If capital accumulation is not the source of sustained growth, it follows that growth must ultimately be due to technological progress. Only by finding more efficient ways of producing goods, or ways of producing new and better goods, can an economy grow at a steady rate.

This conclusion has an obvious but important implication. A country that is able to achieve a higher rate of technological progress than other countries can eventually become much richer than the others. In terms of our earlier example, if country B is able to maintain a higher rate of technological progress than country A, the two paths of output will look as in Figure 22-7. Even if country B is behind to start with, it will eventually overtake country A, and steadily become richer.

This raises the question of what determines the rate of technological progress. What we know about the determinants of technological progress—from the role of spending on fundamental and applied research, to the role of patent laws, to the role of education and training—is the topic of Chapter 24. That chapter ends by returning to the facts of growth in rich countries since the early 1950s—the facts we saw in Section 22-1—and interpreting them in the light of the theory we have developed.

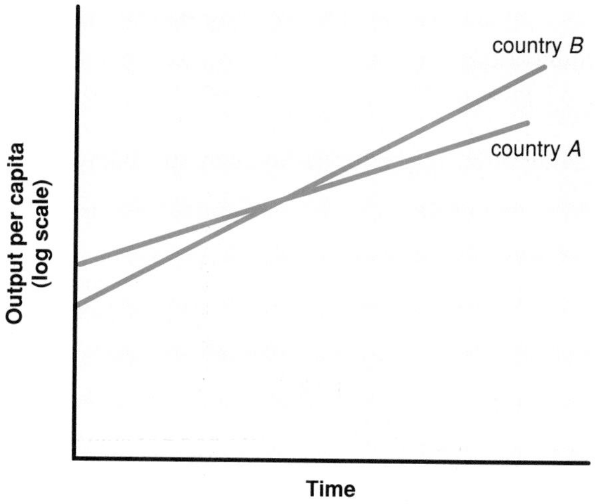

FIGURE 22-7

The Effects of Different Rates of Technological Progress

A higher rate of technological progress leads to a higher growth rate.

SUMMARY

■ Over long periods of time, fluctuations in output are dwarfed by growth, the steady increase of aggregate output over time.

■ Looking at growth in the G7 countries (Canada, France, Germany, Italy, Japan, the United Kingdom, and the United States) since 1950, three main facts emerge:

1. All seven countries have experienced strong growth and a large increase in the standard of living. Growth since 1950 has increased real output per capita by a factor of 2.6 in Canada, by a factor of 4.3 in Germany, and by a factor of 10.6 in Japan.

2. Growth has slowed since the mid-1970s. The average growth rate of output per capita went from 4.2% per year from 1950 to 1973 to 1.9% from 1973 to 1992.

3. The levels of output per capita across the seven countries have converged over time. Put another way, those countries that were behind have grown faster, reducing the gap between them and the world's economic leader, the United States.

■ Looking at the evidence across a broader set of countries and a longer period of time, the following facts emerge:

1. On the scale of human history, sustained output growth is a recent phenomenon. From the end of the Roman Empire to roughly year 1500, there was essentially no growth of output per capita in Europe. Even during the Industrial Revolution, growth rates were not high by current standards. The growth rate of output per capita from 1820 to 1950 in the United States was only equal to 1.5%.

2. Convergence of levels of output per capita is not a worldwide phenomenon. Many Asian countries are rapidly catching up, but most African countries have both very low levels of output per capita and low growth rates.

■ To think about growth, economists start from an aggregate production function relating aggregate output to two factors of production, capital and labour. How much output is produced, given these inputs, depends on the state of technology.

■ The aggregate production function implies that increases in output per worker can come either from increases in capital per worker or from improvements in the state of technology.

■ Capital accumulation by itself cannot sustain growth. Nevertheless, how much a country saves is very important because the saving rate determines the *level* of output per capita, if not its growth rate. The relation of the saving rate to capital and output is the topic of Chapter 23.

■ Sustained growth of output per capita is ultimately due to technological progress. Thus, perhaps the most important question in growth theory is what the determinants of technological progress are. The sources and the effects of technological progress are the topic of Chapter 24.

KEY TERMS

- growth, 430
- logarithmic scale, 430
- output per capita, 432
- purchasing power, 433
- purchasing power parity (PPP), 433
- convergence, 436
- leapfrogging, 439
- four tigers, 440

- aggregate production function, 442
- state of technology, 442
- constant returns to scale, 443
- decreasing returns to capital, 443
- decreasing returns to labour, 443
- capital accumulation, 444
- technological progress, 444
- saving rate, 445

QUESTIONS AND PROBLEMS

1. Use Table 22-1 (page 432) to answer each of the following questions.
   a. If the Canadian growth rate had not slowed down from 1973 to 1992, what would Canadian output per capita have been in 1992?
   b. If Japan's growth rate had not slowed down from 1973 to 1992, what would Japan's output per capita have been in 1992?
   c. Among the OECD countries, was there continued convergence toward the U.S. standard of living between 1973 and 1992?

2. Assume that typical consumers in Mexico and Canada buy the quantities and pay the prices indicated in the table below:

   a. Calculate Canadian consumption per capita in dollars.
   b. Calculate Mexican consumption per capita in pesos.
   c. Suppose that the exchange rate is 0.1 ($0.10 per peso). Using the current exchange rate, calculate Mexico's consumption per capita in dollars.
   d. Using the purchasing power parity method and Canadian prices, calculate Mexico's consumption per capita in dollars.
   e. Under each method, how much lower is the standard of living in Mexico than in Canada? Does the choice of method make a difference?

3. Consider the production function $Y = (\sqrt{K})(\sqrt{N})$.
   a. Calculate output when $K = 64$ and $N = 100$.

	FOOD		CONSUMER DURABLES	
	Price	Quantity	Price	Quantity
Mexico	I peso	1000	10 pesos	500
Canada	$1	2000	$2	4000

b. If both capital and labour double, what happens to output?

c. Is this production function characterized by constant returns to scale? Explain.

d. Write this production function as a relationship between output per worker and capital per worker. (*Hint:* Divide both sides of the production function by N.)

e. As capital per worker increases from 2 to 4 to 6, what happens to output per worker?

f. Will this production function have the same general shape as the one in Figure 22-5? Explain.

4. As discussed in the chapter, sustained growth in output per worker requires sustained technological progress. Table 22-1 seems to contradict this: From 1950 to 1973, Japan had higher growth than the United States, yet there were many more important technological discoveries in the United States than in Japan. Can you resolve this apparent contradiction?

5. In the Focus box titled "The Construction of PPP Numbers," we used U.S. prices to compare U.S. and Russian consumption per capita under the PPP method. Now, use *Russian* instead of U.S. prices to estimate consumption in both countries. What happens to our estimate of relative consumption per capita in Russia?

FURTHER READING

A broad presentation of facts about growth is given by Angus Maddison in *Phases of Economic Development* (New York: Oxford University Press, 1982).

Chapter 3 in *Productivity and American Leadership* by William Baumol, Sue Anne Batey Blackman, and Ed-ward Wolff (Cambridge, MA: MIT Press 1989) gives a vivid description of how life has been transformed by growth in the United States since the mid-1880s.

SAVING, CAPITAL ACCUMULATION, AND OUTPUT

Since 1950 the Canadian **saving rate,** defined as the ratio of saving to GDP, has averaged 22.4%, compared with 24.8% in Germany and 33.8% in Japan. Can this fact explain why the Canadian growth rate has been lower than in most OECD countries in the last 40 years? Would increasing the Canadian saving rate lead to sustained higher Canadian growth in the future?

We have already given the basic answer to these questions at the end of Chapter 22. And the basic answer is no. Over long periods of time (an important qualification to which we shall return), an economy's growth rate does not depend on its saving rate. It does not appear that lower Canadian growth in the last 40 years comes primarily from a low saving rate. Nor should one expect that an increase in the saving rate would lead to sustained higher Canadian growth.

However, this conclusion does not imply that concerns about the low Canadian saving rate are misplaced. Even if it does not permanently affect the growth rate, the saving rate can permanently affect the level of output and the standard of living. A higher saving rate would eventually lead to a higher standard of living in Canada.

The effects of the saving rate on capital and output per capita are the topics of this chapter. We proceed in four steps. In the first two sections we look at the interactions between output and capital accumulation, and the effects of the saving rate. In the third section we plug in numbers to get a better sense of the magnitudes involved. In the fourth section we extend the initial model to allow not only for physical capital but also for human capital.

23-1 INTERACTIONS BETWEEN OUTPUT AND CAPITAL

The amount of capital in the economy determines the level of output that can be produced. The level of output determines in turn the level of saving and investment, and thus how much capital is accumulated. Together, these interactions, which are represented in Figure 23-1, determine the movements in output and capital. Let's look at each of them in turn.

THE EFFECTS OF CAPITAL ON OUTPUT

We discussed the first of these two relations, the effect of capital on output, in Section 22-3. There we introduced the aggregate production function and saw that, under the assumption of constant returns to scale, we can write the following relation between output and capital per worker:

$$\frac{Y}{N} = F\left(\frac{K}{N}, 1\right)$$

Output per worker (Y/N) is an increasing function of capital per worker (K/N). Under the assumption of decreasing returns to capital, however, the *rate* at which output per worker increases will decline as we increase capital per worker. When capital is very low, an increase in capital will produce a large increase in output. When capital is already very high, a further increase will yield only a very small increase in output.

To simplify notation, it will be convenient to rewrite the relation between output and capital per worker simply as

$$\frac{Y}{N} = f\left(\frac{K}{N}\right)$$

where the function f captures the same relation between output and capital per worker as the function F: $f(K/N) \equiv F(K/N, 1)$.

If we ignore fluctuations, we can think of N as the number of workers employed in the economy when unemployment is at its natural rate. Thus we can think of N as equal to total population times the participation rate (the proportion of people who are in the labour force) times the proportion of the labour force that is

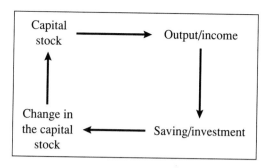

FIGURE 23-1

Capital, Output, and
Saving/Investment

employed (a proportion that is itself equal to 1 minus the unemployment rate). In this chapter, we shall assume that population, the participation rate, and the natural unemployment rate are all constant. These assumptions imply that the number of workers, N, is also constant. Also, under these assumptions, output per worker, output per capita, and output all move proportionately. Although we shall usually refer to movements in output or capital *per worker,* to lighten the text we shall often just talk about movements in output or capital.

Next, to focus on the interaction between capital and output, we shall ignore technological progress in this chapter, and thus take the production function f (or equivalently F) as given and unchanging through time. The basic conclusions we derive will still hold when we introduce population growth and technological progress in Chapter 24.

Under these assumptions, our first relation between output and capital per worker, from the production side, is given by

$$\frac{Y_t}{N} = f\left(\frac{K_t}{N}\right) \qquad (23.1)$$

where we have introduced time indexes for output and capital because we want to focus on their dynamics below. (Labour, N, is constant and thus does not have a time index.) In short, higher capital per worker leads to higher output per worker.

THE EFFECTS OF OUTPUT ON CAPITAL ACCUMULATION

How much capital per worker will the economy accumulate? The answer to this question clearly depends on how much people save. Assume that private saving is proportional to income, so that

$$S = sY$$

S is private saving. The parameter s is the saving rate, and has a value between zero and 1. Thus, we assume the saving rate to be constant. This assumption does not do justice to our discussion of consumption and saving behaviour in Chapter 8. But it does capture two basic facts about saving. The saving rate does not appear to increase or decrease systematically as a country becomes richer. And richer countries do not appear to have systematically higher or lower saving rates than poorer ones.

Assume that the economy is closed and that the budget deficit is equal to zero, so that in equilibrium investment equals private saving:[1]

$$I = S$$

Combining the two relations above gives investment as a function of output:

$$I = sY$$

We need to relate investment, which is a flow (the new machines and plants produced in the economy during a given period), to capital, which is a stock (the existing machines and plants in the economy at a point in time). Assume, as we did in Chapter 8, that capital depreciates at rate δ: Every period, a proportion δ of the capital stock becomes useless. Capital accumulation is then given by

$$K_{t+1} = (1 - \delta)K_t + I_t$$

Because we are looking at dynamics, we have introduced time indexes explicitly. You can think of the unit period as being a year. The capital stock next year is equal to the capital stock this year, adjusted for depreciation, plus investment this year.

Replacing investment by saving, and dividing both sides by N (the number of workers in the economy), gives

$$\frac{K_{t+1}}{N} = (1 - \delta)\frac{K_t}{N} + s\frac{Y_t}{N}$$

How much capital per worker the economy has in year $t + 1$ is equal to how much capital per worker it had in year t, adjusted for depreciation, plus investment per worker in year t. Investment per worker is in turn equal to the saving rate times output per worker.

Moving K_t/N to the left-hand side of the equation and reorganizing, we can rewrite the preceding equation as

$$\frac{K_{t+1}}{N} - \frac{K_t}{N} = s\frac{Y_t}{N} - \delta\frac{K_t}{N}$$

$$\dot{k} = sy - dK \quad (Jones) \tag{23.2}$$

The change in the capital stock per worker—the term on the left-hand side—is equal to saving per worker (the first term on the right) minus depreciation per worker (the second term on the right). This equation gives us our second relation between output and capital per worker.

23-2 IMPLICATIONS OF ALTERNATIVE SAVING RATES

We have derived two relations. From the production side, equation (23.1) tells us that capital determines output. From the saving side, equation (23.2) tells us that output in turn determines capital accumulation. Let's now put them together.

---

[1]As you will recall from Chapter 12, saving and investment need not be equal in an open economy. A country may save more than it invests, and lend the difference to the rest of the world. This is the case for Japan, which is running a large trade surplus and thus lending part of its saving to the rest of the world (see the discussions of the Japanese trade surplus in Sections 1-2 and 14-3). We shall, however, ignore open-economy issues here, and thus use saving and investment interchangeably.

Replacing output per worker (Y_t/N) in equation (23.2) by its expression in terms of capital per worker from equation (23.1) gives

$$\frac{K_{t+1}}{N} - \frac{K_t}{N} = sf\left(\frac{K_t}{N}\right) - \delta\frac{K_t}{N} \qquad (23.3)$$

change in capital = investment – depreciation

This relation fully describes what happens to capital per worker over time. The capital stock (let us omit "per worker" for the rest of the paragraph) this year determines output this year. Given the saving rate, this output determines in turn the amount of saving and thus of investment this year, the first term on the right. The capital stock also determines the amount of depreciation, the second term on the right. If investment exceeds depreciation, capital increases. If investment is less than depreciation, capital decreases.

Given the evolution of capital per worker, we can then use equation (23.1) to find the evolution of output per worker:

$$\frac{Y_t}{N} = f\left(\frac{K_t}{N}\right)$$

The best way to understand the implications of equations (23.1) and (23.3) for the dynamics of capital and output is to use a graph. We do this in Figure 23-2, where output per worker is measured on the vertical axis, and capital per worker on the horizontal axis.

The figure first draws output per worker, $f(K_t/N)$, as a function of capital per worker. The relation is the same as the production function in Figure 22-5 (page 444). Output per worker increases with capital per worker, but the effect is smaller the higher the level of capital per worker.

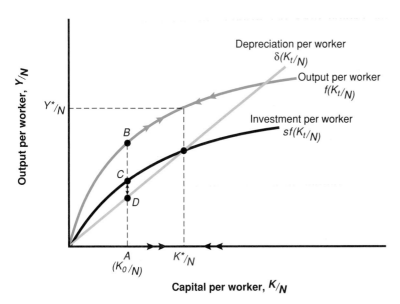

FIGURE 23-2
Capital and Output Dynamics
When capital and output are low, investment exceeds depreciation, and capital increases. When capital and output are high, investment is less than depreciation and capital decreases.

The figure then plots the two components of the right-hand side of equation (23.3). The first is investment per worker, $sf(K_t/N)$. This curve has the same shape as the production function, except that it is lower by a factor s. At the level of capital per worker K_0/N, for example, output per worker is given by the distance AB, and investment per worker is given by the distance AC, which is equal to s times the distance AB.[2] Thus, investment increases with capital, but by less and less as capital increases. When capital is already very high, the effect of a further increase in capital on output, and thus in turn on investment, is very small.

The second component is depreciation per worker, $\delta K_t/N$. Depreciation per worker increases in proportion to capital per worker; the relation is thus represented by a line with slope equal to δ. At the level of capital per worker given by K_0/N, depreciation is given by the distance AD.

The change in capital per worker is given by the difference between investment and depreciation per worker. Thus, at K_0/N it is positive and given by the distance $CD = AC - AD$. As we move to the right along the horizontal axis and look at higher and higher levels of capital per worker, investment increases by less and less, while depreciation keeps increasing in proportion to capital. Thus, for some level of capital per worker, K^*/N in Figure 23-2, investment is just enough to cover depreciation, and capital per worker remains constant. To the left of K^*/N, investment exceeds depreciation and capital per worker increases. To the right of K^*/N, depreciation exceeds investment and capital per worker decreases.

Characterizing the evolution of capital per worker over time is now easy. Consider an economy that starts with a low level of capital per worker, say K_0/N in Figure 23-2. Because investment exceeds depreciation, capital per worker increases. And because output moves with capital, output per worker increases as well. Capital per worker eventually reaches K^*/N, the level at which investment is equal to depreciation. From that time on, output and capital per worker remain constant at Y^*/N and K^*/N, their long-run equilibrium levels.

Think, for example, of a country that loses part of its capital stock, say as a result of a war. The mechanism we have just seen suggests that, if it has suffered much larger capital than human losses, it will come out of the war with a low level of capital per worker, thus at a point to the left of K^*/N. It will then experience a large increase in both capital and output per worker for some time. This appears to describe well what happened after World War II to countries that had proportionately larger destructions of capital than of human lives (see the Global Macro box entitled "Capital Accumulation and Growth in France in the Aftermath of World War II").

Similarly, if a country starts with a high level of capital per worker, a point to the right of K^*/N, then capital and output per worker will decrease. The initial level of capital per worker is just too high to be sustained given the saving rate. This decrease in capital per worker will continue until the economy again reaches the point where investment is equal to depreciation, where capital per worker is equal to K^*/N. From that point on, capital and output per worker will remain constant.

---

[2] *To make the graph easier to read, we have assumed an unrealistically high saving rate.*

CAPITAL ACCUMULATION AND GROWTH IN FRANCE IN THE AFTERMATH OF WORLD WAR II

When World War II ended in 1945, France had suffered some of the heaviest losses of all European countries. The losses in lives were large; more than 550 000 people had died, out of a population of 42 million. The losses in capital were larger. Estimates are that the French aggregate capital stock in 1945 was about 30% below its prewar value. A more vivid picture of the destruction of capital is provided by the numbers in Table 1.

Our model makes a clear prediction about what will happen to a country that loses a large part of its capital stock. It predicts that the country will experience fast capital accumulation and output growth for some time. In terms of Figure 23-2, a country with capital per worker initially far below K^*/N will grow rapidly as it converges to K^*/N and output converges to Y^*/N.

This prediction fares well in the case of France. There is plenty of anecdotal evidence that small increases in capital led to large increases in output. Minor repairs to a major bridge would lead to the reopening of the bridge. Reopening of the bridge would lead in turn to large reductions in the travel distance between two cities, and thus a large reduction in transport costs. A large reduction in transport costs would then allow a plant to get much-needed inputs to increase production, and so on.

The more convincing evidence, however, comes directly from the numbers on growth of aggregate output itself. From 1946 to 1950, the annual growth rate of French GDP was a very high 9.6% per year, leading to an increase in real GDP of about 60% over five years.

Was all the increase in French GDP due to capital accumulation? The answer is no. There were other forces in addition to the mechanism in our model. Much of the remaining capital stock in 1945 was old. Investment had been low in the 1930s (a decade dominated by the Great Depression) and nearly nonexistent during the war. Thus much of the postwar capital accumulation was associated with the introduction of more modern production techniques. Technological progress was another reason for the high growth rates of the postwar period.

Source: Gilles Saint-Paul, "Economic Reconstruction in France, 1945–1958," in Rudiger Dornbusch, Willem Nolling, and Richard Layard, eds., *Postwar Economic Reconstruction and Lessons for the East Today* (Cambridge, MA: MIT Press, 1993), 83–114.

TABLE 1

PROPORTION OF THE FRENCH CAPITAL STOCK DESTROYED AT THE END OF WORLD WAR II

Railways:		Rivers:	
Tracks	6%	Waterways	86%
Stations	38%	Canal locks	11%
Engines	21%	Barges	80%
Hardware	60%	Buildings:	
Roads:		Dwellings	1 229 000
Cars	31%	Industrial	246 000
Trucks	40%		

Source: See source note for this box.

STEADY-STATE CAPITAL AND OUTPUT

It will be useful for later to characterize the levels of output and capital per worker to which the economy converges in the long run. The state in which output and capital per worker are no longer changing is called—quite logically—the **steady state** of the economy. Putting the left-hand side of equation (23.3) equal to zero (in the steady state, by definition, the change in capital per worker is zero), the steady-state value of capital per worker, K^*/N, is given by

$$sf\left(\frac{K^*}{N}\right) = \delta\frac{K^*}{N} \tag{23.4}$$

The steady-state value of capital per worker is such that the amount of saving (the left-hand side) is just sufficient to cover depreciation of the existing capital stock (the right-hand side).

Given capital per worker (K^*/N), the steady-state value of output per worker (Y^*/N) is given in turn by the production function relation:

$$\frac{Y^*}{N} = f\left(\frac{K^*}{N}\right) \tag{23.5}$$

We now have the elements we need to discuss the effects of the saving rate on output per worker, both over time and in the steady state.

THE SAVING RATE AND OUTPUT

How does the saving rate affect the level of output per worker? Our analysis leads to a three-part answer:

(1) *The saving rate has no effect on the growth rate of output in the long run, which is* [*steady-state*] *equal to zero.*

There is a way of thinking about this result that will prove very useful when we introduce technological progress in Chapter 24. Think of what would be needed to sustain a constant positive growth rate of output per worker in the long run. Capital per worker would have to increase. And because of decreasing returns to capital, it would have to increase faster than output per worker. This implies that each year the economy would have to save a larger and larger fraction of output and put it toward capital accumulation. At some point, even saving all the output would not be enough to sustain growth. This is why it is impossible to sustain a constant positive growth rate forever. In the long run, capital per worker must be constant, and so must be output per worker.

(2) Nonetheless, *the saving rate determines the level of output per worker in the long run.* Other things equal, countries with a higher saving rate will achieve higher output per worker in the long run.

Figure 23-3 illustrates this point. Consider two countries with the same production function, the same level of employment, and the same depreciation rate, but different saving rates, say s_0 and $s_1 > s_0$. Figure 23-3 draws their common production function, $f(K_t/N)$, and the functions giving saving/investment as a function of capital for each of the two countries, $s_0f(K_t/N)$ and $s_1f(K_t/N)$. In the long run, the country with saving rate s_0 will reach the level of capital per worker K_0/N and output Y_0/N. The country with saving rate s_1 will reach the higher levels K_1/N and Y_1/N.

(3) *An increase in the saving rate will lead to higher growth for some time, but not forever.*

This conclusion follows from the two propositions we just discussed. From the first, we know that an increase in the saving rate does not affect the long-run *growth*

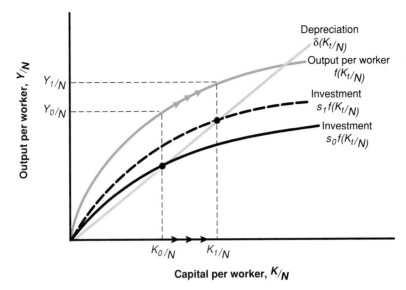

FIGURE 23-3

The Effects of Different Saving Rates

A country with a higher saving rate achieves a higher level of output in the steady state.

rate of output per worker, which remains equal to zero. From the second, we know that an increase in the saving rate leads to an increase in the long-run *level of output per worker.* It follows that as output increases to its new higher level in response to the increase in the saving rate, the economy will go through a period of positive growth. This period of growth will come to an end when the economy reaches its new steady state.

We can use Figure 23-3 again to illustrate this point. Consider a country that has an initial saving rate of s_0. Assume that capital per worker is initially equal to K_0/N, with associated output per worker Y_0/N. Now consider the effects of an increase in the saving rate from s_0 to s_1. (You can think of this increase as coming from tax changes that make it more attractive to save or from reductions in the budget deficit; the origin of the increase in the saving rate does not matter here.) The function giving saving/investment per worker as a function of capital per worker shifts upward from $s_0 f(K_t/N)$ to $s_1 f(K_t/N)$.

At the initial level of capital per worker, K_0/N, investment now exceeds depreciation, so capital per worker increases. As capital per worker increases, so does output per worker, and the economy goes through a period of positive growth. When capital eventually reaches K_1/N, investment is again equal to depreciation and growth ends. The economy remains from then on at K_1/N, with associated output per worker Y_1/N. The movement of output per worker is plotted against time in Figure 23-4. Output per worker is initially constant at level Y_0/N. After the increase in the saving rate at, say, time t, output per worker increases for some time until it reaches the higher level Y_1/N and the growth rate returns to zero.

We have derived these three results under the assumption of no technological progress and thus no growth in the long run. But, as we shall see in Chapter 24, the three results extend straightforwardly to an economy in which there is technological progress. Let us briefly indicate how. An economy where there is technological progress has a positive growth rate of output per worker even in the long run. This growth rate is independent of the saving rate—the extension of the first result

FIGURE 23-4

The Effects of an Increase in the Saving Rate on Output per Worker

An increase in the saving rate leads to a period of higher growth until output reaches its new, higher steady-state level.

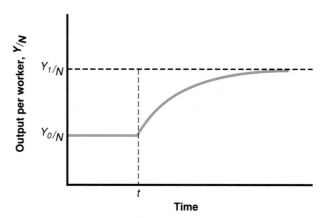

FIGURE 23-5

The Effects of an Increase in the Saving Rate on Output per Worker in an Economy with Technological Progress

An increase in the saving rate leads to a period of higher growth until output reaches a new, higher path.

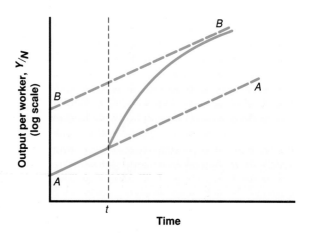

above. However, the saving rate does affect the level of output per worker—the extension of the second result. An increase in the saving rate leads to growth in excess of the steady-state growth rate for some time until the economy reaches its new higher path—the extension of our third result.

These three results are illustrated in Figure 23-5, which extends Figure 23-4 by plotting the effect of an increase in the saving rate in an economy with positive technological progress. At the initial saving rate, s_0, the economy moves along AA. If, at time t, the saving rate increases to s_1, the economy experiences higher growth for some time until it reaches its new higher path, BB. On path BB, the growth rate is again the same as before the increase in the saving rate (that is, the slope of BB is the same as the slope of AA).[3]

THE SAVING RATE AND THE GOLDEN RULE

Governments can use various instruments to affect the saving rate. They can run budget deficits or surpluses. They can give tax breaks to savers, making it more attractive for them to save. What aggregate saving rate should governments aim for?

---

[3]Note that Figure 23-5 is drawn with a log scale. For an explanation, see footnote 1 in Chapter 22.

To think about this question, we must shift our focus from the behaviour of *output* to the behaviour of *consumption*. What matters to people is not output per se, but rather, how much they consume.

It is clear that an increase in saving initially decreases consumption. (To make for lighter prose, we shall drop the "per worker" in this subsection and just refer to consumption rather than consumption per worker, capital rather than capital per worker, and so on.) A change in the saving rate this year has no effect on output and income *this year*. (By assumption, it takes a year for the additional saving/investment to show up in capital and thus in output.) Thus an increase in saving comes initially with an equal decrease in consumption.

Does an increase in saving lead to an increase in consumption in the long run? The answer is: not necessarily. Consumption may decrease not only in the short run, but also in the long run. You may find this surprising. After all, we know from Figure 23-3 that an increase in the saving rate always leads to an increase in the level of *output* per worker in the long run. But output is not the same as consumption. To see why, consider two extreme values of the saving rate.

An economy in which the saving rate is (and has always been) equal to zero is an economy in which capital is equal to zero. In this case, output is also equal to zero, and so is consumption. Thus a saving rate equal to zero implies zero consumption in the long run.

Now consider instead an economy in which the saving rate is equal to 1. People save all of their income. The level of capital, and thus output, will be very high. But because people save all of their income, consumption is again equal to zero. The economy is carrying an excessive amount of capital. Simply maintaining that level requires that all output be devoted to replacing depreciation! Thus a saving rate equal to 1 also implies zero consumption in the long run.

These two extreme cases imply that there is some value of the saving rate between zero and 1 at which the steady-state level of consumption reaches a maximum value. Increases in the saving rate beyond that value decrease consumption not only in the short run but also in the long run. This happens because the increase in capital associated with the increase in the saving rate leads to only a small increase in output, an increase that is too small to cover the increased depreciation. In effect, the economy carries too much capital. The level of capital associated with this critical value of the saving rate is known as the **golden-rule level of capital.** It is the level of capital at which long-run consumption is maximized. Increases in capital beyond the golden level can only reduce consumption.

If an economy already has so much capital that it is operating beyond the golden rule, then increasing saving further will decrease consumption not only now, but also later. Is this a relevant worry? Where do countries stand in practice? The empirical evidence indicates that most countries are actually far below their golden-rule level of capital. Thus increased saving now would lead to higher consumption in the future.

This conclusion implies that, in practice, governments face a tradeoff. An increase in the saving rate implies lower consumption for some time, higher consumption later. What should they do? The answer is that how close to the golden rule governments should try to get depends on how much weight they put on the welfare of current generations—who are more likely to lose from policies aimed at increasing the saving rate—versus that of future generations, who are more likely to gain. A point often made in politics is that future generations do not vote. This implies that governments are unlikely to ask current generations for large sacrifices, and thus that capital may stay far below its golden-rule level.

23-3 GETTING A SENSE OF MAGNITUDES

How large is the effect of a change in the saving rate on output in the long run? For how long and by how much does an increase in the saving rate affect growth? How far is Canada from the golden-rule level of capital? To get a better sense of the answers to these questions, let us now make more specific assumptions, plug in some numbers, and see what they imply.

Assume that the production function is given by

$$Y = \sqrt{K}\sqrt{N} \tag{23.6}$$

Output is equal to the product of the square root of capital and the square root of labour.[4]

Note that this production function satisfies constant returns to scale. Doubling both capital and labour leads to a doubling of output:

$$\sqrt{2K}\sqrt{2N} = (\sqrt{2})^2\sqrt{K}\sqrt{N} = 2Y$$

Note that it also implies <u>decreasing</u> returns to capital. For example, a quadrupling of capital, keeping labour constant, leads only to a doubling of output:

$$\sqrt{4K}\sqrt{N} = \sqrt{4}\sqrt{K}\sqrt{N} = 2Y$$

Dividing both sides of equation (23.6) by N gives

$$\frac{Y}{N} = \frac{\sqrt{K}\sqrt{N}}{N} = \frac{\sqrt{K}}{\sqrt{N}} = \sqrt{\frac{K}{N}}$$

where the second equality follows from the fact that $\sqrt{N}/N = 1/\sqrt{N}$. Output per worker is equal to the square root of capital per worker.

Substituting output per worker from the previous equation into the equation for capital accumulation (23.2) gives

$$\frac{K_{t+1}}{N} - \frac{K_t}{N} = s\sqrt{\frac{K_t}{N}} - \delta\frac{K_t}{N} \tag{23.7}$$

Let's now look at what this equation implies.

---

[4]***Digging deeper.*** *Consider the class of production functions given by* $Y = K^\alpha N^{1-\alpha}$, *where* α *is a number between zero and 1. The production function we use in the text takes* $\alpha = 0.5$, *giving equal weights to capital and labour. A more realistic production function would give relatively more weight to labour and less to capital, for example* $\alpha = 0.35$. *There are two reasons why we use* $\alpha = 0.5$ *in the text. The first is that it makes the algebra much simpler. The second is based on a broader interpretation of capital than just physical capital. As we shall see in the Section 23-4, we can think of the accumulation of skills, say through education or on-the-job training, as a form of capital accumulation as well. Under this broader view of capital, a coefficient of 0.5 for capital is roughly appropriate.*

What is the effect of the saving rate on the steady-state level of output per worker? In the steady state the amount of capital per worker is constant, so the left-hand side of equation (23.7) is equal to zero. This implies that

$$s\sqrt{\frac{K}{N}} = \delta\frac{K}{N}$$

when $\dot{K} = 0$

We have dropped time indexes, which are no longer needed because in the steady state K/N is constant. Square both sides to get

$$s^2\frac{K}{N} = \delta^2\left(\frac{K}{N}\right)^2$$

Divide both sides by K/N and reorganize to get

$$\frac{K}{N} = \left(\frac{s}{\delta}\right)^2 \tag{23.8}$$

This gives us an equation for steady-state capital per worker. From equations (23.6) and (23.8), steady-state output per worker is given by

$$\frac{Y}{N} = \sqrt{\frac{K}{N}} = \sqrt{\left(\frac{s}{\delta}\right)^2} = \frac{s}{\delta} \tag{23.9}$$

panel a $y = k$ | *Same idea as:* $\frac{s}{(n+g+d)}$

Output per worker is equal to the ratio of the saving rate to the depreciation rate; capital per worker is equal to the square of that ratio. A higher saving rate and a lower depreciation rate both lead to higher capital and output per worker in the long run.

Suppose that the depreciation rate is equal to 10% per year, and take the initial saving rate to be 10% as well. Then, using equations (23.8) and (23.9), we see that capital and output per worker are both equal to 1. Now suppose that the saving rate doubles, from 10% to 20%. It follows from equation (23.8) that, in the new steady state, capital per worker increases from 1 to 4. And from equation (23.9) $\left(\frac{20}{10}\right)^2 = t$ output per worker doubles, from 1 to 2. Thus a doubling of the saving rate leads, in the long run, to a doubling of output.

After an increase in the saving rate, how long does it take for the economy to reach the higher level of output? To answer this question, we must use equation *(speed of convergence)* (23.7) and solve it for capital at time 0, time 1, and so on. Assume that the saving rate increases from 0.1 to 0.2 at time 0. At time 0, $K_0/N = 1$. At time 1, equation (23.7) gives

$$\frac{K_1}{N} - \frac{K_0}{N} = s\sqrt{\frac{K_0}{N}} - \delta\frac{K_0}{N}$$

With a depreciation rate equal to 0.1 and a saving rate now equal to 0.2, this equation implies that $(K_1/N) - 1 = [(0.2)(\sqrt{1})] - [(0.1)1]$ so that $K_1/N = 1.1$. In the same way, we can solve for K_2/N, and so on. We can then use equation (23.6) to solve for output per worker at time 0, time 1, and so on.

The results of this computation are presented in Figure 23-6. Figure 23-6(a) plots the *level* of output per worker against time. Y/N increases over time from its initial value of 1 at time 0 to its steady-state value of 2 in the long run. Figure 23-6(b) gives the same information in a different way, plotting instead the *growth rate* of output per worker against time. As Figure 23-6(a) shows, output per worker

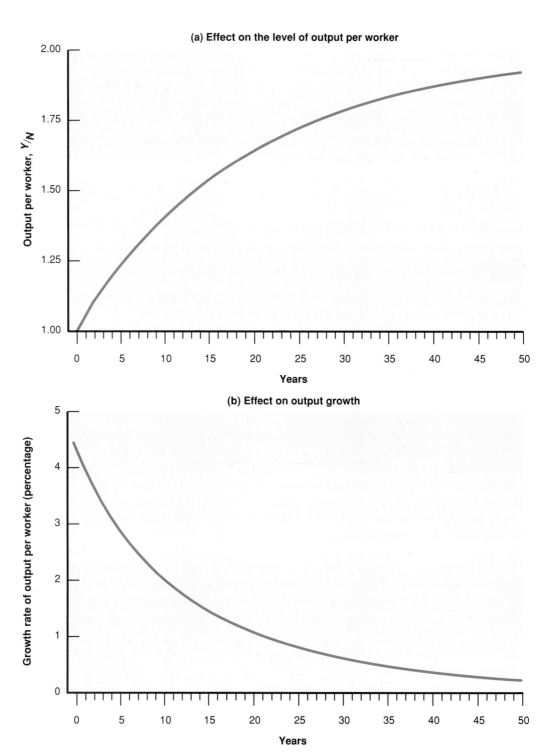

FIGURE 23-6

Dynamic Effects of an Increase in the Saving Rate from 10% to 20% on the Level and the Growth Rate of Output per Worker

It takes a long time for output to adjust to its new higher level after an increase in the saving rate. Put another way, an increase in the saving rate leads to a long period of higher growth.

increases fastest at the beginning. Thus, growth of output per worker is highest at the beginning and then decreases over time. In the new steady state, growth of output per worker is again equal to zero.

What comes out clearly from Figure 23-6 is that the adjustment to the new, higher, long-run equilibrium takes a long time. It is only 40% complete after 10 years, 63% complete after 20 years. Put another way, the increase in the saving rate increases the growth rate of output per worker for a long time. The average growth rate is equal to 3.1% for the first 10 years, 1.5% for the next ten. Thus, while changes in the saving rate have no effect on growth in the long run, they do nevertheless lead to higher growth for quite some time.

To go back to the question raised at the beginning of the chapter, can the low saving/investment rate in Canada explain the relatively low Canadian growth since 1950? The answer would be yes if Canada had had a higher saving rate in the past, and *if this saving rate had decreased substantially in the last 40 years.* If this were the case, it could explain the period of lower growth in Canada in the last 40 years along the lines of the mechanism in Figure 23-6 (with the sign reversed). But this is not the case; although the Canadian saving rate has declined slightly in the 1990s, the Canadian saving rate has been low for a long time. Thus, low saving cannot explain lower Canadian growth over the last 40 years.

On which side of the golden rule is Canada? In the steady state, consumption per worker is equal to output per worker minus depreciation per worker:

$$\frac{C}{N} = \frac{Y}{N} - \delta\frac{K}{N}$$

Using equations (23.8) and (23.9) for the steady-state values of output and capital per worker, consumption per worker is given by

$$\frac{C}{N} = \frac{Y}{N} - \delta\frac{K}{N}$$

$$= \frac{s}{\delta} - \delta\left(\frac{s}{\delta}\right)^2 \qquad \frac{s}{\delta}\left[1 - \delta\left(\frac{s}{\delta}\right)\right]$$

$$= \frac{s(1-s)}{\delta} \qquad = \frac{s\,(1-s)}{\delta}$$

Using this equation, together with equations (23.8) and (23.9), Table 23-1 gives the steady-state values of capital, output, and consumption per worker for different values of the saving rate (and for a depreciation rate equal to 10%).

Steady-state consumption is largest when $s(1-s)$ is largest, thus when s is equal to 50%. Thus the golden-rule level of capital is associated with a saving rate of 50%. Below that level, increases in the saving rate lead to an increase in long-run consumption. Above that level, they lead to a decrease. Few economies in the world today have saving rates above 40%, and (as we saw at the beginning of the chapter) the Canadian saving rate is about 20%. As rough as it is, our computation therefore suggests that, in most economies, an increase in the saving rate would increase both output and consumption levels in the long run.

THE SAVING RATE AND THE STEADY-STATE LEVELS OF CAPITAL, OUTPUT, AND CONSUMPTION PER WORKER

SAVING RATE s	CAPITAL PER WORKER K/N	OUTPUT PER WORKER Y/N	CONSUMPTION PER WORKER C/N
0.0	0.0	0.0	0.0
0.1	1.0	1.0	0.9
0.2	4.0	2.0	1.6
0.3	9.0	3.0	2.1
0.4	16.0	4.0	2.4
0.5	25.0	5.0	2.5
0.6	36.0	6.0	2.4
⋮	⋮	⋮	⋮
1.0	100.0	10.0	0.0

23-4 PHYSICAL VERSUS HUMAN CAPITAL

We have concentrated so far on physical capital—on machines, plants, office buildings, and so on. But economies have another type of capital: the set of skills possessed by the workers in the economy, what economists call **human capital.** An economy with many highly skilled workers is likely to be much more productive than an economy in which most workers cannot read or write.

The increase in human capital has been as dramatic as the increase in physical capital over the last two centuries. At the beginning of the Industrial Revolution, only 30% of the population knew how to read. Today, the literacy rate in rich countries is above 95%. Schooling was not compulsory prior to the Industrial Revolution. Today it is, usually until the age of 16. Still, there are large differences across countries. Today, in rich countries, 100% of children get a primary education, 90% get a secondary education, and 38% get a higher education. The corresponding numbers in poor countries, countries with GDP per capita below U.S.$400 in 1985, are 95%, 32%, and 4% respectively.

How should we think about the effect of human capital on output? How does the introduction of human capital change our earlier conclusions? These are the questions we take up in this section.

EXTENDING THE PRODUCTION FUNCTION

The most natural way of extending our analysis to allow for human capital is to modify the production function relation (23.1) to read

$$\frac{Y}{N} = f\left(\frac{K}{N}, \frac{H}{N}\right)$$ (23.10)

$$(+, \ +)$$

The level of output per worker now depends on both the level of physical capital per worker (K/N) and the level of human capital per worker (H/N).[5] As before, an increase in capital per worker leads to an increase in output per worker. And an increase in the average level of skill also leads to more output per worker. More skilled workers can use more complex machines; they can deal more easily with unexpected complications; they can adapt faster to new tasks. All of these lead to higher output per worker.

We assumed earlier that increases in physical capital per worker increased output per worker, but that the effect became smaller as the level of capital per worker increased. The same assumption is likely to apply to human capital per worker as well. Think of increases in H/N as coming from increases in the number of years of education. The evidence is that the returns to increasing the proportion of children acquiring a primary education are very large. At the very least, the ability to read and write allows people to use more sophisticated equipment. For rich countries, however, primary education—and, for that matter, secondary education—are no longer the relevant margins: Most children now get both (although the content of secondary education, it is often argued in Canada, could be improved substantially). The relevant margin is higher education. The evidence here—and we are sure this will come as good news to most of you—is that higher education indeed increases skills, at least as measured by the increase in wages for those who acquire it.[6] But, to take an extreme example, it is not clear that forcing everybody to acquire an advanced university degree would increase aggregate output very much. Many people would end up overqualified and probably more frustrated rather than more productive.

How should we construct the measure for human capital, H? The answer is: very much the same way we construct the measure for physical capital, K. In constructing K, we just add the values of the different pieces of capital, so that a machine that costs $2000 gets twice the weight of a machine that costs $1000. Similarly, we construct the measure of H so that workers who are paid twice as much get twice the weight.[7] Take, for example, an economy with 100 workers, half of them unskilled and half of them skilled. Suppose that the relative wage of skilled workers is twice that of unskilled workers. We can then construct H as $[(50 \times 1) + (50 \times 2)] = 150$. Human capital per worker, H/N, is equal in turn to $150/100 = 1.5$.

HUMAN CAPITAL, PHYSICAL CAPITAL, AND OUTPUT

How does the introduction of human capital change the analysis of the previous sections?

---

[5] Note that we are using the same letter, H, to denote the monetary base in Chapter 5 and human capital in this chapter. Both uses are traditional, and no confusion should arise.

[6] We return to this issue in Chapter 25.

[7] **Digging deeper.** The logic for using relative wages as weights is that they are supposed to capture the relative marginal products of different workers, so that a worker who is paid three times as much as another has a marginal product that is three times higher. This would indeed be true if labour markets were perfectly competitive. But, as we discussed in Chapter 15 and its appendix, labour markets are not perfectly competitive, and one may doubt that relative wages always accurately reflect relative marginal products.

Our conclusions about physical capital accumulation remain true: An increase in the saving rate increases steady-state physical capital per worker, and thus output per worker. But our conclusions now extend to human capital accumulation as well. An increase in how much society "saves" in the form of human capital—through education and on-the-job training—increases steady-state human capital per worker, and thus output per worker.

Thus, our extended model gives us a richer picture of the determination of output per worker. In the long run, it tells us, output per worker depends on both how much society saves and how much it spends on education. This conclusion raises in turn a question: What is the relative importance of human and physical capital in the determination of output per worker?

A place to start is to compare how much is spent on formal education to how much is invested in physical capital. In Canada, spending on formal education is about 8.1% of GDP. This number includes both government expenditures on education and private expenditures by people on education. This number is thus between one-third and one-half of the gross investment rate for physical capital (which is around 20%). But this comparison is only a first pass. Consider the following complications:

■ Education, especially higher education, is partly consumption—done for its own sake—and partly investment. We should include only the investment part for our purposes. However, the number in the preceding paragraph (8.1%) includes both.

■ At least for postsecondary education, the opportunity cost of an education is also foregone wages while one is acquiring the education. Spending on education should thus include not only the actual cost of education but also the opportunity cost. The 8.1% number does not include the opportunity cost.

■ Formal education is only part of education. Much of what we learn comes from on-the-job training, formal or informal. Both the actual costs and the opportunity costs of on-the-job training should also be included. The 8.1% number does not include them.

Investment in human capital—whether through education or on-the-job training—increases steady-state human capital per worker, and thus output per worker.

■ One should compare investment rates net of depreciation. Depreciation of physical capital, especially of machines, is likely to be higher than depreciation of human capital. Skills deteriorate, but do so slowly. Indeed, unlike physical capital, skills deteriorate more slowly, indeed often improve, the more they are used.

For all these reasons, it is difficult to come up with reliable numbers for investment in human capital. A recent study concludes that investment in physical capital and in education play roughly similar roles in the determination of output.[8] This conclusion seems to be a good benchmark. It implies that output per worker depends roughly equally on the amount of physical capital and the amount of human capital in the economy. Countries that save more, and/or spend more on education, can achieve substantially higher steady-state levels of output per worker.

ENDOGENOUS GROWTH

Note what our conclusion in the preceding paragraph did and did not say. It did say that a country that saves more and/or spends more on education will achieve a *higher level* of output per worker in the steady state. It did not say that by doing so a country can sustain permanently *higher growth* of output per worker.

This conclusion, however, has been challenged in the past decade. Following the lead of Robert Lucas[9] and Paul Romer, researchers have explored the possibility that the combination of physical and human capital accumulation may actually be enough to sustain growth. The question they have asked is the following. Given human capital, increases in physical capital will run into decreasing returns. And given physical capital, increases in human capital will also run into decreasing returns. But consider the case where both physical and human capital increase in tandem. Can't an economy grow forever just by having steadily more capital and more skilled workers?

The models that these researchers have explored are called **models of endogenous growth** to reflect the fact that in those models—in contradiction to the model we saw in earlier sections of this chapter—growth depends, even in the long run, on variables such as the saving rate and the rate of spending on education. The jury is still out, but the indications so far are that the conclusions we drew earlier need not be abandoned. There is no evidence that countries can sustain higher growth just from capital accumulation and skill improvements.

To end this chapter, let us restate our earlier conclusions, modified to take into account the presence of human capital.

Output per worker depends on the level of both physical and human capital per worker. Both forms of capital can be accumulated, one through investment, the other through education and training. Increasing either the saving rate and/or the fraction of output spent on education and training can lead to much higher levels of output per worker in the long run. However, given the rate of technological progress, such measures are unlikely to lead to a permanently higher growth rate.

---

[8]*See N. Gregory Mankiw, David Romer, and David Weil, "A Contribution to the Empirics of Economic Growth,"* Quarterly Journal of Economics, *May 1992, 407–437.*

[9]*We have mentioned Lucas twice already: in connection with rational expectations in Chapter 10, and in connection with the Lucas critique in Chapter 18.*

Note the qualifier in the last proposition: *given the rate of technological progress*. But is technological progress unrelated to the level of human capital in the economy? Can't a more educated and more skilled labour force lead to a higher rate of technological progress? These questions take us to the topic of the next chapter, the sources and the effects of technological progress.

SUMMARY

■ How much output is produced depends on the amount of existing capital.* Capital accumulation depends in turn on the level of output, which determines saving and investment.

■ These interactions between capital and output imply that starting from any level of capital (and ignoring technological progress, the topic of Chapter 24), an economy converges in the long run to a *steady-state* level of capital. Associated with this level of capital is a steady-state level of output. In the steady state, the levels of capital and output are not changing.

■ The steady-state level of capital, and thus the steady-state level of output, depend positively on the saving rate. A higher saving rate leads to a higher steady-state level of output; during the economy's transition to the new steady state, a higher saving rate leads to positive output growth. But (again ignoring technological progress) in the long run the growth rate of output is equal to zero, and is thus independent of the saving rate.

■ An increase in the saving rate requires an initial decrease in consumption. In the long run, the increase in the saving

rate may lead to an increase or a decrease in consumption. The long-run outcome depends on whether the economy is below or above the *golden rule of capital,* the level of capital at which long-run consumption is maximized.

■ Most countries appear to have a level of capital below the golden-rule level. Thus, an increase in the saving rate will lead to a decrease in consumption in the short run but an increase in consumption in the long run. In thinking about whether or not to take policy measures aimed at changing the saving rate, policy makers must decide how much weight to put on the welfare of current generations versus the welfare of future generations.

■ While most of the analysis of this chapter focusses on the effects of physical capital accumulation, output depends on the levels of both physical *and* human capital. Both forms of capital can be accumulated, one through investment, the other through education and training. Increasing the saving rate and/or the fraction of output spent on education and training can lead to large increases in output in the long run.

―――――
*To make the reading of this summary lighter, we shall omit "per worker" in what follows.

KEY TERMS

■ saving rate, 449
■ steady state, 455
■ golden-rule level of capital, 459

■ human capital, 464
■ models of endogenous growth, 467

QUESTIONS AND PROBLEMS

1. "An increase in saving per worker implies an increase in capital per worker, an increase in output per worker, and therefore a higher growth rate of output per worker." Discuss.

2. Suppose that the government, which was running a balanced budget, now decides to increase spending without increasing taxes and thus to run a budget deficit equal to 2% of GDP forever. Assume that the private saving rate does not change.

   a. On a diagram similar to Figure 23-2, show the impact of the rise in the budget deficit.

   b. How will the new steady state be different from the old one? In what way will it be similar to the old one?

   c. Does the change in the budget deficit affect the economy's growth rate? Explain carefully.

3. Discuss the likely impact of the following changes on the level of output per worker in the long run:

a. A decrease in the retirement age

b. A new discovery that prolongs the life of capital equipment

*4. Suppose that the economy's production function is $Y = K^{1/4}N^{3/4}$.

a. Is this production function characterized by constant returns to scale? Explain.

b. Are there decreasing returns to capital? Demonstrate.

c. Are there decreasing returns to labour? Demonstrate.

d. Transform the production function into a relationship between output per worker and capital per worker.

e. For a given saving rate (s) and a depreciation rate (δ), give an expression for capital per worker in the steady state.

f. Give an expression for output per worker in the steady state.

g. Solve for the steady-state level of output per worker when the depreciation rate (δ) and the saving rate (s) are both equal to 0.10.

h. Suppose the depreciation rate remains equal to 0.10, while the saving rate doubles to 0.20. What happens to steady-state output per worker? Does a doubling of the saving rate in this problem have the same effect on output as in the example in the text?

5. Suppose the production function in an economy is given by $Y = \sqrt{K}\sqrt{N}$, and both the saving rate (s) and the depreciation rate (δ) are equal to 0.10.

a. What is the steady-state level of capital per worker?

b. What is the steady-state level of output per worker?

c. Suppose that output and capital per worker have reached their steady states in period t, and then, in period $t + 1$, the depreciation rate increases to 0.20.

(i) What will be the new steady-state levels of capital per worker and output per worker?

(ii) Calculate the path of capital per worker and output per worker over the first three periods after the change in the depreciation rate.

FURTHER READING

The classic treatment of the relation between the saving rate and output is by Robert Solow, *Growth Theory: An Exposition* (New York: Oxford University Press, 1970).

An important (but harder) recent article, extending the Solow model to assess the relative roles of physical and human capital, is by N. Gregory Mankiw, David Romer, and David Weil, "A Contribution to the Empirics of Economic Growth," *Quarterly Journal of Economics,* May 1992, 107-2, 407–437.

Two (harder) articles presenting endogenous growth models are (1) Robert Lucas, "On the Mechanics of Economic Development," *Journal of Monetary Economics,* 22, 1988, 3–42, and (2) Paul Romer, "Capital Accumulation in the Theory of Long-Run Growth," in *Modern Business Cycle Theory,* Robert Barro, ed. (Cambridge, MA: Harvard University Press, 1989), 51–127.

24

TECHNOLOGICAL PROGRESS AND GROWTH

Our conclusion in Chapter 23 that capital accumulation cannot by itself sustain growth has a straightforward implication: Sustained growth *requires* technological progress. This chapter therefore focusses on technological progress, its determinants and its implications for growth.

24-1 THE DETERMINANTS OF TECHNOLOGICAL PROGRESS

Technological progress conjures up images of major discoveries: the invention of the microchip, the discovery of the structure of DNA, and so on. These discoveries in turn suggest a process driven largely by scientific research and chance rather than by economic forces. But the truth is that most technological progress in modern economies is the result of a much more humdrum process: the outcome of firms' **research and development (R&D)** activities. Industrial R&D expenditures account for between 1.4% and 3% of GDP in six of the major rich countries we looked at in Chapter 22 (the United States, Canada, France, Germany, Japan, and the United Kingdom).[1] About 75% of the roughly 1 000 000 U.S. scientists and researchers

---

[1]*We were unable to find comparable data for Italy.*

working in R&D are employed by firms. U.S. firms' R&D spending is equal to more than 20% of their spending on gross investment, and to more than 60% of their spending on net investment. And there is little doubt that how much firms spend on R&D is based on economic considerations, on each firm's assessment of the costs and benefits of R&D.

R&D SPENDING DECISIONS

Firms spend on R&D for the same reason they buy new machines or build new plants: to increase expected profits.

Earlier in this book (Section 8-2) we examined firms' investment decisions. We argued that, in deciding whether to purchase a new machine, a firm compares the expected present value of profits from the machine to the cost of purchasing and installing the machine. If the expected present value of profits exceeds the cost, the firm invests; otherwise, it does not.

This logic gives us a good starting point to think about R&D decisions. By increasing R&D spending, a firm increases the probability that it will discover and develop a new product. (We shall use *product* as a generic term to denote new goods or new techniques of production.) If the new product is successful, the firm's future profits will increase. If the expected present value of profits exceeds the expected cost of research, the firm will embark on a new R&D project; otherwise, it will not.

There is, however, an important difference between the purchase of a machine and R&D spending, and this is why the theory of technological progress is not just an application of general investment theory. The difference is that the outcome of R&D is fundamentally *ideas*. And, unlike a specific machine, ideas can potentially be used by many firms at the same time.

R&D, IDEAS, AND PATENT LAWS

A firm that has just acquired a new machine does not have to worry that another firm will use that machine. A firm that has discovered and developed a new product can make no such assumption.

How much a firm should worry about others appropriating its idea depends in large part on the legal protection given to the new product. Without legal protection, expected profits from developing a new product are likely to be small. Except in rare cases where the product is based on a trade secret, it will generally not take long for other firms to produce the same product, eliminating any advantage the innovating firm may have had initially. This is why countries have patent laws. **Patents** give a firm that has discovered a new product—usually a new technique or device—the right to exclude anyone else from the production or use of the new product for some time.

But even in the presence of patent laws, protection is far from complete. By looking at the new product and at the research process that led to its development, other firms may learn ways of making another product not covered by the patent, and in this way compete with the original product. They may even learn how to make a better product, thus eliminating the market for the original product altogether.

Thus, how the discovery and development of new products translate into higher profits depends on both the legal system and the nature of the research process. If protection of new products is weak—either because of poor patent protection or because the ideas in the new products can be easily used or adapted—expected profits from new products will be small, and so will the incentives for firms to engage in R&D.

R&D SPENDING AND TECHNOLOGICAL PROGRESS

What ultimately determines the level of R&D and the rate of technological progress? Our previous discussion suggests that two main factors are at work: the *fertility* of research and the *appropriability* of research results. Let's consider each one in turn.

The fertility of research. The **fertility of research** tells us how spending on R&D translates into new ideas and new products. If research is very fertile—if R&D spending leads to many new products—then, other things equal, firms will have more incentives to do R&D, and R&D and technological progress will be higher.

The determinants of the fertility of research lie largely outside the realm of economics. Many factors interact here:

■ Applied research and development ultimately depend on basic research. Much of the computer industry's development can be traced to a few breakthroughs, from the invention of the transistor to the invention of the microchip. Some worry that most major discoveries have already taken place and that technological progress will now slow down. This fear is based on analogies with mining, where higher-grade mines were exploited first and where we have had to turn to lower- and lower-grade mines. But this analogy is far from convincing, and there is no evidence that it is right.

■ It takes many years, and often many decades, for the full potential of major discoveries to be realized. The usual sequence is one in which a major discovery leads to the exploration of potential applications, then to the development of new products, then to the adoption of these new products. The Focus box entitled "The Diffusion of New Technology: Hybrid Corn" shows the results of one of the first studies of this process of diffusion of ideas. Closer to us is the example of personal computers. Many years after the commercial introduction of personal computers, it often feels as if we have just started discovering their uses.

The interactions among basic research, applied research, and development are both important and poorly understood. Some countries appear more successful at basic research; others are more successful at applied research and development. Many studies point out the importance of the education system. For example, it is often said that the French higher education system, with its strong emphasis on abstract thinking, produces researchers who are better at basic research than at applied research and development. Studies also point to the importance of entrepreneurs, of a "culture of entrepreneurship"; an important part of technological progress comes from the entrepreneurs' ability to organize the successful development and marketing of new products.

New technologies are not developed or adopted overnight. One of the first studies of the diffusion of new technologies was carried out in 1957 by Zvi Griliches, who looked at the diffusion of hybrid corn in different states in the United States.

Hybrid corn is, in the words of Griliches, "the invention of a method of inventing." Producing hybrid corn entails crossing different strains of corn to develop a type of corn adapted to local conditions. Introduction of hybrid corn can increase the corn yield by up to 20%.

Although the idea was first developed at the beginning of the twentieth century, the first commercial application of hybridization on a substantial scale did not take place until the 1930s in the United States. Figure I shows the rate at which hybrid corn was adopted in a number of U.S. states from 1932 to 1956.

The figure clearly shows two dynamic processes at work. The first is the process through which appropriate hybrid corns were discovered for each state. Hy-

brid corn became available in southern states (Texas, Alabama) more than 10 years after it had become available in northern states (Iowa, Wisconsin, Kentucky). The second is the speed at which hybrid corn was adopted within each state. Within eight years of introduction, practically all corn in Iowa was hybrid corn. The process was much slower in the south. More than 10 years after its introduction, hybrid corn accounted for only 60% of total acreage in Alabama.

Why was the speed of adoption higher in Iowa than in the South? One of the important contributions of Griliches' article was to show that the reason was an economic one. The speed of adoption in each state was a function of the profitability of introducing hybrid corn. And profitability was higher in Iowa than in the southern states.

Source: Zvi Griliches, "Hybrid Corn: An Exploration in the Economics of Technological Change," *Econometrica,* October 1957, 25-4.

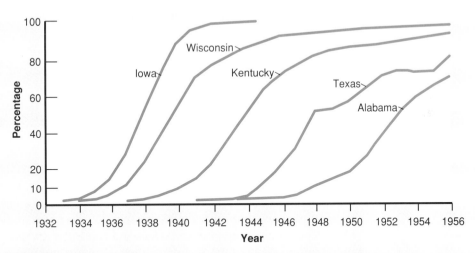

FIGURE I

Percentage of Total Corn Acreage Planted with Hybrid Seed, Selected U.S. States, 1932–1956

Source: See source note for this box.

The appropriability of research results. **Appropriability** captures the extent to which firms benefit from the results of their R&D. If firms cannot appropriate the profits from the development of new products, they will not engage in R&D and technological progress will be slow. Here again many factors are at work:

■ The first is the nature of the research process itself. For example, if it is widely believed that the discovery of a new product will lead in short order to the discovery of better products, there may be little payoff to being first. Thus a highly fertile field of research may not generate high levels of R&D. This example is extreme, but revealing.

■ A second factor is the degree of protection given to new products by patent laws. How should governments design patent laws? On the one hand, protection is needed to provide firms with the incentives to spend on R&D. On the other, once firms have discovered new products, it would be best for society if the knowledge embodied in those new products were made available to other firms without restrictions. A patent law must strike a difficult balance. Too little protection will lead to little R&D. Too much protection will make it difficult for new R&D to build on the results of past R&D, and may also lead to little R&D.

Countries that are less technologically advanced often have poorer patent protection. Our discussion helps explain why. The poorer countries are typically users rather than producers of new technologies. Much of technological progress comes, not from inventions within the poor country, but rather from the adaptation of foreign technologies. In this case the costs of weak patent protection are small, because there were few domestic inventions to start with. But the benefits of low patent protection are clear: They allow domestic firms to use and adapt foreign technology without having to pay royalties to the foreign firms that developed the technology.[2]

24-2 TECHNOLOGICAL PROGRESS AND THE PRODUCTION FUNCTION

In an economy in which there is both capital accumulation and technological progress, at what rate will output grow? Can an increase in the saving rate increase growth, and for how long? Answering these questions requires extending the model we developed in Chapter 23 to allow for technological progress. The first step is to revisit the aggregate production function, which is what we do in this section.

Technological progress has many dimensions. It may mean larger quantities of output for given quantities of capital and labour; think of a new type of lubricant that allows a machine to run at a higher speed. It may mean better products; think of the steady improvement in car safety and comfort over time. It may mean new products; think of the introduction of the CD player and the fax machine. It may mean more types of products; think of the steady increase in the number of products available at your local supermarket.[3]

---

[2]*Think about the pharmaceutical industry, where this issue has been hotly debated. Should Canadians give the large drug companies long patent lives to encourage R&D or should we exploit research elsewhere so that we can benefit from having low-cost generic drug producers?.*

[3]*The average number of items carried by a supermarket in the United States increased from 2200 in 1950 to 17 500 in 1985. To get a sense of what this means, watch Robin Williams (who plays an immigrant from the Soviet Union) in the supermarket scene in* Moscow on the Hudson.

These dimensions are in fact more similar than they first appear. If we think of consumers as caring, not about goods per se, but rather about the services these goods provide, then all these examples have something in common: In each case, technological progress provides more services to consumers. A better car provides more security, a new product such as the fax machine provides more information services, and so on.[4]

Thus, if we think of output as the set of underlying services provided by the goods produced in the economy, we can think of technological progress as leading to increases in output for given amounts of capital and labour. We can then think of the *state of technology* as a variable that tells us how much output can be produced from capital and labour at any time. Let's denote the state of technology by the letter A and rewrite the production function as

$$Y = F(K, N, A)$$
$$(+, +, +)$$

This is our extended production function. Output depends on both capital and labour (K and N) and on the state of technology (A).[5] Given capital and labour, an improvement in the state of technology, A, leads to an increase in output.

It will prove convenient to use a slightly more restrictive form of the preceding equation, namely:

$$Y = F(K, NA) \qquad (24.1)$$

This equation states that production depends on capital and on labour multiplied by the state of technology. This way of introducing the state of technology makes it easier to think about the effect of technological progress on the relation among output, capital, and labour. Under equation (24.1), we can think of technological progress in two equivalent ways.

1. Given the existing capital stock, technological progress reduces the number of workers needed to achieve a given amount of output. A doubling of A produces the same quantity of output with only half the original number of workers, N.

2. Technological progress increases NA, the amount of **effective labour** in the economy. If the state of technology doubles, it is as if the economy had twice as many workers. We can then think of output being produced by two factors: capital (K) on the one hand, and effective labour (NA) on the other.[6]

What restrictions should we impose on the extended production function (24.1)? We can build directly here on our discussion in Chapter 22.

It is again reasonable to assume constant returns to scale. *For a given state of technology (A)*, doubling both the amount of capital (K) and the amount of labour (N) is likely to lead to a doubling of output:

---

[4]*As we saw in the Focus box entitled "Real GDP, Technological Progress, and the Price of Computers" in Chapter 2, thinking of products as providing a number of underlying services is indeed the method used to construct the price index for computers.*

[5]*For simplicity, we ignore accumulation of human capital here.*

[6]*NA is also sometimes called **labour in efficiency units**. The use of "efficiency" here and for efficiency wages in Chapter 15 is a coincidence; the two notions are unrelated.*

$$2Y = F(2K, 2NA)$$

More generally, for any number λ,

$$\lambda Y = F(\lambda K, \lambda NA)$$

It is also reasonable to assume decreasing returns to each of the two factors, capital and effective labour. Given effective labour, an increase in capital is likely to increase output, but at a decreasing rate.

It was convenient in Chapter 23 to think in terms of output and capital *per worker*. The reason was that the steady state of the economy was a state where output and capital *per worker* were constant. It is convenient here to use a similar approach, and to look at output and capital *per effective worker*. The reason is the same: As we shall soon see, the steady state of the economy will be a state where output and capital *per effective worker* are constant.

To get a relation between output and capital per effective worker, take $\lambda = 1/NA$ in the preceding equation, to get

$$\frac{Y}{NA} = F\left(\frac{K}{NA}, 1\right)$$

Or, if we define the function f so that $f(K/NA) \equiv F(K/NA, 1)$,

$$\frac{Y}{NA} = f\left(\frac{K}{NA}\right) \tag{24.2}$$

This equation gives us a relation between *output per effective worker* and *capital per effective worker*. Output per effective worker increases if and only if capital per effective worker increases. The relation between output and capital per effective worker is drawn in Figure 24-1. It looks very much the same as the relation we drew in Figure 23-2 between output and capital per worker in the absence of technological progress. Increases in K/NA lead to increases in Y/NA, but at a decreasing rate.

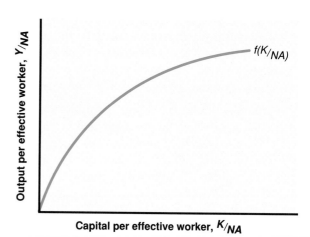

FIGURE 24-1
Output per Effective Worker versus Capital per Effective Worker
Increases in capital per effective worker lead to smaller and smaller increases in output per effective worker.

24-3 TECHNOLOGICAL PROGRESS AND THE RATE OF GROWTH

We now have the elements we need to think about the determinants of growth. Our analysis will parallel that of Chapter 23. There we looked at output and capital *per worker;* here we look at the dynamics of output and capital *per effective worker.*

INTERACTIONS BETWEEN OUTPUT AND CAPITAL

Let's first think of investment in the economy. Under the same assumptions as before—namely that the saving rate is constant and equal to *s,* and that investment and saving are equal—investment is given by

$$I = S = sY$$

Divide both sides by the number of effective workers, $NA,$ to get

$$\frac{I}{NA} = s\frac{Y}{NA}$$

Replacing output per effective worker by its expression from equation (24.2) gives

$$\frac{I}{NA} = sf\left(\frac{K}{NA}\right)$$

The relation between investment per effective worker and capital per effective worker is drawn in Figure 24-2. The upper curve replicates the relation between output and capital per effective worker drawn in Figure 24-1. The relation between investment and capital per effective worker is equal to the upper curve times *s.*

Let's now ask what investment has to be just to maintain a given level of capital per effective worker. In Chapter 23 the answer was simple; investment had to be equal to depreciation of the existing capital stock. Here the answer is slightly more complicated. The reason is as follows. Now that we allow for technological progress, the number of effective workers (NA) is increasing over time. Thus, maintaining the same ratio of capital to effective workers (K/NA) requires an increase in the capital stock (K) proportional to the increase in the number of effective workers (NA). Let's look at this condition more closely.

Assume that population is growing at rate $g_N.$[7] If we assume that the ratio of workers to the total population remains constant, the number of workers (N) also grows at rate $g_N.$ Assume also that the rate of technological progress is equal to $g_A.$ We discussed the determinants of g_A in Section 24-1. Here, we shall take g_A as given.

Together, these two assumptions imply that the growth rate of effective labour (NA) is equal to $g_N + g_A.$[8] If the number of workers is growing at 1% per year and the rate of technological progress is 2%, then the growth rate of effective labour is equal to 3%.

---

[7] *Our focus is on technological progress, not on population growth. But once we allow for technological progress introducing population growth is straightforward. Thus, we allow for both here.*

[8] *We are using here the property that the growth rate of the product of two variables is equal to the sum of the growth rates of the two variables. See Proposition 7 in Appendix 3 at the end of the book.*

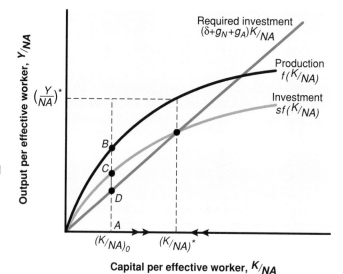

FIGURE 24-2

Dynamics of Capital and Output per Effective Worker

Capital and output per effective worker converge to constant values in the long run.

Let δ be, as before, the depreciation rate of capital. Then the level of investment needed to maintain a given level of capital per effective worker is given by

$$(\delta + g_N + g_A)K$$

An amount δK is needed just to keep the capital stock constant. If the depreciation rate is 10%, then investment must be equal to 10% of the capital stock just to maintain the capital stock. And an additional amount $(g_N + g_A)K$ is needed to ensure that the capital stock increases at the same rate as effective labour. If effective labour increases by 3%, then capital must increase by 3% to maintain the same level of capital per effective worker. Putting the two together, if the depreciation rate is 10% and the growth rate of effective labour is 3%, then investment must be equal to 13% of the capital stock to maintain a constant level of capital per effective worker.

Dividing the preceding expression by the number of effective workers to get the amount of investment per effective worker needed to maintain a constant level of capital per effective worker gives

$$(\delta + g_N + g_A)\frac{K}{NA}$$

The level of investment per effective worker needed to maintain a given level of capital per effective worker is represented by the line denoted "Required investment" in Figure 24-2. The line has a slope equal to $\delta + g_N + g_A$.

DYNAMICS OF CAPITAL AND OUTPUT

We can now give a graphical description of the dynamics of capital and output per effective worker. Consider in Figure 24-2 a given level of capital per effective worker, say $(K/NA)_0$. At that level, output per effective worker is equal to the distance AB. Investment per effective worker is given by AC. The amount of investment required to maintain that level of capital per effective worker is given by the distance AD. Because investment exceeds what is required to maintain the existing level of capital per effective worker, K/NA increases.

Thus, starting from $(K/NA)_0$, the economy moves to the right, with the level of capital per effective worker increasing over time. This process continues until investment is just sufficient to maintain the existing level of capital per effective worker, until capital per effective worker reaches $(K/NA)^*$. Thus, in the long run, capital per effective worker reaches a constant level, and so does output per effective worker. Put another way, the steady state of this economy is such that capital and output per effective worker are constant, and equal to $(K/NA)^*$ and $(Y/NA)^*$ respectively.

Note what this conclusion implies. *In the steady state, in this economy, what is constant is not output but rather output per effective worker.*[9] This implies that in the steady state output (Y) is growing at the same rate as effective labour (NA) (so that the ratio of the two is indeed constant). Because effective labour grows at rate $g_N + g_A$, output growth in the steady state is thus also equal to $g_N + g_A$. The same reasoning applies to capital. Because capital per effective worker is constant in the steady state, capital is also growing at the rate $g_N + g_A$.

These conclusions give us our first important result. *In the steady state, the growth rate of output is equal to the rate of population growth (g_N) plus the rate of technological progress (g_A). The growth rate is independent of the saving rate.*

The best way to strengthen the intuition for this result is to recall the logic we used in Chapter 23 to show that, without technological progress and population growth, the economy could not sustain positive growth forever. The argument went as follows. Suppose that the economy tried to achieve positive output growth. Because of decreasing returns to capital, capital would have to grow faster than output. The economy would have to devote a larger and larger proportion of output to capital accumulation. At some point there would be no more output to devote to capital accumulation.

Exactly the same logic is at work here. Effective labour now grows at rate $g_N + g_A$. Suppose that the economy tried to achieve output growth in excess of $g_N + g_A$. Because of decreasing returns to capital, capital would again have to increase faster than output. The economy would have to devote a larger and larger proportion of output to capital accumulation. At some point this would prove impossible. Thus the economy cannot permanently grow faster than $g_N + g_A$.

We have focussed on the behaviour of aggregate output. To get a sense of what happens, not to aggregate output, but rather to the standard of living over time, we can look instead at the behaviour of output per worker (not output per *effective* worker). Because output grows at rate $g_N + g_A$ and the number of workers grows at rate g_N, output per worker grows at rate g_A.[10] In other words, *in the steady state, output per worker grows at the rate of technological progress.*

---

[9]*If the number of effective workers is constant, constancy of output per effective worker obviously implies constant output as well. This was the case in Chapter 23, where there was neither population growth nor technological progress, and where, therefore, output was constant in the steady state. But this is not the case here.*

[10]*We are using here the property that the growth rate of Y/N is equal to the growth rate of Y minus the growth rate of N. See Proposition 8 in Appendix 3 at the end of the book.*

TABLE 24-1

THE CHARACTERISTICS OF BALANCED GROWTH

GROWTH RATE OF:	
1. Capital per effective worker	0
2. Output per effective worker	0
3. Capital per worker	g_A
4. Output per worker	g_A
5. Labour	g_N
6. Capital	$g_N + g_A$
7. Output	$g_N + g_A$
8. Effective labour	$g_N + g_A$

Because output, capital, and effective labour all grow at the same rate ($g_N + g_A$) in the steady state, the steady state of this economy is also called a state of **balanced growth**: In the steady state, output and the two inputs (capital and effective labour) grow in balance (at the same rate). The characteristics of balanced growth will be helpful later in the chapter, and are summarized in Table 24-1.

On the balanced growth path (equivalently, in steady the state; equivalently, in the long run):

■ Capital and output per effective worker are constant; this is the result we derived in Figure 24-2.

■ Equivalently, capital and output per worker are growing at the rate of technological progress, g_A.

■ Or, in terms of labour, capital, and output: Labour is growing at the rate of population growth, g_N; capital, output, and effective labour are growing at a rate equal to the sum of population growth and the rate of technological progress, $g_N + g_A$.

THE EFFECTS OF THE SAVING RATE

In the steady state, the growth rate of output depends *only* on the rate of population growth and the rate of technological progress. Thus changes in the saving rate do not affect the steady-state growth rate. But changes in the saving rate do increase the steady-state *level* of output per effective worker.

This result is best seen in Figure 24-3, which shows the effect of an increase in the saving rate from s_0 to s_1. The increase in the saving rate shifts the investment relation from $s_0 f(K/NA)$ to $s_1 f(K/NA)$. It follows that the steady-state level of capital per effective worker increases from $(K/NA)_0$ to $(K/NA)_1$, with a corresponding increase in the level of output per effective worker from $(Y/NA)_0$ to $(Y/NA)_1$.

Thus, following the increase in the saving rate, capital and output per effective worker increase for some time as they converge to their new higher level. Figure 24-4 plots the evolution of output and capital against time. Both output and capital are measured on logarithmic scales.[11] The economy is initially on the balanced

---

[11] *See Chapter 22, footnote 1.*

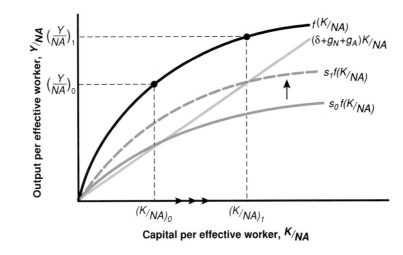

FIGURE 24-3
The Effects of an Increase in the Saving Rate: I
An increase in the saving rate leads to an increase in the steady-state levels of output and capital per effective worker.

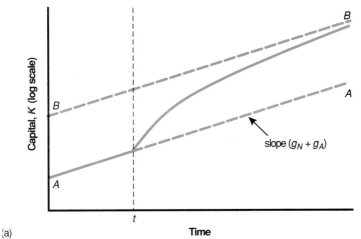

(a)

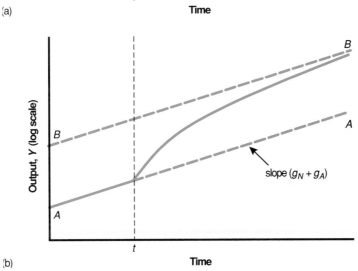

(b)

FIGURE 24-4
The Effects of an Increase in the Saving Rate: II
An increase in the saving rate leads to higher growth until the economy reaches its new, higher, balanced growth path.

growth path AA: Capital and output are growing at rate $g_N + g_A$ (the slope of AA is equal to $g_N + g_A$). After the increase in the saving rate at time t, output and capital grow faster for some period of time. Eventually, capital and output end up at higher levels than they would have without the increase in saving. But their growth rate returns to $g_N + g_A$. In the new steady state, the economy grows at the same rate, but on a higher growth path, BB. Line BB, which is parallel to AA, also has a slope equal to $g_N + g_A$.[12]

24-4 THE FACTS OF GROWTH REVISITED

In Chapter 22, we looked at growth in rich countries since 1950 and identified three main facts:

- Sustained growth, especially from 1950 to the mid-1970s

- A slowdown in growth since the mid-1970s

- Convergence, or the fact that countries that are further behind have been growing faster, closing the gap between them and the United States

Let us now use the theory we have developed and see what light it sheds on these facts.

CAPITAL ACCUMULATION VERSUS TECHNOLOGICAL PROGRESS

Suppose that we see an economy growing unusually fast—either in relation to its own growth in the past, or in relation to growth in other countries. Our theory suggests that this fast growth may be due to one of two causes:

- It may be due to a higher rate of technological progress, with faster output growth reflecting faster balanced growth. In other words, if g_A is higher, then balanced output growth—$g_N + g_A$—will also be higher.

- Or it may reflect the adjustment to a higher level of capital per effective worker. As we saw in Figure 24-4, such an adjustment leads to a period of higher growth, even if the rate of technological progress has not increased.

How can we tell which is the cause? If high growth reflects high balanced growth, output per worker should be growing at a rate *equal* to the rate of technological progress (see Table 24-1, line 4). If high growth reflects instead the adjustment to a higher level of capital per effective worker, this adjustment should be reflected in a growth rate of output per worker that *exceeds* the rate of technological progress.

These facts suggest a simple strategy, that of computing the growth of output per worker and the rate of technological progress for our six countries since 1950, then comparing the numbers. Angus Maddison recently implemented this strategy; his results are summarized in Table 24-2. (What Table 24-2 actually gives is the growth rate of output *per capita* rather than the growth rate of output *per worker;* this is to make comparison with the numbers in Table 22-1 easier. The numbers for

---

[12] *Figure 24-4 is the same as Figure 23-5, which anticipated the derivation presented here.*

TABLE 24-2

AVERAGE RATES OF OUTPUT GROWTH PER CAPITA AND OF TECHNOLOGICAL PROGRESS IN
SIX RICH COUNTRIES, 1950–1987 (PERCENTAGE)

	GROWTH OF OUTPUT PER CAPITA			RATE OF TECHNOLOGICAL PROGRESS		
	1950–1973 (1)	1973–1987 (2)	Change (3)	1950–1973 (4)	1973–1987 (5)	Change (6)
Canada	3.0	2.5	−0.5	3.0	2.0	−1.0
France	4.0	1.8	−2.2	4.9	2.3	−2.6
Germany	4.9	2.1	−2.8	5.6	1.9	−3.7
Japan	8.0	3.1	−4.9	6.4	1.7	−4.7
United Kingdom	2.5	1.8	−0.7	2.3	1.7	−0.6
United States	2.2	1.6	−0.6	2.6	0.6	−2.0
Average	4.1	2.2	−1.9	4.1	1.7	−2.4

Average is unweighted. Germany is West Germany only.

Source: Constructed from Tables 3-3, 5-3, 5-4, and 5-19 in Angus Maddison, *Dynamic Forces in Capitalist Development* (New York: Oxford University Press, 1991). Canadian capital stock data are from CANSIM matrices 6572, 6593, and 6832.

growth of output per worker and growth of output per capita are nearly identical.[13])

The first two columns correspond roughly to the first two columns of Table 22-1. They give the average annual growth rates of output per capita during 1950–1973 and 1973–1987 respectively.[14] The third column gives the change in the growth rate from the first to the second period. Columns (4) and (5) give the average annual rates of technological progress during 1950–1973 and 1973–1987 respectively. The sixth column gives the change in the rate of technological progress from the first period to the second. The method of construction of the rate of technological progress—which is not directly observable—is presented in the appendix at the end of this chapter.

Let's now return to our three main facts. The table suggests the following conclusions:

(1) *The period of high growth, from 1950 to 1973, was due to rapid technological progress, not to unusually high capital accumulation.*

Look at columns (1) and (4) of the table. In all six countries, the growth rate of output per capita from 1950 to 1973 was roughly equal to the rate of technological progress. This is exactly what we would expect when countries are growing along their balanced growth path; the main cause of growth was technological progress.

This is an important conclusion, because it rejects one hypothesis for why growth was so high from 1950 to 1973. The hypothesis is that fast growth was

---

[13] *If the ratio of employment to population had remained constant, the two growth rates would be identical. Because the participation rate has increased over time (as more women joined the work force), the growth rate in output per capita is slightly higher than that of output per worker.*

[14] *There are two minor differences between the two tables, due to differences in sources and construction of the numbers. The first is that the numbers for France and Japan for 1950–1973 are slightly different from those in Table 22-1, leading to a slightly different average growth rate for 1950–1973. The second is that the second period in this table ends in 1987, instead of 1992 as in Table 22-1.*

partially the result of the destruction of capital during World War II, leading thus to rapid rates of capital growth after the war. As we saw in the Global Macro box in Chapter 23, this explanation does explain some of the high growth in the immediate postwar period in France, and probably in other countries as well. But it is not the reason for the sustained growth of the 1950s and 1960s in the six countries we are looking at.

(2) *The slowdown in growth since 1973 has come from a decrease in the rate of technological progress, not from unusually low capital accumulation.*

This conclusion is strongly supported by columns (3) and (6) of Table 24-2. If lower capital accumulation were to blame for the growth slowdown, we would see a larger decline in the growth rate of output per capita than in the rate of technological progress. But this is not what the table shows. In all six countries, the decrease in technological progress has been roughly equal to the decrease in the growth rate of output.

Thus, contrary to some popular beliefs, the slowdown in growth since the mid-1970s is not due to a sharp drop in the saving rate, to the "disappearance of thrift." It is due to the decrease in the rate of technological progress, which declined from an average of 4.1% per year during 1950–1973 to only 2.2% per year from 1973–1987. This is potentially bad news for the future. In contrast to a decline in the saving rate—which, as we have seen, leads to only a temporary decline in growth—lower technological progress implies a permanently lower rate of growth.

(3) *Convergence of output per capita across countries has come from higher technological progress, rather than from faster capital accumulation, in the countries that started behind.*

Look at column (4) of Table 24-2. During 1950–1973, the annual rate of technological progress in Japan was 3.8% higher than that in the United States. The German rate was 3.0% higher, the French rate 2.3% higher. Only the U.K. rate was slightly below that of the United States. During 1973–1987, the differences narrowed to 1.1% for Japan, 1.3% for Germany, and 1.7% for France. Canada is a bit of an exception. The rate of technological progress slowed after 1973, but the gap between our rate and that of the United States actually widened in 1973–1987.

These facts yield an important conclusion. In thinking about the sources of convergence, one thinks of two possible causes. The first is that the poorer countries are poorer because they have less capital to start with. Thus, over time, they accumulate capital faster than the others, generating convergence. The second is that the poorer countries are poorer because they are less technologically advanced than the others. Over time, they become more sophisticated, by either importing advanced countries' technology or developing their own. As technological levels converge, so do countries. The conclusion we can draw from Table 24-2 is that the more important mechanism in this case has clearly been the second one. For example, Japan's output per worker has increased relative to that of the United States, not so much because Japan has accumulated capital extremely quickly, but rather because the state of technology has improved very quickly in Japan over the last 40 years.

The conclusions we drew in the preceding section represent intellectual progress. But they raise in turn a new set of issues. One obvious issue is *why* technological progress has slowed down since the mid-1970s. Much research has been devoted to answering this question. Many hypotheses have been proposed. The most common centre around measurement error, the rise of the service sector, and decreased spending on R&D.

Measurement error. The first hypothesis is that, in fact, there has been no slow-down in technological progress, and that the measured slowdown is solely the result of measurement error. The fact that measurement error could indeed be important is obvious to anybody who looks at how measures of output (such as GDP) are actually constructed.[15] In a number of sectors, productivity is not easily measurable. How do you measure the evolution of the productivity of doctors or lawyers over time? Because of the difficulties involved, the National Income and Expenditure Accounts make simple assumptions about productivity growth in those sectors, and these assumptions may well be wrong. To take an example, technological progress in financial services is assumed equal to zero. But there is plenty of evidence that, in fact, there has been substantial technological progress in financial services.

There is thus no question that there is measurement error and that we may be systematically understating technological progress and output growth. This is an important point. Our standard of living may in fact be increasing faster than the official statistics suggest. To explain the slowdown, however, one must show that the error has become larger since the mid-1970s, so that technological progress is *more* understated now than it was earlier in time. There is so far little evidence that this is the case.

The rise of the service sector. The second hypothesis is that the technological slowdown reflects the fact that the rich countries have become **post-industrial economies** in which manufacturing's share of GDP is steadily declining. The share of services is steadily increasing. And, the argument goes, the scope for technological progress is much more limited in services than in manufacturing. How much technological progress can take place in haircuts?

This argument is plausible. However, the facts suggest that the shift toward services has played a limited role in the slowdown. We can see why this is so in Figure 24-5, which plots the change in labour productivity growth in the United States, from 1948–1973 to 1973–1987, by industry. (For reasons of data availability, the figure uses labour productivity growth—that is, the growth rate of output per worker. Based on what we know for specific industries, the results would be very similar if one used instead estimates of the rate of technological progress.)

What is striking about Figure 24-5 is how the slowdown has affected nearly all sectors in the United States. Only farming and nonelectrical machinery (which is mainly computers) have seen increases in labour productivity growth from the first period to the second. The decline has been largest in mining (reflecting the depletion of the most easily available reserves) and in utilities (where it is in large part

---

[15]*This is what Martin N. Baily and Robert Gordon do in "The Productivity Slowdown, Measurement Issues, and the Explosion of Computer Power,"* Brookings Papers on Economic Activity, *1988:2, 347–431.*

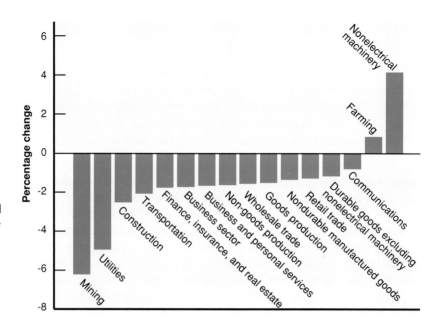

FIGURE 24-5

Changes in Annual Labour Productivity Growth, 1948–1973 to 1973–1987, by Industry

Most sectors of the U.S. economy have experienced a slowdown in productivity growth.

Source: Martin N. Baily and Robert Gordon, "The Productivity Slowdown, Measurement Issues, and the Explosion of Computer Power," *Brookings Papers on Economic Activity,* 1988:2, 347–431.

the result of more stringent environmental regulations). More directly relevant for our purposes here, the decline has been roughly the same in manufacturing sectors and in service sectors. Thus the shift in composition toward services cannot account for the slowdown in overall productivity growth.

Decreased R&D spending. The third hypothesis focusses on R&D. Because of the roughly equal decline in the productivity of the manufacturing and service sectors, the search for explanations must centre on factors that can explain why there has been a slowdown in most sectors. It seems natural to hypothesize that there was a general decline in R&D, and thus in technological progress. It turns out that the facts do not support this hypothesis. Table 24-3 shows the evolution of spending on R&D in each of our six countries. In all six countries, spending on R&D remained constant or increased as a percentage of GDP between 1963 and 1989.

The facts therefore suggest that the proximate reason for the decline in the rate of technological progress is a decline in the *fertility* of R&D. While rich countries are

....

TABLE 24-3
SPENDING ON R&D AS A PERCENTAGE OF GDP

	1963	1975	1989
Canada	1.0	1.1	1.4
France	1.6	1.8	2.3
Germany	1.4	2.2	2.9
Japan	1.5	2.0	3.0
United Kingdom	2.3	2.0	2.3
United States	2.7	2.3	2.8

Source: Kumiharu Shigehara, "Causes of Declining Growth in Industrialized Countries," in *Policies for Long-Run Economic Growth* (Kansas City, MO: Kansas City Fed, 1993), Table 4, p. 22. Canadian data from Statistics Canada, CANSIM Matrix 7943.

spending as much as or more than they used to on R&D, measured technological progress has slowed down. This is unfortunately the extent of our knowledge at this point. Some economists argue that we are in a phase where there has not been a truly major discovery for some time. Others argue that different sectors have developed sector-specific technologies, with the result that discoveries affect a smaller number of sectors than in the past, leading to smaller spillovers of research across sectors. At this stage, these hypotheses are largely untested.

24-5 EPILOGUE: THE SECRETS OF GROWTH

Why the rate of technological progress has declined since the mid-1970s is not the only unanswered question in the economics of growth. Indeed, many questions remain.

For example, we understand the basic mechanisms of growth in rich countries. But we are not very good at answering more specific questions, such as which measures could be taken to increase growth. Are governments spending the right amount on basic research? Should patent laws be modified? Is there a case for an **industrial policy,** a policy aimed at helping specific sectors of the economy (for example, those sectors with the potential for high technological progress, and thus the potential for large spillovers for the rest of the economy)? What can we expect from increasing the average number of years of education by another year?

We know even less about growth in poorer countries, and in particular why some countries seem unable to grow. The basic question economists face here is simple. Many countries have levels of output per capita that are less than one-tenth that of the G7 average. What prevents these countries from simply adapting much of the advanced countries' technology, thus quickly closing a good part of the **technology gap?** The elements of the answer are clear, and they include many of the factors we left aside in thinking about the determinants of the production function in Chapter 22. They include poorly established property rights, political instability, the lack of entrepreneurs, poorly developed financial markets, and a low average level of education. But the specific role of each of these factors is much harder to pinpoint. Many are as much the effect as they are the cause of low growth.

Looking at the poor countries that have grown rapidly in the last 20 years (such as the "four tigers"—Hong Kong, Taiwan, Singapore, and South Korea) or at the even more recent fast growers (such as China, Indonesia, Malaysia, and Thailand) would seem to be the best way of uncovering the secrets of growth.[16] But here again the lessons are proving anything but simple. In all these countries, growth has come with the rapid accumulation of both physical and human capital. And in all these countries it has come with an increase in the importance of foreign trade, an increase in exports and imports. But beyond these two factors, clear differences emerge. Some countries, such as Hong Kong, have relied mostly on free markets and limited government intervention. Others, such as Korea and Singapore, have relied instead on government intervention and an industrial policy aimed at fostering the growth of specific industries.

---

[16]*As the crisis in Asia in 1997 shows, even these countries have something to learn.*

(The cases of Hong Kong and Singapore are discussed in detail in the In Depth box entitled "Hong Kong and Singapore: A Tale of Two Cities.") The bottom line is clear: We have not yet unravelled the secrets of growth.[17]

---

[17]*These few paragraphs do not do justice to what is in effect a whole field of economics,* development economics. *Much progress has been made from studies of individual countries, as well as from systematic comparisons across countries. Much is known—even if we do not yet have the secrets of growth.*

▶ ## *HONG KONG AND SINGAPORE: A TALE OF TWO CITIES*

Between 1960 and 1985, the average growth rate of output in both Hong Kong and Singapore was a high 6.1% per year.* What were the causes of this high growth rate?

THE SIMILARITIES

Hong Kong and Singapore have several things in common. Both are former British colonies. Both are essentially cities that served initially as trading ports with little manufacturing activity. The postwar population of both countries was composed primarily of immigrant Chinese from southern China. During the course of their rapid growth, they have gone through a similar sequence of industries, with Singapore starting later than Hong Kong by 10 to 15 years. The respective sequences are summarized in Table 1.

THE DIFFERENCES

A closer look, however, shows major differences in the way the two countries have grown.

These numbers are computed using PPP measures of GDP, from Heston and Summers (see the Focus box on PPP measures in Chapter 22).

Hong Kong has grown under a policy of minimal government intervention. For the most part, the government has limited its intervention to providing infrastructure and selling land as it became required for further growth. In contrast, growth in Singapore has been dominated by government intervention. Through budget surpluses, as well as forced saving through pension contributions, the government has achieved a very high national saving rate. Singapore's share of gross investment in GDP increased from 9% in 1960 to 43% in 1984, one of the highest investment rates in the world. The development of specific industries has been the result of systematic government targeting, implemented through large tax incentives, mostly for foreign investors.

These differences in strategies are reflected in the relative roles of capital accumulation and technological progress. In Hong Kong, the annual growth rate of output per worker from 1970 to 1990 was 2.4%; the growth rate of technological progress over the same period was 2.3%. Thus, using the interpretation provided by the model we developed in this chapter, growth in Hong Kong has been roughly balanced. In

....
TABLE 1

THE SEQUENCE OF ACTIVITIES IN HONG KONG AND SINGAPORE SINCE THE EARLY 1950s

HONG KONG		SINGAPORE	
Early 1950s	Textiles	Early 1960s	Textiles
Early 1960s	Clothing, plastics	Late 1960s	Electronics, refining
Early 1970s	Electronics	Early 1970s	Electronics, refining, textiles, clothing
1980s	Trade, banking	1980s	Banking, electronics

Source: See source note for this box.

Singapore, the growth rate of output per worker from 1971 to 1990 was 1.5%; the growth rate of technological progress was a surprisingly low 0.1%. In other words, Singapore has grown nearly entirely through unusually high capital accumulation, not technological progress. Singapore's growth has been unbalanced.

Why has Singapore achieved so little technological progress? In the article on which this box is based, Alwyn Young, an economist at Boston University, argues that in effect Singapore has gone too fast from one industry to the next. By going so fast, it has not had time to learn how to produce in any of them very efficiently. And, by relying largely on foreign investment, it has not allowed a class of domestic entrepreneurs to learn and replace foreign investment in the future.

What lies in store for Singapore? The model we have developed in this chapter suggests that a slowdown in growth is inevitable. High saving and investment rates can lead to high growth only for a while. The numbers would appear brighter for Hong Kong, which seems to be growing on a balanced growth path. But major changes are probably in store for Hong Kong as well, as it adapts to becoming part of China.

Source: Alwyn Young, "A Tale of Two Cities: Factor Accumulation and Technical Change in Hong Kong and Singapore," *NBER Macroeconomics Annual*, 1992, 13–63.

SUMMARY

■ Technological progress depends on both (1) the fertility of research and development and (2) the appropriability of R&D results (that is, the extent to which firms benefit from the results of their R&D).

■ In designing patent laws, governments must trade off protection for future discoveries with a desire to make existing discoveries available to potential users without restrictions.

■ In thinking about the implications of technological progress for growth, it is useful to think of technological progress as increasing the amount of effective labour available in the economy (that is, labour multiplied by the state of technology). We can then think of output as being produced with capital and effective labour.

■ In the steady state, output and capital *per effective worker* are constant. Put another way, output and capital *per worker* grow at the rate of technological progress. Put yet another way, output and capital grow at the same rate as effective labour, thus at a rate equal to the growth rate of labour plus the rate of technological progress (that is, the rate of growth of the state of technology). When the economy is in the steady state, it is said to be on a balanced growth path.

■ The rate of output growth in the steady state is independent of the saving rate. However, the saving rate affects the steady-state *level* of output per effective worker. And increases in the saving rate lead to a period of growth in excess of the steady-state output growth rate.

■ Canada, Germany, France, Japan, the United Kingdom, and the United States have had roughly balanced growth since 1950. The slowdown in growth since the mid-1970s comes from a decrease in the rate of technological progress. Convergence of output appears to have come primarily from a convergence in the levels of technological progress.

■ There is no good explanation for the decline in the rate of technological progress since the mid-1970s. More generally, our understanding of the determinants of technological progress—and its relation to factors such as property rights, political instability, and education—remains limited.

KEY TERMS

■ research and development (R&D), 470
■ patents, 471
■ fertility of research, 472
■ appropriability, 474
■ effective labour, 475

■ balanced growth, 480
■ post-industrial economies, 485
■ industrial policy, 487
■ technology gap, 487

QUESTIONS AND PROBLEMS

1. Name two discoveries or inventions (aside from discoveries or innovations in the computer industry) that have not yet fully diffused in the economy.

2. How has technological change improved (or added to) the services provided by each of the following?
   a. Telephone calls
   b. Chequing accounts
   c. Visits to the doctor
   d. College textbooks

3. How would each of the following policy changes affect (1) the appropriability, (2) the fertility, and (3) the amount of R&D spending?
   a. Increasing patent protection from 17 to 25 years
   b. Allowing corporations a full credit against taxes for their R&D spending, rather than just a deduction from revenue
   c. Government-sponsored conferences between university scientists and corporate managers to discuss the latest discoveries

4. Why is it more "pessimistic" to attribute the recent decline in growth in the richer countries to a decline in the rate of technological progress rather than to a decline in the saving rate?

5. Suppose the economy's production function is $Y = \sqrt{K}\sqrt{NA}$ and both the saving rate (s) and the depreciation rate (δ) are equal to 0.10. Further, suppose that the number of workers grows at 1.5% per year ($g_N = 0.015$) and the rate of technological progress is 3.5% per year ($g_A = 0.035$).
   a. Find the steady-state values of
      (i) the capital stock per effective worker
      (ii) output per effective worker
      (iii) the growth rate of output per effective worker
      (iv) the growth rate of output per worker
   b. Suppose that the rate of technological progress increases to 7% per year. Recalculate the answers to (i)–(iv) in part **a** above.
   c. Now suppose that the rate of technological progress is 7% per year, and the number of workers grows at 3% per year. Recalculate the answers to (i)–(iv) above.

6. Use your answers in parts **b** and **c** in question 5 above to evaluate each of the following statements:
   a. "In the steady state, the capital stock per effective worker is independent of the rate of technological progress."
   b. "In the steady state, the level of output per effective worker is independent of the rate of technological progress."
   c. "In the steady state, growth in the standard of living depends on the rate of technological progress."
   d. "In the steady state, growth in the standard of living depends on how fast the population is growing."

FURTHER READING

Two classic books on the nature of invention and its role in growth:

Joseph Schmookler, *Invention and Economic Growth* (Cambridge, MA: Harvard University Press, 1966) looks at the precise nature of invention and inventions.

Joseph Schumpeter, *Capitalism, Socialism, and Democracy* (New York: Harper & Row, 1942) builds a general theory of fluctuations and growth, giving central roles to innovations and to entrepreneurs.

A thorough assessment of what we know about technological progress in rich countries is given in William Baumol, Sue Anne Batey Blackman, and Edward Wolff, *Productivity and American Leadership: The Long View* (Cambridge, MA: MIT Press, 1989).

For more on rapid growth in Asian countries, read John Page, "The East Asia Miracle: Four Lessons for Development Policy," *NBER Macroeconomics Annual*, 1994, 219–281.

For an issue we have not explored in the text, growth and the environment, read the World Bank's *Development and the Environment, World Development Report* (Oxford: Oxford University Press, 1992).

APPENDIX

CONSTRUCTING A MEASURE OF TECHNOLOGICAL PROGRESS*

In 1957, Robert Solow suggested a way of constructing an estimate of technological progress. The method, which is still used today, relies on one important assumption: that each factor of production is paid its marginal product, its marginal contribution to output.

Under this assumption, it is easy to compute the contribution of an increase in any factor of production to the increase in output. For example, if a worker is paid $30 000 a year, the assumption implies that her contribution to output is equal to $30 000. Now suppose that this worker increases the number of hours she works by 10%. The increase in output coming from the increase in her hours will therefore be equal to $30 000 \times 10\%$, or $3000.

The argument can be applied more generally. Assume that there are two factors of production in the economy, capital and labour. Also assume that the share of total income paid to labour is α, and that the growth rate of labour is g_N. Then the increase in output attributable to the increase in labour is equal to α times g_N.

Similarly, we can compute the increase in output attributable to capital. Because we assume that there are only two factors of production, labour and capital, the share of capital in income is $(1 - \alpha)$. If the rate of growth of capital is equal to g_K, then the increase in output attributable to the increase in capital is equal to $(1 - \alpha)$ times g_K.

Putting the two together, the increase in output due to increases in both labour and capital is equal to $\alpha g_N + (1 - \alpha)g_K$. Thus we can measure the effects of technological progress by computing what Solow called the residual, the excess of actual output growth over what can be accounted for by the growth in capital and labour:

$$\text{residual} \equiv g_Y - [\alpha g_N + (1 - \alpha)g_K]$$

This measure is called the **Solow residual.** It is easy to compute; all it requires knowing are the growth rates of output, labour, and capital, as well as the shares of labour and capital.

The Solow residual is sometimes called the **rate of growth of multifactor productivity.** This is to distinguish it from the *rate of growth of labour productivity,* which is defined as $g_Y - g_N$, the rate of output growth minus the rate of labour growth.

The Solow residual is related to the rate of technological progress in a simple way. The residual is equal to the share of labour times the rate of technological progress:

$$\text{residual} = \alpha g_A$$

We shall not derive this result here. But the intuition for this relation comes from the fact that what matters in the production function $Y = F(K,NA)$ [equation (24.1)] is the product of labour times the state of technology, NA. We saw that to get the contribution of labour growth to output growth, we must multiply the growth rate of labour by its share. Because N and A enter in the same way in the production function, it is clear that to get the contribution of technological progress to output growth we must also multiply it by the share of labour.

If the Solow residual is equal to zero, so is technological progress. To construct an estimate of g_A, one must construct the Solow residual and then divide it by the share of labour. This is how the estimates of g_A presented in the text were constructed.

***Source**: Robert Solow, "Technical Change and the Aggregate Production Function," Review of Economics and Statistics, *1957, 312–320.*

KEY TERM

....

■ Solow residual, or rate of growth of multifactor productivity, 491

TECHNOLOGICAL PROGRESS, UNEMPLOYMENT, AND WAGES

We spent much of the preceding three chapters celebrating the merits of technological progress. In the long run, we argued, technological progress is the key to steady increases in output per capita, to increases in the standard of living.

Popular and political discussions on technological progress are often more ambivalent. Since the beginning of the Industrial Revolution, workers have worried that technological progress will eliminate their jobs and throw them into unemployment. In early nineteenth-century England, groups of workers in the textile industry, known as the Luddites, destroyed the new machines that they saw as a direct threat to their jobs. Similar movements took place in other countries as well. The word *saboteur* comes from one of the ways French workers destroyed machines: by putting their sabots (their heavy wooden shoes) in the machines.

The theme of **technological unemployment** typically comes back to the fore whenever unemployment is high. During the Great

Depression, adherents to the *technocracy movement* argued that high unemployment came from the introduction of machinery, and that things would only get worse if technological progress were allowed to continue. Today in Europe—where, as we saw in Chapter 20, unemployment is also very high—there is widespread support in many countries for a shorter workweek, 30 or 35 hours. In Canada, various groups have proposed "job sharing" as a solution to the unemployment problem. Because of technological progress, the argument goes, there is no longer enough work for all workers to have full-time jobs. The solution is thus to have each worker work fewer hours so that more workers can be employed.

In its crudest form, the argument that technological progress must lead to unemployment is patently false. The very large improvements in the standard of living that advanced countries have witnessed over the twentieth century have been associated with large *increases* in employment and no systematic increase in the unemployment rate. In Canada, output per capita has increased by a factor of 4 since 1926, and so has the number of workers employed. Nor, looking across countries, is there any evidence of a systematic positive relation between the unemployment rate and the level of productivity. Japan and the United States, two of the countries with the highest levels of productivity, have two of the lowest unemployment rates among OECD countries.

Do these facts mean that the fears reflected in popular perceptions are groundless? No, or at least not necessarily. To organize the discussion here, it is useful to distinguish between two related but separate dimensions of technological progress:

■ Technological progress allows the production of larger quantities of existing goods using the same number of workers.

■ Technological progress leads to the production of new goods and the disappearance of old ones.

Consider the first dimension, and note that the relation between output and the number of workers can be stated in two ways. First, technological progress allows the economy to produce *more and more* output with the *same* number of workers. Second, technological progress allows the economy to produce the *same* amount of output with *fewer and fewer* workers. Those who emphasize the role of technological progress in increasing output and the standard of living think in terms of the first formulation. Those who worry about technological unemployment think in terms of the second. The evidence we saw in previous chapters clearly shows that, in the long run, the adjustment is through increases in output. But how long is the long run? Does output increase quickly enough in response to an increase in productivity to avoid a prolonged period of unemployment? This is the question we take up in the first two sections of this chapter. In Section 25-1 we look at the short-run response of output to increases in productivity. In Section 25-2 we look at the medium- and long-run responses. As we shall see, neither theory nor evidence supports the fear that faster technological progress leads to more unemployment. If anything, the effect seems to go the other way: Productivity slowdowns, not increases, appear to be associated for some time with more unemployment.

Consider now the second dimension. With technological progress comes a complex process of job creation and job destruction. This theme was central to the work of Joseph Schumpeter, a Harvard economist who in the 1930s emphasized that the growth process was fundamentally a process of **creative destruction.** For those who lose their jobs and have to find new ones, or for those who have skills that are no longer in demand, technological progress can indeed be a curse rather than a blessing. As consumers, they benefit from the availability of new goods. As workers, however, they may suffer from prolonged unemployment and settle for lower wages when taking a new job. This concern is of particular relevance today. The last 15 to 20 years have been characterized by a decline, both relative and absolute, in the wages of low-skill workers. Many signs point to technological progress as the main cause. The distribution effects of technological progress, and in particular the change in the structure of wages, are the topics taken up in Section 25-3.

25-1 PRODUCTIVITY, OUTPUT, AND UNEMPLOYMENT IN THE SHORT RUN

In Chapter 24 we formalized technological progress as an increase in A, the *state of technology,* in the production function:

$$Y = F(K, NA)$$

Capital accumulation is not central to the issues we shall be discussing here. So we shall ignore capital altogether and assume that output is produced according to the following production function:

$$Y = NA \qquad (25.1)$$

Output is produced using only labour, N, and each worker produces A units of output. Increases in A represent technological progress. Note that A has two interpretations here. The first is indeed as the state of technology. The second, which follows from the fact that $A = Y/N$, is as labour productivity (output per worker).[1] Thus, when referring to increases in A below, we shall feel free to use *technological progress* or (labour) *productivity growth* interchangeably.

Rewrite equation (25.1) as

$$N = \frac{Y}{A} \qquad (25.2)$$

Employment is equal to output divided by productivity. Given output, the higher the level of productivity, the lower the level of employment. This naturally leads to the following question: When productivity increases, does output increase enough to avoid a decrease in employment—equivalently, an increase in unemployment? In this section we look at the short-run responses of output, employment, and unemployment. In the next section we look at their medium- and long-run responses.

TECHNOLOGICAL PROGRESS, AGGREGATE SUPPLY, AND AGGREGATE DEMAND

The right model to use when thinking about the short-run response of output to a change in productivity is the model of aggregate supply and aggregate demand that we developed in Chapter 16. Let's recall its basic structure.

Output is determined by the intersection of the aggregate supply curve and the aggregate demand curve. The *aggregate supply* relation captures the effects of output on the price level. The aggregate supply curve is upward-sloping: An increase in the level of output leads to an increase in the price level. Behind the scenes, the mechanism is the following: An increase in output leads to a decrease in unemployment. The decrease in unemployment leads to an increase in wages and, in turn, to an increase in prices—an increase in the price level.

The *aggregate demand* relation captures the effects of the price level on output. The aggregate demand curve is downward-sloping: An increase in the price level leads to a decrease in the demand for output. Behind the scenes, the mechanism is the following: An increase in the price level leads to a decrease in the real money stock. The decrease in real money leads in turn to an increase in the interest rate. The increase in the interest rate leads to a decrease in the demand for goods, and thus to a decrease in output.

Aggregate supply is drawn as AS in Figure 25-1. Aggregate demand is drawn as AD. Their intersection gives the level of output Y consistent with equilibrium in

---

[1] **Digging deeper.** *Output per worker and the state of technology are in general not the same. Remember from Chapter 24 that an increase in output per worker may come from an increase in capital per worker, even if the state of technology has not changed. They are equal here because, in writing the production function as equation (25.1), we ignore the role of capital in production.*

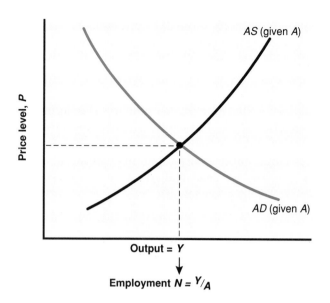

FIGURE 25-1

Aggregate Supply and Aggregate Demand for a Given Level of Productivity

Aggregate supply is upward-sloping: An increase in output leads to an increase in the price level. Aggregate demand is downward-sloping: An increase in the price level leads to a decrease in output.

labour, goods, and financial markets. Given output, the level of employment is determined in turn by the relation $N = Y/A$.

Suppose now that the level of productivity increases from A to A'. What happens to output, and in turn to employment and unemployment? The answer depends on how the increase in productivity shifts the aggregate supply curve and the aggregate demand curve.

Take aggregate supply first. The effect of an increase in productivity is to decrease the amount of labour needed to produce a unit of output, and thus to reduce cost for firms. This leads firms to reduce the price they charge at any level of output; they are now willing to supply the same level of output at a lower price. Thus, aggregate supply shifts down, from AS to AS' in Figure 25-2.

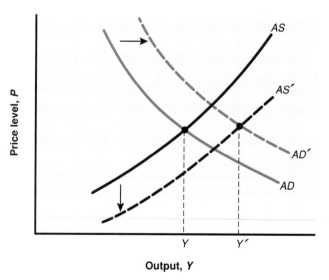

FIGURE 25-2

The Effects of an Increase in Productivity on Output in the Short Run

An increase in productivity shifts the aggregate supply curve down. It has an ambiguous effect on the aggregate demand curve, which may shift to the left or to the right. (In this figure, we assume a rightward shift.)

Now take aggregate demand. Does an increase in productivity increase or decrease the demand for goods at a given price level? The answer is that there is no general answer. The reason is that productivity increases do not appear in a vacuum, and what happens to aggregate demand depends on what triggered the productivity increase in the first place.

Take the case where productivity increases come from the widespread implementation of a major technological breakthrough. It is easy to see how this change may be associated with an increase in demand at a given price level. The prospect of higher growth in the future leads consumers to feel more optimistic about future income, and thus to increase their consumption, given current income. The prospect of higher profits in the future, as well as the need to put the new technology in place, may also lead to a boom in investment. In this case, the demand for goods increases at a given price level; the aggregate demand curve shifts to the right.

Good egs ↓

Now take instead the case where productivity growth comes not from the introduction of new technologies but rather from the more efficient use of existing technologies. One of the implications of increased international trade has been an increase in foreign competition. This competition has forced many firms to cut costs by reorganizing production and eliminating jobs (a process euphemistically referred to as "reducing fat" or "downsizing"). When such reorganizations are the source of productivity growth, there is no presumption that aggregate demand will increase; reorganization of production may require little or no new investment. Increased uncertainty and worries about job security may well lead workers to want to save more, and thus reduce consumption spending. In this case, aggregate demand may well shift to the left rather than to the right.

Let's assume the more favourable case (more favourable from the point of view of output and employment), namely the case where aggregate demand shifts to the right. The effects of the increase in productivity are thus to shift the aggregate supply curve from AS to AS' and the aggregate demand curve from AD to AD'. These shifts are drawn in Figure 25-2. Both shifts contribute to an increase in equilibrium output, from Y to Y'. In this case, the increase in productivity leads to an increase in output.

Without more information, however, we cannot tell what happens to employment. To see why, note that equation (25.2) implies the following relation:[2]

% change in employment = % change in output − % change in productivity

Thus, what happens to employment depends on whether output increases proportionately more or less than productivity. For example, if productivity increases by 2%, it takes an increase in output of at least 2% to avoid a decrease in employment—that is, an increase in unemployment. And without a lot more information about the slopes and the size of the shifts of the two curves, we just cannot tell whether this condition is satisfied in Figure 25-2. In short, increases in productivity may or may not lead to an increase in unemployment. Theory alone cannot settle the issue.

---

[2] *This follows from Proposition 8 in Appendix 3.*

THE EMPIRICAL EVIDENCE

Can empirical evidence help us reach a conclusion? At first glance, it would seem to. Look at Figure 25-3, which plots the behaviour of labour productivity and the behaviour of output for Canada from 1960 to 1996. The coloured line gives the annual rate of change of output; the black line gives the annual rate of change of labour productivity. The figure shows a strong positive relation between output growth and productivity growth. Furthermore, the movements in output are typically larger than the movements in productivity. Figure 25-3 might thus seem to imply that when productivity growth is high output increases by enough to avoid any adverse effect on employment.

   Can we thus conclude from Figure 25-3 that increases in productivity lead to increases in output large enough to increase employment? Unfortunately not. The reason is that the causal relation actually runs the other way, from output growth to productivity growth. That is, output growth leads to productivity growth, not the other way around. We saw why when we discussed Okun's law in Chapter 18. In bad times, firms hoard labour—that is, they keep more workers than is absolutely necessary for production. When the demand for goods increases for any reason, firms respond partly by increasing employment and partly by having existing workers work harder. Thus increases in output lead to increases in productivity. This is what we

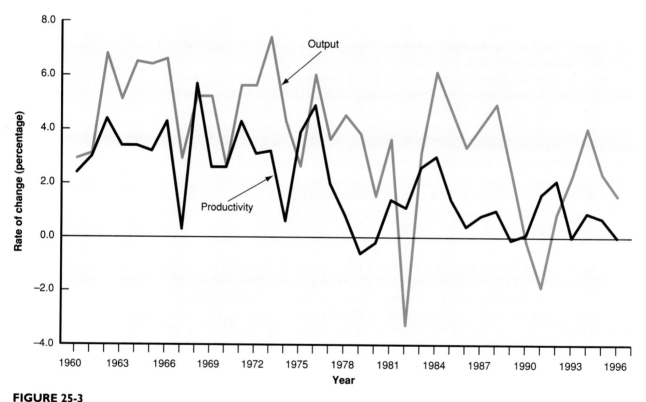

FIGURE 25-3

Labour Productivity and Output Growth in Canada, 1960–1996

There is a strong positive relation between output growth and productivity growth. But the causality runs from output growth to productivity growth, not the other way around.

Source: Statistics Canada, CANSIM Matrices 0603 and 2075 and Series D14442.

see in Figure 25-3, and it is clearly not the effect we are interested in. Rather, we want to know what happens to output and unemployment when there is an *exogenous* change in productivity—that is, a change in productivity that comes from a change in technology, not from the response of firms to movements in output. Figure 25-3 does not help us here. And the conclusion from the research that has looked at the effects of exogenous movements in productivity growth on output is that the data give an answer just as ambiguous as the answer given by the theory: Sometimes increases in productivity lead to increases in output sufficient to maintain or even increase employment in the short run; sometimes they do not, and unemployment increases in the short run.

25-2 PRODUCTIVITY AND THE NATURAL RATE OF UNEMPLOYMENT

We have looked so far at the *short-run* effects of productivity on output, employment, and unemployment. In the longer run, we know that the economy returns to the natural level of output, the level of output consistent with the natural rate of unemployment. Thus, the next question we must ask is: Is the natural rate of unemployment itself affected by changes in productivity?

Recall from Chapter 15 that the natural rate of unemployment is determined by two relations, price setting and wage setting. Our first step must be to think about how changes in productivity affect each of these two relations.

PRICE AND WAGE SETTING REVISITED

Consider price setting first. Recall from equation (25.1) that each worker produces A units of output; equivalently, producing 1 unit of output requires $1/A$ workers. If the nominal wage is equal to W, the cost of producing 1 unit of output is thus equal to $(1/A)W = W/A$. Assuming that firms set prices as a markup over cost, μ, the price level is thus given by

$$\text{Price setting:} \qquad P = (1 + \mu)\frac{W}{A} \qquad (25.3)$$

The only difference between this equation and equation (15.3) is the presence of the productivity term, A (which we had implicitly set to 1 in that chapter). An increase in productivity decreases cost, and thus decreases the price level given the nominal wage.

Turn now to wage setting. The evidence suggests that, other things equal, wages are typically set to reflect the increase in productivity over time. If productivity has been growing at 3% a year on average for some time, then wage contracts will build in a wage increase of 3% a year. This suggests the following extension of our earlier wage-setting equation:

$$\text{Wage setting:} \qquad W = A^e P^e F(u, z) \qquad (25.4)$$

Let's look at the three terms on the right-hand side of equation (25.4).

The last two are familiar from Chapter 15 and equation (15.4). Wages depend on the unemployment rate (u) and on institutional factors captured by the catch-all variable z. Workers care about real wages, not nominal wages. Thus, wages depend on the expected price level, P^e.

The new term is the first: Wages now also depend on the expected level of productivity, A^e. If workers and firms expect productivity to increase, they incorporate those expectations into the wages set in bargaining.[3]

THE NATURAL RATE OF UNEMPLOYMENT

Let's now characterize the natural rate graphically. Recall that the natural rate of unemployment is determined by the price- and wage-setting relations, and the additional condition that expectations be correct. In this case, this condition requires that expectations of *both* prices *and* productivity be correct.

The price-setting equation determines the real wage paid by firms. Reorganizing equation (25.3), we can write

$$\frac{W}{P} = \frac{A}{1+\mu} \tag{25.5}$$

The real wage paid by firms, W/P, depends on both productivity and the markup. The higher the level of productivity, the lower will be the price set by firms given the nominal wage, and thus the higher the real wage paid by firms. Equation (25.5) is drawn in Figure 25-4 for a given level of productivity, A. The real wage is measured on the vertical axis. The unemployment rate is measured on the horizontal axis. Equation (25.5) is represented by the solid horizontal line at $A/(1 + \mu)$: The real wage implied by price setting is independent of the unemployment rate.

Under the condition that expectations be correct—so that both $P^e = P$ and $A^e = A$—the wage-setting equation (25.4) becomes

$$\frac{W}{P} = A\,F(u, z) \tag{25.6}$$

The real wage implied by wage bargaining depends on both the level of productivity and the unemployment rate. The higher the level of productivity, the higher will be the real wage. The higher the unemployment rate is, the lower the real wage. For a given level of productivity, equation (25.6) is represented by the solid downward-sloping curve in Figure 25-4: The real wage implied by wage setting is a decreasing function of the unemployment rate.

Equilibrium in the labour market is given by point B, and the natural rate of unemployment is equal to u_n. Let's now ask what happens to the natural rate in response to an increase in productivity. Suppose that A increases by 5%, and that the new level of productivity A' is therefore equal to 1.05 times A.

From equation (25.5) we see that the real wage implied by price setting is now higher by 5%. The price-setting curve shifts up. From equation (25.6) we see that, at a given unemployment rate, the real wage implied by wage setting is also higher by 5%. The wage-setting curve shifts up. Note that at the initial unemployment rate u_n

---

[3]***Digging deeper.*** *How economy-wide increases in productivity affect wage setting is one of the main questions examined in the book by Edmund Phelps,* Structural Slumps *(Cambridge, MA: Harvard University Press, 1994), already mentioned in footnote 7 in Chapter 15.*

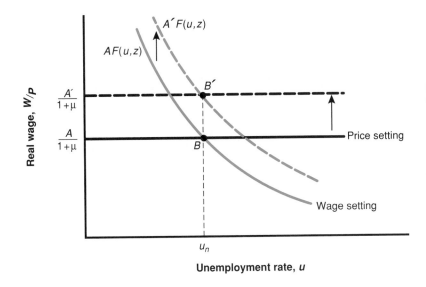

FIGURE 25-4

The Effects of an Increase in Productivity on the Natural Rate of Unemployment

An increase in productivity shifts both the wage-setting curve and the price-setting curve in the same proportion and thus has no effect on the natural rate of unemployment.

both curves shift up by the same amount, namely 5% of the initial real wage. Thus the new equilibrium is at B': The real wage is higher by 5%, and the natural rate remains the same.

The intuition for this result is straightforward. A 5% increase in productivity leads firms to reduce prices by 5% given wages, leading to a 5% increase in real wages. This increase exactly matches the increase in real wages from wage bargaining at the initial unemployment rate. Thus, real wages increase by 5% and the natural unemployment rate remains the same.

We have looked at a one-time increase in productivity, but the argument we have developed also applies to productivity growth. Suppose that productivity steadily increases, so that each year A increases by 5%. Then, each year, real wages will increase by 5%, and the natural rate will remain unchanged.

THE EMPIRICAL EVIDENCE

We have derived two strong results: The natural rate of unemployment should depend neither on the level nor on the rate of productivity growth. How do these two results fit the facts?

An obvious difficulty in answering this question is that we do not observe the natural rate of unemployment. But we can work around this problem by looking at the relation between average productivity growth and the average unemployment rate over decades. Because the actual unemployment rate moves around the natural rate, looking at the average unemployment rate over a decade should give us a good estimate of the natural rate for that decade. Looking at average productivity growth over a decade also takes care of another problem we discussed earlier: While changes in labour hoarding can have a large effect on yearly changes in labour productivity, they are unlikely to make much difference when we look at average productivity growth over a decade.

Figure 25-5 plots average Canadian labour productivity growth and the average unemployment rate for each decade from 1927 to 1996. Surprisingly, the relation that appears between the two is the opposite of that predicted by those who believe in technological unemployment. Periods of *high productivity growth,* such as the decade following World War II, were associated with *a lower unemployment rate.* Periods of *low productivity growth,* such as Canada has seen since the mid-1970s, have been associated with *a higher unemployment rate.*

Can the theory we just developed be extended to explain this inverse relation between productivity growth and unemployment? The answer is yes. To do so, we must look more closely at the formation of expectations of productivity in wage setting.

We have looked at the rate of unemployment that prevails when *both* price expectations *and* expectations of productivity are correct. However, one of the lessons of the 1970s and 1980s is that it takes a very long time for expectations of productivity to adjust to the reality of lower productivity growth. When productivity growth slows down for any reason, it takes a long time for society in general, and workers in particular, to adjust their expectations. In the meantime, workers keep asking for wage increases that are no longer consistent with the new lower rate of productivity growth.

To see what this implies, let's look at what happens to the unemployment rate when price expectations are correct (that is, $P^e = P$), but expectations of productiv-

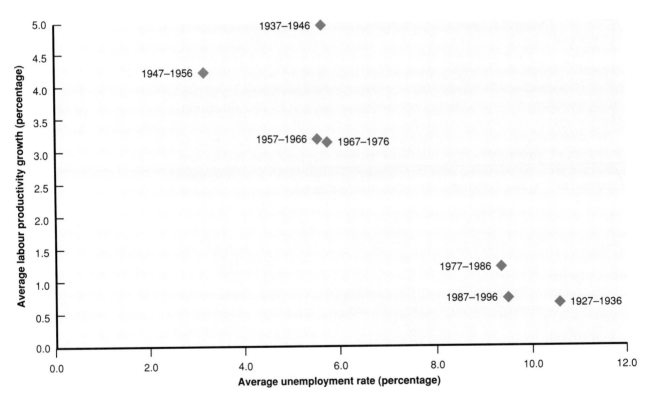

FIGURE 25-5

Productivity Growth and Unemployment in Canada: Averages by Decade, 1927–1996

There is a clear negative relation between decadal averages of productivity growth and unemployment. Higher productivity growth is associated with lower unemployment.

Source: Statistics Canada, CANSIM Matrices 0603 and 2075 and Series D14442, and *Historical Statistics of Canada,* 2nd edition; also, *Historical Statistics of Canada,* M.C. Urquhart and K.A.H. Buckley, eds. (Toronto: Macmillan, 1965).

ity (A^e) are not. In this case, the relations implied by price and wage setting are given by:

$$\text{Price setting:} \quad \frac{W}{P} = \frac{A}{1 + \mu}$$

$$\text{Wage setting:} \quad \frac{W}{P} = A^e F(u, z)$$

If expectations of productivity growth adjust slowly, then A^e will increase by more than A when productivity growth declines. What will then happen to unemployment is shown in Figure 25-6. If A^e increases by more than A, the wage-setting relation will shift up by more than the price-setting relation. The equilibrium will move from B to B', and the natural rate of unemployment will increase from u_n to u'_n. The natural rate will remain higher until expectations of productivity have adjusted to the new reality, until A^e and A are again equal.

Let's summarize what we have learned in this and the preceding section. Put simply, we have not found much support, either in theory or in the data, for the idea that faster technological progress leads to higher unemployment. In the short run, there is no reason to expect, nor does there appear to be, a systematic relation between movements in productivity and movements in unemployment. The relation between the two in the longer run appears to be an inverse relation. Lower productivity growth appears to lead to higher unemployment, and higher productivity growth appears to lead to lower unemployment. One plausible explanation is that high unemployment is what it takes to reconcile workers' wage aspirations with lower productivity growth until those aspirations have adjusted to this new reality.

So, where do the fears of technological unemployment come from? They are likely to come from the dimension of technological progress we have neglected so far, the dimension of structural change. And for some workers, those with the wrong skills or those in the wrong industries, structural change may indeed mean unemployment and lower wages.

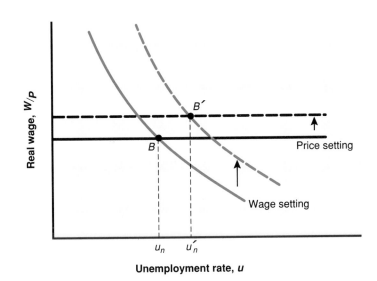

FIGURE 25-6
The Effects of a Decrease in Productivity Growth on the Natural Unemployment Rate When Expectations of Productivity Growth Adjust Slowly

If it takes time for workers to adjust their expectations of productivity growth, a slowdown in productivity growth will lead to an increase in the natural rate of unemployment for some time.

25-3 TECHNOLOGICAL PROGRESS AND DISTRIBUTION EFFECTS

Technological progress is a process of structural change. New goods are developed, making old ones obsolete. New techniques of production appear, requiring new skills and making some old skills less useful. The essence of this **churning** process is nicely captured in the following quote from the president of the Federal Reserve Bank of Dallas in his introduction to a report entitled *The Churn:*[4]

> *My grandfather was a blacksmith, as was his father. My dad, however, was part of the evolutionary process of the churn. After quitting school in the seventh grade to work for the sawmill, he got the entrepreneurial itch. He rented a shed and opened a filling station to service the cars that had put his dad out of business. My dad was successful, so he bought some land on the top of a hill, and built a truck stop. Our truck stop was extremely successful until a new interstate went through 20 miles to the west. The churn replaced US 411 with Interstate 75, and my visions of the good life faded.*

Many professions, from blacksmiths to harness makers, have indeed vanished. Canada was largely an agrarian society at the beginning of the twentieth century. At its peak in 1881, almost half of total employment was in agriculture. Because of high productivity growth in agriculture, only about 3% of workers are now employed in jobs directly related to agriculture; almost as many workers are now employed in art, culture, recreation, and sport. The Focus box entitled "The Changing Composition of Employment by Industry in Canada since 1986" discusses recent and predicted future shifts in the composition of employment in Canada.

---

[4]The Churn: The Paradox of Progress *(Dallas, TX: Federal Reserve Bank of Dallas, 1993).*

THE CHANGING COMPOSITION OF EMPLOYMENT BY INDUSTRY IN CANADA SINCE 1986

In 1996, about three-quarters of workers in Canada had jobs in the service sector. Using the 1986 and 1996 census data from Statistics Canada, reported in Table 1 below, we can see the shifts in employment by industry clearly at work. Although total employment grew on average by 1.2% annually over this decade, almost all industries involving natural resources or manufacturing experienced a decline. Agriculture, which has produced tremendous gains in productivity this century, continued a long-established trend of declining employment. Employment in government services declined as governments at all levels worked to reduce their expenditures, but all other service industries expanded. Business services was the sector with the fastest employment growth. The Bank of Montreal issued a forecast of employment growth for the next few years that we have reproduced in the last column of Table 1. The Bank of Montreal's economists expect total employment growth to be slightly higher from 1997 to 2001 than in the previous decade. With the notable exceptions of health, education, and government services, they expect growth in the service sector to continue to be strong. They also predict that many of the jobs lost in natural resources and manufacturing will be restored over the next few years, but employment growth will vary significantly across industries.

FOCUS

TABLE |

EMPLOYMENT BY INDUSTRY IN CANADA

INDUSTRY	EMPLOYMENT 1996	AVERAGE ANNUAL GROWTH RATE 1986–1996	PREDICTED ANNUAL AVERAGE GROWTH RATE 1997–2001
NATURAL RESOURCES:			
Agriculture	485 605	−0.5	−0.3
Fishing	45 696	−0.2	0.6
Forestry	102 715	−0.9	0.2
Mining	124 120	−0.8	2.2
Oil and Gas	44 200	−2.8	2.2
MANUFACTURING:			
Food	227 565	−1.1	1.3
Beverages	26 755	−3.4	2.9
Tobacco	5 185	−3.8	−0.9
Rubber	24 400	−0.5	2.5
Plastic	62 255	2.1	5.2
Leather	14 835	−6.2	−1.9
Textiles	54 265	−2.1	2.1
Clothing	112 580	−2.4	1.1
Wood	151 730	0.7	0.2
Furniture	59 555	−1.1	5.1
Paper	115 045	−1.2	−0.3
Printing and Publishing	175 855	0.6	0.9
Primary Metals	91 875	−3.0	−1.6
Fabricated Metal Products	172 155	0.0	2.8
Machinery	86 905	0.1	4.9
Transportion Equipment	257 025	0.6	1.6
Electrical/Electronic	137 580	−1.8	3.0
Nonmetallic Minerals	55 000	−1.2	2.2
Refined Petroleum and Coal	15 515	−4.8	−2.3
Chemicals	95 940	−0.8	1.7
Other Manufacturing	97 825	0.0	3.7
CONSTRUCTION			
SERVICES:	822 345	0.8	3.1
Transportation/Storage	589 970	0.6	2.7
Pipelines	8 955	1.4	1.0
Communications	302 750	1.0	1.5
Other Utilities	144 020	0.5	−0.4
Wholesale Trade	711 820	2.0	2.6
Retail Trade	1 781 250	1.0	2.2
Financial Real Estate	787 800	1.3	1.3
Business Services	937 635	4.8	4.3
Education Services	1 005 585	1.8	0.4
Health Services	1 409 170	3.1	0.7
Hospitality/Recreation	1 261 870	2.7	3.1
Personal Services	824 750	2.3	3.0
Government Services	887 450	−0.9	−0.5
ALL INDUSTRIES	14 317 545	1.2	2.0

Source: Statistics Canada, Catalogue No. 93F0027XDB96009; Bank of Montreal, *Sectoral Outlook*, March 12, 1998.

For those in the right sectors or with the right skills, technological progress leads to new opportunities and higher wages. But for those in declining sectors, or those with skills that are no longer in demand, technological progress can mean the loss of their jobs, a period of unemployment, and possibly much lower wages.

The past 15 to 20 years have seen the emergence of large changes in the patterns of worker compensation. In both Canada and the United States, the average level of real wages has shown little growth, but important differences have appeared in the patterns of wages for different groups. In the United States, the differences have been closely related to education: College-educated workers have seen their wages increase slightly, but workers with at best a high-school education have seen sharp declines in their wages. This divergence between the wages of low-skilled and higher-skilled workers in the United States has been the focus of much discussion and research; we summarize the main findings in the Focus box entitled "Increasing Wage Inequality in the United States, 1963–1987." Canada has experienced similar patterns, but the quantitative changes have been substantially different.

Table 25-1 summarizes some of the important changes in real income that occurred in Canada from 1972 to 1991, broken down by sex, education, and age. The big story since 1972 is that men and women have fared very differently. Real income for men grew only 3% over the 20-year period; by contrast, women's wages have grown sharply.[5] Both sexes had higher income from 1972 to 1980 than in the next decade; men's real income actually declined from 1980 to 1991. Over the 20-year period, skilled workers (those with a university degree) had below-average income

....

TABLE 25-1

CHANGES IN REAL INCOME, 1972–1991 (PERCENTAGE)

	1972–1980	1980–1991	1972–1991
Men (overall):	8.4	−5.0	3.0
By education:			
University degree	−2.2	0.3	−1.9
By age			
25–34	8.7	−17.5	−10.3
35–44	8.1	−10.7	−3.5
45–54	9.8	−2.3	7.3
Women (overall):	25.6	17.0	47.0
By education:			
University degree	10.5	20.9	33.6
By age			
25–34	27.3	3.6	31.9
35–44	29.2	18.9	53.6
45–54	29.5	19.5	54.8

Source: Charles M. Beach and George A. Slotsve, *Are We Becoming Two Societies?* Toronto: C. D. Howe Institute, 1996, Table 9.

---

[5]*In 1967, the first year for which such data were collected, women earned on average 58 cents to each dollar earned by men. By 1996, the ratio had climbed to 73 cents.*

Table 1 shows the evolution of real wages for various groups of workers in the United States, by education level, sex, and experience (defined as the number of years of employment). While the table gives the evolution of wages only up to 1987, available evidence shows that the evolution since has been similar to that of the period 1979–1987. (Notice that Table 1 refers to *real wages* in the United States; Table 25-1 in the text refers to *real income*; although the difference is important, it does not affect the broad patterns and we shall ignore this distinction.)

There are several similarities between the Canadian and the U.S. experience since 1971. Both coun-tries have seen average real wage growth virtually dis-appear, but some groups have gained while others have lost. Full-time working women have narrowed the gap with their male counterparts; women's wages have in-creased while men's wages have tended to fall. The gap between the real wages of young workers and older workers has widened; young workers are worse off than those born 10 or 20 years earlier, but the rel-ative decline has been less pronounced in the United States. After falling behind in the 1970s, the incomes of university-educated workers increased in both coun-tries, but the turnaround has been much more notice-able in the United States than in Canada.

TABLE 1

REAL WAGE CHANGES FOR FULL-TIME WORKERS, 1963–1987 (PERCENTAGE)

	1963–1971	1971–1979	1979–1987
All workers	19.2	−2.8	−0.3
By education:			
Less than high school	17.1	0.3	−6.6
High school	16.7	1.4	−4.0
Less than four years of college	16.4	1.5	1.5
Four years of college or more	25.5	−10.1	7.7
By sex:			
Men	19.7	−3.4	−2.4
Women	17.6	−0.8	6.1
By experience:			
1–5 years	17.1	−3.5	−6.7
26–35 years	19.4	−0.6	0.0

Source: Lawrence Katz and Keven Murphy, "Changes in Relative Wages, 1963–1987," Quarterly Journal of Economics, February 1992, 35–78.

growth. Although university graduates fared better than the average worker after 1980, the differences were not very large. However, very large differences have emerged based on age. In Canada, the gap between younger and older workers has widened considerably. From 1972 to 1980, the rate of growth of real incomes for both male and female workers did not vary much with age; since 1980, however, young workers have fared much more poorly. For example, while older male work-ers have pretty much held their own, the real income of young male workers de-clined by almost 18%!

How can we explain the shifting patterns of compensation related to education and age since 1972?[6] Some observers point to the Canada-United States Free Trade Agreement (FTA) and the impact of international competition on wages as the cause of important adjustments in the Canadian labour market. The evidence, however, seems to indicate that the main impact of the FTA has been on employment; some industries have grown faster because of the FTA while others have seen employment contract. The FTA appears to have had very little impact on wages.[7]

Economists believe that the most important force at work in shaping the relative compensation of different groups of workers has been **skill-biased technological progress**. New machines and new methods of production, the argument goes, require skilled workers, more so today than in the past. The development of computers requires workers to be increasingly computer-literate. The new methods of production require workers to be more flexible, better able to adapt to new tasks. Greater flexibility requires more skills and more education. In the United States, the shift in the demand for highly skilled workers has shown up as a sharp increase in the return to education. In Canada, however, this shift in demand for highly skilled workers has coincided with a sharp increase in supply. The proportion of young people who choose to go to university or obtain some other post-secondary school training has increased noticeably in the past 25-years; by some measures, it now exceeds that of the United States. Not surprisingly, a relatively large increase in the supply of skilled workers in Canada has led to a muted increase in the relative wages of these workers. Can skill-biased technological progress explain the widening gap between young and older workers? Perhaps, but the evidence is less persuasive. Older workers are indeed more experienced, and experience tends to be a good proxy for being skilled. But it is impossible to distinguish this explanation from others that are cohort-specific. For now, it must be admitted that we don't really understand why young workers are faring so poorly compared with previous cohorts.

Can we expect the reward for education to continue to increase? Will the wages of the least-skilled, least-experienced workers continue to decrease? There are three reasons for some optimism.

1. The trend in relative demand may simply slow down. For example, one can think of computers in the future as being easier and easier to use, even by unskilled workers. One can even think of computers as replacing skilled workers, those whose skills involve primarily the ability to compute or to memorize (accountants, lawyers?).

2. Technological progress is not exogenous; this is a theme we explored in Chapter 24. How much firms spend on R&D and in what directions firms direct their research efforts depend on expected profits. Thus the low relative wage of unskilled workers may lead firms to explore new technologies that take advantage of unskilled, low-wage workers. Market forces may lead technological progress to become less skill-biased in the future.

---

[6]*For those interested in changes in the male-female gap, we recommend Chapter 12 in Dwayne Benjamin, Morley Gunderson, and Craig Riddell,* **Labour Market Economics**, *4th ed.*

[7]*For a detailed discussion of this evidence, see Noel Gaston and Daniel Trefler, "The Labour Market Consequences of the Canada-U.S. Free Trade Agreement,"* Canadian Journal of Economics, *February 1997.*

3. The relative supply of skilled versus unskilled workers is also not exogenous. The large increase in the relative wage of more educated workers implies that the returns to acquiring more education and training are higher than they were one or two decades ago. Higher returns to training and education can in turn increase the relative supply of skilled workers, and thus work to stabilize relative wages. We have seen this process at work in Canada since 1970. Many economists believe that policy has an important role to play here, making sure that the quality of primary and secondary education for the children of low-wage workers does not further deteriorate, and that those who want to acquire more education can borrow to do so.

SUMMARY

■ Popular discussions often reflect fears that technological progress destroys jobs and leads to higher unemployment. Such fears were present during the Great Depression. They have reemerged in Europe today, where there is widespread support for a shorter workweek to allow more workers to have jobs. Theory and evidence suggest that these fears are largely unfounded.

There is not much support, either in theory or in the data, for the idea that faster technological progress leads to higher unemployment.
■ In the short run, there is no reason to expect, nor does there appear to be, a systematic relation between changes in productivity and movements in unemployment.
■ If there is a relation between changes in productivity

KEY TERMS

■ technological unemployment, 492
■ creative destruction, 494

■ churning, 504
■ skill-biased technological progress, 508

QUESTIONS AND PROBLEMS

1. Suppose an economy is characterized by the following equations:

$$\text{Price setting: } P = (1 + \mu) \frac{W}{A}$$

$$\text{Wage setting: } W = A^e P^e (1 - u)$$

where

A = productivity
u = unemployment rate
μ = markup
and the superscript ^{e} indicates an expected variable

Let the markup be 10% over labour costs.
a. Solve for the rate of unemployment. Assume initially that expectations of prices and productivity are accurate.
b. What does your answer suggest about the relationship between the rate of unemployment and productivity in the long run?

c. Now suppose that productivity is always expected to be equal to last year's level, $A_t^e = A_{t-1}$, while price expectations are still accurate.
  (i) Find an expression for the rate of unemployment.
  (ii) If productivity has been constant for some time and then rises by 2%, what will be the impact on the unemployment rate? Explain your result briefly.
2. How might each of the following affect the wage gap between skilled and unskilled workers in Canada?
  a. Increased government subsidies for college education
  b. High tariffs on imported goods and services
  c. Increased use of industrial robots to manufacture goods
3. Should university professors fear "technological unemployment"? Explain.
4. Suppose the central bank lags behind in recognizing changes in the natural rate of unemployment. If labour productivity begins to grow at a higher rate, what will happen to the inflation rate over time?

5. "Skill-biased technological progress presents countries with a tough choice regarding low-skilled workers: Either let their wages fall, or maintain their wages but accept higher low-skilled unemployment rates. The United States has chosen the first course. Most European countries have chosen the second. Canada has taken a middle road." Discuss.

FURTHER READING

For more on the process of reallocation that characterizes modern economies, read *The Churn: The Paradox of Progress,* a report by the Federal Reserve Bank of Dallas, 1993.

For a modern, but demanding, presentation of Schumpeterian growth theory, read Phillipe Aghion and Peter Howitt, *Endogenous Growth Theory* (Cambridge, MA: MIT Press, 1998).

TRANSITION IN EASTERN EUROPE

skip?

In the early 1990s, the countries of Eastern Europe abandoned communism and shifted from central planning to the market. Much has been accomplished, but the economic part of the transition has been painful. Output has declined, and so has employment.

In this chapter we examine the transition process. The first section provides an overview. In the following sections we look at the performance of central planning in the past, the evolution of output since the beginning of transition, and the shape of things to come.

26-1 AN OVERVIEW

For most of the post–World War II period, the economies of Eastern Europe were centrally planned. The government sent production plans to firms, telling them what to produce and how to produce it. Prices were mostly accounting devices and played little or no role in the allocation process.

Starting in the 1970s, some countries, most notably Hungary, introduced reforms aimed at giving firms incentives to produce more efficiently, sometimes allowing for a limited role for prices. But the first fundamental reform was introduced in Poland in January 1990. After winning the elections in 1989, *Solidarity*—the trade union that the communist regime had banned earlier—decided to abandon central planning altogether and shift quickly to a market economy. In January 1990, in what has been called an economic **big bang**, it removed controls on nearly all prices and implemented a macroeconomic stabilization plan.[1] Since then, all the countries of Eastern Europe have moved in the same direction. Central planning is gone. What has taken its place is less clear, however, and varies from country to country.

Looking at the evolution of all the countries in transition would be too large a task: There are too many of them. Fifteen countries have emerged from the breakup of the former Soviet Union alone, each with its own transition dynamics.[2] And transition is not limited to Eastern Europe. China and Vietnam are also moving away from central planning. Indeed, of all the countries that were centrally planned in the 1980s, only Cuba and North Korea have so far resisted the tide. To keep things manageable, we shall focus here on six countries only, four from Central Europe—Bulgaria, the Czech Republic, Hungary, and Poland—and two from the former Soviet Union—Russia and Ukraine. Table 26-1 gives their basic characteristics and where they stood in their transition at the end of 1997.

The first two columns of the table give population and GDP per capita. In terms of GDP per capita (in PPP terms[3]), all six countries are middle-income countries. GDP per capita in the Czech Republic, the highest in the list, is roughly the same as that of Portugal.

The next three columns report on progress in three important dimensions of transition.

····················

TABLE 26-1

COUNTRIES OF EASTERN EUROPE: CHARACTERISTICS AND PROGRESS

	POPULATION (MILLIONS)	GDP PER CAPITA (US$ PPP)	PRICE LIBERALIZATION	PRIVATIZATION	MACRO CONTROL
Bulgaria	8.4	4190	3	3	1
Czech Republic	10.3	9770	3	4	4
Hungary	11.2	6410	3+	4	3
Poland	38.6	5400	3	3+	3
Russia	147.5	4480	3	3+	3
Ukraine	50.9	2400	3	2+	3

GDP per capita is for 1995. The other numbers refer to 1997. The numbers for the last three columns are indexes, constructed as follows:

Price liberalization refers to easing of restrictions on price setting by firms. 1: Most prices still controlled; 2: price controls still important; 3: substantial progress on price liberalization; 4: comprehensive price liberalization and antitrust legislation in place.

Privatization refers to the privatization of large firms. 1: Little done; 2: some program in place, and some sales; 3: more than 25% of large firms privatized; 4: more than 50% of large firms privatized.

Macroeconomic control is measured by the rate of inflation. 1: Inflation above 100% a year in 1997; 2: inflation between 50 and 100%; 3: inflation between 10 and 50%; 4: inflation under 10%.

Source: European Bank for Reconstruction and Development, *Transition Report,* 1997.

---

[1] See Jeffrey Sachs, Poland's Jump to the Market Economy *(Cambridge, MA: MIT Press, 1993). The author, a Harvard economist, played an important role in the design of Polish economic reform.*

[2] Twelve of them—Armenia, Azerbaijan, Belarus, Georgia, Kazakhstan, Kyrgyz, Moldova, Russia, Tajikistan, Turkmenistan, Ukraine, and Uzbekistan—have formed a loose alliance called the Commonwealth of Independent States (CIS). Three Baltic states (Estonia, Latvia, and Lithuania) have gone their own way; they are independent from one another and from the CIS.

[3] See Chapter 22 for a definition and a discussion.

- The first is **price liberalization,** the degree to which prices are decontrolled and allowed to clear markets. Firms need reliable price signals to know what to produce. The table shows that this step has been largely achieved in all six countries; this is indeed a dramatic departure from central planning.

- The second is **privatization,** the transfer of state-owned firms to private owners, thus to owners with the incentives to maximize profit and to respond to price signals. The table shows some differences across countries. Substantial progress has been made in the Czech Republic, Russia, Hungary, and Poland.

- The third is **macroeconomic control.** As we shall see, this term has various dimensions. An important one is that firms cannot systematically expect transfers from the state if they make losses. When firms know that the government will bail them out if they make losses, their incentives are clearly distorted. And the resulting transfers can easily lead to large budget deficits and inflation. The table again shows sharp differences across countries. Macroeconomic control has been largely achieved in all the countries except Bulgaria, where inflation in 1997 exceeded 1000%.

As a result of these measures, these economies are undergoing profound structural change—more on this later in this chapter. But one of the transition's most dramatic outcomes so far has been the decline in output. Figure 26-1 shows the evolution of real GDP in each of the six countries since 1989. In each country, GDP is normalized to equal 1 in 1989.

All six countries suffered a large decline in measured output. Only Poland has climbed above its 1989 value. Note the use of "measured output" here; as we shall soon discuss, some of the decline probably reflects measurement error. But much of the decline is indeed real.

The decline has been steepest in the countries from the former Soviet Union. Russian GDP fell for seven years before increasing modestly in 1997. At the end of 1997, GDP in Poland was above its pre-transition level (110% of its pretransition level). GDP was equal to 90% of its pre-transition level in the Czech Republic, 89% in Hungary, and 63% in Bulgaria. The output decline in Russia and Ukraine was much larger. At the end of 1997, measured output in Russia stood at 57% of its 1989 level; measured output in Ukraine stood at 37% of its 1989 level. These are very large declines indeed; by comparison, at the trough of the Great Depression in Canada in 1933, real GDP stood at 72% of its pre-depression (1929) level.

This overview sets the stage for the three issues on which we focus in the rest of this chapter:

1. What triggered the transition in the first place?

2. Why did the transition start with such a large output decline?

3. What lies ahead? Will growth become stronger? And what are the potential dangers?

In studying each of these issues, we shall use the tools we developed earlier, in particular the study of growth in Chapters 22 to 24, the study of technological progress and unemployment in Chapter 25, and the study of episodes of high inflation in Chapter 21.

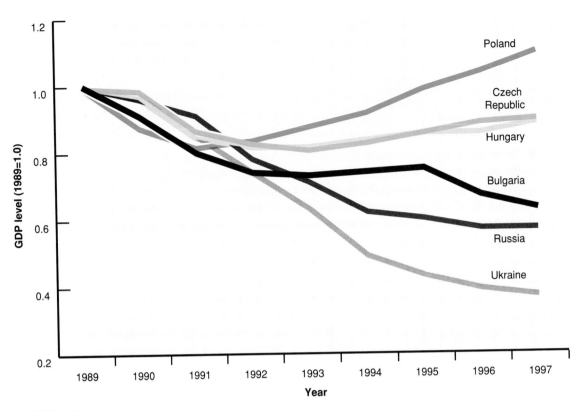

FIGURE 26-1

The Evolution of GDP in Six Eastern European Countries
Transition has been associated with a large decrease in real GDP. The decrease has been larger in Russia and Ukraine than in Central Europe.

Source: European Bank for Reconstruction and Development.

26-2 ECONOMIC GROWTH UNDER CENTRAL PLANNING

Many factors contributed to the fall of communism. The main one was its failure as an economic system. We focus in this section on the economic performance of the Soviet Union, but the lessons apply to the other communist countries as well.

When Russia became communist in 1917, many believed that central planning could lead to a better organization of economic activity than the market, while maintaining a more equal distribution of income. And central planning was indeed able to transform an essentially agricultural economy into a major industrial power. Even as late as the early 1960s, Khrushchev's forecast that the Soviet Union would soon be ahead of the United States was not seen as idle boasting but as a real possibility. But by the 1970s and 1980s, the Soviet Union became increasingly unable to fulfill its own expectations. And, in the early 1990s, a full-fledged economic crisis triggered the fall of communism.

Why did Soviet economic performance deteriorate? Let's first look at the basic numbers. Table 26-2 gives average annual growth rates for output per worker,

TABLE 26-2

ANNUAL AVERAGE GROWTH RATES OF OUTPUT AND CAPITAL PER WORKER, AND
ANNUAL AVERAGE RATE OF TECHNOLOGICAL PROGRESS IN THE SOVIET UNION, 1928–1987
(PERCENTAGE)

	GROWTH RATE OF OUTPUT PER WORKER	GROWTH RATE OF CAPITAL PER WORKER	RATE OF TECHNOLOGICAL PROGRESS
1928–1939	2.9	4.9	1.6
1940–1949	1.9	1.5	1.0
1950–1959	5.8	7.4	2.1
1960–1969	3.0	5.4	4.6
1970–1979	2.1	5.0	0.2
1980–1987	1.4	4.0	−0.3
1928–1987	2.8	4.8	1.5

Source: William Easterly and Stanley Fischer, "The Soviet Economic Decline: Historical and Republican Data," *NBER Working Paper 4735,* May 1994.

capital per worker, and the rate of technological progress for each decade since the late 1920s. These numbers are based on Western estimates of Soviet production, which are thought to be the most reliable, but even these should be taken with a grain of salt. (See the Global Macro box entitled "Growth in the Soviet Union: Can We Trust the Numbers?")

We learn several things from the table. Most important, over the period 1928 to 1987, Soviet economic growth compared favourably with the growth performance of Western countries. During that period, average annual growth of output per worker was 3%—compared, for example, with about 2.2% for Canada. True, growth decreased in the 1970s and 1980s, down to 2.1% in the 1970s, and to 1.4% from 1980 to 1987. But, as we saw in Chapter 22, during that period growth was also lower in Western countries than it had been in the two preceding decades.

So where did the increasing perception of economic crisis come from? The numbers in the second and third columns give us two clues:

■ Over the period 1928 to 1987, the growth rate of capital per worker was typically much higher than the growth rate of output per worker. For the 1928–1987 period as a whole, the annual growth rate of capital was 2% higher than that of output. Put another way, there was a steady increase in the ratio of capital to output at the rate of 2% a year, which roughly implies a threefold increase ($1.02^{60} = 3.28$) in the ratio of capital to output over the six decades.

■ For the last 30 years of its history, the Soviet Union's rate of technological progress steadily declined. The annual rate was roughly equal to zero in the 1970s. It was actually negative in the 1980s.

The growth model we developed in Chapters 22 to 24 tells us how to interpret these two clues. Recall the main two results we derived in those chapters. First, in the long run the growth rate of output per worker is equal to the rate of technological progress (see Table 24-1 on page 480). The second is that an economy can grow faster for a while by accumulating capital at a rate faster than the growth rate of output. A high rate of investment leads, for some time, to output growth in excess of technological progress.

The main problem in assessing past Soviet growth is data reliability. There are many problems with the Soviet Union's official statistics:

■ Under central planning, firms were evaluated according to whether they met the plan's objectives. Thus managers had strong incentives to report output numbers at least equal to the plan's requirements, thus to overstate production if needed. And the evidence is that they did.

■ Under central planning, firms could justify price increases on the basis of quality improvements. To increase prices (and thus increase their profits), they therefore had strong incentives to report fictitious quality improvements. The result was an overstatement of quality-adjusted output—and a corresponding understatement of inflation (what was truly a price increase was treated as a quality improvement and thus an output increase).

■ The Soviet Union as a country had strong incentives to report high growth, in order to show the rest of the world that central planning was a workable economic system. There are good reasons to believe that some numbers, such as agricultural production, were indeed systematically overstated.

■ Prices under central planning do not reflect market prices. Thus, industries that had prices higher than would have been the case in a market economy had a higher weight in output than they would have had in a market economy. This was often the case for industries that were growing the most, thus again leading to an overstatement of output growth.

■ The measure of total output used in the Soviet Union—net material product, or NMP for short—was different from that used in the West. For example, the measure did not include services that were not related to production, such as passenger transportation, housing, and the output of government employees not producing material output.

For decades, Western economists tried to adjust for these problems and produced their own numbers. The numbers considered most reliable were constructed by the CIA, and they are the ones used in Table 26-2. The differences between official Soviet numbers and Western estimates are large. For example, the annual growth rate of industrial output from 1928 to 1987 is 6.3% according to official Soviet numbers, but only 3.4% according to Western estimates. The Soviet estimates imply a fortyfold increase in industrial output ($1.063^{60} = 39.1$) over those 60 years; Western estimates imply only a sevenfold ($1.034^{60} = 7.4$) increase.

The construction of measures of technological progress, such as the measure presented in the last column of Table 26-2, raises further issues. The standard method, described in the appendix to Chapter 24, constructs the Solow residual as the difference between the growth rate of output and a weighted average of the growth rates of capital and labour. The weights are the shares of capital and labour in production. The problem here is that the rationale for using the shares is based on the assumption that the factors of production (capital and labour) are paid their marginal products. This assumption may be very wrong for centrally planned economies, where the shares of labour and capital were determined by the central planning authority. Table 26-2 uses the shares of labour and capital for market economies at roughly the same level of economic development as the Soviet Union, namely 0.6 for labour and 0.4 for capital. This is a very rough way of taking care of the problem, and the numbers in the last column must therefore be considered tentative.

From 1928 to 1987, the Soviet Union sustained growth by accumulating capital at a rate much higher than the rate of output growth. It could do this for so long only by steadily increasing its investment rate. The investment rate, which was equal to roughly 15% in 1950, had increased to 30% in the 1980s.

Investment rates cannot be increased forever. As they become higher and higher, less and less output is left for consumption. At some point, the economy goes beyond the golden-rule level of capital, and past that point further capital accumulation leads to less rather than more consumption.[4] This decrease in consumption is likely to lead to increasing dissatisfaction and political crisis. This is exactly what happened in the 1980s. Achieving growth through higher investment rates became politically impossible, and growth could proceed only at the rate of technological progress, which had come to a halt in the 1970s and 1980s.

Why did technological progress come to a halt? To economists, the more natural question is how the Soviet Union was able to sustain positive technological progress for as long as it did. Recall the discussion of technological progress in Chapter 24. The pace of technological progress, we argued, depends on the firms' incentives to innovate and to implement innovations. These incentives were for the most part absent or even perverse in the Soviet Union. Firms, which could always obtain the inputs they needed to achieve the output goals set in the plan, had little incentive to develop more efficient methods of production. Indeed, firms had incentives *not* to adopt new methods or introduce new products. Changing machines or products leads to temporary disruptions of production, which makes meeting output quotas harder. Why bother, for no reward?

So how, despite all these disincentives, was the Soviet Union able to achieve fairly high rates of technological progress until the 1970s? One answer is that technological progress comes in part from the implementation of techniques developed elsewhere. Thus the central plan could require firms to introduce and use new techniques that had been developed in the West. The high level of education and thus of human capital in the Soviet Union implied that firms had the competence to adopt and implement the new techniques, if ordered to do so. Why did this system no longer work from the 1970s on? There is no fully convincing answer. The most plausible is that the nature of required technological progress changed. The increasing complexity of economic organization and the nature of innovations made central planning a poorer and poorer substitute for the market. Take as an example the development of the computer industry in the rest of the world; the market fostered and coordinated a myriad of interactions among software development, hardware development, and the market signals of what consumers like and do not like. These interactions just could not be replicated or managed by a central planning system.

With the slowdown in output growth, Soviet policy makers found themselves under increasing pressure to try economic reform. The 1980s saw a succession of reforms, all introducing incentives for firms to become more efficient, but within the framework of central planning. Few reforms had any effect; some indeed had perverse effects. For example, increased incentives for firms to maximize profits, but within a price system determined in the plan, led firms to increase production of those products with the highest profit margin—whether or not there was sufficient demand for those products. By the late 1980s, then-Premier Mikhail Gorbachev concluded that wider-ranging reforms were needed and tried to implement a slow transition to a market economy. But by then it was too late. The fall of communism and the shift to the market were going to be much faster than he had anticipated.

---

[4]*See Chapter 23.*

26-3 EXPLAINING THE DECLINE IN OUTPUT

If the previous system functioned so badly, why has transition been characterized by a further decrease in output? This is clearly an important question and economists do not yet fully agree on the answer.

MEASUREMENT, STABILIZATION, AND STRUCTURAL CHANGE

Economists have explored three main lines of explanation: measurement issues, macroeconomic stabilization, and structural change. Let's examine each one in turn.

Measurement issues. It is clear that part of the decline in output reflects measurement issues rather than a true decline in output.

Some of the goods produced under central planning were practically useless. But their prices, which were set by the central planner and did not reflect the forces of demand and supply, were positive. With price liberalization, demand has dropped and the production of these goods has stopped. This phenomenon should not be counted as a decrease in output; but under standard national accounting practices, it is. To understand the issues involved here, consider the following example.

Suppose that, in 1989, a country—we shall call it Poland, and use zlotys (zl) as the monetary unit, but there is no attempt at realism here—was producing 2000 pounds of bread at 10 zl a pound, and 10 TV sets at 1000 zl each. In 1990, Poland liberalizes prices. Domestic TV sets turn out to be so bad as to be useless; their price and their production both fall to zero. Both the quantity of bread and the price of bread remain unchanged. Has true GDP changed? Clearly not. Something that was useless in the first place (and thus would have had a price of zero in a market economy) is now no longer produced. But, as Table 26-3 shows, measured real GDP, using 1989 as the base year, will register a decrease of 33%, from 30 000 zl in 1989 to 20 000 zl in 1990.[5]

···············
TABLE 26-3
DECLINES IN GDP: TRUE DECLINES OR INCORRECT MEASUREMENT?

	QUANTITY	PRICE	Q × P	Q × P$_{1989}$
1989				
Bread	2000	10	20 000	20 000
TV sets	10	1000	10 000	10 000
Real GDP				**30 000**
1990				
Bread	2000	10	20 000	20 000
TV sets	0	0	0	0
Real GDP				**20 000**

---

[5] *For a review of the construction of real GDP, see Chapter 2.*

What is the source of the problem? It is the fact that, under central planning, the price of TV sets did not reflect their true market price (which was equal to zero). More generally, goods that were overpriced under central planning, and thus are most likely to see a large decline in production during the transition, receive too large a weight in national income accounting. This leads conventional measures of GDP to overestimate the true decline in GDP during transition.

Incorrect pretransition prices are not the only measurement problem, however. Part of the transition has involved the emergence of a large number of small private firms. Because the surveys needed to monitor the activity of small firms are only slowly being put in place, the activity of these firms is underreported. And small firms are typically not eager to report their activity to official authorities, because doing so implies paying taxes, from profit taxes to social security contributions—something they would rather avoid if they can.

How important are these measurement errors? The consensus is that they account for some, but by no means all, of the decline in measured output. In Poland, for example, estimates are that the decrease in output in 1990, the first year of transition, was closer to 6% to 8% than to 11.6%, the official estimate. And there is evidence that measurement errors play a more important role in Russia and Ukraine than in Central European countries. But, even allowing for measurement error, there is no question that countries have suffered a large decline in output, so we must search for other explanations.

Macro stabilization.   Another reason for the output decline in some of the countries in transition is likely to have been macroeconomic stabilization. In many countries, transition was associated with high inflation. Inflation has now been reduced through stabilization plans involving both a reduction in budget deficits and in money growth. We saw in Chapter 21 how stabilizations often lead to a reduction in output, as measures taken to control the budget and stop money growth lead to sharp increases in interest rates and decreases in aggregate demand.

Table 26-4 gives the rate of inflation for each of the six countries for each year since 1989 and shows that some countries have indeed experienced large movements in inflation. But, together with Figure 26-1, it does not give strong support to the notion that macroeconomic stabilization is at the root of the output decline. Two countries—Hungary and the Czech Republic—have avoided high inflation and thus the need for drastic macro stabilization; yet they have suffered a large decline in output. Poland had very high inflation in 1989–1990 and has since decreased it,

TABLE 26-4
ANNUAL RATES OF INFLATION (USING THE CPI) 1989–1997 (PERCENTAGE)

	1989	1990	1991	1992	1993	1994	1995	1996	1997
Bulgaria	6	26	334	82	73	96	62	123	1049
Czech Republic	1	11	57	11	21	10	9	9	10
Hungary	17	29	35	23	23	19	28	24	18
Poland	251	586	70	43	36	32	28	20	16
Russia	2	6	93	1526	875	307	198	48	17
Ukraine	2	4	91	1210	4700	891	376	90	20

Source: European Bank for Reconstruction and Development.

yet it has had the best output performance of all six countries. The countries with the largest output declines—Bulgaria, Russia, and Ukraine—have had the highest inflation.

Structural change. The third explanation for the decline in output is **structural change.** Transition has been associated with a sharp shift in relative demand. State firms that accounted for most of production under central planning have suffered a sharp decrease in the demand for their goods, and thus reduced their production. Demand has shifted to new products and to new firms. But these firms, which for the most part did not exist pre-transition, face many constraints and can only expand production slowly. The net result has therefore been a decrease in total output. As new firms expand and old firms shift to new products, output has started increasing in a number of countries. This mechanism is probably the most important factor behind the evolution of output. The rest of this section explores it in more detail.

SHIFTS IN RELATIVE DEMAND AND AGGREGATE EMPLOYMENT

In thinking about aggregate supply so far in this book, we have made a strong simplifying assumption: We have thought of the economy as producing only one type of good, so that we did not have to keep track of the supply of each good. When we look at transition in Eastern Europe, this is an assumption that we must relax because much of what has happened can be understood only by looking at the state and the private sector separately.

The way to think about the determination of output and employment in an economy with two sectors is given by Figure 26-2. For the moment, let's call the two sectors simply 1 and 2. The real wage is measured on the vertical axis. Employment in sector 1 is measured from the origin (point O) on the horizontal axis, going from left to right. The demand for labour in sector 1 is given by the downward-sloping

FIGURE 26-2

Employment and Unemployment in an Economy with Two Sectors

Employment in each sector is a decreasing function of the real wage.

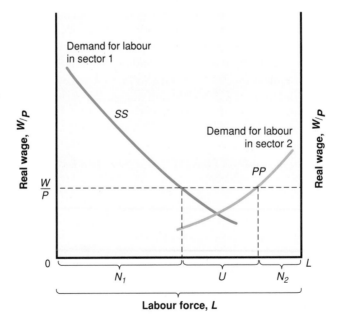

relation *SS*. (We use the letter *S* because we shall interpret sector 1 as the state sector in what follows.) For example, at wage W/P, employment in sector 1 is given by N_1. The lower the real wage, the higher the level of employment in sector 1.[6]

Now draw point *L* on the horizontal axis so that the distance *OL* measures the labour force. Then measure employment in sector 2 by the distance from point *L*, now going from right to left. The demand for labour in sector 2 is given by *PP*. (We use the letter *P* because we shall interpret sector 2 as the private sector in what follows.) It is also downward-sloping, as we go from right to left. The lower the real wage, the higher the level of employment in sector 2. For example, at wage W/P, employment in sector 2 is given by N_2.

To characterize employment and unemployment, we must describe how the real wage is determined. We could assume that W/P is determined in the same way as in our earliest chapters, that it is a function of the unemployment rate. It will simplify matters here to take the real wage as given. So assume that the real wage is the same in both sectors and is given by W/P. Then, Figure 26-2 tells us, employment in sector 1 is given by N_1, and employment in sector 2 is given by N_2. Because the labour force is given by *L*, unemployment is given by $U = L - N_1 - N_2$.

TRANSITION AND THE DECREASE IN EMPLOYMENT

In terms of this framework, where were Eastern European economies before the transition started? In most countries, most employment was in the state sector, the sector composed of the large firms producing under the central plan. The private sector was small, and unemployment was officially equal to zero. Despite the official numbers, there are reasons to believe that the unemployment rate was in fact positive, but the evidence is that it was small indeed.

This initial position is captured in Figure 26-3. Think now of sector 1 as the state sector. Think of sector 2 as the private sector. *SS* gives the demand for labour in the state sector; *PP* gives the demand for labour in the private sector. The real wage is W/P. At that wage, employment in the state sector is equal to N_S, employment in the private sector is equal to N_P, and unemployment is equal to zero.

The initial effects of transition were to sharply reduce the demand for labour in the state sector and to increase it in the private sector. The net effect was a decrease in employment and output, and an increase in unemployment. Let's examine each of these effects in turn.

Many state firms experienced sharp decreases in demand. With price liberalization, many of the goods they produced were no longer in demand. With the decline in defence spending, there was a sharp drop in the demand for defence-related goods, an effect particularly important in the former Soviet Union. Trade arrangements among centrally planned economies collapsed in 1991, leading many state firms to lose their Eastern European export markets. This adverse shift in the

---

[6]***Digging deeper.*** *In drawing the demand for labour in each sector as downward-sloping, we implicitly assume decreasing returns to labour, so that higher employment requires a lower real wage. In Chapter 15 we assumed constant returns to labour, and therefore the corresponding relation was horizontal instead (see in particular the appendix to Chapter 15). This assumption was convenient, but it is important to relax it here. As we will make clear, we want to capture the notion that it may be impossible to employ all workers, even at a very low wage.*

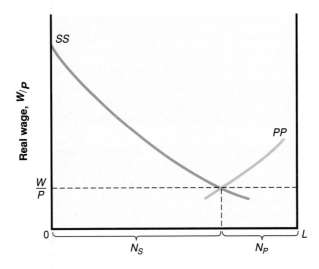

demand for labour by state firms is represented as a shift to the left of the *SS* curve, from *SS* to *SS'* in Figure 26-4.[7]

Why the demand for goods produced by state firms decreased is no great mystery. But why didn't demand shift to goods produced by the new private sector, leading to an increase in private employment sufficient to offset the decrease in state employment? This question is closely related to the issues we discussed in Chapter 25. Modern economies are characterized by churning. Sectors that experience adverse shifts in demand decrease employment; those that experience favourable shifts increase employment. Why is it that, in the case of Eastern Europe, private-sector employment did not increase enough to maintain employment?

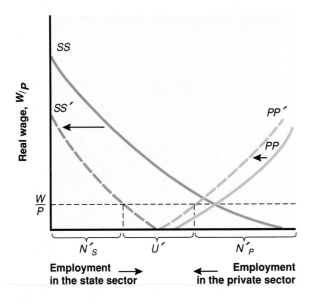

---

[7]*This shift did not happen overnight. Faced with sharp declines in output, state firms only slowly decreased employment. We ignore these dynamics in Figure 26-4.*

The answer is that the private sector could not grow fast enough. Some services, such as small-scale retail trade, can be provided with little or no capital and little expertise. Indeed, within a few weeks of the start of economic reform in Poland in January 1990, thousands of farmers and other small-scale vendors were selling their goods directly to consumers on the sidewalks of the capital, Warsaw. But production of new goods requires capital and expertise, and both were absent at the beginning of transition. Because the banking system also lacked the expertise to make loans, and because entrepreneurs-to-be had no past credit record to be assessed, they could not obtain credit to buy new capital. And because entrepreneurial skills were not taught or useful under central planning, most would-be entrepreneurs lacked the knowledge and the skills needed to create and run new firms. As a result, the increase in employment in the private sector was insufficient to offset the decrease in employment in the state sector.

In terms of Figure 26-4, there was an increase in the demand for labour from the private sector, a shift to the left from PP to PP'. But the net effect of the shifts in SS and PP was to decrease total employment. Thus the initial effect of the transition was to increase unemployment from zero to U'.

This conclusion raises a last issue. Why didn't wages decrease enough so as to maintain the high level of employment? The first answer is that in fact real wages did decrease in most countries at the beginning of transition. The second answer is that, even if wages had decreased much more, there would probably still have been unemployment. This is the case in Figure 26-4: As we have drawn it, the economy would still have some unemployment even at zero wages.

26-4 SCENARIOS FOR THE FUTURE

What we have described thus far is the initial phase of transition. But most Central European countries have already entered the next phase. If everything goes well, this phase will be characterized by (1) a restructuring of state firms and (2) a steady increase in the size of the private sector, both of which will lead to a steady increase in output over time.

A SCENARIO OF SUSTAINED TRANSITION AND GROWTH

We can represent this hoped-for evolution in Figure 26-5. The demand for labour in the new private sector will increase, leading to a steady leftward shift of PP. Restructuring of state firms may lead initially to further plant closings and thus a further decrease in employment in former state firms. But, eventually, restructuring of former state firms will allow them to survive and grow, and lead to a rightward shift of SS. The shifts in SS and PP will both lead to output growth and a decrease in unemployment, from U' to U''.

How much time is this process, and the associated reduction in unemployment, likely to take? The answer depends on a number of factors.

■ The lack of domestic expertise in many aspects of business, the lack of entrepreneurial experience, and the lack of capital are still hindering growth of the new private sector. A crucial factor will therefore be **foreign direct investment (FDI),** the purchase of existing firms or the development of new

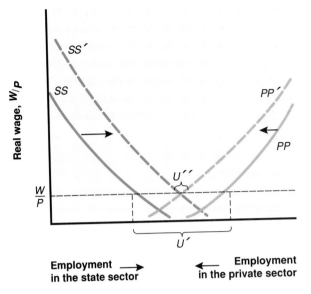

FIGURE 26-5
The Next Phase of Transition, If All Goes Well

If all goes well, the next phase of transition will see an increase in the demand for labour by former state firms, together with further increases in employment in the private sector.

plants by foreign firms. Because many Eastern European countries are geographically close to Western markets, have a highly educated labour force, and very low wages compared to OECD countries, FDI is increasing rapidly. In 1996, FDI to Central European countries was almost 2% of their combined GDP, compared to 0.1% in 1990. To keep this number in perspective, however, total FDI in 1996 to all of Eastern Europe was less than FDI to either Mexico or China.

■ Before they can expand production and employment, most state firms need to undergo drastic restructuring. So far, restructuring has been slow, for three reasons. First, restructuring involves large amounts of capital, since many plants are technologically obsolete. Second, restructuring typically involves an initial further decrease in employment; some plants must be shut down and some redundant workers must be laid off. The prospect of layoffs often leads to strong opposition by workers, who (given the already high unemployment rate) care very much about job security. Third, restructuring is complex and requires that somebody in the firm have both strong incentives and strong authority. In many countries, privatization has proceeded slowly, and in many firms nobody is in a position to make and implement such decisions. (See the Global Macro box entitled "Privatization in the Czech Republic and Russia.")

For these reasons, even under the best of scenarios it will take many years, indeed probably a few decades, before these countries achieve a standard of living comparable to those of the average OECD country. Interestingly, a country that no longer exists—East Germany, which has become the eastern part of a reunified Germany—looks quite different from the countries in Figure 26-1 (page 514). It had a much sharper initial decrease in employment and output, but is also having much faster growth. The reason is that although it was subject to a much larger initial shock, it is also free of many of the constraints on growth that face the other Central European countries. The evolution of Eastern Germany is explored in more detail in the Global Macro box entitled "The Transition in Eastern Germany (page 527)".

When transition started, most firms formally belonged to the state. One of the first tasks governments faced was thus to establish a private ownership structure. In planning privatization, governments had various objectives.

1. To establish a fair distribution of ownership rights. After all, these firms belonged implicitly to all citizens before transition; it was fair that they belong to them explicitly after the transition.

2. To establish an efficient structure of ownership. Study of firms in Western economies suggests that it is important for firms to have at least a few large owners. When a firm is owned by only small shareholders, managers are freer to do what they want, which may not always be in the owners' best interest.

3. To proceed quickly, so as not to leave the state firms in ownership limbo, which would prevent restructuring.

4. To obtain revenues from the sale of state firms, so as to make it easier to finance new spending and avoid budget deficits.

The problem is that these objectives are incompatible. For example, a fair distribution leads to many small owners; an efficient structure requires a few large owners. Getting substantial revenues from privatization is incompatible with proceeding quickly: selling firms at a good price is complicated and requires time.

The two countries that have been the most successful in balancing these objectives, the Czech Republic and Russia, have both relied on what is known as

voucher privatization. In the Czech Republic, people were given vouchers allowing them to bid for shares of state firms. They had the choice of either bidding directly or selling their vouchers to privately created investment funds, which then used the vouchers to bid for the shares. About three-fourths of the vouchers were sold to about 260 funds. Each investment fund typically has a large stake in a few firms. The hope is that, as large shareholders, these funds will force managers to restructure the firms. It is too early yet to know whether this will actually happen. But it appears at this time that this privatization method has achieved the goals of fairness, speed, and a potentially efficient ownership structure.

Russia has followed a similar approach. People were given vouchers that they could either sell in a market for vouchers or use to bid for shares of firms. During the 20 months in which vouchers were traded, their price varied from $4 to $20. One important difference from the process used in the Czech Republic was that the government gave firms' managers and workers a large proportion of the shares to induce them to accept privatization. From 1992 to 1997, the number of wholly state-owned enterprises was reduced from 205 000 to 88 000. At the beginning of 1997, the private sector in Russia accounted for 70% of GDP, over 60% of employment, and 74% of investments.

Reference

For more on privatization in Russia, see Maxim Boycko, Andrei Shleifer, and Robert Vishny, *Privatizing Russia* (Cambridge, MA: MIT Press, 1995).

THE DANGERS AHEAD

If things go well, the process of transition will continue, eventually leading to lower unemployment, a modern economic structure, and sustained growth. But can things go wrong? They always can, and the transition has not been smooth so far, but one danger is particularly clear at this point.

Part of the transition process involves the closing of large parts of state firms, most often in heavy industry and defence. Closing firms even in market economies is always politically difficult. It is particularly difficult in Eastern Europe, for two reasons.

First, under central planning, Central European firms traditionally relied on subsidies if and when they were making losses. This is what Janos Kornai, a

Hungarian economist and a long-time advocate of economic reform in centrally planned economies, has called the **soft budget constraint.** Part of the transition involves convincing firms that the budget constraint is now *hard,* that they will not receive subsidies from the state to cover their losses.

Second, it is politically harder to close large portions of the economy than it is to close one or two firms. For firms in difficulty, there is strength in numbers. And it is even more difficult to close firms when unemployment is already high and when workers are politically powerful.

While there are good arguments for getting governments involved in making the transition easier, through training and conversion programs or temporary subsidies, a substantial risk is involved. Governments may start subsidizing firms on a large scale, allowing them to maintain employment indefinitely even in the face of sharply lower sales. The result is then likely to be twofold. First, managers are likely to devote their energy to obtaining subsidies rather than to restructuring. Second, large subsidies may lead to a large budget deficit, to money creation, and to high inflation.

This mechanism is indeed the basic cause of the high inflation numbers we saw earlier in this chapter. In 1989 in Poland, for example, the still-communist government attempted to appease workers by offering them large nominal wage increases. These were in turn financed by subsidies to firms, a budget deficit, and money creation. The result was a budget deficit of 10% of GDP in 1989, entirely financed by money creation. The monthly rate of inflation in August 1989 was 30% and had increased to 54% in October. This high level of inflation is why macroeconomic stabilization, along with the elimination of both subsidies and the budget deficit, was an essential part of transition in Poland.

Subsidies to firms have also been the cause of high inflation in Russia, Ukraine, and the other former republics of the Soviet Union. As a result of price liberalization, the decrease in defence spending, and the collapse of trade among the former republics, most state firms have experienced very large decreases in sales. They were able to maintain employment, in part by extracting subsidies from the government—subsidies financed in turn through money creation. This is why, in 1993, inflation was close to 1000% in Russia and 5000% in Ukraine. Inflation has been reduced sharply in these two countries (but exploded again in Bulgaria). But large government deficits, caused by an inability to collect taxes, threatens to lead to high money creation and high inflation once again.

At this point, it appears that the soft budget constraint in most countries in transition is becoming harder. But the risk that adverse shocks, higher unemployment, or weaker governments will lead to larger subsidies, and thus to inflation, will be present for some time to come.

In November 1989, the Berlin Wall fell. The full economic and monetary integration of West and East Germany followed less than eight months later, in July 1990.

The immediate economic result was a collapse of the economy of what is now called Eastern Germany. Industrial production decreased from an index value of 100 in January 1990 to a value of 33 in January 1991! Table 1 gives the behaviour of output, productivity, and employment growth, together with the unemployment rate, in Eastern Germany from 1990 to 1994. Eastern German GDP declined by 13% in 1990, and by another 20% in 1991. The unemployment rate reached 15.5% at the end of 1992.

From 1992 on, however, the output picture improved. From 1992 to 1994, annual GDP growth averaged 7.7%. GDP growth came with very large increases in productivity, so that employment declined by a further 12.3% over these three years. Migration to Western Germany and decreases in participation limited the increase in unemployment, which stood at 16.8% in 1994.

Compared with Central European countries, the decrease in Eastern German production was larger, but lasted for less time. And output is now growing faster in Eastern Germany than in Central Europe. Two main factors account for these differences.

1. With unification came considerable political pressure to increase Eastern German wages to levels comparable to those in Western Germany. By the beginning of 1992, average wages in Eastern Germany stood at 60% of Western German wages—six times their level in 1990, and about eight times wages in Poland or the Czech Republic. With a productivity level estimated to be about one-third of the West German level, most East German firms found themselves unable to compete. This explains why the decline in production was so sharp. In contrast, other Central European countries have been able to maintain much lower wages and have not been exposed to the same level of competition from Western firms.

2. With unification, however, also came large transfers and large flows of capital and expertise. Transfers from Western to Eastern Germany amounted to 72% of Eastern German GDP in 1991, and to an amazing 90% of Eastern German GDP in 1992. Eastern German firms, sold by the privatization agency called the Treuhandt, were bought by Western German firms. These Western German firms quickly introduced expertise, new techniques, knowledge about markets, and capital. Investment has increased rapidly. Firms are being restructured. These factors explain both the rapid increase in output and the large gains in productivity since 1992. Thus, while the initial fall was larger, the rebound was greater. The prospects for future growth are better than they are in Central Europe, which does not have access to the same amount of expertise and capital.

Reference

For more on German unification, see Gerlinde Sinn and Hans Werner Sinn, *Jumpstart: The Economic Unification of Germany* (Cambridge, MA: MIT Press, 1994).

....

TABLE 1

OUTPUT GROWTH, EMPLOYMENT GROWTH, PRODUCTIVITY GROWTH, AND UNEMPLOYMENT IN EASTERN GERMANY, 1990–1994 (PERCENTAGE)

	1990	1991	1992	1993	1994
GDP growth	−13.0	−20.0	6.8	7.1	9.1
Employment growth	−9.0	−19.0	−7.0	−3.4	−1.9
Productivity growth	−4.0	−1	13.8	10.5	11
Unemployment rate	2	11	15.5	15.8	16.8

Source: OECD Economic Outlook, various issues, 1992–1994.

SUMMARY

■ Most countries that operated under central planning have moved toward becoming market economies. The first to break away from central planning was Poland in 1990, followed by the other countries of Eastern Europe.

■ The governments of the countries in transition had to take three types of measures to establish a market economy: price liberalization (allowing prices to clear markets), privatization (transferring previously state-owned firms to private owners), and macroeconomic control (including the elimination of subsidies to loss-making firms). Most countries have achieved price liberalization. Progress on privatization continues, but many large firms are still state owned. Progress on macroeconomic control has been uneven. As of 1997, only Bulgaria had hyperinflation, but high deficits and money growth were a problem for many countries.

■ The origin of transition is in large part the economic failure of central planning as an economic system. Growth in the Soviet Union was quite high for a long time, compared with that of its Western counterparts. But it was achieved largely by a steady increase in the ratio of capital to output, and thus by higher and higher investment rates, rather than by high technological progress. When the increase in investment rates came to an end, growth slowed down, leading to increasing dissatisfaction with the economic system.

■ Transition has been associated in all countries with a large decrease in output (although the true decrease is smaller than the measured decrease). This output decline is mostly the result of a process of structural change. The state sector, which was too large and produced many goods that are no longer in demand, has suffered a large decline in output. Demand for new goods, and thus for the output of the new private sector, is high. But growth in the private sector is hampered by many factors, particularly a lack of capital, a poorly functioning banking system, and a lack of expertise and experience. The output increase in the new private sector has been insufficient to offset the decrease in the state sector, resulting in a net decrease in aggregate output and employment.

■ The evolution of transition from here on depends largely on the speed with which state firms are restructured. Faster restructuring may lead to a further decline in employment in the short run, but will lead to higher productivity, output, and employment in the longer run. High unemployment in some countries is creating opposition to restructuring, slowing down transition.

■ One of the dangers facing the transition process is that political pressure to maintain employment in state firms will lead the state to subsidize these firms, leading not only to a slower transition but also to a large budget deficit, high money growth, and high inflation. Some countries, such as Russia and Ukraine, have so far made limited progress in reducing subsidies.

KEY TERMS

- big bang, 511
- price liberalization, 513
- privatization, 513
- macroeconomic control, 513

- structural change, 520
- foreign direct investment (FDI), 523
- voucher privatization, 525
- soft budget constraint, 526

QUESTIONS AND PROBLEMS

1. In addition to the factors mentioned in the text (lack of expertise and entrepreneurship and the inability to get loans), what other constraints do you think have prevented rapid expansion of the private sector in the countries in transition? Are any of these forces still operating, slowing down economic recovery? Explain.

2. Can you think of examples (historical or current) of soft budget constraints in Canada? Explain.

3. "The Soviet economy was doing very well, even through the 1980s. Capital per worker was steadily rising, and this could have continued indefinitely, leading to higher and higher living standards under central planning." Comment.

4. Why might the Soviet Union have had a more difficult time taking advantage of Western technological progress in the 1980s compared with in earlier periods? [*Hint:* What were the most significant changes in technology in the 1980s?]

5. What has happened to the Russian economy since the end of 1997? Has inflation come down? Has output

recovered? [*Suggestion:* Find data on output and inflation in the back of a recent issue of *The Economist* magazine.]

6. Poland's GDP per capita in 1995 was $5400 (in PPP terms, 1995 U.S. dollars), compared with an average of $19 300 for the EU countries. If growth in Poland exceeds growth in the EU by 3% per year, how long will it take for Poland to have a GDP per capita

a. One-half that of the EU countries?
b. Two-thirds that of the EU countries?
c. Equal to that of the EU countries?

7. Explain in simple terms how going from an inefficient economic system (central planning) to a more efficient one (a market economy) can lead to a temporary decline in output. Can you think of other examples in life where reorganization may lead to a temporary decline in output?

FURTHER READING

In addition to the books by Jeffrey Sachs on Poland (see footnote 1) and by Maxim Boycko et al. on privatization in Russia (see the Global Macro box entitled "Privatization in the Czech Republic and Russia" on page 525), the following references may be useful.

On the Soviet Union under central planning, read Padma Desai, *The Soviet Economy: Problems and Prospects* (New York: Basil Blackwell, 1987).

In *Gorbachev's Struggle for Economic Reform* (London: Pinter Publishers, 1991), Anders Aslund gives a first-hand description of Mikhail Gorbachev's failed attempts to reform the Soviet economy. In *How Russia Became a Market Economy* (Washington, DC: Brookings Institution, 1995), he gives a description of reforms since Gorbachev.

A very useful document, in terms of both analysis and basic numbers, is the *Annual Transition Report* of the European Bank for Reconstruction and Development, a bank created for the sole purpose of helping the transition in Eastern Europe.

SHOULD POLICY MAKERS BE RESTRAINED?

A recurrent theme of this book has been that macroeconomic policy has an important role to play. The right mix of fiscal and monetary policy can, we have argued, help a country out of a recession, improve its trade position without increasing activity and igniting inflation, slow down an overheating economy, stimulate investment and capital accumulation, and so on.

This theme, however, is clearly at odds with growing demands that policy makers be tightly restrained. Having fought hard to eliminate the deficit, some Canadians want to see a law passed that would prevent any future federal government from running a deficit. The requirement that the budget be balanced year by year would all but prevent the use of fiscal policy as an instrument of stabilization policy.

Monetary policy is also under fire. For example, the new charter of the central bank of New Zealand, written in 1989, defines monetary policy's role as the maintenance of price stability, to the exclusion of any other macroeconomic goal.

Arguments for restraints on policy makers fall into two general categories. The first is that policy makers may have good intentions, but end up doing more harm than good. The second is that policy makers do what is best for themselves, which is not necessarily what is best for the country. This chapter develops and examines these arguments in the context of macroeconomic policy in general. Chapters 28 and 29 then examine aspects of monetary policy and fiscal policy in more detail.

27-1 UNCERTAINTY AND POLICY

A blunt way of stating the first argument in favour of policy restraints is that those who know little should do little. The argument has two parts: first, that macroeconomists, and by implication the policy makers who rely on their advice, know little; and second, that policy makers should therefore do little. Let's discuss both parts in turn.

HOW MUCH DO MACROECONOMISTS ACTUALLY KNOW?

Macroeconomists are very much like doctors treating cancer. They know a lot, but there is also a lot they don't know.

Take the case of an economy with high unemployment, where the central bank is considering the use of monetary policy to increase economic activity. Think of the sequence of links between an increase in money and an increase in output, and thus of all the questions the central bank faces when deciding whether and by how much to increase the money supply:

■ By how much will the change in the money supply decrease short-term interest rates (an issue we discussed in Chapter 5)? By how much will a decrease in short-term interest rates lead to a decrease in long-term interest rates (Chapter 9)? By how much will stock prices increase (Chapter 9)? By how much will the currency depreciate (Chapters 13 and 14)?

■ How long will it take for lower long-term interest rates and higher stock prices to affect investment and consumption spending (Chapter 8)? How long will it take for the J-curve effects to work themselves out and for the trade balance to improve (Chapter 12)? What is the risk that the effects will come too late, when the economy has already recovered?

■ Is the current high rate of unemployment a sign that unemployment is above the natural rate, or a sign that the natural rate has increased (Chapters 15, 18, and 20)? If the economy is too close to the natural rate, isn't there a risk that monetary expansion will lead to a decrease in unemployment below the natural rate and thus to an acceleration of inflation (Chapter 18)?

When assessing these questions, central banks—or macroeconomic policy makers in general—do not operate in a vacuum. They rely in particular on macroeconometric models. The equations in these models give estimates of how these individual links have looked in the past. But different models give different answers. This is because they have different structures, different lists of equations, and different lists of variables.

Figure 27-1 shows a good example of this diversity. In the early 1990s, researchers at the Brookings Institution—a research institution in Washington, DC—asked the builders of the 12 main macroeconometric multicountry models in use today to answer a similar set of questions. (The models aredescribed in the Focus box entitled "Twelve Macroeconometric Models.") The goal was to see how different the answers would be across models.

One of the questions was the following. Consider a case where the U.S. economy is growing at its normal growth rate, and where unemployment is at its natural rate; call this the *baseline* case. Suppose now that, over the period of a year, the Fed increases money faster than in the baseline, so that after a year, nominal money is 4% higher than it would have been in the baseline case. From then on, nominal money grows at the same rate as in the baseline case, so that the level of nominal money remains 4% higher than it would have been without the change in monetary

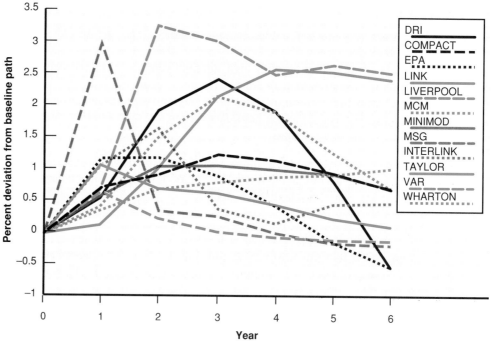

FIGURE 27-1

The Response of Output to a Monetary Expansion: 12 Predictions from 12 Models

While all 12 models predict that output will increase for some time in response to a monetary expansion, the range of answers regarding the size and the length of the output response is large.

Source: See footnote 1.

FOCUS

Together, the set of models used in the Brookings project is representative of the different types of macroeconomic models used for forecasting and policy in the world today.

■ Two, DRI (which stands for Data Resources Incorporated) and WHARTON, are commercial models. They are used regularly to generate and sell economic forecasts to firms and financial institutions.

■ Five are used for forecasting and to help in the design of policy. MCM (for multicountry model) is used by the Federal Reserve Board in Washington for the conduct of U.S. monetary policy; INTERLINK is used by the OECD in Paris; COMPACT is used by the Commission of the European Union in Brussels; EPA is used by the Japanese Planning Agency. Each of these four models was constructed by one team of researchers doing all the work—that is, building submodels for countries or groups of countries and linking them through trade and financial flows. In contrast, the fifth, LINK, is composed of individual country models—models constructed in each country by researchers from that country and then linked together by trade and financial relations.

■ Four models incorporate rational expectations explicitly: the LIVERPOOL model, based in England; MINIMOD, used at the International Monetary Fund; MSG, developed by Warwick McKibbin and Jeffrey Sachs at Harvard University; and the TAYLOR model—which we saw in Section 18-3—developed by John Taylor of Stanford University. These are typically smaller models, with less detail than those listed above. But they are better at capturing the expectation effects of various policies.

■ The last one, VAR (for Vector AutoRegression, the technique of estimation used to build the model), developed by Christopher Sims and Robert Litterman at the University of Minnesota, is very different from the others. It is not a structural model but rather a statistical summary of the relations between the different variables, without an explicit economic interpretation. Its strength is in its fit of the data. Its weakness is that it is, essentially, a black box.

policy. Suppose further that interest rates in the rest of the world remain unchanged. What will happen to U.S. output?[1]

Figure 27-1 shows the deviation of output from the baseline predicted by each of the 12 models. All 12 models predict that output will increase for some time after the increase in money. After one year, the average deviation of output from the baseline is positive. But the range of answers is large, from near 0% to close to 3%; even leaving out the most extreme prediction, the range is still more than 1%. Two years out, the average deviation is 1.2%; again leaving out the most extreme prediction, the range is still 2%. Six years out, the average deviation is 0.6%, and the answers range from −0.3% to 2.5%, with no obvious outlier. In short, if we measure uncertainty by the range of answers from this set of models, there is indeed substantial uncertainty about the effects of policy.

SHOULD UNCERTAINTY LEAD POLICY MAKERS TO DO LESS?

Should uncertainty about the effects of policy lead policy makers to do less? In general, the answer is yes. Borrowing from the simulations that we've just looked at, we

---

[1] *A detailed description of the Brookings study is given in Ralph Bryant et al.,* Empirical Macroeconomics for Interdependent Economies *(Washington, DC: Brookings Institution, 1988). The simulation described here is simulation E in the supplemental volume.*

shall assume that a 4% increase in money is expected to increase output next year by 0.85% with a range of 0% to 3%.

Suppose that the Canadian economy is in recession. The unemployment rate stands at 10% and the Bank of Canada is considering using monetary policy to expand output. To concentrate on uncertainty about the effects of policy, let's assume that the Bank knows everything else for sure. Based on its forecasts, it *knows* that, in the absence of changes in monetary policy, unemployment will still be at 10% next year. It knows that the natural rate of unemployment is 7.6%, and that the unemployment rate is therefore 2.4% above the natural rate. It also knows from Okun's law that 1% more output growth for a year leads to a reduction in the unemployment rate of 0.4%.[2] Under these assumptions, the Bank knows that, if it could achieve 6% more output growth over the coming year, the unemployment rate a year from now would be lower by 0.4 times 6% = 2.4%, thus down to its natural rate of 7.6%. By how much should the Bank increase the money supply?

In our calculation, we assume an increase in the money supply of 4% leads to a 0.85% increase in output in the first year. Equivalently, a 1% increase in the money supply leads to a 0.85/4 = 0.21% increase in output.

Suppose that the Bank took this average relation as holding with *certainty*. What it should then do is straightforward. To return the unemployment rate to the natural rate in one year requires 6% more output growth. Six percent output growth in turn requires the Bank to increase money by 6%/0.21 = 28.2%. The Bank should therefore increase the money supply by 28.2%. Under our assumptions, this increase in money will return the economy to the natural rate of unemployment at the end of the year.

Suppose that the Bank actually increases money by 28.2 %. But let's now take into account uncertainty. Suppose that the range of responses of output to a 4% increase in money after one year varies from 0% to 3%; equivalently, a 1% increase in money leads to a range of increases in output from 0% to 0.75%. These ranges imply that an increase in money of 28.2% leads to an output response anywhere between 0% and (28.2% × 0.75 =) 21.2%. These output numbers imply in turn a decrease in unemployment anywhere between 0% and 8.5%, or values of the unemployment rate a year hence anywhere between 10% and 1.5%!

The conclusion is clear. Given the range of uncertainty about the effects of monetary policy on output, increasing money by 28.2% would be irresponsible. If the effects of money on output are as strong as suggested by one of the 12 models, unemployment by the end of the year could be 6% below the natural rate, leading to enormous inflationary pressures. Given this uncertainty, the Bank should increase money by much less than 28.2%. For example, increasing money by 10% leads to a range for unemployment a year hence of 10% to 7%, clearly a safer range of outcomes.[3]

---

[2] *In the real world, of course, the Bank does not know any of these numbers with certainty. It has only forecasts of what will happen; it does not know the exact value of the natural rate or the coefficient in Okun's law. Introducing these sources of uncertainty would only reinforce our basic conclusion.*

[3] **Digging deeper.** *The theory behind this example relies on the notion of* multiplicative uncertainty—*the notion that, because the effects of policy are uncertain, more active policies lead to more uncertainty. See William Brainard, "Uncertainty and the Effectiveness of Policy,"* American Economic Review, *May 1967, 411–425.*

Let's summarize our conclusions so far.

There is indeed substantial uncertainty about the effects of macroeconomic policies. This uncertainty should lead policy makers to be more cautious, to use policies which are less active. Policies should be broadly aimed at avoiding prolonged recessions, slowing down booms, and avoiding inflationary pressure. The higher the level of unemployment or inflation, the more active the policies should be. But they should stop well short of **fine-tuning,** of trying to achieve constant unemployment or constant output growth.

These conclusions would have been controversial 25 years ago. Back then, there was a heated debate between two groups of economists. One group was headed by Milton Friedman from Chicago, who argued that, because of long and variable lags, activist policy is likely to do more harm than good. The other group, headed by Franco Modigliani from MIT, had just built the first generation of large macroeconometric models and believed that economists' knowledge of the economy was becoming good enough to allow for increasingly fine tuning of the economy.[4] Today, most economists recognize that there is substantial uncertainty about the effects of policy. They also accept the implication that this uncertainty should lead to less-active policies.

Note that what we have developed is an argument for *self-restraint* by policy makers, not for *restraints on* policy makers. If policy makers understand the implications of uncertainty—and there is no reason to think that they don't—they will, on their own, follow less-active policies. There is no reason to impose further restraints, such as the requirement that money growth be constant or that the budget be balanced. Let's now turn to arguments for restraints *on* policy makers.

27-2 EXPECTATIONS AND POLICY

One of the reasons why the effects of macroeconomic policy are uncertain is the interaction of policy and expectations. How a policy works, and sometimes whether it works at all, depends not only on how it affects current variables but also on how it affects expectations about the future (this was indeed the main theme of Chapter 10). But the importance of expectations for policy goes beyond uncertainty about the effects of policy. This conclusion brings us to a discussion of *games*.

Until 25 years ago, macroeconomic policy was seen in the same way as the control of a complicated machine. Indeed, methods of **optimal control,** developed initially to control and guide rockets, were being increasingly used to design macroeconomic policy.

Economists no longer think in terms of **optimal control theory**. It has become clear that the economy is fundamentally different from a machine. Unlike a machine, the economy is composed of people and firms who try to anticipate what policy makers will do, who react not only to current policy but also to expectations of future policy. In this sense, macroeconomic policy can be thought of as a **game**

---

[4]*Friedman and Modigliani are the same two economists who independently developed the modern theory of consumption, presented in Chapter 8.*

between policy makers and the economy. Thus, when thinking about policy, what we need is not optimal control theory but rather **game theory.**

Let's clarify semantics. When economists use the word "game," they do not mean "entertainment" but rather **strategic interactions** between **players.** In the context of macroeconomic policy, the players are the policy makers and "the economy"—more concretely, people and firms. The strategic interactions are clear: What people and firms do depends on what they expect policy makers to do. In turn, what policy makers do depends on what is happening in the economy.

Game theory has given economists many insights, often explaining how some apparently strange behaviour makes sense when one understands the nature of the game being played.[5] Among these insights is one that is particularly important for our discussion of restraints here: Sometimes you can do better in a game by giving up some of your options. To see why this is so, let's start with an example from outside economics, governments' policies toward hijackers.

HIJACKINGS AND NEGOTIATIONS

Most governments have a stated policy that they will not negotiate with airplane hijackers. The reason for this stated policy is clear: to deter hijacking by making it unattractive to hijack planes.

Suppose that, despite the stated policy, a hijacking takes place. Now that the hijacking has taken place anyway, why not negotiate? Whatever compensation the hijackers demand is likely to be less costly than the alternative, the likelihood that lives will be lost if the plane has to be taken by force. So the best policy would appear to be: Announce that you will not negotiate, but if a hijacking happens, negotiate.

But upon reflection it is clear that this is a terrible policy. Hijackers' decisions depend not so much on the stated policy as on what they expect will actually happen if they hijack a plane. If they know that negotiations will actually take place, they will rightly consider the stated policy to be irrelevant. And hijackings will take place.

So what is the best policy? Despite the fact that negotiations typically lead to a better outcome once hijackings have taken place, the best policy is for governments to commit *not* to negotiate. By giving up the option to negotiate, they are likely to prevent hijackings in the first place.

Let's now turn to a macroeconomic example, based on the relation between inflation and unemployment.[6] As you will see, exactly the same logic is involved.

INFLATION AND UNEMPLOYMENT REVISITED

Recall the relation between inflation and unemployment we derived in Chapter 17 [equation (17.7), with the time indexes omitted]:

$$\pi = \pi^e - \alpha(u - u_n) \tag{27.1}$$

Inflation (π) depends on expected inflation (π^e) as embodied in wages set in labour

---

[5] *Game theory is becoming an important tool in all branches of economics. The 1994 Nobel Prize was given to three game theorists, John Nash from Princeton, John Harsanyi from Berkeley, and Reinhard Selten from Germany.*

[6] *This example was first developed by Finn Kydland, from Carnegie-Mellon, and Edward Prescott, from Minnesota, in "Rules Rather Than Discretion: The Inconsistency of Optimal Plans,"* Journal of Political Economy, *85(3), June 1977, 473–492.*

contracts, and on the difference between the actual unemployment rate and the natural unemployment rate $(u - u_n)$. The coefficient α captures the effect of unemployment on inflation, given expected inflation. When unemployment is above the natural rate, inflation is lower than expected; when it is below the natural rate, inflation is higher than expected.

Suppose that the Bank of Canada announces that it will follow a monetary policy consistent with zero inflation. On the assumption that wage setters believe the announcement, expected inflation (π^e) as embodied in wage contracts is thus equal to zero, and the Bank now faces the following relation:

$$\pi = -\alpha(u - u_n) \tag{27.2}$$

If the Bank follows through on its announced policy of zero inflation, expected and actual inflation are both equal to zero, and unemployment is equal to the natural rate.[7]

Zero inflation and unemployment equal to the natural rate are not a bad outcome. But it would seem that the Bank can actually do better. Now that wages are set, the Bank faces an attractive tradeoff between inflation and unemployment. Recall from Chapter 17 that in Canada α in equation (27.1) is roughly equal to 0.5. Thus, by accepting just 2% inflation, the Bank can achieve an unemployment rate of 1% below the natural rate. Suppose that the Bank—and everybody else in the economy—finds the tradeoff attractive, and decides to decrease unemployment by 1% in exchange for an inflation rate of 2%.[8]

This incentive to deviate from the announced policy once the other player has moved—in this case, once wage setters have set the wage—is known in game theory as the **time inconsistency** of optimal policy. In our example, the Bank can indeed improve the outcome this period by deviating from its announced policy of zero inflation. By accepting some inflation, it can achieve a substantial reduction in unemployment.

Unfortunately, this is not the end of the story. Seeing that the Bank of Canada has increased money more than it announced it would, wage setters are likely to wise up and now to expect positive inflation of 2%. If the Bank still wants to achieve an unemployment rate 1% below the natural rate, it now has to accept 4% inflation. However, if it does so, wage setters are likely to increase their expectations of inflation further, and so on.

The eventual outcome is likely to be high inflation. Because wage setters understand the Bank's motives, expected inflation catches up with actual inflation, and the Bank must eventually be unsuccessful in its attempt to achieve unemployment below the natural rate. In short, attempts by the Bank of Canada to make things better lead in the end to things being worse. The economy ends up with the *same unemployment rate* as would have prevailed if the Bank had followed its announced policy, but with *much higher inflation.*

How relevant is this example? A rereading of Chapter 17 suggests that it is very relevant indeed. One can read the history of the Phillips curve in the 1970s as coming from the Bank's attempts to sustain unemployment below the natural rate.

---

[7]*For simplicity, we assume that the Bank can choose the rate of inflation exactly. In doing so, we ignore uncertainty about the effects of policy, which was the topic of Section 27-1 but is not central here.*

[8]*As we saw in Chapters 15 and 17, despite its name, the natural rate of unemployment has no claim to being best in any sense.*

In that light, the shift of the original Phillips curve can be seen as the adjustment of wage setters' expectations to the central bank's behaviour.

What is therefore the best policy? It is for the Bank of Canada to make a credible commitment that it will not try to decrease unemployment below the natural rate. By giving up the option of deviating from its announced policy, the Bank can achieve unemployment equal to the natural rate and zero inflation. The analogy with the hijacking example is clear: By credibly committing not to do something that would appear desirable at the time, policy makers can achieve a better outcome—no hijackings in our earlier example, no inflation here.

ESTABLISHING CREDIBILITY

How can the Bank of Canada credibly commit not to deviate from its announced policy?

One way for a central bank to establish its credibility is to give up—or to be stripped by law of—its policy-making power. For example, the mandate of the bank can be defined by law in terms of a simple rule, such as setting money growth at 0%.[9]

Such a law surely takes care of the problem of time inconsistency. But such a tight restraint comes close to throwing the baby out with the bathwater. We want to prevent the central bank from pursuing too high a rate of money growth in an attempt to lower unemployment below the natural rate. But—subject to the restrictions discussed in Section 27-1—we still want the central bank to be able to expand the money supply when unemployment is far above the natural rate and contract the money supply when unemployment is far below the natural rate. Such actions become impossible under a constant-money-growth rule.

There are indeed better ways to deal with time inconsistency. In the case of monetary policy, our discussion suggests one way that this can be done:

1. Make the central bank independent. Appointing central bankers for longer terms and making it harder to fire them will make them more likely to resist political pressure to decrease unemployment below the natural rate.

2. Choose a "conservative" central banker, somebody who dislikes inflation and is unwilling to accept inflation for less unemployment when unemployment is at the natural rate. When the economy is at the natural rate, such a central banker will simply not be tempted to embark on a monetary expansion. Thus the problem of time inconsistency will disappear altogether.

Appointing as the central bank's head somebody who does not have the same preferences as the people as a whole might seem like a solution that only game theorists would concoct. But this is actually the way many countries have been responding to the problem of time inconsistency in monetary policy. In the last two decades, many countries have given their central banks more independence. And government leaders typically have appointed central bankers who are more "conservative" than themselves, who appear to care more about inflation and less about unemployment than the governments themselves. (See the Focus box entitled "Was Alan Blinder Wrong in Speaking the Truth?")

Figure 27-2 suggests that this approach has been quite successful. The vertical axis gives the average annual inflation rate in 18 OECD countries for the period

[9]*In the early discussions leading up to the failed Meech Lake Accord, one of the issues discussed was a constitutional amendment that would commit the Bank of Canada to zero inflation.*

1960–1992. The horizontal axis gives the value of an index of "central bank independence," constructed by looking at a number of legal provisions in the bank's charter—for example, whether and how the government can remove the head of the bank.[10] There is a striking inverse relation, as summarized by the regression line. More central bank independence is indeed systematically associated with lower inflation.

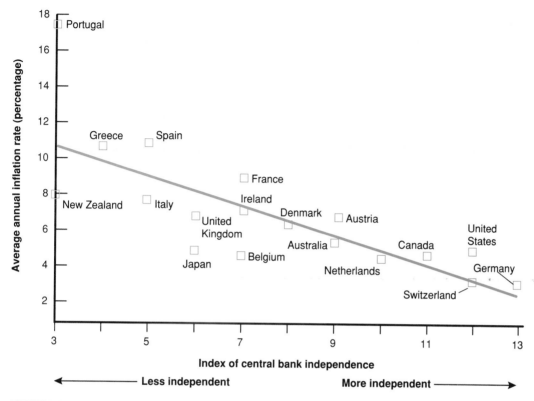

FIGURE 27-2

Inflation and Central Bank Independence

Across OECD countries, the higher the degree of central bank independence, the lower is the inflation rate.

Source: See footnote 10.

---

[10] *Vittorio Grilli, Donato Masciandaro, and Guido Tabellini, "Political and Monetary Institutions and Public Financial Policies in the Industrial Countries,"* Economic Policy, *October 1991, 341–392.*

■ **WAS ALAN BLINDER WRONG IN SPEAKING THE TRUTH?**

FOCUS

In the summer of 1994, President Clinton appointed Alan Blinder, an economist from Princeton, vice-chairman (in effect, second in command) of the Fed-

eral Reserve Board. A few weeks later Blinder, speaking at an economic conference, indicated his belief that the Fed has both the responsibility and the ability, when unemployment is high, to use monetary policy to help the economy recover. This statement was badly received. Bond prices decreased, and most newspapers ran editorials critical of Blinder.

Why was the reaction of markets and newspapers so negative? It was surely not that Blinder was wrong. There is no doubt that monetary policy can help the economy out of a recession. It is also the case that the Federal Reserve Bank Act of 1978 requires the Fed to pursue full employment as well as low inflation.

The reaction was negative because, in terms of the argument we developed in the text, Blinder revealed by his words that he was not a conservative central banker—that he cared about unemployment as well as inflation. With the unemployment rate at the time equal to 6.1%, close to the natural rate, markets interpreted Blinder's statements as suggesting that he might want to decrease unemployment below the natural rate. Interest rates increased because of higher expected inflation, and thus bond prices decreased.

The moral of the story is clear. Whatever views they may hold, central bankers should try to look and sound conservative. . . . This is indeed why many heads of central banks will not, for example, admit in public the existence of any tradeoff between unemployment and inflation, even in the short run.

TIME CONSISTENCY AND RESTRAINTS ON POLICY MAKERS

Let's summarize what we have learned in this section. We have examined a second line of argument for putting restraints on policy makers, based on the issue of time inconsistency. We have looked at the case of monetary policy. But similar issues arise in the context of fiscal policy as well. We shall discuss, for example, the issue of debt repudiation—the option for the government to cancel its debt obligations—in Chapter 29.

When issues of time inconsistency are relevant, tight restraints on policy makers—such as a fixed-money-growth rule in the case of monetary policy—can indeed provide a coarse solution. But the solution may have large costs if it prevents the use of macroeconomic policy altogether. Better ways typically involve designing better institutions (such as an independent central bank) that can reduce the problem of time inconsistency without eliminating monetary policy as a macroeconomic stabilization tool.

27-3 POLITICS AND POLICY

We have so far assumed *benevolent* policy makers, policy makers who try to do what is best for the economy. However, much of the current public discussion challenges that assumption. Politicians or policy makers, the arguments go, do what is best for themselves, and this is not always what is best for the country.

You have heard the arguments. Politicians avoid the hard decisions, and they pander to the electorate. Discussing the flaws of democracy goes far beyond the scope of this book. What we *can* do here is review how these arguments apply to macroeconomic policy, then look at the empirical evidence and see what light it sheds on the issue of policy restraints.

GAMES BETWEEN POLICY MAKERS AND VOTERS

Many macroeconomic measures involve trading off short-run losses against long-run gains—or, conversely, short-run gains against long-run losses.

Take, for example, tax cuts. By definition, tax cuts lead to lower taxes today. They are also likely to lead to an increase in activity, and thus in pretax income, for some time. But, unless they are matched by equal decreases in government spending, they lead to a larger budget deficit and to the need for an increase in taxes in the future.[11]

If voters are shortsighted, the temptation for politicians to cut taxes may prove irresistible. Politics may lead to systematic deficits, at least until the level of government debt has become so high that politicians are scared into action.

Now move from taxes to macroeconomic policy in general. Suppose again that voters are shortsighted. If the politicians' main goal is to please voters and get re-elected, what better policy than to expand aggregate demand before an election, leading to higher growth and lower unemployment? True, growth in excess of the normal growth rate cannot be sustained, and eventually the economy must return to the normal level of output. Higher growth must be followed later by lower growth.[12] But with the right timing and shortsighted voters, higher growth can win the elections. Thus, one might expect a clear **political business cycle,** with higher growth on average before elections than after elections.

The arguments we have just laid out are familiar; in one form or another, you surely have heard them before. And their logic is convincing. Thus it may come as a surprise that they do not fit the facts very well.

Our discussion of taxes might lead you to expect that large budget deficits, and therefore increasing levels of government debt, have always been the norm and are unavoidable. Figure 27-3, which gives the evolution of the ratio of gross federal debt to GDP in Canada since 1926, shows that this is simply not the case. Note how the sharp buildups in debt prior to the mid-1970s happened under very special circumstances—the Great Depression and World War II, times of unusual declines in output or unusually high military spending. Note also how, from the end of World War II to the middle of the 1970s, the ratio of debt to GDP steadily *decreased*. By 1976, the ratio of gross federal debt to GDP stood at 32%, down from 147% in 1946.

True, the steady increase in debt since the mid-1970s fits the argument of shortsighted politicians quite well. The ratio of debt to GDP has steadily increased and in March 1996 was back up to 80%. But explaining the last 20 years through the behaviour of shortsighted politicians raises the issue of why things were different earlier. The broader historical record suggests that, by itself, shortsightedness does not explain much of the past evolution of deficits and debt.[13]

Return to the political-business-cycle argument that policy makers try to get high output growth before the elections so that they will be reelected. If the political

---

[11] We shall look at the relation between current and future taxes more formally when we examine the implications of government budget constraint in Chapter 29.

[12] See Chapter 18, in particular Table 18-1.

[13] Chapter 29 will examine alternative—and empirically more successful—explanations for the evolution of government debt, both over time and across countries.

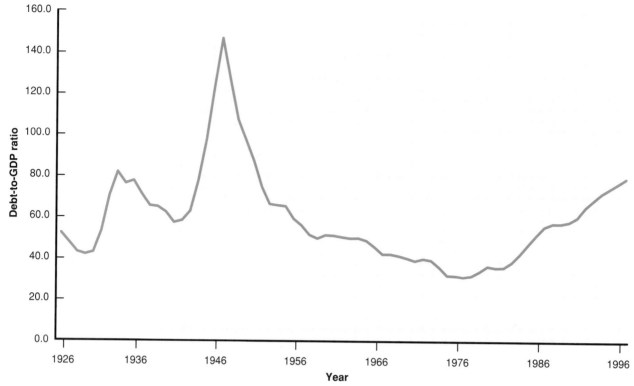

FIGURE 27-3

The Evolution of the Ratio of Canadian Debt to GDP, 1926–1996

The major buildups of federal government debt in Canada since 1926 have been associated with the Great Depression and World War II.

Source: Statistics Canada, CANSIM Series D11011 and D469409.

business cycle were indeed important, one would expect to see faster growth before elections than after. Table 27-1 presents evidence from the United States. It gives GDP growth rates for each of the four years of each U.S. administration since President Truman in 1948 up to the end of the first Clinton administration in 1996. It is indeed the case that growth has been highest on average in the last year of an administration. But the average difference across years is small: 3.7% in the last year of an administration versus 3.3% in the first year. There is thus little evidence of manipulation—or at least of successful manipulation—of the macroeconomy to win elections.

GAMES BETWEEN POLICY MAKERS

Another line of argument focusses not on games between politicians and voters, but rather on games between political parties. For example, take the issue of budget deficit reduction in the United States. Progress was slow; it took more than a decade after deficits were first recognized as a major U.S. macroeconomic issue to get them under control. Some of the delays are part and parcel of the normal democratic process. Deficit reductions involve making painful decisions, and forging a consensus takes time. But other factors seemed to be at work too. While agreeing on the need for deficit reduction, the two political parties differed on how it should be

TABLE 27-1

U.S. GDP GROWTH DURING DEMOCRATIC AND REPUBLICAN
ADMINISTRATIONS (PERCENTAGE PER YEAR)

	YEAR			
	First	Second	Third	Fourth
Democratic				
Truman	0.0	8.5	10.3	3.9
Kennedy/Johnson	2.6	5.3	4.1	5.3
Johnson	5.8	5.8	2.9	4.1
Carter	4.7	5.3	2.5	−0.2
Clinton	4.0	4.1	2.7	2.8
Average: Democratic	3.4	5.8	4.5	3.2
Republican				
Eisenhower I	4.0	−1.3	5.6	2.1
Nixon	2.4	−0.3	2.8	5.0
Nixon/Ford	5.2	−0.5	−1.3	4.9
Reagan I	1.9	−2.5	3.6	6.4
Reagan II	3.6	3.0	2.7	3.0
Bush	2.5	1.2	−0.7	2.6
Average: Republican	3.3	−0.1	2.1	4.0
Average	3.3	2.6	3.2	3.7

Source: Alberto Alesina, "Macroeconomics and Politics," *NBER Macroeconomics Annual,* 1988, 13–61. Table 4.
Updated.

done. Because they believe in a smaller role for government, Republicans focussed on decreases in spending. In contrast, Democrats were more open to increases in taxes. For a long time, each side held out, hoping the other party would give in.

Game theorists refer to these situations as *wars of attrition.* The hope that the other side will give in leads to long and often costly delays. Wars of attrition are not limited to fiscal policy—witness the 1994 baseball strike. But they are endemic in fiscal policy. Deficit reduction often takes place long after it would be advisable. This is particularly visible during episodes of hyperinflation. As we saw in Chapter 21, hyperinflation comes from monetary finance of large budget deficits. While the need to reduce those deficits is usually recognized early on, support for a stabilization program—which includes the elimination of those deficits—typically comes only when inflation has reached such high levels that economic activity is severely affected.

BACK TO THE BALANCED-BUDGET AMENDMENT

Let's end this chapter with the issue we started with, the case for and against a balanced-budget amendment. What have we learned?

First, despite common beliefs, the picture of politicians pandering to short-sighted voters does not fit the broad evidence on the evolution of deficits and debt. The large peacetime federal budget deficits from 1980 to 1996 are the exception rather than the rule. Fiscal policy is not typically characterized by chronic deficits.

This is not to say that all is well, or that the political process always delivers the best macroeconomic policy decisions. Hard decisions are often delayed. Deficit reductions often come late, only after debt has increased to high levels.

Do these problems justify a law that forces the federal government to run a balanced budget? To most economists, the answer is no. Their arguments go as follows.

The case against a balanced-budget amendment. A balanced-budget amendment would eliminate the problem of deficits. But it would also eliminate the use of fiscal policy as a macroeconomic policy instrument. This is too high a price to pay.

The evidence suggests that the problem is not that politicians systematically want deficits. Rather it appears to be that politicians find it difficult to agree on and implement a deficit-reduction plan. Deficit control and reduction can be achieved with looser constraints than a constitutional amendment.

The case for a balanced-budget amendment. To some economists, however, the answer is: yes, a balanced-budget amendment is necessary. These economists are typically more skeptical of the usefulness of macroeconomic policy in general, and of fiscal policy in particular. They worry that running deficits during recessions may have adverse effects on financial markets and thus hinder rather than help the recovery (a potential perverse policy effect we discussed in Section 10-3). Because of the lags involved in the legislative process, they are also skeptical of the federal government's ability to change fiscal policy in time to stabilize the economy. Thus they are willing to give up fiscal policy as a macroeconomic instrument.

They are also skeptical of any rules that a government may impose upon itself but can undo by a later vote. They point to the mixed record that the United States had with such rules in the 1980s. (We examine this record in the In Depth box entitled "Self-Restraints and Fiscal Policy in the United States in the 1980s: Gramm-Rudman-Hollings.") This leads them to conclude that nothing short of a constitutional amendment can do the job of ending deficit spending.

We have made clear what our views are. We see the costs of a balanced-budget amendment as far exceeding the benefits.

▶ **SELF-RESTRAINTS AND FISCAL POLICY IN THE UNITED STATES IN THE 1980s: GRAMM-RUDMAN-HOLLINGS**

In 1985, increasing pressure to reduce the deficit—which stood at 5.4% of GDP—led to two major proposals in Congress. The first, by Senate Republicans, was built around a freeze in defence spending and the elimination of inflation adjustments (called *COLAs, for cost of living adjustments*) for Social Security benefits for one year. The other, by House Democrats, was built around cuts in defence spending and the maintenance of COLAs for Social Security. The result was deadlock.

Frustrated by Congress's inability to achieve deficit reduction, two Republican senators, Senators Gramm and Rudman, and one Democratic senator, Senator Hollings, jointly introduced a bill aimed at forcing deficit reduction through restraints on the budget process.

The bill easily passed the Senate and the House. Its principle was simple. The bill set ceilings for the deficit in each fiscal year, with the goal of eliminating the deficit by 1991. If the budget proposed by Congress implied a deficit above the ceiling, a procedure known as *sequestration* automatically went into effect, with spending on all programs cut by the same percentage in order to achieve the target deficit.

A number of spending programs were excluded from the cuts, mainly interest payments on the debt,

Social Security benefits, and some low-income transfer programs. There were also *escape clauses* to prevent deficit reduction from standing in the way of macro-economic stabilization. For example, if projected growth was below 3%, the deficit ceiling was relaxed in proportion to the difference between projected growth and 3%.

How did Gramm-Rudman-Hollings (GRH) work in practice? It had a short and checkered history.

The first obstacle was a constitutional challenge on the grounds that GRH took too much power away from the legislative bodies. The bill was indeed declared unconstitutional by the Supreme Court in 1986. But the ruling was based largely on technicalities, and a second GRH bill, which avoided the problems mentioned by the Court, was passed in 1987. The occasion was used, however, to increase the ceilings, with the target date for zero deficit moved to 1993.

The later history of GRH was marked by the use of loopholes, optimistic forecasts, and other gimmicks:

■ Because the GRH ceiling applied only to the coming year's budget, Congress systematically shifted spending to the previous year's budget. This creative accounting made the deficit in the previous fiscal year look worse. But it allowed the current year's budget to satisfy the GRH ceilings more easily.

■ Public assets were sold, and proceeds from sales counted as revenues, thus reducing the measured deficit but doing little to reduce the true deficit.

■ With the complicity of the administration, which was officially in charge of giving the economic forecasts needed to compute projected deficits, optimistic projections for economic activity were used. The result was optimistic projections for revenues and thus optimistically low projections for deficits.

In view of all these gimmicks, how effective was GRH in reducing deficits? Table 1 gives the initial ceilings (GRH I), the revised ceilings (GRH II), and the actual deficits—both in current dollars and in ratio to GDP—for each fiscal year from 1986 to 1990, the year in which GRH was replaced by another set of rules (more on this later in this box).

In each of the fiscal years 1987 to 1989, the budget adopted by Congress satisfied the GRH ceiling. In each of those years, however, the realized deficit was larger than the GRH ceiling, by anywhere between $6 billion in 1987 and $16 billion in 1989. Nevertheless, the period was characterized by a steady decrease in the ratio of the deficit to GDP, from 5.0% in 1986 to 2.9% in 1989.

In 1990, however, the actual deficit was $121 billion above the GRH ceiling. There were two main reasons for this. First, economic activity was lower than forecast when the budget had been passed; GDP growth was only 1.2% in 1990. And the government was faced with a savings-and-loan crisis: Many S&L institutions had become insolvent, and the government had to make good on its promise to insure depositors against losses.

Both phenomena—unusually low growth and the S&L crisis—justified a higher deficit. Nevertheless, because the actual deficit was so much higher than the ceiling, GRH had lost its credibility. A new system of rules, known as the *Budget Enforcement Act of 1990,*

TABLE 1

GRAMM-RUDMAN-HOLLINGS CEILINGS AND ACTUAL DEFICITS

FISCAL YEAR*	DEFICIT CEILING UNDER:		ACTUAL DEFICIT (BILLIONS OF CURRENT DOLLARS)	ACTUAL DEFICIT GDP (%)
	GRH I (BILLIONS OF CURRENT DOLLARS)	GRH II		
1986	172		221	−5.0
1987	144		150	−3.4
1988	108	144	155	−3.2
1989	72	136	152	−2.9

*The fiscal year runs from October 1 of the preceding calendar year to September 30 of the current calendar year.

was introduced. Still largely in place in 1996, it differed from GRH by putting caps on specific spending programs and limiting Congress's role to reallocating spending within those programs, a system known as the *flexible freeze*.

What is the lesson of GRH? That good design of restraints is important and very difficult. It is important both to limit loopholes and to allow for realistic escape clauses. Some loopholes may actually make restraints more flexible and thus more acceptable. One can read the 1986–1989 outcomes in that light: Despite the use of creative accounting and optimistic forecasts, the ratio of the deficit to GDP was steadily reduced. Realistic escape clauses or exceptions are also important. In view of low growth and the S&L crisis, the deficit of 1990 was largely justified. But because it was so much larger than the GRH ceiling, GRH's credibility was destroyed, and another system had to be put in its place.

Reference

For further reading, read Steven Sheffrin, *The Making of Economic Policy* (New York: Blackwell, 1989).

SUMMARY

■ There is substantial uncertainty about the effects of macroeconomic policies. This uncertainty should lead policy makers to be more cautious, to use less-active policies. Policies must be broadly aimed at avoiding prolonged recessions, slowing down booms, and avoiding inflationary pressure. The higher the level of unemployment or inflation, the more active the policies should be. But they should stop well short of fine-tuning, of trying to maintain constant unemployment or constant output growth.

■ Using macroeconomic policy to control the economy is fundamentally different from controlling a machine. Unlike a machine, the economy is composed of people and firms who try to anticipate what policy makers will do, who react not only to current policy but also to expectations of future policy. In this sense, macroeconomic policy can be thought of as a game between policy makers and the economy.

■ When playing a game, it is sometimes better for a player to give up some options. Take the example of hijackings. When a hijacking occurs, it is best to negotiate with hijackers. But a government that credibly commits to not negotiating with hijackers—that gives up the option of negotiation—is actually more likely to deter hijackings in the first place.

The same argument applies to various aspects of macroeconomic policy. By credibly committing not to use monetary policy to decrease unemployment below its natural rate, a central bank can alleviate fears that money growth will be high, and in the process decrease both expected and actual inflation. When issues of time inconsistency are relevant, tight restraints on policy makers—such as a fixed-money-growth rule in the case of monetary policy—can indeed provide a coarse solution. But the solution may have large costs if it prevents the use of macroeconomic policy altogether. Better methods typically involve designing better institutions (such as an independent central bank) that can reduce the problem of time inconsistency without eliminating monetary policy as a macroeconomic stabilization tool.

■ Another argument for putting restraints on policy makers is that they may play games either with the public or among themselves, and that these games may lead to undesirable outcomes. Politicians may try to fool a short-sighted electorate by choosing policies with short-run benefits but large long-term costs—for example, large budget deficits. Political parties may delay painful decisions, hoping that the other party will make the adjustment and take the blame. These problems exist, although they are less prevalent than is often perceived. In such cases, tight restraints on policy, such as a constitutional amendment to balance the budget, again provide a coarse solution. Better ways typically involve better institutions and better ways of designing the process through which policy and decisions are made.

KEY TERMS

- fine-tuning, 535
- optimal control, 535
- optimal control theory, 535
- game, 535
- game theory, 536
- strategic interactions, 536
- players, 536
- time inconsistency, 537
- political business cycle, 541

QUESTIONS AND PROBLEMS

1. Briefly discuss an example where the problem of "time inconsistency" has occurred in your personal life. Who were the "players" in this "game"?
2. Section 27-1 discusses the uncertainties in the timing and potency of monetary policy. Are there also uncertainties in the timing and potency of fiscal policy? Explain.
3. In the example in Section 27-1, a more conservative increase in the money supply—10% instead of 28.2%—is recommended, resulting in an unemployment rate between 7% and 10% in one year. Suppose, however, that in addition to the uncertainty in the text, there is *also* uncertainty about the value of Okun's coefficient during that year. Specifically, suppose that

Okun's coefficient is known to be somewhere between 0.3 and 0.5. What is the *new* range for unemployment one year hence when the money supply is increased by 10%?

4. It has been suggested that monetary policy be more closely supervised by the Minister of Finance. Discuss the implications of this proposal.
5. Suppose that there were a *strict* balanced-budget amendment, with no escape clauses: a deficit equal to zero, regardless of economic conditions. Would this produce a more stable, or more unstable, economy? Explain.

FURTHER READING

A leading proponent of the view that governments misbehave and should be tightly restrained is James Buchanan, from George Mason University. Buchanan received the Nobel Prize in 1986 for his work on public choice. Read, for example, his book with Richard Wagner, *Democracy in Deficit: The Political Legacy of Lord Keynes* (New York: Academic Press, 1977).

For a survey of the politics of fiscal policy, read Alberto Alesina and Roberto Perotti, "The Political Economy of Budget Deficits," *IMF Staff Papers*, 1995.

A nice discussion of fiscal rules is given in the debate between William Niskanen, "The Case for a New Fiscal Constitution," and Charles Schultze, "Is There a Bias towards Excess in U.S. Government Budgets or Deficits?" *Journal of Economic Perspectives,* Spring 1992, 13–44.

For more on the politics of monetary policy, read Alberto Alesina and Lawrence Summers, "Central Bank Independence and Macroeconomic Performance: Some Comparative Evidence," *Journal of Money, Credit and Banking,* May 1993, 289–297.

For evidence regarding the absence of political business cycles in Canada, see Apostolos Serletis and Panos Afxentiou, "Electoral and partisan cycle regularities in Canada," *Canadian Journal of Economics,* February 1998, 28–46.

MONETARY POLICY: A SUMMING UP

Barely a chapter has gone by without a reference to monetary policy. The purpose of this chapter is to put together what we have learned, and tie up the remaining loose ends.

Let's first briefly review what we have learned (the Focus box entitled "Monetary Policy: What We Have Learned and Where" gives a more detailed summary):

■ In the short and the medium run, monetary policy affects output as well as its composition. An increase in money leads to a decrease in interest rates and a depreciation of the currency. These in turn lead to an increase in the demand for goods and an increase in output.

■ In the long run, however, monetary policy is neutral. Changes in the level of money eventually lead to proportional increases in prices, leaving output and unemployment unaffected. Changes in the rate of money growth lead to corresponding changes in the inflation rate.

We can therefore think of monetary policy as involving two basic decisions. The first is the choice of an average inflation rate and thus a corresponding average rate of money growth. The second is the choice of how much to deviate from this average to reduce fluctuations in output.

In this context, there are two issues we want to explore in this chapter. The first is the issue of the optimal inflation rate. There is little doubt that high inflation is costly. But how low should inflation be?

Should central banks aim for an average inflation rate of, say, 3%, or aim for price stability, or even aim for deflation (negative inflation)? The answer clearly depends on the costs and benefits of different inflation rates. We study these options in Section 28-1 and discuss what the optimal inflation rate might be.

The second issue is the relation between money growth and inflation. We have assumed so far that there was a stable relation between money growth and inflation in the long run. As we shall see in Section 28-2, this assumption must be qualified. Discussing this issue will force us to think about the relation between money and liquidity.

Having explored these issues, we end our study of monetary policy by looking in Section 28-3 at what the Bank of Canada actually does, how it designs and carries out monetary policy, and how well it has done in the recent past.

MONETARY POLICY: WHAT WE HAVE LEARNED AND WHERE

■ In Chapter 5 we looked at the determination of money demand, money supply, and the effects of monetary policy on the interest rate. We saw how an increase in the money supply leads to a decrease in the interest rate.

■ In Chapter 6 we looked at the short-run effects of monetary policy on output. We saw how an increase in money leads, through a decrease in the interest rate, to an increase in spending and output.

■ In Chapters 7 and 9 we introduced two distinctions, the first between the nominal interest rate and the real interest rate, the second between short-term and long-term interest rates. We saw that the real interest rate is equal to the nominal rate minus expected inflation. We saw how the long-term interest rate depends on current and expected future short-term rates.

■ In Chapter 10 we returned to the short-run effects of monetary policy on output, taking into account monetary policy's effect on expectations. We saw that monetary policy affects the short-term nominal interest rate, but that spending depends primarily on the long-term real interest rate. We saw how, as a result,

monetary policy's effects on output depend on how expectations respond to policy.

■ In Chapter 13 we looked at the effects of monetary policy when goods and financial markets are open. We saw how, in an open economy, monetary policy affects spending and output not only through interest rates but also through the exchange rate. An increase in money leads to both a decrease in the interest rate and a depreciation, and both increase spending and output. We discussed the implications of two different régimes for monetary policy, flexible versus fixed exchange rates.

■ In Chapter 14 we looked further at the effects of monetary policy on interest rates and exchange rates, taking into account the role of expectations in financial and foreign-exchange markets. We saw how the real exchange rate depends on the difference between current and expected future domestic and foreign real interest rates. We saw how an increase in the money supply can lead initially to a large depreciation, and thus have a large effect on output through the exchange rate.

■ In Chapter 16 we looked at the effects of changes in money on output and prices, not only in the short run but also in the medium and long run. We established the proposition that in the long run, money is neutral; changes in money are fully reflected in changes in prices.

■ In Chapter 18 we looked at the relation among money growth, inflation, and unemployment. We saw that in the long run money growth is reflected one for one in inflation, leaving the unemployment rate unaffected. We looked at alternative disinflation strategies. Based on the Canadian disinflation in the early 1980s and other disinflations around the world, we concluded that disinflations typically come at a cost of higher unemployment for some time.

■ In Chapter 19 we returned to the relation between money growth and interest rates. We saw how higher money growth decreases nominal interest rates in the short run but increases them in the long run. We also returned to fixed exchange rates and examined the pros and cons of adopting a common currency, as is now being considered in Europe.

■ In Chapter 20 we looked at monetary policy in the Great Depression. We saw how a contraction in nominal money and deflation were important sources of the large decline in output in the United States

from 1929 to 1933. Then, looking at unemployment in Europe, we examined the argument that disinflation may not only increase the unemployment rate above the natural rate, but also increase the natural rate itself.

■ In Chapter 21 we studied hyperinflations and looked at the conditions under which such episodes arise and eventually end. We focussed on the relation among budget deficits, money growth, and inflation, and how large budget deficits can lead to rapidly increasing money growth and to hyperinflation.

■ In Chapter 27 we looked at the problems facing macroeconomic policy in general and monetary policy in particular. We saw that uncertainty about the effects of policy should lead to more cautious policies. We saw that even well-intentioned policy makers may sometimes not do what is best, and that there is indeed a case for restraints on policy makers. We also saw that there is a case for making the central bank independent and appointing a conservative central banker.

■ In this chapter we discuss the issues of the optimal inflation rate, the relation between money and liquidity, and the implications for monetary policy. We conclude the chapter by looking at how the Bank of Canada actually conducts monetary policy in Canada today.

28-1 THE OPTIMAL INFLATION RATE

As Table 28-1 shows, inflation has steadily gone down in rich countries since the early 1980s. Average inflation in the OECD stood in 1995 at 4.1%, and all but two OECD countries (out of 25) had less than 10% inflation. Inflation in the OECD countries continued to decline in 1996 and 1997.

Does this mean that most central banks have now achieved their goal, or should they aim for price stability—zero inflation? The answer depends on the costs and benefits of inflation. Let's discuss each in turn.

THE COSTS OF INFLATION

Money serves three functions. First, it serves as a *unit of account:* Setting the prices of goods in the same units makes it easier for people and firms to assess relative prices, and thus to make the right decisions. Second, money serves as a *medium of exchange;* the fact that it is accepted in payment for all goods eliminates the need for complicated barter. Third, money serves as a *store of value*—even though there are many other assets, such as bonds, that pay a positive nominal interest rate and thus are better than money in that respect (money typically pays no interest or at most a small nominal interest rate).

TABLE 28-1

INFLATION RATES IN THE OECD, 1980–1995

YEAR	1980	1985	1990	1995
OECD average[a] (%)	11.8	6.6	6.0	4.1
Number of countries with inflation below 10%[b]	10	19	21	23

[a]Average using relative GDPs at PPP prices as weights.
[b]Out of 25 countries. The two countries with inflation above 10% in 1995 were Turkey (90%) and Mexico (28%).

Source: OECD *Economic Outlook*, December 1995.

When inflation is very high, such as during the episodes of hyperinflation we studied in Chapter 21, money serves all three functions badly. Because prices change so often, it becomes very difficult to assess relative prices and make informed decisions. Transactions require carrying larger and larger amounts of money, making transactions increasingly cumbersome; recall the example of people using wheelbarrows to carry currency during the German hyperinflation after World War I. Because money loses its value so fast, people and firms keep as little money as they can, engaging in frequent and time-consuming transactions. At very high rates of inflation, say 50% a month or more, economic activity is seriously disrupted and the cost in terms of lost output is large.

The debate in OECD countries today, however, is not about the costs of inflation rates of 50% a month. Rather, it centres on the advantages of, say, 0% versus 3% inflation a year. Within that range, economists identify four main types of costs: shoe-leather costs, tax distortions, and costs coming from money illusion and inflation variability.

Shoe-leather costs. In the long run, a higher inflation rate leads to higher nominal interest rates,[1] thus to a higher opportunity cost of holding money. As a result, people decrease their money balances by making trips to the bank more often; thus the expression "shoe-leather costs." These trips could be avoided if inflation were lower, and people could be doing other things instead, working more or enjoying leisure.

During hyperinflations, shoe-leather costs can become quite large. But their importance in times of moderate inflation is limited. If an inflation rate of 3% leads people to go to the bank one more time every month, or to do one more transaction between their money market fund and their chequing account every month, this hardly qualifies as a major cost of inflation.

Tax distortions. The second cost of inflation comes from the interaction between the tax system and inflation. To see why, consider the example of a capital-gains tax. Taxes on capital gains are typically based on the change in the asset's dollar price between the time it was purchased and the time it is sold. This method implies that the higher the rate of inflation, the higher will be the tax. An example will make this clear.

Suppose that inflation has been running at π% a year for the last 10 years. Suppose that you (or your family) own a summer cottage and the value of your cottage has also increased by π% a year, so its real value has not changed. More

---

[1]*This is the Fisher effect, which we studied in Chapter 19.*

concretely, suppose that you bought your cottage for $50 000 ten years ago and are thus selling it today for $50 000 times $(1 + \pi\%)^{10}$. If the capital-gains tax is 30%, the *effective tax rate* on the sale of your cottage—defined as the ratio of the tax you pay to the price for which you sell your cottage—is equal to

$$(30\%) \frac{50\,000(1 + \pi\%)^{10} - 50\,000}{50\,000(1 + \pi\%)^{10}}$$

Because you are selling your cottage for the same real price for which you bought it, your real capital gain is zero and you should not be paying any tax. Indeed, if $\pi = 0$—if there has been no inflation—then the effective tax rate is 0. But if $\pi = 5\%$, then the effective tax rate is 11.6%. If $\pi = 10\%$, the effective tax rate is 18%! Despite the fact that you have made no real capital gain, you may end up paying a high tax.

The problems extend beyond capital-gains taxes. Although the real rate of return on an asset is the real interest rate, not the nominal interest rate, income for the purpose of income taxation includes nominal interest payments, not real interest payments. Or, to take yet another example, since the early 1990s in Canada, the income levels corresponding to different income-tax rates have not increased systematically with inflation. As a result, people have been pushed into higher tax brackets as their nominal income—but not necessarily their real income—increased over time, an effect known as *bracket creep*.[2]

One may argue that this cost is not a cost of inflation per se, but rather the result of a badly designed tax system. If the government wanted to avoid the problems we've discussed, it could *index* the purchase price to the price level—that is, adjust the purchase price for inflation since the time of purchase—and compute the tax on the difference between the sale price and the adjusted purchase price. Then, in our example, there would be no capital gains and thus no capital-gains tax to pay. But because tax codes rarely allow for such systematic adjustment, the inflation rate matters.

Money illusion. The third cost comes from *money illusion,* an issue we discussed in Chapter 19—the notion that people appear to make systematic mistakes in assessing nominal versus real changes. A number of computations that would be simple under price stability become more complicated when there is inflation. In comparing their incomes this year with their incomes in the past, people have to keep track of the history of inflation. In choosing between different assets or deciding how much to consume or save, they have to keep track of the difference between the real interest rate and the nominal interest rate. Casual evidence suggests that many people find these computations difficult and often fail to make the relevant distinctions. Economists and psychologists have gathered more formal evidence, and, as we saw in Chapter 19, this evidence suggests indeed that inflation leads people and firms to make incorrect decisions.

Inflation variability. The last cost comes from the fact that higher inflation is typically associated with *more variable inflation.* And more variable inflation means that financial assets such as bonds, which promise fixed nominal payments in the future, become riskier.

---

[2]*Some economists argue that the costs of bracket creep have been very large indeed. They argue that as tax revenues increased over time, there was little pressure on the government to control spending, and that the result has been an excessively large and inefficient government.*

Take, for example, a bond that pays $1000 in 10 years. With constant inflation over the next 10 years, the real value of the bond is known with certainty. But with variable inflation the real value of $1000 in 10 years becomes uncertain. Saving for retirement becomes more difficult. For those who have invested in bonds, lower inflation than expected means a better retirement; but higher inflation may mean poverty. This is one of the reasons why retirees, for whom part of income is fixed in dollar terms, typically worry more about inflation than do other groups in the population.

As in the case of taxes, one may argue that these costs are not due to inflation per se, but rather to the financial markets' inability to provide assets that protect their holders against inflation. Rather than issuing only nominal bonds (bonds that promise a fixed nominal amount in the future), governments or firms could also issue *indexed bonds,* bonds that promise a nominal amount adjusted for inflation. This is precisely what some governments, such as those of Canada and the United Kingdom, have done, as we saw in the first Focus box in Chapter 9. The indexed bonds they are selling allow savers to avoid the risks associated with inflation and have been a great success.

THE BENEFITS OF INFLATION

Inflation is not all bad. One can identify three benefits of inflation: (1) seigniorage, (2) the option of negative real interest rates for macroeconomic policy, and (3) paradoxically, the interaction of money illusion and inflation in making it easier to achieve decreases in real wages when required.

Seigniorage. Money creation—the ultimate source of inflation—is one of the ways in which the government can finance its spending. Put another way, money creation is an alternative to borrowing from the public or raising taxes.

Technically, the government does not "create" money to pay for its spending. Rather, it issues bonds and spends the proceeds. Some of the bonds are bought by the central bank, which then creates money to pay for them. But the result is the same: Other things equal, the revenues from money creation—that is, *seigniorage*—allow the government to borrow less from the public or to lower taxes.

How large is seigniorage in practice? When looking at hyperinflation in Chapter 21, we saw that seigniorage was often an important source of government finance in countries with very high inflation rates. But its importance in OECD economies today, and for the range of inflation rates we are considering, is much more limited. Take, for example, the case of Canada. The ratio of the monetary base—the money issued by the Bank of Canada (see Chapter 5)—to GDP is equal to about 4%. An increase in money growth of 5% per year (which eventually leads in turn to a 5% increase in inflation) would therefore lead to an increase in seigniorage of 5% × 4%, or 0.2 percent of GDP. This is a small amount of revenue to get in exchange for 5% more inflation.

Thus, while the seigniorage argument is sometimes relevant (for example, in economies that do not yet have a good fiscal system in place), it hardly seems relevant in the discussion of whether OECD countries today should have, say, 0% versus 5% inflation.

The option of negative real interest rates. A positive inflation rate allows the monetary authority to achieve *negative real interest rates,* an option that may be useful when an economy is in recession. Let's look at this argument more closely.

The nominal interest rate on a bond cannot be negative. If it were, bondholders would be better off holding money rather than bonds. Thus the lowest possible nominal interest rate is zero. This was nearly the case in 1946, when the interest rate on three-month T-bills went down to 0.36%.

The real interest rate is equal to the nominal rate minus expected inflation (see Section 7-1). If expected inflation is positive, then the real interest rate can be negative. But if it is equal to zero, the lowest value the real interest rate can take is zero. And if there is expected deflation, the real rate must remain positive.

As we discussed in Chapter 20, this was indeed one of the adverse implications of deflation during the Great Depression: Expected deflation put a floor on real interest rates, limiting the role of monetary policy. And now that inflation is very low, the issue is resurfacing in a number of countries. Take Japan at the end of 1995. Faced with a sharp decrease in activity, the Japanese central bank decreased the short-term nominal interest rate to an extremely low 0.5% (at an annual rate). If expected inflation had been, say, 5%, such a low nominal rate would have led to a negative real interest rate of −4.5% (= 0.5% − 5%), and would have made investment spending very attractive. But because inflation was actually negative in Japan in 1995, expected inflation, using available forecasts, was also negative. Thus the real interest rate was above the nominal interest rate, and the effect of low nominal interest rates on spending was limited. Put another way, the ability of monetary policy to help the Japanese economy recover was very limited.

Money illusion revisited.   There is a reasonable argument that money illusion actually provides an argument *for* having a positive inflation rate. To see why, consider two situations. In the first, inflation is equal to 0% and somebody's wage is cut by 3% in dollar terms. In the second, inflation is equal to 5% and the wage is increased by only 2% in dollar terms. Both lead to the same decrease in the real wage, namely 3%. There is some evidence, however, that workers will accept the real wage cut more easily in the second case than in the first.[3]

Why is this example relevant to our discussion? Because the constant process of change that characterizes modern economies (see Chapter 25) means that some workers must sometimes take a real pay cut. Thus, the argument goes, inflation allows for these downward real-wage adjustments more easily than no inflation.

This argument is plausible. And because many economies now have very low inflation, we may soon be in a position to test it more thoroughly. If the argument is valid, we should see a smaller proportion of workers experiencing real wage decreases (as this would imply decreases in nominal wages) than was the case when inflation was higher (when decreases in real wages were consistent with increases in nominal wages). So far the evidence is mixed, but it may be too early to tell.

THE OPTIMAL INFLATION RATE: THE CURRENT DEBATE

At this stage, the debate in OECD countries is between those who think that some inflation (say 3%) is fine and those who want to achieve price stability—that is, 0% inflation.

---

[3]*See, for example, the results of a survey of managers by Alan Blinder and Don Choi, in "A Shred of Evidence on Theories of Wage Rigidity,"* Quarterly Journal of Economics, *1990, 1003–1016.*

Those who are happy with an inflation rate around 3% emphasize that the costs of 3% versus 0% inflation are small and that the benefits of some inflation are worth keeping. They argue that some of the costs of inflation could be avoided by indexing the tax system and issuing indexed bonds. They also emphasize that going from 3% to 0% is likely to involve an increase in unemployment for some time, and that this transition cost will exceed the eventual benefits.

Those who want to aim at 0% argue that, if inflation is bad, it should be eliminated, and that there is no reason to stop at 3% inflation just because we happen to be there. They make the point that 0% is a very different target rate from all others: It corresponds to price stability. This is desirable in itself. And, given the time inconsistency problem facing central banks that we discussed in Chapter 27, credibility and simplicity of the target inflation rate are important. Price stability achieves these goals better than a target inflation rate of 3%.

28-2 MONEY VERSUS LIQUIDITY

We have so far maintained a sharp distinction between money and other assets. But innovations in financial markets make the distinction increasingly fuzzy. As we shall now see, these changes make the work of the central bank more difficult.

MONEY AND OTHER LIQUID ASSETS

Table 28-2 shows the components of one of the most commonly used definitions of money. The first line, currency held outside the banks (both bank notes and coins), requires little explanation. The second line, chartered bank net demand deposits,[4] includes all deposits against which the owner can write cheques without providing notice. In December 1997, the value of such deposits was just over $51 billion. Company current accounts made up about three-quarters of the total; the remainder were personal chequing accounts. The first two lines are the assets that we have in mind when talking about "money." Their sum, which in December 1997 stood at about $80 billion, is called **M1**, or sometimes **narrow money**.

TABLE 28-2
COMPONENTS OF M1 IN CANADA, DECEMBER 1997
(MILLIONS OF DOLLARS)

Currency outside banks	29 390
Chartered bank net demand deposits	51 113
Adjustments to M1	−881
= M1	79 622

Source: Statistics Canada, CANSIM Series B2001, B478, B2050, and B2033

---

[4]*Net demand deposits excludes some items, such as interbank deposits and Government of Canada deposits, and adjusts for the value of cheques in transit (also called "the float"). Each spring, the* Bank of Canada Review *contains an article that reviews recent developments in the monetary aggregates and gives more detailed descriptions of these aggregates than we do here.*

The property that distinguishes money from other assets is its role as a **medium of exchange**; that is, it is immediately exchangeable for goods and services. While everyone agrees that currency and chequing deposits are money, there are other assets that are very similar. For example, some savings accounts at the chartered banks pay daily interest and allow cheques to be written. In theory, the bank could demand notice before allowing the cheque to be written; but this never happens. Near-banks (trust companies, credit unions, and caisses populaires) also offer accounts with chequing privileges. Most of these institutions have to make arrangements with the banks (more precisely, the *directly clearing members* of the Canadian Payments Association) in order to offer chequing privileges. But their clients see very little difference. Many other assets are not exchanged directly for goods and services, but are very liquid because they can be exchanged quickly and at very little cost into narrow money. Because it is difficult to distinguish between these assets and those that make up $M1$, a number of notions of **broad money** are also used:

■ The sum of $M1$ plus personal savings deposits and nonpersonal notice deposits at chartered banks is called **$M2$**. In December 1997, $M2$ totalled $404 billion.

■ The sum of $M2$ and nonpersonal term deposits and foreign currency deposits of residents is called **$M3$**. In December 1997, $M3$ totalled $550 billion.

■ The sum of $M2$ plus all deposits at trust and mortgage loan companies, credit unions and caisses populaires, and government savings institutions, plus the value of money market mutual funds, plus the value of individual annuities held by life insurance companies is called **$M2+$**. In December 1997, $M2+$ totalled $630 billion.

We could go on and consider other **monetary aggregates** as well. Where should we draw the line? Rather than split hairs about definitions, economists prefer to say that the best definition of money is the one that is most useful. Among the various definitions about which reasonable people can disagree a priori, we should pick those that display a stable and reliable relationship with other macro aggregates such as nominal income and interest rates. So far, there appears little would be gained by using aggregates other than $M1$ and $M2$, or their close cousins. This may change as new assets and payment instruments are introduced. If they are introduced, we'll adopt new definitions of money.

You may have noticed that you (or your parents) are far more likely to "pay" for a restaurant dinner with a credit card than with a life insurance annuity or even a cheque. Why do none of the various measures of money make any reference to credit cards? One answer is that a line of credit is not an asset in the usual sense. When you hand over your card, the credit card company is actually guaranteeing a loan for you. It guarantees to pay the merchant directly and will come after you to settle the account. The true payment comes when you write your cheque to the credit card company. This answer may strike you as semantic quibbling and, if so, we are somewhat sympathetic: Although the introduction of credit cards has not changed the quantity of money, it has affected profoundly the way we manage our money balances. More on this below.

THE DEMAND FOR MONEY REVISITED

Let's take stock. Until now, we thought of money as liquid and paying no interest, and of bonds as not liquid but paying interest. What we have just seen is that there is, in fact, a continuum of assets between money and bonds. These assets differ in two dimensions: liquidity and the interest rate they pay. Typically, the most liquid assets pay the lowest interest rate. $M1$ includes only the most liquid assets, those that can be used in transactions directly. $M2$ uses a more generous definition of liquidity, including assets that can be used for some transactions or exchanged for money at low cost.

In this more complex world, it remains true that the Bank of Canada controls the *supply of M1*. The analysis we developed in Chapter 5 remains applicable: The Bank directly controls the monetary base. And because cheques are cleared by exchanging deposits held at the Bank, the Bank indirectly controls the supply of $M1$.

What we need to reexamine, however, is the *demand for M1*. In choosing among the assets that go into $M3$ and $M2+$, people tradeoff more liquidity for a lower interest rate. This tradeoff has a straightforward implication: The demand for $M1$ depends on how attractive $M1$'s components are in comparison with other liquid assets. Thus, changes in the menu of assets can lead to large movements in the demand for $M1$. Two examples will make the point:

■ The widespread adoption of credit cards over the past three decades has made it possible to economize on $M1$ balances. Rather than carrying currency around in our pockets or holding on to large balances in our chequing accounts so that we can pay for our purchases, it is now easy to arrange our affairs so that we can make most of our payments on one day of the month. The result has been a gradual but important decline in the demand for $M1$.

■ With the introduction of money market funds in the late 1970s, people were able to hold a very liquid asset while receiving an interest rate close to that on T-bills. Money market funds grew from almost nothing in 1976 to \$1 billion in 1988, and reached \$33 billion in 1997. It is clear that the rapid growth in the 1990s has been financed to some extent by a shift away from chequing and savings deposits.

IMPLICATIONS FOR MONETARY POLICY

We have just seen that the demand for $M1$ should be thought of in the larger context of the choice between assets based on their liquidity and their return. This tradeoff has an important implication for monetary policy: Trying to maintain a constant rate of $M1$ growth, either in the short or the long run, would be a major mistake.

The short run. Consider the short run first, and take the simplest version of the *IS-LM*, without expectations or open-economy considerations (the version we saw in Chapter 6):

$$IS: \qquad Y = C(Y - T) + I(i, Y) + G$$

$$LM: \qquad \frac{M}{P} = YL(i)$$

The equations are familiar. M in the LM relation stands here for $M1$. The equation says that the real supply of $M1$ must be equal to the real demand for $M1$. The IS and the LM curves are drawn in Figure 28-1. The LM curve is drawn for a given price level; because we are looking at the short run, we take the price level as given.

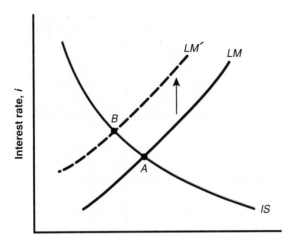

FIGURE 28-1

Shifts in Money Demand and the *IS-LM*

If the Bank of Canada does not increase the money supply, an upward shift in the demand for money will lead to an increase in the interest rate and a decrease in output.

Suppose that a new regulation puts tighter restrictions on cheque writing on money market fund accounts, leading to a shift back toward chequable deposits. The demand for $M1$ thus increases, leading to an upward shift in the LM curve; at a given level of output and given an unchanged money supply, the increase in money demand leads to an increase in the interest rate.

Should the Bank keep $M1$ constant or adjust it in response to the shift in demand? If it keeps $M1$ constant, the economy moves from point A to point B: The interest rate increases and output decreases. It is hard to see why it would be desirable to let output decrease in response to people's desire to shift out of money market funds into chequable deposits. It is thus clearly more appropriate for the Bank to prevent this portfolio shift from affecting output. It can do so by increasing the money supply, $M1$, in line with the shift in demand. By increasing $M1$ enough to leave the interest rate unchanged, it can leave the economy at point A, with unchanged output and an unchanged interest rate.

The lesson is therefore that the central bank must be ready to adjust $M1$ in response to shifts in the demand for money. One way it can identify such shifts is by looking at the evolution of different monetary aggregates. For example, if there is a large increase in $M1$, but not much change in $M2$, the Bank may reasonably conclude that it is indeed facing a shift in the demand for $M1$, which it may want to accommodate through an increase in the supply of $M1$.

The long run. The same issues are relevant in the long run. We argued earlier, in Section 28-1, that the central bank should aim at maintaining a low average rate of inflation. Our discussion implies that a stable inflation rate may not be associated with stable $M1$ growth.

Why? New assets are continually introduced in financial markets. Many of these offer new combinations of liquidity and return. We discussed the introduction of money market funds earlier in this section. But in addition to such large changes are hundreds of small ones, which continually offer new combinations of liquidity and return. In addition, new payment mechanisms arise that allow $M1$ balances to be economized. As people and firms reallocate their portfolios in response to these

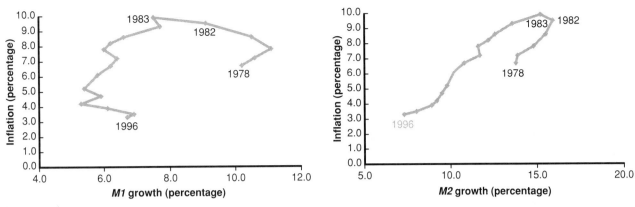

FIGURE 28-2

*M*1 and *M*2 Growth versus Inflation: Ten-Year Averages, 1978–1996

Even when we look at 10-year averages, there is no tight relation between *M*1 growth and inflation. The relation is stronger between *M*2 growth and inflation.

Source: Statistics Canada, CANSIM Series B2031, B2033, and P700000.

changes, the demand for *M*1 continually shifts. This phenomenon suggests that, even in the long run, there is unlikely to be a very tight relation between *M*1 growth and inflation. And, as Figure 28-2 shows, this is indeed the case.

Figure 28-2(a) plots 10-year averages of the inflation rate (using the CPI as the price index) against 10-year averages of the growth rate of *M*1, for each year from 1978 to 1996 (for example, the point corresponding to 1978 gives the average inflation and *M*1 growth rates for 1969 to 1978). The reason for using these 10-year averages should be clear. In the short run, changes in money growth affect mostly output, not inflation. It is only in the longer run that a relation between money growth and inflation may emerge.[5] Taking 10-year averages of both money and inflation is a way of looking for the presence of such a long-run relation.

Figure 28-2(a) shows that there is no tight relation between *M*1 growth and inflation. Since 1983, inflation has decreased with no consistent corresponding decrease in *M*1 growth.

Our discussion suggests that *M*2 might be a better measure of liquidity, so there might be a tighter relation between the inflation rate and the growth rate of *M*2. Thus, Figure 28-2(b) carries out the same exercise as Figure 28-2(a), but this time with 10-year averages of the inflation rate and of the growth rate of *M*2. The relation is indeed tighter. Inflation and *M*2 growth increased together from 1978 to 1982, and then have decreased together since 1983.

However, even for *M*2 the relation is not always reliable. As the points corresponding to the 1990s in Figure 28-2(b) show, average *M*2 growth has been unusually low in the 1990s compared to inflation. The reason appears to be a shift from savings accounts to other assets such as money market funds.

The preceding paragraph points to the shortcomings of *M*2 as a measure of liquidity: Some assets not included in *M*2 are very similar in terms of liquidity to

---

[5] *See Chapter 18.*

some assets included in *M2*. This problem has led researchers to try to construct better measures of liquidity. However, at this point—and somewhat surprisingly—none appears to be more reliable than *M2*.

28-3 THE BANK OF CANADA IN ACTION

Let's end this chapter by looking at how the Bank of Canada actually designs and carries out monetary policy.

THE BANK'S MANDATE

The Bank was formed in 1935.[6] The Bank of Canada Act has been amended many times since then, but the preamble still gives the Bank the responsibility: *to regulate credit and currency in the best interests of the economic life of the nation, to control and protect the external value of the national monetary unit and to mitigate by its influence fluctuations in the general level of production, trade, prices and employment, so far as may be possible within the scope of monetary action, and generally to promote the economic and financial welfare of Canada.*

Such a vague mandate leaves a lot of room for interpretation. What are the best interests of the nation? Should the Bank focus on obtaining stable income growth and high employment, low and stable inflation, some particular value for the Canadian dollar, low interest rates, or some undetermined combination of each? Since its founding, the Bank has at one time or another focussed on each of these goals. During the constitutional debate that led up to the Meech Lake Accord of 1987, some economists, including the governor of the Bank of Canada, pressed for a statement in the new Constitution that would narrow the Bank's mandate to maintaining price stability. Although interesting, this suggestion was quickly shunted to the side by more pressing issues for renewing the confederation. Undaunted, the Bank has since 1989 chosen to focus almost exclusively on controlling inflation. The Bank has adopted this position because it has determined (largely by trial and error) that providing price stability is the best thing it can do to encourage a well-functioning economy.

THE INSTRUMENTS OF MONETARY POLICY

In Chapter 5, we emphasized the link between money and reserves [equation (5.11)]:

$$M = (1 + c)/(1 + \theta)H \tag{28.1}$$

M is the money supply, where money is defined as *M1*, the sum of currency and chequable deposits. *H* is the monetary base—the sum of currency and reserves held by banks. *Reserves* refers to deposits held by the banks at the Bank of Canada. In some countries, such as the United States, and Canada before 1994, banks must hold reserves that equal a certain fraction of their deposits. In Canada since 1994, reserves are not held to meet legal requirements but to facilitate the clearing of cheques and other transfers; that's why they are often called **settlement balances**.

---

[6]*A brief (and very readable) history of the Bank is available on its web page.*

The parameter c is the ratio of currency to chequable deposits; θ denotes the ratio of reserves held by banks to chequable deposits. The expression $(1 + c)/(1 + \theta)$ is called the **money multiplier**. The public chooses c and the banks choose θ. The main influence of monetary policy is on the quantity of reserves held by the banks.

Only a small number (currently 13) of the financial institutions in Canada hold deposits at the Bank of Canada. Strictly speaking, we should refer to them as "direct clearers" rather than banks, but we shall stick with the simpler label. Every night,[7] transfers drawn on these banks are matched up and the balance is settled by transferring deposits at the Bank of Canada. For example, suppose bank A holds claims such as cheques drawn on bank B of $10 billion, while bank B holds claims on bank A of $11 billion. The difference of $1 billion is paid by transferring that amount from the account held at the Bank of Canada from bank A to the account held by bank B. Deposits at the Bank of Canada pay low interest, so the banks have an incentive to minimize these balances. What if bank A doesn't have enough in its account to cover this transfer? It can borrow the amount from the Bank of Canada overnight. By controlling the rate of interest on these overnight advances, the Bank effectively controls the interest rate in a much larger overnight market involving a wider number of financial players. The Bank of Canada fixes a band of 50 basis points (one-half of 1%) for such overnight loans. That is, banks can always earn the lowest interest rate in the Bank's range by holding reserves above the minimum needed for settlement. Also banks can borrow at the highest interest rate in the Bank's range by reducing their reserves below the minimum need for settlement. The upper bound of its interest rate band is called the *Bank rate*.[8]

In a world of perfect certainty, it would not matter if the Bank of Canada fixed the level of reserves to achieve a given market-clearing interest rate in the overnight market, or if it fixed the market-clearing interest rate to obtain a given level of reserves. In practice, because of uncertainties in the links between its actions and the subsequent path of output and prices, the Bank follows the second strategy[9]. Having determined a desired path for the Bank rate (following a procedure to be described below), the Bank of Canada then adjusts reserves in order to keep the overnight interest rate in its band. The main tool in managing reserves is transfers of federal government accounts. The federal government maintains fairly large cash balances to finance its activities. These balances are held in accounts at the Bank of Canada and private banks. Managing these balances provides an effective way to change the supply of reserves. For example, a transfer of $1 billion from the government's account at the Bank of Canada to bank A creates an increase of $1 billion in the reserves held by bank A. An opposite transfer reduces reserves by the same amount. Other tools for influencing the interest rate in the overnight market are open-market operations such as short-term loans called *Special Purchase and*

---

[7]*Beginning in late 1998, large-value transfers (amounts of over $50 000) will be cleared in real time. Large-value items represent a very small fraction of the settlements exchanged, but account for over 90% of their value. Same-day settlement of large-value items allows the banks to more accurately estimate their settlement needs and allows them to reduce their balances held at the Bank.*

[8]*From March 1980 to February 1996, the Bank rate was set equal to the average interest rate established at the auction of three-month Government of Canada Treasury bills, plus 25 basis points. The new definition of the Bank rate went into effect February 22, 1996.*

[9]*The Bank experimented with trying to control monetary aggregates directly from 1975 to 1982 but determined that the links between these aggregates and the goals of monetary policy were too unreliable.*

Resale Agreements (SPRA)[10] and short-term borrowing called *Sale and Repurchase Agreements (SRA)*. Since 1985, traditional open-market operations involving the purchase or sale of Treasury bills have rarely been used to "fine-tune" the quantity of reserves. However, the Bank uses purchases of Treasury bills to expand its asset holdings and determine the trend in monetary expansion.

THE PRACTICE OF POLICY

How does the Bank decide what policy to follow?

Because the actions of the monetary authority feed into the growth of nominal income and prices with long and variable lags, the Bank of Canada needs to keep its sights on what it expects will happen to the economy six months to several years ahead. The Bank goes through a semiannual cycle. After the release of the National Accounts in the second quarter (Q2) and the fourth quarter (Q4), the Bank's staff prepares forecasts and simulations of the effects of different monetary policies. A path for the Bank rate is sketched out for the next few years that will generate inflation consistent with the Bank's goals. Within this semiannual cycle, the Bank frequently updates its estimates of the path of the economy and reassesses its desired path for the Bank rate. Information about the path of short-term interest rates plays a large role. If short-term rates are lower than the Bank expected, it is often a signal that aggregate demand in the economy has not grown as quickly as the Bank had forecast. The Bank takes this as a signal that it should reduce the Bank rate in order to stimulate growth in aggregate demand. Because of the importance of international trade, the Bank also keeps an eye on the multilateral exchange rate; it views sharp declines in the exchange rate as stimulating aggregate demand and as a potential signal to raise the Bank rate. Of course, both interest rates and the exchange rate can move in response to shocks that have no consequence for inflation. The Bank tries to use a variety of information sources to isolate the source of the shocks that appear and to determine if any response is warranted. Experience has taught us that it is unwise to react to every little wiggle that appears in financial markets. Changes in the Bank rate are infrequent.

THE ROLE OF TARGET RANGES

Since 1991, the Bank of Canada has announced explicit targets for the rate of inflation. The target was set at 3% for 1992 and was reduced slowly to 2% in 1995 where it has been since. In fact, the Bank announces a range that is its target, plus or minus 1%. How close has the Bank come to hitting its inflation targets? The answer is given in Figure 28-3 which plots the rate of inflation for each year since 1988 and gives the Bank's target range since 1992. The Bank prefers to use a measure of the CPI that excludes volatile components such as food, energy, and the effect of indirect taxes. Although the Bank doesn't always hit its target, the errors since 1992 have all been on the low side. That is, either inflation has been in the Bank's target range or it has fallen below.

Why should the Bank announce such a range? There are two reasons:

1. The target serves as a signal of the Bank's intentions. By adopting a target of gradually declining inflation after 1992, the Bank tried to signal its commitment to achieving a lower inflation rate.

---

[10]*The Bank also makes loans called* Purchase and Resale Agreements (PRA). *While an SPRA is made on the initiative of the Bank, a PRA is made on the initiative of a qualified borrower such as an investment dealer.*

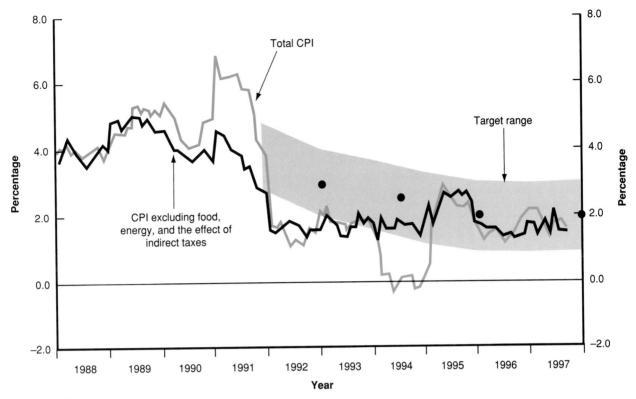

FIGURE 28-3

Inflation in Canada, 1988–1997: Realized Inflation Rate and Target Inflation Range

Inflation has been on the low side of the target range since 1992 and has fallen below the target range on several occasions.

Source: Bank of Canada Monetary Policy Report, November 1997

2. The target serves as a benchmark for judging the Bank's behaviour. If actual inflation ends up higher than the Bank's range, then it has some explaining to do. Similarly, raising the Bank rate when inflation is below its target range (as the Bank did in January 1998) forces the Bank to explain its actions. Has it forgotten its focus on inflation and chosen to worry about the exchange rate? Unless it can provide a convincing explanation, the Bank risks losing its credibility and leaves financial markets worrying about how to interpret monetary policy.

With these two functions in mind, the Bank sets the target ranges for inflation and then adjusts monetary policy during the year. The width of the range indicates the degree of flexibility that the Bank thinks it may need. The Bank then tries to keep inflation within these bounds. But, when faced with either unexpectedly large special factors or macroeconomic shocks, the Bank is willing to tolerate a rate of inflation outside its range. It knows that as long as it can convince the markets that such a deviation is justified it will not lose credibility.

Does this way of running monetary policy work? The answer is that it has worked well since 1992. The decline in the Bank's targets has been accompanied by a similar decline in inflation around the OECD, although Canada's performance on inflation has been better than average. It remains to be seen how well the Bank's targets will work if inflation starts to challenge the upper bound of its target range.

SUMMARY

On the Optimal Rate of Inflation

■ Inflation is down to very low levels in most OECD countries. One question facing central banks is whether they should try to achieve zero inflation—that is, price stability.

■ The main arguments for zero inflation are the following:
- The combination of inflation and an imperfectly indexed tax system leads to tax distortions.
- Because of money illusion, inflation leads people and firms to make incorrect decisions.
- Higher inflation typically comes with higher inflation variability, creating more uncertainty and making it more difficult for people and firms to make the right decisions about the future.
- As a target, price stability has a simplicity and a credibility that a positive inflation target does not have.

■ There are also arguments for maintaining low but positive inflation.
- Revenues from money growth (seigniorage) allow for lower taxes elsewhere. However, this argument is quantitatively unimportant when comparing inflation rates of 0% versus, say, 3%.
- When real wage cuts are needed, positive inflation allows firms to achieve real wage cuts without requiring nominal wage cuts.
- Positive actual and expected inflation allows the central bank to achieve negative real interest rates, an option that may be useful in fighting a recession.
- A further decrease from the current positive rate of inflation to zero would imply an increase in unemployment for some time, and this transition cost may exceed whatever benefits come from zero inflation.

On Money versus Liquidity

■ Rather than a sharp distinction between "money" and "bonds," there is a continuum of financial assets, with different degrees of liquidity and different rates of return. The menu of liquid assets ranges from currency to chequable deposits, to money market funds and time deposits. Thus, the demand for $M1$—currency and chequable deposits—depends on the attractiveness of close substitutes such as money market funds. As new types of interest-bearing assets are introduced, the demand for $M1$ shifts, complicating the task of the central bank.

■ Given the shifts in money demand, it would be unwise for the central bank to maintain a constant growth rate of $M1$. Monitoring monetary aggregates such as $M1$ and $M2$ helps the central bank identify and respond to shifts in the demand for money.

On the Bank of Canada

■ The Bank of Canada was founded in 1935 *"to regulate credit and currency in the best interests of the economic life of the nation."* The Bank influences the level of short-term interest rates by controlling reserves. Since 1994 (and in practice for several years before that), these reserves have been deposits held to settle transfers between the direct clearers. The main instrument of monetary policy is the transfer of federal government balances between the Bank and the direct clearers. The Bank also acts to influence interest rates directly in the overnight market through loans and borrowings that go by the acronyms SPRA and SRA respectively.

■ The Bank conducts a semiannual exercise to sketch out the path for its Bank rate that will allow it to hit its inflation targets approximately nine months to several years ahead. Within the six-month cycle, the Bank constantly monitors its assumptions and forecasts. On a week-to-week basis, the Bank pays particular attention to the movements in short-term interest rates and the multilateral exchange rate. Unexpected movements in these variables sometimes prompt a change in the Bank rate.

■ Every few years, the Bank announces a target range for inflation. Since 1995, the target has been 2% with a band of ±1%.

■ The Bank has done a good job of hitting its inflation targets since 1992, although inflation has sometimes fallen below the target range.

KEY TERMS

- $M1$, or narrow money, 555
- medium of exchange, 556
- broad money, 556
- $M2$, 556
- $M3$, 556
- monetary aggregates, 556
- settlement balances, 560

QUESTIONS AND PROBLEMS

1. How would each of the following, *ceteris paribus,* affect the demand for *M*1 and *M*2+ as currently defined?
   a. New restrictions specifying a $5000 minimum on each cheque written from a money market account
   b. Lower penalties on early withdrawal from a time deposit (guaranteed investment certificate)
   c. The new technology permitting you to pay for your purchases at the store by electronically and instantaneously withdrawing from your savings or money market account
   d. Because of increasing fraud, store owners stop accepting personal cheques.

2. Suppose that asset holdings and interest rates in the economy are as follows:

TYPE OF ASSET	QUANTITY (BILLIONS OF DOLLARS)	INTEREST RATE (%)
Cash	300	0
Chequable deposits	800	1
Money market fund shares	250	5
Money market accounts	600	4
Savings accounts	400	3
Time deposits (under $100 000)	500	6
Three-month Treasury bills	1000	8

   a. What is the value of *M*1?
   b. What is the value of *M*2?
   c. What is the value of *M*2+?

3. Suppose that your job is advising the Bank of Canada on monetary policy for the coming month. One day, on the way to work, you hear on the radio that a change in the earth's magnetic field has rendered all credit cards and ABM cards permanently useless by destroying the data on their magnetic strips. What would you advise the governor of the Bank to do?

4. Suppose you have a mortgage of $100 000 under one of the following three cases:
   (i) Expected inflation is 0% and the nominal interest rate on your mortgage is 4% per year.
   (ii) Expected inflation is 5% and the nominal interest rate on your mortgage is 9% per year.
   (iii) Expected inflation is 10% and the nominal interest rate on your mortgage is 14% per year.
   a. Compute the annual nominal interest payment you would make on your mortgage in each case.
   b. Compute the annual real interest payment you would make on your mortgage in each case.
   c. Suppose you can deduct nominal mortgage interest payments from your income before paying the income tax (as in the United States). Assume your tax rate is 30%. Thus, for each dollar you pay in mortgage interest, you can deduct 30 cents from your gross income, in effect getting a subsidy from the government for your mortgage costs. Compute in each case the real interest payment on your mortgage, adjusted for this subsidy.
   d. "In the United States, inflation is good for home owners." Discuss.

FURTHER READING

For more details on how the Bank of Canada operates, download "The Transmission of Monetary Policy in Canada" from the Bank of Canada's web site. It contains articles from the *Bank of Canada Review* that outline the details of monetary policy during the first half of the 1990s.

   "Modern Central Banking," written by Stanley Fischer for the 300th anniversary of the Bank of England—published in *The Future of Central Banking*, Forrest Capie, Stanley Fischer, Charles Goodhart, and Norbert Schnadt, eds. (Cambridge: Cambridge Univeristy Press, 1995)—provides a very nice discussion of the current issues in central banking.

   As its title indicates, "Central Bank Behavior and the Strategy of Monetary Policy: Observations from Six Industrialized Countries," by Ben Bernanke and Frederic Mishkin, *NBER Macroeconomics*, 1992, 183-237, describes how monetary policy is conducted in six major OECD countries.

FISCAL POLICY: A SUMMING UP

The purpose of this chapter is to do for fiscal policy what we did for monetary policy in Chapter 28: to review what we have learned and tie up remaining loose ends.

Let's first review what we have learned (the Focus box entitled "Fiscal Policy: What We Have Learned and Where" gives a more detailed summary).

- In the short run, a budget deficit (triggered, say, by a decrease in taxes) is likely to increase demand and thus output. The strength of the initial impact on output depends very much on expectations. For example, if financial markets worry that deficits will remain large in the future, the result may be a large increase in long-term interest rates that offsets much of the direct expansionary effect of lower taxes.

- In the long run, output returns to its natural level. But the natural level of output itself may be affected by fiscal policy. If deficits lead to lower investment, a lower capital stock in the long run implies lower output as well.

In deriving these conclusions, however, we did not pay close attention to the *government budget constraint*—that is, to the relation among debt, deficits, government spending, and taxes. Thus, this chapter's first order of business is to look at the government's budget constraint and its implications. Having done so, we examine a number of fiscal policy issues where this constraint plays an important role, from the proposition that deficits do not really matter to the dangers of accumulating very high levels of public debt.

FISCAL POLICY: WHAT WE HAVE LEARNED AND WHERE

■ In Chapters 3 and 4 we looked at the role of government spending and taxes in determining demand and output in the short run. We saw how, in the short run, increases in government spending and decreases in taxes both increase spending and output.

■ In Chapter 6 we looked at the short-run effects of fiscal policy on output and the interest rate. We saw how a fiscal contraction leads to decreases in both output and the interest rate. We also saw how fiscal and monetary policy can be used to affect both the level and the composition of output.

■ In Chapter 10 we looked at the short-run effects of fiscal policy, taking into account not only its direct effects through spending but also its effects through expectations. We saw how the effects of a deficit reduction on output depend on expectations of future fiscal and monetary policy. We also saw how a deficit reduction may, in some circumstances, be expansionary.

■ In Chapter 12 we looked at the effects of fiscal policy when the economy is open to trade. We saw how fiscal policy affects both output and the trade balance, and examined the relation between budget deficits and trade deficits. We saw how fiscal policy and exchange-rate adjustments can be used to affect both the level and the composition of output.

■ In Chapter 13 we looked at the role of fiscal policy in an economy with open goods and financial markets. We saw how, in the presence of international capital mobility, the effects of fiscal policy depend on the exchange-rate régime. Fiscal policy has a much stronger effect on output under fixed exchange rates than under flexible exchange rates.

■ In Chapter 16 we looked at the effects of fiscal policy not only in the short run, but also in the medium run and the long run. We established the proposition that in the long run (ignoring the effects of policy on capital accumulation) changes in fiscal policy have no effect on output, but do affect the price level and the interest rate. In the long run, changes in fiscal policy are also reflected in a different composition of spending.

■ In Chapter 21 we looked at the relations among fiscal policy, money growth, and inflation. We saw how budget deficits must be financed either by borrowing or by money creation. We saw that when money creation becomes the main source of finance, the result of large deficits is high money growth and hyperinflation.

■ In Chapter 23 we looked at how saving, and thus budget deficits, affect the level of capital accumulation, and thus affect the standard of living in the long run. We saw how, once capital accumulation is taken into account, larger deficits lead to lower capital accumulation and thus to a lower level of output in the long run.

■ In Chapter 27 we looked at the problems facing fiscal policy makers, from uncertainty about the effects of policy to issues of time in consistency and credibility. We discussed the pros and cons of restraints on the conduct of fiscal policy, such as a constitutional amendment to balance the budget.

■ In this chapter we look further at the implications of the budget constraint facing the government and discuss current issues of fiscal policy in Canada.

29-1 THE GOVERNMENT BUDGET CONSTRAINT

Suppose that, starting from a balanced budget, the government cuts taxes, thus creating a deficit. What will happen to debt over time? Will the government need to increase taxes later? If so, by how much?

To answer these questions, let us start with the definition of the budget deficit. We can write the budget deficit in year t as

$$\text{deficit} = rB_{t-1} + G_t - T_t \qquad (29.1)$$

All variables are in real terms. B_{t-1} is government debt at the end of year $t - 1$, or equivalently at the beginning of year t; r is the real interest rate, which we shall take to be constant here. Thus rB_{t-1} is equal to the real interest payments on the existing government debt. G_t is government spending on goods and services in year t. T_t is equal to taxes minus transfers in year t. Thus the budget deficit is equal to spending, inclusive of interest payments, minus taxes net of transfers.

Note two characteristics of equation (29.1):

1. We measure interest payments as real interest payments—that is, the product of the *real* interest rate times existing debt—rather than as actual interest payments—that is, the product of the nominal interest rate times existing debt. As we discuss in the Focus box entitled "Inflation Accounting and the Measurement of Deficits," this is indeed the correct way of measuring interest payments. However, official measures of the deficit include actual (nominal) interest payments and are therefore incorrect. The correct measure of the deficit is sometimes called the **inflation-adjusted deficit.**

2. For consistency with our definition of G as spending on goods and services earlier in the book, G does not include transfer payments. Transfers are instead subtracted from T, so that T stands for taxes minus transfers. Official measures of government spending add transfers to spending on goods and services, and define revenues as taxes, not taxes net of transfers. These are only accounting conventions. Whether transfers are added to spending or subtracted from taxes makes a difference to the measurement of G and T, but clearly does not affect the measure of the deficit.

The **government budget constraint** then simply states that the *change in government debt during year t* is equal to the *deficit in year t:*

$$B_t - B_{t-1} = \text{deficit}$$

Thus, if the government runs a deficit, government debt increases. If the government runs a surplus, government debt decreases.

Using our definition of the deficit, we can rewrite the government budget constraint as

$$B_t - B_{t-1} = rB_{t-1} + G_t - T_t \qquad (29.2)$$

The government budget constraint thus links the change in debt to the initial level of debt (which affects interest payments) and to current government spending and taxes.

Official measures of the budget deficit are constructed as (dropping the time indexes, which are not needed here) nominal interest payments, iB, plus spending on goods and services, G, minus taxes net of transfers, T:

$$\text{official measure of deficit} = iB + G - T$$

This measure is indeed an accurate measure of the *change in nominal debt*. If it is positive, the government is spending more than it receives, and must therefore issue new debt. If it is negative, the government retires debt.

But it is not an accurate measure of the *change in real debt*, the change in how much the government owes, expressed in terms of goods rather than dollars. To see why, suppose that the official measure of the deficit is equal to zero, so the government neither issues nor retires debt. Suppose that inflation is positive and equal to 10%. Then, at the end of the year, the real value of the debt has decreased by 10%. Thus, if we define—as we should—the deficit as the change in the real value of debt, the government is, in fact, running a budget surplus equal to 10% times the initial level of debt.

More generally, if B is debt and π is inflation, the official measure of the deficit overstates the correct measure by an amount equal to πB. The correct measure of the deficit is thus equal to

$$
\begin{aligned}
\text{correct measure of the deficit} &= iB + G - T - \pi B \\
&= (i - \pi)B + G - T \\
&= rB + G - T
\end{aligned}
$$

where $r = i - \pi$ is the real interest rate.* The correct measure of the deficit is thus equal to real interest payments plus government spending minus taxes net of transfers, the measure we have used in the text.

The difference between the official and the correct measures of the deficit is equal to πB. Thus, the higher the rate of inflation (π) or the higher the level of debt (B), the more inaccurate the official measure is. In countries in which both inflation and debt are high, the official measure may record a very large budget deficit, when in fact real government debt is actually decreasing. Thus you should always make sure that you do the inflation adjustment before deriving conclusions about the position of fiscal policy.

Today, with inflation running at 1% to 2% a year, and the Canadian ratio of debt to GDP equal to roughly 70%, the difference between the two measures is roughly equal to 1% to 2% times 70%, or 0.7% to 1.4% of GDP. Put another way, an official deficit of 0.7% to 1.4% of GDP corresponds in fact to budget balance.

Note that r, which is equal here to the nominal interest rate minus actual inflation, should be more accurately called the "ex post real interest rate," to distinguish it from the real interest rate, which we defined in Chapter 7 as equal to the nominal interest rate minus expected inflation.

For reasons that will be clear later in this chapter, it is often convenient to decompose the deficit into the sum of two terms: (1) interest payments on the debt, rB_{t-1}, and (2) the difference between spending and taxes, $G_t - T_t$. This second term is called the **primary deficit** (or **primary surplus** if taxes exceed spending). Using this decomposition, we can rewrite equation (29.2) as

$$B_t - B_{t-1} = \underbrace{rB_{t-1}}_{\text{interest payments}} + \underbrace{G_t - T_t}_{\text{primary deficit}}$$

Moving B_{t-1} to the right-hand side of the equation and reorganizing yields

$$B_t = (1 + r)B_{t-1} + \underbrace{G_t - T_t}_{\text{primary deficit}} \qquad (29.3)$$

Debt at the end of year t is equal to $(1 + r)$ times debt at the end of year $t - 1$, plus the primary deficit, which is equal to $G_t - T_t$. This relation will prove very useful in what follows.

CURRENT VERSUS FUTURE TAXES

Let's now look at the implications of a one-year decrease in taxes for the path of debt and future taxes. Start from a situation where, until year 0, the government has balanced its budget, so that debt is equal to zero. In year 0, the government decreases taxes by $1 billion (1 for short in what follows) for one year. Thus, debt at the end of year 0, B_0, is equal to 1. What happens thereafter? Let's consider different cases.

Full repayment in year 1. Suppose that the government decides to repay the debt fully in year 1. From equation (29.3), the budget constraint in year 1 is given by

$$B_1 = (1 + r)B_0 + (G_1 - T_1)$$

If debt is fully repaid in year 1, then debt at the end of year 1 is equal to zero: $B_1 = 0$. Replacing B_0 by 1, and B_1 by 0, in the preceding equation gives

$$T_1 - G_1 = 1 + r$$

To repay the debt fully in year 1, the government must therefore run a primary surplus equal to $(1 + r)$ billion dollars. It can do so in one of two ways: through a decrease in spending or through an increase in taxes. We shall assume here and in what follows that the adjustment comes through taxes, so that the path of spending is unaffected. It follows that the decrease in taxes by 1 below normal in year 0 must be offset by an increase in taxes by $(1 + r)$ billion dollars above normal in year 1. The path of taxes and debt corresponding to this case is given in Figure 29-1(a) (assuming a value for r of 10%). The blue bars represent the deviation of taxes from their initial level, and the black bars represent the level of debt.

Repayment after t years. Now suppose that the government instead waits t years to increase taxes and repay the debt. Thus, until year t, the primary deficit is equal to zero. Let's work out what this implies for the level of debt at the beginning of year t (equivalently, the end of year $t - 1$).

In year 1, the primary deficit is equal to zero. Thus, from equation (29.3), debt at the end of year 1 is equal to

$$B_1 = (1 + r)B_0 + 0 = 1 + r$$

where the final equality follows from our original assumption that $B_0 = 1$. In year 2, with the primary deficit still equal to zero, debt at the end of the year is given by

$$B_2 = (1 + r)B_1 + 0 = (1 + r)(1 + r) = (1 + r)^2$$

Solving for debt in year 3 and so on, it is clear that, as long as the government keeps a primary deficit equal to zero, debt grows at a rate equal to the interest rate, and thus debt at the end of year $t - 1$ is given by

$$B_{t-1} = (1 + r)^{t-1} \tag{29.4}$$

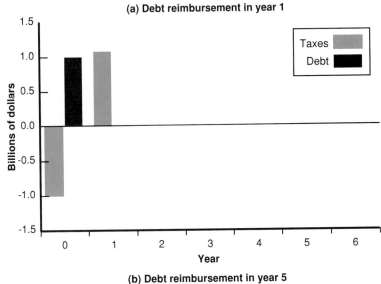

(a) Debt reimbursement in year 1

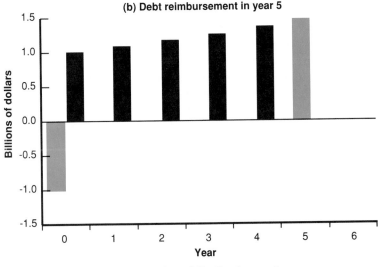

(b) Debt reimbursement in year 5

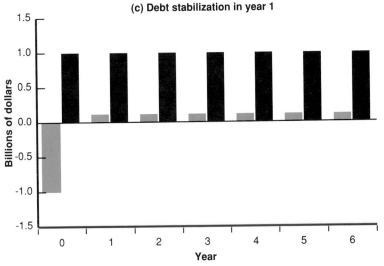

(c) Debt stabilization in year 1

FIGURE 29-1
Tax Cuts, Debt Repayment, and Debt Stabilization
(a) If debt is fully repaid in year 1, the decrease in taxes of 1 in year 0 requires an increase in taxes equal to $(1 + r)$ in year 1. (b) If debt is fully repaid in year t, the decrease in taxes of 1 in year 0 requires an increase in taxes equal to $(1 + r)^t$ in year t. (c) If debt is stabilized from year 1 on, then taxes must be permanently higher by r from year 1 on.

Despite the fact that taxes are lower only in year 0, debt steadily increases from year 0 on, at a rate equal to the interest rate. The reason is simple: While the primary deficit is equal to zero, debt is now positive, and so are interest payments on the debt. Each year, the government must issue more debt to pay the interest on existing debt.

In year t, the year in which the government decides to repay the debt, the budget constraint is given by

$$B_t = (1+r)B_{t-1} + (G_t - T_t)$$

If debt is fully repaid in year t, then B_t (debt at the end of year t) is equal to zero. Replacing B_t by zero, and B_{t-1} by its expression from equation (29.4), gives

$$0 = (1+r)(1+r)^{t-1} + (G_t - T_t)$$

Reorganizing and bringing $G_t - T_t$ to the left-hand side (and keeping in mind that one has to add exponents—see Appendix 3 at the end of the book) implies that

$$T_t - G_t = (1+r)^t$$

To pay back the debt, the government must run a primary surplus equal to $(1+r)^t$ billion dollars. If the adjustment is done through taxes, the initial decrease in taxes of 1 in year 0 leads, t years later, to a one-year increase in taxes of $(1+r)^t$. The path of taxes and debt corresponding to this case is given in Figure 29-1(b).

This example yields our first basic but important conclusion. If spending is unchanged, a decrease in taxes must eventually be offset by an increase in taxes in the future. The longer the government waits to increase taxes or the higher the real interest rate, the higher will be the eventual increase in taxes.

DEBT AND PRIMARY SURPLUSES

We have assumed so far that the government fully repays the debt. Let's now look at what happens to taxes if the government only *stabilizes* the debt—takes measures so that debt remains constant.

Suppose that the government decides to stabilize the debt from year 1 on. From equation (29.3), the budget constraint in year 1 is given by

$$B_1 = (1+r)B_0 + (G_1 - T_1)$$

Stabilizing the debt means that debt at the end of year 1 is equal to debt at the end of year 0, that $B_1 = B_0 = 1$. Replacing in the preceding equation gives

$$1 = (1+r) + (G_1 - T_1)$$

Reorganizing, and bringing $G_1 - T_1$ to the left-hand side, we have

$$T_1 - G_1 = (1+r) - 1 = r$$

To avoid a further increase in debt in year 1, the government must run a primary surplus equal to real interest payments on the existing debt. It must do so in following years as well: Each year, the primary surplus must be sufficient to cover interest payments, and thus leave the debt level unchanged. The path of taxes and debt is shown in Figure 29-1(c): Debt remains at 1 from year 1 on; taxes are permanently higher from year 1 on, by an amount equal to r.

The logic of this argument extends straightforwardly to the case where the government waits instead t years to stabilize. Whenever the government stabilizes, it must from then on run a primary surplus sufficient to pay interest on the debt.

This example yields our second conclusion. The legacy of past deficits is higher debt. To stabilize the debt, the government must eliminate the deficit. To do so, it must run a primary surplus equal to the interest payments on the existing debt.

THE EVOLUTION OF THE DEBT-TO-GDP RATIO

We have focussed so far on the evolution of the level of debt. But in an economy in which output grows over time it makes more sense to focus instead on the ratio of debt to GDP. To see how this change in focus modifies our conclusions, let us go from equation (29.3) to an equation that gives the evolution of the **debt-to-GDP ratio**—the **debt ratio,** for short. To do this, a few manipulations are needed.

Divide both sides of equation (29.3) by real output, Y_t, to get

$$\frac{B_t}{Y_t} = (1+r)\frac{B_{t-1}}{Y_t} + \frac{G_t - T_t}{Y_t}$$

Rewrite part of the first term on the right, B_{t-1}/Y_t, as $(B_t/Y_{t-1})(Y_{t-1}/Y_t)$ (in other words, multiply top and bottom by Y_{t-1}). The relation becomes

$$\frac{B_t}{Y_t} = (1+r)\left(\frac{Y_{t-1}}{Y_t}\right)\frac{B_{t-1}}{Y_{t-1}} + \frac{G_t - T_t}{Y_t}$$

We are nearly where we want to be; all the terms are now in terms of ratios to GDP. However, we can simplify a bit further. Denote the growth rate of output by g, so that Y_{t-1}/Y_t can be written as $1/(1+g)$. And use the approximation $(1+r)/(1+g) = 1 + r - g$.[1] We can then rewrite the equation as

$$\frac{B_t}{Y_t} = (1+r-g)\frac{B_{t-1}}{Y_{t-1}} + \frac{G_t - T_t}{Y_t}$$

Finally, move B_{t-1}/Y_{t-1} to the left-hand side of the equation to get

$$\frac{B_t}{Y_t} - \frac{B_{t-1}}{Y_{t-1}} = (r-g)\frac{B_{t-1}}{Y_{t-1}} + \frac{G_t - T_t}{Y_t} \qquad (29.5)$$

The change in the debt ratio is equal to the sum of two terms. The first is the difference between the real interest rate and the growth rate times the initial debt ratio. The second is the ratio of the primary deficit to GDP.

Compare equation (29.5), which gives the evolution of the ratio of debt to GDP, with equation (29.2), which gives the evolution of debt itself. The difference is

---

[1] *See Proposition 6 in Appendix 3.*

the presence of $r - g$ in equation (29.5) compared with r in the equation (29.2). The reason for the difference is simple. Suppose that the primary deficit is zero. Debt will then increase at a rate equal to the real interest rate, r. But if GDP is growing as well, the ratio of debt to GDP will grow more slowly; it will grow at a rate equal to $r - g$, the real interest rate minus the growth rate of output.[2]

The evolution of the debt ratio in the OECD. Equation (29.5) implies that the increase in the debt ratio will be larger the higher the real interest rate, or the lower the growth rate, or the higher the initial debt ratio, or the higher the ratio of the primary deficit to GDP. As such, it provides a useful guide to the evolution of the debt-to-GDP ratio over the last three decades in the OECD countries.

The 1960s was a decade of strong growth in rich countries, so strong that the average growth rate exceeded the average real interest rate in most countries. As a result, $r - g$ was actually negative, and most countries were able to decrease their debt ratios without having to run large primary surpluses.

The 1970s was a period of slower growth, but of very low (often negative) real interest rates. Thus, $r - g$ was again negative on average, and the result was a further decrease in the debt ratio in most OECD countries.

The situation has changed drastically since the early 1980s. Real interest rates have increased and growth rates have decreased. Thus, to avoid an increase in their debt ratios, rich countries would have had to run large primary surpluses. Most have not, and, as Table 29-1 shows, the result has been a large increase in those countries' debt ratios.

The first line of Table 29-1 shows that the debt-to-GDP ratio for the OECD as a whole increased by almost 25% of GDP between 1978 and 1997. And the next four lines show how large the increase has been in some countries. Canada, Italy, Belgium, and Greece have all had increases in their debt ratios of more than 50% of GDP; in 1997, the last three had debt ratios in excess of 100%. All four countries are now trying to decrease these debt ratios slowly. To do so, in 1997 all were running large primary surpluses, as the first column shows.

....
TABLE 29-1
DEBT AND PRIMARY SURPLUSES IN THE OECD AS A WHOLE, AND FOR SELECTED COUNTRIES, 1978–1997 (PERCENT OF GDP)

COUNTRY	PRIMARY SURPLUS/GDP 1997 (MINUS SIGN: DEFICIT)	DEBT/GDP 1978	DEBT/GDP 1997
OECD	1.4	20.3	44.6
Canada	5.4	11.6	66.7
Italy	5.2	57.4	109.2
Belgium	5.1	58.2	118.8
Greece	5.0	24.5	107.3

"Debt" is net debt—that is, financial liabilities of the government minus financial assets held by the government.

Source: OECD Economic Outlook, December 1993 and December 1997,. Tables 32, 34, and 35.

_____

[2]*If two variables (here debt and GDP) grow at rates* r *and* g *respectively, then their ratio (here the ratio of debt to GDP) will grow at rate* (r − g). *See Proposition 8 in Appendix 3.*

29-2 FOUR ISSUES IN FISCAL POLICY

Having looked at the mechanics of the government budget constraint, we can now take up four issues in which this constraint plays an important role.

RICARDIAN EQUIVALENCE

How does taking into account the government budget constraint affect the way we should think of the effects of deficits on output?

One view is that once this constraint is taken into account, neither deficits nor debt have any effect on economic activity! This argument is known as the **Ricardian equivalence** proposition. Its name comes from David Ricardo, a nineteenth-century English economist, who was the first to articulate its logic. But the argument was further developed and given prominence in the 1970s by Robert Barro from Harvard University. For this reason, the argument is also often known as the **Ricardo-Barro proposition.**

The best way to understand the proposition's logic is to use the first example of tax changes from Section 29-1. Suppose that the government decreases taxes by 1 this year. Also suppose that as it does so it announces that, in order to repay the debt, it will increase taxes by $1 + r$ next year.

What will be the effect of the initial tax cut on consumption? A plausible answer is: no effect at all. Why? Because consumers realize that the tax cut is not much of a gift: Lower taxes this year are exactly offset by higher taxes next year. Indeed, their human wealth—the present value of after-tax labour income—is unaffected.[3] Current taxes go down by 1, but the present value of next year's taxes goes up by $(1 + r)/(1 + r) = 1$, and the net effect of the two changes is exactly equal to zero.

We can look at the same result another way, by looking at saving rather than consumption. To say that consumers do not change consumption in response to the tax cut is the same as saying that *private saving increases one for one with the deficit.* Thus the Ricardian equivalence proposition says that, if a government finances a given path of spending through deficits, private saving will increase one for one with the decrease in public saving, leaving total saving unchanged. As a result, the total amount left for investment will not be affected. Over time, the mechanics of the government budget constraint imply that government debt will increase. But this increase will not come at the expense of capital accumulation.

Under the Ricardian equivalence proposition, the long sequence of deficits and the increase in OECD debt since the 1980s are no cause for worry. As governments were dissaving, the argument goes, people were saving more in anticipation of the higher taxes to come. The decrease in public saving was offset by an equal increase in private saving. Total saving was therefore unchanged, and so was investment. These economies have the same capital stock today that they would have had if there had been no increase in debt. High debt is thus no cause for concern.

How seriously should one take the Ricardian equivalence proposition? Most economists would answer: "Seriously, but not seriously enough to think that deficits and debt are irrelevant." A major theme of this book has been that expectations matter, that consumption decisions depend not only on current income but also on

---

[3] *See Chapter 8 for a definition of human wealth and a discussion of its role in consumption.*

future income. Indeed, if it were widely believed that a tax cut this year is going to be followed by an offsetting increase in taxes *next year,* the effect on consumption would probably be small. Many consumers would save most or all of the tax cut in anticipation of higher taxes next year. (Replace "year" by "month" or "week" and the argument becomes even more convincing.)

However, tax cuts rarely come with the announcement of tax increases a year later. Consumers have to guess when and how taxes will eventually be increased. This fact does not by itself invalidate the Ricardian equivalence argument. No matter when taxes will be increased, the government budget constraint still implies that the present value of future tax increases must always be equal to the decrease in taxes today. Take, for example, the second example we looked at in Section 29-1—drawn in Figure 29-1(b)—in which the government waits t years to increase taxes, and thus increases them by $(1 + r)^t$. The present value in year 0 of this expected tax increase is equal to $(1 + r)^t/(1 + r)^t = 1$—exactly equal to the original tax cut, so the change in human wealth from the tax cut is still equal to zero.

But insofar as future tax increases appear more distant and their timing more uncertain, consumers are more likely to ignore them. This may be the case because they expect to die before taxes go up, or, more likely, because they just do not think that far into the future. In either case, Ricardian equivalence is likely to fail.

There is little evidence that government deficits are offset by increases in private spending. Figure 29-2 plots Canadian private and public saving for the years 1950–1996. *Public saving* is defined as the sum of federal, provincial, and local budget surpluses (a negative value indicates a deficit). The figure shows that the deficits in the 1970s and early 1980s were matched by increases in private saving. Outside of this episode, however, both public and private saving have tended to move together. The figure shows that the large increase in deficits since the early 1980s has been associated for the most part with a *decrease* rather than an increase in private saving. Such a figure must clearly be accompanied by a warning. Other factors may have been at work that would hide a positive relation between deficits and private saving. Nonetheless, sophisticated empirical research has not so far been able to find much positive response of private saving to the higher deficits over the period.

Thus, a safe conclusion is that deficits have an important effect on activity. In the short run, larger deficits are likely to lead to higher demand and higher output. In the long run, however, higher government debt lowers capital accumulation and thus lowers output.

DEFICITS, OUTPUT STABILIZATION, AND THE CYCLICALLY ADJUSTED DEFICIT

The fact that deficits have long-run adverse effects on capital accumulation and output does not imply that deficits should not be used for output stabilization. Rather, it implies that deficits during recessions should be offset by surpluses during booms, so as not to lead to a steady increase in debt.[4]

To help assess whether fiscal policy is on track, economists have constructed deficit measures that tell them what the deficit would be, under the existing tax and

---

[4]*For simplicity, we shall ignore output growth in this section, and thus ignore the difference between stabilizing the debt and stabilizing the debt-to-GDP ratio. The arguments extend straightforwardly to the case where output is growing.*

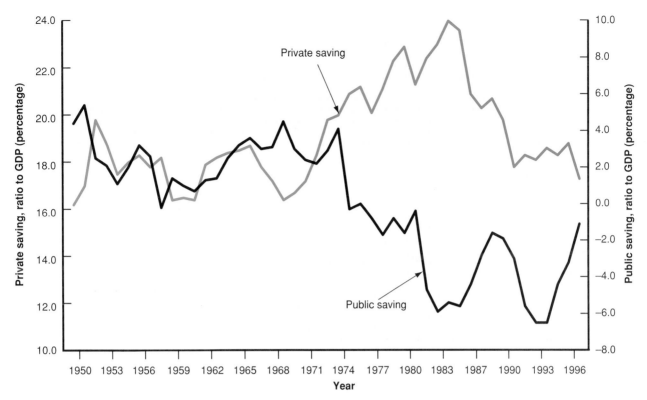

FIGURE 29-2
Canadian Private and Public Saving, Ratios to GDP (Percentage), 1950–1996
There is little evidence that increased public dissaving since the early 1980s has been off-set by increased private saving.

Source: Statistics Canada, CANSIM Matrix 6705.

spending rules, if output were at its natural level. Such measures come under many names, from **full-employment deficit,** to **midcycle deficit,** to **standardized employment deficit,** to **structural deficit** (the term used by the OECD). We shall use **cyclically adjusted deficit,** the term we find the most intuitive.

Such a measure gives a simple benchmark by which to judge the direction of fiscal policy. If the actual deficit is large but the cyclically adjusted deficit is equal to zero, then current fiscal policy is consistent with no systematic increase in debt over time. Debt will increase as long as output is below its natural level; but as output returns to its natural level, deficits will eventually disappear and the debt will stabilize.

This does not imply that the goal of policy should be to maintain a cyclically adjusted deficit equal to zero at all times. In a recession, the government may want to run a deficit large enough that even the cyclically adjusted deficit is positive. In that case, the fact that the cyclically adjusted deficit is positive provides a warning that the return of output to its natural level will not be enough to stabilize the debt, a warning that the government will have to take specific measures to decrease the deficit at some point in the future.

The theory underlying the cyclically adjusted deficit is a simple one. The practice has proven more tricky. To see why, we need to look at how measures of the cyclically adjusted deficit are constructed. Their construction requires two steps.

The first is to establish how much lower the deficit would be if output were, say, 1% higher. The second is to assess how far away output is from its natural level.

The first step is relatively straightforward. As a rule of thumb, a 1% decrease in output leads automatically to a deficit increase of 0.5% of GDP. This decrease occurs because most taxes are proportional to output, while most government spending does not depend on the level of output. Thus a decrease in output, which leads to a decrease in revenues and not much change in spending, naturally leads to a larger deficit. If output is, say, 5% below its natural level, the deficit as a ratio to GDP will be about 2.5% larger than it would be if output were at its natural level. (This effect of activity on the deficit has been called an **automatic stabilizer:** A recession naturally generates a deficit, and thus a fiscal expansion that partly counteracts the recession.)

The second step has proven more difficult. Recall from Chapter 15 that the natural level of output is the output level that would be produced if the economy were operating at the natural rate of unemployment. Too low an estimate of the natural rate of unemployment will lead to too high an estimate of the natural level of output, and thus to too optimistic a measure of the cyclically adjusted deficit. This phenomenon explains in part what happened in Europe in the 1980s. Based on the assumption of a constant natural unemployment rate, the cyclically adjusted deficits did not look bad in the 1980s. If European unemployment had indeed returned to its level of the late 1970s, the output increase would have been sufficient to reestablish budget balance in most countries. But, as we saw in Chapter 1 and in more detail in Chapter 21, unemployment remained very high during the 1980s. As a result, most of the decade was characterized by high deficits and a large increase in debt-to-GDP ratios.

WARS AND DEFICITS

Wars typically bring about large budget deficits. As we saw in Chapter 27, the largest increases in Canadian government debt in the twentieth century occurred during World War II. We examine this case at greater length in the Focus box entitled "Deficits, Consumption, and Investment in Canada during World War II."

Is it right for governments to rely so much on deficits to finance wars? After all, war economies are usually operating at low unemployment, so the output stabilization reasons for running deficits we examined earlier are irrelevant. The answer, nevertheless, is yes. In fact, there are two good reasons to run deficits during wars. The first is distributional: Deficit finance is a way to pass some of the burden of the war to those alive after the war, and it seems only fair for future generations to share in the sacrifices the war requires. The second is more narrowly economic: Deficit spending helps reduce tax distortions. Let's look at each reason in turn.

Passing on the burden of the debt. Wars lead to large increases in government spending. Consider the implications of financing this increased spending either through increased taxes or through higher deficits. To distinguish this case from our earlier discussion of output stabilization, let's also assume that output is fixed at its natural level.

Suppose that the government relies on deficit finance. With government spending sharply up, there will be a very large increase in the demand for goods.

Given our assumption that output cannot increase, interest rates will have to increase enough so as to maintain equilibrium. Investment, which depends on the interest rate, will thus decrease sharply.

Suppose instead that the government finances the spending increase through an increase in taxes. Consumption will decline sharply. Exactly how much depends on consumers' expectations. The longer they expect the war to last, the more they will decrease consumption and the less they will decrease saving. In any case, the increase in government spending will be partly offset by a decrease in consumption. Interest rates will increase by less than they would have under deficit spending. Investment will thus decrease by less.

In short, for a given output, the increase in government spending requires either a decrease in consumption and/or a decrease in investment. Whether the government relies on tax increases or deficits determines whether consumption or investment does more of the adjustment.

How does all this affect who bears the burden of the war? The more the government relies on deficits, the smaller will be the decrease in consumption during the war and the larger the decrease in investment. Lower investment means a lower capital stock after the war, and thus lower output after the war. By reducing capital accumulation, deficits become a way of passing some of the burden of the war onto future generations.

FOCUS

DEFICITS, CONSUMPTION, AND INVESTMENT IN CANADA DURING WORLD WAR II

In 1938, the share of Canadian government spending on goods and services in GDP was 9.7%, and the federal government's share was just 2.7%. By 1944, the figures were 40.8% and 36.9% respectively. Not surprisingly, the bulk of the increase was due to increased spending on national defence, which went from 0.7% of GDP in 1938 to 35.6% in 1944.

Faced with such a massive increase in spending, the federal government reacted with large tax increases. For the first time in Canadian history, income taxes became a major source of revenue; income tax revenues which were less than 1% of GDP in 1938 increased to 7.2% in 1944. But the tax increase was still far less than the increase in expenditures. The increase in federal revenues, from 7.4% of GDP in 1938 to 35.6% in 1944, was only about two-thirds of the increase in expenditures.

The result was a sequence of large deficits. By 1944, the federal deficit reached 22% of GDP. The ratio of debt to GDP, already high at 65% in 1938 because of the deficits the government had run during the Great Depression, climbed to almost 150% by the end of the war.

Was the increase in government spending achieved at the expense of consumption or private investment? [In principle, it could have come from higher imports and a current account deficit (see Chapter 11). But Canada's trading partners were in no better shape than we were; net exports remained about 6% of GDP throughout the war.]

The 30% increase in the share of GDP going to government purchases was met in large part by a 19% drop in consumptions's share of GDP from 70.3% to 51.9%. Part of the decrease in consumption may have been in anticipation of higher taxes after the war; part was also the result of the unavailability of many consumer durables; and patriotism probably played a role in leading people to save more and buy the war bonds issued by the government to finance the war. But the increase in government purchases was also met by a 3% decrease in the share of (private) investment in GDP—a decrease from 10% to 7%. Thus, part of the burden of the war was indeed passed on in the form of lower capital accumulation to those living after the war.

Reducing tax distortions. There is another argument for running deficits not only during wars but, more generally, in times when government spending is exceptionally high. Think, for example, of reconstruction after an earthquake or of the costs involved in the reunification of Germany in the early 1990s.[5]

The argument is as follows: If the government were to increase taxes in line with the increase in spending, tax rates would have to be very high. Very high tax rates can lead to very high distortions. Faced with very high income tax rates, people work less or engage in illegal, untaxed activities. Thus, rather than moving the tax rate up and down so as to always balance the budget, it is better (from the point of view of reducing distortions) to maintain a relatively constant tax rate, to *smooth taxes*. **Tax smoothing** implies running large deficits when government spending is exceptionally high and small surpluses the rest of the time.

THE DANGERS OF VERY HIGH DEBT

High government debt leads to lower capital accumulation. It also means higher taxes, and thus higher distortions. Canada's recent experience with general government debt ratios above 100% points to yet another cost: High debt can lead to vicious circles and makes the conduct of fiscal policy extremely difficult.

To see why this is so, return to equation (29.5), which gives the evolution of the debt ratio:

$$\frac{B_t}{Y_t} - \frac{B_{t-1}}{Y_{t-1}} = (r - g)\frac{B_{t-1}}{Y_{t-1}} + \frac{G_t - T_t}{Y_t}$$

Take a country with a high debt ratio, say 100%. Suppose that the real interest rate is equal to 3%, the growth rate equal to 2%. The first term on the right-hand side of the equation is thus equal to (3% – 2%) times 100% = 1% of GDP. Suppose further that the government is running a primary surplus of 1%, thus just enough to keep the debt ratio constant [the right-hand side of the equation is equal to 1% + (–1%) = 0%].

Now suppose that financial investors start requiring a risk premium to hold domestic bonds. This may be because fear of Quebec separation increases political uncertainty, or for any other reason. Suppose further that the central bank wants to maintain the exchange rate and, to do so, has to increase the domestic interest rate from 3% to 6%.[6] Finally, suppose that the high interest rate triggers a recession and the growth rate plummets to 0%.

Now assess the fiscal situation: $r - g$ is now 6% – 0% = 6%. With the increase in $r - g$ from 1% to 6%, the government must increase its primary surplus from 1% to 6% of GDP just to keep the debt-to-GDP ratio constant. Now come the potential vicious circles.

Suppose that the government takes steps to avoid an increase in the debt ratio. The spending cuts or tax increases prove politically costly, generating even more political uncertainty and the need for an even higher interest rate to maintain the exchange rate. The sharp fiscal contraction leads to an even worse recession, further decreasing the growth rate. Both the increase in the interest rate and the decrease in growth further increase $r - g$, making it even harder to stabilize the debt ratio.

---

[5] See the In Depth box in Chapter 6.

[6] See Chapter 14 for a review of the underlying mechanisms.

Alternatively, suppose that the government proves unable or unwilling to increase the budget surplus by 5% of GDP. Debt increases, leading financial markets to become even more worried and require an even higher interest rate. There is strong pressure to finance the budget through money creation.[7] The prospect of potentially higher inflation from money creation further worries financial markets, requiring even higher interest rates to maintain the exchange rate, and so on.

These are not idle intellectual speculations. In Italy, corruption scandals and serious political crises since the early 1990s have led to attacks on the lira, requiring increases in interest rates and thus making it harder for the government to balance the budget. So far, the crises have been more limited than in the story we told above. But it is a reasonable forecast that more crises will come, and that they may indeed be worse.

There is therefore little question that high-debt countries should aim at reducing their debt ratios. How and how fast should debt ratios be reduced? The answer is: through many years, indeed many decades, of surpluses. The historical reference here is that of England in the nineteenth century. By the end of its war against Napoleon in the early 1800s, England had run up a debt ratio in excess of 200% of GDP. It spent most of the nineteenth century reducing the ratio, so that by 1900 it stood at only 30% of GDP.

The prospect of many decades of fiscal austerity is unpleasant. Thus, when debt ratios are very high, an alternative solution keeps coming up, that of **debt repudiation.** The argument is a simple one. Repudiating the debt—that is, cancelling it in part or in full—is good for the economy. It allows for a decrease in taxes and thus a decrease in distortions. It decreases the risk of vicious circles. The problem with repudiation is the problem of time inconsistency we studied in Chapter 27. If the government reneges on its promises to repay the debt, it may find it very difficult to borrow again for a long time in the future; financial markets will remember what happened and be reluctant to lend again. Thus, what seems best today may in fact be unappealing in the long run. Debt repudiation is very much a last resort, and is probably not the solution even for high-debt countries such as Italy and Belgium.

SUMMARY

■ The government budget constraint gives the evolution of government debt as a function of spending and taxes. One way of expressing the constraint is that the change in debt (the deficit) is equal to the primary deficit plus interest payments on the debt. The primary deficit is the difference between government spending on goods and services, G, and taxes net of transfers, T.

■ If government spending is unchanged, a decrease in taxes must eventually be offset by an increase in taxes in the future. The longer the government waits to increase taxes or the higher the real interest rate, the higher the eventual increase in taxes.

■ The legacy of past deficits is higher debt. To stabilize the debt, the government must eliminate the deficit. To do so, it must run a primary surplus equal to the interest payments on the existing debt.

■ Under the Ricardian equivalence proposition, a larger deficit is offset by an equal increase in private saving. Thus deficits have no effect on demand and output. The accumulation of debt does not affect capital accumulation. When Ricardian equivalence fails, larger deficits lead to higher demand and higher output in the short run. The accumulation of debt leads to lower capital accumulation, and thus to lower output in the long run.

[7]See Chapter 21.

■ To stabilize the economy, the government should run deficits during recessions, and surpluses during booms. The cyclically adjusted deficit tells what the deficit would be, under existing tax and spending rules, if output were at its natural level.

■ Deficits are justified in times of high spending, such as wars. Compared with tax finance, they lead to higher consumption and lower investment during wars. They there-fore shift some of the burden of the war from people living during the war to those living after the war. They also help smooth taxes and reduce tax distortions.

■ Canada and a number of European countries have very high debt-to-GDP ratios. In addition to reducing capital and requiring higher taxes and thus tax distortions, high debt ratios increase the risk of fiscal crisis.

KEY TERMS

■ inflation-adjusted deficit, 568
■ government budget constraint, 568
■ primary deficit (surplus), 569
■ debt-to-GDP ratio, or debt ratio, 573
■ Ricardian equivalence, or Ricardo-Barro proposition, 575

■ full-employment deficit, midcycle deficit, standardized employment deficit, structural deficit, or cyclically adjusted deficit, 577
■ automatic stabilizer, 578
■ tax smoothing, 580
■ debt repudiation, 581

QUESTIONS AND PROBLEMS

1. Consider the following data for an economy (all figures as a percentage of GDP):

Tax revenues/GDP	15%
Expenditures/GDP (including transfers, excluding interest payments)	20%
Nominal interest rate	10%
Annual inflation rate	4%
Annual growth rate of GDP	0%
Debt-to-GDP ratio (at beginning of year)	50%
Unemployment rate	7%

Give all answers to the following as a fraction of GDP.
   a. What is the official deficit?
   b. What is the correct or "inflation-adjusted" deficit?
   c. What is the primary deficit?
   d. By how much will the real national debt increase over the year?

2. Using the data in problem 1, suppose that the natural rate of unemployment is 7.5%, Okun's coefficient is 0.4, and the deficit responds to changes in GDP according to the rule of thumb given in the text. What is the value of the *official* cyclically adjusted deficit (as a fraction of GDP)? The inflation-adjusted and cyclically adjusted deficit?

3. Suppose that a nation's debt, inflation rate, and GDP growth rate are all equal to zero, and its interest rate is 5%. In year t, the nation runs a deficit equal to 10% of GDP, and then eliminates the *primary* deficit in year $t + 1$ and afterward. Calculate the deficit (as a fraction of GDP) in years $t + 1$ and $t + 2$.

4. Have you ever thought about taxes you might have to pay in future years when deciding how much of your paycheque to spend? Do you think you are typical or atypical in this respect? What do your answers have to do with the effects of fiscal policy and the Ricardian equivalence proposition?

5. "A deficit during a war can be a good thing. First, the deficit is temporary, so after it's over the government can go right back to its old level of spending and its old level of taxes. Second, since the evidence supports the Ricardian equivalence proposition, the deficit should help stimulate the economy during wartime, helping to keep the unemployment rate low." Identify four distinct mistakes in this statement. Is anything in this statement correct?

FURTHER READING

The modern statement of the Ricardian equivalence proposition is Robert Barro, "Are Government Bonds Net Wealth?" *Journal of Political Economy,* December 1974, 1095–1117.

THE STORY OF MACROECONOMICS

We have spent the previous 29 chapters presenting the core of macro-economics—the framework that most economists use to think about macroeconomic issues and the major conclusions they draw, as well as the issues on which they disagree. How this core has been built over time is a fascinating story. And this is the story we want to tell in this chapter.

30-1 KEYNES AND THE GREAT DEPRESSION

John Maynard Keynes

The history of modern macroeconomics starts in 1936, with the publication of Keynes's *General Theory of Employment, Interest and Money.* As he was writing the *General Theory,* Keynes confided to a friend:[1] "I believe myself to be writing a book on economic theory which will largely revolutionise—not I suppose at once but in the course of the next ten years, the way the world thinks about economic problems."

Keynes's prediction was right. The book's timing was surely one of the reasons for its immediate success. The Great Depression was not only an economic catastrophe, but also an intellectual failure for the economists working

<hr>

[1]*The friend was the writer George Bernard Shaw. J. M. Keynes,* Collected Writings, *Volume 13 (New York: Macmillan Press, 1973), 492.*

on **business cycle theory**—as macroeconomics was then called. Few economists had a coherent explanation for the Depression, for either its depth or its length. The economic measures taken by the Roosevelt administration in the New Deal had been based on instinct rather than on economic theory. The *General Theory* offered an interpretation of events, an intellectual framework, and a clear argument for government intervention.

The centrepiece of the *General Theory* was the emphasis on **effective demand**—what we now call aggregate demand. In the short run, Keynes argued, effective demand determines output. Even if output eventually returns to its natural level in the long run, the process is a slow one at best. Indeed, one of Keynes's most famous quotes is: "In the long run, we are all dead."

In the process of deriving effective demand, Keynes introduced many of the building blocks of modern macroeconomics: the multiplier, which explains how shocks to demand can be amplified and lead to larger shifts in output; the notion of **liquidity preference** (the term Keynes gave to the demand for money), which explains how monetary policy can affect interest rates and effective demand; the importance of expectations in affecting consumption and investment; and the idea that *animal spirits* (shifts in expectations) are a major factor behind shifts in demand and output.

Finally, the *General Theory* was much more than just a treatise for economists. It offered clear policy implications, and they were in tune with the times. Waiting for the economy to return by itself to its natural level was irresponsible. In the midst of a depression, trying to balance the budget was not only stupid, it was dangerous. Active use of fiscal policy was essential to return the country to high employment.

30-2 THE NEOCLASSICAL SYNTHESIS

Within a few years, the *General Theory* had transformed macroeconomics. Not everybody was converted, and few agreed with it all. But most discussions became organized around it.

By the early 1950s a large consensus had emerged, based on an integration of many of Keynes's ideas and the ideas of earlier economists. This consensus was called the **neoclassical synthesis.** To quote from Paul Samuelson, in the 1955 edition of his textbook, *Economics* (New York: McGraw-Hill)—the first modern economics textbook:

> *In recent years, 90 per cent of American economists have stopped being "Keynesian economists" or "Anti-Keynesian economists." Instead, they have worked toward a synthesis of whatever is valuable in older economics and in modern theories of income determination. The result might be called neo-classical economics and is accepted, in its broad outlines, by all but about five per cent of extreme left-wing and right-wing writers.*

The neoclassical synthesis was to remain the dominant view for another 20 years. Progress was astonishing, and the period from the early 1940s to the early 1970s can be called the golden age of macroeconomics.

PROGRESS ON ALL FRONTS

The first order of business after publication of the *General Theory* was to formalize mathematically what Keynes meant. While Keynes knew mathematics, he had avoided using math in the *General Theory*. One unfortunate result was endless controversies about what Keynes meant and whether there were logical flaws in some of his arguments.

Franco Modigliani

The *IS-LM* model. A number of formalizations of Keynes's ideas were offered. The most influential one was the *IS-LM* model, developed by John Hicks and Alvin Hansen in the 1930s and early 1940s. The initial version of the *IS-LM* model—which was actually very close to the one presented in Chapter 6 of this book—was criticized for emasculating many of Keynes's insights: Expectations played no role, and the adjustment of prices and wages was altogether absent. Yet the *IS-LM* model provided a basis from which to start building, and as such it was immensely successful. Discussions became organized around the slopes of the *IS* and *LM* curves, what variables were missing from the two relations, what equations for prices and wages should be added to the model, and so on.

Theories of consumption, investment, and money demand. Keynes had emphasized the importance of consumption and investment behaviour, and of the choice between money and other financial assets. Major progress was soon made along all three fronts.

In the 1950s, Franco Modigliani (then at Carnegie Mellon and now at MIT) and Milton Friedman (at the University of Chicago) independently developed the theory of consumption. Both insisted on the importance of expectations in determining current consumption decisions.

James Tobin, from Yale, developed the theory of investment, based on the relation between the present value of profits and investment. The theory was further developed and tested by Dale Jorgenson, from Harvard.

Tobin also developed the theory of the demand for money, and more generally the theory of the choice between different assets based on liquidity, return, and risk. His work has become the basis not only for an improved treatment of financial markets in macroeconomics, but also for the theory of finance in general.

James Tobin

Growth theory. In parallel with the work on short-run fluctuations, there was a renewed focus on growth. In contrast with the stagnation in the pre-World War II era, most countries were growing fast in the 1950s and 1960s. Even if they experienced fluctuations, their standard of living was increasing rapidly. The growth model developed by MIT's Robert Solow in 1956, which we saw in Chapter 23, provided a framework to think about the determinants of growth. It was followed by an explosion of work on the roles of saving and technological progress in growth.

Macroeconometric models. All these contributions were integrated into larger and larger macroeconometric models. The first U.S. macroeconometric model, developed by Lawrence Klein from the University of Pennsylvania in the early 1950s, was an extended *IS* relation with 16 equations. With the availability of better data and the development of econometrics and computers, the models quickly grew in size. The most important effort was the construction of the MPS model (MPS stands for MIT-Penn-SSRC, for the two universities and the research institution in-

Robert Solow

volved in its construction), developed during the 1960s by a group of people led by Franco Modigliani. Its structure was still the *IS-LM* model writ large, plus a Phillips curve mechanism. But its components—consumption, investment, and money demand—all reflected the tremendous theoretical and empirical progress made since Keynes.

KEYNESIANS VERSUS MONETARISTS

With such rapid progress, many macroeconomists came to believe that the future was bright. The nature of deviations of output from its natural level was steadily better understood; the development of models allowed for the sophisticated use of policy. The time when the economy could be fine-tuned, and recessions all but eliminated, seemed not far in the future.

This optimism was met with skepticism by a small but influential minority, the **monetarists.** Their intellectual leader was Milton Friedman. While Friedman saw much progress being made—and indeed was himself the father of one of the major contributions, the theory of consumption—he did not share in the general enthusiasm. He believed that the understanding of the economy remained very limited. He questioned the motives of governments as well as the notion that they actually knew enough to improve macroeconomic outcomes.

In the 1960s, debates between "Keynesians" and "monetarists" dominated the economic headlines. The debates centred around three issues: the effectiveness of monetary versus fiscal policy, the Phillips curve, and the role of policy.

Monetary versus fiscal policy. Keynes had emphasized *fiscal* rather than *monetary* policy as the key to fighting recessions. And this had remained the prevailing wisdom. The *IS* curve, many argued, was quite steep; changes in the interest rate had little effect on demand and output. Thus, monetary policy did not work very well. Fiscal policy, which affects demand directly, could affect output faster and more reliably.

Friedman strongly challenged this conclusion. In a 1963 book titled *A Monetary History of the United States, 1867–1960,* Friedman and Anna Schwartz painstakingly reviewed the evidence on monetary policy and the relation between money and output in the United States over a century. Their conclusion was not only that monetary policy was very powerful, but that it could explain most of the fluctuations in output. They interpreted the Great Depression in the United States as the result of a tragic mistake in monetary policy, a decrease in the money supply due to bank failures—a decrease that the Fed could have avoided by increasing the monetary base, but had not.

Friedman and Schwartz's challenge was followed by a vigorous debate and by intense research on the respective effects of fiscal and monetary policy. In the end, a consensus was reached. The truth was in between: Both fiscal and monetary policies clearly had effects. And if one cared about the composition of output and took into account the openness of the economy, the best policy was typically a mix of the two.

The Phillips curve. The second debate focussed on the Phillips curve. The Phillips curve was not part of the initial Keynesian model. But, because it provided such a convenient (and apparently reliable) way of explaining the movements of wages and prices over time, it had become part of the neoclassical synthesis. In the

Milton Friedman

1960s, based on the empirical evidence up to then, many Keynesian economists believed that there was a reliable tradeoff between unemployment and inflation, even in the long run.

Milton Friedman and Edmund Phelps (from Columbia University) strongly disagreed. They argued that the existence of such a long-run tradeoff flew in the face of basic economic theory. They argued that the apparent tradeoff would quickly vanish if policy makers actually tried to exploit it—that is, if they tried to achieve low unemployment by accepting higher inflation. As we saw in Chapter 17 when we studied the evolution of the Phillips curve, Friedman and Phelps were most definitely right. By the mid-1970s, the consensus was indeed that there was no long-run tradeoff between inflation and unemployment.

Edmund Phelps

The role of policy.   The third debate centred on the role of policy. Much less certain that economists knew enough to stabilize output and that policy makers could be trusted to do the right thing, Friedman argued for the use of simple rules, such as steady money growth. Here is what he said in 1958:[2]

> A steady rate of growth in the money supply will not mean perfect stability even though it would prevent the kind of wide fluctuations that we have experienced from time to time in the past. It is tempting to try to go farther and to use monetary changes to offset other factors making for expansion and contraction. . . . The available evidence casts grave doubts on the possibility of producing any fine adjustments in economic activity by fine adjustments in monetary policy—at least in the present state of knowledge. There are thus serious limitations to the possibility of a discretionary monetary policy and much danger that such a policy may make matters worse rather than better.
>
> Political pressures to "do something" in the face of either relatively mild price rises or relatively mild price and employment declines are clearly very strong indeed in the existing state of public attitudes. The main moral to be drawn from the two preceding points is that yielding to these pressures may frequently do more harm than good.

As we saw in Chapter 27, this debate has not been settled. The nature of the arguments has changed somewhat, but they are still with us today.

30-3   THE RATIONAL-EXPECTATIONS CRITIQUE

Despite the battles between Keynesians and monetarists, macroeconomics around 1970 looked like a successful and mature field. It appeared successful at explaining events, at guiding policy choices. Most debates were framed within a common intellectual framework. But within a few years the field was in crisis. The crisis had two sources.

The first was events. By the mid-1970s, most countries were experiencing *stagflation,* a new word coined at the time to denote the simultaneous existence of

---

[2]Milton Friedman, *"The Supply of Money and Changes in Prices and Output,"* testimony to Congress, 1958.

Robert Lucas

high unemployment and high inflation. Macroeconomists had not predicted stagflation. After the fact and after a few years of research, a convincing explanation was provided, based on the effects of adverse supply shocks on both prices and output. But it was too late to undo the damage to the discipline's image.

The second was ideas. In the early 1970s, a small group of economists—Robert Lucas from Chicago; Thomas Sargent, then from Minnesota and now at Chicago; and Robert Barro, then from Chicago and now at Harvard—led a strong attack against mainstream macroeconomics. They did not mince words. In a 1978 paper, Lucas and Sargent stated:[3]

> *That the predictions [of Keynesian economics] were wildly incorrect, and that the doctrine on which they were based was fundamentally flawed, are now simple matters of fact, involving no subtleties in economic theory. The task which faces contemporary students of the business cycle is that of sorting through the wreckage, determining what features of that remarkable intellectual event called the Keynesian Revolution can be salvaged and put to good use, and which others must be discarded.*

THE THREE IMPLICATIONS OF RATIONAL EXPECTATIONS

Lucas and Sargent's main argument was that Keynesian economics had ignored the full implications of the effect of expectations on behaviour. The way to proceed, they argued, was to assume that people formed expectations as rationally as they could, based on the information they had. Thinking of people as having *rational expectations* had three major implications, all highly damaging to Keynesian macroeconomics.

Thomas Sargent

The Lucas critique.   The first implication was that existing macroeconomic models could not be used to help design policy. While these models recognized that expectations affect behaviour, they did not incorporate expectations explicitly. All variables were assumed to be functions of current and past values of other variables, including policy variables. Thus, what the models captured was the set of relations between economic variables as they had held in the past, under past policies. Were these policies to be changed, Lucas argued, the way people formed expectations would change as well, making estimated relations—and, by implication, simulations generated using existing macroeconometric models—poor guides to what would happen under these new policies. This critique of macroeconometric models became known as the **Lucas critique.** To take again the history of the Phillips curve as an example, the data up to the early 1970s had suggested a tradeoff between unemployment and inflation. As policy makers tried to exploit that tradeoff, it disappeared.

Rational expectations and the Phillips curve.   The second implication was as follows. When rational expectations were introduced in Keynesian models, these models actually delivered the very un-Keynesian conclusion that deviations of output from its natural level were short-lived, much more so than Keynesian economists claimed. This argument was based on a reexamination of aggregate supply.

Robert Barro

---

[3]*"After Keynesian Economics,"* in After the Phillips Curve: Persistence of High Inflation and High Unemployment *(Boston: Federal Reserve Bank of Boston, 1978).*

In Keynesian models, the slow return of output to its natural level came from the slow adjustment of prices and wages through the Phillips curve mechanism. An increase in money, for example, led first to higher output and to lower unemployment. Lower unemployment then led to higher nominal wages and to higher prices. The adjustment continued until wages and prices had increased in the same proportion as nominal money, until unemployment and output were back at their natural levels.

This adjustment, Lucas pointed out, was highly dependent on wage setters' backward-looking expectations of inflation. In the MPS model, for example, wages responded only to current and past inflation, and to current unemployment. But once one made the assumption that wage setters had rational expectations the adjustment was likely to be much faster. Indeed, changes in money, to the extent that they were anticipated, might have no effect on output. For example, anticipating an increase in money of 5% over the coming year, wage setters would increase the nominal wages set in contracts for the coming year by 5%. Firms would in turn increase prices by 5%. The result would be no change in the real money supply, and thus no change in demand or output.

Within the logic of the Keynesian models, Lucas therefore argued, only *unanticipated changes in money* should affect output. Predictable movements in money should have no effect on activity. More generally, if wage setters had rational expectations, shifts in aggregate demand were likely to have effects on output for only as long as wages were set in nominal terms, a year or so. Thus, even on its own terms, the Keynesian model did not deliver a convincing theory of the long-lasting effects of demand on output.

Optimal control versus game theory. The third implication of rational expectations was as follows. If people and firms had rational expectations, it was wrong to think of policy as the control of a complicated but passive system. Rather, the right way was to think of policy as a game between policy makers and the economy. The right tool was thus not *optimal control,* but rather *game theory.* And game theory led to a different vision of policy. A striking example was the issue of *time inconsistency* discussed by Finn Kydland and Edward Prescott (then at Carnegie Mellon, now at the University of Minnesota), an issue that we discussed in Chapter 27: Good intentions on the part of policy makers could actually lead to disaster.

In summary: When rational expectations were introduced, (1) Keynesian models could not be used to determine policy, (2) Keynesian models could not explain long-lasting deviations of output from its natural level, and (3) the theory of policy needed to be redesigned, using the tools of game theory.

THE INTEGRATION OF RATIONAL EXPECTATIONS

As you might guess from the tone of Lucas and Sargent's quote, the intellectual atmosphere in macroeconomics was tense in the early 1970s. But within a few years a process of integration (of ideas, not people, because tempers remained high) had started, and it was to dominate the 1970s and the 1980s.

Fairly quickly, the idea that rational expectations was the right working assumption gained wide acceptance. This did not happen because all macroeconomists believe that people, firms, and participants in financial markets always form

Rudiger Dornbusch

Stanley Fischer

John Taylor

expectations rationally. But rational expectations appears to be a natural benchmark, at least until economists have made more progress in understanding whether and how actual expectations systematically differ from rational expectations.

Work then started on the challenges raised by Lucas and Sargent.

The implications of rational expectations.   First, there was a systematic exploration of the role and the implications of rational expectations in goods, financial, and labour markets. Much of what was discovered has been presented in this book already. Thus we shall limit ourselves to two examples here:

■ Robert Hall, then from MIT and now at Stanford, showed that if consumers are very foresighted (in the sense we defined in Chapter 8), then changes in consumption will be unpredictable. The best forecast of consumption next year will be consumption this year! This result came as a surprise to most macroeconomists at the time, but it is in fact based on a simple intuition: If consumers are very foresighted, they will change their consumption only when they learn something new about the future; but, by definition, such news cannot be predicted. This consumption behaviour, known as the **random walk** of consumption, has served as a benchmark in consumption research ever since.

■ Rudiger Dornbusch from MIT showed that the large swings in exchange rates under flexible exchange rates, which had previously been thought of as the result of speculation by irrational investors, were in fact fully consistent with rationality. We saw this result in Chapter 14. Changes in monetary policy can lead to long-lasting changes in interest rates; changes in current and expected interest rate differentials between two countries can lead in turn to large changes in the exchange rate. Dornbusch's model, known as the *overshooting* model of exchange rates, has become the benchmark in discussions of exchange rate movements.

Wage and price setting.   Second, there was a systematic exploration of the determination of wages and prices, going far beyond the Phillips curve relation. Two important contributions were made by MIT's Stanley Fischer and John Taylor, then from Columbia University and now at Stanford. Both showed that the adjustment of prices and wages in response to changes in unemployment can be slow *even under rational expectations.*

They pointed to an important characteristic of both wage and price setting, the **staggering of wage and price decisions.** In contrast with the simple story we told earlier, where all wages and prices increased simultaneously in anticipation of an increase in money, actual wage and price decisions are staggered over time. Thus there is not one sudden synchronized adjustment of all wages and prices to an increase in money. Rather, the adjustment is likely to be a slow one in which wages and prices adjust, through a process of leapfrogging, to the new level of money, but do so only over time. Fischer and Taylor thus showed that the second issue raised by the rational-expectations critique could be resolved, that a slow return of output to its natural level is in fact consistent with rational expectations in the labour market.

The theory of policy.   Third, thinking about policy in terms of game theory led to an explosion of research on the nature of the games being played, not only between policy makers and the economy but also between policy makers—between political parties, or between the central bank and the government, or between the governments of different countries. One of the major achievements of this research has been the development of a way of thinking more rigorously about such fuzzy notions as "credibility," "reputation," and "commitment." At the same time, there has been a distinct shift in focus from "what governments should do" to "what governments actually do," and thus a focus on the political constraints that economists should take into account when advising policy makers.

In summary: By the end of the 1980s, the challenges raised by the rational-expectations critique had led to an overhaul of macroeconomics. The basic structure had been extended to take into account the implications of rational expectations—or, more generally, of forward-looking behaviour by people and firms. Indeed, what we have presented in this book is what we see as the synthesis that has emerged and that now constitutes the core of macroeconomics. In the last section of this chapter, we shall summarize what we see as the core and its basic propositions. But before we do so, we want to turn briefly to current research. Much of it is still too speculative to have made it into the core, but no doubt some of it will.

30-4 CURRENT DEVELOPMENTS

Today, three groups dominate the research headlines: the new classicals, the new Keynesians, and the new growth theorists. (Note the generous use of the word "new." Unlike producers of laundry detergents, economists stop short of using "new and improved." But the subliminal message is the same.)

NEW CLASSICAL ECONOMICS AND REAL BUSINESS CYCLE THEORY

The rational-expectations critique was more than just a critique of Keynesian economics. It also offered its own interpretation of fluctuations. Instead of relying on imperfections in labour markets, on the slow adjustment of wages and prices, and so on, to explain fluctuations, Lucas argued, macroeconomists should see how far they could go in explaining fluctuations as the effects of shocks in competitive markets with fully flexible prices and wages.

This is the research agenda that has been pursued by the **new classicals.** The intellectual leader is Edward Prescott, and the models he and his followers have developed are known as **real business cycle (RBC) models.** These models assume that output is always at its natural level. Thus, all fluctuations in output are movements of the natural level of output, as opposed to movements away from the natural level of output.

Where do these movements come from? The answer proposed by Prescott is technological progress. As new discoveries are made, productivity increases, leading to an increase in output. The increase in productivity leads to an increase in the wage, which makes it more attractive to work and thus leads workers to work more. Productivity increases therefore lead to increases in both output and employment, as we indeed observe in the real world.

The RBC approach has been criticized on many fronts. As we discussed in Chapter 24, technological progress is the result of very many innovations, each of which takes a long time to diffuse. It is hard to see how this process could generate anything like the large short-run fluctuations in output that we observe in practice.

Edward Prescott

It is also hard to think of recessions as times of technological *regress,* times in which productivity and output both go down. Finally, as we have seen, there is very strong evidence that changes in money, which have no effect on output in RBC models, in fact have strong effects on output in the real world.

Thus, at this point, most economists do not believe that the RBC approach provides a convincing explanation of major fluctuations in output. The approach has nevertheless proven useful. It has drilled in the correct point that not all fluctuations in output are deviations of output from its natural level. At a more technical level, it has provided a number of new techniques for building stochastic dynamic general equilibrium (SDGE) models, which are widely used in research today. It is likely to evolve rather than disappear. Already the label seems misleading, as some recent RBC models have introduced nominal rigidities (which we discussed in Chapter 18), thus allowing for the effects of money on output.

New Keynesian Economics

The term **new Keynesians** denotes a loosely connected group of researchers who share a common belief that the synthesis that has emerged in response to the rational-expectations critique is basically correct. But they also share the belief that much remains to be learned about the nature of imperfections in different markets, and about the implications of those imperfections for macroeconomic evolutions.

One line of research has focussed on the determination of wages in the labour market. We discussed in Chapter 15 the notion of *efficiency wages*—the idea that wages, if perceived by workers as too low, may lead to shirking, problems of morale within the firm, difficulties in recruiting or keeping good workers, and so on. One influential researcher in this area has been George Akerlof from Berkeley, who has explored the role of "norms," the rules that develop in any organization—in this case, the firm—to assess what is fair or unfair. This research has led him and others to explore issues previously left to research in sociology and psychology, and to examine their macroeconomic implications.

Another line of new Keynesian research has explored the role of imperfections in credit markets. Except for a few remarks and a box in Chapter 20, we have assumed that the effects of monetary policy work through interest rates, and that firms and people can borrow freely at the quoted interest rate. In practice, most people and many firms can borrow only from banks. And banks often turn down potential borrowers, despite their willingness to pay the posted interest rate. Why this happens, and how it affects our view of how monetary policy works, has been the subject of much research, in particular by Ben Bernanke of Princeton.

Yet another direction of research is **nominal rigidities.** As we saw earlier in this chapter, Fischer and Taylor have shown that, with staggering of wage or price decisions, output can deviate from its natural level for a long time. This conclusion raises a number of issues. If staggering is indeed responsible, at least in part, for fluctuations, why don't wage setters/price setters synchronize decisions? Why aren't prices and wages adjusted more often? Why aren't all prices and wages changed, say, on the first of each month? In tackling these issues, Akerlof and N. Gregory Mankiw (from Harvard University) have derived a surprising and important result, often referred to as the **menu cost** explanation of output fluctuations.

In the menu cost explanation, each wage or price setter is largely indifferent as

George Akerlof

to when and how often he changes his own wage or price (for a retailer, changing the prices on the shelf every day or every week does not make much difference to profits). Thus, even small costs of changing prices—such as those involved in printing a new menu, for example—may lead to infrequent and staggered price adjustment. This staggering leads in turn to slow adjustment of the price level, and thus to large aggregate output fluctuations in response to movements in aggregate demand. In short, decisions that do not matter much at the individual level (how often to change prices or wages) lead to large aggregate effects (slow adjustment of the price level, and thus large effects of shifts in aggregate demand on output).

NEW GROWTH THEORY

After being one of the most active topics of research in the 1960s, growth theory went into an intellectual slump. Since the mid-1980s, however, growth theory has made a strong comeback. The set of new contributions goes under the name of **new growth theory.**

Two economists, Robert Lucas (the same Lucas who spearheaded the rational-expectations critique) and Paul Romer (from Berkeley), have played an important role in defining the issues. When growth theory faded in the late 1960s, two issues were left largely unresolved. The first was the determinants of technological progress. The second was the role of increasing returns to scale—whether, say, doubling capital and labour may actually lead to more than a doubling of output. These are the two major issues on which new growth theory has concentrated. The discussions of technological progress in Chapter 24, and of the interaction between technological progress and unemployment in Chapter 25, reflect some of the advances that economists have made on this front. The work of Alwyn Young (from Boston University) on growth in fast-growing Asian countries, which we discussed in the In Depth box in Chapter 24, is a good example of this new research.

30-5 THE CORE

As we come to the end of this book, let us state what we see as the basic set of propositions that define the core of macroeconomics today.

- In the short run, shifts in aggregate demand affect output. Higher consumer confidence, a larger budget deficit, and faster growth of money are all likely to increase output and employment, and thus to decrease unemployment.

- Expectations play a major role in determining the behaviour of the economy. How people and firms respond to a change in policy determines the size and even sometimes the direction of the economy's response to the change.

- In the long run, output returns to its natural level. This natural level depends on the natural rate of unemployment (which, together with the size of the labour force, determines the level of employment), on the capital stock, and on the state of technology.

- Monetary policy affects output in the short and medium run, but not in the long run. A higher rate of money growth eventually translates one for one into a higher rate of inflation.

■ Fiscal policy has both short- and long-run effects on output. Higher deficits are likely to increase activity in the short run. However, they are likely to decrease capital accumulation and output in the long run.

Clearly, these propositions leave a lot of room for disagreement.

The major area of disagreement is in the length of the "short run," the period of time over which aggregate demand affects output. At one extreme, real business cycle theorists start from the assumption that output is always at its natural level: The short run is very short indeed. At the other, theories of hysteresis in unemployment (which we explored in Chapter 20) imply that the effects of demand may be extremely long-lasting, that the short run may in fact be very long.

The other area of disagreement is in the role for policy. While conceptually distinct, it is largely related to the first. Those who believe that output returns quickly to its natural level are typically willing to impose tight rules on both monetary and fiscal policy, from constant money growth to the requirement of a balanced budget. Those who believe that the adjustment is slow typically believe in the need for more flexible stabilization policies.

Despite these disagreements, this core provides a framework in which to conduct and organize research. More important, it provides a framework to interpret events and discuss policy. This is what we have done in this book.

SUMMARY

■ The history of modern macroeconomics starts in 1936, with the publication of Keynes's *General Theory of Employment, Interest and Money*. Keynes's contribution was formalized in the *IS-LM* model by John Hicks and Alvin Hansen in the 1930s and early 1940s.

■ The period from the early 1940s to the early 1970s can be called the golden age of macroeconomics. Among the major developments were the development of the theories of consumption, investment, money demand, and portfolio choice; the development of growth theory; and the development of large macroeconometric models.

■ The main debate during the 1960s was between Keynesians and monetarists. Keynesians believed that developments in macroeconomic theory allowed for better and better control of the economy. Monetarists, led by Milton Friedman, were more skeptical of the ability of governments to help stabilize the economy.

■ In the 1970s, macroeconomics experienced a crisis. There were two reasons. The first was the appearance of stagflation, which came as a surprise to most economists. The second was a theoretical attack led by Robert Lucas. Lucas and his followers showed that when rational expectations were introduced (1) Keynesian models could not

be used to determine policy, (2) Keynesian models could not explain long-lasting deviations of output from its natural level, and (3) the theory of policy needed to be redesigned, using the tools of game theory.

■ Much of the 1970s and 1980s was spent integrating rational expectations into macroeconomics. As is reflected in this book, macroeconomists are now much more aware of the role of expectations in determining the effects of shocks and policy, and of the complexity of policy, than they were two decades ago.

■ Current research in macroeconomic theory is proceeding along three lines. New classical economists are exploring the extent to which fluctuations can be explained as movements in the natural level of output, as opposed to movements away from the natural level of output. New Keynesian economists are exploring more formally the role of market imperfections in fluctuations. New growth theorists are exploring the role of R&D and of increasing returns to scale in growth.

■ Despite the differences, there exists a core of macroeconomic beliefs to which most macroeconomists adhere. The two basic propositions are: In the short run, shifts in aggregate demand affect output. In the long run, output returns to its natural level.

KEY TERMS

- business cycle theory, 611
- effective demand, 611
- liquidity preference, 611
- neoclassical synthesis, 611
- monetarists, 613
- Lucas critique, 615
- random walk, 617

- staggering of wage and price decisions, 617
- new classicals, 618
- real business cycle (RBC) models, 618
- new Keynesians, 619
- nominal rigidities, 619
- menu cost, 619
- new growth theory, 620

QUESTIONS AND PROBLEMS

1. What role do you think political beliefs—particularly one's belief that government's role in our lives should be relatively limited or relatively expansive—have played in the development of different macroeconomic theories?

2. Of the various macroeconomic theories presented in this chapter, which seemed to arise primarily from *events*, and which primarily from *ideas?*

3. Can the schools of thought discussed in this chapter be classified as either "optimistic" or "pessimistic"? Explain.

4. Look up the change in real GDP, the unemployment rate, and the inflation rate for Canada over the most recent two years. How would this change be explained by a proponent of:
   a. Real business cycle theory?
   b. New Keynesian theory?

5. In which areas do you think macroeconomists have a particularly good understanding of the basic mechanisms at work? A particularly poor understanding of the basic mechanisms at work?

FURTHER READING

The two classics are J. M. Keynes, *The General Theory of Employment, Money and Interest* (London: Macmillan Press, 1936), and Milton Friedman and Anna Schwartz, *A Monetary History of the United States, 1867–1960* (Princeton, NJ: Princeton University Press, 1963). Be warned: The first makes for hard reading, and the second is a heavy volume.

In the introduction to *Studies in Business Cycle Theory* (Cambridge, MA: MIT Press, 1981), Robert Lucas develops his approach to macroeconomics and gives a guide to his contributions.

The paper that launched real business cycle theory is Edward Prescott, "Theory Ahead of Business Cycle Measurement," *Federal Reserve Bank of Minneapolis Review,* Fall 1986, 9–22. A recent survey is by George Stadler, "Real Business Cycles," *Journal of Economic Literature,* December 1994, 1750–1783. Neither of these papers makes for easy reading.

For more on new Keynesian economics, read David Romer, "The New Keynesian Synthesis," *Journal of Economic Perspectives,* Winter 1993, 5–22, or N. Gregory Mankiw, "A Quick Refresher Course in Macroeconomics," *Journal of Economic Literature,* December 1990.

For more on new growth theory, read Paul Romer, "The Origins of Endogenous Growth," *Journal of Economic Perspectives,* Winter 1994, 3–22.

In a lighter mode, for an extremely well-written set of essays on many economists and their ideas, read David Warsh, *Economic Principals: Masters and Mavericks of Modern Economics* (New York: Free Press, 1993).

More about macroeconomic issues and theory

Most economics journals are heavy on mathematics and are hard to read. But a few make an effort to be more friendly. The *Journal of Economic Perspectives* in particular has nontechnical articles on current economic research and issues.

Canadian Public Policy is a very readable journal that tries to present economic perspectives on policy for non-economists.

The *Brookings Papers on Economic Activity,* published twice a year, analyze current U.S. macroeconomic problems. So does *Economic Policy,* published in Europe, which focusses more on European issues.

More advanced treatments of current macroeconomic theory—roughly at the level of a first graduate course in macroeconomics—are given by David Romer, *Advanced Macroeconomics* (New York: McGraw-Hill, 1995), and by Olivier Blanchard and Stanley Fischer, *Lectures on Macroeconomics* (Cambridge, MA: MIT Press, 1989).

APPENDICES

Appendix 1: Where to Find the Numbers

The purpose of this appendix is to help you find numbers, from inflation in Canada last quarter, to consumption in the United States in 1952, to youth unemployment in Ireland in the 1980s, and so on. Although it is evolving rapidly, we decided to emphasize resources that can be accessed via the World Wide Web.

FOR A QUICK LOOK AT CURRENT NUMBERS

■ The best source for the most recent numbers on production, unemployment, inflation, exchange rates, interest rates, and stock prices for a large number of countries is the last four pages of *The Economist,* published every week.

■ A good summary of current developments in the Canadian economy is provided by the *Canadian Economic Observer*, published each month by Statistics Canada. If you can't wait, the information is released every day as it becomes available in *The Daily* on their Website.

FOR MORE DETAIL ABOUT THE CANADIAN ECONOMY

■ Statistics Canada provides a number of sources. Their Web page (**www.statcan.ca**) provides free access to all the main economic indicators for the past few years. In addition, you can also download various guides detailing the construction of the CPI, the Labour Force Survey, etc.

The main database of economic time series available from Statistics Canada is called CAN-SIM. Finding what you want in CANSIM is not easy. Related series are grouped in what are called *CANSIM matrices*, but there is very little relationship between adjacent matrices, so you'll need some patience (and a good search engine!). Access to the CANSIM database is on a subscription basis only (ask if your University has such access). CANSIM gives yearly data as far back as 1926 and quarterly data as far back as 1947 on hundreds of thousands of economic time series.

■ The Department of Finance publishes *Economic Reference Tables,* which gives annual data on the main economic and financial aggregates over the past few years. Their Website (**www.fin.gc.ca**) has a number of useful (and free) publications including a detailed description of the federal budget and monthly statements about federal government revenues and expenditures *(The Fiscal Monitor). The Economy in Brief*, published quarterly, gives a succinct summary of major developments in the Canadian economy. The various provincial ministries also make similar information available on-line.

■ On a quarterly basis, the Bank of Canada publishes *Bank of Canada Review*. In addition to two or three articles related to monetary policy, it gives data at various frequencies on financial variables and monetary aggregates (the most recent data are reported by day or week; older data are reported by month or year). Their Website (**www.bank-banque-canada.ca**) provides access to much of this data in *Weekly Financial Statistics*. In addition, there is now a small library of historical data series that can be downloaded, as well as *Bank of Canada Working Papers* and articles on monetary policy.

■ Several of the large chartered banks provide daily and weekly summaries of developments in financial markets, including current term structures for Canada and the United States. A few also provide current values of the main economic aggregates. Read their explanations of what is currently happening and what they expect to occur in the near future in financial markets. Ask yourself how it compares with what you have learned in this class.

■ Recent data is much easier to obtain than earlier data. A good source for the latter is F. H. Leacy (ed.), *Historical Statistics of Canada*, 1983.

FOR MORE DETAIL ABOUT THE U.S. ECONOMY

■ Unlike Canada, the United States has no central statistical agency. A guide to the more than 70 agencies than produce statistical data is available at **www.fedstats.gov**.

■ For a detailed presentation of the most recent numbers, look at *Survey of Current Business*, published monthly by the U.S. Department of Commerce, Bureau of Economic Analysis.

■ The U.S. Federal Reserve System (**www.bog. frb.us**) has excellent data on financial and monetary variables. The regional banks also provide on-line access to data. A particularly useful source is the Federal Reserve Bank of St. Louis's FRED (Federal Reserve Economic Data) database, **www.stls.frb.org**.

■ Once a year, *Economic Report of the President*, written by the Council of Economic Advisers and published by the U.S. government Printing Office in Washington, gives a description of current evolutions, as well as numbers for most major macroeconomic variables, often going back to the 1950s. (**www.access.gpo.gov**)

■ The authoritative source for statistics going back as far as data have been collected is *Historical Statistics of the United States, Colonial Times to 1970*, Parts 1 and 2, published by the U.S. De-

partment of Commerce, Bureau of the Census. 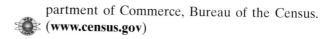 (**www.census.gov**)

NUMBERS FOR OTHER COUNTRIES

The OECD, located in Paris, publishes three extremely useful publications that give data for most of the rich countries in the world (see the Focus box in Chapter 1 for the list). (**www.oecd.org**)

■ The first OECD publication is *OECD Economic Outlook*, published twice a year. In addition to describing current macroeconomic issues, it includes data for many macroeconomic variables. The data typically go back to the 1970s, and are reported consistently both across time and across countries.

■ The second is *OECD Employment Outlook*, published annually. It focusses more specifically on labour-market issues and numbers.

■ Occasionally, the OECD puts together current and past data, and publishes *OECD Historical Statistics*. At this time, the most recent is *Historical Statistics, 1960–1993*, published in 1995.

The main strength of the publications of the International Monetary Fund (IMF), located in Washington, DC, is that they cover most of the countries of the world. (**www.imf.org**)

The IMF issues four particularly useful publications:

■ *International Financial Statistics* (IFS), published monthly. It has data for member countries, usually going back a few years, mostly on financial variables, but also on some aggregate variables (such as GDP, employment, and inflation).

■ *International Financial Statistics Yearbook*, published annually. It has the same coverage of countries and variables as the IFS, but gives annual data going back up to 30 years.

■ *Government Finance Statistics Yearbook*, published annually, gives data on each country's budget, typically going back 10 years. (Because of delays in the construction of the numbers, data for the most recent years are often unavailable.)

■ *World Economic Outlook*, published twice a year, describes major evolutions in the world and in specific member countries.

Many Eastern European countries are not yet included in IMF publications (although most of them are now members of the IMF). Two good sources of data for these countries are:

■ *Short-term Indicators: Central and Eastern Europe*, published by the OECD, quarterly.

■ The *Annual Transition Report*, published by the European Bank for Reconstruction and Development (EBRD) in London. The EBRD was set up in 1990 to help finance investments in Central and Eastern Europe.

For long-term historical statistics for a number of countries, a precious new data source is *Monitoring the World Economy, 1820–1992*, Development Centre Studies, OECD, Paris, 1995. This study gives data going back to 1820 for 56 countries.

Finally, check your department's Web page. It will probably have links to various resources (data, journals, working papers, etc.) available via the Web.

Appendix 2: An Introduction to National Income and Expenditure Accounts

The purpose of this appendix is to introduce the basic structure as well as the terms used in the national income and expenditure accounts. The basic measure of aggregate activity is gross domestic product, or GDP. The **national income and expenditure accounts (**or simply **national accounts)** are organized around two decompositions of GDP. The first looks at *income:* Who receives what? The other looks at *expenditures:* What is produced, and who buys it?

THE INCOME SIDE

Table A-1 looks at the income side of GDP, at who receives what. The top half of the table (lines 1 to 10) goes from GDP to national income, the sum of the incomes received by the different factors of production.

■ The starting point, in line 1, is **gross domestic product,** or **GDP.** It is defined as *the market value of the final goods and services produced by labour and property located in Canada.* (Alternatively, it is the value of the income *produced* in Canada.)

■ The next two lines take us from GDP to **GNP**, the **gross national product** (line 3). GNP is an alternative measure of aggregate output. It is defined as the market value of the final goods and services produced by labour and property supplied by Canadian residents (alternatively, it is the income *accruing* to Canadian residents).

For many years, most countries used GNP rather than GDP as the main measure of aggregate activity. The emphasis in the Canadian national accounts shifted from GNP to GDP in 1986. The difference between the two comes from the distinction between "located in Canada" (used to define GDP) and "supplied by Canadian residents" (used to define GNP). For example, profit from a Canadian-owned plant in the United States is not included in Canadian GDP, but is included in Canadian GNP. Thus, to go from GDP to GNP, we must first add the difference between factor payments received from foreign residents and factor payments made to foreign residents. In the national accounts this is called **net investment income received from non-**

Source: *Statistics Canada, CANSIM Matrices 6547 and 2360.*

TABLE A-1
GDP: THE INCOME SIDE, 1996 (MILLIONS OF DOLLARS)

From gross domestic product to net domestic income at factor cost:	
1 Gross domestic product (GDP) at market prices	820 323
2　Add: Net investment income received from nonresidents	−27 705
3 Equals: Gross national product (GNP) at market prices	792 618
4　Deduct: Capital consumption allowances	105 935
5 Equals: Net national product (NNP) at market prices	686 683
6　Deduct: Indirect taxes less subsidies	109 944
7　Deduct: Statistical discrepancy	211
8 Equals: Net national income at factor cost	576 528
9　Deduct: Net investment income received from nonresidents	−27 705
10 Equals: Net domestic income at factor cost	604 233
The decomposition of (net) domestic income at factor cost:	
11 Wages, salaries, and supplementary labour income	429 601
12 Corporate profits before taxes	67 988
13 Government business enterprise profits before taxes	6 476
14 Interest and miscellaneous investment income	48 789
15 Accrued net income of farm operators from farm production	3 457
16 Net income of non-farm unincorporated businesses including rents	49 491
17 Inventory valuation adjustment	−1 569

residents and appears on line 2 (net factor payments to labour are small and are ignored). In 1996, payments to the rest of the world exceeded receipts from the rest of the world by about $27 billion, so GNP was smaller than GDP by the same amount.

■ The next step takes us from GNP to **net national product**, or **NNP** (line 5). The difference between GNP and NNP is the depreciation of capital, which is called "capital consumption allowances" in the national accounts. Statistics Canada estimates the economic depreciation of capital from capital depreciation reported by companies on their tax forms.

■ Lines 6 to 8 take us from NNP to **net national income** at factor cost (line 8). This gives us the value of income that is received by labour and capital supplied by Canadian residents.

The main step in going from NNP to national income is subtracting indirect taxes (or sales taxes) less subsidies (line 6). Indirect taxes decrease the amount left for factor income, while subsidies increase it.

National income is actually constructed in two independent ways. One is from the top down, starting from GDP constructed from the expenditure side and going through the steps we have just gone through in Table A-1. The other way is from the bottom up, by adding the different components of factor income (wages and salaries, profits, and so on). The two measures typically differ and the difference is split between the two estimates and called the **statistical discrepancy**. In 1996, national income computed from the top down exceeded national income computed from the bottom up by $422 billion, so half this amount, $211 billion, was subtracted, before arriving at line 8.

■ Finally, to look at the components of income, we need to add back the net payments to foreign factors and arrive at **net domestic income at factor cost.**

Various keywords appear in the national accounts often enough that it pays to give them a brief discussion. *Factor cost* measures prices before indirect taxes (GST and other sales taxes) are added on to obtain *market prices*. Aggregates that have been adjusted for depreciation are given the modifier *net*; otherwise they are called *gross*. Aggregates that refer to Canadian residents are called *national*; if an aggregate refers to something that took place in Canada it is called *domestic*. Using these distinctions, you can construct your own aggregates rather easily (or at least quickly see how to get from some published number to what you want). For example, we can compute gross domestic income at market prices by starting with line 8 and adding back lines 4 and 6.

Statistics Canada produces monthly estimates of GDP at factor cost, including the contributions by industry, with a lag of about 60 days. The monthly estimates are based on a number of clever tricks to proxy for unavailable data. Although useful, they tend to be rather noisy. Estimates of GDP at market prices (both income and expenditure accounts) are available on a quarterly and on an annual basis. The quarterly estimates arrive with a lag of about two months, and are updated regularly for several years as better data become available. Some historical data are also available on an annual basis, but in December 1997 Statistics Canada announced that it would shift its emphasis to its quarterly series and many annual series would no longer be updated.

The bottom half of Table A-1 (lines 11 to 17) decomposes domestic income into different types of income:

■ Wages, salaries, and supplementary labour income (line 11) is by far the largest component, accounting for 71% of domestic income.

■ Corporate profits before taxes (line 12) is a highly volatile component of domestic income. In 1996, it accounted for almost 11%.

■ Interest and miscellaneous investment income (line 14) is the interest paid by firms minus the interest received by firms, plus interest received from the rest of the world. In 1996, this component was approximately equal to 8% of domestic income.

■ The income of farm operators (line 15) is now a very small component of domestic income. It is measured on an *accrued* basis (rather than a *cash flow* basis). That is, the value of farm production is counted in the year in which it is produced, rather than the year in which it is sold.

■ The income of other small businesses (line 16) accounted for about 8% of domestic income. Rental income includes the rents received by landlords and an *imputed* value for owner-occupied housing. Owner-occupied housing generates a flow of services. To capture the value of this flow in domestic income, Statistics Canada estimates how much homeowners would have to pay themselves to rent their own homes and adds this into line 16.

■ Finally, the inventory valuation adjustment (line 17) measures capital gains (or losses) from business inventories. These show up in corporate profits, but because they do not reflect economic production they are subtracted before arriving at domestic income.

Table A-2 shows how to construct **personal disposable income**, the income available to consumers after they have received transfers and paid taxes.

■ Personal income, the sum of income earned by households and transfers received from the other sectors in the economy, totalled $680 billion in 1996. Many of the components of personal income are already familiar from our discussion of Table A-1. Wages, salaries, and supplementary labour income (line 2) is by far the largest component of personal income. Unincorporated business net income (line 3) is just the sum of lines 15 and 16 in Table A-1. Line 4 in Table A-2 is obtained by adding the value of the dividends received from corporations to line 14 in Table A-1. In addition to their earnings from supplying labour and capital, households also receive transfers such as

employment insurance benefits. In 1996, transfers amounted to 16% of personal income.

■ From personal income, we subtract direct taxes (income taxes) and other transfers to government (some small things including fines).

■ The net result is **personal disposable income** (line 9), the amount of income households have to spend after controlling for taxes and transfers. In 1996, personal disposable income was $518 billion, or about 63% of GDP.

THE EXPENDITURE SIDE

Table A-3 looks at the expenditure side of the national accounts, at who buys what. Let us start with the three components of domestic demand: consumption, investment, and government spending.

■ Consumption, called personal expenditure on consumer goods and services (line 2), is by far the largest component of demand, accounting for 58% of GDP. It is defined as *the sum of goods and services purchased by persons resident in Canada.*

In the same way as they include rental income on the income side, the national accounts include imputed housing services as part of consumption. Owners of a house are assumed to consume housing services, for a price equal to the imputed rental income.

Consumption is decomposed into four components: purchases of durable goods such as cars (line 3), semi-durable goods such as shoes (line 4), non-durable goods such as strawberries (line 5), and services (line 6). Non-durable goods and services account for just over three-quarters of total consumption.

■ Investment is called business investment in fixed capital. It is the sum of two very different components: (1) residential construction (line 8) is the purchase of new houses or apartments by persons; (2) non-residential construction (line 9) and machinery and equipment (line 10) are the investment decisions of firms related to plant and equipment.

■ Government purchases of goods and services (line 11) are mostly current expenditures, but

....

TABLE A-2

PERSONAL INCOME AND PERSONAL DISPOSABLE INCOME, 1996 (MILLIONS OF DOLLARS)

1	Personal income	680 412
2	Wages, salaries and supplementary labour income	429 601
3	Unincorporated business income	52 948
4	Interest, dividends and miscellaneous investment receipts	89 085
5	Transfers from government, corporations, and nonresidents	108 778
6	Less: Income taxes and other transfers to government	162 245
7	Equals: Personal disposable income	518 167

Source: Statistics Canada, CANSIM Matrix 6524.

GDP: THE EXPENDITURE SIDE, 1996 (MILLIONS OF DOLLARS)

1	Gross domestic product	820 323
2	Personal expenditure on consumer goods and services	477 927
3	Durable goods	58 500
4	Semi-durable goods	43 117
5	Non-durable goods	118 720
6	Services	257 590
7	Business investment in fixed capital	121 907
8	Residential construction	40 083
9	Nonresidential construction	35 437
10	Machinery and equipment	46 387
11	Government purchases	188 234
12	Net exports	31 420
13	Exports of goods and services	320 739
14	Imports of goods and services	289 319
15	Changes in business inventories	1 045
16	Statistical discrepancy	−210

Source: Statistics Canada, CANSIM Matrix 6548.

also include government investment (e.g., buildings) and a very small component for changes in government inventories. They do *not* include transfers to persons (such as employment insurance or Canada Pension Plan payments) or interest on the public debt. In 1996, government purchases accounted for almost 23% of GDP.

■ The sum of consumption, investment, and government purchases gives the demand for goods by Canadian firms, Canadian persons, and Canadian governments. If Canada were a closed economy, this would be the same as the demand for Canadian goods. But because Canada is open the two numbers are different. To get to the demand for Canadian goods, we must first add the foreign purchases of domestic goods, exports (line 13). Second, we must subtract Canadian purchases of foreign goods, imports (line 14). In 1996, exports exceeded imports by just over $31 billion. Thus net exports (or, equivalently, the **trade balance**) was equal to $31 billion (line 12).

■ Adding consumption, investment, government purchases, and net exports gives the total purchases of Canadian goods. However, production may be less than those purchases if firms satisfy the difference by decreasing inventories; or production may be greater than purchases, in which case firms accumulate inventories. Thus line 15 in Table A-3 gives changes in business inventories (line 15), also called (rather misleadingly) "business investment in inventories." This term is defined as the value of the change in the physical volume of inventories held by business. The change in business inventories can be positive or negative. In 1996, it was positive: Canadian production exceeded total purchases of Canadian goods by $1.05 billion. Although relatively small, the change in business inventories is one of the most cyclically sensitive components of aggregate expenditure.

A WARNING

National accounts give an internally consistent description of aggregate activity. But underlying these accounts are many choices of what to include and what not to include, where to put some types of income or spending, and so on. Here are three examples:

■ Work within the home is not counted in GDP. Thus, to take an extreme example, if two women decide to babysit each other's child rather than take care of their own child and pay each other for the babysitting services, measured GDP will go up, while true GDP clearly does not change. The solution would be to count work within the home in GDP, in the same way that we impute rents for owner-occupied housing. But so far this has not been done.

■ The purchase of a house is treated as an investment, and housing services are then treated as part of consumption. Contrast this with the treatment of automobiles. Despite the fact that they provide services for a long time—although not as long a time as houses do—purchases of automobiles are not treated as investment. They are treated as consumption and appear in the national accounts only in the year in which they are bought.

■ Physical investment and education are treated

asymmetrically. Firms' purchases of machines are treated as investment. The purchase of education is treated as consumption of education services. But education is clearly in part an investment: People acquire it in part to increase their future income.

The list goes on. However, the purpose of these examples is not to make you conclude that national accounts are wrong. Most of the choices we just saw were made for good reasons, often because of data availability or for simplicity of treatment. Rather, the point is that, to use the national accounts best, you should understand not only their logic, but also their choices and thus their limitations.

KEY TERMS

....

- national income and expenditure accounts, or national accounts, A4
- gross domestic product (GDP), A4
- gross national product (GNP), A4
- net investment income received from nonresidents, A4
- net national product (NNP), A4
- net domestic income at factor cost, A5
- personal disposable income, A6
- trade balance, A7

FURTHER READING

....

For more details on national income accounting, see Statistics Canada, *Guide to Income and Expenditure Accounts*, Cat. No. 13-603E, No. 1, 1990.

Appendix 3: A Math Refresher

The purpose of this appendix is to present in a simple way the mathematical tools and the mathematical results that are used in this book.

GEOMETRIC SERIES

Definition. A geometric series is a sum of numbers of the form

$$1 + x + x^2 + \cdots + x^n$$

where x is a number that may be greater or smaller than 1 and x^n denotes x to the power n, that is, x times itself n times.

Examples of such series are:

■ The sum of spending in each round of the multiplier (Chapter 4). If c is the marginal propensity to consume, then the sum of increases in spending after n rounds is given by

$$1 + c + c^2 + \cdots + c^{n-1}$$

■ The present discounted value of a sequence of payments of 1 each year for n years (Chapter 7), when the interest rate is equal to i:

$$1 + \frac{1}{1+i} + \frac{1}{(1+i)^2} + \cdots + \frac{1}{(1+i)^{n-1}}$$

We usually have two questions we want to answer when encountering such a series. The first one is what the sum is equal to. The second is whether the sum explodes as we let n increase, or whether it reaches a finite limit and what that limit is. The following propositions tell you what you need to know to answer these questions.

Proposition 1 tells you how to compute the sum:

Proposition 1:

$$1 + x + x^2 + \cdots + x^n = \frac{1 - x^{n+1}}{1-x} \qquad \text{(A.1)}$$

The proof is as follows. Multiply the sum by $(1 - x)$, and use the fact that $x^a x^b = x^{a+b}$ (that is, one must add exponents when multiplying):

$$(1 + x + x^2 + \cdots + x^n)(1-x) = 1 + x + x^2 + \cdots + x^n$$
$$- x - x^2 - \cdots - x^n - x^{n+1}$$
$$= 1 - x^{n+1}$$

All the terms on the right except for the first and last cancel. Dividing both sides by $(1 - x)$ gives equation (A.1).

This formula can be used for any x and any n. If, for example, x is 0.9 and n is 10, then the sum is equal to 6.86. If x is 1.2 and n is 10, then the sum is equal to 32.15.

Proposition 2 tells you what happens as n becomes large.

Proposition 2: If x is less than 1, the sum goes to $1/(1 - x)$ as n becomes large. If x is equal to or greater than 1, the sum explodes as n becomes large.

The proof is as follows. If x is less than 1, then x^n goes to zero as n becomes large. Thus, from equation (A.1), the sum goes to $1/(1 - x)$. If x is greater than 1, then x^n becomes larger and larger as n increases, $1 - x^n$ becomes a larger and larger negative number, and the ratio $(1 - x^n)/(1 - x)$ becomes a larger and larger positive number. Thus, the sum explodes as n becomes large.

Application from Chapter 7: Consider the present value of a payment of \$1 forever, starting next year, when the interest rate is equal to i. The present value is given by

$$\frac{1}{1+i} + \frac{1}{(1+i)^2} + \cdots \qquad \text{(A.2)}$$

Factoring out $1/(1 + i)$, rewrite this present value as

$$\frac{1}{1+i}\left(1 + \frac{1}{1+i} + \cdots\right)$$

The term in parentheses is a geometric series, with $x = 1/(1 + i)$. As the interest rate i is positive, x is less than 1. Applying Proposition 2, when n becomes large, the term in parentheses is thus equal to

$$\frac{1}{1 - \dfrac{1}{(1+i)}} = \frac{1+i}{1+i-1} = \frac{1+i}{i}$$

Replacing the term in large parentheses in the previous equation by $(1 + i)/i$ gives

$$\frac{1}{1+i}\left(\frac{1+i}{i}\right) = \frac{1}{i}$$

The present value of a sequence of payments of $1 a year forever, starting next year, is thus equal to 1 over the interest rate. If i is equal to 5%, the present value is equal to $20.

USEFUL APPROXIMATIONS

Throughout the book, we use a number of approximations that make computations easier. These approximations are most reliable when the variables x, y, z used below are small, say between 0% and 10%. The numerical examples in Propositions 3–8 below are based on the values $x = 0.05$ and $y = 0.03$.

Proposition 3:

$$(1 + x)(1 + y) \approx 1 + x + y \qquad (A.3)$$

The proof is as follows. Expanding $(1 + x)(1 + y)$ gives $(1 + x)(1 + y) = 1 + x + y + xy$. If x and y are small, then the product xy is very small and can be ignored as an approximation (for example, if $x = 0.05$ and $y = 0.03$, then $xy = .0015$). So $(1 + x)(1 + y)$ is approximately equal to $(1 + x + y)$.

For the values x and y above, for example, the approximation gives 1.08 compared with an exact value of 1.0815.

Application from Chapter 11: Arbitrage between domestic and foreign bonds leads to the following relation:

$$1 + i_t = (1 + i_t^*)\left(1 + \frac{E_{t+1}^e - E_t}{E_t}\right)$$

Using Proposition 3 on the right-hand side of the equation gives

$$(1 + i_t^*)\left(1 + \frac{E_{t+1}^e - E_t}{E_t}\right) \approx 1 + i_t^* + \frac{E_{t+1}^e - E_t}{E_t}$$

Replacing in the arbitrage equation gives

$$1 + i_t \approx 1 + i_t^* + \frac{E_{t+1}^e - E_t}{E_t}$$

Subtracting 1 from both sides gives

$$i_t \approx i_t^* + \frac{E_{t+1}^e - E_t}{E_t}$$

The domestic interest rate is approximately equal to the foreign interest rate plus the expected rate of depreciation of the domestic currency.

Proposition 4:

$$(1 + x)^2 \approx 1 + 2x \qquad (A.4)$$

The proof follows directly from Proposition 3, with $y = x$. For the value of $x = 0.05$, the approximation gives 1.10, compared with an exact value of 1.1025.

Application from Chapter 9: From arbitrage, the relation between the two-year interest rate and the current and expected one-year rates is given by

$$(1 + i_{2t})^2 = (1 + i_{1t})(1 + i_{1t+1}^e)$$

Using Proposition 4 for the left-hand side of the equation gives

$$(1 + i_{2t})^2 \approx 1 + 2i_{2t}$$

Using Proposition 3 for the right-hand side of the equation gives

$$(1 + i_{1t})(1 + i_{1t+1}^e) \approx 1 + i_{1t} + i_{1t+1}^e$$

Replacing in the original relation gives

$$1 + 2i_{2t} \approx 1 + i_{1t} + i_{1t+1}^e$$

Reorganizing, we obtain

$$i_{2t} \approx \frac{i_{1t} + i_{1t+1}^e}{2}$$

The two-year rate is approximately equal to the average of the current and expected one-year rates.

Proposition 5:

$$(1 + x)^n \approx 1 + nx \qquad (A.5)$$

The proof follows by repeated application of Propositions 3 and 4. For example, $(1 + x)^3 = (1 + x)^2 (1 + x) \approx (1 + 2x)(1 + x)$ by Proposition 4, $\approx (1 + 2x + x) = 1 + 3x$ by Proposition 3.

The approximation becomes worse as n increases, however. For example, for $x = 0.05$ and $n = 5$, the approximation gives 1.25, compared to an exact value of 1.2763. For $n = 10$, the approximation gives 1.50, compared with an exact value of 1.63.

Application: In Chapter 14 arbitrage between n-year U.S. bonds and n-year Canadian bonds implies that

$$(1 + r_{nt})^n = \left(\frac{\varepsilon_{t+n}^e}{\varepsilon_t}\right)(1 + r_{nt}^*)^n$$

From Proposition 5, it follows that

$$(1 + r_{nt})^n \approx 1 + nr_{nt}$$

and

$$(1 + r_{nt}^*)^n \approx 1 + nr_{nt}^*$$

Note also that we can rewrite the first term on the right in the arbitrage equation as

$$\frac{\varepsilon_{t+n}^e}{\varepsilon_t} = 1 + \frac{\varepsilon_{t+n}^e - \varepsilon_t}{\varepsilon_t}$$

Replacing these three expressions in the arbitrage relation gives

$$1 + nr_{nt} \approx \left(1 + \frac{\varepsilon_{t+n}^e - \varepsilon_t}{\varepsilon_t}\right)(1 + nr_{nt}^*)$$

From Proposition 3, it follows that

$$1 + nr_{nt} \approx \left(1 + nr_{nt}^* + \frac{\varepsilon_{t+n}^e - \varepsilon_t}{\varepsilon_t}\right)$$

Simplifying yields

$$n(r_{nt} - r_{nt}^*) \approx \frac{\varepsilon_{t+n}^e - \varepsilon_t}{\varepsilon_t}$$

The difference between the nth year Canadian and U.S. real interest rates, multiplied by n, is approximately equal to the expected rate of real depreciation of Canadian goods vis-à-vis American goods over the next n years.

Proposition 6:

$$\frac{1 + x}{1 + y} \approx 1 + x - y \qquad (A.6)$$

The proof is as follows. Consider the product of $(1 + x - y)(1 + y)$. Expanding this product gives $(1 + x - y)(1 + y) = 1 + x + xy - y^2$. If both x and y are small, then xy and y^2 are very small, so $(1 + x - y)(1 + y) \approx (1 + x)$. Dividing both sides of this approximation by $(1 + y)$ gives the proposition in equation (A.6).

For the values of $x = 0.05$ and $y = 0.03$, the approximation gives 1.02, while the correct value is 1.019.

Application from Chapter 7: The real interest rate is defined by

$$1 + r_t = \frac{1 + i_t}{1 + \pi_t^e}$$

Using Proposition 6 gives

$$1 + r_t \approx 1 + i_t - \pi_t^e$$

Simplifying yields

$$r_t \approx i_t - \pi_t^e$$

This gives us the approximation we use at many points in the book: The real interest rate is approximately equal to the nominal interest rate minus the expected inflation rate.

These approximations are also very convenient when dealing with growth rates. Define the rate of growth of x by $g_x \equiv \Delta x / x$, and similarly for z, g_z and y, g_y. The numerical examples in Propositions 7 and 8 below are based on the values $g_x = 0.05$ and $g_y = 0.03$.

Proposition 7: If $z = xy$, then

$$g_z \approx g_x + g_y \qquad (A.7)$$

The proof is as follows. Let Δz be the increase in z when x increases by Δx and y increases by Δy. Then, by definition,

$$z + \Delta z = (x + \Delta x)(y + \Delta y)$$

Divide both sides by z, so that

$$\frac{z + \Delta z}{z} = \left(\frac{x + \Delta x}{x}\right)\left(\frac{y + \Delta y}{y}\right)$$

where we have used on the right-hand side the fact that dividing by z is the same as dividing by xy. Simplifying gives

$$1 + \frac{\Delta z}{z} = \left(1 + \frac{\Delta x}{x}\right)\left(1 + \frac{\Delta y}{y}\right)$$

or, equivalently,

$$1 + g_z = (1 + g_x)(1 + g_y)$$

From Proposition 3, $1 + g_z \approx 1 + g_x + g_y$, or, equivalently, $g_z \approx g_x + g_y$.

For the values of $g_x = 0.05$ and $g_y = 0.03$, the approximation gives $g_z = 8\%$, while the correct value is 8.15%.

Application from Chapter 25: Let the production function be of the form $Y = NA$, where Y is production, N is employment, and A is productivity. Denoting the growth rates of Y, N, and A by g_Y, g_N, and g_A respectively, Proposition 7 implies $g_Y \approx g_N + g_A$: The rate of output growth is approximately equal to the rate of employment growth plus the rate of productivity growth.

Proposition 8: If $z = x/y$, then

$$g_z \approx g_x - g_y \qquad (A.8)$$

The proof is as follows. Let Δz be the increase in z, when x increases by Δx and y increases by Δy. Then, by definition,

$$z + \Delta z = \frac{x + \Delta x}{y + \Delta y}$$

Dividing both sides by z and using the fact that $z = x/y$ gives

$$1 + \left(\frac{\Delta z}{z}\right) = \frac{1 + \dfrac{\Delta x}{x}}{1 + \dfrac{\Delta y}{y}}$$

Or, substituting:

$$1 + g_z = \frac{1 + g_x}{1 + g_y}$$

From Proposition 6, $1 + g_z \approx 1 + g_x - g_y$, or, equivalently, $g_z \approx g_x - g_y$.

For the values of $g_x = 0.05$ and $g_y = 0.03$, the approximation gives $g_z = 2\%$, while the correct value is 1.9 percent.

Application from Chapter 18: Let aggregate demand be given by $Y = \gamma M/P$, where Y is output, M is nominal money, P is the price level, and γ is a constant parameter. It follows from Propositions 7 and 8 that

$$g_Y \approx g_\gamma + g_M - \pi$$

where π is the rate of growth of prices, equivalently the rate of inflation. Because γ is constant, g_γ is equal to zero. Thus

$$g_Y \approx g_M - \pi$$

The rate of output growth is approximately equal to the rate of growth of nominal money minus the rate of inflation.

FUNCTIONS

We use functions informally in this book, as a way of denoting how a variable depends on one or more other variables.

In some cases, we look at how a variable Y moves with a variable X. We write this relation as

$$Y = f(X)$$
$$+$$

A plus sign below X indicates a positive relation: An increase in X leads to an increase in Y. A minus sign indicates a negative relation: An increase in X leads instead to a decrease in Y.

In some cases, we allow the variable Y to depend on more than one variable. For example, we allow Y to depend on X and Z:

$$Y = f(X, Z)$$
$$(+, -)$$

The signs indicate that an increase in X leads to an increase in Y, and that an increase in Z leads to a decrease in Y. An example of such a function is the investment function in Chapter 6:

$$I = I(Y, i)$$
$$(+, -)$$

This equation says that investment, I, increases with production, Y, and decreases with the interest rate, i.

In some cases, it is reasonable to assume that the relation between two or more variables is a **linear relation**. A given increase in X always leads to the same increase in Y. In that case, the function is given by

$$Y = a + bX$$

The parameter a is called the **intercept:** It gives the value of Y when X is equal to zero. The parameter b is called the **slope:** It tells us by how much Y increases when X increases by 1.

The simplest linear relation is the relation $Y = X$, which is represented by the 45-degree line and has a slope of 1. Another example of a linear relation is the consumption function introduced in Chapter 3:

$$C = c_0 + c_1 Y_D$$

where C is consumption and Y_D is disposable income. c_0 tells us what consumption would be if disposable income were equal to zero. c_1 tells us by how much consumption increases when income increases by 1 unit. c_1 is called the *marginal propensity to consume.*

KEY TERMS

· ·

▪ linear relation, A12
▪ intercept, A12
▪ slope, A12

GLOSSARY

above the line, below the line In the balance of payments, the items in the *current account* are above the line drawn to divide them from the items in the *capital account,* which appear below the line.

accelerationist Phillips curve See *modified Phillips curve.*

accommodation The central bank's increase in the money supply in order to avoid an increase in the interest rate.

adaptive expectations A backward-looking method of forming expectations by adjusting for past expectational mistakes.

adjusted nominal money growth Nominal money growth minus normal output growth.

aggregate demand relation The demand for output at a given price level. It is derived from equilibrium in goods and financial markets.

aggregate private spending The sum of all nongovernmental spending. Also called *private spending.*

aggregate production function The relation between the quantity of aggregate output produced and the quantities of inputs used in production.

aggregate supply relation The price level at which firms are willing to supply a given level of output. It is derived from equilibrium in the labour market.

animal spirits A term introduced by Keynes to refer to movements in investment that could not be explained by movements in current variables.

anticipated money Movements in nominal money that could have been predicted based on the information available at some time in the past.

appreciation (nominal) An increase in the price of the domestic currency in terms of a foreign currency. Corresponds to a decrease in the exchange rate.

appropriability (of research results) The extent to which firms benefit from the results of their research and development efforts.

arbitrage The proposition that the expected rates of return on two financial assets must be equal. Also called *risky arbitrage* to distinguish it from *riskless arbitrage,* the proposition that the actual rates of return on two identical financial assets must be the same.

automatic stabilizer The fact that a decrease in output leads, under given tax and spending rules, to an increase in the budget deficit. This increase in the budget deficit in turn increases demand and thus stabilizes output.

autonomous spending That component of the demand for goods that does not depend on the level of output.

balance of payments A set of accounts that summarize a country's transactions with the rest of the world.

balanced budget A budget in which taxes are equal to government spending.

balanced growth The situation in which output, capital, and effective labour all grow at the same rate.

band (for exchange rates) The limits within which the exchange rate is allowed to move under a fixed exchange rate system.

Bank of Canada Canada's central bank

bank run Simultaneous attempts by depositors to withdraw their funds from a bank.

bargaining power The relative strength of each side in a negotiation or a dispute.

barter The exchange of goods for other goods rather than for money.

behavioural equation An equation that captures some aspect of behaviour.

big bang The simultaneous implementation of many reforms at once.

bilateral exchange rate The real exchange rate between two countries.

bond A financial asset that promises a stream of known payments over some period of time.

bond rating The assessment of a bond based on its default risk.

broad money See *M2.*

budget deficit The excess of government expenditures over government revenues.

budget surplus See *public saving.*

business cycles See *output fluctuations.*

Canada bonds Bonds with a maturity of 1 to 30 years when issued that pay interest semiannually.

Canada Deposit Insurance Corporation A government body that insures bank deposits in order to avoid bank runs.

capital accumulation Increase in the capital stock.

capital controls Restrictions on the foreign assets domestic residents can hold and on the domestic assets foreigners can hold.

cash flow The net flow of cash a firm is receiving.

causality A relation between cause and effect.

central bank money Money issued by the central bank. Also known as the *monetary base* and *high-powered money.*

central parity The reference value of the exchange rate around which the exchange rate is allowed to move under a fixed exchange rate system. The centre of the *band.*

chequable deposits Deposits at banks and other financial institutions against which cheques can be written.

churning The concept that new goods make old goods obsolete, that new production techniques make older techniques and worker skills obsolete, and so on.

civilian noninstitutionalized population The relevant population available for civilian employment.

confidence band When estimating the dynamic effect of one variable on another using econometrics, the range of values where we can be confident the true dynamic effect lies.

constant returns to scale The proposition that a proportional increase (or decrease) of all inputs leads to the same proportional increase (or decrease) in output.

consumer confidence index An index computed monthly that estimates consumer confidence regarding current and future economic conditions.

Consumer Price Index (CPI) The cost of a given list of goods and services consumed by a typical household.

consumption (C) Goods and services purchased by consumers.

consumption function A function that relates consumption to its determinants.

contractionary open-market operation An open-market operation in which the central bank sells bonds to decrease the money supply.

controlled experiment A set of test conditions in which one variable is altered while the others are kept constant.

convergence The tendency for countries with lower output per capita to grow faster, leading to convergence of output per capita across countries.

coordination (of macroeconomic policies between two countries) The joint design of macroeconomic policies to improve the economic situation in the two countries.

corporate bond A bond issued by a corporation.

correlation A measure of the way two variables move together. A positive correlation indicates that the two variables tend to move in the same direction. A negative correlation indicates that the two variables tend to move in opposite directions. A correlation of zero indicates that there is no apparent relation between the two variables.

cost of living The average price of a consumption bundle.

coupon bond A bond that promises multiple payments before maturity and one payment at maturity.

coupon payments The payments before maturity on a coupon bond.

coupon rate The ratio of the coupon payment to the face value of a coupon bond.

crawling peg An exchange rate mechanism in which the exchange rate is allowed to move over time according to a pre-specified formula.

creative destruction The concept that growth simultaneously creates and destroys jobs.

credibility The degree to which people and markets believe that a policy announcement will actually be instituted and followed through.

currency Coins and banknotes.

current account In the balance of payments, the summary of a country's payments to and from the rest of the world.

current yield The ratio of the coupon payment to the price of a coupon bond.

cyclically adjusted deficit A measure of what the government deficit would be under existing tax and spending rules, if output were at its natural level. Also called a *full-employment deficit, midcycle deficit, standardized employment deficit,* or *structural deficit.*

debt finance Financing based on loans or the issuance of bonds.

debt monetization The practice of creating money to finance a deficit.

debt ratio See *debt-to-GDP ratio.*

debt repudiation A unilateral decision by a debtor not to repay its debt.

debt-to-GDP ratio The ratio of debt to gross domestic product. Also called simply the *debt ratio.*

decreasing returns to capital The property that increases in capital lead to smaller and smaller increases in output as the level of capital increases.

decreasing returns to labour The property that increases in labour lead to smaller and smaller increases in output as the level of labour increases.

default risk The risk that the issuer of a bond will not pay back the full amount promised by the bond.

deflation Negative inflation.

degrees of freedom The number of useable observations in a *regression* minus the number of parameters to be estimated.

demand deposit A bank account that allows depositors to write cheques or get cash on demand, up to an amount equal to the account balance.

demand for domestic goods The demand for domestic goods by people, firms, and governments, both domestic and foreign. Equal to the domestic demand for goods plus net exports.

dependent variable A variable whose value is determined by one or more other variables.

depreciation (nominal) A decrease in the price of the domestic currency in terms of a foreign currency. Corresponds to an increase in the exchange rate.

depreciation rate A measure of how much usefulness a piece of capital loses from one period to the next.

depression A deep and long-lasting recession.

devaluation An increase in the exchange rate in a fixed exchange-rate system.

discount bond A bond that promises a single payment at maturity.

discount factor The value today of a dollar (or other national currency unit) at some time in the future.

discount rate The interest rate used to discount a sequence of future payments. Equal to the nominal interest rate when discounting future nominal payments, to the real interest rate when discounting future real payments.

discouraged worker A person who has given up looking for employment.

disinflation A decrease in inflation.

disposable income The income that remains once consumers have received transfers from the government and paid their taxes.

dividends The portion of a corporation's profits that the firm pays out each period to shareholders.

dollarization The use of U.S. dollars in domestic transactions in a country other than the United States.

domestic demand for goods The sum of consumption, investment, and government spending.

dual labour market A labour market that combines a *primary labour market* and a *secondary labour market.*

duration of unemployment The period of time during which a worker is unemployed.

dynamics Movements of one or more economic variables over time.

econometrics Statistical methods applied to economics.

effective demand Synonym for *aggregate demand.*

effective labour The number of workers in an economy times the state of technology.

effective exchange rate See *multilateral exchange rate.*

efficiency wage That wage at which a worker is performing a job most efficiently or productively.

employment rate Ratio of employment to the labour force.

endogenous variable A variable that depends on other variables in a model and is thus explained within the model.

entitlement programs Programs that require the payment of benefits to all who meet the eligibility requirements established by law.

equilibrium The equality between demand and supply.

equilibrium equation An equation that represents an equilibrium condition.

equity finance Financing based on the issuance of shares.

European Monetary System (EMS) A fixed exchange rate system adopted by most of the countries of the European Union prior to 1999.

European Union A political and economic organization of 15 European nations. Formerly called the European Community.

Eurosclerosis A term coined to reflect the belief that Europe suffers from excessive rigidities, especially in the labour market.

exogenous variable A variable that is not explained within a model but rather is taken as given.

expansion A period of positive GDP growth.

expansionary open-market operation An open-market operation in which the central bank buys bonds to increase the money supply.

expectations-augmented Phillips curve See *modified Phillips curve.*

expected present discounted value The value today of an expected sequence of future payments. Also called *present value.*

experiment A test carried out under controlled conditions to assess the validity of a model or hypothesis.

exports (X) The purchases of domestic goods and services by foreigners.

ex post real interest rate The difference between the nominal interest rate and the realized rate of inflation.

face value (on a bond) The single payment at maturity promised by a discount bond.

fad A period of time during which, for reasons of fashion or overoptimism, financial investors are willing to pay more than the fundamental value of a stock.

Federal Reserve Bank (Fed) The U.S. central bank.

fertility of research The degree to which spending on research and development translates into new ideas and new products.

financial intermediary A financial institution that receives funds from people and/or firms, and uses these funds to make loans or buy financial assets.

financial investment The purchase of financial assets.

financial markets The markets in which financial assets are bought and sold.

financial wealth The value of all of one's financial assets minus all financial liabilities. Sometimes called *wealth* for short.

fine-tuning A macroeconomic policy aimed at precisely hitting a given target, such as constant unemployment or constant output growth.

fiscal contraction A policy aimed at reducing the budget deficit through a decrease in government spending or an increase in taxation.

fiscal expansion An increase in government spending or a decrease in taxation, which leads to an increase in the budget deficit.

fiscal policy A government's choice of taxes and spending.

Fisher effect or **Fisher hypothesis** The proposition that in the long run an increase in nominal money growth is reflected in an identical increase in both the nominal interest rate and the inflation rate, leaving the real interest rate unchanged.

Fisher hypothesis See *Fisher effect*.

fixed exchange rate An exchange rate between the currencies of two or more countries that is fixed at some level and adjusted only infrequently.

fixed investment See *investment (I)*.

floating exchange rate An exchange rate determined in the foreign-exchange market without central bank intervention.

flow A variable that can be expressed as a quantity per unit of time (such as income).

forecast error The difference between the actual value of a variable and a forecast of that variable.

foreign direct investment The purchase of existing firms or the development of new firms in an economy by foreign investors.

foreign exchange Foreign currency; all currencies other than the domestic currency of a given country.

foreign-exchange reserves Foreign assets held by the central bank.

full-employment deficit See *cyclically adjusted deficit*.

fundamental value (of a stock) The present value of expected dividends.

G7 The seven major economic powers in the world: the United States, Japan, France, Germany, the United Kingdom, Italy, and Canada.

games *Strategic interactions* between *players*.

game theory The prediction of outcomes from *games*.

GDP deflator The ratio of nominal GDP to real GDP; a measure of the overall price level. Gives the average price of the final goods produced in the economy.

GDP growth The growth rate of real GDP in year t; equal to $(Y_t - Y_{t-1})/Y_{t-1}$.

general equilibrium A situation in which there is equilibrium in all markets (goods, financial, and labour) at the same time.

geometric series A mathematical sequence in which the ratio of one term to the preceding term remains the same. A math sequence of the form $1 + c + c^2 + \ldots + c^n$.

gold standard A system in which a country fixed the price of its currency in terms of gold and stood ready to exchange gold for currency at the stated parity.

golden-rule level of capital The level of capital at which long-run consumption is maximized.

government bond A bond issued by a government or a government agency.

government budget constraint The budget constraint faced by the government. The constraint implies that an excess of spending over revenues must be financed by borrowing, and thus leads to an increase in debt.

government spending (G) The goods and services purchased by federal, state, and local governments.

government transfers Payments made by the government to individuals that are not in exchange for goods or services. Example: Canada Pension Plan payments.

Great Depression The severe worldwide depression of the 1930s.

gross domestic product (GDP) A measure of aggregate output in the national income accounts. (The market value of the goods and services produced by labour and property located in Canada.)

gross national product (GNP) A measure of aggregate output in the national income accounts. (The market value of the goods and services produced by labour and property supplied by Canadian residents.)

growth The steady increase in aggregate output over time.

hedonic pricing An approach to calculating real GDP that treats goods as providing a collection of characteristics, each with an implicit price.

heterodox stabilization program A stabilization program that includes incomes policies.

high-powered money See *central bank money*.

hires Workers newly employed by firms.

human capital The set of skills possessed by the workers in an economy.

human wealth The labour-income component of wealth.

hyperinflation Very high inflation.

hysteresis In general, the proposition that the equilibrium value of a variable depends on its history. With respect to unemployment, the proposition that a long period of sustained actual unemployment leads to an increase in the equilibrium rate of unemployment.

identification problem In econometrics, the problem of finding whether correlation between variables X and Y indicates a causal relation from X to Y, or from Y to X, or both. This problem is solved by finding exogenous variables, called *instruments*, that affect X and do not affect Y directly, or affect Y and do not affect X directly.

identity An equation that holds by definition, denoted by the sign $\equiv$.

imports (Q) The purchases of foreign goods and services by domestic consumers, firms, and government.

income The flow of revenue from work, rental income, interest, and dividends.

incomes policies Government policies that set up wage and/or price guidelines or controls.

independent variable A variable that is taken as given in a relation or in a model.

index number A number, such as the GDP deflator, that has no natural level and is thus set to equal some value (typically 1 or 100) in a given period.

indexed bond A bond that promises payments adjusted for inflation.

industrial policy A policy aimed at helping specific sectors of an economy.

industrial product price index (IPPI) An index of prices of domestically produced goods that excludes transportation costs and indirect taxes.

inflation A sustained rise in the general level of prices.

inflation rate The rate at which the price level increases over time.

inflation tax The product of the rate of inflation and real money balances.

inflation-adjusted deficit The correct economic measure of the budget deficit: the sum of the *primary deficit* and real interest payments.

instrumental variable methods In econometrics, methods of estimation that use *instruments* to estimate causal relations between different variables.

instruments In econometrics, the exogenous variables that allow the identification problem to be solved.

intercept In a relation between two variables, the value of the first variable when the second variable is equal to zero.

interest parity condition See *uncovered interest parity relation*.

intermediate good A product used in the production of a final good.

International Monetary Fund (IMF) The principal international economic organization. Publishes *World Economic Outlook* twice a year and *International Financial Statistics (IFS)* monthly.

inventory investment (I_S) The difference between production and sales.

investment (I) Purchases of new houses and apartments by people, and purchases of new capital goods (machines and plants) by firms.

investment income In the current account, income received by residents of one country from their holdings of foreign assets.

IS curve A downward-sloping curve relating output to the interest rate. The curve corresponding to the *IS relation*, the equilibrium condition for the goods market.

IS relation An equilibrium condition stating that the demand for goods must be equal to the supply of goods, or equivalently that investment must be equal to saving. The equilibrium condition for the goods market.

J-curve A curve depicting the initial deterioration in the trade balance caused by a real depreciation, followed by an improvement in the trade balance.

junk bond A bond with a high risk of default.

labour force The sum of those employed and those unemployed.

labour force survey (LFS) A large survey of households used to compute the unemployment rate in Canada.

labour hoarding The practice of retaining workers during a period of low product demand rather than laying them off.

labour-market rigidities Restrictions on firms' ability to adjust their level of employment.

labour productivity The ratio of output to the number of workers.

Laffer curve A curve showing the relation between tax revenues and the tax rate.

lagged value The value of a variable in the preceding time period.

layoffs Workers who lose their jobs either temporarily or permanently.

leapfrogging Advancing on and then overtaking the leader. Used to describe the process by which economic leadership passes from country to country.

life-cycle theory of consumption The theory of consumption, developed initially by Franco Modigliani, that emphasizes that the planning horizon of consumers is their lifetime.

linear relation A relation between two variables such that a one-unit increase in one variable always leads to an increase of n units in the other variable.

liquidity preference The term introduced by Keynes to denote the demand for money.

LM curve An upward-sloping curve relating the interest rate to output. The curve corresponding to the *LM relation*, the equilibrium condition for financial markets.

LM relation An equilibrium condition stating that the demand for money must be equal to the supply of money, or equivalently the demand for bonds must be equal to the supply of bonds. The equilibrium condition for financial markets.

logarithmic scale A scale in which the same proportional increase represents the same distance on the scale, so that a variable that grows at a constant rate is represented by a straight line on the scale.

long run A period of time extending over decades.

long-term bond A bond with maturity of 10 years or more.

Lucas critique The proposition, put forth by Robert Lucas, that existing relations between economic variables may change when policy changes. An example is the apparent tradeoff between inflation and unemployment, which may disappear if policy makers try to exploit it.

M1 The sum of currency, and demand deposits—assets that can be used directly in transactions. Also called *narrow money*.

M2 *M*1 plus personal savings deposits, and nonpersonal time deposits. Also called *broad money*.

M3 A *monetary aggregate* constructed by the Bank of Canada and broader than *M*2.

macroeconomic control In the context of transition, the reduction of subsidies to firms, the reduction of budget deficits, and moderate money growth.

macroeconomics The study of aggregate economic variables, such as production for the economy as a whole, or the average price of goods.

Maastricht treaty A treaty signed in 1991 that defines the steps involved in the transition to a common currency for the European Union.

marginal propensity to consume (*mpc*, or c_1) The effect on consumption of an additional dollar of disposable income.

marginal propensity to import The effect on imports from an additional dollar in income.

marginal propensity to save The effect on saving of an additional dollar of disposable income (equal to one minus the marginal propensity to consume).

Marshall-Lerner condition The condition under which a real depreciation leads to an increase in net exports.

maturity The length of time over which a financial asset (typically a bond) promises to make payments to the holder.

medium of exchange An asset that is immediately exchangeable for goods and services.

medium run A period of time between the *short run* and the *long run*.

medium-term bond A bond with maturity of one to 10 years.

menu cost The cost of changing a price.

merchandise trade Exports and imports of goods.

microeconomics The study of production and prices in specific markets.

midcycle deficit See *cyclically adjusted deficit*.

model A simple structure used to think about and interpret an economic phenomenon.

models of endogenous growth Models in which accumulation of physical and human capital can sustain growth even in the absence of technological progress.

modified Phillips curve The curve that plots the change in the inflation rate against the unemployment rate. Also called an *expectations-augmented Phillips curve* or an *accelerationist Phillips curve*.

monetarists A group of economists, led by Milton Friedman, who emphasized the importance of monetary policy in fluctuations.

monetary aggregate The market value of a sum of liquid assets. *M*1 is a monetary aggregate that includes only the most liquid assets.

monetary base See *central bank money*.

monetary contraction A decrease in the money supply. Also called *monetary tightening*.

monetary expansion An increase in the money supply.

monetary-fiscal policy mix The combination of monetary and fiscal policies in effect at a given time.

monetary tightening See *monetary contraction*.

money Those financial assets that can be used directly to buy goods.

money illusion The failure to perceive that the dollar, or any other unit of account, expands or shrinks in value over time.

money-market funds Financial institutions that receive funds from people and use them to buy short-term bonds.

money multiplier The increase in the money supply resulting from a one-dollar increase in central bank money.

multilateral exchange rate The real exchange rate between a country and its trade partners, computed as a weighted average of bilateral real exchange rates. Also called the *trade-weighted exchange rate* or *effective exchange rate*.

multiplier The ratio of the change in an *endogenous variable* to the change in an *exogenous variable* (for example, the ratio of the change in output in response to a change in autonomous spending).

Mundell-Fleming model A model of simultaneous equilibrium in both goods and financial markets for an open economy.

narrow money See *M1*.

national income and expenditure accounts The system of accounts used to describe the evolution of the sum, the composition, and the distribution of aggregate output.

natural experiment A real-world event or period that can be used to test an economic theory.

natural level of employment The level of employment that prevails when unemployment is equal to its natural rate.

natural level of output The level of production that prevails when employment is equal to its natural level.

natural rate of unemployment The unemployment rate at which price and wage decisions are consistent.

neoclassical synthesis A consensus in macroeconomics, developed by the early 1950s, based on an integration of Keynes's ideas and the ideas of earlier economists.

net capital flows Capital flows from the rest of the world to the domestic economy minus capital flows to the rest of the world from the domestic economy.

net domestic income at factor cost The result of adding back net payments to foreign factors.

net exports The difference between exports and imports. Also called the *trade balance*.

net investment income received from nonresidents Income earned from claims to foreign capital by Canadian residents less income earned from claims to domestic capital by foreign residents.

net national product (NNP) Gross national product minus capital depreciation.

net transfers received In the current account, the net value of the foreign aid that a country gives and receives.

neutrality of money The proposition that an increase in nominal money has no effect on output or the interest rate but is reflected entirely in a proportional increase in the price level.

new classicals A group of economists who interpret fluctuations as the effects of shocks in competitive markets with fully flexible prices and wages.

New Deal The set of programs put in place by the Roosevelt administration to get the U.S. economy out of the Great Depression.

new growth theory Recent developments in growth theory that explore the determinants of technological progress and the role of increasing returns to scale in growth.

new Keynesians A group of economists who believe in the importance of nominal rigidities in fluctuations, and are exploring the role of market imperfections in explaining fluctuations.

nominal exchange rate The number of units of domestic currency you can get for one unit of foreign currency; the price of foreign currency in terms of domestic currency.

nominal GDP The sum of the quantities of final goods produced in an economy times their current price.

nominal interest rate Interest rate in terms of the national currency (in terms of dollars in Canada). Tells us how many dollars one has to repay in the future in exchange for one dollar today.

nominal rigidities The slow adjustment of nominal wages and prices to changes in economic activity.

nonaccelerating inflation rate of unemployment (NAIRU) The unemployment rate at which inflation neither decreases nor increases. See *natural rate of unemployment*.

nonhuman wealth The financial and housing component of wealth.

normal growth rate The rate of output growth needed to maintain a constant unemployment rate.

North American Free Trade Agreement (NAFTA) An agreement signed by the United States, Canada, and Mexico in which the three countries agreed to establish all of North America as a free-trade zone.

not in the labour force Those persons in a population who are neither working in the marketplace nor looking for work.

n-year interest rate See *yield to maturity*.

Okun's law The relation between GDP growth and the change in the unemployment rate.

open-market operation The purchase or sale of government bonds by the central bank for the purpose of increasing or decreasing the money supply.

openness in factor markets The opportunity for firms to choose where to locate production and for workers to choose where to work and whether or not to migrate.

openness in financial markets The opportunity for financial investors to choose between domestic and foreign financial assets.

openness in goods markets The opportunity for consumers and firms to choose between domestic and foreign goods.

optimal control The control of a system (a machine, a rocket, an economy) by means of mathematical methods.

optimal control theory The set of mathematical methods used for *optimal control*.

ordinary least squares A statistical method to find the best fitting relation between two or more variables.

Organization for Economic Cooperation and Development (OECD) An international organization. Collects and studies economic data for many countries. Most of the world's rich countries belong to the OECD.

orthodox stabilization program A stabilization program that does not include incomes policies.

output fluctuations Movements in output around its trend.

output per capita A country's gross domestic product divided by its population.

overnight interest rate The interest rate charged for lending and borrowing overnight.

overshooting The large movement in the exchange rate triggered by a monetary expansion.

panel data set A data set that gives the values of one or more variables for many individuals or many firms over some period of time.

paradox of saving The result that people's attempts to save more may lead both to a decline in output and to unchanged saving.

parameter A coefficient in a behavioural equation.

participation rate The ratio of the labour force to the noninstitutional population.

patent The legal right granted by a government to a person or firm to exclude anyone else from the production or use of a new product or technique for a certain period of time.

peg The exchange rate to which a country commits under a fixed exchange rate system.

permanent income theory of consumption The theory of consumption, developed by Milton Friedman, that emphasizes that people make consumption decisions based not on current income, but on their notion of permanent income.

personal disposable income *Personal income* minus personal tax and nontax payments. The income available to consumers after they have received transfers and paid taxes.

personal income The income actually received by persons.

Phillips curve The curve that plots the relation between (1) movements in inflation and (2) unemployment. The original Phillips curve captured the relation between the inflation rate and the unemployment rate. The *modified Phillips curve* captures the relation between (1) the change in the inflation rate and (2) the unemployment rate.

players The participants in *games*. Depending on the context, players may be people, firms, governments, and so on.

point-year of excess employment A difference between the actual unemployment rate and the natural unemployment rate of one percentage point for one year.

policy mix See *monetary-fiscal policy mix.*

political business cycle Fluctuations in economic activity caused by the manipulation of the economy for electoral gain.

postindustrial economies Economies in which the manufacturing sector's share of gross domestic product is in steady decline.

present value See *expected present discounted value.*

price level The general level of prices in an economy.

price liberalization The process of eliminating subsidies, and decontrolling prices and allowing them to clear markets.

primary deficit Government spending, excluding interest payments on the debt, minus government revenues. (The negative of the *primary surplus.*)

primary labour market A labour market where jobs are good, wages are high, and turnover is low. Contrast with the *secondary labour market.*

primary surplus Government revenues minus government spending, excluding interest payments on the debt.

private saving (S) Saving by consumers. The value of consumers' disposable income minus their consumption.

private spending See *aggregate private spending.*

privatization The transfer of state-owned firms to private ownership.

production function The relation between the quantity of output and the quantities of inputs used in production.

profitability The expected present discounted value of profits.

propagation mechanism The dynamic effects of a *shock* on output and its components.

public saving Saving by the government; equal to government revenues minus government spending. Also called the *budget surplus.* (A *budget deficit* represents public dissaving.)

purchasing power The level of purchases that can be attained given income.

purchasing power parity (PPP) A method of adjustment used to allow for international comparisons of GDP.

quits Workers who leave their jobs in search of better alternatives.

quotas Restrictions on the quantities of goods that can be imported.

R^2 A measure of fit, between zero and one, from a *regression.* An R^2 of zero implies that there is no apparent relation between the variables under consideration. An R^2 of 1 implies a perfect fit: All the *residuals* are equal to zero.

random walk The path of a variable whose changes over time are unpredictable.

rate of growth of multifactor productivity See *Solow residual.*

rational expectations The formation of expectations based on rational forecasts, rather than on simple extrapolations of the past.

rational speculative bubble An increase in stock prices based on the rational expectation of further increases in prices in the future.

real appreciation An increase in the relative price of domestic goods in terms of foreign goods.

real business cycle (RBC) models Economic models that assume that output is always at its natural level. Thus all output fluctuations are movements of the natural level of output, as opposed to movements away from the natural level of output.

real depreciation A decrease in the relative price of domestic goods in terms of foreign goods.

real exchange rate The relative price of foreign goods in terms of domestic goods.

real GDP The sum of quantities produced in an economy times their price in a base year. Also known as "GDP in terms of goods," "GDP in constant dollars," "GDP adjusted for inflation," and "GDP in 1986 dollars" (for the 1986 base year).

real interest rate Interest rate in terms of goods. Tells us how many goods one has to repay in the future in exchange for one good today.

realignment Adjustment of parities in a fixed exchange-rate system.

recession A period of negative GDP growth. Usually refers to at least two consecutive quarters of negative GDP growth.

reference week An arbitrary week chosen by the Labour Force Survey (LFS) for the purpose of determing labour force status.

regression The output of *ordinary least squares.* Gives the equation corresponding to the estimated relation between variables, together with information about the degree of fit and the importance of the different variables.

regression line The best-fitting line corresponding to the equation obtained by using *ordinary least squares.*

rental cost of capital See *user cost of capital.*

research and development (R & D) Spending aimed at discovering and developing new ideas and products.

reservation wage The wage that would make a worker indifferent to working or becoming unemployed.

residual The difference between the actual value of a variable and the value implied by the *regression line.* Small residuals indicate a good fit.

revaluation A decrease in the exchange rate in a fixed exchange-rate system.

Ricardian equivalence The proposition that neither government deficits nor government debt have an effect on economic activity. Also called the *Ricardo-Barro proposition.*

Ricardo-Barro proposition See *Ricardian equivalence.*

risk premium The difference between the interest rate paid on a bond and the interest rate paid on a bond with the highest rating.

riskless arbitrage See *arbitrage*.

risky arbitrage See *arbitrage*.

sacrifice ratio The number of point-years of excess unemployment needed to achieve a decrease in inflation of 1%.

saving The sum of private and public saving, denoted by S.

saving rate The proportion of income that is saved in an economy.

savings The accumulated value of past saving. Also called *wealth*.

scatter diagram A graphic presentation that plots the value of one variable against the value of another variable.

secondary labour market A labour market where jobs are poor, wages are low, and turnover is high. Contrast with the *primary labour market*.

seigniorage The revenues from the creation of money.

separations Workers who are leaving or losing their jobs.

settlement balances Reserves held by banks, not to meet legal requirements but to facilitate the clearing of cheques and other transfers.

severance payments Payments made by firms to laid-off workers.

share A financial asset issued by a firm that promises to pay a sequence of payments, called dividends, in the future. Also called *stock*.

shock A movement in the factors that affect aggregate demand and/or aggregate supply.

short run A period of time extending over one year or a few years at most.

short-term bond A bond with maturity of one year or less.

simulation The use of a model to look at the effects of a change in an exogenous variable on the variables in the model.

skill-biased technological progress The proposition that new machines and new methods of production require skilled workers to a greater degree than in the past.

slope In a relation between two variables, the amount by which the first variable increases when the second increases by one unit.

soft budget constraint The granting of subsidies to firms that make losses, thus decreasing the incentives for these firms to take the measures needed to generate profits.

Solow residual The excess of actual output growth over what can be accounted for by the growth in capital and labour.

stabilization program A government program aimed at stabilizing the economy (typically stopping high inflation).

stagflation The combination of stagnation and inflation.

staggering of wage and price decisions The fact that different wages and prices are adjusted at different times, making it impossible to achieve a synchronized decrease in nominal wage and price inflation.

standardized employment deficit See *cyclically adjusted deficit*.

state of technology The degree of technological development in a country or industry.

statistical discrepancy A difference between two numbers that should be equal, based on differences in sources or methods of construction.

steady state In an economy without technological progress, the state of the economy where output and capital per worker are no longer changing. In an economy with technological progress, the state of the economy where output and capital per effective worker are no longer changing.

stock A variable that can be expressed as a quantity at a point in time (such as wealth). Also a synonym for *share*.

stocks An alternative term for *inventories*.

strategic interactions An environment in which the actions of one player depend on and affect the actions of another player.

structural change In the context of transition: The sharp shifts in the structure of demand, leading to sharp shifts in the structure of production as transition takes place.

structural deficit See *cyclically adjusted deficit*.

structural rate of unemployment See *natural rate of unemployment*.

supply sider An economist who believes that a cut in tax rates will lead people and firms to work much harder and more productively, and that the resulting increase in activity will lead to an increase in tax revenues.

Tanzi-Olivera effect The adverse effect of inflation on tax revenues and in turn on the budget deficit.

tariffs Taxes on imported goods.

tax smoothing The principle of keeping tax rates roughly constant, so that the government runs large deficits when government spending is exceptionally high and small surpluses the rest of the time.

technological progress An improvement in the state of technology.

technological unemployment Unemployment brought about by technological progress.

technology gap The differences between states of technology across countries.

term structure of interest rates See *yield curve*.

time inconsistency In game theory, the incentive for one player to deviate from his previously announced course of action once the other player has moved.

total wealth The sum of human wealth and nonhuman wealth.

tradeable goods Goods that compete with foreign goods in either domestic or foreign markets.

trade balance The difference between exports and imports. Also called *net exports*.

trade deficit A negative trade balance; that is, what exists when imports exceed exports.

trade surplus A positive trade balance; that is, what exists when exports exceed imports.

trade-weighted exchange rate See *multilateral exchange rate*.

transfers See *government transfers*.

Treasury bill (T-bill) A Canadian government discount bond with a maturity of up to one year.

t-**statistic** A statistic associated with an estimated coefficient in a regression that expresses the level of confidence that the true coefficient differs from zero.

unanticipated money Movements in nominal money that could not have been predicted based on the information available at some time in the past.

uncovered interest parity relation An arbitrage relation stating that domestic and foreign bonds must have the same expected rate of return, expressed in terms of the domestic currency.

underground economy That part of a nation's economic activity that is not measured in official statistics, either because the activity is illegal or because people and firms are seeking to avoid taxes.

unemployment rate The ratio of the number of unemployed to the labour force.

union density The proportion of the work force that is unionized.

useable observation An observation for which the values of all the variables under consideration are available for *regression* purposes.

user cost of capital The cost of using capital over a year, or a given period of time. The sum of the real interest rate and the depreciation rate. Also called the *rental cost of capital*.

value added The value a firm adds in the production process, equal to the value of its production minus the value of the intermediate inputs it uses in production.

velocity The ratio of nominal income to money; the number of transactions for a given quantity of money, or the rate at which money changes hands.

voucher privatization A system of privatization in which the government grants vouchers to private citizens allowing them to bid for shares in state-owned firms.

wage indexation A rule that automatically increases wages in response to an increase in prices.

wage-price spiral The mechanism by which increases in wages lead to increases in prices, which lead in turn to further increases in wages, and so on.

wealth See *financial wealth*.

yield curve The relation between yield and maturity for bonds of different maturities. Also called the *term structure of interest rates*.

yield to maturity The constant interest rate that makes the price of an *n*-year bond today equal to the present value of future payments. Also called the n-*year interest rate*.

INDEX

When referring to a table, page numbers are followed by a lowercase t. When referring to a figure, page numbers are followed by a lowercase f. When referring to a note, page numbers are followed by a lowercase n.

908